Lecture Notes in Mathematics

Edited by A. Dold, Heidelberg and B. Eckmann, Zürich

344

A. S. Troelstra
(Editor)

Universiteit van Amsterdam, Amsterdam/Nederland

Metamathematical Investigation of Intuitionistic Arithmetic and Analysis

Springer-Verlag
Berlin · Heidelberg · New York 1973

MS Subject Classifications (1970): 02C15, 02D05, 02D99, 02H10

3N 3-540-06491-5 Springer-Verlag Berlin · Heidelberg · New York
3N 0-387-06491-5 Springer-Verlag New York · Heidelberg · Berlin

setdruck: Julius Beltz, Hemsbach/Bergstr.

Dedicated to

GEORG KREISEL

who has contributed so much to the subject of this volume

Preface

The present volume found its origin in a course on functional and realizability interpretations on intuitionistic formal systems, presented at the Rijksuniversiteit Utrecht (Netherlands) in the spring of 1970, and a course on the metamathematics of intuitionistic formal systems at the University of Amsterdam in 1971 - 1972. The literature on the subject was widely scattered, the connection between certain rules was often not made explicit in the literature, and some obvious questions were not answered there.

Therefore I thought it would be useful to give a coherent presentation of the principal methods for metamathematical investigation of intuitionistic formal systems and the results obtained by these methods, connecting results in the literature, filling gaps and adding some new material. A first attempt (for realizability and functional interpretations) was made in Troelstra 1971, which, however, because of a rather terse style, was not readily assimilated by readers new to the field. (It still provides a useful survey of the applications to first-order systems however.) Therefore a more elaborate presentation, including other techniques of metamathematical research, seemed to be called for.

Having learnt of the unpublished Ph.D. work of C.Smorynski on applications of Kripke-models to intuitionistic arithmetic, and of Dr Zucker's thesis on the intuitionistic theory of higher-order generalized inductive definitions, subjects which both fitted very well into the scope of the planned volume, I asked them to contribute a chapter each ; their contributions appear as chapters V, and VI respectively. The models for intuitionistic arithmetic of finite type, functional and realizability interpretations, and normalization for natural deduction systems, and also the general editing of the volume I undertook myself.

Finally, W.A. Howard contributed an Appendix supplementing discussions in § 2.7 and § 3.5.

The organization of the volume is primarily method-centered, i.e. the material presented is grouped mostly around methods and techniques, and not arranged according to the results obtained. Hence some results, obtainable by different methods, appear at various places in the book. This will enable the reader to compare the relative merits of the various methods.

As regards intuitionistic arithmetic and closely related systems, the treatment is almost wholly self-contained ; some experience with classical metamathematics, and the elements of intuitionism, such as may be gleaned from Kleene's Introduction to metamathematics and Heyting's book on Intuitionism suffices. The parts dealing with arithmetic can therefore be used in a course for graduate students or a seminar.

The sections dealing with analysis are not self-contained, and serve more or less as a running commentary on the literature, connecting and comparing various approaches and adding new results besides. This part was thought of primarily as a help to the beginning researcher, to help him to find his way in the subject. For use in a seminar, these sections should usually be supplemented by the reading of other papers.

In keeping with this set-up, the listing of applications for intuitionistic arithmetic and closely related systems is rather extensive, but in the case of analysis we have often restricted ourselves to some typical examples ; further applications can easily be made by the reader himself once he has understood the method, and its applications to arithmetic.

No special attention has been given to intuitionistic propositional logic and predicate logic, because as formal systems they exhibit many properties which do not generalize to arithmetic and analysis, and therefore would require a separate treatment.

Speedy publication was thought more useful than final polish, so as not to make the material outdated at the moment of its appearance. Hence also the choice for publication in the "Lecture Notes in Mathematics". Even while refraining from a completely self-contained treatment of all parts, it was not possible to take all relevant work into account, not even on arithmetic ; for example, N. Goodman's work on the theory of constructions was left out altogether, since it would not easily be fitted into the framework of the other developments and so would consume too much space.

We have no doubt that there are still many imperfections in this presentation ; it hardly needs saying that the authors will be grateful for errors, misprints, additions to the bibliography being brought to their attention.

The contents of the present volume are primarily technical in character ; but it is to be hoped that the material will not inspire a thought- and mind-less multiplication of metamathematical results, without a thought spent on their possible significance for an analysis of intuitionistic basic notions and for foundations of mathematics in general. On the other hand, the "philosophical interest" of the subject is not promoted by uncritical analysis. (A single example : the interest of the well known disjunction property $\vdash A \vee B \Rightarrow \vdash A$ or $\vdash B$, and the explicit definability for existential statements are frequently overrated, especially as a criterion for the "constructive character" of the system considered. See e.g. the discussion in Troelstra A.) As regards potential "philosophical interest", it seems to me to be more promising (but also more difficult) to look for new results for well-known systems (possibly different in kind from the results discussed in this volume), instead of trying to extend known results to stronger and stronger systems. Of course, to be potentially interesting, the new results should also have a clear intuitive meaning in terms of the intended interpretation of the systems considered.

Directions for use. In order to help the reader find his way, there is an analytical table of contents at the beginning, a bibliography, and lists of notions and notations at the end. Reference to the bibliography are self-explanatory. § 3.5 refers (except in the appendix) to chapter III, § 5, etc.

The parts on arithmetic and closely related systems are more or less self-contained. As such we mention especially: Chapter I, §§ 1 - 8, §§ 10, 11; chapter II, §§ 1 - 4 (2.4.18 excepted), § 5, § 7 (except where results of § 6 are used); chapter III, § 1 (3.1.1 - 18), § 2 (3.2.1 - 28; 3.2.33), § 4 (3.4.1 - 14; 3.4.29), § 5 (3.5.1 - 11; 3.5.16 (i), (iii)); § 6 (3.6.1-3.6.16), § 7 (3.7.1 - 8), § 8 (except 3.8.7), § 9; chapter IV, §§ 1 - 4; chapter V, §§ 1 - 6.

Chapter I contains all generalities, and should usually be consulted when needed only.

Acknowledgements. As regards my own contribution to this volume, I am especially indebted to G. Kreisel, who permitted the use of unpublished material in his course notes (apart from the general indebtedness expressed by the dedication), to J.I. Zucker, for his patient and careful reading of drafts of my chapters, suggesting many stylistic, expository and mathematical improvements and corrections, and to Miss Judith van Witsen, who undertook the seemingly endless task of typing the manuscript. Some other acknowledgements have been made in footnotes.

Amsterdam, June 1973.

A. S. Troelstra

TABLE OF CONTENTS

Chapter I

INTUITIONISTIC FORMAL SYSTEMS

§ 1. Intuitionistic logic.

1.1.1. Contents. In the present section we describe logical notation and systems for intuitionistic predicate logic to be used in the sequel. The reader already acquainted with these subjects may skip them and use them for reference only. We shall presuppose acquaintance with classical predicate logic and the treatment of "elementary" metamathematics of classical systems, and elementary recursion theory (for example, as found in Kleene 1952). In the sequel, proofs in formal systems are usually not set out in a completely formalized form, but we compromise between readability and rigour ; i.e. the proof is described in sufficient detail so as to make full formalization a routine matter ; but we try to avoid an excess of detail which obscures the underlying idea.

In view of this aim, we usually freely employ theorems and rules of intuitionistic predicate logic, whose proof is not to be found in this monograph. For the reader with little previous acquaintance with intuitionistic reasoning, we recommend Heyting 1956, Chapter VII and Troelstra 1969, § 2 for intuitive background, and Kleene 1952 for formal details. We remark here only that in order to convert an intuitive proof of an intuitionistic logical theorem into a formal argument, the system of natural deduction described in 1.1.7 is usually very convenient.

In agreement with the attitude towards formalization described above, the description of formal systems for intuitionistic predicate logic below does not serve as a basis for deductions in the formal systems to be studied, but as a reference for metamathematical arguments proceeding by induction on the length of deductions. Nevertheless, the discussion is fairly detailed, to enable a reader without experience with intuitionistic formal systems to get accustomed to them. In later sections and chapters the development gradually becomes more condensed.

1.1.2. Some notational conventions.

(i) As logical symbols we use $\&$, $\vee$, $\exists$, $\forall$, $\rightarrow$, $\curlywedge$ (absurdity) ; as metamathematical abbreviations we use $\Rightarrow$, $\Leftrightarrow$, $\underset{\sim}{\forall}$, $\underset{\sim}{\exists}$, $\in$, $\subseteq$, etc.

Definitions (or abbreviating expressions) of a more or less permanent character are usually indicated by $\equiv_{def}$; $\equiv$ is used for definitions or abbreviations of a more local character (i.e. within a certain argument), and to express syntactical identity.

(ii) x, y, z, u, v, w (provided with sub- or superscripts if necessary) will be used as syntactical variables for variables; in systems containing arithmetic they are usually reserved for numerical variables. In the sequel we shall often have to introduce other categories of symbols as syntactical variables for certain sorts of variables in the formal systems studied.

Usually we do not use separate sets of symbols for free and bound variables, with the exception of systems of natural deduction, where we feel the notational distinction between (bound) variables and parameters (free variables) to be a definite advantage. In this case, lower case letters $a, b, c, \ldots$ from the beginning of the alphabet are used to indicate parameters.

(iii) Capitals (primarily from the beginning of the alphabet) $A, B, C, \ldots$ are used as syntactical variables for formulae. t, s will be used to denote terms (with an exception in chapter IV, where s is reserved for successor).

Variables $x, y, z, \ldots$ occurring free (perhaps only as dummy variables) in a term t are indicated by the use of square brackets $t[x]$, $t[x,y]$, etc.

(iv) In all categories of variables introduced, sub- and superscripts may be added to create more variables of the same category.

(v) Syntactical descriptions of the classes of formulae and terms, in our various formal systems, are as usual; if we wish, we may assume the actual notation to be bracket-free ("Polish" notation, which is convenient in gödelization) and think of the usual notations with brackets as "abbreviations" for better readability. Our bracketing conventions are us usal: unary operators bind stronger than binary ones, $\vee$, & bind stronger than $\rightarrow$. In general, we shall omit brackets whenever we can do so without impairing readability.

(vi) Some abbreviations:

$\neg A \equiv_{def} A \rightarrow \curlywedge$, $\quad A \leftrightarrow B \equiv_{def} (A \rightarrow B) \,\&\, (B \rightarrow A)$,

$\forall x_1 \ldots x_n A \equiv_{def} \forall x_1 \forall x_2 \ldots \forall x_n A$, $\quad \exists x_1 \ldots x_n A \equiv_{def} \exists x_1 \ldots \exists x_n A$,

$\forall x \in A(B) \equiv_{def} \forall x(Ax \rightarrow B)$, $\quad \exists x \in A(B) \equiv_{def} \exists x(Ax \,\&\, B)$.

Also in formulae, we sometimes use $x \in Q$ as an alternative to Qx, for unary predicates Q.

(vii) For substitution of a term t for a variable x (t, x of the same sort) in an expression (generally a term or a formula) or in a deduction (especially in chapter IV) we write $[x/t]E$, where E is the expression.

Quite frequently, when there is no danger of confusion, we shall also use the more imprecise convention that whenever an expression $E(x)$ has been introduced, $E(t)$ denotes $[x/t]E$. For variables occurring in terms we use square brackets: $t[x,y]$, etc.; in contrast, if φ is a function,

φt or $\varphi(t)$ stand for φ applied to the argument t.

(viii) Formal systems will be indicated by capitals or combination of capitals with a wavy underlining (e.g. $\underset{\sim}{HA}$, $qf\text{-}\underset{\sim}{N}\text{-}\underset{\sim}{NA}^{\omega}$, $\underset{\sim}{H}$ etc.).

The language of a formal system $\underset{\sim}{H}$ is denoted by $\mathcal{L}(\underset{\sim}{H})$, the set of well-formed formulae by $Fm(\underset{\sim}{H})$ or Fm_H, the set of theorems by $Thm(\underset{\sim}{H})$ or Thm_H.

Deducibility in $\underset{\sim}{H}$ is indicated by $\underset{\sim}{H} \vdash$ or, rarely, as $\vdash_H$.

$A \in \underset{\sim}{H}$ means the same as $\underset{\sim}{H} \vdash A$, i.e. A is a theorem of $\underset{\sim}{H}$. $\underset{\sim}{H} \subseteq \underset{\sim}{H}'$ is also interpreted as usual. If we add to $\underset{\sim}{H}$ a set of axioms Γ, we write either $\underset{\sim}{H} \cup \Gamma$ or $\underset{\sim}{H} + \Gamma$.

If $\mathcal{L}$ denotes a given language, and P a predicate letter not occurring in $\mathcal{L}$, we write $\mathcal{L}[P]$ for the language obtained by adding P to the constants of $\mathcal{L}$.

Similarly, if $\underset{\sim}{H}$ is a formal system, presented by giving a set of rules, axioms and axiom schemata, $\underset{\sim}{H}[P]$ is the system with language $\mathcal{L}(\underset{\sim}{H})[P]$, with the same rules, axioms and axiom schemata (i.e. the schematic letters in an axiom schema now stand for formulae of the extended language).

(ix) Church's λ-notation will sometimes be used informally to indicate functions or predicates.

1.1.3. Spector's system.

The systems for intuitionistic predicate logic described in this and the next section are "Hilbert-type systems", i.e. based on logical axioms and inference rules. The present system, taken from Spector 1962 (leaving out his A2, in view of footnote 7 on page 10 of Spector 1962), is given by the following axioms and rules:

PL 1) $A \to A$

PL 2) $A, A \to B \Rightarrow B$

PL 3) $A \to B,\ B \to C \Rightarrow A \to C$

PL 4) $A \& B \to A,\ A \& B \to B,\ A \to A \vee B,\ B \to A \vee B$

PL 5) $A \to C,\ B \to C \Rightarrow A \vee B \to C$

PL 6) $A \to B,\ A \to C \Rightarrow A \to B \& C$

PL 7) $A \& B \to C \Rightarrow A \to (B \to C)$

PL 8) $A \to (B \to C) \Rightarrow A \& B \to C$

PL 9) $\wedge \to A$.

and for predicate logic (x a variable of sort i, t a term of sort i, C not containing x free)

Q 1^i) $B \to A(x) \Rightarrow B \to \forall x A(x)$

Q 2^i) $\forall x Ax \to At$

Q 3^i) $At \to \exists x Ax$

$Q\,4^i$) $Ax \to B \Rightarrow \exists x Ax \to B$.

In applications of $Q\,1^i$ and $Q\,4^i$ the premiss is supposed not to depend on assumptions comtaining x free, i.e. has been derived without the use of such assumptions.

1.1.4. Gödel's system.

For the purpose of verifying the soundness of the so-called Dialectica interpretation (see chapter III), Gödel suggested another system, based on Q1 - Q4, PL2, 3, 7, 8, 9 and

PL 10) $A \vee A \to A$, $A \to A \,\&\, A$

PL 11) $A \to A \vee B$, $A \,\&\, B \to A$

PL 12) $A \vee B \to B \vee A$, $A \,\&\, B \to B \,\&\, A$

PL 13) $A \to B \Rightarrow C \vee A \to C \vee B$.

This system has the advantage of keeping complexities down to a minimum (i.e. in the rules and axioms there appear fewer logical symbols than in the previous system).

1.1.5. Equivalence of Spector's and Gödel's system.

In Hilbert - type systems, we may either suppose deductions from assumptions be represented as finite sequences of formulae, each formula of the sequence being an axiom, an assumption, or obtained from formulae appearing earlier in the sequence by means of a rule of the system. (This form is often quite convenient for actual arithmetization; however, it should be noted that in some cases it is more natural to suppose the rule or axiom involved to be indicated explicitly at each element of the sequence). A more pictorial representation is by means of deduction trees, which we shall use below.

It is perhaps useful to remark already here, that in case the proof trees themselves are objects of study (as in chapter IV) we must think of them as being completely presented by a tree of formulae together with an indication which rule or axiom is applied at each node. However, in presenting proof trees pictorially below, we shall not always explicitly indicate the rules used, so as not to encumber typography.

A proof given as a sequence may be thought of as being obtained by consistently extending the partial order of the tree to a linear order.

If $\vdash_S$ denotes deducibility in Spector's system, $\vdash_G$ in Gödel's system, Γ a set of assumption formulae, then the two systems are equivalent in the sense that

$$\Gamma \vdash_S A \Leftrightarrow \Gamma \vdash_G A$$

(for first- and higher-order languages, one- or many-sorted)

$\Gamma \vdash_G A \Rightarrow \Gamma \vdash_S A$ follows by the following deductions :

PL 1)

$$\text{(PL 3)}\ \frac{A \rightarrow A \& A \quad \text{(PL10)} \quad A \& A \rightarrow A \quad \text{(PL11)}}{A \rightarrow A}$$

PL 4) $A \& B \rightarrow A$ is the second half of PL11

$$\text{(PL 3)}\ \frac{A \& B \rightarrow B \& A \quad \text{(PL12)} \quad B \& A \rightarrow B \quad \text{(PL11)}}{A \& B \rightarrow B}$$

$A \rightarrow A \lor B$ is the first half of PL11

$$\text{(PL 3)}\ \frac{B \rightarrow B \lor A \quad \text{(PL11)} \quad B \lor A \rightarrow A \lor B \quad \text{(PL12)}}{B \rightarrow A \lor B}$$

PL 5)

$$\frac{\text{(PL13)}\ \dfrac{B \rightarrow C \ \text{(ass)}}{A \lor B \rightarrow A \lor C} \qquad \dfrac{\text{(PL12)}\ A \lor C \rightarrow C \lor A \qquad \dfrac{\text{(PL 3)}\ \dfrac{\text{(PL13)}\ \dfrac{A \rightarrow C \ \text{(ass)}}{C \lor A \rightarrow C \lor C} \qquad C \lor C \rightarrow C \ \text{(PL10)}}{}}{C \lor A \rightarrow C}\ \text{(PL 3)}}{A \lor C \rightarrow C}\ \text{(PL 3)}}{A \lor B \rightarrow C}$$

PL 6)

$$\frac{\text{(PL10)}\ A \rightarrow A \& A \qquad \dfrac{\dfrac{\text{(ass)}\ A \rightarrow C \qquad \dfrac{\dfrac{\text{(PL12)}\ C \& A \rightarrow A \& C \qquad \dfrac{\dfrac{\text{(ass)}\ A \rightarrow B \qquad \dfrac{\overset{\text{(PL 1 deduction)}}{B \& C \rightarrow B \& C}}{B \rightarrow (C \rightarrow B \& C)}\ \text{(PL 7)}}{A \rightarrow (C \rightarrow B \& C)}\ \text{(PL 3)}}{A \& C \rightarrow B \& C}\ \text{(PL 8)}}{C \& A \rightarrow B \& C}\ \text{(PL 3)}}{C \rightarrow (A \rightarrow B \& C)}\ \text{(PL 7)}}{A \rightarrow (A \rightarrow B \& C)}\ \text{(PL 3)}}{A \& A \rightarrow B \& C}\ \text{(PL 8)}}{A \rightarrow B \& C}\ \text{(PL 3)}$$

Conversely, we verify that $\Gamma \vdash_S A \Rightarrow \Gamma \vdash_G A$ by the following deductions :

PL10)

$$\frac{A \rightarrow A \qquad A \rightarrow A}{A \lor A \rightarrow A}\ \text{(PL 5)} \qquad \frac{A \rightarrow A \qquad A \rightarrow A}{A \rightarrow A \& A}\ \text{(PL 6)}$$

PL11) is part of PL 4.

PL12)

$$\text{(PL 5)}\ \frac{B \rightarrow A \lor B \quad \text{(PL 4)} \quad A \rightarrow A \lor B \quad \text{(PL 4)}}{B \lor A \rightarrow A \lor B}$$

$$\text{(PL 6)}\ \frac{A \& B \rightarrow B \quad \text{(PL 4)} \quad A \& B \rightarrow A \quad \text{(PL 4)}}{A \& B \rightarrow B \& A}$$

PL13)

$$\frac{(PL\ 4)\ C \to C \vee B \qquad \dfrac{(ass)\ A \to B \qquad B \to C \vee B\ (PL\ 4)}{A \to C \vee B}\ (PL\ 3)}{C \vee A \to C \vee B}\ (PL\ 5)$$

1.1.6. Equivalence of Spector's and Kleene's formalization.

Yet another Hilbert-type system is described in Kleene 1952, chapters IV, V. The equivalence with Spector's system is proved in Spector 1962.

Warning. In one respect our conventions differ from Kleene's : Kleene 1952 permits application of Q 1, Q 4 also when the variable occurs in assumption formulae (i.e. the assumption formulae are treated as if they are universally closed) ; if Kleene wishes to indicate that certain variables are to be treated as parameters, and hence are not permitted as proper variables of an application of Q 1, Q 4 ("variables held constant") he uses the notation $\vdash^{x_1 \ldots x_n}$.

1.1.7. A natural deduction system.

We now distinguish between parameters and (bound) variables. Below, we shall use a, b, c, ... for the parameters, x, y, z, ... for the variables. We describe the system briefly (a detailed and rigorous description is in Prawitz 1965 ; more briefly in Prawitz 1971).

The rules may be schematically described as follows :

&I) $\dfrac{A \qquad B}{A \& B}$ &E$_l$) $\dfrac{A \& B}{A}$ &E$_r$) $\dfrac{A \& B}{B}$

$\vee$I$_r$) $\dfrac{A}{A \vee B}$ $\vee$I$_l$) $\dfrac{B}{A \vee B}$ $\vee$E) $\dfrac{A \vee B \qquad \begin{matrix} \not{A} \\ \vdots \\ C \end{matrix} \qquad \begin{matrix} \not{B} \\ \vdots \\ C \end{matrix}}{C}$

$\to$I) $\dfrac{\begin{matrix} \not{A} \\ \vdots \\ B \end{matrix}}{A \to B}$ $\to$E) $\dfrac{A \qquad A \to B}{B}$

$\forall$I) $\dfrac{Aa}{\forall x Ax}$ $\forall$E) $\dfrac{\forall x Ax}{At}$

$\exists$I) $\dfrac{At}{\exists x Ax}$ $\exists$E) $\dfrac{\exists x Ax \qquad \begin{matrix} \not{Aa} \\ \vdots \\ C \end{matrix}}{C}$

$\bot_I$ $\dfrac{\bot}{A}$.

In the explanations below, it should be taken into account that we are primarily concerned with formula occurrences (fo's), i.e. a formula together with a position in a tree-like arrangement of formulas. "A formula occurrence A" means "an occurrence of the formula A". We shall sometimes loosely use "formula", when, as is apparent from the context, formula occurrences are meant.

The I - rules are called introduction rules, the E - rules elimination rules, since a logical constant is introduced in the conclusion, respectively eliminated from a premiss. So we sometimes write "$\to$ introduction " for "$\to$I" etc.

We suppose deductions to be represented in tree form; the top formulas of the tree are then the assumptions, and the (uniquely determined) end formula (occurrence) is the conclusion of the deduction. Each formula occurrence is either a top formula, or the conclusion of an application of one of the inference rules, its immediate predecessors being the premisses in the application of the rule. At applications of $\vee$E, $\exists$E, $\to$I certain assumptions (of the form indicated by the formulae crossed out in our list of rules) are discharged; a discharged assumption is said to be <u>closed</u> (by the inference). Only assumptions which have not been discharged previously (i.e. at a node of the proof tree above the one considered) can be discharged. It should also be stressed that not necessarily <u>all</u> assumptions (possibly even none) of the same form occurring above the application of $\vee$E, $\exists$E, $\to$I are discharged.

We shall think of the assumptions to be grouped into assumption classes; each assumption class consists of a number of occurrences of the <u>same</u> formula. All formulas of an assumption class are always treated simultaneously, i.e. at each application of a rule in the deduction either all formulas of the class are discharged or none of them is discharged.

A formula occurrence A is said to <u>depend</u> on the assumptions standing above A that have not been closed by some inference above A.

In the applications of $\forall$I the premiss Aa must not depend on assumptions containing a, and in an application of $\exists$E of the form $\frac{\exists x Ax \quad C}{C}$, the upper occurrence of C must not depend on assumptions other than Aa containing the parameter a. In applications of $\forall$I, $\exists$I a is called the <u>proper</u> parameter of the inference; a parameter is a <u>proper</u> parameter of a deduction if it is used as the proper parameter of an $\forall$I, $\exists$E inference.

The <u>open</u> assumptions of a deduction are the assumptions on which the end formula of the deduction depends. A deduction is said to be <u>closed</u> if there are no open assumptions.

We shall always assume a completely described deduction to have specified at each node which rule is being applied, and for the assumptions (top formulae), at which inference (if any) they are discharged.

With respect to the rules, we wish to introduce some further terminology. In an application of an E- rule, the premiss containing the occurrence of the constant to be eliminated, is called the <u>major</u> premiss of the inference; the other premisses, if any, are called <u>minor</u> premisses. So, in our list of inferences above, $A \& B$, $A \vee B$, $A \to B$, $\forall x Ax$, $\exists x Ax$ are the major premisses of $\&E$, $\vee E$, $\to E$, $\forall E$, $\exists E$ respectively. It is convenient to call any premiss of an application of an I-rule or $\curlywedge_I$ also a major premiss.

It will be obvious that deductions which only differ in the naming of their proper parameters may be regarded as assentially the same. It is easy to verify that we may always select our proper parameters so as to satisfy the following requirements, for a given deduction Π of A from assumptions Γ (cf. Prawitz 1965, chapter I, § 3).

(i) The proper parameter of an application α of $\forall I$ in Π occurs in Π only in formula occurrences above the consequence of α.

(ii) The proper parameter of an application α of $\exists E$ in Π occurs in Π only in formula occurrences above the minor premiss of α.

(iii) Every proper parameter in Π is a proper parameter of exactly one application of the $\forall I$-rule or the $\exists E$-rule in Π.

1.1.8. Examples. We give some examples of deductions in the system of natural deduction; the theorems derived will be used later on (§ 10).

In the examples, "(ass)" marks an assumption which is not discharged (i.e. "open") in the deduction. Assumptions which are discharged are marked by a number "(1)", "(2)", etc., all assumptions in the same class getting the same number. This number is then repeated at the application of a rule where the assumption is discharged.

I)

$$\dfrac{\dfrac{\dfrac{\dfrac{\dfrac{(1)\ A \qquad A \to B\ (\text{ass})}{B} \qquad B \to \curlywedge}{\curlywedge}\ (1)}{A \to \curlywedge}}{\neg B \to \neg A}}{}$$

II)

$$\dfrac{\dfrac{\dfrac{(3)\ \neg\neg\forall x Ax \qquad \dfrac{\dfrac{(2)\ \neg Aa \qquad \dfrac{\forall x Ax\,(1)}{Aa}}{\curlywedge}\,{\to}E}{\neg\forall x Ax}\,{\to}I\ (1)}{\dfrac{\curlywedge}{\neg\neg Aa}\,{\to}I\ (2)}\,{\to}E}{\forall x\, \neg\neg Ax}\,\forall I}{\neg\neg\forall x Ax \to \forall x\, \neg\neg Ax}\,{\to}I\ (3)$$

III)

$$\cfrac{\cfrac{\cfrac{(3)\ \neg\neg\neg A \qquad \cfrac{\cfrac{(1)\ A \quad \neg A\ (2)}{\curlywedge}}{\neg\neg A}\ (2)}{\curlywedge}}{\neg A}\ (1)}{\neg\neg\neg A \to \neg A}\ (3)$$

Since also

$$\cfrac{\cfrac{\cfrac{(1)\ \neg A \quad \neg\neg A\ (2)}{\curlywedge}}{\neg\neg\neg A}\ (2)}{\neg A \to \neg\neg\neg A}\ (1)$$

it follows that

$$\neg A \leftrightarrow \neg\neg\neg A\ .$$

IV)

$$\cfrac{\cfrac{\cfrac{\cfrac{\cfrac{\cfrac{\cfrac{(2)\ A \quad A \to B\ (1)}{B} \qquad \neg B\ (3)}{\curlywedge}}{\neg(A \to B)}\ (1) \qquad \neg\neg(A \to B)\ (4)}{\curlywedge}}{\neg\neg B}\ (3)}{A \to \neg\neg B}\ (2)}{\neg\neg(A \to B) \to (A \to \neg\neg B)}\ (4)$$

$$\cfrac{\cfrac{\cfrac{\cfrac{\cfrac{\cfrac{\cfrac{\cfrac{\cfrac{(1)\ A \quad \neg A\ (2)}{\curlywedge}}{B}}{A \to B}\ (1) \qquad \neg(A \to B)\ (4)}{\curlywedge}}{\neg\neg A}\ (2) \qquad \neg\neg A \to \neg\neg B\ (5)}{\neg\neg B} \qquad \cfrac{\cfrac{\cfrac{\cfrac{(3)\ B}{A \to B} \qquad \neg(A \to B)\ (4)}{\curlywedge}}{\neg B}\ (3)}{}}{\curlywedge}}{\neg\neg(A \to B)}\ (4)}{(\neg\neg A \to \neg\neg B) \to \neg\neg(A \to B)}\ (5)\ .$$

An application of $\to$I in (I) yields $(C \to D) \to (\neg D \to \neg C)$.

Hence

$$\begin{aligned} (A \to \neg\neg B) &\to (\neg\neg\neg B \to \neg A) \\ &\to (\neg\neg A \to \neg\neg\neg\neg B) \\ &\to (\neg\neg A \to \neg\neg B) \qquad \text{(III)} \end{aligned}$$

and thus

$$\neg\neg(A \to B) \leftrightarrow (\neg\neg A \to \neg\neg B) \leftrightarrow (A \to \neg\neg B)$$

V) $\neg\neg(A \& B) \to \neg\neg A \& \neg\neg B$ may be proved very similarly to II.

Further

$$\dfrac{\dfrac{\dfrac{\dfrac{(4)\ \neg\neg A \,\&\, \neg\neg B}{\neg\neg A} \qquad \dfrac{\dfrac{(3)\ \neg A \,\&\, B \qquad \dfrac{(1)\ A \quad B\ (2)}{A \,\&\, B}}{\Lambda}}{\neg A}\,(1)}{\dfrac{\Lambda}{\neg B}\,(2)} \qquad \dfrac{\neg\neg A \,\&\, \neg\neg B\ (4)}{\neg\neg B}}{\dfrac{\Lambda}{\neg\neg(A \,\&\, B)}\,(3)}}{\neg\neg A \,\&\, \neg\neg B \to \neg\neg(A \,\&\, B)}\,(4)\ .$$

Hence $\neg\neg(A \,\&\, B) \leftrightarrow \neg\neg A \,\&\, \neg\neg B$.

1.1.9. Lemma. The following schemata and rules are derivable in Spector's system:

(a) $A \to (B \to A)$

(b) $A \Rightarrow B \to A$

(c) $A \to B,\ A \to (B \to C) \Rightarrow A \to C$

(d) $[(A \to (B \to C)) \,\&\, (A \to B)] \,\&\, A \to C$

(e) $(A \to (B \to C)) \to [(A \to B) \to (A \to C)]$

(f) $[(A \to B) \to (A \to C)] \to (A \to (B \to C))$

(g) $A \to (B \to C),\ A \to (C \to D) \Rightarrow A \to (B \to D)$

(h) $A \to (B \to C) \Rightarrow B \to (A \to C)$

(i) $(A \to C) \,\&\, (B \to C) \to (A \vee B \to C)$

(j) $(A \to B) \to [(A \to C) \to (A \to (B \,\&\, C))]$.

(k) $A \to (B \,\&\, C \to D) \Rightarrow A \to (B \to (C \to D))$

(l) $A \to (B \to (C \to D)) \Rightarrow A \to (B \,\&\, C \to D))$.

Proof.

(a) From $A \,\&\, B \to A$ (PL4), $A \to (B \to A)$ by (PL7).

(b) Immediate from (a) with PL2.

(c)
$$\dfrac{\dfrac{A \to A\ (\mathrm{PL1}) \qquad A \to B\ (\mathrm{ass})}{A \to A \,\&\, B} \qquad \dfrac{A \to (B \to C)\ \mathrm{ass}}{A \,\&\, B \to C}}{A \to C}\ .$$

(d) We put $\bar{B} \equiv [A \to (B \to C)] \,\&\, (A \to B)$,
$\bar{A} \equiv (\bar{B} \,\&\, A)$.

$$\dfrac{\dfrac{\bar{A} \to A \qquad \dfrac{(\mathrm{PL4})\ \bar{A} \to \bar{B} \qquad \bar{B} \to (A \to B)\ (\mathrm{PL4})}{\bar{A} \to (A \to B)}\,(\mathrm{PL3})}{\bar{A} \to B}\,(c) \qquad \dfrac{\bar{A} \to A\,(\mathrm{PL3}) \qquad \dfrac{\bar{A} \to \bar{B}\ (\mathrm{PL4}) \qquad \bar{B} \to (A \to (\dot{B} \to C))\ (\mathrm{PL4})}{\bar{A} \to (A \to (B \to C))}}{\bar{A} \to (B \to C)}\,(c)}{\bar{A} \to C}\,(c)\ .$$

Here we have used an abbreviated notation for the proof tree ; (c) next to a horizontal line indicates that the line represents a part of the proof tree of the same form as for the part of (c) above. In other words, we use (c) as a derived rule to abbreviate our proof trees. Similar conventions are used below.

(e) Apply PL7 twice to (d).

(f) We put $D \equiv (A\to B)\to(A\to C)$, $E \equiv (D\,\&\,A)\,\&\,B$.

$$\dfrac{E\to A\ (PL4,\ PL4,\ PL3)\qquad \dfrac{E\to D\ (PL4,\ PL4,\ PL3)\qquad D\to(A\to C)\ (PL4)}{E\to(A\to C)}}{E\to C}\ (c)\ .$$

(f) is obtained by two applications of PL7 to $E\to C$.

(g)

$$\dfrac{\dfrac{\dfrac{\dfrac{A\to(B\to C)\ (ass)}{(A\to B)\to(A\to C)}\ ((e),\ PL2)\qquad \dfrac{A\to(C\to D)\ (ass)}{(A\to C)\to(A\to D)}\ ((e),\ PL2)}{(A\to B)\to(A\to D)}\ (PL3)}{A\to(B\to D)}\ (f)}{}\ .$$

(h)

$$\dfrac{\dfrac{\dfrac{B\&A\to A\qquad B\&A\to B}{B\&A\to A\&B}\qquad \dfrac{A\to(B\to C)\ \ ass}{A\&B\to C}}{B\&A\to C}\ (PL3)}{B\to(A\to C)}\ .$$

(i) Let $D \equiv (A\to C)\,\&\,(B\to C)$, $E \equiv A\vee B\to C$

$$\dfrac{\dfrac{\dfrac{D\to(A\to C)}{A\to(D\to C)}\ (h)\qquad \dfrac{D\to(B\to C)}{B\to(D\to C)}\ (h)}{A\vee B\to(D\to C)}\ (h)}{D\to E}\ .$$

(j) Let $D \equiv (A\to B)\,\&\,(A\to C)$, $E \equiv D\,\&\,A$.

$$\dfrac{\dfrac{\dfrac{\dfrac{E\to A\qquad E\to(A\to B)}{E\to B}\ (c)\qquad \dfrac{E\to A\qquad E\to(A\to C)}{E\to C}\ (c)}{E\to B\,\&\,C}}{D\to(A\to B\,\&\,C)}}{(A\to B)\to[(A\to C)\to(A\to B\&C)]}\ .$$

(k) By repeated use of PL4, PL3 $(A\,\&\,B)\,\&\,C\to A\,\&\,(B\,\&\,C)$

$$\dfrac{\dfrac{\dfrac{(A\,\&\,B)\,\&\,C\to A\,\&\,(B\,\&\,C)\qquad \dfrac{A\to(B\,\&\,C\to D)\ (ass)}{A\,\&\,(B\,\&\,C)\to D}}{(A\,\&\,B)\,\&\,C\to D}}{(A\,\&\,B)\to(C\to D)}}{A\to(B\to(C\to D))}$$

(l) Similar to (k), but arguing in the inverse direction.

1.1.10. <u>Deduction theorem for Spector's system</u>. $\Gamma, A \vdash_S B \Rightarrow \Gamma \vdash_S A \to B$.

<u>Proof</u>. We write simply $\vdash$ for $\vdash_S$. We show, by induction on the length of deductions , that a deduction of B from $\Gamma \cup \{A\}$ can be transformed into a deduction of $A \to B$ from Γ.

<u>Basis</u>. The deduction has length 1; then the deduction consists of B itself and therefore either $B \in \Gamma$, or $B \equiv A$, or B is an axiom.

In the first case, $\Gamma \vdash B$, hence by 1.1.9 (b) $\Gamma \vdash A \to B$.

In the second case, $\Gamma \vdash A \to A$ by PL1.

In the third case, $\Gamma \vdash A \to B$ by 1.1.9 (b).

<u>Induction step</u>. Assume the assertion to have been proved for all deductions of length $\leq k$, and let $A_1, \ldots, A_k, B$ be a deduction. By induction hypothesis, we have already shown $\Gamma \vdash A \to A_i$, $1 \leq i \leq k$.

If B is an axiom, or belongs to $\Gamma \cup \{A\}$, we proceed as for the basis step.

If B is obtained from the A_i by application of a rule, we must distinguish various cases.

<u>Case</u> PL2: $A_j \equiv A_i \to B$. We have $\Gamma \vdash A \to A_i$, $\Gamma \vdash A \to (A_i \to B)$; also (1.1.9 (e)) $\Gamma \vdash [A \to (A_i \to B)] \to [(A \to A_i) \to (A \to B)]$, hence $\Gamma \vdash A \to B$.

<u>Case</u> PL3: $A_i \equiv A' \to A''$, $A_j \equiv A'' \to A'''$, $B \equiv A' \to A'''$. We have $\Gamma \vdash (A \to (A' \to A''))$, $\Gamma \vdash (A \to (A'' \to A'''))$; application of 1.1.9 (g) yields $\Gamma \vdash A \to (A' \to A''')$.

<u>Case</u> PL5: Assume, by hypothesis $\Gamma \vdash A \to (B \to D)$, $\Gamma \vdash A \to (C \to D)$.
Then $\Gamma \vdash A \to [(B \to D) \& (C \to D)]$ (1.1.9 (j)), and $\vdash [(B \to D) \& (C \to D)] \to [(B \vee C) \to D]$ (1.1.9 (i)). So $\Gamma \vdash A \to [(B \vee C) \to D]$ (PL3).

<u>Case</u> PL6: Similarly, using 1.1.9 (j).

<u>Case</u> PL7: Use 1.1.9 (k).

<u>Case</u> PL8: Use 1.1.9 (l).

<u>Case</u> Q1: Assume $\Gamma \vdash A \to (C \to Bx)$, Γ, A not containing x free.
Use PL8: $\Gamma \vdash A \& C \to Bx$; then apply Q1: $\Gamma \vdash A \& C \to \forall x Bx$, and thus by PL7 $\Gamma \vdash A \to (C \to \forall x Bx)$.

<u>Case</u> Q4: Assume $\Gamma \vdash A \to (Bx \to C)$. Use 1.1.9 (h), so $\Gamma \vdash Bx \to (A \to C)$; apply Q4: $\Gamma \vdash \exists x Bx \to (A \to C)$; apply 1.1.9 (h) again to obtain $\Gamma \vdash A \to (\exists x Bx \to C)$.

1.1.11. <u>Theorem</u>. (Equivalence between natural deduction and Spector's system.) $\Gamma \vdash_N A$ iff $\Gamma \vdash_S A$ ($\Gamma \vdash_N A$ indicates that A can be deduced from assumptions Γ in the natural deduction system).

<u>Proof</u>. First we show that if $\Gamma \vdash_S A$, then $\Gamma \vdash_N A$. This is a routine matter: we have to show that (a) for the axioms A of Spector's system,

$\vdash_N A$, and (b) that for an instance of a rule $F_1, F_2 \Rightarrow F_3$ in Spector's system there is a deduction $F_1, F_2 \vdash_N F_3$ (in fact, as we shall see, the deductions in $\vdash_N$ for axioms and rules are uniform in the formula variables used to describe the axiom (schemata) and rules). For example, $A \to B \vee C$ is represented by the natural deduction proof

$$\dfrac{\dfrac{A}{A \vee B}\,\vee I}{A \to A \vee B}\,\to I$$

and the rule (PL5) by

$$\dfrac{\dfrac{A \vee B \qquad \dfrac{A \qquad A \to C}{C}\,\to E \qquad \dfrac{B \qquad B \to C}{C}\,\to E}{C}\,\vee E}{A \vee B \to C}\,\to I$$

etc., etc.

It remains to be shown that if $\Gamma \vdash_N A$, then $\Gamma \vdash_S A$. To see this, we have to show that each rule of the natural deduction calculus corresponds to a derived rule in Spector's system. For example, for $\vee E$ we have to show:

If $\Gamma, A \vdash_S C$, $\Gamma, B \vdash_S C$, then $\Gamma, A \vee B \vdash_S C$.

By the deduction theorem, $\Gamma \vdash_S A \to C$, $\Gamma \vdash_S B \to C$, and with PL5, $\Gamma \vdash_S A \vee B \to C$, so with PL2 $\Gamma, A \vee B \vdash_S C$.
The only crucial case is $\to I$, but this is provided by the deduction theorem.

1.1.12. Remark on the equivalence proofs in 1.1.11 and 1.1.5 under additional axioms.

The equivalence proofs remain obviously valid if we add further axioms (in the case of the natural deduction system, axioms may appear as top formulas but are not counted as assumptions). In the case of additional rules, however, the equivalence proofs have to be extended ; e.g. the proof of the deduction theorem for Spector's system, essential in 1.1.11, has to be extended with the consideration of further cases corresponding to the additional rules.

1.1.13. Sequent calculi.

We do not make use of Gentzen's calculus of sequents and its variants (cf. Gentzen 1935, Kleene 1952, § 77) ; for an equivalence proof the reader is referred to Prawitz 1965, Appendix A.

1.1.14. Convention (for indicating the classical equivalent of an intuitionistic system). If $\underset{\sim}{H}$ is any formal system based on intuitionistic (many-sorted) predicate logic, $\underset{\sim}{H}^c$ denotes the corresponding system with classical (many-sorted) predicate logic ("^{c}" for "classical").

§ 2. Conservative and definitional extensions, expansions.

1.2.1. Contents of the section. In this section we have brought together some theorems on definitional extensions, which are not emphasized in most text books, but which will be used quite frequently in this volume, either explicitly or implicitly. For the proofs, we must refer to Kleene 1952, § 74.

Under an intuitionistic (many-sorted) predicate calculus with equality we shall understand a system of intuitionistic predicate logic with equality = satisfying the axiom

$$\forall x(x = x)$$

and the schema

$$x = y \rightarrow (Ax \rightarrow Ay) .$$

It readily follows that = is symmetric and transitive.
(In what follows equality need not be given for all sorts of variables, if one makes some obvious stipulations in the theorems ; but we shall leave the formulation of the theorems in this more general situation to the reader.)

In this section, a formal system $\underset{\sim}{H}$ is said to be based on intuitionistic predicate logic with equality, if $\underset{\sim}{H}$ is based on the rules of intuitionistic (many-sorted) predicate logic, the equality axiom and schema, and possibly additional axioms and axiom schemata (the non-logical axioms).

1.2.2. Definition. Let $\underset{\sim}{H}'$, $\underset{\sim}{H}$ be formal systems based on (many-sorted) intuitionistic predicate logic, and let $\underset{\sim}{H} \subseteq \underset{\sim}{H}'$ (i.e. the language of $\underset{\sim}{H}'$ is an extension of the language of $\underset{\sim}{H}$, and the set of theorems of $\underset{\sim}{H}$ is contained in the set of theorems of $\underset{\sim}{H}'$). Then $\underset{\sim}{H}'$ is said to be a conservative extension of $\underset{\sim}{H}$ (or conservative over $\underset{\sim}{H}$), if

$$\mathrm{Thm}(\underset{\sim}{H}') \cap \mathrm{Fm}(\underset{\sim}{H}) = \mathrm{Thm}(\underset{\sim}{H}) .$$

Let $\Gamma \subseteq \mathrm{Fm}(\underset{\sim}{H})$. Then $\underset{\sim}{H}'$ is said to be conservative over $\underset{\sim}{H}$ relative to Γ (or w.r.t. Γ) if

$$\mathrm{Thm}(\underset{\sim}{H}') \cap \Gamma = \mathrm{Thm}(\underset{\sim}{H}) \cap \Gamma$$

(we shall sometimes abbreviate this as $\underset{\sim}{H}' \cap \Gamma = \underset{\sim}{H} \cap \Gamma$).

1.2.3. Definition. Let $\underset{\sim}{H}'$, $\underset{\sim}{H}$ be formal systems based on many-sorted intuitionistic predicate logic, and let $\underset{\sim}{H} \subseteq \underset{\sim}{H}'$. $\underset{\sim}{H}'$ is said to be an expansion of $\underset{\sim}{H}$, if there is a mapping φ of those formulae of $\mathrm{Fm}(\underset{\sim}{H}')$ where the only free variables are of sorts occurring in $\mathrm{Fm}(\underset{\sim}{H})$, such that

(i) $\underset{\sim}{H}' \vdash A \leftrightarrow \varphi A$

(ii) $\underset{\sim}{H}' \vdash A \Leftrightarrow \underset{\sim}{H} \vdash \varphi A$

(iii) $\varphi A \equiv A$ for $A \in Fm(\underset{\sim}{H})$.

We say that $\underset{\sim}{H}'$ is an <u>expansion</u> of $\underset{\sim}{H}$ relative to Γ, $\Gamma \subseteq Fm(\underset{\sim}{H})$ if (iii) is weakened to

(iii)' $\varphi A \equiv A$ for $A \in \Gamma$.

1.2.4. <u>Definition</u>. Let $\underset{\sim}{H}'$, $\underset{\sim}{H}$ be formal systems based on many-sorted predicate logic, and let the language of $\underset{\sim}{H}'$ be obtained by adding non-logical constants (i.e. constants assumed to be in the range of certain sorts of variables, and predicate constants). Then $\underset{\sim}{H}'$ is said to be a <u>definitional extension</u> of $\underset{\sim}{H}$, if there exists a mapping φ such that (i), (ii), (iii) hold and

(iv) $\varphi(\curlywedge) = \curlywedge$, $\varphi(A \circ B) = \varphi A \circ \varphi B$ for $\circ \equiv \vee, \&, \rightarrow$

(v) $\varphi((Qx)A) \equiv (Qx)\varphi A$ for $Q \equiv \forall, \exists$.

(i.e. φ is a homomorphism w.r.t. logical operations)

1.2.5. <u>Remark</u>. A definitional extension is an expansion, and an expansion is a conservative extension.

1.2.6. <u>Theorem</u> (<u>Addition of symbols for definable predicates</u>). Let $\underset{\sim}{H}$ be any theory based on (many-sorted) intuitionistic predicate logic, and let $A(x_1,\ldots,x_n) \in Fm(\underset{\sim}{H})$, where all the free variables of A are among $x_1, \ldots, x_n$. Let $\underset{\sim}{H}'$ be obtained by addition of a predicate symbol M, with axiom

$$A(x_1,\ldots,x_n) \leftrightarrow M(x_1,\ldots,x_n).$$

Then $\underset{\sim}{H}'$ is a definitional extension of $\underset{\sim}{H}$. (<u>Kleene</u> 1952, § 74, Example 1)

<u>Proof</u>. Trivial; see <u>Kleene</u> 1952, loc. cit. For future reference we describe the mapping φ_0 required by the definition of definitional extension:

(a) If P is a prime formula, not of the form $Mt_1\ldots t_n$, $\varphi_0 P \equiv P$

(b) $\varphi_0(Mt_1\ldots t_n) \equiv A(t_1,\ldots,t_n)$

(c) φ_0 is a homomorphism w.r.t. the logical operations.

1.2.7. <u>Theorem</u> (<u>Addition of symbols for definable functions</u>). Let $\underset{\sim}{H}$ be a theory based on (many-sorted) intuitionistic predicate calculus with equality. Assume, for a formula A containing free only $x_1,\ldots,x_n,y$:

$$\underset{\sim}{H} \vdash \exists! y\, A(x_1,\ldots,x_n, y)$$

where $\exists! y\, By$ abbreviates, as usual, $\exists y[By \,\&\, \forall z(Bz \rightarrow z=y)]$.

Let $\underset{\sim}{H}'$ be obtained from $\underset{\sim}{H}$ by addition of a new function symbol f, together with an axiom

$$A(x_1,\ldots,x_n,\ f(x_1,\ldots,x_n))$$

and extension of the axiom schemata to formulae in the extended language.

Let φ_1 be the mapping of $Fm(\underset{\sim}{H}')$ into $Fm(\underset{\sim}{H})$ defined as follows. Let us call a term not containing occurrences of f, f-<u>less</u>; and a term of the form $ft_1\ldots t_n$, an f-<u>term</u>; if $t_1,\ldots,t_n$ are f-less, $ft_1\ldots t_n$ is a <u>plain</u> f-term.

We define φ_1 for prime formulae P by induction on the number q of occurrences of f-terms in P. $\varphi_1 P \equiv P$, if $q = 0$.

Assume $q > 0$ for P; let (on some standard enumeration of term occurrences in prime formulae) $ft_1\ldots t_n$ be the first occurrence of a plain f-term in P; let v be a variable not occurring in P; let C(v) be obtained from P by replacing the occurrence $ft_1\ldots t_n$ by v.
Then $\varphi_1 P \equiv \exists v[F(t_1\ldots t_n, v) \,\&\, \varphi_1 C(v)]$.
(The variable v may be assumed to be chosen in a standard manner.)

φ_1 is defined for logically compound formulas by the requirement that it is a homomorphism w.r.t. the logical operations. $\varphi_1 A$ is called the f-less transform of A.

Then the assertion of the theorem is as follows:
$\underset{\sim}{H}'$ is a definitional extension of $\underset{\sim}{H}$ (with φ_1 as the mapping required by the definition of definitional extension), provided the additional axiom schemata satisfy the condition
(a) if A is an axiom by an additional axiom schema, then $\underset{\sim}{H} \vdash \varphi_1 A$.
<u>Proof</u>. <u>Kleene</u> 1952, § 74 (Theorem 42).

1.2.8. <u>Theorem</u> (<u>Replacement of function symbols by predicate symbols</u>). Let $\mathcal{L}$ be a language containing an n-ary function symbol f and an (n+1)-ary predicate symbol F. Let φ_0, φ_1 be mappings of the formulae of $\mathcal{L}$ into the formulae of $\mathcal{L}$ such that $\varphi_0(A)$ is obtained by replacing every occurrence of $F(t_1,\ldots,t_n,t)$ in A by $f(t_1,\ldots,t_n) = t$ (cf. proof of 1.2.6), and $\varphi_1(A)$ is the f-less transform of A (see 1.2.7).
Let $\underset{\sim}{H}$, $\underset{\sim}{H}'$ be two formal systems, based on intuitionistic predicate logic with equality, such that

(i) $\mathcal{L}(\underset{\sim}{H})$ is obtained from $\mathcal{L}$ by omitting f, $\mathcal{L}(\underset{\sim}{H}')$ is obtained by omitting F;

(ii) $\underset{\sim}{H}$ contains an axiom $\exists! y\, F(x_1,\ldots,x_n,y)$

(iii) If A is a non-logical axiom of $\underset{\sim}{H}$ (resp. of $\underset{\sim}{H}'$) then $\underset{\sim}{H}' \vdash \varphi_0(A)$ (resp. $\underset{\sim}{H} \vdash \varphi_1(A)$).

Then

$$\underset{\sim}{H} \vdash \varphi_1\varphi_0(A) \leftrightarrow A,\quad \underset{\sim}{H}' \vdash \varphi_0\varphi_1(A) \leftrightarrow A,$$

$$\underset{\sim}{H} + \Gamma \vdash A \Leftrightarrow \underset{\sim}{H}' + \varphi_0 \Gamma \vdash \varphi_0 A ,$$
$$\underset{\sim}{H}' + \Gamma \vdash A \Leftrightarrow \underset{\sim}{H} + \varphi_1 \Gamma \vdash \varphi_1 A .$$

Note that if $\underset{\sim}{H}''$ is the common extension of $\underset{\sim}{H}$ and $\underset{\sim}{H}'$ in the language $\mathcal{L}$, containing the axioms and axiom schemata of both, then $\underset{\sim}{H}''$ is a definitional extension of $\underset{\sim}{H}$ as well as $\underset{\sim}{H}'$.

Proof. See Kleene 1952, § 74, theorem 43.

1.2.9. Theorem (Addition of defined sorts of variables).

Let $\underset{\sim}{H}$ be a system based on intuitionistic (many-sorted) predicate logic; let $M(x)$ be a formula of $Fm(\underset{\sim}{H})$, containing free only x, and $\underset{\sim}{H} \vdash \exists x M(x)$.

Let $\underset{\sim}{H}'$ be obtained by addition of a new sort of variables (say $\underline{x}, \underline{y}, \underline{z}, \ldots$), with rules for term and formula construction also extended to the new variables, with the axiom schemata and rules of $\underset{\sim}{H}$ (but where in an axiom or axiom schema involving quantified variables, the axiom or axiom schema is not to be generalized by replacing quantification over the original variables by quantifiers over the new variables), and with the new axiom and schemata and rules

$M\underline{x}$, $Mt \rightarrow (\forall \underline{x} A\underline{x} \rightarrow At)$, $Mt \rightarrow (At \rightarrow \exists \underline{x} A\underline{x})$,

$A \rightarrow B\underline{x} \Rightarrow A \rightarrow \forall \underline{x} B\underline{x}$, $B\underline{x} \rightarrow A \Rightarrow \exists \underline{x} B\underline{x} \rightarrow A$.

Then $\underset{\sim}{H}'$ is an expansion of $\underset{\sim}{H}$.

Proof. (Kleene 1952, § 74, Example 13). We describe the correlation φ_2 as follows. Let A be a formula containing a set V of free variables. We define a mapping φ_V for the subformulae of A by induction on their complexity:

$$\begin{array}{lll}
\varphi_V(P(\underline{x}_1,\ldots,\underline{x}_n)) \equiv P(y_1,\ldots,y_n) & & \text{for prime formulae } P, \\
\varphi_V(B_1 \circ B_2) \equiv \varphi_V(B_1) \circ \varphi_V(B_2) & & \text{for } \circ \equiv \rightarrow, \vee, \&, \\
\varphi_V((Qx)B) \equiv (Qx)\, \varphi_V(B) & & \text{for } Q \equiv \forall, \exists, \\
\varphi_V(\forall \underline{x}_i B\underline{x}_i) \equiv \forall y_i (My_i \rightarrow \varphi_V B x_i) & & \\
\varphi_V(\exists \underline{x}_i B\underline{x}_i) \equiv \exists y_i (My_i \,\&\, \varphi_V B x_i) & &
\end{array}$$

Here $y_1,\ldots,y_n$ are variables not in V, and $\underline{x}_1,\ldots,\underline{x}_n$ may be assumed to be a complete list of the new variables occurring in A.

Then we put $\varphi_2 A = \varphi_V A$.

1.2.10. Alternative approach to defined sorts of variables.

In 1.2.9, the defined sort of variables was treated as a subset of the original set of individual variables, with respect to the formation rules. If we wish afterwards to introduce symbols for functions defined on n-tuples of elements of $\{x \mid Mx\}$, it is preferable to treat the new sort of variables as completely disjoint, and state the formation rules separately.

We then need axioms $\forall \underline{x} \exists y (\underline{x} = y)$ (with a primitive or defined $=$), $\forall x \in M \exists \underline{y} (x = \underline{y})$. Cf. e.g. Kreisel - Troelstra 1970, 3.3.4).

§ 3. Intuitionistic first-order arithmetic.

1.3.1. Contents of the section. In this section we describe intuitionistic first-order arithmetic and notational conventions, choice of pairing and sequence codings, and the formalization of elementary recursion theory.

1.3.2. Language of HA. The language of Heyting's arithmetic HA is a first-order language, with logical constants $\forall, \exists, \rightarrow, \&, \vee, \curlywedge$ (which may be identified with $0 = 1$), numerical variables (indicated by x, y, z, u, v, w) a constant 0 (zero), a unary function constant S (successor), constant function symbols for all primitive recursive functions (see below in 1.3.4), and a single binary predicate constant $=$ (equality between numbers). Terms and formulae are defined as usual.

1.3.3. Axioms and rules of HA.

HA is based on intuitionistic first-order predicate logic and contains in addition the following axioms:

$$\mathrm{ES}\quad\begin{cases} x = x \\ x = y \;\&\; z = y \rightarrow x = z \\ x_i = x_i' \rightarrow \varphi(x_1,\ldots,x_i,\ldots,x_n) = \varphi(x_1,\ldots,x_i',\ldots,x_n) \\ \quad\text{for any } n\text{-ary function constant } \varphi,\ 1 \leq i \leq n \\ Sx \neq 0, \\ Sx = Sy \rightarrow x = y, \end{cases}$$

the definining axioms for the primitive recursive functions (see 1.3.4) and all instances of the schema of induction

$$\mathrm{IND}\qquad A0 \;\&\; \forall x(Ax \rightarrow A(Sx)) \rightarrow \forall x Ax .$$

1.3.4. Definining axioms for primitive recursive functions.

The precise selection of initial functions and defining schemata is not relevant to our discussion of intuitionistic arithmetic in this volume. A very simple set is as follows:

Initial functions are the zero-place 0, 1-place successor S, and the n-place projection function I_n^i, $1 \leq i \leq n$, for all n, satisfying

$$I_n^i(x_1,\ldots,x_i,\ldots,x_n) = x_i .$$

Our defining schemata are composition and recursion.

Composition is described as follows. If $\varphi_1, \ldots, \varphi_m$ are n-place functions and ψ an m-place function which have been defined before, then we may introduce a new n-place function ξ with axiom

$$\xi(x_1,\ldots,x_n) = \psi(\varphi_1(x_1,\ldots,x_n), \ldots, \varphi_m(x_1,\ldots,x_n)) .$$

ξ is said to be defined by composition from ψ, $\varphi_1,\ldots,\varphi_m$.

Recursion : if φ is an n-place function and ψ a n+2-place function which have been defined before, then we may introduce a new (n+1)-place function ξ with axioms

$$\begin{cases} \xi(0, x_1,\ldots,x_n) = \varphi(x_1,\ldots,x_n) \\ \xi(Sy,x_1,\ldots,x_n) = \psi(\xi(y,x_1,\ldots,x_n), y,x_1,\ldots,x_n) . \end{cases}$$

In this case, ξ is said to be defined by recursion from φ, ψ.

1.3.5. Rule and axiom schema of induction.

Instead of using IND, we might also have added the rule of induction

Rule-IND $\qquad B0 , \; Bx \rightarrow B(Sx) \Rightarrow By$

(x not occurring in assumptions on which $Bx \rightarrow B(Sx)$ depends).

A minor variant :

$$B0 , \; \forall x(Bx \rightarrow B(Sx)) \Rightarrow By .$$

It is obvious that IND implies Rule-IND. For the converse, we must apply the rule to

$$Bx \equiv A0 \,\&\, \forall y(Ay \rightarrow A(Sy)) \rightarrow Ax .$$

1.3.6. Natural deduction variant of HA.

The description in 1.3.2 - 1.3.4 of HA is independent of the particular formalization of intuitionistic predicate logic which is used. However, to obtain a natural deduction variant of HA which is especially suited to the proof-theoretic researches in chapter V, we have to make some changes in the non-logical part also.

As in the discussion of intuitionistic predicate logic, we distinguish between parameters and variables. We have one individual constant, 0, a single unary function constant, denoted by S (successor), a binary predicate constant = (equality), and a denumerable sequence of predicate constants F_1, F_2, F_3, ... for the graphs of primitive recursive functions.

To the rules of predicate logic we add rules (the basic rules)

$$\begin{cases} t = t \qquad \dfrac{t = t'}{t' = t} \qquad \dfrac{t = t' \quad t' = t''}{t = t''} \\[2ex] \dfrac{t_i = t_i' \qquad F_k t_1 \ldots t_i \ldots t_n}{F_k t_1 \ldots t_i' \ldots t_n} \end{cases}$$

$$\frac{0 = St}{\curlywedge} \qquad \frac{t = t'}{St = St'} \qquad \frac{St = St'}{t = t'}$$

and, for example, parallel to the first set of initial functions and defining schemata given in 1.3.4, we introduce F_k's such that

$$F_k x_1 \ldots x_i \ldots x_n x_i$$

and if $F_{k_0}, F_{k_1}, \ldots, F_{k_m}$ have already been introduced, we introduce an F_k $(k > k_o, k_1, \ldots, k_m)$

$$\frac{F_{k_1} t_1 \ldots t_n t_1' \quad , \; F_{k_2} t_1 \ldots t_n t_2', \ldots, F'_{k_m} t_1 \ldots t_n t_m', F_{k_o} t_1' \ldots t_m' t}{F_k \; t_1 \ldots t_n t}$$

and if F_m, F_n have already been introduced, we introduce an F_k $(k > m,n)$ with two rules:

$$\frac{F_m \; t_1 \ldots t_n t}{F_k 0 t_1 \ldots t_n t} \qquad \frac{F_k t_o t_1 \ldots t_n t \quad F_n t t_o t_1 \ldots t_n t'}{F_k (St_o) t_1 \ldots t_n t'} \quad .$$

(Thus to each n-ary primitive recursive function φ there corresponds an F such that, intuitively, $\varphi(x_1, \ldots, x_n) = y \leftrightarrow Fx_1 \ldots x_n y$.)

Finally, we add a rule of induction * in the form

$$\text{IND}' \qquad \frac{A0 \qquad \begin{matrix} \not{A}a \\ \vdots \\ A(Sa) \end{matrix}}{At}$$

where a is a parameter not occurring in assumptions on which A(Sa) depends except of the form Aa.

a is called the <u>proper parameter</u> of the IND-application.

A proper parameter of a deduction may now be a proper parameter of an application of $\forall$I, $\exists$E, IND.

We shall also agree to call the premisses of an application of IND <u>minor</u> premisses; the premiss of the form A0 is called the <u>zero premiss</u>, the premiss of the form A(Sa) the <u>inductive premiss</u>; the premisses of the basic rules are regarded as <u>major premisses</u>.

The conditions on variables given in 1.1.7 may now be sharpened by reformulating (iii) as

(iii) Every proper parameter in Π is a proper parameter of exactly one application of $\forall$I, $\exists$E or IND in Π, and adding

---

*) In the sequel, IND will be used indiscriminately to refer to the formulation in 1.3.3 and the one given here.

(iv) The proper parameter of an application of IND in Π occurs in Π only in formula occurrences above the inductive premiss of IND.

The present formulation of HA (say Nat - HA) is equivalent to the one described in 1.3.2 - 1.3.4 (to be called simply HA below), in the following sense: obviously, Nat - HA is equivalent to Nat - HA* obtained by replacing the basic rules by corresponding implications, and IND' by the schema of 1.3.3. Addition of symbols for the primitive recursive functions with their axioms is then an expansion H of Nat - HA*; also, H is an expansion of HA (cf. 1.2.7, 1.2.8). So there exist mappings φ, φ' such that

$$(1)\quad \begin{cases} \mathrm{HA} \vdash A \Leftrightarrow \mathrm{Nat\text{-}HA} \vdash \varphi A \\ \mathrm{Nat\text{-}HA} \vdash A \Leftrightarrow \mathrm{HA} \vdash \varphi' A \\ \mathrm{H} \vdash (\varphi A \leftrightarrow A) \,\&\, (\varphi' A \leftrightarrow A). \end{cases}$$

1.3.7. Eliminability of disjunction in systems containing arithmetic.

In intuitionistic arithmetic, we have

$$A \vee B \leftrightarrow \exists x[(x=0 \rightarrow A) \,\&\, (x \neq 0 \rightarrow B)].$$

In order to show that this may be taken as a definition for $\vee$, we have to show that the axioms and rules for $\vee$ are derivable for this defined connective from the rules and axioms for the other logical operators.
In order to see this for the natural deductive formulation, let us first introduce the following notational convention: $\overset{[A]}{\Sigma}$ stands for a (finite sequence of) deductions, where $[A]$ indicates a set of open assumptions in Σ of the form A. If Π is a deduction with conclusion A,

$$\begin{matrix}\Pi \\ [A] \\ \Sigma\end{matrix}$$

denotes the result of substituting Π for the occurrences of the class $[A]$ in Σ.

Now $\vee I$ can be shown to be a derived rule as follows:

$$\dfrac{\dfrac{\dfrac{A}{0=0 \rightarrow A}\rightarrow I \qquad \dfrac{\dfrac{\dfrac{\overset{(1)}{0\neq 0} \quad 0=0}{\Lambda}\Lambda_I}{B}}{0\neq 0 \rightarrow B}(1)}{(0=0\rightarrow A) \,\&\, (0\neq 0 \rightarrow B)}}{\exists x[(x=0\rightarrow A) \,\&\, (x\neq 0 \rightarrow B)]} \qquad \dfrac{\dfrac{\dfrac{\dfrac{\dfrac{S0\neq 0 \quad S0=0}{\Lambda}}{A}}{S0=0 \rightarrow A} \qquad \dfrac{B}{S0\neq 0 \rightarrow B}}{(S0=0 \rightarrow A) \,\&\, (S0\neq 0 \rightarrow B)}}{\exists x[(x=0 \rightarrow A) \,\&\, (x\neq 0 \rightarrow B)]}\;.$$

$\vee E$ is obtained as follows:

Assume $\begin{array}{c}[A]\\ \Sigma \\ \hline C\end{array}$, $\begin{array}{c}[B]\\ \Sigma' \\ \hline C\end{array}$ to be given.

$$\dfrac{\exists x[(x=0\to A)\ \&\ (x\neq 0\to B)] \qquad \dfrac{b=b \qquad \dfrac{\dfrac{\begin{array}{c}\dfrac{b=0 \quad \dfrac{(b=0\to A)\&(b\neq 0\to B)}{b=0\to A}}{[A]}\\ \Sigma \\ \hline C\end{array}}{b=0\to C} \qquad \dfrac{\begin{array}{c}\dfrac{\dfrac{\dfrac{\dfrac{\dfrac{b=0}{0=b}\quad b=Sa}{0=Sa}}{\curlywedge}}{b\neq 0}\quad \dfrac{(b=0\to A)\&(b\neq 0\to B)}{b\neq 0\to B}}{[B]}\\ \Sigma' \\ \hline C\end{array}}{b=Sa\to C}}{b=b\to C}}{C}}{C}$$

Further we note that we obtain corresponding results for the Spector system and the Gödel system since the proof of equivalence between these systems also applies if we restrict ourselves to the fragments not involving $\vee$.

<u>In practice</u>, <u>however</u>, we shall usually treat $\vee$ as a primitive, since this requires only very little additional effort in our proofs, and moreover many developments then apply to predicate logic also, almost without change or additions.

1.3.8. <u>Formulation of HA without any function symbols</u>.

It is of course possible to carry the step of eliminating function symbols in favour of predicate constants one step further than has been done in the natural deduction formulation of HA , and to replace the successor function by a binary predicate $S(x,y)$ such that

$$x=y \ \ \& \ S(x,z) \to S(y,z)$$
$$S(x,z)\ \&\ S(y,z) \to x=y$$
$$S(x,y) \to 0 \neq y$$

and the inductive clause for a predicate constant F representing a function introduced by the recursion schema now appears as

$$F_k x_0 \ldots x_n x \ \&\ F_n x x_0 \ldots x_n y \ \&\ S(x_0,z) \to F_k z x_0 \ldots x_n y$$

(where F_n has been introduced before F_k).
Induction appears as

$$A(0)\ \&\ \forall xy(Ax\ \&\ Sxy \to Ay) \ \to\ \forall x Ax\ .$$

This formulation is equivalent to our original formulations (again by using § 1.2) in the same manner as indicated for Nat - HA in 1.3.6 (formulas (1)). A formulation of this type is used in chapter V.

1.3.9. Notational conventions. We list a number of notational conventions and abbreviations to be used in the sequel.

A) For frequently used primitive recursive functions and predicates we adopt notations in common use, such as "prd" for the predecessor function, satisfying

$$\text{prd}\ 0 = 0\,, \quad \text{prd}(Sx) = x\,;$$

cut-off subtraction is denoted by $\dot{-}$,

$$x \dot{-} 0 = x\,, \quad x \dot{-} Sy = \text{prd}(x \dot{-} y)\,;$$

signature: "sg", satisfies

$$\text{sg}\ 0 = 0\,, \quad \text{sg}(Sx) = S0\,;$$

absolute difference " $|\cdot - \cdot|$ ":

$$|x - y| = (x \dot{-} y) + (y \dot{-} x)\,;$$

maximum and minimum "max", "min":

$$\max(x,y) = x + (y \dot{-} x)$$
$$\min(x,y) = \max(x,y) \dot{-} |x - y|\,.$$

For Kleene's T-predicate and the result-extracting function we use T, U respectively.

B) Pairing.

j, j_1, j_2 are assumed to be a pairing function from $N \times N$ onto N, with its inverses:

(1) $$j_1 j(x,y) = x\,, \quad j_2 j(x,y) = y\,, \quad j(j_1 z, j_2 z) = z\,.$$

The use of a pairing function onto the natural numbers is not essential; e.g. we might have used Kleene's $2^x 3^y$ to encode the pair (x,y). However, it is often convenient to assume the pairing to be onto N. In nearly all cases, the only properties which matter are given by (1), together with the information that j, j_1, j_2 are primitive recursive. For definiteness, we may fix on some definite pairing function, e.g. by requiring

(2) $$2j(x,y) = (x+y)(x+y+1) + 2x\,.$$

The pairing function represented by (2) has the additional advantages that

(3) $$x < j(x,y)\,, \quad y < j(x,y) \quad \text{if } x+y > 0\,; \quad j(0,0) = 0$$

(4) $$x < x' \rightarrow j(x,y) < j(x',y)\,, \quad y < y' \rightarrow j(x,y) < j(x,y')$$

which we shall assume from now on.

C) Coding of finite sequences. For a coding of finite sequences of natural numbers we also prefer to have a coding onto N (as used in Kreisel and Troelstra 1970, 2.5.3, in order not to have to specify that a certain variable is to range over code numbers of sequences only (as in Kleene and Vesley 1965). Of course, an elegant solution would be to introduce a separate sort of variables running over finite sequences (obviously a conservative extension, since they can be coded by natural numbers ; cf. 1.2.9) ; but this is rather a heavy draw on our typographical resources, which we wish to avoid.
For the sake of definiteness, we may assume our coding to be constructed from the standard pairing function as follows : we first introduce codings ν_u of u - tuples

$$\nu_1(x) = x$$
$$\nu_2(x_1,x_2) = j(x_1,x_2)$$
$$\nu_{u+1}(x_0,x_1,\ldots,x_u) = j(x_0,\nu_u(x_1,\ldots,x_u)) ;$$

there exist inverses j_i^u such that

$$j_i^u\nu_u(x_1,\ldots,x_u) = x_i , \quad \nu_u(j_1^u z,\ldots,j_u^u z) = z ;$$

now we fix our coding of finite sequences by

$$\langle\ \rangle = 0 , \quad \langle x_0\rangle \equiv_{def} Sj(0,x_0) ,$$
$$\langle x_0,\ldots,x_u\rangle = Sj(u,\nu_{u+1}(x_0,\ldots,x_u)) ,$$

where $\langle x_0,\ldots,x_u\rangle$ denotes the code number for $x_0,\ldots,x_u$, $\langle\ \rangle$ the code number for the empty sequence. The present choice of coding implies :

$$x_u < \langle x_0,\ldots,x_u,\ldots\rangle ,$$
$$\langle x_0,\ldots,x_u\rangle < \langle x_0,\ldots,x_u,\ldots,x_{u+v}\rangle \quad \text{for } v > 0 .$$

As an abbreviation we introduce

$$\hat{x} \equiv_{def} \langle x\rangle .$$

$lth(n)$ is used to denote the length of the sequence coded by n, so

$$lth\langle\ \rangle = 0 , \quad lth\langle x_0,\ldots,x_{u-1}\rangle = u .$$

$*$ denotes the concatenation, so

$$\langle x_0,\ldots,x_{u-1}\rangle * \langle x_u,\ldots,x_{u+v}\rangle = \langle x_0,\ldots,x_{u+v}\rangle .$$

We put

$$n \preceq m \equiv_{def} \exists n'(n * n' = m)$$
$$n \prec m \equiv_{def} n \preceq m \ \&\ n \neq m .$$

($\succ$, $\succeq$ are used in different meanings in this volume : 1°) as the natural partial ordering between finite sequences, as just defined ; 2°) as a symbol for an arbitrary primitive recursive well-ordering, in the discussion of the principle of transfinite induction $TI(<)$; 3°) as a metamathematical symbol for "reduces to". In all cases the meaning will be clear from the context.) Let us write $(n)_x$ for a primitive recursive function of n, x such that, for $n = \langle x_0, \ldots, x_{u-1} \rangle$

$$(n)_i = x_i \quad \text{for} \quad i < u$$
$$(n)_i = 0 \quad \text{for} \quad i \geq u.$$

$tl(n)$ ("tail of n") is a primitive recursive function such that

$$tl(0) = 0, \quad tl(\hat{x}) = 0, \quad tl(\hat{x} * n) = n.$$

The derivation of elementary properties of the codings (a tedious affair, which the reader might wish to skip) can be found in Kreisel and Troelstra 1970, § 2.

D) Proof-predicates, gödelnumbers, gödel- and rossersentences, numerals.
We shall frequently have to use formalized proof-predicates and formalized provability. We use

$$\mathrm{Proof}_{\underset{\sim}{H}}(x,y), \quad \text{or} \quad \mathrm{Proof}_H(x,y)$$

for any "canonical" proof-predicate for the formal system $\underset{\sim}{H}$, with intuitive interpretation: x is the gödelnumber of a proof of a formula with gödelnumber y. "Proof_H" may be assumed to be primitive recursive.

"Canonical" will mean that it satisfies some natural derivability conditions, so as to make it possible to prove Gödel's second incompleteness theorem for them. (For derivability conditions, see Hilbert and Bernays 1970, Vol. II, pp. 294 - 295, where they are described in detail.)
We put

$$\mathrm{Pr}_H(y) \equiv_{def} \exists x\, \mathrm{Proof}_H(x,y).$$

"Pr_H" is the "provability" - predicate, and may be assumed to be of Σ^0_1-form. (In our versions of $\underset{\sim}{HA}$, "Proof(x,y)" may be represented by a prime formula.)

A gödelsentence for a system $\underset{\sim}{H}$ (containing, say $\underset{\sim}{HA}$) is a Π^0_1 - sentence (i.e. of the form $\forall x Ax$, Ax primitive recursive) such that on assumption of consistency of $\underset{\sim}{H}$, $\not\vdash \forall x Ax$, and on assumption of ω - consistency of $\underset{\sim}{H}$, $\not\vdash \neg \forall x Ax$.

A rosser-sentence is described like a gödelsentence, but now consistency of $\underset{\sim}{H}$ is sufficient for $\not\vdash \neg \forall x Ax$. Gödel- or rosser-sentences are sometimes called sentences independent w.r.t. $\underset{\sim}{H}$.

In our standard formulation of $\underset{\sim}{HA}$ (the first one described in this section), Ax may be supposed to be a prime formula.

Numerals. As syntactical variables for numerals we use $\bar{x}, \bar{y}, \bar{z}, \bar{u}, \bar{v}, \bar{w}$ and especially $\bar{n}, \bar{m}$ (in chapter V there is a deviating local convention concerning numerals).

If A is a formula, $\ulcorner A \urcorner$ denotes its gödelnumber ; if t is a term, then t denotes its gödelnumber. If all the free variables of $A(x_1,\dots,x_n)$ are among $x_1,\dots,x_n$, we shall use the convention (unless indicated otherwise) that $A(\bar{x}_1,\dots,\bar{x}_n)$ stands for the gödelnumber of $A(\bar{x}_1,\dots,\bar{x}_n)$ as a function of $x_1,\dots,x_n$. (More precisely, if $s(\ulcorner A(x_1,\dots,x_n)\urcorner, y_1,\dots,y_n)$ is the gödelnumber of the formula obtained by substitution of the numerals $\bar{y}_1,\dots,\bar{y}_n$ for $x_1,\dots,x_n$ in A, then $s(\ulcorner A(x_1,\dots,x_n)\urcorner, y_1,\dots,y_n) = \ulcorner A(\bar{x}_1,\dots,\bar{x}_n)\urcorner$). The notation may cause problems in more complicated contexts, but suffices for our purposes. Note that in view of the preceding conventions, $\ulcorner A(\bar{x}_1,\dots,\bar{x}_n)\urcorner$, $\mathrm{Proof}_{HA}(y, \ulcorner A(\bar{x}_1,\dots,\bar{x}_n)\urcorner)$, etc. are in HA represented by formulae containing $x_1,\dots,x_n$ free.

1.3.10. Formalization of elementary recursion theory.

For some of our researches, notably in the case of formalized realizability (chapter III, § 2) it is necessary to know that the principal theorems such as the s-m-n-theorem and the recursion theorem can be formalized in HA. The reader may take this on faith, or better rely on the detailed formalization of recursion theory in Part I of Kleene 1969 ; by omitting there everything pertaining to function arguments (Kleene 1969 discusses formalized recursive functionals) one obtains a formalization of elementary recursion theory in HA. (The use of different pairing functions and encodings of finite sequences is completely irrelevant in this context ; cf. remarks in Kreisel and Troelstra 1970, 2.4.15 , 2.5.3.)

Below we shall list the principal facts needed.

About the T-predicate we may assume

$$\mathrm{HA} \vdash T(x, y, z) \,\&\, T(x, y, z') \rightarrow z = z'.$$

In our first version of HA, where symbols for the primitive recursive functions are present, we may suppose Txyz to be represented by a prime formula $C_T xyz = 0$.

We shall freely use Kleene-brackets $\{.\}$ as a notation for partial recursive functions and partially defined terms, and also the equality $\simeq$ between partially defined terms ; in Kleene 1969, Part I it is shown in great detail how these notations can be used as systematic abbreviations.

Let us, following Kleene, denote as p-terms the class of expressions satisfying the formation rules for terms and in addition :

If $t_0, t_1, \dots, t_n$ are p-terms then so is $\{t_0\}^n(t_1,\dots,t_n)$.

(Actually, we have no need for Kleene-brackets with more than one argument, but it is no trouble including them.) Instead of $\{t_0\}^1(t_1)$ we usually write simply $\{t_o\}(t_1)$.

Each p-term represents a partial recursive function of its free variables. We use furthermore

$$!t \equiv_{def} \exists x(t \simeq x) , \quad t = s \equiv_{def} !t \ \& \ !s \ \& \ t \simeq s .$$

Note that $!t$ and $t = s$ can be expressed as Σ_1^0-formulae.

The s-m-n-theorem may now be stated as follows: There exists a primitive recursive function s_n^m such that

$$\{z\}^{m+n}(x_1,\ldots,x_{m+n}) = \{s_n^m(z,x_1,\ldots,x_m)\}^n(x_{m+1},\ldots,x_{m+n}) .$$

This makes it easy to prove the recursion theorem:

$$\forall x \exists y \forall z_1 \ldots z_n(\{x\}^{n+1}(y,z_1,\ldots,z_n) \simeq \{y\}^n(z_1,\ldots,z_n)) .$$

(Consider $\{x\}^{n+1}(s_n^1(u,u),z_1,\ldots,z_n)$; we can find a v such that $\{v\}^{n+1}(u,z_1,\ldots,z_n) \simeq \{x\}^{n+1}(s(u,u),z_1,\ldots,z_n)$; now take $y = s(v,v)$, then $\{y\}^n(z_1,\ldots,z_n) \simeq \{s(v,v)\}^n(z_1,\ldots,z_n) \simeq \{v\}^{n+1}(v,z_1,\ldots,z_n)$ $\simeq \{x\}^{n+1}(s(v,v),z_1,\ldots,z_n) \simeq \{x\}^{n+1}(y,z_1,\ldots,z_n)$.)

We shall make frequent use of the recursion theorem.

A convenient abbreviation is

$\{t\}(t_1,\ldots,t_n) \equiv_{def} \{\ldots\{\{t\}(t_1)\}(t_2)\ldots\}(t_n)$.

(Actually, $\{x\}^n(y_1,\ldots,y_n)$ can be defined in such a way that $\{x\}^n(y_1 \ldots y_n) \simeq \{x\}(y_1 \ldots y_n)$.)

§ 4. Inductive definitions in HA

1.4.1. We intend to show in this section how certain inductive definitions of sets of natural numbers may be replaced by explicit definitions. The result is used repeatedly, especially in § 4.4. The reading of this section may be postponed until needed.

1.4.2. Definition. Let $\mathcal{L}[X] \equiv \mathcal{L}(HA)[X]$ denote the language of HA extended by a single additional unary predicate symbol X.
Γ is the least class of formulae of $\mathcal{L}[X]$ such that
(i) the formulae of HA are contained in Γ;
(ii) formulae Xt', t' a term of HA, are in Γ;
(iii) if $A, B \in \Gamma$ then $A \& B$, $A \vee B \in \Gamma$
(iv) if $A \in \Gamma$, then $\exists x A \in \Gamma$, $\forall x \leq t A \in \Gamma$ (t not containing x free, t a term of HA).

Our next aim is to show that formulae $A(X, x) \in \mathcal{L}[X]$ can be shown to be equivalent to a certain type of standard formula of $\mathcal{L}[X]$ (see the statement of 1.4.4 below); we first need a simple lemma:

1.4.3. Lemma. $\forall x_0 \leq t_0[y]\, \forall x_1 \leq t_1[x_0,y]\, A(x_0, x_1, y)$ is equivalent to $\forall x \leq t[y]\, A(\varphi_0(x,y), \varphi_1(x,y), y)$, for suitable t, φ_0, φ_1, primitive recursive in x, y, t_0, t_1.

Proof. For our standard pairing function j onto the natural numbers, and with its inverses j_1, j_2 we shall assume $x < x' \rightarrow j(x,y) < j(x',y)$, $y < y' \rightarrow j(x,y) < j(x,y')$.

We define

$$\psi(y) = \sup\{t_1[x_0,y] \mid x_0 \leq t_0[y]\}$$
$$t(y) = j(t_0[y], \psi(y))$$
$$\varphi_0(x,y) = \begin{cases} j_1x & \text{if } j_2x \leq t_1[j_1x,y] \;\&\; j_1x \leq t_0[y] \\ 0 & \text{otherwise}, \end{cases}$$
$$\varphi_1(x,y) = \begin{cases} j_2x & \text{if } j_2x \leq t_1[j_1x,y] \;\&\; j_1x \leq t_0[y] \\ 0 & \text{otherwise}. \end{cases}$$

Now assume

$$(1) \qquad \forall x_0 \leq t_0[y]\forall x_1 \leq t_1[x_0,y]\; A(x_0, x_1, y)$$

and let $x \leq t[y]$.
Now either $j_1x \leq t_0[y]$, $j_2x \leq t_1[j_1x, y]$, and then $\varphi_0(x, y) = j_1x$, $\varphi_1(x, y) = j_2x$, and by (1) $A(\varphi_0(x,y), \varphi_1(x,y), y)$, or $j_1x > t_0[y] \vee j_2x > t_1[j_1x, y]$; in this case, $\varphi_0(x,y) = \varphi_1(x,y) = 0$, and since by (1) $A(0, 0, y)$, once again $A(\varphi_0(x,y), \varphi_1(x,y), y)$.

Conversely, let

(2) $\forall x \le t[y]\ A(\varphi_0(x,y),\ \varphi_1(x,y),\ y)$

and assume $x_0 \le t_0[y]$, $x_1 \le t_1[x_0,y]$. If we put $j(x_0,x_1) = x$, then $x \le t[y]$; $\varphi_0(x,y) = x_0$, $\varphi_1(x,y) = x_1$, so by (2) $A(x_0, x_1, y)$.

1.4.4. <u>Lemma</u>. Each formula $A(X,z)$ of Γ is provably equivalent in $\underset{\sim}{HA}[X]$ (i.e. $\underset{\sim}{HA}$ extended to the language $\mathcal{L}[X]$) to a formula of the following general form

(1) $\exists x \forall y \le t(x,z)\ [P(x, y, z) \vee (Q(x, y, z)\ \&\ X\,t'[x, y, z])]$.

<u>Proof</u>. To prove the lemma we have to show that formulae equivalent to a formula of type (1) satisfy the closure conditions (i) - (iv) in the definition of Γ.

Below we shall omit all variables which are redundant in the context of the argument.

(i) Let Pz be an arbitrary formula of $\underset{\sim}{HA}$. Then obviously

$$Pz \leftrightarrow \exists x \forall y \le 0\ (Pz \vee [0 = 1\ \&\ X\,0])$$

(x,y are assumed not to occur free in P).

(ii) $X\,t'$ is equivalent to

$$\exists x \forall y \le 0\ (0 = 1 \vee [0 = 0\ \&\ X\,t'])$$

(t' does not contain x,y).

(iv) We note the following equivalences

(2) $\exists xy \forall z \le t[x,y]\ A(x,y,z) \leftrightarrow \exists x \forall z \le t[j_1x,\ j_2x]\ A(j_1x,\ j_2x,\ z)$

(3) $\forall x \le t\ \exists y\ A(x,y) \leftrightarrow \exists n \forall x \le t\ A(x,\ (n)_x)$

(t not containing x).

The closure of Γ under existential quantification is immediate by (2). The closure under bounded universal quantification follows from (3) and the previous lemma :

$$\forall u \le t^* \exists y\ \forall x \le t[u,y]\ A(u, x, y) \leftrightarrow$$
$$\leftrightarrow \exists n\ \forall u \le t^*\ \forall x \le t[u, (n)_u]\ A(u, x, (n)_u) \leftrightarrow$$
$$\leftrightarrow \exists n\ \forall v \le t'[n]\ A(\psi_0(v,n),\ \psi_1(v,n),\ (n)_{\psi_0(v,n)})\ .$$

(iii) Let

$$A \equiv \exists x\ \forall y \le t_0(x)\ \ [P_0 \vee (Q_0\ \&\ Xt_0')],$$
$$B \equiv \exists x\ \forall y \le t_1(x)\ \ [P_1 \vee (Q_1\ \&\ Xt_1')].$$

After contraction of two existential quantifiers, we obtain

$$A\ \&\ B \equiv$$
$$\exists x\ \forall y_0 \le t_0[j_1x] \forall y_1 \le t_1[j_2x]\ \{[P_0(j_1x,y) \vee (Q_0(j_1x,y)\ \&\ Xt_0'[j_1x,y])]\ \&$$
$$\&\ [P_1(j_2x,y) \vee (Q_1(j_2x,y)\ \&\ Xt_1'[j_2x,y])]\}.$$

We put

$$P(x, y, u) \equiv (P_0(j_1x,y,u) \& u=0) \vee (P_1(j_2x,y,u) \& u\neq 0)$$
$$Q(x, y, u) \equiv (Q_0(j_1x,y,u) \& u=0) \vee (Q_1(j_2x,y,u) \& u\neq 0)$$
$$t[x, u] \equiv (1 \dot{-} u)t_0[x] + sg(u) . t_1[x]$$
$$t'[x,y,u] \equiv (1 \dot{-} u)t'_0[x,y] + sg(u).t'_1[x,y] .$$

Then

$$A \& B \equiv \exists x\, \forall u \leq 1\, \forall y \leq t[x,u]\, [P \vee (Q \& Xt')] .$$

By the previous lemma, this is equivalent to a formula of type (1).

$$A \vee B \longleftrightarrow \exists x\, \{\forall y \leq t_0[x]\, (P_0 \vee (Q_0 \& Xt'_0)) \vee \vee \forall y \leq t_1[x]\, (P_1 \vee (Q_1 \& Xt'_1))\} .$$

We put

$$P'(x,y,u) \equiv (P_0 \& u=0) \vee (P_1 \& u\neq 0)$$
$$Q'(x,y,u) \equiv (Q_0 \& u=0) \vee (Q_1 \& u\neq 0)$$
$$t'''[x,y,u] \equiv (1 \dot{-} u)t'_0 + sg(u) . t'_1$$
$$t''[x, u] \equiv (1 \dot{-} u)t_0 + sg(u) . t_1 .$$

Then $A \vee B \longleftrightarrow \exists x \exists u\, \forall y \leq t''[x,u]\, [P' \vee (Q' \& Xt''')]$; by (2) this is equivalent to a formula of the form (1).

1.4.5. Theorem. Let $A(X,z)$ be a formula of the class Γ; then there is an arithmetical predicate (i.e. definable in HA) $P_A z$ such that

(1) $$A(P_A, z) \rightarrow P_A z$$

and for each arithmetical predicate Q

(2) $$\forall z[A(Q,z) \rightarrow Qz] \rightarrow \forall z[P_A z \rightarrow Qz]$$

are derivable in HA.

Proof. By lemma 1.4.4, we may restrict our attention to a predicate $A(X,z)$ of the form

(3) $$\exists x \forall y \leq t[x,z][P(x,y,z) \vee (Q(x,y,z) \& Xt'[x,y,z])] .$$

A proof that z satisfies (3) may be supposed to be in tree form: the proof Π provides an x, and for each $y \leq t[x,z]$ Π contains a sub-proof Π_z which either establishes $P(x,y,z)$ (and then represents an end node of the tree T associated with Π) or establishes $Q(x,y,z)$ and $Xt'[x,y,z]$; in the latter case, with Π_z is associated a subtree T_z of T, which establishes

$$\exists x' \forall y' \leq t[x', t'[x,y,z]](P(x',y',t'[x,y,z]) \vee \vee (Q(x',y',t'[x,y,z]) \& Xt'[x',y',t'[x,y,z]])) ,$$

and so on.

It is this intuitive idea which suggests the explicit definition which will be given below.

Any natural number may be supposed to code a finite tree ; $\langle\ \rangle$ codes the empty tree, and if $n = \langle x_o,\dots,x_u\rangle$, then the tree T_n represented by n has the structure

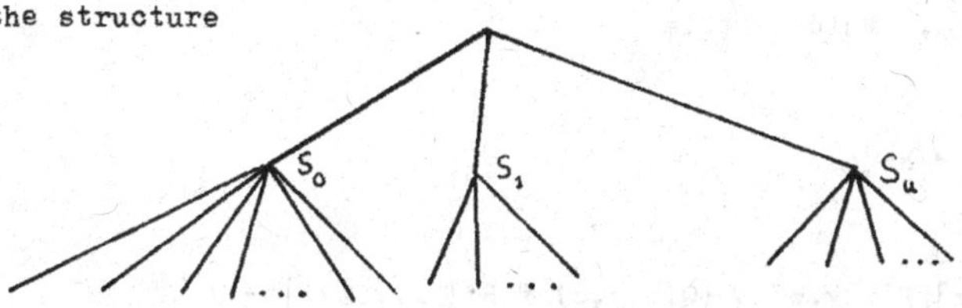

where $S_o,\dots,S_u$ are the trees coded by $x_o,\dots,x_u$, respectively. Below we adopt the following notations : we write n_u for $(n)_u$, and define $[n]_m$ inductively on $\mathrm{lth}(m)$ by

$$[n]_o = n,\ [n]_{\langle u\rangle} = (n)_u = n_u,\ [n]_{m*\langle u\rangle} = [[n]_m]_{\langle u\rangle} = ([n]_m)_u .$$

Now we put for $P_A z$:

$$P_A z \equiv \exists n\exists m\exists p[n\neq 0 \,\&\, m_o = z \,\&\, \forall u\forall y\leq t[p_u,m_u]\{([n]_{u*\langle y\rangle}\neq 0 \rightarrow$$
$$\rightarrow Q(p_u,y,m_u)\,\&\, t'[p_u,y,m_u] = m_{u*\langle y\rangle})\ \&$$
$$\&\ ([n]_{u*\langle y\rangle} = 0 \,\&\, [n]_u\neq 0 \rightarrow P(p_u,y,m_u))\}] .$$

Part I. We have to show $A(P_A,z)\rightarrow P_A z$.

$$\exists x\forall y\leq t[x,z][P(x,y,z)\ \vee\ (Q(x,y,z)\ \&\ \exists n\exists m\exists p\ B(n,m,p,t'[x,y,z]))] ,$$

where $\exists n\exists m\exists p\ B(n,m,p,z) \equiv P_A z$, implies that for a suitable x_o

$$\forall y\leq t[x_o,z][P(x_o,y,z)\ \vee\ (Q(x_o,y,z)\,\&\,\exists n\exists m\exists p\ B(n,m,p,t'[x_o,y,z]))] .$$

Hence we can find w, n, m, p such that

$$\forall y\leq t[x_o,z][(w_y\neq 0\rightarrow P(x_o,y,z))\ \&$$
$$\&\ (w_y = 0\rightarrow Q(x_o,y,z)\,\&\,B(n_y,m_y,p_y,t'[x_o,y,z]))] .$$

Now we define n', m', p' such that

$$n' = \langle \tilde{n}_o,\dots,\tilde{n}_{t(x_o,z)}\rangle$$

with $\tilde{n}_y = 0$ for $w_y\neq 0$, $\tilde{n}_y = n_y$ for $w_y = 0$.

$$m' \text{ satisfies } \begin{cases} m'_o = z \\ m'_{\langle y\rangle * u} = (m_y)_u \end{cases}$$

$$p' \text{ satisfies } \begin{cases} p'_o = x_o \\ p'_{\langle y\rangle * u} = (p_y)_u \end{cases} .$$

Then obviously

$$n' \neq 0 \;\&\; [m'_o = z \;\&\; \forall u \forall y \leq t(p'_o, z)\; \{([n']_{u*\langle y\rangle} \neq 0 \rightarrow$$
$$\rightarrow Q(p'_u, y, m'_u) \;\&\; t'[p'_u, y, m'_u] = m'_{u*\langle y\rangle}) \;\&$$
$$\&\; ([n']_u \neq 0 \;\&\; [n']_{u*\langle y\rangle} = 0 \rightarrow P(p'_u, y, m'_u))\}]$$

i.e. $B(n', m', p', z)$, which implies $P_A z$.

<u>Part</u> II. Assume

(4) $$\forall z[A(R,z) \rightarrow Rz],$$

i.e.

$$\exists x \forall y \leq t[x,z][P(x,y,z) \vee (Q(x,y,z) \;\&\; Rt'[x,y,z])] \rightarrow Rz.$$

We shall prove by induction on n

(5) $$\forall m \forall p \forall z[B(n,m,p,z) \rightarrow Rz].$$

Basis: For $n = 0$ (5) is trivially fulfilled, because the antecedent is then false.

Induction step: Now assume (5) to have been proved for all $n < n'$, hence for all x such that $[n']_{\langle x\rangle} \neq 0$.

We may rewrite $B(n', m, p, z)$ as

$$n' \neq 0 \;\&\; m_o = z \;\&$$
$$\&\; \forall w \forall u \forall y \leq t[p_{\langle w\rangle*u}, m_{\langle w\rangle*u}]\{([n']_{\langle w\rangle*u*\langle y\rangle} \neq 0 \rightarrow$$
$$\rightarrow Q(p_{\langle w\rangle*u}, y, m_{\langle w\rangle*u}) \;\&\; t'[p_{\langle w\rangle*u}, y, m_{\langle w\rangle*u}] = m_{\langle w\rangle*u*\langle y\rangle})$$
$$\&\; ([n']_{\langle w\rangle*u*\langle y\rangle} = 0 \;\&\; [n']_{\langle w\rangle*u} \neq 0 \rightarrow P(p_{\langle w\rangle*u}, y, m_{\langle w\rangle*u}))\} \;\&$$
$$\&\; \forall y \leq t[p_o, m_o]\{([n']_{\langle y\rangle} \neq 0 \rightarrow Q(p_o, y, m_o) \;\&\; t'[p_o, y, m_o] = m_{\langle y\rangle}) \;\&$$
$$\&\; ([n']_{\langle y\rangle} = 0 \;\&\; [n']_o \neq 0 \rightarrow P(p_o, y, m_o))\}.$$

It follows that

$$[n']_{\langle w\rangle} \neq 0 \rightarrow B([n']_{\langle w\rangle}, m, p, m_{\langle w\rangle})$$

hence

(6) $$[n']_{\langle w\rangle} \neq 0 \rightarrow R(m_{\langle w\rangle}) \qquad \text{(induction hypothesis)}$$

and

(7) $$\begin{cases} \forall y \leq t[p_o, z]\{([n']_{\langle y\rangle} \neq 0 \rightarrow Q(p_o, y, z) \;\&\; t'[p_o, y, z] = m_{\langle y\rangle}) \;\& \\ \&\; ([n']_{\langle y\rangle} = 0 \;\&\; [n']_o \neq 0 \rightarrow P(p_o, y, z))\}, \end{cases}$$

therefore, combining (6) and (7):

$$\forall y \leq t[p_o, z](P(p_o, y, z) \vee (Q(p_o, y, z) \;\&\; R\, t'[p_o, y, z])).$$

By assumption (4), Rz.

§ 5. Partial reflection principles.

1.5.1. Contents of the section. Let $\mathrm{Proof}_n(x,\ulcorner A\urcorner)$ indicate the proof-predicate of $\underset{\sim}{HA}$ (e.g. for Spector's system), restricted to derivations containing formulae of logical complexity $\leq n$ only. In other words, if the logical complexity of a proof is described as the maximum of the logical complexities of the formulae occurring in it, then $\mathrm{Proof}_n(x,\ulcorner A\urcorner)$ holds iff $\mathrm{Proof}_{HA}(x,\ulcorner A\urcorner)$ and the logical complexity of the deduction represented by x is $\leq n$.

The principal aim of this section is to establish in $\underset{\sim}{HA}$ reflection principles for the subsystems of $\underset{\sim}{HA}$ obtained by putting a bound on the complexity of the formulae considered, i.e.

(1) $\underset{\sim}{HA} \vdash \exists x\, \mathrm{Proof}_n(x,\ulcorner A\urcorner) \to A$.

The main step towards establishing (1) is the construction of a "valuation-function" which assigns to (the gödelnumber of) a closed term its intended value (and which can be shown to do so in $\underset{\sim}{HA}$).

1.5.2. Gödelnumbering of function constants and terms.

For definiteness in the formal description, we specify some details of the gödelnumbering.

We put $\tilde{\varphi}$ for the code number of the function constant φ, where

$\tilde{0} \equiv \langle 0\rangle$

$\tilde{S} \equiv \langle 1\rangle$

$\tilde{I}_n^i \equiv \langle 2, n, i\rangle$

$\tilde{\xi} \equiv \langle 3, \tilde{\psi}, \tilde{\varphi}_1, \ldots, \tilde{\varphi}_m\rangle$ if ξ is defined by composition from $\psi, \varphi_1, \ldots, \varphi_m$.

$\tilde{\xi} \equiv \langle 4, \tilde{\varphi}, \tilde{\psi}\rangle$ if ξ is defined by recursion from φ, ψ.

The gödelnumber of an arbitrary closed term is defined as follows. Each closed term is of the form $\xi t_1 \ldots t_n$ (n possibly 0), where ξ is a function constant of our language. We put $\ulcorner \xi t_1 \ldots t_n \urcorner = \langle \tilde{\xi}, \ulcorner t_1\urcorner, \ldots, \ulcorner t_n\urcorner \rangle$. (Note that 0 as a function constant has number $\langle 0\rangle$, as a term number $\langle\langle 0\rangle\rangle$.)

The sequence of gödelnumbers of numerals is primitive recursive; if νx is to denote the gödelnumber assigned to the numeral $\bar{x}$, ν is given by $\nu 0 = \langle\langle 0\rangle\rangle$, $\nu(Sx) = \langle\langle 1\rangle, \nu x\rangle$.

1.5.3. Evaluation of closed terms.

Any closed term t in $\underset{\sim}{HA}$ can be evaluated by a standard procedure, and has a standard deduction of $t = \bar{y}$ for a numeral $\bar{y}$ in $\underset{\sim}{HA}$. This procedure may be described as follows.

A <u>contractible</u> term is a term of the form $\varphi\bar{x}_1\ldots\bar{x}_n$, φ a constant introduced by composition or recursion, or a projection.

For closed terms t, we define the "<u>right most contractible subterm occurrence of</u> t" (abbreviated: $\mathrm{rcso}(t)$) inductively as follows:

1°) 0 does not have an $\mathrm{rcso}(t)$;

2°) $\mathrm{rcso}(St)$ is the occurrence in St corresponding to $\mathrm{rcso}(t)$, if this exists;

3°) if $t \equiv \varphi(t_1,\ldots,t_i,\bar{x}_{i+1},\ldots,\bar{x}_n)$, φ a function constant, t_i not a numeral, then $\mathrm{rcso}(t)$ is the occurrence corresponding to $\mathrm{rcso}(t_i)$;

4°) if $t \equiv \varphi(\bar{y}_1,\ldots,\bar{y}_n)$, φ a function constant, not S, 0, then $\mathrm{rcso}(t)$ is t itself.

A <u>contraction</u> is the replacement of $I^i_n\bar{x}_1\ldots\bar{x}_n$ by $\bar{x}_i$, or of a term of the form $\varphi(\bar{x}_1,\ldots,\bar{x}_n)$, if φ defined by composition from $\psi, \varphi_1, \ldots, \varphi_m$, by $\psi(\varphi_1(\bar{x}_1,\ldots,\bar{x}_n), \ldots, \varphi_m(\bar{x}_1,\ldots,\bar{x}_n))$, and if φ is defined by recursion from ψ, χ, by $\psi(\bar{x}_1,\ldots,\bar{x}_n)$ if $\bar{x}_1 \equiv \bar{0}$, and by $\chi(\varphi(\bar{y},\bar{x}_2,\ldots,\bar{x}_n), \bar{y}, \bar{x}_2,\ldots,\bar{x}_n)$ if $\bar{x}_1 \equiv S\bar{y}$.

A "<u>standard reduction sequence for</u> t" is a sequence $t_1, \ldots, t_n$, $t_1 \equiv t$, t_n a numeral, such that t_{i+1} results from t_i by a contraction applied to the $\mathrm{rcso}(t_i)$. We then call y, if $t_n \equiv \bar{y}$, a value for t. Since the construction of t_{i+1} out of t_i is uniquely determined, the value is obviously unique.

Let $\mathrm{SRED}(z, z')$ be the Σ^0_1-arithmetical predicate expressing: the term with gödelnumber z has a standard reduction sequence to the term with gödelnumber z'.
Let us write, for each n-ary function constant φ:

(1) $\quad \mathrm{Val}(\varphi) \equiv_{\mathrm{def}} \forall x_1\ldots x_n \mathrm{SRED}(\ulcorner\varphi\,\bar{x}_1\ldots\bar{x}_n\urcorner, \nu(\varphi(x_1,\ldots,x_n)))$.

For $n=0$, we may put $\mathrm{Val}(\varphi) \equiv_{\mathrm{def}} \mathrm{SRED}(\ulcorner\varphi\urcorner, \nu(\varphi))$.
Now we establish, for each function constant φ

(2) $\quad \mathbf{HA} \vdash \mathrm{Val}(\varphi)$.

a) For $\varphi \equiv 0$, this is immediate.

b) For $\varphi \equiv S$ also, from the definition of ν.

c) Suppose φ is defined by composition from $\psi, \varphi_1, \ldots, \varphi_m$, and assume $\mathbf{HA} \vdash \mathrm{Val}(\psi)$, $\mathbf{HA} \vdash \mathrm{Val}(\varphi_i)$ $(i = 1, \ldots, m)$.
Then we easily verify $\mathbf{HA} \vdash \mathrm{Val}(\varphi)$.

d) Suppose φ is defined by recursion from ψ and χ, and assume $\mathrm{Val}(\psi)$, $\mathrm{Val}(\chi)$ in $\mathbf{HA}$. We now establish (1) by induction on x_1.

In the verification of (a) - (d) we have to use repeatedly that a standard reduction sequence for $\varphi t_1\ldots t_n$ always "contains" standard reduction sequences for $t_1, \ldots, t_n$.

We have established (2), and now we readily prove, by induction on the logical complexity of t, for any term $t[x_1,\ldots,x_n]$ whose free variables are among $x_1,\ldots,x_n$:

(3) $\quad \underset{\sim}{HA} \vdash \mathrm{SRED}(\ulcorner t[\bar{x}_1,\ldots,\bar{x}_n]\urcorner, \nu t[x_1,\ldots,x_n])$.

1.5.4. Construction of partial truth definitions.

By induction on n we construct truth definitions T_n, such that in $\underset{\sim}{HA}$, for all formulae A of logical complexity $\leq n$,

(1) $\quad \underset{\sim}{HA} \vdash T_n(\ulcorner A(\bar{x}_1,\ldots,\bar{x}_m)\urcorner) \leftrightarrow A(x_1,\ldots,x_m)$

(the variables free in A are among $x_1,\ldots,x_m$).

For prime formulae we put

$$T_o(\ulcorner t(\bar{x}_1,\ldots,\bar{x}_n) = s(\bar{x}_1,\ldots,\bar{x}_n)\urcorner) \leftrightarrow \exists y[\mathrm{SRED}(\ulcorner t(\bar{x}_1,\ldots)\urcorner,\ulcorner\bar{y}\urcorner) \;\& \;\mathrm{SRED}(\ulcorner s(\bar{x}_1,\ldots)\urcorner,\ulcorner\bar{y}\urcorner)].$$

Assume T_n to have been defined. Then we define T_{n+1} such that for A, B, $(Qx)Cx$ closed (where Q stands for $\exists$ or $\forall$)

(2) $$\begin{cases} T_{n+1}(\ulcorner A \circ B\urcorner) \leftrightarrow T_n(\ulcorner A\urcorner) \circ T_n(\ulcorner B\urcorner) & (\text{for } \circ \equiv \rightarrow, \&, \vee) \\ T_{n+1}(\ulcorner (Qx)Cx\urcorner) \leftrightarrow (Qx)T_n(\ulcorner C\bar{x}\urcorner) & (\text{for } Q \equiv \exists, \forall) \end{cases}.$$

Such a definition is possible, since for the usual standard gödelnumberings it is primitive recursively decidable what the main logical operator of a formula is; so we may define T_{n+1} by cases in agreement with (2). Now (1) is readily proved by induction over n. For $n=0$, (1) is immediate by the result (3) of 1.5.3, and the induction step is given by (2).

1.5.5. Lemma.

(a) (For Spector's or Gödel's version of $\underset{\sim}{HA}$.)

$\underset{\sim}{HA} \vdash \mathrm{Proof}_n(y, \ulcorner A(x_1,\ldots,x_n)\urcorner) \rightarrow \forall x_1\ldots x_n\; T_n(\ulcorner A(\bar{x}_1,\ldots,\bar{x}_n)\urcorner)$.

(b) (For the natural deduction version of $\underset{\sim}{HA}$.) Let us write $\Gamma(a_1,\ldots,a_n) \Rightarrow A(a_1,\ldots,a_n)$, where $\Gamma = \{C_1(a_1,\ldots,a_n),\ldots, C_p(a_1,\ldots,a_n)\}$, for: $A(a_1,\ldots,a_n)$ can be deduced from assumptions Γ; let $a_1,\ldots,a_n$ be a list containing all parameters free in Γ, A. Then

$\underset{\sim}{HA} \vdash \mathrm{Proof}_n(x, \ulcorner\Gamma(a_1,\ldots,a_n) \Rightarrow A(a_1,\ldots,a_n)\urcorner) \rightarrow$
$\rightarrow \forall x_1\ldots x_n(T_n(\ulcorner C_1(\bar{x}_1,\ldots,\bar{x}_n)\urcorner) \;\&\; \ldots \;\&\; T_n(\ulcorner C_p(\bar{x}_1,\ldots,\bar{x}_n)\urcorner) \rightarrow$
$\rightarrow T_n(\ulcorner A(\bar{x}_1,\ldots,\bar{x}_n)\urcorner)$.

Proof. We consider case (a); the treatment of case (b) is entirely similar. Let us write $\varphi(z,x)$ for the function which is such that

$$\varphi(\ulcorner A(v_{i_o},\ldots,v_{i_m})\urcorner, x) = \ulcorner A(\overline{(x)}_{i_o},\ldots,\overline{(x)}_{i_m})\urcorner.$$

Then we prove by induction on $\mathrm{lth}\,x$ that

$$\mathrm{Proof}_n(x,z) \to \forall y T_n(\varphi(z,y)) .$$

To give an example of the argument for the induction step, assume

$$\forall x \forall v < u(\mathrm{lth}(x) = v \,\&\, \mathrm{Proof}_n(x,z) \to \forall y\, T_n(\varphi(z,y))) .$$

Now assume

$$\mathrm{lth}(x) = u \,\&\, \mathrm{Proof}_n(x,z) .$$

We have to distinguish various cases, depending on the last rule applied in the proof with number x. For example, assume x to be the number of a proof obtained by application of Q 1 to some subproof x_1 of x of an assertion of the form

$$A(v_{i_0},\ldots,v_{i_m}) \equiv B(v_{i_1},\ldots,v_{i_m}) \to C(v_{i_0},v_{i_1},\ldots,v_{i_m}) .$$

Note that x_1, $\ulcorner A\urcorner$, $\ulcorner B\urcorner$, $\ulcorner C\urcorner$ and the numbers $i_0,\ldots,i_m$ can be found recursively (in fact, for the usual gödelnumberings, primitive recursively) from x. So by induction hypothesis

$$\forall y\, T_n(\varphi(\ulcorner A\urcorner, y))$$

in $\underset{\sim}{\mathrm{HA}}$, which implies in turn (in $\underset{\sim}{\mathrm{HA}}$):

$$\forall y\{T_n(\ulcorner B(\overline{(y)}_{i_1}, \ldots, \overline{(y)}_{i_m})\urcorner) \to T_n(\ulcorner C(\overline{(y)}_{i_0}, \overline{(y)}_{i_1}, \ldots, \overline{(y)}_{i_m})\urcorner)\},$$

i.e.

$$\forall y \forall x_0\{T_n(\ulcorner B(\overline{(y)}_{i_1}, \ldots, \overline{(y)}_{i_m})\urcorner) \to T_n(\ulcorner C(\bar{x}_0, \overline{(y)}_{i_1}, \ldots, \overline{(y)}_{i_m})\urcorner)\}$$

hence

$$\forall y\{T_n(\ulcorner B(\overline{(y)}_{i_1},\ldots,\overline{(y)}_{i_m})\urcorner) \to \forall x_0\, T_n(\ulcorner C(\bar{x}_0, \overline{(y)}_{i_1}, \ldots, \overline{(y)}_{i_n})\urcorner)\},$$

hence

$$\forall y\{T_n(\varphi(\ulcorner B\urcorner,y)) \to T_n(\varphi(\ulcorner \forall v_{i_0} C\urcorner,y))\},$$

i.e. by the properties of T_n

$$\forall y\, T_n(\varphi(\ulcorner B \to \forall v_{i_0} C\urcorner,y))$$

as desired.

1.5.6. <u>Theorem</u> (Partial reflection principles).

(a) For Gödel's or Spector's formulation of $\underset{\sim}{\mathrm{HA}}$:
$\underset{\sim}{\mathrm{HA}} \vdash \mathrm{Proof}_n(x,\ulcorner A(\bar{x}_1,\ldots,\bar{x}_n)\urcorner) \to A(x_1,\ldots,x_n)$.

(b) For the natural deduction formulation of $\underset{\sim}{\mathrm{HA}}$:
$\mathrm{HA} \vdash \mathrm{Proof}_n(x, \ulcorner \Rightarrow A(\bar{x}_1,\ldots,\bar{x}_n)\urcorner) \to A(x_1,\ldots,x_n)$.

<u>Proof</u>. We consider (a); (b) is treated similarly.
Assume $\mathrm{Proof}_n(x,\ulcorner A(\bar{x}_1,\ldots,\bar{x}_n)\urcorner)$, then by 1.5.5

$$T_n(\ulcorner A(\bar{x}_1,\dots,\bar{x}_n)\urcorner)$$

and therefore by 1.5.4

$$A(x_1,\dots,x_n) .$$

Q. e. d.

1.5.7. Remark on refinements. We may introduce a more refined measure of complexity, e.g. by contracting conjunctions, disjunctions and successive occurrences of the same type of quantifiers. I.e. we define a degree d by

$$d(\ulcorner A\urcorner) = 0 \quad \text{for prime formulae}$$
$$d(\ulcorner A \to B\urcorner) = \max(d(\ulcorner A\urcorner), d(\ulcorner B\urcorner)) + 1$$
$$d(\ulcorner A_1 \circ \dots \circ A_n\urcorner) = \max(d(\ulcorner A_1\urcorner), \dots, d(\ulcorner A_n\urcorner)) + 1$$

where $A_1 \circ \dots \circ A_n$ stands for any formula obtained from $A_1,\dots,A_n$ by insertion of brackets and the binary operator $\circ$ (either $\circ$ is everywhere &, or everywhere $\vee$), $A_1,\dots,A_n$ not being of the form $B \circ B'$.

$$d(\ulcorner (Qx_1) \dots (Qx_n)A(x_1,\dots,x_n)\urcorner) = d(\ulcorner A(x_1,\dots,x_n)\urcorner) + 1 ,$$

where A is not of the form $(Qy)B$, and either all Q are $\forall$ or all Q are $\exists$. It is easy to adapt the definition of T_n to this new measure.

1.5.8. Remark on quantifier-free systems.

In the various quantifier-free systems discussed here and in the sequel, (namely qf - HA, being described as arithmetic restricted to quantifier-free formulae, with a rule of induction, and the systems qf - N - HA^ω, qf - WE - HA^ω, qf - I - HA^ω, qf - HA^ω described in 1.6.13 - 1.6.15) there are two possible variants in the formulation:

(i) We state axioms such as e.g. $x = x$ with free variables, and have a rule of substitution of terms for free variables; the induction rule may then be stated as $A0$, $A(x) \to A(sx) \Rightarrow A(y)$, or

(ii) we state the axioms for arbitrary terms (as schemata), e.g. $t = t$, and formulate induction as $A(0)$, $A(x) \to A(Sx) \Rightarrow A(t)$; then we may omit the substitution rule.

1.5.9. Theorem. Let qf - HA be HA restricted to quantifier-free formulae, with Rule - IND instead of induction. Then

$$HA \vdash \mathrm{Proof}_{qf\text{-}HA}(x, \ulcorner A(\bar{x}_1,\dots,\bar{x}_n)\urcorner) \to A(x_1,\dots,x_n) .$$

Proof. We may reformulate qf - HA as an equational calculus (i.e. each quantifier-free formula corresponds to an equation $t = s$, cf. 1.6.14), or what amounts to the same, we can adapt the definition of T_o. We then prove as in 1.5.6 the present theorem.

1.5.10. Corollary to 1.5.3. Let $\underset{\sim}{H}$ be qf-$\underset{\sim}{HA}$ with Rule-IND left out.

(i) $\underset{\sim}{HA} \vdash t[x_1,\ldots,x_n] = y \rightarrow Pr_H(\ulcorner t[\bar{x}_1,\ldots,\bar{x}_n] = \bar{y}\urcorner)$.

(ii) There exists a (primitive) recursive φ_t such that

$\underset{\sim}{HA} \vdash t[x_1,\ldots,x_n] = y \rightarrow Proof_H(\varphi_t(x_1,\ldots,x_n,y), \ulcorner t[\bar{x}_1,\ldots,\bar{x}_n] = \bar{y}\urcorner)$.

Proof. (i) is immediate from (3) making use of

$$\underset{\sim}{HA} \vdash SRED(\ulcorner t\urcorner, \ulcorner t'\urcorner) \rightarrow Pr_H(\ulcorner t=t'\urcorner) .$$

(ii) A more detailed inspection of the proof of (2), (3) in 1.5.3 yields (ii). One first establishes

(4) $\underset{\sim}{HA} \vdash \xi(x_1,\ldots,x_n) = y \rightarrow Proof_{qf-HA}(\lambda_\xi(x_1,\ldots,x_n,y), \ulcorner \xi(\bar{x}_1,\ldots,\bar{x}_n) = \bar{y}\urcorner)$

for the constants ξ of $\underset{\sim}{HA}$, suitable recursive λ_ξ, and then establishes (ii) by induction on the complexity of t.

For example, in the proof of (4), we have to consider the case that ξ is defined by recursion from χ, ψ, and assume as induction hypotheses

$$\underset{\sim}{HA} \vdash \chi(x_1,\ldots,x_n) = y \rightarrow Proof_H(\lambda_\chi(x_1,\ldots,x_n,y), \ulcorner \chi(\bar{x}_1,\ldots,\bar{x}_n) = \bar{y}\urcorner)$$

$$\underset{\sim}{HA} \vdash \psi(z,x_0,x_1,\ldots,x_n) = y \rightarrow Proof_H(\lambda_\psi(z,x_0,\ldots,x_n,y), \ulcorner \psi(\bar{z},\bar{x}_0,\ldots,\bar{x}_n) = \bar{y}\urcorner) .$$

Now we note

$$\xi(0, x_1, \ldots, x_n) = y \rightarrow \chi(x_1,\ldots,x_n) = y ,$$

hence $Proof_H(\lambda_\chi(x_1,\ldots,x_n, y), \ulcorner \chi(\bar{x}_1,\ldots,\bar{x}_n) = \bar{y}\urcorner)$.

There exists a primitive recursive ξ_1 which transforms the gödelnumber of a proof of $\chi(x_1,\ldots,x_n) = y$ into a number of a proof of $\xi(0, \bar{x}_1,\ldots,\bar{x}_n) = \bar{y}$ (since the new proof is obtained by adding $\xi(0,\bar{x}_1,\ldots,\bar{x}_n) = \chi(\bar{x}_1,\ldots,\bar{x}_n)$ (instantiation of an axiom) and applying the equality axioms). Now take

$$\lambda_\xi(0, x_1,\ldots,x_n, y) = \xi_1\lambda_\chi(x_1,\ldots,x_n, y).$$

Suppose $\lambda_\xi(z, x_1,\ldots,x_n, y)$ to be defined.

Let $\xi(Sz, x_1,\ldots,x_n) = y$. Then

$$\psi(\xi(z, x_1,\ldots,x_n), z, x_1,\ldots,x_n) = y$$

and (abbreviating $\overline{\xi(z, x_1,\ldots,x_n)}$ as $\bar{\xi}$)

$$Proof_H(\lambda_\psi(\xi(z,x_1,\ldots,x_n), z, x_1,\ldots,x_n, y) , \ulcorner \psi(\bar{\xi}, \bar{z}, \bar{x}_1,\ldots,\bar{x}_n) = \bar{y}\urcorner)$$

$$Proof_H(\lambda_\xi(z, x_1,\ldots,x_n, \xi(z, x_1,\ldots,x_n)) , \ulcorner \xi(\bar{z}_1, \bar{x}_1,\ldots,\bar{x}_n) = \bar{\xi}\urcorner) .$$

Combining the proofs of $\psi(\bar{\xi}, \bar{z}, \bar{x}_1,\ldots,\bar{x}_n) = y$ and of $\xi(\bar{z}, \bar{x}_1,\ldots,\bar{x}_n) = \bar{\xi}$ and the instantiation $\xi(S\bar{z},\bar{x}_1,\ldots,\bar{x}_n) = \psi(\bar{\xi},\bar{z},\bar{x}_1,\ldots,\bar{x}_n)$, we obtain, primitive recursively in the numbers of these proofs, the proof number $\lambda_\xi(Sz, x_1,\ldots,x_n, y)$ of a proof of $\xi(S\bar{z}, \bar{x}_1,\ldots,\bar{x}_n) = \bar{y}$, etc. etc.

§ 6. Intuitionistic arithmetic in all finite types.

1.6.1. Contents of the section. This section deals with extensions of HA to theories involving objects of finite type. The type structure is defined inductively in 1.6.2.

Subsections 3 - 7 describe the basic system $N - HA^\omega$ of intuitionistic arithmetic in all finite types ("N" from "neutral"). Subsection 8 introduces the combinatorially defined λ - operator, in subsection 9 it is shown that HA is properly contained in $N - HA^\omega$.

In subsections 10 - 14 the extensions of $N - HA^\omega$ to an intensional variant $I - HA^\omega$ ("I" from "intensional") and extensional variants $WE - HA^\omega$, $E - HA^\omega$ ("E" from "extensional", "WE" from "weakly extensional") and the quantifier-free fragments $qf - N - HA^\omega$, $qf - I - HA^\omega$, $qf - WE - HA^\omega$ are described.

Subsection 15 describes a weak system HA^ω, with its quantifier-free part $qf - HA^\omega$, in a language with only equations between type zero terms as prime formulae. The system is of interest in connection with the Dialectica interpretation. The reader who is not interested in the Dialectica interpretation (§ 3.5) may decide to skip this section, and also the material in § 1.7 dealing with HA^ω.

Subsection 16 discusses pairing operators and simultaneous recursion in general, and subsection 17 describes a pairing for $qf - WE - HA^\omega$, which can even be used in a suitable version of $N - HA^\omega$ with the λ - operator instead of combinators as a primitive. More material on these subjects in § 1.7.

Subsection 18 gives historical notes and a discussion of the literature.

Directions for use. The reader who is primarily interested in HA, may skip §§ 1.6, 1.7, 1.8 altogether. If the reader has no previous acquaintance with the intuitionistic theory of finite types, he may find it enlightening, after a brief glance at subsection 1 - 14, to have a look at the models HRO, HEO of these theories described in chapter II.

1.6.2. Type structure T.

The type structure T is defined inductively by the following two clauses:

T1) $0 \in T$

T2) $\sigma, \tau \in T \Rightarrow (\sigma)\tau \in T$.

Remarks. (i) Intuitively, each type represents a class of objects: type 0 represents the natural numbers, and if σ, τ are types, then $(\sigma)\tau$ represents a class of mappings from objects of type σ to objects of type τ.

(ii) There are many alternative notations for $(\sigma)\tau$ in the literature, such as (σ,τ), $\sigma \to \tau$, τ^σ, $(\tau)\sigma$, etc.

(iii) Each $\sigma \in T$ is of the form $(\sigma_1)\ldots(\sigma_n)0$, as is readily verified by induction over T.

1.6.3 - 1.6.7. Description of the neutral theory $\underset{\sim}{N}$ - $\underset{\sim}{HA}^{\omega}$.

1.6.3. Language of $\underset{\sim}{N}$ - $\underset{\sim}{HA}^{\omega}$.

The language contains variables (indicated by x^σ, y^σ, z^σ, u^σ, v^σ, w^σ) for each type $\sigma \in \underset{\approx}{T}$. (Type superscripts are often omitted.)
There is a symbol $=_\sigma$ for equality between objects of type σ, for each $\sigma \in \underset{\approx}{T}$; we usually omit the type subscript.

Furthermore, there are constants for objects of certain types: a constant 0 of type 0 (zero); S of type (0)0 (successor), and constants $\Pi_{\sigma,\tau}$, $\Sigma_{\rho,\sigma,\tau}$, R_σ for all $\rho,\sigma,\tau \in \underset{\approx}{T}$, whose types will be described below.

As logical constants we use $\&$, $\rightarrow$, $\vee$, $\forall x^\sigma$, $\exists x^\sigma$ (for each variable x^σ, $\sigma \in \underset{\approx}{T}$).

1.6.4. Terms.

Let Tm_σ denote the class of terms of type σ, $Tm = \bigcup\{Tm_\sigma \mid \sigma \in \underset{\approx}{T}\}$.
Terms are defined inductively as follows:

Tm 1) Constants and variables of type σ belong to Tm_σ

Tm 2) $t \in Tm_{(\sigma)\tau}$, $t' \in Tm_\sigma \Rightarrow (tt') \in Tm_\tau$.

1.6.5. Notational conventions.

We will reserve s, t, T as syntactical variables for terms, if necessary provided with primes or subscripts to create more variables, and provided with a type superscript: s^σ, t^σ for clarity.
We abbreviate $(\ldots((t_1t_2)t_3)\ldots t_n)$ as $t_1t_2t_3\ldots t_n$. So $t_1t_2t_3$ abbreviates $((t_1t_2)t_3)$, but $t_1(t_2t_3)$ stands for $(t_1(t_2t_3))$.
We shall use $\underline{\underline{x}}$, $\underline{\underline{y}}$, $\underline{\underline{z}}$, $\underline{\underline{u}}$, $\underline{\underline{v}}$, $\underline{\underline{w}}$, $\underline{\underline{X}}$, $\underline{\underline{Y}}$, $\underline{\underline{Z}}$, $\underline{\underline{U}}$, $\underline{\underline{V}}$, $\underline{\underline{W}}$ for finite (possibly empty) strings of variables. So if $\underline{\underline{x}} \equiv (x_1,\ldots,x_n)$, $\forall \underline{\underline{x}} A$, $\exists \underline{\underline{x}} A$ abbreviate $\forall x_1\ldots\forall x_n A$, $\exists x_1\ldots\exists x_n A$ respectively.
$\underline{\underline{s}}$, $\underline{\underline{t}}$, $\underline{\underline{T}}$ will be used to denote finite (possibly empty) strings of terms.

If we wish to indicate that a term is of type σ, we often simply write $t \in \sigma$. We shall often omit type superscripts; but it is always assumed in writing down an expression, that the terms are well-formed (i.e. the types are fitting). So, for example, if we write $xyz = u$, and if we assume $y \in \sigma$, $u \in \tau$, $z \in \rho$, then $x \in (\sigma)(\rho)\tau$.

Let $\underline{\underline{s}} \equiv s_1,\ldots,s_n$, $\underline{\underline{t}} \equiv t_1,\ldots,t_m$, $s_i \in (\tau_1)\ldots(\tau_m)\sigma_i$, $t_j \in \tau_j$ $(1 \leq i \leq n,\ 1 \leq j \leq m)$, then

$$\underline{\underline{s}}\,\underline{\underline{t}} \equiv_{def} s_1t_1\ldots t_m,\ \ldots,\ s_nt_1\ldots t_m.$$

Immediately after variable-binding operators $(\forall, \exists, \lambda)$ juxtaposition indeed indicates concatenation however, so $\forall \underline{\underline{x}}\,\underline{\underline{y}}$ where $\underline{\underline{x}} \equiv (x_1,\ldots,x_n)$, $\underline{\underline{y}} \equiv (y_1,\ldots,y_m)$ stands for $\forall x_1\ldots\forall x_n\ \forall y_1\ldots\forall y_m$ etc.

1.6.6. Formulae.

Let us denote the class of formulae as Fm. Prime formulae are expressions of the form $t^\sigma =_\sigma s^\sigma$. Fm is defined as usual by two inductive clauses:

Fm 1) Prime formulae belong to Fm

Fm 2) If $A,B \in$ Fm, then also $(A \& B)$, $(A \vee B)$, $(A \to B)$, $(\forall x^\sigma A)$, $(\exists x^\sigma A)$.

In bracketing we follow the usual conventions.

1.6.7. Axioms and rules.

(a) Axioms and rules for many-sorted intuitionistic predicate logic.

(b) Axioms for equality:

$$x^\sigma = x^\sigma ,$$
$$x^\sigma = y^\sigma \;\&\; z^\sigma = y^\sigma \to x^\sigma = z^\sigma ,$$
$$x^\sigma = y^\sigma \to z^{(\sigma)\tau}x^\sigma = z^{(\sigma)\tau}y^\sigma ,$$
$$x^{(\sigma)\tau} = y^{(\sigma)\tau} \to x^{(\sigma)\tau}z^\sigma = y^{(\sigma)\tau}z^\sigma ,$$

and the usual equality axioms for successor

$$Sx^0 \neq 0, \quad x^0 = y^0 \leftrightarrow Sx^0 = Sy^0 .$$

(c) The rule or axiom schema of induction (for arbitrary formulae of the language).

(d) Defining axioms for $\Pi_{\rho,\sigma}$, $\Sigma_{\rho,\sigma,\tau}$, R_σ:

$$\Pi_{\rho,\sigma}x^\rho y^\sigma = x^\rho , \qquad \Pi_{\rho,\sigma} \in (\rho)(\sigma)\rho$$

$$\Sigma_{\rho,\sigma,\tau}xyz = xz(yz) , \qquad x \in (\rho)(\sigma)\tau,\ y \in (\rho)\sigma,\ z \in \rho ,$$
$$\Sigma_{\rho,\sigma,\tau} \in ((\rho)(\sigma)\tau)((\rho)\sigma)(\rho)\tau .$$

$$\left.\begin{array}{l} R_\sigma xy\,0 = x \\ R_\sigma xy(Sz) = y(R_\sigma xyz)z \end{array}\right\} \quad \begin{array}{l} x \in \sigma,\ y \in (\sigma)(0)\sigma,\ z \in 0 , \\ R_\sigma \in (\sigma)((\sigma)(0)\sigma)(0)\sigma . \end{array}$$

1.6.8. Theorem (Definition of the λ-operator).

To each term $t^\tau[x^\sigma]$ we can construct a term $\lambda x^\sigma.t^\tau[x^\sigma]$ such that

(i) $(\lambda x^\sigma.t^\tau[x^\sigma])(t') = t^\tau[t']$ $\quad (t' \in \sigma)$,

(ii) $\lambda x^\sigma.\, tx^\sigma = t$ for t not containing x^σ,

(iii) if $x^\sigma \notin t', t''$, then
$t' = t'' \to \lambda x^\sigma.[y/t']t = \lambda x^\sigma.[y/t'']t$, y a variable different from x.

Proof. $\lambda x^\sigma.t$ is defined by induction on the complexity of t:

(a) $x^\sigma \notin t^\tau \Rightarrow \lambda x^\sigma.t^\tau \equiv_{def} \Pi_{\tau,\sigma}t^\tau$

(b) $\lambda x^\sigma.x^\sigma \equiv_{def} \Sigma_{\sigma,(o)\sigma,\sigma}\Pi_{\sigma,(o)\sigma}\Pi_{\sigma,o}$

(c) $x^\sigma \notin t$; then $\lambda x^\sigma.tx^\sigma \equiv_{def} t$

(d) $x^\sigma \in t$, or ($x^\sigma \in t'$ and $t' \neq x^\sigma$); then $\lambda x^\sigma.tt' \equiv_{def}$
$\equiv \Sigma(\lambda x^\sigma.t)(\lambda x^\sigma.t')$.

Now (ii) is immediate by clause (c) of the definition. (i) and (iii) can be proved simultaneously by induction on the complexity of t.
As an example, we prove (i).

(a) Let $x^\sigma \notin t^\tau$. Then $(\lambda x^\sigma . t^\tau)(t_1^\sigma) = \Pi_{\tau,\sigma} t^\tau t_1^\sigma = t^\tau$,

(b) $\Sigma_{\sigma,(o)\sigma,\sigma} \Pi_{\sigma,(o)\sigma} \Pi_{\sigma,o} x^\sigma = \Pi_{\sigma,(o)\sigma} x^\sigma (\Pi_{\sigma,o} x^\sigma) = x^\sigma$.

(c) Let $x^\sigma \notin t$, then $(\lambda x^\sigma . t x^\sigma)(t') = tt'$.

(d) Let $x^\sigma \in t$, or ($x^\sigma \in t'$ and $t' \neq x^\sigma$);

$$(\lambda x^\sigma . t[x^\sigma] t'[x^\sigma]) t'' = \Sigma(\lambda x^\sigma . t[x^\sigma])(\lambda x^\sigma . t'[x^\sigma]) t'' =$$
$$= (\lambda x^\sigma . t[x^\sigma]) t'' ((\lambda x^\sigma . t'[x^\sigma]) t'') = t[t''] t'[t''] \quad \text{(induction hypothesis).}$$

Etc., etc.

Remark. In combinatory logic, the defined λ-operator is usually written with square brackets: $[x]t$ for $\lambda x.t$. We find it more suggestive to use the λ-notation instead; but if one has to discuss in a single context defined and primitive λ-operators, there should be a notational distinction.

Abbreviation: $\lambda \underline{\underline{x}}.t$, where $\underline{\underline{x}} \equiv (x_1, \ldots, x_n)$, stands for $\lambda x_1.(\lambda x_2.(\ldots(\lambda x_n.t)\ldots))$.

1.6.9. HA as a subsystem of N-HA$^\omega$.

Let us associate to each function constant φ of HA, a term T_φ of N-HA$^\omega$, as follows.

(i) $T_0 \equiv 0$, $T_S \equiv S$, and if U_n^i is the function such that $U_n^i(x_1, \ldots, x_n) = x_i$, we put $T_{U_n^i} \equiv \lambda x_1 \ldots x_n . x_i$.

(ii) If ψ is explicitly defined from $\psi_0, \varphi_1, \ldots, \varphi_m$ such that $\psi(x_1, \ldots, x_n) = \psi_0(\varphi_1(x_1, \ldots, x_n), \ldots, \varphi_m(x_1, \ldots, x_n))$, we put $T_\psi \equiv \lambda x_1 \ldots x_n . T_{\psi_0}(T_{\varphi_1} x_1 \ldots x_n) \ldots (T_{\varphi_m} x_1 \ldots x_n)$.

(iii) If φ is defined by primitive recursion from ψ_1, ψ_2 such that

$$\varphi(0, x_1, \ldots, x_n) = \psi_1(x_1, \ldots, x_n)$$
$$\varphi(Sz, x_1, \ldots, x_n) = \psi_2(\varphi(z, x_1, \ldots, x_n) z, x_1, \ldots, x_n)$$

we put $T_\varphi \equiv RT_{\psi_1}(\lambda \underline{\underline{x}} yz . T_{\psi_2}(y\underline{\underline{x}}) z\underline{\underline{x}})$.

Then, for any n-ary function constant φ of HA, $T_\varphi x_1 \ldots x_n$ "behaves like" $\varphi(x_1, \ldots, x_n)$. More precisely, if we define a mapping Δ on terms and formulae of HA by induction on the complexity:

(iv) $\Delta\varphi \equiv T_\varphi$ for each function constant φ of HA, $\Delta x^o \equiv x^o$,

(v) $\Delta\varphi(t_1, \ldots, t_n) \equiv (\Delta\varphi)(\Delta t_1) \ldots (\Delta t_n)$,

(vi) $\Delta(t = s) \equiv (\Delta t = \Delta s)$,

(vii) Δ is a homomorphism w.r.t. logical operators,

then HA translates under Δ into a subsystem of N-HA$^\omega$.
Note that Δ is bi-unique.

1.6.10. Intensional identity or equality.

A basic feature of intuitionism is that we have to deal with mathematical objects as they are given to us ; for example, a species (set) of natural numbers is given by a description (definition) of a property of natural numbers ; the extension of the set (in the classical sense) may then be conceived either as a mode of speech, to avoid speaking about equivalence w.r.t. membership, or as an equivalence class, i.e. a species of higher type; the latter point of view makes sense if we accept the concept of a power species.

Similarly, a (lawlike) function is given as a rule ; its extension (graph) in the classical sense is a derived notion. From a foundational point of view, it is therefore natural to pay attention to the concept of definitional or intensional equality : two objects are said to be definitionally or intensionally equal, if they are given to us as the same object. We do not a priori suppose the concept of "definition" or "description" to be restricted to definition in a given language.

Whatever the precise content of the concept of intensional identity, it seems clear that it should be decidable whether two objects are definitionally equal or not. This is expressed in the extension $\underset{\sim}{\mathrm{I}}-\underset{\sim}{\mathrm{HA}}^{\omega}$ of $\underset{\sim}{\mathrm{N}}-\underset{\sim}{\mathrm{HA}}^{\omega}$ described below.

1.6.11. Description of $\underset{\sim}{\mathrm{I}}-\underset{\sim}{\mathrm{HA}}^{\omega}$.

In $\underset{\sim}{\mathrm{I}}-\underset{\sim}{\mathrm{HA}}^{\omega}$, the intended interpretation of $=_{\sigma}$ is : intensional equality between objects of type σ ; $\underset{\sim}{\mathrm{I}}-\underset{\sim}{\mathrm{HA}}^{\omega}$ is obtained from $\underset{\sim}{\mathrm{N}}-\underset{\sim}{\mathrm{HA}}^{\omega}$ by adding a constant $E_{\sigma} \in (\sigma)(\sigma)0$ for each type $\sigma \in \underset{\sim}{\mathrm{T}}$ such that

$$E_{\sigma}x^{\sigma}y^{\sigma} = 0 \leftrightarrow x^{\sigma} = y^{\sigma} ,$$

$$E_{\sigma}x^{\sigma}y^{\sigma} = 0 \vee E_{\sigma}x^{\sigma}y^{\sigma} = 1 ,$$

which implies decidability of equality at all types :

$$x^{\sigma} = y^{\sigma} \vee x^{\sigma} \neq y^{\sigma} .$$

1.6.12. Description of $\underset{\sim}{\mathrm{E}}-\underset{\sim}{\mathrm{HA}}^{\omega}$, $\underset{\sim}{\mathrm{WE}}-\underset{\sim}{\mathrm{HA}}^{\omega}$.

Another way of interpreting equality in $\underset{\sim}{\mathrm{N}}-\underset{\sim}{\mathrm{HA}}^{\omega}$ is assuming it to be extensional equality : two objects of type $(\sigma)\tau$ are equal if for every argument of type σ they yield equal values. This amounts to

(1) $$\forall z^{(\sigma)\tau} u^{(\sigma)\tau}(z = u \leftrightarrow \forall y^{\sigma}(zy = uy)) .$$

Adding (1) to $\underset{\sim}{\mathrm{N}}-\underset{\sim}{\mathrm{HA}}^{\omega}$ yields an extensional version of intuitionistic arithmetic in all finite types, which we may denote by $\underset{\sim}{\mathrm{E}}-\underset{\sim}{\mathrm{HA}}_{0}^{\omega}$.

For some purposes (notably the study of the Dialectica - interpretation in Chapter III) it is more convenient to use another version of $\underset{\sim}{\mathrm{E}}-\underset{\sim}{\mathrm{HA}}_{0}^{\omega}$, to be denoted by $\underset{\sim}{\mathrm{E}}-\underset{\sim}{\mathrm{HA}}^{\omega}$. Here equality between objects of type 0 is the only primitive concept, equality for all other types is a defined notion (inductively on the complexity of the types)

$$z^{(\sigma)\tau} = u^{(\sigma)\tau} \equiv_{def} \forall y^{\sigma}(zy = uy) .$$

The axioms for equality of $\underset{\sim}{N} - \underset{\sim}{HA}^{\omega}$ are retained (but some of them become redundant, such as $z = u \rightarrow zy = uy$, which now holds by definition); the full force of extensionality is now in

$$(2) \qquad x^{\sigma} = y^{\sigma} \rightarrow z^{(\sigma)\tau}x = z^{(\sigma)\tau}y .$$

Note that $\underset{\sim}{E} - \underset{\sim}{HA}_0^{\omega}$ may be interpreted as a definitional extension of $\underset{\sim}{E} - \underset{\sim}{HA}^{\omega}$, obtained by addition of new symbols $=_{\sigma}$ for definable equality of type σ to $\underset{\sim}{E} - \underset{\sim}{HA}^{\omega}$.

$\underset{\sim}{WE} - \underset{\sim}{HA}^{\omega}$ is obtained from $\underset{\sim}{E} - \underset{\sim}{HA}^{\omega}$ weakening (2) to the following <u>rule of extensionality</u>

EXT - R. $\vdash tx_1 \ldots x_n = sx_1 \ldots x_n, \ \vdash A(t) \Rightarrow \vdash A(s)$.

($A(t)$ quantifier-free, $x_1, \ldots, x_n$ a sequence of variables not occurring in A, t, s, such that $tx_1 \ldots x_n$ and $sx_1 \ldots x_n$ are of type 0. The notation $\vdash F$ indicates that F has been derived <u>without</u> assumptions.) A slightly stronger rule is

EXT - R'. $\vdash P \rightarrow tx_1 \ldots x_n = sx_1 \ldots x_n, \ \vdash A(t) \Rightarrow \vdash P \rightarrow A(s)$.

(P quantifier-free, other conditions as before.) In certain analogues of $\underset{\sim}{WE} - \underset{\sim}{HA}^{\omega}$ (theories with bar recursion of higher type, or induction over trees; cf. 1.9.25 - 1.9.27) EXT - R must be replaced by EXT - R'.

Note also that EXT - R is equivalent to the following rule

$$(3) \qquad \vdash tx_1 \ldots x_n = sx_1 \ldots x_n \Rightarrow \vdash F[t] = F[s],$$

where $F[x^{\sigma}]$ is a term in the language of $\underset{\sim}{E} - \underset{\sim}{HA}^{\omega}$. For, if $A[x^{\sigma}]$ is a formula, we can always find a term $F_A[x^{\sigma}]$ such that $F_A[x^{\sigma}] = 0 \longleftrightarrow A(x^{\sigma})$ (see 1.6.14); therefore, from (3)

$$\vdash tx_1 \ldots x_n = sx_1 \ldots x_n \Rightarrow \vdash F_A[t] = F_A[s],$$

hence

$$\vdash tx_1 \ldots x_n = sx_1 \ldots x_n, \quad \vdash F_A[t] = 0 \Rightarrow \vdash F_A[s] = 0 .$$

Conversely, apply EXT - R with $F[t] = F[x^{\sigma}]$ for $A(x^{\sigma})$, then $A(t)$ obviously holds, and $A(s) \equiv F[t] = F[s]$.

If we would have formulated EXT - R as

$$(4) \qquad A(t), \ tx_1 \ldots x_n = sx_1 \ldots x_n \Rightarrow A(s)$$

(conditions as for EXT - R, but $x_1, \ldots, x_n$ not free in assumptions on which $tx_1 \ldots x_n = sx_1 \ldots x_n$ depends), then the resulting system does not satisfy the deduction theorem. For (4) implies the axiom of extensionality (2), since (4) yields

$$x^\sigma = y^\sigma,\ z^{(\sigma)\tau}x^\sigma = z^{(\sigma)\tau}x^\sigma \vdash z^{(\sigma)\tau}x^\sigma = z^{(\sigma)\tau}y^\sigma$$

hence the deduction theorem would yield (2). From results in Howard B however, it follows that (2) cannot be proved in $\underset{\sim}{\text{WE}} - \underset{\sim}{\text{HA}}^\omega$ (with (4) replacing EXT - R) and that therefore $\underset{\sim}{\text{WE}} - \underset{\sim}{\text{HA}}^\omega$ in this version does not satisfy the deduction theorem.

1.6.13. Description of $\text{qf} - \underset{\sim}{\text{N}} - \underset{\sim}{\text{HA}}^\omega$, $\text{qf} - \underset{\sim}{\text{I}} - \underset{\sim}{\text{HA}}^\omega$, $\text{qf} - \underset{\sim}{\text{WE}} - \underset{\sim}{\text{HA}}^\omega$.

The quantifier-free part of $\underset{\sim}{\text{N}} - \underset{\sim}{\text{HA}}^\omega$, $\underset{\sim}{\text{I}} - \underset{\sim}{\text{HA}}^\omega$, $\underset{\sim}{\text{WE}} - \underset{\sim}{\text{HA}}^\omega$, denoted as $\text{qf} - \underset{\sim}{\text{N}} - \underset{\sim}{\text{HA}}^\omega$ etc., may be obtained from the corresponding theories with quantifiers as follows.

(i) From logic we drop quantifier rules and - axioms. In discussing deductions in $\text{qf} - \underset{\sim}{\text{WE}} - \underset{\sim}{\text{HA}}^\omega$, $t^\sigma = s^\sigma$ is then to be conceived as an abbreviation for $t^\sigma x_1 \ldots x_n = s^\sigma x_1 \ldots x_n$, $x_1, \ldots, x_n$ variables not occurring in t, s or (open) assumptions of the deduction, such that $tx_1 \ldots x_n$ is of type 0.

(ii) The induction schema is replaced by the induction rule:
$A(0),\ A(x) \to A(Sx^0) \Rightarrow Ax^0$, for A quantifier free, x not occurring free in assumptions of the deduction.

(iii) $Ax^\sigma \Rightarrow At^\sigma$, if x does not occur in (open) assumptions.

(Cf. our remark on quantifier-free systems in 1.5.8.)

Note that for $\text{qf} - \underset{\sim}{\text{WE}} - \underset{\sim}{\text{HA}}^\omega$ with EXT - R' instead of EXT - R, the deduction theorem holds even if we liberalize EXT - R' by deleting $\vdash$.

1.6.14. $\text{qf} - \underset{\sim}{\text{I}} - \underset{\sim}{\text{HA}}^\omega$, $\text{qf} - \underset{\sim}{\text{WE}} - \underset{\sim}{\text{HA}}^\omega$ as equational calculi.

Since in $\text{qf} - \underset{\sim}{\text{I}} - \underset{\sim}{\text{HA}}^\omega$, $\text{qf} - \underset{\sim}{\text{WE}} - \underset{\sim}{\text{HA}}^\omega$ prime formulae are decidable, all propositional formulae are decidable, and therefore the propositional operators may be represented by certain constant terms expressing the classical truth functions.

We consider the case of $\text{qf} - \underset{\sim}{\text{I}} - \underset{\sim}{\text{HA}}^\omega$: Let con, dis, imp be primitive recursive functions (of type (0)(0)0) such that

$$\begin{array}{llll} \text{con}(Sx,y) = \text{con}(x,Sy) = S0, & \text{con}(0,0) = 0, \\ \text{dis}(0,x) = \text{dis}(0,x) = 0, & \text{dis}(Sx,Sy) = S0, \\ \text{imp}(Sx,y) = \text{imp}(x,0) = 0, & \text{imp}(0,Sx) = S0. \end{array}$$

Now we construct, for any propositional A, a term T_A such that $T_A = 0 \leftrightarrow A$:

For $T_{t=s}$ we take Ets. $T_{A\&B} \equiv \text{con}(T_A, T_B)$; $T_{A\vee B} \equiv \text{dis}(T_A, T_B)$; $T_{A\to B} \equiv \text{imp}(T_A, T_B)$.

Similarly for $\text{qf} - \underset{\sim}{\text{WE}} - \underset{\sim}{\text{HA}}^\omega$; here we only need to put $T_{t=s} \equiv |t - s|$.

1.6.15. The systems $\underset{\sim}{HA}^{\omega}$, qf - $\underset{\sim}{HA}^{\omega}$.

In connection with the Dialectica interpretation in § 3.5, the following subsystem $\underset{\sim}{HA}^{\omega}$ of $\underset{\sim}{N}$ - $\underset{\sim}{HA}^{\omega}$ is of interest. (The reader who is not interested in the Dialectica interpretation can omit this section, or postpone it till he arrives at studying the Dialectica interpretation.)

The only prime formulae of $\underset{\sim}{HA}^{\omega}$ are equations between terms of type 0; the constants are those of $\underset{\sim}{N}$ - $\underset{\sim}{HA}^{\omega}$, except that we now only need equality of type 0 as a primitive. The logical basis is many-sorted intuitionistic predicate logic. The nonlogical axioms consist of the induction schema, the usual axioms for type 0 equality and successor

$$x^0 = x^0, \quad x^0 = z^0 \;\&\; y^0 = z^0 \rightarrow x^0 = y^0$$
$$Sx^0 \neq 0, \quad x^0 = y^0 \leftrightarrow Sx^0 = Sy^0,$$

substitutivity for type 0 objects:

$$x^0 = y^0 \rightarrow t[x^0] = t[y^0]$$

and the following schemata (which may be viewed as very special instances of the extensionality rule), for all $t \in 0$:

$$\text{SUB} \quad \begin{cases} t[\Pi xy] = t[x] \\ t[\Sigma xyz] = t[xz(yz)] \\ t[Rxy0] = t[x], \quad t[Rxy(Sz)] = t[y(Rxyz)z]. \end{cases}$$

$\underset{\sim}{HA}^{\omega}$ is clearly a subsystem of $\underset{\sim}{N}$ - $\underset{\sim}{HA}^{\omega}$. (We do not know whether $\underset{\sim}{N}$ - $\underset{\sim}{HA}^{\omega}$ is conservative over $\underset{\sim}{HA}^{\omega}$.) qf - $\underset{\sim}{HA}^{\omega}$ is defined in the obvious way, similarly to qf - $\underset{\sim}{N}$ - $\underset{\sim}{HA}^{\omega}$.

Remarks. (i) The schemata SUB give us for the defined λ-operator: $t[(\lambda x^\sigma.t')t''] = t[[x^\sigma/t'']t']$ $(t \in 0)$, and especially $(\lambda x^\sigma.t)t' = [x^\sigma/t']t$ for $t \in 0$.

(ii) If we use $t_1^\sigma = t_2^\sigma$ as a metamathematical abbreviation for $\vdash F[t_1] = F[t_2]$ for all type 0 terms $F[x^\sigma]$, then we see that $t_1^\sigma = t_2^\sigma$ amounts to $\vdash x^{(\sigma)0}t_1^\sigma = x^{(\sigma)0}t_2^\sigma$. For $t_1^\sigma = t_2^\sigma$ implies the latter assertion, and conversely, taking for $x^{(\sigma)0}$: $\lambda x^\sigma.F[x^\sigma]$, we obtain $\vdash F[t_1] = F[t_2]$, in view of the preceding remark.

(iii) $\underset{\sim}{HA}$ is of course also a subsystem of $\underset{\sim}{HA}^{\omega}$.

1.6.16. Simultaneous recursion and pairing: a comparison of various treatments.

We have formulated our system $\underset{\sim}{N}$ - $\underset{\sim}{HA}^{\omega}$ with a primitive R_σ for each σ. For certain applications, however, (cf. § 3.4, 3.5) one requires constants for simultaneous recursion: for each sequence of types $\sigma^* \equiv \langle\sigma_1,\ldots,\sigma_n\rangle$, $\sigma_i \in \underset{\approx}{T}$ for $1 \leq i \leq n$, one requires a sequence of constants $R^1_{\sigma^*}, \ldots, R^n_{\sigma^*}$ (to be abbreviated as $\underset{=}{R}_{\sigma^*}$ or $\underset{=}{R}$) satisfying

(1) $\underline{\underline{R}}\,\underline{\underline{x}}\,\underline{\underline{y}}\,0 = \underline{\underline{x}}, \quad \underline{\underline{R}}\,\underline{\underline{x}}\,\underline{\underline{y}}(Sz) = \underline{\underline{y}}(\underline{\underline{R}}\,\underline{\underline{x}}\,\underline{\underline{y}}\,z)z$

where $\underline{\underline{x}} \in \sigma^*$; $\underline{\underline{y}} = (y_1, \ldots, y_n)$, $y_i \in \tau_i \equiv (\sigma_1)\ldots(\sigma_n)(0)\sigma_i$, $1 \leq i \leq n$; $R^i_{\sigma^*} \in (\sigma_1)\ldots(\sigma_n)(\tau_1)\ldots(\tau_n)(0)\sigma_i$, $1 \leq i \leq n$.

It has been shown by Schütte (see Hindley - Lercher - Seldin 1972, p.156) that constants for simultaneous recursion satisfying (1) can be defined in $\underset{\sim}{N}$ - $\underset{\sim}{HA}^\omega$ (1.7.7); the proof that they satisfy (1) can be given in the quantifier-free fragment of qf - $\underset{\sim}{N}$ - $\underset{\sim}{HA}^\omega$ (1.5.13).

A more complicated way of constructing constants $\underline{\underline{R}}$, via simultaneous course-of-values recursion is described in Diller - Schütte 1971.

If one wishes to avoid the fairly long and ad hoc argument needed to obtain $\underline{\underline{R}}$ from R_σ, there are various possibilities.

(A) One may include constants $\underline{\underline{R}}_{\sigma^*}$ satisfying (1) as primitives in the descriptions of $\underset{\sim}{N}$ - $\underset{\sim}{HA}^\omega$. Since quite consistently sequences of variables and terms can be dealt with in complete analogy to the treatment of single variables and terms, this causes no particular difficulties apart from a certain awkwardness in notation now and then. (This alternative has been followed in Troelstra 1971.)

(B) One may extend the type structure $\underset{\approx}{T}$ to a structure $\underset{\approx}{T}'$ including Cartesian-product types, adding to T1 , T2 a third closure condition ($\times$ binds stronger than application):

T3): $\sigma, \tau \in \underset{\approx}{T}' \Rightarrow \sigma \times \tau \in \underset{\approx}{T}'$

(and replacing $\underset{\approx}{T}$ by $\underset{\approx}{T}'$ in T1 , T2).

and to the set of constants one adds pairing operators $D_{\sigma,\tau} \in (\sigma)(\tau)\sigma\times\tau$ with inverses $D'_{\sigma,\tau} \in (\sigma\times\tau)\sigma$, $D''_{\sigma,\tau} \in (\sigma\times\tau)\tau$ (for all $\sigma,\tau \in \underset{\approx}{T}'$) with axioms

(2) $D'(Dxy) = x, \quad D''(Dxy) = y,$

(3) $D(D'z)(D''z) = z.$

It is shown in 1.8.2 that this enlargement $\underset{\sim}{N}$ - $\underset{\sim}{HA}^\omega_p$ ("p" for "pairing) is an expansion (and a fortiori a conservative extension) of $\underset{\sim}{N}$ - $\underset{\sim}{HA}^\omega$.

In the presence of pairing operators satisfying (2) (but not necessarily (3)), we are able to define operators D_{σ^*}, $D^i_{\sigma^*}$ for n - tuples $\sigma^* = (\sigma_1, \ldots, \sigma_n)$ such that

$$D^i(Dx_1\ldots x_n) = x_i, \quad 1 \leq i \leq n.$$

By a standard trick well known from recursion theory, simultaneous recursion operators may then be defined; we illustrate the process for double recursion. We put

$$T \equiv R_{\sigma_1\times\sigma_2}(Dx_1x_2)(\lambda uz\,.\,D(y_1(D'u)z)(y_2(D''u)z)).$$

Then we readily prove by induction

$$D'(T0) = x_1, \quad D''(T0) = x_2$$
$$D'(T(Sz)) = y_1(D'(Tz))z, \quad D''T(Sz) = y_2(D''(Tz))z$$

and therefore we may take

$$R^1_{\sigma_1,\sigma_2} \equiv \lambda x_1 x_2 y_1 y_2 z \,.\, D'(Tz), \quad R^2_{\sigma_1,\sigma_2} \equiv \lambda x_1 x_2 y_1 y_2 z \,.\, D''(Tz) .$$

It has been shown in Barendregt A that it is impossible to define in $\underset{\sim}{I}-\underset{\sim}{HA}^{\omega}$ types $\sigma \times \tau \in \underset{\approx}{T}$, and operators $D \in (\sigma)(\tau)\sigma\times\tau$, $D' \in (\sigma\times\tau)\sigma$, $D'' \in (\sigma\times\tau)\tau$ satisfying (2).

In suitable versions of $\underset{\sim}{N}-\underset{\sim}{HA}^{\omega}$ with the λ-operators as a primitive it is possible to construct $\sigma\times\tau$, D, D', D'' such that (2) is satisfied (see end of 1.6.17) ; then, of course, the constants $\underset{=}{R}$ can be defined.

A fortiori, in the extensional system $qf-\underset{\sim}{WE}-\underset{\sim}{HA}^{\omega}$, it is possible to define product types and pairing operators such that (2) is satisfied (cf. 1.6.17), so in extensional contexts simultaneous recursion does not cause any problems.

It should be noted that the methods of Schütte from Hindley - Lercher - Seldin 1972 , or of Diller and Schütte 1971 do not extend automatically to other schemata for defining functionals which have "simultaneous" and "single" versions, such as bar-recursion, or definition by induction over well-founded trees ; in such cases, we are forced to fall back on the methods of treatment described under (A) and (B). For this reason, we have e.g. in § 2.3 indicated a treatment of computability including pairing operators.

1.6.17. Pairing operators in $qf-\underset{\sim}{WE}-\underset{\sim}{HA}^{\omega}$.

A pairing operator for the extensional theory, with inverses, is implicit in the reduction to pure types (1.8.5 - 1.8.8), since in the description of the reduction pairing operators with inverses for the pure types are given.

Assume a product type $0\times 0 \in \underset{\approx}{T}$, and operators $D_{0,0}$, $D'_{0,0}$, $D''_{0,0}$ to be given such that

(1) $\quad D'_{0,0}(D_{0,0}x^0y^0) = x^0, \quad D''_{0,0}(D_{0,0}x^0y^0) = y^0 .$

Then product types $\sigma\times\tau$, and operators $D_{\sigma,\tau} \in (\sigma)(\tau)\sigma\times\tau$, $D'_{\sigma,\tau} \in (\sigma\times\tau)\sigma$, $D''_{\sigma,\tau} \in (\sigma\times\tau)\tau$ (satisfying 1.6.16 (2)) may be constructed relative to $D_{0,0}$, $D'_{0,0}$, $D''_{0,0}$ as follows. Let

$$\sigma \equiv (\sigma_1)\ldots(\sigma_m)0, \quad \tau \equiv (\tau_1)\ldots(\tau_n)0 .$$

We put

$$\sigma\times\tau \equiv_{def} (\sigma_1)\ldots(\sigma_m)(\tau_1)\ldots(\tau_n)\, 0\times 0$$

and define

$$0^0 \equiv_{def} 0, \quad 0^{(\sigma)\tau} \equiv_{def} \Pi_{\tau,\sigma} 0^{\tau} ,$$

$$D_{\sigma,\tau} \equiv \lambda x^\sigma y^\tau x_1^{\sigma_1} \ldots x_m^{\sigma_m} y_1^{\tau_1} \ldots y_n^{\tau_n} . D_{o,o}(xx_1 \ldots x_m)(yy_1 \ldots y_n),$$
$$D'_{\sigma,\tau} \equiv \lambda z^{\sigma\times\tau} x_1^{\sigma_1} \ldots x_m^{\sigma_m} . D'_{o,o} z x_1 \ldots x_m 0^{\tau_1} \ldots 0^{\tau_n},$$
$$D''_{\sigma,\tau} \equiv \lambda z^{\sigma\times\tau} y_1^{\tau_1} \ldots y_n^{\tau_n} . D''_{o,o} z 0^{\sigma_1} \ldots 0^{\sigma_m} y_1 \ldots y_n .$$

Then

$$D'_{\sigma,\tau}(D_{\sigma,\tau} x^\sigma y^\tau) = \lambda x_1^{\sigma_1} \ldots x_m^{\sigma_m} . D'_{o,o}(D_{o,o}(xx_1 \ldots x_m)(y0^{\tau_1} \ldots 0^{\tau_n})) =$$
$$= \lambda x_1^{\sigma_1} \ldots x_m^{\sigma_m} . xx_1 \ldots x_m = x$$

and similarly for $D''_{\sigma,\tau}$.

We may take $0 \times 0 = 0$, and $D_{o,o}$, $D'_{o,o}$, $D''_{o,o}$ to be given by the standard pairing function j with inverses j_1, j_2 (1.3.9)

$$D_{o,o} \equiv \lambda xy . j(x,y), \; D'_{o,o} \equiv j_1, \; D''_{o,o} \equiv j_2, \quad j_1 j(x,y) = x, \; j_2 j(x,y) = y .$$

Alternatively, if we take $0 \times 0 \equiv ((0)(0)0)0$, and we define

$$D_{o,o} \equiv_{def} \lambda x^o y^o z^{(o)(o)o} . zxy ,$$

$$D'_{o,o} \equiv_{def} \Delta_{(o)(o)o,o} \Pi_{o,o}, \quad D''_{o,o} \equiv_{def} \Delta_{(o)(o)o,o} \Pi^*_{o,o}$$

where

$$\Delta_{\sigma,\tau} \equiv_{def} \lambda x^\sigma y^\tau . y^\tau x^\sigma , \qquad \Pi^*_{o,o} \equiv_{def} \lambda xy . y ,$$

then (1) is satisfied again; in fact, we can now establish 1.6.16 (2) in a variant of $\mathrm{qf}\text{-}\underset{\sim}{\mathrm{N}}\text{-}\underset{\sim}{\mathrm{HA}}^\omega$ with the λ-operator as a primitive, with rules $t = t' \Rightarrow \lambda x . t = \lambda x . t'$, $\lambda x . tx = t$ if t does not contain x, but without use of the induction rule (cf. also the discussions in 1.8.4, 2.2.27-34).

1.6.18. Historical notes; variants in the literature.

A quantifier-free theory of primitive recursive functionals (corresponding to $\mathrm{qf}\text{-}\underset{\sim}{\mathrm{I}}\text{-}\underset{\sim}{\mathrm{HA}}^\omega$) was first introduced in Gödel 1958. The system is sketched only, i.e. no detailed list of primitives, axioms and rules is given. From footnote 3 in Gödel 1958, however, it is obvious that an intensional, not an extensional version was intended.

In Kreisel 1959, Spector 1962, and Grzecorczyk 1964 an extensional version of the quantifier-free theory is studied. If we disregard the schema of bar-recursion, Spector's system corresponds to $\mathrm{qf}\text{-}\underset{\sim}{\mathrm{WE}}\text{-}\underset{\sim}{\mathrm{HA}}^\omega$. His primitives for functionals contain 0, S, constants Φ_s (substitution) satisfying $\Phi_s x y z = x(y z)$, constants Φ_p^n (projection) satisfying $\Phi_p^n x_1 \ldots x_n = x_1$, and schemata for defining functionals by composition and (primitive) recursion. Spector's paper also contains the generalized induction rule (1.7.10).

In Kreisel 1959, the type structure includes product types. As noted in section 5 of Spector 1962, Kreisel's schemata are very similar to Spector's schemata, but have to be supplemented by a schema permitting λ-abstraction.

Grzegorczyk 1964 describes two variants of the quantifier-free theory. The first set is based on closure under composition, and contains 0, S, and I, B, C, D, R' satisfying $Ix = x$, $Bxyz = x(yz)$, $Cxzy = xyz$, $Dxy = xyy$, $R'xy0 = x$, $R'xy(Sz) = yz(R'xyz)$. The second set is based on 0, S, I, and five schemata for defining new functionals from previously defined ones. Grzegorczyk also describes pairing operators in detail.

In Tait 1967, the full intensional theory, with logical operators is studied; Tait's formalism corresponds to $\underset{\sim}{I}$-$\underset{\sim}{HA}^{\omega}$, but with $x^{\sigma} = y^{\sigma} \vee x^{\sigma} \neq y^{\sigma}$ as an axiom instead of having E_{σ} as a primitive.

Diller and Schütte 1971 seems to be the first paper where a neutral theory (contained in qf-$\underset{\sim}{N}$-$\underset{\sim}{HA}^{\omega}$) is taken as a starting point.

The pairing operators described in 1.6.17 seem to belong to the "folklore" of the subject; the first variant in 1.6.17 is e.g. found in Luckhardt ~~1970~~ 1973. This variant has the advantage of being consistent with a notion of intensional equality for the typed λ-calculus (cf. end of 1.6.17, Barendregt A).

Other studies of the functionals of $\underset{\sim}{N}$-$\underset{\sim}{HA}^{\omega}$, in the context of a typed theory of combinators, are e.g. Sanchis 1967, Stenlund 1971, 1972.

§ 7. Induction and simultaneous recursion.

1.7.1. Contents of the section. The section discusses various points of detail concerning induction and simultaneous recursion in qf - N - HA$^{\omega}$, HA$^{\omega}$ and extensions of these systems. The reading of the section may be postponed until the need arises.

In 1.7.2 - 1.7.7 we describe a method, due to Schütte, for obtaining constants for definition by simultaneous recursion from the constants for simple recursion. Proofs are carried out in qf - N - HA$^{\omega}$. If the reader is interested say in a discussion of modified realizability, or the Dialectica interpretation (in § 3.4, 3.5 respectively), and wishes to avoid the tedious details of 1.7.7, he may either simply believe the result, or use the type structure extended by Cartesian product of formation (cf. 1.8.2), or simply postulate in the description of N - HA$^{\omega}$ simultaneous recursion outright.

In 1.7.8 - 1.7.10 we obtain an induction lemma, needed for the Dialectica interpretation in § 3.5, taken from Spector 1962.

In 1.7.11 we briefly discuss (following Diller and Schütte 1971) how the iterator may replace the recursor as a primitive.

Subsection 1.7.12 discusses the treatment of simultaneous recursion and the induction lemma for HA$^{\omega}$.

1.7.2 - 1.7.7. Simultaneous recursion in qf - N - HA$^{\omega}$.

1.7.2. Definition. We put

$$\text{prd} \equiv_{\text{def}} \lambda x . R0(\lambda yz . y)x ,$$
$$x \dot{-} y \equiv_{\text{def}} Rx(\lambda uv . \text{prd}(u))y$$
$$x > y \equiv_{\text{def}} x \dot{-} y \neq 0, \quad x < y \equiv_{\text{def}} y > x$$
$$x \geq y \equiv_{\text{def}} (x>y) \vee x=y , \quad x \leq y \equiv_{\text{def}} y \geq x .$$

The following two lemmas will be needed in part in 1.7.7 below, but especially in 1.7.10. The reason for explicitly proving all these elementary properties is that we wish to verify that they indeed have quantifier-free proofs.

1.7.3. Lemma. In qf - N - HA$^{\omega}$:

(i) $\text{prd}(0) = 0$, $\text{prd}(Sx) = x$

(ii) $x \dot{-} 0 = x$, $x \dot{-} Sy = \text{prd}(x \dot{-} y)$

(iii) $\text{prd}(x) \neq 0 \rightarrow x \neq 0$

(iv) $Sx \dot{-} Sy = x \dot{-} y$

(v) $x \dot{-} x = 0$

(vi) $Sx \dot{-} x = 1$

(vii) $x \neq 0 \to x = S\,\mathrm{prd}\,x$

(viii) $Sy \leq x \to y < x$

(ix) $y < x \to x \dot{-} y = S(x \dot{-} Sy)$

(x) $Sx \dot{-} y = 0 \to x \dot{-} y = 0$.

Proof. (i), (ii) immediate from the definition.

(iii): Contraposition of $x = 0 \to \mathrm{prd}(x) = 0$.

(iv) : Induction on y: $Sx \dot{-} S0 = \mathrm{prd}(Sx \dot{-} 0) = \mathrm{prd}(Sx) = x = x \dot{-} 0$; assume $Sx \dot{-} Sz = x \dot{-} z$, then $Sx \dot{-} SSz = \mathrm{prd}(Sx \dot{-} Sz) = \mathrm{prd}(x \dot{-} z) = x \dot{-} Sz$; apply rule of induction.

(v) : By (iv) and induction on x.

(vi) : Induction on x.

(vii): Induction on x: $0 \neq 0 \to 0 = S\,\mathrm{prd}\,0$; assume $z \neq 0 \to z = S\,\mathrm{prd}\,z$, then $Sz \neq 0 \to S\,\mathrm{prd}(Sz) = Sz$.

(viii): To show $x \dot{-} Sy \neq 0 \vee x = Sy \to x \dot{-} y \neq 0$.

$x = Sy \to y = \mathrm{prd}\,x$, so $x \dot{-} y = S\,\mathrm{prd}\,x \dot{-} \mathrm{prd}\,x = 1$ (by vi).

$x \dot{-} Sy \neq 0 \to \mathrm{prd}(x \dot{-} y) \neq 0$; so $x \dot{-} Sy \neq 0 \to x \dot{-} y \neq 0$ (by iii).

(ix) : $y<x \to x \dot{-} y \neq 0$; $x \dot{-} y \neq 0 \to S\,\mathrm{prd}(x \dot{-} y) = S(x \dot{-} Sy)$.

(x) : By induction on y .

1.7.4. Lemma. In qf-N-HA$^{\omega}$ $y<x \to x \dot{-} (x \dot{-} y) = y$.

Proof. Induction on y .

(1) $0 < x \to x \dot{-} (x \dot{-} 0) = x \dot{-} x = 0$ (1.7.3, (v)),

(2) $S0<x \to x \dot{-} (x \dot{-} S0) = x \dot{-} \mathrm{prd}\,x = 1$ (1.7.3, (vii)).

Assume

(3) $Sz<x \to x \dot{-} (x \dot{-} Sz) = Sz$.

If $SSz < x$, then $SSz \leq x$, hence $Sz < x$ (1.7.3, (viii)) ; with 1.7.3(ix) $x \dot{-} Sz = S(x \dot{-} SSz)$.

$\mathrm{prd}(x \dot{-} (x \dot{-} SSz)) = x \dot{-} S(x \dot{-} SSz) = x \dot{-} (x \dot{-} Sz) = Sz \neq 0$, so $x \dot{-} (x \dot{-} SSz) = SSz$ (1.7.3, (vii)).

Hence

(4) (3) $\to [(SSz<x) \to x \dot{-} (x \dot{-} SSz) = SSz]$.

(2), (4) give the induction rule

(5) $Sy<x \to (x \dot{-} (x \dot{-} Sy) = Sy$.

Hence also by induction from (1), (5), the assertion of the lemma.

1.7.5. Simultaneous recursion in qf-N-HA$^{\omega}$.

For each sequence of types $\sigma_1, \ldots, \sigma_n$ we wish to construct a sequence of constants $R^i_{\sigma_1,\ldots,\sigma_n}$, $i = 1,\ldots,n$ (abbreviated as $\underline{R}$) such that for sequences of variables $\underline{\underline{x}} \equiv x_1,\ldots,x_n$, $\underline{\underline{y}} \equiv y_1,\ldots,y_n$, where $x_i \in \sigma_i$, $y_i \in (\sigma_1)\ldots(\sigma_n)(0)\sigma_i$

(1) $\qquad \underline{\underline{R}}\,\underline{\underline{x}}\,\underline{\underline{y}}\,0 = \underline{\underline{x}}, \quad \underline{\underline{R}}\,\underline{\underline{x}}\,\underline{\underline{y}}(Sz) = \underline{\underline{y}}(\underline{\underline{R}}\,\underline{\underline{x}}\,\underline{\underline{y}}\,z)z$.

In order to obtain constants $\underline{\underline{R}}$ as required in (1), it is sufficient to establish the recursion rule :

(2) $\left\{\begin{array}{l}\text{If } \underline{\underline{t}}, \underline{\underline{s}} \text{ are sequences of closed terms (of fitting types), there} \\ \text{is a sequence of closed terms } \underline{\underline{T}} \text{ such that } \underline{\underline{T}}0 = \underline{\underline{t}}, \\ \underline{\underline{T}}(Sz) = \underline{\underline{s}}z(\underline{\underline{T}}z) .\end{array}\right.$

To obtain (1) from (2), we apply (2) with $\lambda \underline{\underline{x}}\,\underline{\underline{y}}\,.\,\underline{\underline{x}}$, $\lambda v^0 \underline{\underline{u}}\,\underline{\underline{x}}\,\underline{\underline{y}}\,.\,\underline{\underline{y}}(\underline{\underline{u}}\,\underline{\underline{x}}\,\underline{\underline{y}})v$ (with $\underline{\underline{u}}$ and $\underline{\underline{x}}$ having the same types) for $\underline{\underline{t}}$, $\underline{\underline{s}}$ respectively. Then $\underline{\underline{T}}$ satisfies the equations for $\lambda z \underline{\underline{x}}\,\underline{\underline{y}}\,.\,\underline{\underline{R}}\,\underline{\underline{x}}\,\underline{\underline{y}}\,z$, i.e.

$$\underline{\underline{T}}0\,\underline{\underline{x}}\,\underline{\underline{y}} = \underline{\underline{x}}, \quad \underline{\underline{T}}(Sz)\,\underline{\underline{x}}\,\underline{\underline{y}} = \underline{\underline{y}}(\underline{\underline{T}}z\,\underline{\underline{x}}\,\underline{\underline{y}})z ,$$

and therefore $\lambda \underline{\underline{x}}\,\underline{\underline{y}}\,z\,.\,\underline{\underline{T}}z\,\underline{\underline{x}}\,\underline{\underline{y}}$ satisfies the equations (1) for $\underline{\underline{R}}$.

The method for establishing (2) below seems to be due to Schütte * (cf. Hindley, Lercher and Seldin 1972, page 156) ; a stronger result (simultaneous course-of-values recursion) is established in Diller and Schütte 1971 ; this implies (2).

1.7.6. Pairing functions for objects of equal type.

Let P_σ be a pairing function of type $(\sigma)(\sigma)(0)\sigma$, satisfying

$$P_\sigma x^\sigma y^\sigma 0 = x^\sigma, \quad P_\sigma x^\sigma y^\sigma (Sz) = y^\sigma .$$

For P_σ we may take

$$P_\sigma \equiv_{def} \lambda x^\sigma y^\sigma z^0 . Rx(\lambda u^\sigma v^0 . y)z .$$

1.7.7. Theorem * (Schütte). The recursion rule (2) of 1.7.5 holds.

Proof. We establish this by induction on the length of the sequences $\underline{\underline{t}}$, $\underline{\underline{s}}$. For length 1 the solution is given by $Rt(\lambda uv . svu)$. Assume (induction hypothesis) (2) to have been established for $\underline{\underline{t}}$, $\underline{\underline{s}}$ of length n. We wish to construct $\tilde{t}_1, \ldots, \tilde{t}_{n+1}$ such that

(3) $\left\{\begin{array}{l}\tilde{t}_i(0) = a_i \\ \tilde{t}_i(Sz) = b_i z(\tilde{t}_1 z)\ldots(\tilde{t}_{n+1}z)\end{array}\right.$

where a_i, b_i $(1 \leq i \leq n+1)$ are constants of the appropriate type.

Let $R' = \lambda xyz . Rx(\lambda uv . yvu)z$, so that $R'xy0 = z$, $R'xy(Sz) = yz(R'xyz)$.

----

* I am indebted to R. Hindley for communicating to me a correction to page 156 of Hindley, Lercher and Seldin 1972, together with the proof given below.

We put

$$t = \lambda u_1 \ldots u_n \,.\, R'a_{n+1}(\lambda v \,.\, b_{n+1} v(u_1 v) \ldots (u_n v))$$
$$t_i = \lambda v u_1 \ldots u_n w \,.\, P(u_i w)(b_i v(u_1 v) \ldots (u_n v)(t u_1 \ldots u_n v))(w \dot{-} v) \,.$$

By induction hypothesis, there are $t_1^*, \ldots, t_n^*$ such that, for $1 \leq i \leq n$

$$t_i^* 0 = \Pi a_i \,, \quad t_i^*(Sz) = t_i z(t_1^* z) \ldots (t_n^* z) \,.$$

We put

$$\tilde{t}_i \equiv \lambda z \,.\, t_i^* zz \,, \quad 1 \leq i \leq n$$
$$\tilde{t}_{n+1} \equiv \lambda z \,.\, t(t_1^* z) \ldots (t_n^* z) z \,.$$

Now (3) is proved by induction.

Case a, $1 \leq i \leq n$.

$$\tilde{t}_i 0 = t_i^* 00 = \Pi a_i 0 = a_i$$
$$\tilde{t}_i(Sz) = t_i^*(Sz)(Sz) = t_i z(t_1^* z) \ldots (t_n^* z)(Sz) =$$
$$= P(t_i^* z(Sz))(b_i z(t_1^* zz) \ldots (t_n^* zz)(t(u_1^* z) \ldots (u_n^* z)z))1$$
$$= b_i z(t_1^* zz) \ldots (t_n^* zz)(t(u_1^* z) \ldots (u_n^* z)z)$$
$$= b_i z(\tilde{t}_i z) \ldots (\tilde{t}_n z)(\tilde{t}_{n+1} z) \,.$$

Case b, $i = n + 1$.

We first establish

(4) $\quad t_i^*(z+w)z = t_i^* zz \quad$ for $1 \leq i \leq n$, all w .

We prove this by induction on w . For $w = 0$ (4) is immediate. For $w > 0$,

$$t_i^*(z + Sw)z = t_i^*(S(z + w))z =$$
$$= t_i(z + w)(t_1^*(z + w)) \ldots (t_n^*(z + w))z =$$
$$= P(t_i^*(z + w)z)\Psi(z \dot{-} (z+w)) = t_i^*(z + w)z = t_i^* zz \,,$$

(the exact form of the expression Ψ is irrelevant here ; we must use $z \dot{-} (z+w) = 0$, derived by induction on w from 1.7.3 (v) as basis). Now we establish, for all z, w

(5) $\quad \tilde{t}_{n+1} z = R'a_{n+1}(\lambda v \,.\, b_{n+1} v(t_1^*(z+w)v) \ldots (t_n^*(z+w)v))z \,.$

We use induction on z .

$$\tilde{t}_{n+1} 0 = t(t_1^* 0) \ldots (t_n^* 0)0 =$$
$$= R'a_{n+1}(\lambda v \,.\, b_{n+1} v(t_1^* 0v) \ldots (t_n^* 0v))0 = a_{n+1}$$

which is equal to the right hand side of (5) for $z = 0$.

$$\tilde{t}_{n+1}(Sz) = t(t_1^*(Sz)) \ldots (t_n^*(Sz))(Sz) =$$
$$= R'a_{n+1}(\lambda v . b_{n+1} v(t_1^*(Sz)v) \ldots (t_n^*(Sz)v)(Sz) =$$
$$= b_{n+1} z(t_1^*(Sz)z) \ldots (t_n^*(Sz)z)\Psi \,,$$

where

$$\Psi \equiv R'a_{n+1}(\lambda v . b_{n+1}\, v\, (t_1^*(Sz)v) \ldots (t_n^*(Sz)v))z .$$

By the induction hypothesis for z, using $w = 1$,

$$\Psi = \tilde{t}_{n+1}z .$$

Also, by the induction hypothesis for z, using Sw for w

$$\tilde{t}_{n+1}z = \Psi' = R'a_{n+1}[\lambda v . b_{n+1}\, v\, (t_1^*(z+Sw)v) \ldots (t_n^*(z+Sw)v)]z ,$$

hence

$$\begin{aligned} \tilde{t}_{n+1}(Sz) &= b_{n+1}z(t_1^*(Sz)z) \ldots (t_n^*(Sz)z)\Psi' \\ &= b_{n+1}z(t_1^*(Sz+w)z) \ldots (t_n^*(Sz+w))\Psi' \qquad \text{(by (4))} \\ &= R'a_{n+1}[\lambda v . b_{n+1}\, v\, (t_1^*(Sz+w)v) \ldots (t_n^*(Sz+w)v)](Sz) . \end{aligned}$$

This establishes (5). Now we can complete the proof :

$$\begin{aligned} \tilde{t}_{n+1}(Sz) &= t(t_1^*(Sz)) \ldots (t_n^*(Sz))Sz = \\ &= R'a_{n+1}[\lambda v . b_{n+1}\, v\, (t_1^*(Sz)v) \ldots (t_n^*(Sz)v)](Sz) = \\ &= b_{n+1}z(t_1^*(Sz)z) \ldots (t_n^*(Sz)z)\Psi'' , \end{aligned}$$

where

$$\begin{aligned} \Psi'' &\equiv R'a_{n+1}(\lambda v . b_{n+1}z(t_1^*(Sz)v) \ldots (t_n^*(Sz)v)z = \\ &= \tilde{t}_{n+1}z , \quad \text{hence} \end{aligned}$$

$$\begin{aligned} \tilde{t}_{n+1}(Sz) &= b_{n+1}z(t_1^*zz) \ldots (t_n^*zz)(\tilde{t}_{n+1}z) \\ &= b_{n+1}z(\tilde{t}_1 z) \ldots (\tilde{t}_n z)(\tilde{t}_{n+1}z) , \end{aligned}$$

and since also $\tilde{t}_{n+1}0 = t(t_1^*0) \ldots (t_n^*0)0 = R'a_{n+1}\xi 0 = a_{n+1}$ (the form of ξ is irrelevant) the proof is completed.

1.7.8 - 1.7.10. <u>The induction lemma for</u> $qf - \underset{\sim}{N} - \underset{\sim\sim}{HA}^{\omega}$.

1.7.8. <u>Lemma.</u> Let $\underline{\underline{T}}$ be a sequence of terms of $\underset{\sim}{N} - \underset{\sim\sim}{HA}^{\omega}$, and let $\underline{\underline{t}}$ be a sequence of terms defined by means of the recursion operator such that

$$\underline{\underline{t}}\, 0x\, \underline{\underline{v}} = \underline{\underline{v}} , \quad \underline{\underline{t}}\, (Sy)\, x\, \underline{\underline{v}} = \underline{\underline{T}}(x \dot{-} Sy)(\underline{\underline{t}}\, yx\, \underline{\underline{v}}) ,$$

where $x, y \in 0$. Let Q be a predicate such that ($x, \underline{\underline{v}}$ not free in Γ) in $qf - \underset{\sim}{N} - \underset{\sim\sim}{HA}^{\omega}$:

$$\Gamma \vdash Q(x, \underline{\underline{T}}x\underline{\underline{v}}) \rightarrow Q(Sx, \underline{\underline{v}}) , \quad \Gamma \vdash Q(0, \underline{\underline{v}}) .$$

Then in $qf - \underset{\sim}{N} - \underset{\sim\sim}{HA}^{\omega}$

$$\Gamma \vdash z < x \rightarrow [Q(z, \underline{\underline{t}}(x \dot{-} z)x\underline{\underline{v}}) \rightarrow Q(Sz, \underline{\underline{t}}(x \dot{-} Sz)x\underline{\underline{v}})] .$$

<u>Proof.</u> In order not to encumber our typography we let $\underline{\underline{t}} \equiv t$, $\underline{\underline{v}} \equiv v$, $\underline{\underline{T}} \equiv T$. Assume $z < z$, Γ. Then $x \dot{-} z = S(x \dot{-} Sz)$ (1.7.3, (ix)) ;

$$t(x \dot{-} z)xv = t(S(x \dot{-} Sz))xv =$$
$$= T(x \dot{-} S(x \dot{-} Sz))(t(x \dot{-} Sz)xv) =$$
$$= T(x \dot{-} (x \dot{-} z))(t(x \dot{-} Sz)xv) =$$
$$= Tz(t(x \dot{-} Sz)xv .$$

Since $\vdash Q(x, Txv) \rightarrow Q(Sx, v)$, it follows that

$$Q(z, Tz\ (t(x \dot{-} Sz)xv)) \rightarrow Q(Sz,\ t(x \dot{-} Sz)xv) .$$

This implies $Q(z,t(x \dot{-} z)xv) \rightarrow Q(Sz,t(x \dot{-} Sz)xv)$.

1.7.9. Lemma. When Q satisfies the conditions of the previous lemma, then in $\mathrm{qf}\text{-}\underset{\sim}{N}\text{-}\underset{\sim}{HA}^{\omega}$

$$\Gamma \vdash y \leq x \rightarrow Q(y, \underline{\underline{t}}(x \dot{-} y)x\underline{\underline{v}}) .$$

Proof. Induction on y.

$$0 \leq x \rightarrow Q(0, \underline{\underline{t}}xx\underline{\underline{v}}) \qquad (\text{since } \Gamma \vdash Q(0,v)).$$

Assume

$$z \leq x \rightarrow Q(z, \underline{\underline{t}}(x \dot{-} z)x\underline{\underline{v}}) .$$

Then

$$Sz \leq x \rightarrow z < x, \quad \text{hence} \quad Sz \leq x \rightarrow (Q(z,\underline{\underline{t}}(x \dot{-} z)x\underline{\underline{v}}) \rightarrow Q(Sz,\underline{\underline{t}}(x \dot{-} Sz)x\underline{\underline{v}})),$$
$$Sz \leq x \rightarrow z \leq x, \quad \text{hence} \quad Sz \leq x \rightarrow Q(z,\underline{\underline{t}}(x \dot{-} z)x\underline{\underline{v}}).$$

hence

$$Sz \leq x \rightarrow Q(Sz,\underline{\underline{t}}(x \dot{-} Sz)x\underline{\underline{v}}) .$$

1.7.10. Induction lemma. In $\mathrm{qf}\text{-}\underset{\sim}{N}\text{-}\underset{\sim}{HA}^{\omega}$, if $\Gamma \vdash Q(0,\underline{\underline{v}})$, $\Gamma \vdash Q(x,Tx\underline{\underline{v}}) \rightarrow Q(Sx,\underline{\underline{v}})$, then $\Gamma \vdash Q(x,\underline{\underline{v}})$ ($x,\underline{\underline{v}}$ not occurring free in Γ).
Proof. By the previous lemma, $\Gamma \vdash x \leq x \rightarrow Q(x,\underline{\underline{t}}(x \dot{-} x)x\underline{\underline{v}})$, hence $\Gamma \vdash Q(x,\underline{\underline{t}}\,0\,x\underline{\underline{v}})$, i.e. $\Gamma \vdash Q(0,\underline{\underline{v}})$.
Note that for $\underline{\underline{v}}$ consisting of a single variable, the proof only requires simple recursion.

1.7.11. Theorem. (Replacement of R_σ by the iterator J_τ; Diller and Schütte 1971.) If we replace in $\mathrm{qf}\text{-}\underset{\sim}{N}\text{-}\underset{\sim}{HA}^{\omega}$ the constants R_σ with its corresponding axioms by a constant J_τ, the iterator of type τ (for each $\tau \in T$) satisfying

$$J_\tau xy0 = x, \quad J_\tau xy(Sz) = y(J_\tau xyz) \qquad (x \in \tau,\ y \in (\tau)\tau),$$

then a constant satisfying the axioms for R_σ becomes definable.
Proof. The λ-operator can be defined as before. In terms of J_σ we may define P_σ as $\lambda x^\sigma y^\sigma . J_\sigma\, x^\sigma(\Pi_{\sigma,\sigma} y^\sigma)$; then obviously

$$P_\sigma x^\sigma y^\sigma 0 = x^\sigma, \qquad P_\sigma x^\sigma y^\sigma (Sz) = y^\sigma .$$

Next we define $U \equiv \lambda x^1 . J_o(x^1(SO))S$. Then

$$Ux^10 \quad = x^1(SO)$$
$$Ux^1(SO) = S(x^1(SO)) .$$

This function enables us to define a predecessor function :

$$prd \equiv \lambda y^o . J(\Pi_{o,o}0)U y^o 0 .$$

We prove by induction on y

$$J(\Pi_{o,o}0)U y (SO) = y .$$

Then it follows that

$$prd0 = J(\Pi_{o,o}0)U 0 0 = \Pi_{o,o}0 0 = 0$$
$$prd(Sy) = J(\Pi_{o,o}0)U(Sy)0 = U(J(\Pi_{o,o}0)Uy)0 =$$
$$= J(\Pi_{o,o}0)Uy(SO) = y .$$

Now define Q_τ as

$$Q_\tau \equiv \lambda xyz[x(y(prd\,z))(prd\,z)] \quad (z \in 0,\ y \in (0)\tau,\ x \in (\tau)(0)\tau) .$$

Then

$$Q_\tau xy(Sz) = x(yz)z .$$

Finally we put

$$R_\tau \equiv \lambda x^\tau y^{(\tau)(o)\tau} z^o[J_{(o)\tau}(\Pi_{\tau,o}x)(Q_\tau y)zz]$$

and then

$$R_\tau x^\tau y^{(\tau)(o)\tau} 0 = x^\tau ,\ R_\tau x^\tau y^{(\tau)(o)\tau}(Sz) = y(Rxyz)z .$$

Q.e.d.

Remark. In the sequel we have usually dealt directly with R_σ as a primitive ; occasionally, in applications, there might be a slight advantage in using the iterator as a primitive.

1.7.12. Simultaneous recursion and the induction lemma in $qf - \underset{\sim}{HA}^\omega$.

We note that the proof of the induction lemma (1.7.10), provided simultaneous recursion is available, can be carried over to $qf - \underset{\sim}{HA}^\omega$ without difficulty. For the case $\underset{=}{v} = v$ (i.e. $\underset{=}{v}$ consists of a single variable) we only need simple recursion.

In the proof of closure under simultaneous recursion, the crucial step is in establishing (5) in 1.7.7. Let us abbreviate (5) as

$$A(z,w) \equiv (\mathfrak{k}_{n+1}z = s[z,w])$$

where $s[z,w]$ represents the left hand side of (5) in 1.7.7.
Assume $\mathfrak{k}_{n+1}z \in \sigma,\ \sigma \equiv (\tau)0$. Inspection of the argument in 1.7.7 shows that by the following sequence of equalities :

$$u^{\sigma}(\tilde{t}_{n+1}(Sz)) = u^{\sigma}(b_{n+1}z(t_1^*(Sz)z)\dots(t_n^*(Sz)z)s[z,w]) =$$
$$= u^{\sigma}(b_{n+1}z(t_1^*(Sz)z)\dots(t_n^*(Sz)z)(\tilde{t}_{n+1}z)) =$$
$$= u^{\sigma}(b_{n+1}z(t_1^*(Sz)z)\dots(t_n^*(Sz)z)s[z,Sw]) =$$
$$= u^{\sigma}s[Sz,w]$$

the assertion $u^{\sigma}(\tilde{t}_{n+1}(Sz)) = u^{\sigma}s[Sz,w]$ can be obtained from

$$(1)\quad \begin{cases} r[u^{\sigma},s[z,w],z] = r[u^{\sigma},\tilde{t}_{n+1}z,z], \\ r[u^{\sigma},\tilde{t}_{n+1}z,\ z] = r[u^{\sigma},s[z,Sw],z], \end{cases}$$

where $r[u^{\sigma},v_0^{\tau},z]$ stands for $u^{\sigma}(b_{n+1}z(t_1^*(Sz)z)\dots(t_n^*(Sz)z)v_0^{\tau})$.

Let us put $T_0 \equiv \lambda u^{\sigma}zv_0^{\tau}.\, r[u^{\sigma},v_0^{\tau},z]$; then (1) is equivalent to

$$T_0u^{\sigma}z(\tilde{t}_{n+1}\ z) = T_0u^{\sigma}z\ s[z,Sw]$$
$$T_0u^{\sigma}z(\tilde{t}_{n+1}\ z) = T_0u^{\sigma}z\ s[z,1].$$

Therefore, intuitively:

$$(2)\quad \begin{cases} \forall w\le Sx(T_0u^{\sigma}z(\tilde{t}_{n+1}z) = T_0u^{\sigma}z\ s[z,w]) \rightarrow \\ \rightarrow \forall w\le x(u^{\sigma}(\tilde{t}_{n+1}(Sz)) = u^{\sigma}s[Sz,w]). \end{cases}$$

In the quantifier-free system bounded quantification can be expressed by introducing a function f by primitive recursion, such that

$$fu^{\sigma}z\ 0 = |u^{\sigma}(\tilde{t}_{n+1}z) - u^{\sigma}s[z,0]|$$
$$fu^{\sigma}z(Sx) = fu^{\sigma}zx + |u^{\sigma}(\tilde{t}_{n+1}z) - u^{\sigma}s[z,Sx]|.$$

Then (2) can be expressed as

$$f(T_0u^{\sigma}z)z(Sx) = 0 \rightarrow fu^{\sigma}(Sz)x = 0.$$

Now let p_{σ}, q_{σ} for all $\sigma \in T$ be defined by

$$p_0x^0 = x^0,\quad p_{(\sigma)\tau}x^{(\sigma)\tau} = p_{\tau}(x^{(\sigma)\tau}0^{\sigma})$$
$$q_0x^0 = x^0,\quad q_{(\sigma)\tau}x^0 = \lambda x^{\sigma}.q_{\tau}x^0.$$

Obviously

$$p_{\sigma}q_{\sigma}x^0 = x^0.$$

So q_{σ} provides an embedding of the natural numbers in type σ.

Now put

$$B(v^{(0)\sigma},z) \equiv_{def} f(v^{(0)\sigma}0)z(p_{\sigma}v^{(0)\sigma}1)) = 0$$

and let

$$T_1 \equiv \lambda v^{(0)\sigma}z.\, p(T_0(v^{(0)\sigma}0)z)(q_{\sigma}(Sp_{\sigma}(v^{(0)\sigma}1))),$$

then

$$B(T_1v^{(0)\sigma}z,z) \rightarrow B(v^{(0)\sigma},Sz).$$

Since obviously $u^\sigma(\xi_{n+1}0) = u^\sigma s[0,w]$, we also have $B(v^{(o)\sigma},0)$. Therefore, by the induction lemma 1.7.10 $B(v^{(o)\sigma},z)$. Substitution of $P_\sigma u^\sigma(q_\sigma w^o)$ for v yields $\vdash u^\sigma(\xi_{n+1}z) = u^\sigma s[z,w]$.

Thus we obtain simultaneous recursion in qf - $\underset{\sim}{\mathrm{HA}}^\omega$, and now we can extend the induction lemma for qf - $\underset{\sim}{\mathrm{HA}}^\omega$ to an arbitrary sequence of variables $\underline{\underline{v}}$.

§ 8. More about $\underset{\sim}{N} - \underset{\approx}{HA}^{\omega}$.

1.8.1. Contents of the section. This section contains further miscella-
neous information of $\underset{\sim}{N} - \underset{\approx}{HA}^{\omega}$, which may be consulted by the reader when the
need arises.

First the extension of the type structure by Cartesian product formation,
together with the addition of pairing operators as primitives is discussed :
in 1.8.2 it is shown that this extension constitutes an expansion of any
theory in the language of $\underset{\sim}{N} - \underset{\approx}{HA}^{\omega}$.

Subsection 1.8.4 is devoted to the discussion of the λ - operator as a
primitive.

Subsections 1.8.5 - 1.8.9 discuss reductions of the type structures.

1.8.2. Theorem (Cartesian product types and pairing operators).
Let the type structure $\underset{\approx}{T}$ be extended to $\underset{\approx}{T}'$ by adding a clause

T 3) $\sigma, \tau \in \underset{\approx}{T}' \Rightarrow \sigma \times \tau \in \underset{\approx}{T}'$

and replacing in T 1 , T 2 $\underset{\approx}{T}$ by $\underset{\approx}{T}'$ (1.6.2).
Furthermore, we assume the existence of constants $D_{\sigma, \tau} \in (\sigma)(\tau)\sigma \times \tau$,
$D'_{\sigma, \tau} \in (\sigma \times \tau)\sigma$, $D''_{\sigma, \tau} \in (\sigma \times \tau)\tau$ satisfying

$$D'(D\,xy) = x , \qquad D''(D\,xy) = y , \qquad D(D'x)(D''x) = x .$$

Let $\underset{\sim}{N} - \underset{\approx}{HA}^{\omega}_p$ denote the extension of $\underset{\sim}{N} - \underset{\approx}{HA}^{\omega}$ thus obtained. Then $\underset{\sim}{N} - \underset{\approx}{HA}^{\omega}_p$
is an expansion of $\underset{\sim}{N} - \underset{\approx}{HA}^{\omega}$, and qf - $\underset{\sim}{N} - \underset{\approx}{HA}^{\omega}_p$ (defined analogously to
qf - $\underset{\sim}{N} - \underset{\approx}{HA}^{\omega}$) is an expansion of $\underset{\sim}{N} - \underset{\approx}{HA}^{\omega}_p$.
Proof. To each type $\sigma \in \underset{\approx}{T}'$ we assign a sequence σ^* of types in $\underset{\approx}{T}$, as
follows :
(i) $0^* \equiv (0)$.
Let $\sigma^* \equiv (\sigma_1, \ldots, \sigma_m)$, $\tau^* \equiv (\tau_1, \ldots, \tau_n)$.
(ii) $(\sigma \times \tau)^* \equiv (\sigma_1, \ldots, \sigma_m, \tau_1, \ldots, \tau_n)$,
(iii) $[(\sigma)\tau]^* \equiv ((\sigma_1) \ldots (\sigma_m)\tau_1, \ldots, (\sigma_1) \ldots (\sigma_m)\tau_n)$.

We define a sequence $\underset{\approx}{\Pi}_{\sigma^*, \tau^*} \equiv (\Pi^1_{\sigma^*, \tau^*}, \ldots, \Pi^m_{\sigma^*, \tau^*})$ such that

$$\underset{\approx}{\Pi}_{\sigma^*, \tau^*} \equiv \lambda \underset{=}{x} \, \lambda \underset{=}{y} . \underset{=}{x} .$$

Further we define a sequence $\underset{\approx}{\Sigma}_{\rho^*, \sigma^*, \tau^*}$ such that

$$\underset{\approx}{\Sigma}_{\rho^*, \sigma^*, \tau^*} \equiv \lambda \underset{=}{x} \, \underset{=}{y} \, \underset{=}{z} . \underset{=}{x} \, \underset{=}{z} (\underset{=}{y} \, \underset{=}{z}) .$$

For $\underset{=}{R}_{\sigma^*}$ we take the sequence as defined in 1.7.5.
$\underset{=}{t} = \underset{=}{s}$, where $\underset{=}{t} \equiv (t_1, \ldots, t_m)$, $\underset{=}{s} \equiv (s_1, \ldots, s_m)$, is interpreted as
$t_1 = s_1 \,\&\, \ldots \,\&\, t_m = s_m$.
Now we define a mapping Γ on terms and formulae of $\underset{\sim}{N} - \underset{\approx}{HA}^{\omega}_p$, as follows.

(i) To each x^σ, $\sigma \in \underset{\approx}{T}'$ we assign a sequence $\underline{\underline{x}} \equiv (x_1^{\sigma_1},\ldots,x_n^{\sigma_n})$, where $(\sigma_1,\ldots,\sigma_n) \equiv \sigma^*$. If x^σ, x'^σ are distinct variables, then Γx^σ and $\Gamma x'^\sigma$ have no element in common.

(ii) $\Gamma\Pi_{\sigma,\tau} \equiv \underline{\underline{\Pi}}_{\sigma^*,\tau^*}$, $\Gamma\Sigma_{\rho,\sigma,\tau} \equiv \underline{\underline{\Sigma}}_{\rho^*,\sigma^*,\tau^*}$, $\Gamma R_\sigma \equiv \underline{\underline{R}}_{\sigma^*}$,
$\Gamma 0 = 0$, $\Gamma(S) = S$.

(iii) If $\Gamma x^\sigma \equiv \underline{\underline{x}}$, $\Gamma y^\tau \equiv \underline{\underline{y}}$, we take for $\Gamma D_{\sigma,\tau}$ the concatenation of the sequences of operators $\underline{\underline{\Pi}}_{\sigma^*,\tau^*}$ and $\lambda \underline{\underline{xy}} . \underline{\underline{y}}$, and for $\Gamma D'_{\sigma,\tau}$, $\Gamma D''_{\sigma,\tau}$ we take

$$\Gamma D'_{\sigma,\tau} \equiv \underline{\underline{\Pi}}_{\sigma^*,\tau^*}, \quad \Gamma D''_{\sigma,\tau} = \lambda \underline{\underline{x}} \lambda \underline{\underline{y}} . \underline{\underline{y}} .$$

(iv) $\Gamma tt' \equiv (\Gamma t)(\Gamma t')$.

(v) $\Gamma(t = s) \equiv (\Gamma t = \Gamma s)$.

(vi) Γ preserves propositional operators and $\wedge$, i.e. $\Gamma(\wedge) \equiv \wedge$,
$\Gamma(A \circ B) \equiv \Gamma(A) \circ \Gamma(B)$ for $\circ = \rightarrow, \vee, \&$.

(vii) $\Gamma(\forall x^\sigma A) \equiv \forall \Gamma x^\sigma(\Gamma(A))$, $\Gamma(\exists x^\sigma A) \equiv \exists \Gamma x^\sigma(\Gamma(A))$.

First note that Γ is the identity on formulae of $\underset{\sim}{N}-\underset{\sim\sim}{HA}^\omega$ (modulo renaming of variables). It remains to be shown, by induction on the length of deductions that $\underset{\sim}{N}-\underset{\sim\sim}{HA}_p^\omega \vdash A \Rightarrow \underset{\sim}{N}-\underset{\sim\sim}{HA}^\omega \vdash \Gamma(A)$. This turns out to be completely trivial.

1.8.3. Remark. The theorem also extends to certain extensions of $\underset{\sim}{N}-\underset{\sim\sim}{HA}^\omega$ obtained by adding definition schemata for functionals, if only the theory without Cartesian product types contains a "simultaneous" variant of the additional definition schemata (examples Ψ_σ for $\underset{\sim}{N}-\underset{\sim\sim\sim}{IDB}^\omega$, B_σ in § 1.9).

1.8.4. The λ-operator as a primitive notion.

Instead of having a theory $\underset{\sim}{E}-\underset{\sim\sim}{HA}^\omega$ with primitives $\Pi_{\rho,\sigma}$, $\Sigma_{\rho,\sigma,\tau}$ we may consider an alternative version $\lambda\underset{\sim}{E}-\underset{\sim\sim}{HA}^\omega$, with λx^σ as primitive operators. The description of $\lambda\underset{\sim}{E}-\underset{\sim\sim}{HA}^\omega$ is similar to the description of $\underset{\sim}{E}-\underset{\sim\sim}{HA}^\omega$, with the following differences:

(a) $\Pi_{\sigma,\tau}$, $\Sigma_{\rho,\sigma,\tau}$ are omitted from the list of constants; the operators λx^σ are added.

(b) The term-definition (1.6.4) is extended with a clause:

Tm 3) If $t \in Tm_\tau$, then $\lambda x^\sigma . t \in Tm_{(\sigma)\tau}$.

(c) The defining axioms for $\Pi_{\rho,\sigma}$, $\Sigma_{\rho,\sigma,\tau}$ are replaced by the rule of λ-conversion:

λ-CON $\quad (\lambda x^\sigma . t[x^\sigma])t'^\sigma = t[t']$ (t' free for x in t).

If we make changes (a), (b), (c) in $\underset{\sim}{E}-\underset{\sim\sim}{HA}_0^\omega$, we obtain a theory $\lambda\underset{\sim}{E}-\underset{\sim\sim}{HA}_0^\omega$ (cf. 2.4.18).

Now $\underset{\sim}{E}-\underset{\sim\sim}{HA}^\omega$ and $\lambda\underset{\sim}{E}-\underset{\sim\sim}{HA}^\omega$ are equivalent in the following sense: as has been shown in 1.6.8, we can define a λ-operator in $\underset{\sim}{E}$-$\underset{\sim}{HA}^{\omega}$ such that the rule of λ-conversion holds; conversely, in $\lambda\underset{\sim}{E}$-$\underset{\sim}{HA}^{\omega}$ we can define operators $\Pi_{\rho,\sigma}$, $\Sigma_{\rho,\sigma,\tau}$ satisfying the defining equations for the corresponding primitives in $\underset{\sim}{E}$-$\underset{\sim}{HA}^{\omega}$, by putting $\Pi_{\rho,\sigma} \equiv_{def} \lambda x^{\rho} y^{\sigma}.x^{\rho}$, $\Sigma_{\rho,\sigma,\tau} \equiv_{def} \lambda x^{(\rho)(\sigma)\tau} y^{(\rho)\sigma} z^{\rho}.xz(yz)$.

Hence the union of $\underset{\sim}{E}$-$\underset{\sim}{HA}^{\omega}$ and $\lambda\underset{\sim}{E}$-$\underset{\sim}{HA}^{\omega}$ is **a definitional** extension of both; and similarly for the quantifier-free theories qf-$\underset{\sim}{WE}$-$\underset{\sim}{HA}^{\omega}$ and λqf-$\underset{\sim}{WE}$-$\underset{\sim}{HA}^{\omega}$ (the latter defined in the obvious way).

The description of a λ-variant of (qf-) $\underset{\sim}{N}$-$\underset{\sim}{HA}^{\omega}$ or (qf-) $\underset{\sim}{I}$-$\underset{\sim}{HA}^{\omega}$ is not such a simple matter however. The problem is this: if we simply include the rule $s = t \Rightarrow \lambda x.s = \lambda x.t$ in our system (say $\underset{\sim}{H}$), then equality between closed terms of $\underset{\sim}{H}$ cannot be recursively decidable. In fact, if

$$S = \{j(\ulcorner s \urcorner, \ulcorner t \urcorner) \mid \underset{\sim}{H} \vdash s = t\}$$
$$T = \{j(\ulcorner s \urcorner, \ulcorner t \urcorner) \mid \underset{\sim}{H} \vdash s \neq t\}$$

(s, t closed terms of $\underset{\sim}{H}$), then S, T are recursively inseparable.

For let A,B be a pair of recursively enumerable, recursively inseparable sets (e.g. as in Rogers 1967, p. 94) such that

(1) $\underset{\sim}{H} \vdash A \cap B = \emptyset$.

Let $C \equiv \{x \mid \underset{\sim}{H} \vdash \bar{x} \notin A\}$.

Then C is recursively enumerable, $A \cap C = \emptyset$, $B \subset C$ (since $x \in B \Rightarrow \underset{\sim}{H} \vdash \bar{x} \in B$ (by completeness of $\underset{\sim}{H}$ for Σ_1^0-predicates) $\Rightarrow \underset{\sim}{H} \vdash \bar{x} \notin A$ (by (1))).

So A,C is a pair of recursively enumerable, recursively inseparable sets.

Now the statement $x \notin A$ is equivalent to $\forall y\, P(x,y)$ for some primitive recursive P. Let t be the closed term of $\underset{\sim}{H}$ representing the characteristic function of P. Then

$$x \in A \Leftrightarrow \exists y \neg Pxy \Leftrightarrow \exists y(\underset{\sim}{H} \vdash t\bar{x}\bar{y} \neq 0)$$
$$\Leftrightarrow \underset{\sim}{H} \vdash t\bar{x} \neq \lambda y.0;$$
$$x \in C \Leftrightarrow \underset{\sim}{H} \vdash P\bar{x}y = 0 \Leftrightarrow \underset{\sim}{H} \vdash t\bar{x} = \lambda y.0$$

(by the proposed rule $s = t \Rightarrow \lambda x.s = \lambda x.t$), and so the pair A,C is 1-1-reducible to S,T respectively, via the mapping $\lambda x.j(\ulcorner t\bar{x} \urcorner, \ulcorner \lambda y.0 \urcorner)$. (Rogers 1967, p. 80). Hence the pair S,T is also recursively inseparable.

The problem of the description of a λ-variant of (qf-) $\underset{\sim}{N}$-$\underset{\sim}{HA}^{\omega}$ or (qf-) $\underset{\sim}{I}$-$\underset{\sim}{HA}^{\omega}$ has therefore to be solved in a different manner, namely by distinguishing "ordinary" provable equality, and equality established by restricted means. A discussion of this possibility is better postponed till after the treatment of computability in Chapter II.

1.8.5 - 1.8.8. <u>Reduction to pure types.</u>

1.8.5. <u>Pure types.</u> The pure types $\underset{\approx}{P}$ are defined inductively by

T 1) $0 \in \underset{\approx}{P}$;

T 4) $\sigma \in \underset{\approx}{P} \rightarrow (\sigma)0 \in \underset{\approx}{P}$.

We introduce an abbreviation using natural numbers to indicate pure types:

$(n)0 \equiv_{def} n+1$.

We now wish to construct a mapping Ω of $\underset{\approx}{T}$ onto $\underset{\approx}{P}$, such that $\Omega\sigma = \sigma$ for $\sigma \in \underset{\approx}{P}$, and such that to each $\sigma \in \underset{\approx}{T}$ there exist mappings $\Gamma_\sigma \in (\sigma)\Omega\sigma$, $\Gamma'_\sigma \in (\Omega\sigma)\sigma$, Γ_σ and Γ'_σ definable in $\underset{\sim}{N}-\underset{\sim}{HA}^\omega$, such that in $qf-\underset{\sim}{WE}-\underset{\sim}{HA}^\omega$

$$\Gamma'_\sigma\Gamma_\sigma f = f, \quad \Gamma_\sigma\Gamma'_\sigma f' = f'.$$

This is done in a number of steps.

1.8.6. <u>Injection in higher types.</u>

We define mappings mp_j, with left-inverses pm_j which map the objects of type j into objects of type $j+1$, as follows:

$$(1)\quad \begin{cases} mp_0 \equiv \lambda x^0 y^0.x^0, \quad pm_0 \equiv \lambda x^1.x^1 0, \text{ and for } j>0 \\ mp_j \equiv \lambda x^j y^j.x^j(pm_{j-1}(y^j)), \\ pm_j \equiv \lambda x^{j+1} y^{j-1}.x^{j+1}(mp_{j-1}(y^{j-1})). \end{cases}$$

One readily verifies that $mp_j \in (j)j+1$, $pm_j \in (j+1)j$. By induction on j we find that

$$pm_j(mp_j(x^j)) = x^j.$$

For we have $pm_0(mp_0 x^0) = pm_0(\lambda y^0.x^0) = (\lambda y^0.x^0)0 = x^0$, and for $j>0$

$$\begin{aligned} pm_j(mp_j(x^j)) &= pm_j(\lambda y^j.x^j(pm_{j-1}(y^j)) = \\ &= \lambda y^{j-1}[\lambda y^j.x^j(pm_{j-1}(y^j))](mp_{j-1}(y^{j-1})) = \\ &= \lambda y^{j-1}.x^j(pm_{j-1}mp_{j-1}(y^{j-1})) = \lambda y^{j-1}.x^j(y^{j-1}) = x^j. \end{aligned}$$

Now we construct type-increasing mappings $mp_j^m \in (j)m$, with left-inverses $pm_j^m \in (m)j$ $(m \geq j)$ by

$$(2)\quad \begin{cases} mp_j^j \equiv \lambda x^j.x^j, \quad pm_j^j \equiv \lambda x^j.x^j, \\ mp_j^{m+1} \equiv \lambda x^j.\, mp_m mp_j^m(x^j), \\ pm_j^{m+1} \equiv \lambda x^{m+1}.\, pm_j^m(x^{m+1}). \end{cases}$$

1.8.7. <u>Coding of</u> <u>n-tuples for all pure types.</u>

Let j, j_1, j_2 denote the standard pairing function with inverses for the natural numbers. We extend these to all pure types by putting

$$j(x^{m+1}, y^{m+1}) \equiv \lambda z^m.\, j(x^{m+1}z^m, y^{m+1}z^m)$$

$$j_1(x^{m+1}) \equiv \lambda z^m . j_1 x^{m+1} z^m$$
$$j_2(x^{m+1}) \equiv \lambda z^m . j_2 x^{m+1} z^m .$$

This may be extended to p - tuples, putting

$$j^2 \equiv j, \quad j_1^2 \equiv j_1, \quad j_2^2 \equiv j_2,$$
$$j^{p+1}(x_1^n, \ldots, x_{p+1}^n) \equiv j(x_1^n, j^p(x_2^n, \ldots, x_{p+1}^n))$$
$$j_1^{p+1} x^n \equiv j_1 x^n, \quad j_{k+1}^{p+1} x^n \equiv j_k^p \, j_2 x^n \qquad \text{for } 1 \leq k \leq p .$$

Now we are able to describe the coding of p - tuples of different pure types:

$$j^p(x_1^{n(1)}, \ldots, x_p^{n(p)}) \equiv j^p(mp_{n(1)}^m x_1^{n(1)}, \ldots, mp_{n(p)}^m x_p^{n(p)}),$$
where $m = \max\{n(1), \ldots, n(p)\}$.

As inverses we have

$$j_{k,1}^{p,m}(x^m) \equiv pm_1^m \, j_k^p(x^m),$$

so that

$$j_{k,n(k)}^{p,m} \, j^p(x_1^{n(1)}, \ldots, x_p^{n(p)}) = x_k^{n(k)} .$$

1.8.8. <u>Description of</u> Ω, Γ_σ, Γ'_σ.

(i) If $\sigma = 0$, $\Omega\sigma \equiv \sigma$, $\Gamma_\sigma f = f$, $\Gamma'_\sigma f = f$.

(ii) If $\sigma = (\sigma_1) \ldots (\sigma_p)0$, $\Omega\sigma_i = n_i$ for $1 \leq i \leq p$, we put

$m = \max\{n_1, \ldots, n_p\}$, then $\Omega\sigma = m+1$, and

$$\Gamma_\sigma x^\sigma = \lambda y^m . x^\sigma(\Gamma'_{\sigma_1} j_{1,n_1}^{p,m} y^m)(\Gamma'_{\sigma_2} j_{2,n_2}^{p,m} y^m) \ldots (\Gamma'_{\sigma_p} j_{1,n_p}^{p,m} y^m),$$

$$\Gamma'_\sigma x^{m+1} = \lambda y_1^{\sigma_1} \ldots y_p^{\sigma_p} . x^{m+1} \, j^p(\Gamma_{\sigma_1} y_1, \ldots, \Gamma_{\sigma_p} y_p).$$

1.8.9. <u>Reduction to numerical types in</u> qf - WE - HA^ω.

Let us consider the following extension T_o of the type structure T:

T_o 1) $\quad 0 \in T_o$,

T_o 2) $\quad \sigma_1, \ldots, \sigma_n, \tau \in T_o \Rightarrow (\sigma_1 \times \ldots \times \sigma_n)\tau \in T_o$

and let T_n (the numerical types) be the substructure of T_o obtained restricting T_o 2) to

$$\sigma_1, \ldots, \sigma_n \in T_n \Rightarrow (\sigma_1 \times \ldots \times \sigma_n)0 \in T_n .$$

Note that in virtue of 1.8.2, qf - N - HA^ω when extended to T_o and with constants $D_{\sigma,\tau}$, $D'_{\sigma,\tau}$, $D''_{\sigma,\tau}$ added such that

$$D'D(x,y) = x, \quad D''D(x,y) = y, \quad D(D'x, D''x) = x$$

is an expansion of the original system qf - N - HA^ω; and similarly for the case which especially interests us, qf - WE - HA^ω.

The type structure $\underset{\approx}{T}_0$ may be reduced to $\underset{\approx}{T}_n$ by mappings Γ_σ, Γ'_σ, Ω as follows:

(a) $\sigma = 0$; then $\Omega\sigma = \sigma$, $\Gamma_\sigma x^\sigma = x^\sigma$, $\Gamma'_\sigma x^\sigma = x^\sigma$.

(b) $\sigma = (\rho_1 \times \ldots \times \rho_n)0$; $\Omega\sigma = (\Omega\rho_1 \times \ldots \times \Omega\rho_n)0$.

$$\Gamma_\sigma x^\sigma = \lambda[y_1,\ldots,y_n]x^\sigma(\Gamma'_{\rho_1}y_1,\ldots,\Gamma'_{\rho_n}y_n)$$

$$\Gamma'_\sigma x^{\Omega\sigma} = \lambda y_1 \ldots y_n . x^{\Omega\sigma}(\Gamma_{\rho_1}y_1,\ldots,\Gamma_{\rho_n}y_n).$$

(c) $\sigma = (\rho_1 \times \ldots \times \rho_n)\tau$, $\Omega\tau = (\mu_1 \times \ldots \times \mu_m)0$. Then $\Omega\sigma = (\Omega\rho_1 \times \ldots \Omega\rho_n \times \mu_1 \times \ldots \times \mu_m)0$, and

$$\Gamma_\sigma x^\sigma = \lambda[y_1,\ldots,y_n, z_1,\ldots,z_m](\Gamma_\tau x^\sigma(\Gamma'_{\rho_1}y_1,\ldots,\Gamma'_{\rho_n}y_n))(z_1^{\mu_1},\ldots,z_m^{\mu_m})$$

$$\Gamma'_\sigma x^{\Omega\sigma} = \lambda y_1 \ldots y_n . \Gamma'_\tau(\lambda z_1^{\mu_1} \ldots z_m^{\mu_m} . x^{\Omega\sigma}(\Gamma_{\rho_1}y_1,\ldots,\Gamma_{\rho_n}y_n, z_1,\ldots,z_m)).$$

Here $\lambda[y_1,\ldots,y_n]\ t[y_1,\ldots,y_n]$ expresses simultaneous abstraction w.r.t. $y_1,\ldots,y_n$; i.e. if t is of type τ, $y_i \in \sigma_i$ $(1 \leq i \leq n)$, then $\lambda[y_1,\ldots y_n]\ t(y_1,\ldots,y_n) \in (\sigma_1 \times \ldots \times \sigma_n)\tau$; but $\lambda y_1 \ldots y_n . t$ abbreviates $\lambda y_1 \lambda y_2 \ldots \lambda y_n . t$, hence is of type $(\sigma_1)(\sigma_2) \ldots (\sigma_n)\tau$. $x(y_1,\ldots,y_n)$ indicates application to the arguments $y_1,\ldots,y_n$ (simultaneously).

We leave it to the reader to verify that the following schemata for defining functionals of $\underset{\approx}{T}_n$ imply the definition schemata of qf - $\underset{\sim}{\text{WE}}$ - $\underset{\sim}{\text{HA}}^{\omega}$ (via the mappings described above) and vice versa.

(i) 0 is a constant of type 0 (with the usual axioms).

(ii) S is a constant of type 1 (with the usual axioms for successor).

(iii) If $t[y_1,\ldots,y_n]$ is a term of type 0, containing $y_1,\ldots,y_n$ free, $y_i \in \sigma_i$ $(1 \leq i \leq n)$, then $\lambda[y_1,\ldots,y_n]t[y_1,\ldots,y_n]$ is a term of type $(\sigma_1 \times \ldots \times \sigma_n)0$, and for $t_i \in \sigma_i$,
$(\lambda[y_1,\ldots,y_n]t[y_1,\ldots,y_n])(t_1,\ldots,t_n) = t[t_1,\ldots,t_n]$.

(iv) There is a term $\varphi \in ((\sigma_1 \times \ldots \times \sigma_p)0 \times ((\sigma_1 \times \ldots \times \sigma_p)0 \times 0 \times \sigma_1 \times \ldots \times \sigma_p) \times \times 0 \times \sigma_1 \times \ldots \times \sigma_p)0$, such that

$$\varphi(x,y,0,y_1^{\sigma_1},\ldots,y_p^{\sigma_p}) = x(y_1,\ldots,y_p)$$

$$\varphi(x,y,Sz,y_1^{\sigma_1},\ldots,y_p^{\sigma_p}) = y(\lambda[y_1,\ldots,y_p].\varphi(x,y,z,y_1,\ldots,y_p),z,y_1,\ldots,y_p).$$

(This reduction may then afterwards be combined with the reduction to pure types as described above.) The proof of the equivalence of the closure conditions first reduces (iv) to comparison with the recursion rule, which, however, is equivalent to asserting the existence of constants R_σ (see 1.7.5).

§ 9. Extensions of arithmetic.

1.9.1. Introduction. The various types of extensions of arithmetic may be divided, according to their language, into three categories:

1°) Extensions of arithmetic w.r.t. the same language or a language obtained by adding to $\mathcal{L}(\mathbf{HA})$ one or more predicate constants. Examples: **HA** with additional reflection principles, or addition of transfinite induction for a certain primitive recursive well-ordering, or **HA** with predicate constants for species or relations introduced by (iterated) generalized inductive definitions (g.i.d's). Such extensions are briefly described and discussed in 1.9.2 below. With respect to the various methods for metamathematical investigation, they behave in most respects like **HA** itself, i.e. as typical first-order systems.

2°) Extensions of arithmetic in a language with variables and quantifiers for species of natural numbers added. In this context full impredicative comprehension can be studied. Pure realizability and pure functional interpretations, as well as normalization theorems for natural deduction systems can be adapted to such systems, but not in such a straightforward way as for the systems under 1°). Variants such as mq-, or q-realizability do not readily extend to these systems. For a more detailed description, see 1.9.3 - 1.9.9 below.

3°) Extensions of arithmetic in a language obtained by addition of function symbols and function quantifiers to $\mathcal{L}(\mathbf{HA})$. Such extensions may (apart from "non-committal" very elementary ones, such as **EL** described in 1.9.10 below) be grouped according to their intended interpretation into two classes:

(A) The function variables are thought of as ranging over "lawlike" sequences (i.e. completely determined objects, given by a "law" or prescription). As long as our concept of "lawlike" has not been analyzed to a degree which prevents identification with "recursive", we may expect systems to be inspired by the idea of lawlike sequences to be consistent with the assumption that lawlike sequences are recursive (Church's thesis, see 1.11.7). To obtain systems which are proof-theoretically stronger than arithmetic, one has also to incorporate additions as under 1°).

Systems for lawlike sequences behave still very much like **HA** with respect to realizability and functional interpretations; with respect to Kripke semantics and natural deduction they have not yet been investigated.

(B) The function variables are thought of as ranging over some kind of choice sequences (i.e. sequences for which it is not assumed that they are a priori completely determined by a law). For detailed discussions of this concept see Troelstra 1968, 1969. Their essential feature is that they enforce certain continuity conditions on operators defined for all choice sequences. I.e., if Φ is a type 2 operator, defined for all choice sequences of a certain "universe", Φ satisfies

$$\forall\alpha \exists x \forall\beta(\bar{\alpha}x = \bar{\beta}x \rightarrow \Phi\alpha = \Phi\beta),$$

where $\bar{\alpha}x = \langle\alpha 0, \ldots, \alpha(x-1)\rangle$.

Continuity conditions do not increase proof-theoretic strength; but it is also possible to postulate the schema of bar induction for choice sequences (which is simply false for the universe of recursive sequences) and which <u>does</u> increase proof-theoretic strength.
Realizability and functional interpretations can be adapted to these systems (replacing partial-recursive-function application $\{.\}(.)$ by continuous-function application).

For a description of the principal systems which fall under this heading, see 1.9.19-24 below.

1.9.2. <u>Extensions of arithmetic expressed in $\mathcal{L}(\underset{\sim}{HA})$ or $\mathcal{L}(\underset{\sim}{HA})$ extended by relation constants.</u>

The most obvious extension is obtained by addition of a local reflection principle to $\underset{\sim}{HA}$:

$$\mathrm{RF}(\underset{\sim}{HA}) \quad \mathrm{Proof}_{HA}(x, \ulcorner A \urcorner) \rightarrow A \quad (A \text{ closed})$$

or a uniform reflection principle

$$\mathrm{RFN}(\underset{\sim}{HA}) \quad \mathrm{Proof}_{HA}(x, \ulcorner A\bar{y} \urcorner) \rightarrow Ay \quad (\forall y Ay \text{ closed}).$$

For a general discussion of such principles, see <u>Kreisel and Levy</u> 1968.
Another method of extension is the addition of the schema of transfinite induction for certain arithmetically definable (in fact, primitive recursive) well-orderings of the natural numbers; i.e. if $<$ is such an ordering, we add

$$\mathrm{TI}(<) \quad \forall x[(\forall y < x)Ay \rightarrow Ax] \rightarrow \forall y Ay.$$

A third, and very interesting possibility is the addition of constants for species introduced by generalized inductive definitions (g.i.d.).

Assume P_A to be a new (unary) predicate constant not occurring in the language $\mathcal{L}$. Let $\underset{\sim}{H}$ be a system with $\mathcal{L}(\underset{\sim}{H}) = \mathcal{L}$, and let $A(P,x) \in \mathcal{L}[P]$, such that A is "monotone":

$$\underset{\sim}{H}[P,P'] \vdash A(P,x) \;\&\; \forall x(Px \rightarrow P'x) \rightarrow A(P'x)$$

(where $\underset{\sim}{H}[P,P']$ is as $\underset{\sim}{H}$, but relative $\mathcal{L}[P,P']$).
Then P_A is said to be introduced by a g.i.d., if we add an axiom and a schema:

$$A(P_A,x) \rightarrow P_A x$$

and for all Q in $\mathcal{L}(P_A)$

$$\forall x[A(Q,x) \rightarrow Qx] \rightarrow \forall x[P_A x \rightarrow Qx] .$$

The best known example is O, the set of recursive ordinals introduced by Kleene (see Kleene 1944, 1955, or Rogers 1967, § 11.7, § 11.8). (To bring the definition in the form described above, we have to rewrite the definition). A simplified version is obtained by taking e.g.

$$A_0(Q,x) \equiv (x=1) \vee (Q(x)_0 \,\&\, x=2^{(x)_0}) \vee [x=3.5^{(x)_2} \,\&\, \forall y(!\{(x)_2\}(y) \,\&\, \&\, Q(\{(x)_2\}(y)))] .$$

Still simpler is

$$A_{P_1}(Q,x) \equiv Q0 \vee \forall y(Q[x](y)) ;$$

here $\lambda xy.[x](y)$ is an enumerating function for all primitive recursive functions, $\lambda y[\bar{n}](y)$ representing the nth primitive recursive function in the enumeration.

We may then define, when P_1 has been introduced, a new predicate P_2, permitting quantification over P_1:

$$A_{P_2}(Q,x) \equiv P_1 x \vee \forall y \in P_1(Q[x](y)) ,$$

etc. (cf. chapter VI). This procedure gives rise to generalized inductive definitions of higher type.

1.9.3. Language of $\underset{\sim}{HAS}_0$.

To the language of $\underset{\sim}{HA}$ we add variables for n-ary relations (species) $(n \geq 0)$, to be denoted by X^n, Y^n, Z^n (when irrelevant to the discussion we shall often omit the superscript), and second-order quantifiers $\forall_2$, $\exists_2$ (we omit the subscript when the context makes it clear that second-order quantifiers are intended). The only second-order terms considered are species variables.

1.9.4. Comprehension principles.

A comprehension principle is a schema of the form

(1) $$\exists X^n \forall x_1 \ldots x_n [A(x_1,\ldots,x_n) \longleftrightarrow X^n x_1 \ldots x_n] ,$$

where A does not contain X^n free.

We call the schema

(i) arithmetical comprehension, if A is a formula of $\underset{\sim}{HA}$ (abbreviation: ACA)

(ii) predicative comprehension, if A is a formula of $\underset{\sim}{HAS}_0$ not containing bound species variables (abbreviation: PCA)

(iii) (full, or impredicative) comprehension, if A is any formula of $\underset{\sim}{HAS}_0$, not containing X^n free (abbreviation: CA).

The system in the language of $\underset{\sim}{HAS}_0$ based on the axioms, rules and axiom schemata of $\underset{\sim}{HA}$ (but with respect to the extended language), together with quantifier rules and axioms for second-order quantifiers, is called $\underset{\sim}{HAS}_0$.

1.9.5. Extensionality.

We denote by EXT the axiom schema

EXT $\qquad \forall xy[Ax \,\&\, x=y \rightarrow Ay]$.

We define $\underset{\sim}{HAS}$ as $\underset{\sim}{HAS}_0 + CA + EXT$.

Note that in $\underset{\sim}{HAS}$, EXT may be replaced by the single axiom

$$\forall X^1 \forall xy\,[X^1x \,\&\, x=y \rightarrow X^1y].$$

1.9.6. Theorem. If $\underset{\sim}{H}$ is one of the systems $\underset{\sim}{HAS}_0$, $\underset{\sim}{HAS}_0 + ACA$, $\underset{\sim}{HAS}_0 + PCA$, $\underset{\sim}{HAS}_0 + CA$, then $\underset{\sim}{H} + EXT$ is conservative over $\underset{\sim}{H}$ w.r.t. formulae of $\underset{\sim}{HA}$.

First proof. Let φ be a mapping of formulae of $\underset{\sim}{HAS}_0$ into formulae of $\underset{\sim}{HAS}_0$, given by: $\varphi(A)$ is obtained from A by replacing each sub-formula of A of the form $Xt_1\ldots t_n$ by $\exists x_1\ldots x_n(t_1=x_1 \,\&\, \ldots \,\&\, t_n=x_n \,\&\, Xx_1\ldots x_n)$. Then one readily verifies by induction on the length of deductions for the systems $\underset{\sim}{H}$ mentioned:

$$\underset{\sim}{H} + EXT \vdash A \Leftrightarrow \underset{\sim}{H} \vdash \varphi(A).$$

Second proof. Let ψ be a mapping of formulae of $\underset{\sim}{HAS}_0$ into formulae of $\underset{\sim}{HAS}_0$, which is defined as the relativization to extensional species, i.e.

$\psi[\forall X^n A] \equiv \forall X^n(Ext(X^n) \rightarrow \psi[A])$,

$\psi[\exists X^n A] \equiv \exists X^n(Ext(X^n) \,\&\, \psi[A])$

and $\psi[A \circ B] \equiv \psi[A] \circ \psi[B]$ for $\circ \equiv \rightarrow, \vee, \&$,

$\psi[(Qx)A] \equiv (Qx)\psi[A]$ for $Q \equiv \forall_1, \exists_1$, and where $Ext(X^n)$ is defined by

$$Ext(X^n) \equiv_{def} \forall x_1\ldots x_n y_1\ldots y_n(X^n x_1\ldots x_n \,\&\, x_1=y_1 \,\&\, \ldots x_n=y_n \rightarrow X^n y_1\ldots y_n).$$

Then also, if all free second-order variables of A are among $X_1,\ldots,X_n$, one proves by induction on the length of deductions

$$\underset{\sim}{H} + EXT \vdash A(X_1,\ldots,X_n) \Leftrightarrow \underset{\sim}{H} + Ext(X_1) + \ldots + Ext(X_n) \vdash \psi[A(X_1,\ldots,X_n)].$$

The verification is quite straightforward and left to the reader.

1.9.7. Lemma. If $\underset{\sim}{H}$ is one of the systems $\underset{\sim}{HAS}_0 + EXT$, $\underset{\sim}{HAS}_0 + EXT + ACA$, $\underset{\sim}{HAS}_0 + EXT + PCA$, $\underset{\sim}{HAS}_0 + EXT + CA$, and $\underset{\sim}{H}'$ is the corresponding system obtained by restriction of the predicate variables to unary ones, then $\underset{\sim}{H}$ is conservative over $\underset{\sim}{H}'$.

Proof. Let $V_0^n, V_1^n, V_2^n, \ldots$ be the list of species variables (i.e. the actual variables in this case, not the syntactical (= metamathematical) variables for variables) with n arguments, for all n.
We define a mapping τ as follows.

$$\tau(t=s) \equiv t=s .$$
$$\tau(V_i^n t_1 \ldots t_n) \equiv V^1_{j(n,i)}(\nu_n t_1 \ldots t_n) . \quad (\nu_n \text{ was defined in 1.3.9 C.})$$
$$\tau(A \circ B) \equiv (\tau A) \circ (\tau B) \quad \text{for } \circ = \rightarrow, \&, \vee .$$
$$\tau((Qx)A) \equiv (Qx)\tau(A) \quad \text{for } Q = \exists_1, \forall_1 ,$$
$$\tau((QV_k^n)A) \equiv (QV^1_{j(n,k)})\tau(A) \quad \text{for } Q = \exists_2, \forall_2 .$$

This mapping transforms each proof in $\underset{\sim}{H}$ into a proof in $\underset{\sim}{H}'$, with a conclusion which only differs in the naming of second-order variables.

If we wish, we might also have kept the variables for one-argument species in the proof unchanged by the translation, by a slightly modified definition of τ (the modification depends on the proof under consideration).
Let m be the maximum index i such that V_i^1 occurs in the given proof.
We then put $\tau'(V_k^1 t) \equiv V_k^1 t$, $\tau'(V_k^n t_1 \ldots t_n) \equiv V^1_{m+j(k,n)}(\nu_n t_1 \ldots t_n)$ for $n \neq 1$, $\tau'((QV_k^1)A) \equiv (QV_k^1)\tau'(A)$, and

$$\tau'((QV_k^n)A) \equiv (QV^1_{m+j(k,n)})\tau'(A) \quad \text{for } Q = \forall_2, \exists_2, \ n \neq 1 .$$

The verification that the mapping has the property stated is quite straightforward; we consider the only case which is not quite trivial, i.e. instances of the comprehension schema for n-argument $(n \neq 1)$ species:

$$\exists V_k^n \forall x_1 \ldots x_n [A(x_1, \ldots, x_n) \leftrightarrow V_k^n x_1 \ldots x_n] .$$

This translates into (under τ)

$$\exists V^1_{j(n,k)} \forall x_1 \ldots x_n [A(x_1, \ldots, x_n) \leftrightarrow V^1_{j(n,k)}(\nu_n x_1 \ldots x_n)] .$$

This follows from

$$\exists V^1_{j(n,k)} \forall z [A(j_1^n z, \ldots, j_n^n z) \leftrightarrow V^1_{j(n,k)} z]$$

together with EXT.

1.9.8. Theorem. $\underset{\sim}{HAS}_0$ + EXT + ACA is conservative over $\underset{\sim}{HA}$.
Proof. By the previous lemma, it suffices to prove $\underset{\sim}{H}$ to be conservative over $\underset{\sim}{HA}$, where $\underset{\sim}{H}$ is the restriction of $\underset{\sim}{HAS}_0$ + EXT + ACA to unary species variables.
We further remark that we may restrict attention to instances of ÁCA containing (at most) one free numerical parameter, since e.g.

$$\forall xy \, \exists X' \forall z [A(x,y,z) \leftrightarrow X' z]$$

is a consequence of EXT and

$$\forall u\, \exists X^1\, \forall z[A(j_1 u, j_2 u, z) \leftrightarrow X^1 z].$$

Let now any proof π in $\underset{\sim}{H}$ of an arithmetical statement be given; we may assume π to use finitely many instances of the comprehension schema, say

$$\forall y\, \exists X^1\, \forall z[A_i(y,z) \leftrightarrow X^1 z], \quad 0 \leq i \leq k \quad (\forall yz\, A_i(y,z) \text{ closed}).$$

We define a predicate $C(x,y,z)$:

$$C(x,y,z) \equiv (x=0 \,\&\, A_0(y,z)) \vee (x=1 \,\&\, A_1(y,z)) \vee \ldots \vee \ldots \vee (x=k \,\&\, A_k(y,z)).$$

Let $v_0, v_1, v_2, \ldots$ be the numerical variables of $\underset{\sim}{H}$, and
let $V_0, V_1, V_2, \ldots$ be the (unary) species variables of $\underset{\sim}{H}$.
Let m be the maximum index i such that v_i occurs free or bound in π.
Now we define a mapping σ:

$$\sigma(t=s) \equiv t=s\,, \quad \sigma(\wedge) = \wedge$$
$$\sigma(A \circ B) \equiv \sigma(A) \circ \sigma(B) \quad \text{for } \circ \equiv \vee, \&, \rightarrow,$$
$$\sigma((Qv_n)A) \equiv (Qv_n)\sigma(A) \quad \text{for } Q \equiv \forall_1, \exists_1;$$
$$\sigma(V_n t) \equiv C(j_1 v_{m+n+1}, j_2 v_{m+n+1}, t)$$
$$\sigma((Q_2 V_n)A) \equiv (Q_1 v_{m+n+1})\sigma(A), \quad \text{where } Q \equiv \exists, \forall.$$

Note that σ is the identity on arithmetical formulae; σ preserves logical inferences, axioms for equality and successor and induction (at least as far as they occur in π).

Now consider an instance of ACA occurring in π; it translates under σ into

$$\forall y\, \exists v_{n+m+i}\, \forall z[A_i(y,z) \leftrightarrow C(j_1 v_{n+m+1}, j_2 v_{m+n+1}, z)].$$

This is obviously derivable in $\underset{\sim}{HA}$, since $A_i(y,z) \leftrightarrow C(i,y,z)$; also note that $\sigma(EXT)$ is derivable in $\underset{\sim}{HA}$. So, after intercalation of some steps, σ transforms π into a proof in $\underset{\sim}{HA}$.

1.9.9. Formulation of $\underset{\sim}{HAS}$ with λ-terms.

Instead of formulating second-order logic with a comprehension schema, it is sometimes more convenient to use a more general class of second-order terms.

So CA is replaced by a rule of term formation: whenever $A(x_1,\ldots,x_n)$ is a formula of $\underset{\sim}{HAS}_0$, then $\lambda x_1\ldots x_n A(x_1,\ldots,x_n)$ is a second-order term, with the rule

$$\{\lambda x_1\ldots x_n . A(x_1,\ldots,x_n)\}(t_1,\ldots,t_n) \leftrightarrow A(t_1,\ldots,t_n).$$

PCA is represented by a similar rule of term formation where A is not permitted to contain bound second-order quantifiers, etc.

1.9.10 - 1.9.11. Intuitionistic analysis with variables for sequences.

1.9.10. Description of EL.

The system EL to be described below is a slight variant of the system EL as described in Kreisel - Troelstra 1970, § 2.5.

For sequence variables we use either x^1, y^1, z^1, u^1, v^1, w^1 (as in the case of $N-HA^\omega$) or we use greek lower case letters α, β, γ, ... $\mathcal{L}(EL)$ is obtained from $\mathcal{L}(HA)$ by the addition of sequence variables and quantifiers, and an application operator "Ap", a recursor R, and abstraction operators λx, and extending the term definition of HA by

(i) function variables are functors (i.e. terms for functions);

(ii) one-argument function constants are functors;

(iii) if φ is a functor, t a (numerical) term, then $Ap\,\varphi t$ (abbreviated as φt) is a term;

(iv) if t, t' are terms, φ a functor, $Rt\varphi t'$ is a term;

(v) if $t[x]$ is a term, $\lambda x.t[x]$ is a functor.

EL is now formalized by adding quantifier rules and axioms for function quantifiers, a rule of λ - conversion λ - CON $(\lambda x.t)t' = [x/t']t$ and defining axioms for R

$$\text{REC} \quad \begin{cases} Rt\varphi 0 = t \\ Rt\varphi(St') = \varphi j(Rt\varphi t', t') \end{cases}$$

and a quantifier-free axiom of choice

$$\text{QF-AC}_{00} \quad \forall x \exists y\, A(x,y) \rightarrow \exists \alpha \forall x\, A(x, \alpha x) \qquad (\text{A quantifier-free}).$$

EL is essentially a subsystem of $N-HA^\omega$, i.e. the primitives of EL are definable in $N-HA^\omega$. (λx only as a syntactical operator, but since EL only deals with equality at lowest type, this makes no difference.)

$QF-AC_{00}$ requires, intuitively speaking, that the universe of functions is closed under "recursive in". EL is easily seen to be a conservative extension of HA, by interpreting all function variables as ranging over total recursive functions.

Sometimes we can get by with a system of elementary analysis with a more restricted language, such as EL_0 in Kreisel - Troelstra 1970; EL is a definitional extension of such a system. In the sequel we shall denote all such systems by EL.

The system H in Howard and Kreisel 1966 only requires the universe of functions to be closed under "primitive recursive in". This is somewhat too weak for the formalization of the elementary theory of recursive functionals, which we need.

1.9.11. Some notations and conventions.

$\bar{\alpha}0 = \langle\ \rangle$,

$\bar{\alpha}x = \langle \alpha 0, \alpha 1, \ldots, \alpha(x-1) \rangle$,

hence $\mathrm{lth}(\bar{\alpha}x) = x$.

$\alpha \in n \equiv_{\mathrm{def}} \forall x < \mathrm{lth}(n)\ (\alpha x = (n)_x)$.

Furthermore we put

$j_1\alpha \equiv_{\mathrm{def}} \lambda x \,.\, j_1\alpha x$, $j_2\alpha \equiv_{\mathrm{def}} \lambda x \,.\, j_2\alpha x$, $j(\alpha, \beta) \equiv_{\mathrm{def}} \lambda x \,.\, j(\alpha x, \beta x)$.

(If we have not included λ as a primitive,, $j_1\alpha$, $j_2\alpha$, $j(\alpha, \beta)$ can nevertheless be used in contexts like $A(j_1\alpha)$, $A(j_2\alpha)$ etc., where $A(j_1\alpha)$ etc. is taken as an abbreviation for the formula A obtained by replacing in $A(\beta)$ every occurrence of βt by $j_1(\alpha t)$ and repeating this process until β has been eliminated.) $(\alpha)_x \equiv_{\mathrm{def}} \lambda y . \alpha j(x,y)$.

1.9.12 - 1.9.16. Formalization of elementary recursion theory in EL.

1.9.12. We have to rely heavily on the development in Kleene 1969. The developments in Part I of Kleene 1969 can be carried out in EL (since $\mathrm{QF}\text{-}\mathrm{AC}_{00}$ includes all instances of Kleene's $^{x}2.1$! needed in Part I of Kleene 1969, cf. footnote 7 in Kleene 1969). That we assume our coding of sequences of natural numbers to be onto the natural numbers (contrary to Kleene's definition) is inessential. We define $\alpha|\beta$, $\alpha(\beta)$ by

$$(\alpha|\beta)(x) \simeq y \equiv_{\mathrm{def}} \alpha(\hat{x} * \bar{\beta}\ \min_z[\alpha(\hat{x} * \bar{\beta}z) \neq 0]) \dot{-} 1 = y$$

$$\alpha(\beta) \simeq y \equiv_{\mathrm{def}} \alpha(\bar{\beta}\ \min_z[\alpha(\bar{\beta}z) \neq 0]) \dot{-} 1 = y .$$

We may use the partially defined expressions constructed from terms and the partially defined application operations $.|.$, $.(.)$ as systematic abbreviations, similar to the use of p-terms (cf. 1.3.10) constructed by means of partial recursive function application. We shall speak of p-terms in this case also. If they are of type 1 (i.e. when they represent a partial function) we may distinguish them as p-functors. We denote p-functors by φ, φ', ...

$(\alpha|\beta)(x)$ and $\alpha(\beta)$ are partial recursive functionals of α, β, x and α, β respectively, hence we can find certain numerals $\bar{n}_0, \bar{n}_1$ such that

$$(1)\quad \begin{cases} \{\bar{n}_0\}(x, \alpha, \beta) \simeq y \leftrightarrow (\alpha|\beta)(x) \simeq y \\ \{\bar{n}_1\}(\alpha, \beta) \quad\ \simeq y \leftrightarrow \alpha(\beta) \simeq y \end{cases} .$$

By virtue of (1), every p-term of type 0 corresponds in a standard way to a p-term in the sense of Kleene 1969, replacing $\varphi|\varphi'$, $\varphi(\varphi')$ by $\lambda x \,.\, \{\bar{n}_0\}(x, \varphi, \varphi')$ and $\{\bar{n}_1\}(\varphi, \varphi')$ respectively.

1.9.13. Coding of sequences and n-tuples.

We put $\nu_u(\alpha_1, \ldots, \alpha_u) = \lambda x \,.\, \nu_u(\alpha_1 x, \ldots, \alpha_u x)$

$$j_i^u \alpha = \lambda x . j_i^u \alpha x .$$

Furthermore we let k_i^n $(0 \leq i \leq n)$ be functions satisfying

$$k_i^n 0 = 0 , \quad k_i^n(m * \hat{x}) = k_i^n m * \langle j_i^n x \rangle .$$

We abbreviate

$$\varphi | (\varphi_1, \ldots, \varphi_u) \equiv_{def} \varphi | \nu_u(\varphi_1, \ldots, \varphi_u)$$
$$\varphi \ (\varphi_1, \ldots, \varphi_u) \equiv_{def} \varphi(\nu_u(\varphi_1, \ldots, \varphi_u)) .$$

1.9.14. Theorem. (i) Let $\varphi[\alpha_1, \ldots, \alpha_n]$ be a p-functor; then there is a (primitive) recursive f_φ such that

$$f_\varphi | (\alpha_1, \ldots, \alpha_n) \simeq \varphi[\alpha_1, \ldots, \alpha_n] .$$

(ii) Let $\varphi[\alpha_1, \ldots, \alpha_n]$ be a p-term of type 0; then there is a (primitive) recursive f'_φ such that

$$f'_\varphi(\alpha_1, \ldots, \alpha_n) \simeq \varphi[\alpha_1, \ldots, \alpha_n] .$$

Proof. We use lemma 41 and § 4 of Kleene 1969 to prove (i). By Kleene 1969, p. 67, lemma 41 we can prove the existence of a numeral $\bar{n}_\varphi$ such that

$$\{\bar{n}_\varphi\}(x, \alpha_1, \ldots, \alpha_n) \simeq \varphi[\alpha_1, \ldots, \alpha_n](x) .$$

By Kleene 1969, *34.1, *34.2 (page 69)

$$\{\bar{n}_\varphi\}(x, \alpha_1, \ldots, \alpha_n) \simeq U \min_y T(\bar{n}_\varphi, x, \bar{\alpha}_1 y, \ldots, \bar{\alpha}_n y)$$

(T, U primitive recursive).

We put

$$f_\varphi 0 = 0$$
$$f_\varphi(\hat{x} * m) = U(\mathrm{lth}(m)) \text{ if } T(\bar{n}_\varphi, x, k_1^n m, \ldots, k_n^n m)$$
$$f_\varphi(\hat{x} * m) = 0 \qquad \text{ if} \neg T(\bar{n}_\varphi, x, k_1^n m, \ldots, k_n^n m) .$$

It is easily verified that f_φ satisfies our requirements.

(ii) is proved similarly.

1.9.15. Theorem (s-m-n theorem analogue).

(i) There exists a primitive recursive function $\wedge_n$ of two arguments such that (writing $\varphi \wedge_n \psi$ for $\wedge_n(\varphi, \psi)$)

$$(\alpha \wedge_n \beta_1) | (\beta_2, \ldots, \beta_n) \simeq \alpha | (\beta_1, \ldots, \beta_n) .$$

(ii) Similarly, there is a primitive recursive function $\wedge'_n$ such that

$$(\alpha \wedge'_n \beta_1)(\nu_{n-1}(\beta_2, \ldots, \beta_n)) \simeq \alpha(\nu_n(\beta_1, \ldots, \beta_n)) .$$

Proof. (i) By Kleene 1969, lemma 41, *34.1 there is a numeral $\bar{m}$ such that

$$(\alpha|(\beta_1,\ldots,\beta_n))(x) \simeq U\,\min_y T(\bar{m}, x, \bar{\alpha}y, \bar{\beta}_1 y, \ldots, \bar{\beta}_n y) .$$

We put (using 1.9.14 (i) implicitly)

$$(\alpha \wedge \beta_1)(0) = 0$$
$$(\alpha \wedge \beta_1)(\hat{x} * u) = y+1 \longleftrightarrow U(\mathrm{lth}(u)=y) \wedge$$
$$\wedge T(\bar{m}, x, \bar{\alpha}(\mathrm{lth}\,u), \bar{\beta}_1(\mathrm{lth}\,u), k_1^{n-1}u, \ldots, k_{n-1}^{n-1}u) ;$$
$$(\alpha \wedge \beta_1)(\hat{x} * u) = 0 \text{ in all other cases.}$$

(ii) Similarly.

1.9.16. Theorem (Recursion theorem analogue).
(i) For each α there exists a β such that

$$\alpha|(\beta, \gamma_1, \ldots, \gamma_n) \simeq \beta|(\gamma_1, \ldots, \gamma_n) .$$

(ii) For each α there exists a β such that

$$\alpha\,(\beta, \gamma_1, \ldots, \gamma_n) \simeq \beta(\gamma_1, \ldots, \gamma_n) .$$

Proof. (i) Consider $\alpha|(\delta \wedge_n \delta,\ \gamma_1, \ldots, \gamma_n)$. There exists an ϵ such that

$$\epsilon|(\delta, \gamma_1, \ldots, \gamma_n) \simeq \alpha|(\delta \wedge_n \delta, \gamma_1, \ldots, \gamma_n) .$$

Take $\beta = \epsilon \wedge_n \epsilon$. Then

$$(\epsilon \wedge_n \epsilon)|(\gamma_1, \ldots, \gamma_n) \simeq \epsilon|(\epsilon, \gamma_1, \ldots, \gamma_n) \simeq$$
$$\simeq \alpha|(\epsilon \wedge_n \epsilon, \gamma_1, \ldots, \gamma_n) \simeq \alpha|(\beta, \gamma_1, \ldots, \gamma_n) .$$

(ii) Similarly.

1.9.17. Definitions of $\Lambda^0 x$, $\Lambda^1 x$, $\Lambda^0 \alpha$, $\Lambda^1 \alpha$.
If t is a p-term of type 0, which is provably defined for all values of x, we put $\Lambda^0 x.t \equiv_{\mathrm{def}} \lambda x.t$.
If t is a p-term of type 0, we take $\Lambda^0 \alpha.t$ to be any φ, primitive recursive in the parameters of t different from α, such that

$$\varphi(\alpha) \simeq t .$$

If t is a p-functor, we take $\Lambda^1 x.t$ to be any φ, primitive recursive in the parameters of t different from α, such that

$$\varphi|\lambda y.x \simeq t .$$

Similarly $\Lambda^1 \alpha.t$ is to be a φ such that

$$\varphi|\alpha \simeq t .$$

According to 1.9.14 and 1.9.15 we can always construct such φ.

1.9.18. Systems of intuitionistic analysis based on the concept of a lawlike sequence ; $\underset{\sim}{IDB}$.

For a universe of lawlike sequences, various forms of axioms of choice seem to be intuitively justified, notably AC_{00}, but also AC_{01}, and even the strongest principle

$$\text{RDC}_1 \qquad \forall\alpha[A\alpha \to \exists\beta(B(\alpha,\beta) \,\&\, A(\beta))]$$
$$\to \forall\alpha[A\alpha \to \exists\gamma[(\gamma)_0=\alpha \,\&\, \forall x B((\gamma)_x, (\gamma)_{Sx})]].$$

RDC_1 implies AC_{01}, hence AC_{00} ; see Kreisel and Troelstra 1970, theorem 2.7.2.

It follows from the results of Goodman 1968, and E, that $\underset{\sim}{EL} + RDC_1$ is in fact conservative over $\underset{\sim}{HA}$; the work of Goodman falls beyond the scope of this book ; the proofs are very long and the method cannot be readily fitted into the framework of the rest of this book.

However, it is not hard to establish, by means of a realizability interpretation, that $\underset{\sim}{EL} + RDC_1$ is consistent relative $\underset{\sim}{HA}$ (cf. 3.6.16, and Kreisel and Troelstra 1970, 3.7). The same interpretation also establishes the consistency of Church's thesis (cf. § 3.2, 1.11.7). Church's thesis acts as a reducibility axiom for systems with function variables ; statements involving functions are reduced to statements involving natural numbers only.

To obtain a proof-theoretic strengthening, we have to add a constant for a species introduced by a generalized inductive definition ; the principal example here being the theories $\underset{\sim}{IDB}$ and $\underset{\sim}{IDB}_1$.
$\underset{\sim}{IDB}$ is obtained by adding a constant K (for a unary predicate of functions) to $\underset{\sim}{EL}$, together with two axioms and a schema

$$\text{K1.} \qquad \alpha = \lambda n.Sx \to K\alpha$$

$$\text{K2.} \qquad \alpha 0 = 0 \,\&\, \forall x\, K(\lambda n.\alpha(\hat{x} * n)) \to K\alpha$$

and if $A_K(Q,\alpha) \equiv_{def} \exists y(\alpha = \lambda x.Sy) \vee (\alpha 0=0 \,\&\, \forall x Q(\lambda n.\alpha(\hat{x} * n)))$ we put

$$\text{K3.} \qquad \forall\alpha[A_K(Q,\alpha) \to Q\alpha] \to \forall\alpha[K\alpha \to Q\alpha]$$

for all Q in the language of $\underset{\sim}{IDB}$.
K1, K2 may be combined into

$$A_K(K,\alpha) \to K\alpha$$

$\underset{\sim}{IDB}_1$ may be defined as $\underset{\sim}{IDB} + AC_{01}$.
($\underset{\sim}{IDB}_1$ in Kreisel and Troelstra 1970 is an expansion of $\underset{\sim}{IDB}_1$ as defined here ; $\underset{\sim}{IDB}$ corresponds to $\underset{\sim}{IDB}_0$ of Kreisel - Troelstra 1970.)

The axioms K1 - K3 are in fact equivalent to the following axiom and schema (cf. Kreisel - Troelstra 1970, 3.2.1):

(1) $Ka \to [\forall n(an \neq 0 \to Qn) \,\&\, \forall n(\forall y Q(n * \hat{y}) \to Qn) \to Q0]$

for all Q of the language, and

(2) $Ka \,\&\, an \neq 0 \to \forall m(an = a(n * m))$.

(1) is called the principle of induction over unsecured sequences.

1.9.19. Systems of intuitionistic analysis based on a concept of choice sequence.

As already remarked in the introduction to this section, universes of choice sequences are supposed to be such that an extensional operator of type 2, defined on the whole universe should be continuous, i.e. satisfy $\forall \alpha \exists x \forall \beta (\bar{\alpha}x = \bar{\beta} \to \Phi\alpha = \Phi\beta)$.

Without introducing higher type objects in the language, the simplest way of expressing this continuity property is by the schema (weak continuity schema)

WC - N $\quad \forall \alpha \exists x A(\alpha,x) \to \forall \alpha \exists x \exists y \forall \beta \in \bar{\alpha}x A(\beta,y)$.

This principle may be conceived as being obtained by combination of the above-mentioned continuity property for type-2 operators with the following "selection principle" (axiom of choice) which is itself not expressible in $\mathcal{L}(\underset{\sim}{EL})$:

$$\forall \alpha \exists x A(\alpha,x) \to \exists \Phi \forall \alpha A(\alpha, \Phi\alpha) .$$

A stronger axiom of continuity C - N is expressed as follows

C - N $\quad \forall \alpha \exists x A(\alpha,x) \to \exists \beta [K_0 \beta \,\&\, \forall n(\beta n \neq 0 \to \forall \alpha \in n A(\alpha, \beta n \dot{-} 1))$

where

$$K_0(\alpha) \equiv_{def} \forall nm(\alpha n \neq 0 \to \alpha n = \alpha(n*m)) \,\&\, \forall \beta \exists x(\alpha(\bar{\beta}x) \neq 0) .$$

(C - N corresponds to the strong continuity of Howard and Kreisel 1966). C - N expresses that there is a modulus-of-continuity functional ; WC - N only expresses local continuity.

1.9.20. Bar induction.

The schema of bar induction, discussed extensively in Howard and Kreisel 1966, appears in various forms. We list some formulae first :

(1) $\forall \alpha \exists x P\bar{\alpha}x$

(2) $\forall nm(Pn \to P(n * m))$

(3) $\forall n(Pn \vee \neg Pn)$

(4) $\forall n(Pn \to Qn)$

(5) $\forall n(\forall y Q(n * \hat{y}) \to Qn)$.

Then bar induction with the monotonicity condition, BI_M, can be expressed as :

BI_M (1) & (2) & (4) & (5) $\to Q0$

and bar induction with the decidability condition, BI_D as

BI_D (1) & (3) & (4) & (5) $\to Q0$.

The weakest version is

BI_{QF} (1) & (4) & (5) $\to Q0$ (with P quantifier-free).

It is shown, in Howard and Kreisel 1966 (Remark 4, page 337) that BI_M can be strengthened to

(1) & (2) & (4) & (5) $\to \forall n\, Qn$.

Similarly, if Q is supposed to be monotone (i.e. $\forall nm(Qn \to Q(n * m))$), then BI_D or BI_{QF} may also be strengthened:

(1) & (3) & (4) & (5) & $\forall nm(Qn \to Q(n*m)) \to \forall n Qn$

(1) & (4) & (5) & $\forall nm(Qn \to Q(n*m)) \to \forall n Qn$ (for P quantifier free).

It should be noted that (Howard and Kreisel 1966, theorem 8C)

$$\underset{\sim}{EL} + AC_{01} + WC-N + BI_M \vdash C-N$$

and also (Howard and Kreisel 1966, theorem 8E)

$$\underset{\sim}{EL} + C-N \vdash AC_{00} .$$

1.9.21. Extended bar induction.

The schema of extended bar induction is described below. Let R be any unary predicate of functions; then we put

$$R^*\beta \equiv_{def} \forall x(R(\beta)_x)$$
$$R^{\cup}\beta \equiv_{def} \exists y((\beta)_0 = \lambda z.y \,\&\, \forall u > y((\beta)_u = \lambda z.0) \,\&\, \forall u < y(R(\beta)_{Su}))$$
$$lth(\beta) = (\beta)_0 0 ; \quad \bar{\beta}_x j(y,z) = \begin{cases} x & \text{if } y = 0 \\ 0 & \text{if } y > x \\ \beta(j(y,z)) & \text{if } 0 < y \le x . \end{cases}$$

$R^{\cup}$ contains all codings of finite sequences of elements of R. Let us denote the coding of a sequence $\beta_0, \ldots, \beta_{x-1}$ by $\langle\beta_0, \ldots, \beta_{x-1}\rangle^1$, $\langle\beta_0, \ldots, \beta_{x-1}\rangle^1$ is a sequence of φ such that $(\varphi)_0 = \lambda z.x$, $(\varphi)_{y+1} = \beta_y$ for $y < x$, $(\varphi)_y = \lambda z.0$ for $y > x$.

$*$ denotes concatenation; we abbreviate $\langle\eta\rangle^1$ as $\hat{\eta}$. $\langle\ \rangle^1$ is $\lambda x.0$.

Now the schema of extended bar induction EBI_D is given by

EBI_D (1) & (2) & (3) & (4) & (5) $\to Q(\langle\ \rangle^1)$

where

(1) $\exists\alpha\, R\alpha$

(2) $\forall\beta \in R^{\smile}\, (P\beta \vee \neg P\beta)$

(3) $\forall\beta \in R^{\smile}\, (P\beta \rightarrow Q\beta)$

(4) $\forall\beta \in R^{*}\, \exists x(P\bar{\beta}_x)$

(5) $\forall\beta \in R^{\smile}\, (\forall\eta \in R\; Q(\beta * \hat{\eta}) \rightarrow Q\beta)$.

1.9.22. Theorem.

$$\underset{\sim}{EL}^c + DC_1 \vdash EBI_D$$

where DC_1 is the schema

DC_1 $\quad \forall\alpha\, \exists\beta\, A(\alpha,\beta) \rightarrow \exists\gamma[(\gamma)_0 = \alpha \,\&\, \forall z A((\gamma)_z, (\gamma)_{Sz})]$.

Proof. We first show that DC_1 implies the following more general schema (a form of RDC_1)

(6) $\quad \forall\alpha \in S\, \exists\beta \in S\; A(\alpha,\beta) \rightarrow \forall\alpha \in S\; \exists\gamma[(\gamma)_0 = \alpha \,\&\, \forall z \{A((\gamma)_z, (\gamma)_{Sz}) \,\&\, (\gamma)_z \in S\}]$.

To see this, assume

$$\forall\alpha \in S\;\; \exists\beta \in S\; A(\alpha,\beta) .$$

Let

$$A^*(\alpha,\beta) \equiv [\alpha \in S \,\&\, A(\alpha,\beta) \,\&\, \beta \in S] \vee \alpha \notin S .$$

Then obviously (using classical logic!)

$$\forall\alpha\, \exists\beta\, A^*(\alpha,\beta) .$$

Let $\alpha \in S$; by DC_1, there is a γ such that

$$[(\gamma)_0 = \alpha \,\&\, \forall z A^*((\gamma)_z, (\gamma)_{Sz})] .$$

Then we prove by induction on z

$$\forall z[(\gamma)_z \in S \,\&\, A((\gamma)_z, (\gamma)_{Sz})] .$$

Thus (6) follows.

Now apply (6), with

$$S \equiv R^{\smile}\alpha \,\&\, \neg Q\alpha,\; A(\alpha,\beta) \equiv \exists\eta \in R(\beta = \alpha * \hat{\eta}) .$$

Assume (1) - (5), $\neg Q(\langle\, \rangle^1)$, ((1) - (5) as in 1.9.21).

Obviously, if $R^{\smile}\alpha \,\&\, \neg Q\alpha$, contraposition of (5) yields

$$\neg\, \forall\eta \in R\; Q(\alpha * \hat{\eta}) .$$

Hence by classical logic $\exists\eta \in R(\neg Q(\alpha * \hat{\eta}))$ and therefore also $\exists\beta \in R^{\smile}(\exists\eta \in R(\beta = \alpha * \hat{\eta}) \,\&\, \neg Q\beta)$.

Thus

$$\forall\alpha \in S\;\; \exists\beta \in S\; A(\alpha,\beta) .$$

By (6) there is a γ such that

$$(\gamma)_0 = \langle\ \rangle^1 \ \&\ \forall z((\gamma)_z \in S \ \&\ \forall z A((\gamma)_z,(\gamma)_{Sz})),$$

hence, by induction on z

(7) $$\forall z(R^{\vee}(\gamma)_z \ \&\ lth(\gamma)_z = z \ \&\ \exists \eta \in R\ ((\gamma)_{Sz} = (\gamma)_z * \eta) \ \&\ \neg Q(\gamma)_z).$$

Now (7) implies

(8) $$(\gamma)_{Sz} = (\gamma)_z * \langle((\gamma)_{Sz})_{Sz}\rangle \ \&\ ((\gamma)_{Sz})_{Sz} \in R.$$

We wish to construct δ such that $\bar{\delta}_z = (\gamma)_z$. This is achieved by taking $(\delta)_z = ((\gamma)_{Sz})_{Sz}$, i.e. $\delta = \lambda x((\gamma)_{Sj_1x})_{Sj_1x}(j_2x)$.
Now we see from (7) that $\forall z \neg Q(\bar{\delta}_z)$; but on the other hand, $((\gamma)_{Sz})_{Sz} \in R$ for all z (by (8)), therefore $\delta \in R^*$. This contradicts (4), since $\forall z \neg Q(\bar{\delta}_z)$ implies by (3) $\forall z(\neg P(\bar{\delta}_z))$.

1.9.23. Remark. Attention has been drawn to the schema EBI_D by recent work of Luckhardt (Luckhardt 1973, and Scarpellini 1972 A). They showed how to construct models for the theory of bar recursion of higher type which could be shown to be models in a theory corresponding to $\underset{\sim}{EL} + EBI_D$. We show how EBI_D can be applied to show that the so-called extensional continuous functionals are a model for the theory of bar-recursive functionals (2.9.10). By the preceding theorem, this also gives a modelling of the bar-recursive functionals which can be shown to be a model in $\underset{\sim}{EL}^c + DC_1$; but this is already explicitly in the literature, e.g. Kreisel 1968, pp. 146-147

1.9.24. Fan theorem. The so-called "fan-theorem" in its simplest form may be stated as follows:

FAN $$\forall \alpha \in B\ \exists x A(\alpha,x) \rightarrow \exists z\ \forall \alpha \in B\ \exists y\ \forall \beta \in B(\bar{\alpha}z = \bar{\beta}z \rightarrow A(\beta,y)).$$

Here $\alpha \in B$ is an abbreviation for $\forall x(\alpha x \leq 1)$ ($\alpha \in B$ may be read as: "α belongs to the binary spread").

Kleene's *27.7 in Kleene and Vesley 1965, page 75 is a generalization of FAN; it is shown there that

$$\underset{\sim}{EL} + WC\text{-}N + BI_D \vdash FAN$$

by first showing that

$$\underset{\sim}{EL} + BI_D \vdash FAN'$$

where

FAN' $$\forall n(Rn \vee \neg Rn) \ \&\ \forall \alpha \in B\ \exists x\, R(\bar{\alpha}x) \rightarrow \exists z\ \forall \alpha \in B\ \exists x \leq z\ R(\bar{\alpha}x)$$

(FAN' corresponds to Kleene's *26a).

Here we have restricted our attentions to functions in the binary spread (i.e. satisfying $\alpha \in B$); but it is not hard to show that FAN and FAN' are equivalent to similar principles for arbitrary finitary spreads.

Let us denote by FAN* the more general principle

FAN* $\forall \alpha \in S \exists y A(\alpha,y) \rightarrow \exists z \forall \alpha \in S \exists y \forall \beta \in S(\beta \in \bar{\alpha}z \rightarrow A(\beta,y))$

where $\alpha \in S$ abbreviates: α belongs to the fan S.
For every fan S, we can find a function α such that $\beta \in S \rightarrow \forall x(\beta x \leq \alpha x)$. The species $S_\alpha = \{\beta \mid \forall x(\beta x \leq \alpha x)\}$ is itself a fan.

Furthermore, if $S \subseteq S'$ for two fans S, S' we can find a projection φ such that $\varphi[S'] = S$, $\varphi\alpha = \alpha$ for $\alpha \in S$. Therefore FAN* w.r.t. S' implies FAN w.r.t. S, as is seen by applying FAN* w.r.t. S' to $\forall \alpha \in S' \exists y A(\varphi\alpha,y)$.

The following mapping transforms every function of natural numbers into a sequence of 0's and 1's, i.e. an element of the binary spread:

$$\alpha \rightarrow 1^{\alpha 0+1}, 0, 1^{\alpha 1+1}, 0, 1^{\alpha 2+1}, \ldots$$

where 1^x stands for $1, 1, \ldots, 1$ (x times).
It is easy to see this mapping is bi-unique, and transforms a fan into a subfan of the binary spread. Thus we may derive FAN* from FAN.

1.9.25 - 1.9.27. Extensions of $\underset{\sim}{N}$ - $\underset{\sim}{HA}^\omega$.

1.9.25. Extensions to theories in all finite types with sets introduced by generalized inductive definitions; $\underset{\sim}{IDB}^\omega$.

A first example is the theory $\underset{\sim}{T}^2$ with objects of finite type over three basic types: the natural numbers, trees of the first class, and trees of the second class (trees of trees of the first class). This theory is discussed at length in chapter VI. $\underset{\sim}{T}^2$ contains as a sub-theory $\underset{\sim}{T}^1$, the theory obtained from $\underset{\sim}{T}^2$ by deleting all reference to trees of the second class.

Another example, equivalent to $\underset{\sim}{T}^1$ as regards proof-theoretic strength, is obtained by extending $\underset{\sim}{N}$ - $\underset{\sim}{HA}^\omega$ to a theory of finite types over two basic types 0 and K (the natural numbers and the Brouwer-operations) thereby extending (a variant of) $\underset{\sim}{IDB}$ in the same manner as $\underset{\sim}{N}$ - $\underset{\sim}{HA}^\omega$ extends $\underset{\sim}{HA}$. A first example of such an extension appears in Howard 1963, a completely reworked version in Howard 1972.

In Troelstra 1971A systems $\underset{\sim}{I}$ - $\underset{\sim}{IDB}^\omega$, E - $\underset{\sim}{IDB}^\omega$, WE - $\underset{\sim}{IDB}^\omega$ are described. We briefly outline these systems by first introducing $\underset{\sim}{N}$ - $\underset{\sim}{IDB}^\omega$ similar to $\underset{\sim}{N}$ - $\underset{\sim}{HA}^\omega_p$. The type-structure is now extended to a structure $\underset{\sim}{T}_K$:

$$0, K \in \underset{\sim}{T}_K;\ \sigma, \tau \in \underset{\sim}{T}_K \Rightarrow (\sigma)\tau,\ \sigma \times \tau \in \underset{\sim}{T}_K.$$

There are constants 0, S, $\Pi_{\sigma,\tau}$, $\Sigma_{\rho,\sigma,\tau}$, R_σ as before, pairing constants $D_{\sigma,\tau}$, $D'_{\sigma,\tau}$, $D''_{\sigma,\tau}$, and moreover constants Φ_1, Φ_2, Φ_3, Ψ_σ, $I(\rho,\sigma,\tau \in T_{aK})$. Furthermore, there is equality $=_\sigma$ as a primitive constant for each type $\sigma \in T_{aK}$. The axioms and rules contain the axioms and rules of $\underset{\sim}{N}-\underset{\sim}{HA}^\omega_p$, I is an "injection"-functional from K into $(0)0$, so $I \in (K)(0)0$. $\Phi_1 \in (0)K$, $\Phi_2 \in (K)(0)K$; the connection between I, Φ_1, Φ_2 is given by:

$$I(\Phi_1 x)y = Sx \qquad (x,y \in 0)$$
$$I(\Phi_2 ex)y = Ie(\langle x\rangle * y) \quad (x,y \in 0,\ e \in K).$$

(We use e, f, e', e'', $e_1, \ldots, f', f'', \ldots$ for K-variables.)
$\Phi_3 \in ((0)K)K$ is a "sup"-operator for sequences of elements of K and satisfies

$$I(\Phi_3 y)0 = 0, \quad I(\Phi_3 y)(\hat{z} * n) = I(yz)n \quad (y \in (0)K,\ z,n \in 0).$$

Ψ_σ is a constant for definition by recursion over K, with axioms

$$Ie0 = Su \rightarrow \Psi_\sigma exy = xu,$$
$$Ie0 = 0 \rightarrow \Psi_\sigma exy = y(\lambda v . \Psi_\sigma(\Phi_2 ev)xy)e$$
$$(u \in 0,\ x \in (0)\sigma,\ y \in ((0)\sigma)(K)\sigma).$$

Here $\lambda v . \Psi_\sigma(\Phi_2 ev)xy$ is assumed to be syntactically defined in terms of Π's and Σ's (cf. 1.6.8).
This completes the description of $\underset{\sim}{N}-\underset{\sim}{IDB}^\omega$. $\underset{\sim}{E}-\underset{\sim}{IDB}^\omega$ is then obtained by requiring

$$x^K = y^K \leftrightarrow \forall z^0(Ixz = Iyz)$$

and also hereditary extensionality, i.e.

$$x^{(\sigma)\tau} = y^{(\sigma)\tau} \leftrightarrow \forall z^\sigma(xz = yz).$$

In $\underset{\sim}{WE}-\underset{\sim}{IDB}^\omega$ extensionality is weakened to

$$\text{If } \vdash P \rightarrow t = s, \text{ then } \vdash P \rightarrow F[t] = F[s],$$

where P is a propositional equation between terms of type 0, and $t,s \in \sigma$.
$\underset{\sim}{I}-\underset{\sim}{IDB}^\omega$ is obtained by adding to $\underset{\sim}{N}-\underset{\sim}{IDB}^\omega$ a constant E_σ for each type $\sigma \in T_{aK}$ with axioms

$$E_\sigma xy = 0 \vee E_\sigma xy = 1$$
$$x = y \leftrightarrow E_\sigma xy = 0,$$

thereby making equality decidable for all types.

There is still an intermediate type of theory possible, where equality is neither extensional nor intensional (but also not neutral). Let us indicate these extensions of $\underset{\sim}{N}-\underset{\sim}{HA}^\omega$, $\underset{\sim}{N}-\underset{\sim}{IDB}^\omega$ resp. by $\underset{\sim}{Int}-\underset{\sim}{HA}^\omega$, $\underset{\sim}{Int}-\underset{\sim}{IDB}^\omega$. We only add to $\underset{\sim}{N}-\underset{\sim}{HA}^\omega$

$$x^1 = y^1 \leftrightarrow \forall z^0 (xz = yz) .$$

For a model of such a theory, see ICF in § 2.6.

1.9.26. Theories with bar recursion of higher type; $\underset{\sim}{N}-\underset{\sim}{HA}^{\omega} + BR$.

Theories of bar recursion of higher type are extensions of $\underset{\sim}{N}-\underset{\sim}{HA}_p^{\omega}$ based on (essentially) the same type structure, with a new schema for certain constants B_σ (the bar-recursion constant for type σ). Such a theory has been first introduced in Spector 1962. (The most important further references are Howard 1968, Kreisel 1968, Girard 1972. Furthermore Scarpellini 1972 A, Luckhardt 1973.)

For the most convenient formulation, assume that we wish to describe the theory of bar recursion as an extension of $\underset{\sim}{N}-\underset{\sim}{HA}_p^{\omega}$. We note that the addition of a type $\sigma^{\smile}$ of finite sequences of objects of type σ is an expansion, since such sequences may be identified with special sequences of type $(0)\sigma$, e.g. as follows.

Let the natural numbers n be coded into higher types as n_σ as follows:

$$n_0 = n, \quad n_{(\sigma)\tau} = \lambda x^\sigma . n_\tau, \quad n_{\sigma\times\tau} = Dn_\sigma n_\tau .$$

Then $\langle x_0^\sigma, \ldots, x_{u-1}^\sigma \rangle$ may be coded as $z^{(0)\sigma}$ with

$$z^{(0)\sigma}0 = u_\sigma, \quad z^{(0)\sigma}(i+1) = x_i^\sigma \quad \text{for } i < u,$$
$$z^{(0)\sigma}i = 0^\sigma \quad \text{for } i > u.$$

Now let c be a variable of type $\sigma^{\smile}$, ranging over finite sequences of objects of type σ, and let us adopt the same notations as for sequences of natural numbers, such as $c_1 * c_2$, $\hat{u}$ for $\langle u \rangle$ $(u \in \sigma)$, $\mathrm{lth}(c)$.
Let $[c]$ denote a sequence of type $(0)\sigma$, where if $c = \langle u_0, \ldots, u_{x-1} \rangle$, $[c](i) = u_i$ for $i < x$, $[c](i) = 0^\sigma$ for $i \geq x$.
The bar-recursion constant B_σ then satisfies

$$\mathrm{BR}_\sigma \quad \begin{cases} y[c] < \mathrm{lth}(c) \to B_\sigma yzuc = zc \\ y[c] \geq \mathrm{lth}(c) \to B_\sigma yzuc = u(\lambda v . B_\sigma yzu(c * \hat{v}))c . \end{cases}$$

To see, intuitively, that B_σ defines a functional, we must think of y as being continuous (i.e. yz, $z \in (0)\sigma$ depending on a finite initial segment of z).
If $y[c] < \mathrm{lth}(c)$, $B_\sigma yzuc$ is determined; the computation of $B_\sigma yzuc$ for $y[c] \geq \mathrm{lth}(c)$ is reduced to the computation of $B_\sigma yzu(c * \hat{v})$ for all $v \in \sigma$.
If the set of c such that $y[c] \geq \mathrm{lth}(c)$ is well-founded, (classically a consequence of continuity), the computation will be eventually reduced to cases with $y[c] < \mathrm{lth}(c)$.
BR is the set of axioms BR_σ for all σ of our type structure.

1.9.27. Girard's theory of functionals.

This theory is introduced to be able to give a direct Dialectica interpretation of the theory of species (i.e. not via a theory with variables for functions, as in Spector 1962 (Girard 1971, Girard 1972).

The type structure $\underset{\approx}{T}_S$ is defined by

(i) $0 \in \underset{\approx}{T}_S$

(ii) $\alpha, \beta, \alpha', \beta', \ldots$ belong to $\underset{\approx}{T}_S$ ($\alpha, \beta, \alpha', \beta', \ldots$ are called variable types)

(iii) $\sigma,\tau \in \underset{\approx}{T}_S \Rightarrow \sigma\times\tau,\ (\sigma)\tau \in \underset{\approx}{T}_S$

(iv) if $\sigma[\alpha] \in \underset{\approx}{T}_S$, then $\forall\alpha.\sigma[\alpha],\ \exists\alpha.\sigma[\alpha] \in \underset{\approx}{T}_S$.

The functional constants contain S, $\Pi_{\sigma,\tau}$, $\Sigma_{\rho,\sigma,\tau}$, R_σ, $D_{\sigma,\tau}$, $D'_{\sigma,\tau}$, $D''_{\sigma,\tau}$, 0_σ for all $\rho,\tau,\sigma \in \underset{\approx}{T}_S$, and two injection functionals $I_{\forall\alpha\sigma,\tau}$ and $I'_{\exists\alpha\sigma,\tau}$, satisfying

$$I_{\forall\alpha\sigma[\alpha],\tau} \in (\forall\alpha.\sigma[\alpha])(\sigma[\tau]),$$

$$I'_{\exists\alpha\sigma[\alpha],\tau} \in (\sigma[\tau])(\exists\alpha.\sigma[\alpha]).$$

Finally, there are two operators DT, ST (not functionals with a type, but corresponding to schemata for introducing new functionals).

Let $t \in \sigma$ be a term, not containing free variables containing in their type α free. Then $DT\alpha t$ is a term of type $\forall\alpha.\sigma$ with axioms

$$I_{\forall\alpha\sigma[\alpha],\tau}(DT\alpha t^{\sigma[\alpha]}) = t^{\sigma[\tau]} .$$

Let $t \in (\sigma[\alpha])\tau$, α not occurring free in τ, and not occurring in a type of a variable free in t; then

$$ST\,\alpha\, t \in (\exists\alpha\sigma[\alpha])\tau,$$

with the axiom

$$ST\,\alpha\, t(I'_{\exists\alpha\sigma[\alpha],\rho}\, s^{\sigma[\rho]}) = t^{(\sigma[\rho])\tau}\, s^{\sigma[\rho]} .$$

Also

$$0_{(\sigma)\tau}\, x^\sigma = 0_\tau,\quad D_{\sigma,\tau}\, 0_\sigma 0_\tau = 0_{\sigma\times\tau}$$

$$I_{\forall\alpha\sigma[\alpha],\tau}\, 0_{\forall\alpha.\sigma[\alpha]} = 0_{\sigma[\tau]}$$

$$ST\,\alpha\, t(0_{\exists\alpha.\sigma[\alpha]}) = t(0_{\sigma[\alpha]}) \quad (t \in (\sigma[\alpha])\rho).$$

For this theory also intensional and extensional versions are possible. For example, we may add the equality functionals E_σ as in $\underset{\sim}{I}$-$\underset{\sim}{HA}^\omega$.

§ 10. Relations between classical and intuitionistic systems: translation into the negative fragment.

1.10.1. Contents of the section. For classical predicate logic and arithmetic there exist a number of mappings into the "negative" fragment of the corresponding intuitionistic systems in the literature; the definition of these translations can be readily extended to higher order languages. A survey is given in Luckhardt 1973, chapter III. References are Gödel 1933, Gentzen 1933, Kuroda 1951; cf. also Kleene 1952, § 81.

For definiteness, let $\underset{\sim}{H}$ denote intuitionistic (many-sorted) predicate logic or $\underset{\sim}{HA}$, and let $\underset{\sim}{H}^c$ be obtained by addition of the excluded third. All translations φ have the following properties:

(i) $\underset{\sim}{H}^c \vdash \varphi A \leftrightarrow A$ for all $A \in Fm(\underset{\sim}{H})$

(ii) $\underset{\sim}{H}^c \vdash A \Leftrightarrow \underset{\sim}{H} \vdash \varphi A$, for all $A \in Fm(\underset{\sim}{H})$

(iii) For all $A \in Fm(\underset{\sim}{H})$ there exists a B, constructed from doubly negated prime formulae by means of $\forall, \&, \rightarrow, \curlywedge$ s.t. $\underset{\sim}{H} \vdash \varphi A \leftrightarrow B$.

As remarked in Luckhardt 1973, all these translations are equivalent, in the sense that, for any two translations φ, φ' satisfying (i) - (iii)

$$\underset{\sim}{H} \vdash \varphi A \leftrightarrow \varphi' A .$$

Below we primarily discuss the variant due to Gentzen.

1.10.2. Definition of the mapping'. Let $\mathcal{L}$ be any many-sorted (first-or higher-order) language, based on the logical primitives $\forall, \exists$ (for any sort of variables), $\&, \vee, \rightarrow, \curlywedge$. Then we define the mapping ' (the "negative translation") by induction on the formula complexity as follows:

(i) $P' \equiv \neg\neg P$ for prime formulae P; $\curlywedge' \equiv \curlywedge$.

(ii) $(A \& B)' \equiv A' \& B'$

(iii) $(A \rightarrow B)' \equiv A' \rightarrow B'$

(iv) $(\forall x A)' \equiv \forall x A'$, for variables x of all sorts

(v) $(A \vee B)' \equiv \neg(\neg A' \& \neg B')$

(vi) $(\exists x A)' \equiv \neg \forall x \neg A'$, for variables x of all sorts.

1.10.3. Remarks.

(i) In a system $\underset{\sim}{H}$ with a language $\mathcal{L}$ where prime formulae are decidable (e.g. $\underset{\sim}{HA}$, $\underset{\sim}{HA}^\omega$, $\underset{\sim}{I}$-$\underset{\sim}{HA}^\omega$) clause (i) may be simplified to $P' \equiv P$; the resulting translation is then obviously logically equivalent to the one given by (i) - (vi). In systems with two types of prime formulae, such as $\underset{\sim}{HAS}$, we may use $P' \equiv P$ for the prime formulae which are decidable (such as $t = s$ in $\underset{\sim}{HAS}$) and $P' \equiv \neg\neg P$ for the other prime formulae.

(ii) If we use $P' \equiv P$ for prime formulae, we have $A'' \equiv A'$; and we always have $A'' \leftrightarrow A'$ in intuitionistic predicate logic. We may also let the treatment of prime formulae depend on the context: if they appear unnegated in A we put a double negation in front, otherwise we do not change them; then A is further defined by (ii) - (vi). Then also $A'' \equiv A'$.

1.10.4. Convention. We shall give our proofs below under the assumption that $'$ is defined by 1.10.2, (i) - (vi).

We sometimes find it convenient, however, to assume $P' \equiv P$ for decidable prime formulae P (cf. our remark 1.10.3 (ii) above).

1.10.5. Definitions.

(a) We define the strictly positive parts (s.p.p.) of A, for A in a given language, inductively as follows:

(i) A is a s.p.p. of A;

(ii) If $B \& C$ or $B \vee C$ are s.p.p. of A, then so are B, C;

(iii) If $B \rightarrow C$ is a s.p.p. of A, then so is C;

(iv) If $\forall x Bx$, $\exists x Bx$ is a s.p.p. of A, then so is Bt for any term t.

(b) A Harrop formula is a formula which does not contain a s.p.p. with $\vee$ or $\exists$ as a principal operator.

Alternatively, the class of Harrop formulae Δ can be defined inductively by the clauses (i) prime formulae belong to Δ,
(ii) $A, B \in \Delta \Rightarrow A \& B \in \Delta$, (iii) $A \in \Delta \Rightarrow \forall x A \in \Delta$,
(iv) $B \in \Delta \Rightarrow A \rightarrow B \in \Delta$.

1.10.6. Definition. In any language $\mathcal{L}$ (as intended in 1.10.2), a formula A is said to be negative, if it is constructed from negated prime formulae by means of $\forall$, $\&$, $\rightarrow$, $\wedge$. (In systems with decidable categories of prime formulae we shall assume that a negative formula may also contain unnegated prime formulae out of the decidable categories.)

1.10.7. Remark. In $\underset{\sim}{HA}$ there are Harrop formulae which are not provably equivalent to a negative formula. An example is $\neg \forall x [\neg \neg \exists y T(x,x,y) \rightarrow \exists y T(x,x,y)]$ (see 3.8.2). But conversely, every negative formula is a Harrop formula.

1.10.8. Lemma. Let $\underset{\sim}{H}$ be a formal system based on many-sorted intuitionistic predicate logic, and let A be a Harrop formula constructed from decidable or doubly negated prime formulae. Then

$$\underset{\sim}{H} \vdash A \leftrightarrow \neg\neg A .$$

Proof. By induction on the complexity of A.

(i) The assertion holds for double negations of prime formulae and decidable prime formulae. $\neg\neg\curlywedge \leftrightarrow \curlywedge$.

(ii) If $A \leftrightarrow \neg\neg A$, $B \leftrightarrow \neg\neg B$, then $(A \& B) \leftrightarrow (\neg\neg A \& \neg\neg B) \leftrightarrow (A \& B)$ (1.1.8, V)

(iii) If $Ax \leftrightarrow \neg\neg Ax$, then
$\forall x Ax \to \neg\neg \forall x Ax \to \forall x \neg\neg Ax \to \forall x Ax$.

(iv) If $\neg\neg B \leftrightarrow B$, then
$\neg\neg(A \to B) \leftrightarrow (A \to \neg\neg B) \leftrightarrow (A \to B)$ (1.1.8, IV).

1.10.9. <u>Lemma</u>. Let $\underset{\sim}{H}$ be a formal system based on intuitionistic (many-sorted) predicate logic; let $\underset{\sim}{H}'$ be obtained from $\underset{\sim}{H}$ by adding the schema

DNS $\quad \forall x \neg\neg Ax \to \neg\neg \forall x Ax$

(for all sorts of variables x). Then

(i) For A not containing $\forall$:

$$\underset{\sim}{H} \vdash \neg\neg A \leftrightarrow A'.$$

(ii) For all A

$$\underset{\sim}{H}' \vdash \neg\neg A \leftrightarrow A'.$$

(iii) If all subformulae of A are stable (B is called stable if $B \leftrightarrow \neg\neg B$) in $\underset{\sim}{H}$, then

$$\underset{\sim}{H} \vdash \neg\neg A \leftrightarrow A'.$$

(<u>Note</u>: this lemma is used in 1.11.4.)

<u>Proof</u>. (i), (ii), (iii) can be proved simultaneously by induction on the complexity of A. We consider two typical cases:

(a) Let $\neg\neg Bx \leftrightarrow (Bx)'$. $(\exists x Bx)' \leftrightarrow \neg \forall x \neg (Bx)' \leftrightarrow \neg \forall x \neg Bx \leftrightarrow \neg\neg \exists x Bx$.

(b) (For (ii) or (iii) only.) Let $\neg\neg Bx \leftrightarrow (Bx)'$.
$\neg\neg \forall x Bx \leftrightarrow \forall x \neg\neg Bx \leftrightarrow \forall x B'x \leftrightarrow (\forall x Bx)'$.

1.10.10. <u>Lemma</u>. Let $\underset{\sim}{H}$ denote many-sorted intuitionistic predicate logic for a first- or higher-order language, and let $\underset{\sim}{H}^c$ be obtained by addition of the principle of the excluded third. Then

(i) $H^c \vdash A \leftrightarrow A'$

(ii) $H^c \vdash A \Leftrightarrow H \vdash A'$.

<u>Proof</u>. (i) is obvious. (ii) from left to right can be proved by induction on the length of derivations in $\underset{\sim}{H}^c$. (The implication from right to left is trivial.) We take for example Gödel's system as a basis for our verification.

(a) Basis: $\underset{\sim}{H}^c \vdash A$, A is an axiom.

If $A \equiv B \vee B \to B$, then $A' \equiv \neg(\neg B' \& \neg B') \to B'$

which is equivalent in $\underset{\sim}{H}$ to $\neg\neg B' \to B'$. This holds in $\underset{\sim}{H}$ because of lemma 1.10.8 and remark 1.10.7.

If $A \equiv B \to B \vee C$, then $A' \equiv B' \to \neg(\neg B' \& \neg C')$; this is derivable in $\underset{\sim}{H}$ since $\neg B' \& \neg C' \to \neg B'$, so by contraposition (1.1.8, I) $\neg\neg B' \to \neg(\neg B' \& \neg C')$ etc.

If $A \equiv Bt \to \exists x Bx$, $A' \equiv B't \to \neg \forall x \neg B'x$. Then as before, since $\forall x \neg B'x \to \neg B't$, by contraposition $\neg\neg Bt' \to \neg \forall x \neg B'x$ etc.

If $A \equiv B \vee \neg B$, $A' \equiv \neg(\neg B' \& \neg\neg B')$, which obviously holds in $\underset{\sim}{H}$.
The other axiom schemata are even less difficult.
(b) Induction step. Assume that, for any formula A, if $\underset{\sim}{H}^c \vdash A$ by a deduction of length $\leq k$, then $\underset{\sim}{H} \vdash A'$. Now suppose $\underset{\sim}{H}^c \vdash A$ by a deduction of length $k+1$. We have to distinguish various cases according to the final rule applied in the derivation. The cases PL2, PL3, PL7, PL8, and Q1 are completely trivial.

Assume the last rule to be applied is PL13, so $A \equiv C \vee B_1 \to C \vee B_2$, and by induction hypothesis

$$\underset{\sim}{H} \vdash B_1' \to B_2'.$$

$A' \equiv \neg(\neg C' \& \neg B_1') \to \neg(\neg C' \& \neg B_2')$. A' follows from $\neg C' \& \neg B_2' \to \neg C' \& \neg B_1'$ by contraposition. Since $\neg C' \& \neg B_2' \to \neg C'$, $\neg C' \& \neg B_2' \to \neg B_2'$ and $\neg B_2' \to \neg B_1'$ by contraposition from our induction hypothesis, $\neg C' \& \neg B_2' \to \neg B_1'$ hence $\neg C' \& \neg B_2' \to \neg C' \& \neg B_1'$.

Assume the last rule to be applied to be Q4, so $A \equiv \exists x Bx \to C$. By induction hypothesis, $\underset{\sim}{H} \vdash B'x \to C'$. $A' \equiv \neg \forall x \neg B'x \to C'$.
By contraposition, $\underset{\sim}{H} \vdash \neg C' \to \neg B'x$, so with Q1 $\underset{\sim}{H} \vdash \neg C' \to \forall x \neg B'x$.
By contraposition again, $\underset{\sim}{H} \vdash \forall x \neg B'x \to \neg\neg C'$. By lemma 1.10.8 and remark 1.10.7 $\neg\neg C' \to C'$, so $\underset{\sim}{H} \vdash (\exists x Bx \to C)'$.

1.10.11. Theorem. Let $\underset{\sim}{H}$ be one of the systems $\underset{\sim}{HA}$, $\underset{\sim}{N}$-$\underset{\sim}{HA}^{\omega}$, $\underset{\sim}{HA}^{\omega}$, $\underset{\sim}{I}$-$\underset{\sim}{HA}^{\omega}$, $\underset{\sim}{E}$-$\underset{\sim}{HA}^{\omega}$, $\underset{\sim}{HAS}$ + PCA, $\underset{\sim}{HAS}$. Then, if $\underset{\sim}{H}^c$ denotes the corresponding classical system,

(i) $\underset{\sim}{H}^c \vdash A \leftrightarrow A'$

(ii) $\underset{\sim}{H}^c \vdash A \Leftrightarrow \underset{\sim}{H} \vdash A'$.

Proof. We have only to add to the proof of 1.10.10 a discussion of the additional axioms and rules.
The equality axioms are trivial to deal with, as are the defining axioms for the constants.
Further we have to verify, for any instance A of the induction schema, that $\underset{\sim}{H} \vdash A'$; this is obvious.
In the case of $\underset{\sim}{HAS}$ + PCA or $\underset{\sim}{HAS}$ finally, we have to check that for an instance

A of WCA, resp. CA, $\underset{\sim}{H} \vdash A'$.
Let e.g.

$$A \equiv \exists X^n \forall x_1 \ldots x_n [B(x_1 \ldots x_n) \leftrightarrow X^n x_1 \ldots x_n].$$

We have to show

$$(1) \qquad \underset{\sim}{H} \vdash \neg \forall X^n \neg \forall x_1 \ldots x_n [B'(x_1 \ldots x_n) \leftrightarrow \neg\neg X^n x_1 \ldots x_n].$$

We note that in $\underset{\sim}{H}$ for some Y^n

$$\forall x_1 \ldots x_n [B'(x_1 \ldots x_n) \leftrightarrow Y^n x_1 \ldots x_n].$$

Hence also

$$\forall x_1 \ldots x_n \neg\neg [B'(x_1 \ldots x_n) \leftrightarrow Y^n x_1 \ldots x_n],$$

and thus using $\neg\neg(D_1 \& D_2) \leftrightarrow \neg\neg D_1 \& \neg\neg D_2$, and $\neg\neg(D_1 \to D_2) \leftrightarrow (\neg\neg D_1 \to \neg\neg D_2)$ (1.1.8, V, IV),

$$\forall x_1 \ldots x_n [\neg\neg B'(x_1 \ldots x_n) \leftrightarrow \neg\neg Y^n x_1 \ldots x_n]$$

and with 1.10.8, 1.10.7

$$\forall x_1 \ldots x_n [B'(x_1 \ldots x_n) \leftrightarrow \neg\neg Y^n x_1 \ldots x_n].$$

Hence (1) follows, by $\exists Y^n D(Y) \to \neg \forall Y^n \neg D(Y)$.

1.10.12. Corollary. Let $\underset{\sim}{H}$ be one of the systems indicated in 1.10.11. Then $\underset{\sim}{H}^c$ is conservative over $\underset{\sim}{H}$ w.r.t. negative formulae.

1.10.13. Some further information is given in Kreisel 1962C, where also lemma 1.10.9 (ii) is stated (Theorem 1, Corollary). The concept of a Harrop formula (i.e. a formula without strictly positive occurrences of $\vee$, $\exists$) appears first in Harrop 1960.

§ 11. General discussion of various schemata and proof-theoretic closure conditions.

1.11.1. Introduction. Certain schemata and rules will appear frequently in the statements of metamathematical results in the sequel. In this section we briefly discuss the principal ones.

Let us define a rule as a set of (n+1) - tuples of formulae ; an element of this set is an instance of the rule. If $\langle F_1,\ldots,F_n,F_{n+1}\rangle$ is an instance of a rule, $F_1,\ldots,F_n$ are called the premisses, F_{n+1} the conclusion.

A rule is said to be a derived rule for a system $\underset{\sim}{H}$, if for each instance $\langle F_o,\ldots,F_n\rangle$ of the rule,

$$\underset{\sim}{H}+\Gamma\vdash F_o,\ldots,\underset{\sim}{H}+\Gamma\vdash F_{n-1}\Rightarrow\underset{\sim}{H}+\Gamma\vdash F_n .$$

A rule is said to be an admissible rule for a system $\underset{\sim}{H}$ (or : "derivable from null assumptions") if for any instance $\langle F_o,\ldots,F_n\rangle$ of the rule

$$\underset{\sim}{H}\vdash F_o,\ldots,\underset{\sim}{H}\vdash F_{n-1}\Rightarrow\underset{\sim}{H}\vdash F_n .$$

Admissible and derivable rules are instances of proof-theoretic closure conditions of a rather crude kind : they only involve the set of provable theorems. By a study of normalization for systems of natural deduction, one can obtain more delicate proof-theoretic closure conditions involving the deductions themselves.

For reference in the discussion below, we now briefly recapitulate the intended interpretation of the intuitionistic logical constants.

1°) A proof of $A\&B$ consists of a proof of A and a proof of B ;

2°) A proof of $A\vee B$ consists of a proof of A or a proof of B ;

3°) A proof of $\forall xA$ consists of a construction Π which, applied to a proof c of the fact that d is in the domain of the variable x, yields a proof Πc of Ad, together with a proof Π' of this property of Π.

4°) A proof of $\exists xA$ consists of a c in the domain of x and a proof of Ac, and a proof that c belongs to the domain of x.

5°) A proof of $A\rightarrow B$ consists of a construction Π which transforms any proof c of A into a proof Πc of B (together with a proof Π' that Π satisfies this condition).

Intuitively, the predicate $\Pi_A(c)$ ("c proves A") is assumed to be decidable.

For a more detailed discussion of this interpretation, see Kreisel 1965, Troelstra 1969, §2.

1.11.2. Disjunction and explicit definability property.

Nearly all intuitionistic formal systems discussed in this monograph satisfy the so-called "explicit definability property" (ED)

ED $\vdash \exists x\, Ax \Rightarrow \underset{\sim}{\exists} n\, (\vdash A\bar{n})$ ($\exists x\, Ax$ closed)

(x a numerical variable). In virtue of $\exists x((x=0 \to A)\ \&\ (x \neq 0 \to B)) \leftrightarrow A \vee B$, ED implies the disjunction property

DP $\vdash A \vee B \Rightarrow \vdash A$ or $\vdash B$ ($A \vee B$ closed).

With respect to other sorts of variables, we often encounter a generalization of ED of the form

ED' $\vdash \exists x\, Ax \Rightarrow \underset{\sim}{\exists} t(\vdash At)$ ($\exists x\, Ax$ closed)

where x is now any sort of variable, and t ranges over terms (definable elements of the range of the variable x) of the same sort as x.

ED, ED', DP have often been presented as criteria for the "constructivity" (constructive character) of a formal system. Of course, if we consider e.g. DP, there is a property of the class of informal proofs which parallels DP; a proof of $A \vee B$ should contain either a proof of A or a proof of B; establishing DP means that the set of formal proofs of the system satisfies a similar closure condition. (Similar, but not the same condition: as we stated DP, the formal proof of A or the formal proof of B whose existence follows from the existence of a formal proof of $A \vee B$, need not be contained as "sub-proof" in the proof of $A \vee B$. For a closure condition on formal proofs which more closely resembles the condition on informal proofs, we must establish more!)

Also, ED, ED', DP are neither sufficient nor necessary for the "constructive character" of the system studied, i.e. they do not enforce uniquely the intended interpretation, since there are divergent extensions $\underset{\sim}{HA}'$, $\underset{\sim}{HA}''$ of $\underset{\sim}{HA}$ (so $\underset{\sim}{HA}' \cup \underset{\sim}{HA}''$ is inconsistent) which both possess ED, DP; and on the other hand, there are systems $\underset{\sim}{H}$, $\underset{\sim}{HA} \subseteq \underset{\sim}{H}$, which are obviously intuitionistically acceptable, on the intended interpretation of the logical constants, but which do not possess DP. As an example of the diverging systems, one may take $\underset{\sim}{HA} + M$, $\underset{\sim}{HA} + CT_0 + IP$ (cf. 3.7.4 (i), 3.7.4 (ii), 3.2.27).

We present an example of an intuitionistically justified system which does not satisfy ED (from Troelstra A ; the example is due to Kreisel). Let $\mathrm{Proof} \equiv \mathrm{Proof}_{HA}$, and

$$Ax \equiv_{def} \mathrm{Proof}(x, \ulcorner 0 = 1 \urcorner) \vee \forall y \neg \mathrm{Proof}(y, \ulcorner 0 = 1 \urcorner).$$

Since $\underset{\sim}{HA}$ is intuitionistically consistent (on the intended interpretation),

$\forall y \neg \text{Proof}(y, \ulcorner 0 = 1 \urcorner)$ is intuitionistically true, hence also $\exists x\, Ax$. Further note that because of $\forall y \neg \text{Proof}(y, \ulcorner 0 = 1 \urcorner)$, we must have $\vdash \neg \text{Proof}(\bar{n}, \ulcorner 0 = 1 \urcorner)$ for any numeral $\bar{n}$. Therefore

$$\vdash A\,\bar{n} \;\longleftrightarrow\; \forall y \neg \text{Proof}(y, \ulcorner 0 = 1 \urcorner)$$

for any numeral $\bar{n}$.

Also

$$\vdash \exists x\, Ax \;\longleftrightarrow\; [\exists y\, \text{Proof}(y, \ulcorner 0 = 1 \urcorner) \;\vee\; \forall y \neg \text{Proof}(y, \ulcorner 0 = 1 \urcorner)].$$

Now assume $\vdash \exists x\, Ax \to A\,\bar{n}$, then it would follows that
$\vdash [\exists y\, \text{Proof}(y, \ulcorner 0=1 \urcorner) \vee \forall y \neg \text{Proof}(y, \ulcorner 0=1 \urcorner)] \to \forall y \neg \text{Proof}(y, \ulcorner 0=1 \urcorner)$, hence
$\vdash \exists y\, \text{Proof}(y, \ulcorner 0=1 \urcorner) \to \forall y \neg \text{Proof}(y, \ulcorner 0=1 \urcorner)$, and thus $\vdash \forall y \neg \text{Proof}(y, \ulcorner 0=1 \urcorner)$;
this contradicts Gödel's second incompleteness theorem. Therefore $\underset{\sim}{HA} + \exists x\, Ax$ does not satisfy ED, since $\underset{\sim}{HA} + \exists x\, Ax \vdash \exists x\, Ax$, but not $\underset{\sim}{HA} + \exists x\, Ax \vdash A\,\bar{n}$.

1.11.3. <u>Schema</u> D: $\forall x(A \vee Bx) \to (A \vee \forall x\, Bx)$, x not free in A.

This schema has attracted attention because intuitionistic predicate logic with this schema added, is semantically characterized by Kripke-models with constant domain (see e.g. <u>Görnemann</u> 1971).
In <u>Görnemann</u> 1971 it is also noted that intuitionistic predicate logic + schema D possesses the disjunction property. However, this property is lost as soon as we go to arithmetic.

Actually, $\underset{\sim}{HA} + D = \underset{\sim}{HA}^c$, as is easily seen; by induction on the logical complexity of A one verifies the provability of $\forall x(A \vee \neg A)$. E.g. if $A \equiv \exists y\, By$, then $(\forall x \neg Bx \vee \exists y\, By) \longleftrightarrow \forall x(\neg Bx \vee \exists y\, By)$; and if $\neg Bx$, then $\neg Bx \vee \exists y\, By$; if Bx, then $\exists y\, By$ hence also $\neg Bx \vee \exists y\, By$. By induction hypothesis, $Bx \vee \neg Bx$, therefore $\neg Bx \vee \exists y\, By$ for all x, and thus $\forall x \neg Bx \vee \exists y\, By$.

If $A \equiv \forall y\, By$, note that $\forall y(By \vee \neg By) \to \forall y(By \vee \neg \forall x\, Bx) \longleftrightarrow$
$\longleftrightarrow \forall x\, Bx \vee \neg \forall x\, Bx$.

Even closed instances of D spoil the disjunction property. Let D^c indicate D restricted to closed instances.

Let for example $\forall x\, Cx$ be a rossersentence for $\underset{\sim}{HA} + D^c$, and let $A \equiv \neg \forall x\, Cx$, $Bx \equiv Cx$. Then $\forall x(A \vee Bx)$ holds in $\underset{\sim}{HA}$; by D^c, $\underset{\sim}{HA} + D^c \vdash A \vee \forall x\, Bx$; hence, if $\underset{\sim}{HA} + D^c$ would satisfy DP,

$$\underset{\sim}{HA} + D^c \vdash \neg \forall x\, Cx \quad \text{or} \quad \underset{\sim}{HA} + D^c \vdash \forall x\, Cx$$

which is impossible. At the same time it follows that D^c is not derivable in $\underset{\sim}{HA}$, not even the rule for sentences:

(1) $\qquad \vdash \forall x(A \vee Bx) \Rightarrow \vdash A \vee \forall x\, Bx$.

This contrasts with the case for intuitionistic predicate logic, where cut elimination for a calculus of sequents or normalization for natural deduction readily yields (1). (In a closed normal deduction, $\forall x(A \vee Bx)$ must be obtained from $\forall$-introduction from $A \vee Ba$, and this in turn from A or Ba by $\vee$-introduction; hence A or $\forall x\, Bx$ can be derived.)

Since D is obviously invalid for our realizability interpretations, and the rule is not admissible even for arithmetic, we shall not spend further attention on it.

1.11.4. The schema $\forall x \neg\neg A \rightarrow \neg\neg \forall x A$. (DNS = Double Negation Shift).

To this schema attaches considerable technical interest, since, as we have seen in 1.10.9, in theories based on intuitionistic logic + DNS, the negative translation satisfies $\neg\neg A \leftrightarrow A'$ for all A.
Also it permits us to derive, in intuitionistic analysis, (AC)' from AC: Assume $\forall x \neg \forall y \neg A'(x,y)$, then $\forall x \neg\neg \exists y A'(x,y)$, so $\neg\neg \forall x \exists y A(x,y)$ by DNS; hence $\neg\neg \exists z \forall x A'(x,zx)$, i.e. $\neg \forall z \neg \forall x A'(x,zx)$. In other words, the '-translation interprets classical analysis, formulated with sequence variables and the axiom of choice, in the corresponding intuitionistic theory + DNS. This fact constitutes the starting point of Spector 1962.

1.11.5. Markov's schema and rule.

Markov's schema in its most general form can be stated as

M $\qquad \forall x[A \vee \neg A] \;\&\; \neg\neg \exists x A \rightarrow \exists x A$.

A simpler and weaker form is

M_{PR} $\qquad \neg\neg \exists x A \rightarrow \exists x A \qquad$ (A primitive recursive).

Let us use M^c_{PR} for M_{PR} restricted to closed instances. Intuitively, for x ranging over natural numbers, M expresses: if we have a property A which can be tested for each x (i.e. $\forall x(A \vee \neg A)$) and an indirect proof of $\exists x A$ (i.e. $\neg\neg \exists x A$) then this amounts to a direct proof of $\exists x A$. In other words, Markov's principle enables us to assert, for a computer testing A for all x, that $\neg\neg \exists x A$ guarantees that the computer will find an x such that A. This is obviously an enlargement of the concept of "constructive".

M^c_{PR} is not derivable in HA (Kreisel 1958A). Consider e.g. the following instance

(1) $\qquad \neg \forall x \neg Bx \rightarrow \exists x\, Bx$

where $\forall x \neg Bx$ is a rossersentence for HA.
Making use of the closure of HA under

$$\vdash \neg A \rightarrow \exists x B \Rightarrow \exists n(\vdash \neg A \rightarrow B\bar{n}) \qquad \text{(cf. e.g. 3.1.7)}$$

it would follow that if $\underset{\sim}{HA} \vdash (1)$,

$$\vdash \neg \forall y \neg By \to B\bar{n} .$$

Bx is primitive recursive, hence $\vdash B\bar{n}$ or $\vdash \neg B\bar{n}$.
In the first case $\underset{\sim}{HA} \vdash \neg \forall x \neg Bx$, in the second case $\underset{\sim}{HA} \vdash \neg\neg \forall x \neg Bx$, i.e. $\underset{\sim}{HA} \vdash \forall x \neg Bx$. Both cases conflict with the assumption that $\forall x \neg Bx$ is a rossersentence for $\underset{\sim}{HA}$.

$\underset{\sim}{HA}$, and many other intuitionistic formal systems, have been shown to be closed under the rule

MR $\quad \vdash \forall x(A \vee \neg A),\ \vdash \neg\neg \exists x A \Rightarrow \vdash \exists x A$,

and a fortiori under

MR_{PR} $\quad \vdash \neg\neg \exists x A \Rightarrow \vdash \exists x A$ for A primitive recursive,

and its specialization MR^c_{PR} to closed formulae.

Using ω-consistency and classical metamathematics freely, we can establish MR^c_{PR} as follows. Assume $\vdash \neg\neg \exists x Ax$, A primitive recursive, $\exists x A$ closed. For all $\bar{n}$, $\vdash A\bar{n}$ or $\vdash \neg A\bar{n}$. Suppose $\underset{\sim}{\forall} n(\vdash \neg A\bar{n})$; then, assuming ω-consistency of $\underset{\sim}{HA}$, we obtain a conflict with $\vdash \neg \forall x \neg A$. Hence $\neg \underset{\sim}{\forall} n(\vdash \neg A\bar{n})$; arguing classically, $\underset{\sim}{\exists} n(\vdash A\bar{n})$; hence $\vdash \exists x Ax$.

For more details on MR, see § 3.8, § 5.4.
Technical interest of Markov's schema also derives from the fact that it is validated by Gödel's Dialectica interpretation, and that by a result of Gödel (Kreisel 1962, § 3) completeness w.r.t. intuitionistic validity (conceived as the analogue of classical set-theoretic validity) implies the validity of Markov's schema in intuitionistic arithmetic.

In recursion theory, there is a rather special application of Markov's schema in the proof of Post's theorem (a set $X \subseteq N$ is recursive if X and $N \setminus X$ are both r.e.). The application involved corresponds to a schema of predicate logic of the following form:

$$\forall xy(A \vee \neg A)\ \&\ \forall xy(B \vee \neg B)\ \&\ \forall x \neg \exists y \exists z(A(x,y)\ \&\ B(x,z))\ \&$$
$$\&\ \forall x \neg\neg(\exists y Axy \vee \exists y Bxy) \to \forall x(\exists y Axy \vee \neg \exists y Axy) .$$

Or with function variables:

$$\neg \exists y \exists z(ay = bz)\ \&\ \forall x\ \neg\neg(\exists y(ay = x) \vee \exists y(by = x)) \to \forall x(\exists y(ay = x) \vee \neg \exists y(ay = x)) .$$

This schema is presumably weaker than MR, but the precise logical relationships are unknown.

1.11.6. Independence-of-premiss schemata.

An "independence-of-premiss schema" is a schema of the form (x not free in A)

(1) $(A \to \exists x B) \to \exists x(A \to B)$

where A in general must satisfy additional restrictions $\mathcal{R}$, either syntactical or logical.
The principal instances of independence-of-premiss schemata which we shall encounter are

IP $(\neg C \to \exists x B) \to \exists x(\neg C \to B)$

(so A must be of the form $\neg C$ here), and the weaker

IP_0 $\forall x(A \vee \neg A) \& (\forall x A \to \exists y B) \to \exists y(\forall x A \to B)$

and the still weaker

IP_{PR} $(\forall x A \to \exists y B) \to \exists y(\forall x A \to B)$ (A primitive recursive).

On the intended interpretation of the logical constants, if we assert $A \to \exists x B$, the "x for which B" may depend essentially on the proof of A (and not only on A being true).
An independence-of-premiss schema (1) expresses, that for A satisfying the restriction $\mathcal{R}$, the x does not depend on the proof of A at all: we can indicate a priori an x which should satisfy B if A holds. So independence-of-premiss schemata do affect the intended (constructive) interpretation of the logical constants: they restrict the type of mappings from proofs to proofs which can be used to establish implications of the form $A \to \exists x B$.
The corresponding admissible rules of the form

(2) $\vdash A \to \exists x B \Rightarrow \vdash \exists x(A \to B)$

(A under an additional restriction) show more or less the same as the consistency of the schema relative to the same system: that the system discussed permits the interpretation of intuitionistic implication enforced by the independence-of-premiss schemata.

A certain technical interest of IP is in the fact that it is validated by modified realizability interpretations (see § 3.4). Not even IP^c_{PR} (i.e. IP_{PR} restricted to closed formulae) is derivable in intuitionistic arithmetic; see e.g. 3.1.11).

1.11.7. Church's thesis and rule.

In a formal system with function symbols, (the intuitionistic version of) Church's thesis can be expressed as

CT $\forall \alpha \exists x \forall y \exists z [Txyz \& \alpha x = Uz]$.

Combined with a choice principle AC_{00}, we obtain a version which can be expressed in the language of arithmetic:

CT_0 $\forall x \exists y A(x,y) \to \exists z \forall x \exists u(Tzxu \& A(x,Uu))$.

The conceptual interest of CT and CT_0 is in their bearing on the question: do the concepts of "humanly computable function" and "mechanically computable function" coincide? (here "humanly computable" should mean "computable by an idealized mathematician in the intuitionistic sense" : for a discussion of these matters, see Kreisel 1970, and especially Kreisel 1972); and secondly in the fact that CT implies the incompleteness of intuitionistic predicate logic (sketched in Kreisel 1970, Technical Note I: for a more detailed exposition see van Dalen A).

Church's thesis turns out to be consistent with all intuitionistic formal systems not involving the concept of choice sequence, and especially not containing the fan theorem or bar induction (cf. § 3.2).

The underivability of CT_0 is obvious, since CT_0 is false in $\underset{\sim}{HA}^c$. "Church's rule" takes the forms:

$$\vdash \exists\alpha A\alpha \Rightarrow \vdash \exists\alpha \in GR(A\alpha)$$

where $\alpha \in GR$ abbreviates $\exists x \forall y \exists u(Txyu \ \& \ \alpha y = Uu)$ and

CR_0 $\quad \vdash \forall x \exists y A(x,y) \Rightarrow \vdash \exists z \forall x \exists u(T zxu \ \& \ A(x,Uu))$.

A weaker version of Church's rules takes the form

WCR $\quad$ If $\vdash \forall x \exists y Axy$, then there is a recursive function f such that $\underset{\sim}{\forall} n \vdash A(\bar{n}, \overline{fn})$.

WCR is closely connected with ED: for systems with a recursive axiomatization, ED is equivalent to WCR (Kreisel 1972). This is seen as follows. Obviously, WCR implies ED. Conversely, if ED holds, we may construct f as

$$fn = \min_m \text{Proof}(j_1 m, \ulcorner A(\bar{n}, \overline{j_2 m})\urcorner)$$

which makes f recursive, in view of the recursiveness of "Proof". As remarked in Kreisel 1972, this result tells us where not to look for a conflict with Church's thesis; all the usual systems satisfying ED cannot be expected to yield a refutation.

Chapter II

MODELS AND COMPUTABILITY

§ 1. Definitions by induction over the type structure.

2.1.1. **Definition over the type structure.** In the sequel we shall meet
repeatedly with definitions over applicative type structures, i.e. type
structures, obtained from certain basic types by closure under the condition:
if σ, τ are types, then so is $(\sigma)\tau$, our principal example of such a
structure being $\underset{\approx}{T}$. In the discussion below, we restrict our attention to
$\underset{\approx}{T}$ for simplicity.

An <u>applicative</u> set of terms M is a set of terms such that if $t \in (\sigma)\tau$,
$t' \in \sigma$, $t,t' \in M$ then $tt' \in M$. A <u>basis</u> for an applicative set M is a
subset $M' \subseteq M$ such that the closure of M' under application yields M
(i.e. M is the smallest applicative set containing M'). Examples of
applicative sets :

(a) The set of closed terms CTM of $\underset{\sim}{N} - \underset{\approx}{HA}^{\omega}$, with the constants of $\underset{\sim}{N} - \underset{\approx}{HA}^{\omega}$
as a basis.
(b) The applicative set generated by the basis consisting of the constants
and type 0 variables of $\underset{\sim}{N} - \underset{\approx}{HA}^{\omega}$ (CTM_0).
(c) The applicative set generated by the basis consisting of the constants,
the type 0 variables and a single fixed type 1 variable ($CTM_0(x^1)$,
if x^1 is the type 1 variable).

Let us define the <u>type level</u> as follows: $l(0) = 0$, $l((\sigma)\tau) = l(\sigma) + l(\tau) + 1$.
Further examples of applicative sets are now provided by restriction of the
types to types σ with $l(\sigma) \leq n$ in examples (a), (b), (c).

In its simplest form, a definition of an n-ary relation over the type
structure $\underset{\approx}{T}$ for an applicative set M of terms takes the following form:

(i) $P_0(t_1,\ldots,t_n)$ if $t_1,\ldots,t_n \in 0$ and $A(t_1,\ldots,t_n)$ (A being any given
predicate); $t_1,\ldots,t_n \in M$.

(ii) $P_{(\sigma)\tau}(t_1,\ldots,t_n)$ if $\forall t_1' \in M \ldots \forall t_n' \in M(P_\sigma(t_1',\ldots,t_n') \Rightarrow P_\tau(t_1 t_1',\ldots,t_n t_n'))$.

A slight generalization (for an example see 2.2.5) takes the following form
(M an applicative set):

(i)' $P_0(t)$ if $t \in 0$, $A_0(t)$ and $t \in M$

(ii)' $P_{(\sigma)\tau}(t)$ if $A_{(\sigma)\tau}(t)$ and $\forall t' \in M(P_\sigma t' \Rightarrow P_\tau tt')$.

Such a definition may be viewed as a superposition of a sequence of definitions
over a type structure, where each type σ is not only viewed as obtained
applicatively from type 0 but also acts as a "ground type" (or "basic type")
for more complex types w.r.t. the property A_σ.

A definition according to (i), (ii) or (i)', (ii)' is a definition over the type structure of metamathematical properties of metamathematical objects (i.c. terms). Of course, similar definitions are possible within the system itself, i.e. we may define properties within the system according to a schema:

(i)" $P_0(x_1^0,\ldots,x_n^0) \equiv_{def} A(x_1^0,\ldots,x_n^0)$

(ii)" $P_{(\sigma)\tau}(x^{(\sigma)\tau},\ldots,x_n^{(\sigma)\tau}) \equiv_{def} \forall y_1^\sigma\ldots y_n^\sigma(P_\sigma(y_1^\sigma,\ldots,y_n^\sigma) \rightarrow P_\tau(x_1y_1,\ldots,x_ny_n))$

and such a schema similarly permits generalizations and variants (cf. 2.3.13 for an example).

2.1.2. Establishing properties for applicative sets of terms.

We consider three types of definitions of a property Q for the terms of an applicative set M:

(A) Q is defined as a unary predicate over the type structure, according to clauses (i) and (ii) in 2.1.1.

(B) $Qt \equiv_{def} P(t,\ldots,t)$, where P is an n-ary predicate defined over the type structure according to clauses (i), (ii) in 2.1.1, and $A(t,\ldots,t)$ holds for $t \in M$.

(C) Q is defined inductively over the type structure according to (i)' and (ii)' in 2.1.1.

In each of these cases, establishing $\forall t \in M[Qt]$ may be reduced to establishing $\forall t \in M'[Qt]$ for a basis M' of M, because of the following lemma:

Lemma. If $t_1,\ldots,t_n$ are terms such that $Q(t_i)$ for $1 \leq i \leq n$, and Q is defined by a definition of type (A), (B), (C), then for any t obtained by repeated application from $t_1,\ldots,t_n$, Qt holds.

Proof. Quite straightforward; we have to show that if $Q(t_1)$, $Q(t_2)$ and $t_1 \in (\sigma)\tau$, $t_2 \in \sigma$, then $Q(t_1t_2)$.

Similar lemmas reduce the establishment of properties Q defined via definitions of type (i)", (ii)", or one of the many variants, to the establishment of Q for a basis of the set of applicative terms considered.

2.1.3. Definability aspects.

Suppose Q to be an n-ary predicate defined by a definition of type (A), (B), (C) in 2.1.2. If we wish to consider an arithmetical version Q^* of Q (so Q^* is a predicate of gödelnumbers of terms, not of the terms themselves) then if there is no bound on the type level, we cannot in general expect Q^* to be arithmetically definable, since Q^* should satisfy e.g. for a definition of type (A):

$$Q^*(m) \leftrightarrow \exists\sigma\ (\text{type}\ (m) = \ulcorner\sigma\urcorner \,\&\, Q^*_\sigma(m)),$$
$$Q^*_{(\sigma)\tau}(m) \leftrightarrow \forall n\ (Q^*_\sigma n \rightarrow Q^*_\tau(\text{app}(m,n)))$$

(where $\ulcorner\sigma\urcorner$ denotes the code number of type σ, app(n,m) the arithmetical representation of the application operation), therefore with increasing type level the logical complexity of the arithmetical formula $Q^*_\sigma(m)$ increases indefinitely; for an actual counterexample see 2.3.11.

So for an arithmetized version, the applicative set of terms considered will have a bound on the type level.
As we shall see, in our applications, the definability of the arithmetized predicates usually ensures formalizability of the proof of Q^*n for all gödel-numbers n of terms in the applicative set of terms considered.

2.1.4. Sets of terms closed under λ-abstraction.

If we consider sets of terms not only closed under application but also under λ-abstraction, the reduction effected by the lemma in 2.1.2 might not be sufficient since the effect of closure under λ-abstraction might force us to consider a very "large" basis in the sense of 2.1.1.

Let us call a λ-set of terms any set closed under application and λ-abstraction w.r.t. variables of the set. A λ-basis for M is a subset which yields M by closure under application and λ-abstraction.
The appropriate trick for establishing a property Q defined according to (A), (B), (C) in 2.1.2 for a λ-set is then to prove a stronger property Q^* ("Q-under-substitution"):
$Q^*(t)$ holds if Qt_1 holds for any t_1 obtained by substituting for some (not necessarily all) occurrences of variables in t terms t' for which Q holds (and possibly renaming bound variables in t so as to avoid variables free in t' becoming bound after substitution), (see e.g. 2.2.27 - 2.2.31).

§ 2. Computability of terms in $\underset{\sim}{N}-\underset{\sim}{HA}^{\omega}$.

2.2.1. In Gödel 1958, the concept of a "berechenbare Funktion" (= computable function) is regarded as a primitive concept, and it is considered evident that each primitive recursive functional definable in Gödel's theory (i.e. a functional represented by a term of $\underset{\sim}{N}-\underset{\sim}{HA}^{\omega}$) is "computable".

In Tait 1965 this concept is made the subject of a formal analysis, and it is shown that each constant term of type 0 "reduces to" a numeral, which implies that each term of type 0 can be formally proved (in a suitable version of $\underset{\sim}{N}-\underset{\sim}{HA}^{\omega}$) to be equal to a certain numeral. As a by-product, Tait's analysis yields more: all terms can be brought into a standard form ("normal form").

This is exploited in Tait 1967, to show that the closed terms of $\underset{\sim}{N}-\underset{\sim}{HA}^{\omega}$ yield a model for $\underset{\sim}{I}-\underset{\sim}{HA}^{\omega}$, if equality between terms is interpreted in the model as: reducing to the same normal form. Tait 1967 also introduces the inductively defined formal computability predicates ("Comp"), a device which has been extensively used since.

In the present section we discuss computability predicates and use them to prove normalization and strong normalization theorems for the terms of $\underset{\sim}{N}-\underset{\sim}{HA}^{\omega}$ and extensions.

The main novelty is the simplification of the treatment and strengthening of the strong normalization theorem in 2.2.19.

We first discuss computability for the terms of $\underset{\sim}{N}-\underset{\sim}{HA}^{\omega}$; then we deal with classes of terms with λ-operators instead of Π, Σ as primitives.

In 2.2.35, the various proofs of normalization and normal form theorems occurring in the literature are discussed and compared.

Additional material on computability is given in the next section.

We feel the notions of computability and strong computability have a certain intrinsic interest, because of their intuitive simplicity; hence the rather extensive discussion, with description of different approaches, below. It should be noted, however, that for all applications of computability given in the sequel of this chapter, in proofs of results which do not require the notion of computability for their formulation, we may restrict our attention to standard computability of terms of type 0; the remainder is a luxury.

2.2.2. Definition of reduction and standard reduction for terms of $\underset{\sim}{N}-\underset{\sim}{HA}^{\omega}$.

We say that a term t contracts to a term t' (t contr. t', or t' is a contraction of t), if one of the following clauses is satisfied:

(a) $t \equiv \Pi t_1 t_2$, $t' \equiv t_1$

(b) $t \equiv \Sigma t_1 t_2 t_3$, $t' \equiv t_1 t_3 (t_2 t_3)$

(c) $t \equiv R t_1 t_2 0$, $t' \equiv t_1$

(d) $t \equiv R t_1 t_2 (S t_3)$, $t' \equiv t_2 (R t_1 t_2 t_3) t_3$.

We adopt the terminology of Curry - Feys 1958 and call t in the clauses (a) - (d) a redex (and similarly for other types of contraction introduced in the sequel).

If t' is obtained from t contracting a single subterm (occurrence) of t (i.e. a subterm of t is replaced by its contraction) then we write $t' <_1 t$ or $t >_1 t'$.

A sequence (finite or not) $t_0, t_1, t_2, \ldots$ with $t_{i+1} <_1 t_i$ for all i is said to be a reduction sequence of t_0 (starting from t_0).

A term which does not admit any contractions, is said to be in normal form. A finite reduction sequence ending in a term in normal form is said to terminate.

We say that $t \geq t'$ (t reduces to t') if there is a reduction sequence of t ending with t' (a reduction sequence from t to t').

A reduction sequence is said to be strict, if the contractions (a) - (d) are applied only in case t_1, t_2, t_3 are normal. Attention to strict reduction sequences implies prescribing a certain (partial) order for the contractions. The order in which contractions have to be executed can be made completely deterministic by introducing the concept of the leftmost minimal redex (lmr). The lmr of t is a subterm of t when t is not normal, otherwise undefined.

We define the lmr by induction on the complexity of t (t assumed to be non-normal).

(i) If $t \equiv \varphi t_1 \ldots t_n$, φ a constant or variable, and $t_1, \ldots, t_{i-1}$ are normal, t_i not, then the lmr of t is the lmr of t_i.

(ii) If t is a redex and (i) does not apply, then lmr(t) is t itself.

A standard reduction sequence $t_0, t_1, t_2, \ldots$ is a reduction sequence such that t_{i+1} is obtained from t_i by contraction of the lmr of t_i. We write $t \geq' t'$ if there is a standard reduction sequence from t to t'.

2.2.3. Comparison of standard and strict reduction.

Intuitively we feel that there is little essential difference between standard and strict reduction: strict reduction corresponds to the natural idea of contracting ("computing") starting "from the inside" (i.e. starting with redexes not containing other redexes); standard reductions make in a convenient but arbitrary way (since not directly related to the partial ordering of the tree of subterms) the procedure completely deterministic. We show

Proposition. If t strictly reduces to t', t' normal, then there is a standard reduction sequence from t to t'.

Proof. By induction on the length of strict reduction sequences. Suppose (1) the assertion to hold if t strictly reduces to t' in less than k steps; we now prove the assertion for strict reduction sequences of length k by a sub-induction on the complexity of the first term of the sequence. So assume (2) also the assertion to have been proved for all strict reduction sequences of length k starting with a term of complexity < 1.

Let $t_1, \ldots, t_k$ be a strict reduction sequence from t_1 to t_k, and let the complexity of t_1 be 1. Then either t_2 is obtained by contracting t_1, and then the assertion readily follows from induction hypothesis (1). If t_2 is not obtained from contracting t_1, $t_1, \ldots, t_k$ starts with an initial segment

$$t_1 \equiv \varphi s_1^{(1)} \ldots s_m^{(1)}, \varphi s_1^{(2)} \ldots s_m^{(2)}, \ldots, \varphi s_1^{(n)} \ldots s_m^{(n)}$$

where φ is a constant or variable of $\underset{\sim}{N}-\underset{\sim}{HA}^\omega$, and $s_1^{(n)}, \ldots, s_m^{(n)}$ are normal. Then the sequences $s_i^{(1)}, \ldots, s_i^{(n)}$ $(1 \le i \le m)$ become strict reduction sequences after omission of repetitions. Then either

(i) all sequences $s_i^{(1)}, \ldots, s_i^{(n)}$ have length $< k$ after omission of repetitions. Then by induction hypothesis (1), there are standard reduction sequences from $s_i^{(1)}$ to $s_i^{(n)}$ $(1 \le i \le m)$ and from $\varphi s_1^{(n)} \ldots s_m^{(n)} \equiv t_n$ to t_k which may be combined into a standard reduction sequence from t_1 to t_k; or

(ii) there is a sequence $s_i^{(1)}, \ldots, s_i^{(n)}$ of length k, without repetitions and $s_j^{(1)} \equiv s_j^{(n)}$ for $1 \le j \le m$, $j \ne i$, and $\varphi s_1^{(n)} \ldots s_m^{(n)} \equiv t_k$. Then the assertion follows by the sub-induction hypothesis (2).

(Cf. Tait 1967, II on page 203.)

Corollary. All terminating strict reduction sequences starting from a given term terminate in the same term.

In the sequel we shall establish a much stronger result (see 2.2.23).

2.2.4. Alternative definition of $\ge$.

$\ge$ may also be defined as a relation between terms, inductively generated by the following closure conditions:

$t \ge t$, $t \ge t' \Rightarrow tt'' \ge t't''$, $t \ge t' \Rightarrow t''t \ge t''t'$, $t \ge t'$ and $t' \ge t'' \Rightarrow t \ge t''$, $\Pi tt' \ge t$, $\Sigma tt't'' \ge tt''(t't'')$, $Rtt'0 \ge t$, $Rtt'(St'') \ge t'(Rtt't'')t''$.

The equivalence of this definition with the one given in 2.2.2 is intuitively obvious, and in fact formally provable in $\underset{\sim}{HA}$, e.g. by an appeal to the theory of § 1.4.

2.2.5. Definition of computability, strict computability, standard computability. We define a predicate $\text{Comp} = \bigcup \{\text{Comp}_\sigma \mid \sigma \in \underset{\approx}{T}\}$, defining Comp_σ

by induction over the type structure :

(i) $\mathrm{Comp}_0(t) \equiv_{def} t \in 0$ and t reduces to normal form ;

(ii) $\mathrm{Comp}_{(\sigma)\tau}(t) \equiv_{def} \forall t'(\mathrm{Comp}_\sigma(t') \Rightarrow \mathrm{Comp}_\tau(tt'))$ and t reduces to normal form.

Similarly we define Comp', Comp'_σ and Comp", Comp''_σ, replacing "reduces to" in the definition of Comp_σ by "strictly reduces to" and "reduces to ... by a standard reduction sequence" respectively.

2.2.6. Theorem. All terms t of $\mathbf{N}$-$\mathbf{HA}^\omega$ satisfy Comp"(t), and hence have a terminating standard reduction sequence.

Proof. We note that if $\mathrm{Comp}''(t_1), \ldots, \mathrm{Comp}''(t_n)$, then Comp"(t) for any term t constructed by repeated application from $t_1, \ldots, t_n$. (Cf. the lemma in 2.1.2.) This is an immediate consequence of the definition. Hence it is sufficient to prove $\mathrm{Comp}''(\varphi)$ for φ a constant or variable of our theory.

(i) Comp"(0) is immediate.

(ii) $\mathrm{Comp}''(t^0) \Rightarrow \mathrm{Comp}''(St^0)$ is also immediate, hence Comp"(S).

(iii) $\mathrm{Comp}''(x^0)$ is immediate.

(iv) Let $\sigma = (\sigma_1) \ldots (\sigma_m)0$. If $\mathrm{Comp}''_{\sigma_1}(t_1), \ldots, \mathrm{Comp}''_{\sigma_m}(t_m)$, there are terminating standard reduction sequences for $t_1, \ldots, t_m$, which are readily combined into a terminating standard reduction sequence for $x^\sigma t_1 \ldots t_m$. Hence $\mathrm{Comp}''(x^\sigma)$.

(v) If $\mathrm{Comp}''_0(t_3)$, then t_3 has a standard reduction sequence terminating in a term in normal form t_3'. There is a uniquely determined k such that $t_3' \equiv S^k t_3''$, $t_3'' \not\equiv St$ for any t.
We put $\nu(t_3) = k$. We now establish $\mathrm{Comp}''(R_\sigma)$ by proving

(1) $\forall t_1 \forall t_2 \forall t_3\, (\mathrm{Comp}''(t_1)$ and $\mathrm{Comp}''(t_2)$ and $\mathrm{Comp}''(t_3)$ and $\nu(t_3) = k \Rightarrow \mathrm{Comp}''(R_\sigma t_1 t_2 t_3))$

by induction w.r.t. k. Let $\sigma \equiv (\sigma_1) \ldots (\sigma_m)0$.

(a) Basis. Let $\mathrm{Comp}''(t_i)$, $1 \leq i \leq m+3$, $\nu(t_3) = 0$.
There are standard reduction sequences from t_i to t_i', t_i' normal, $1 \leq i \leq m+3$. If $t_3' \equiv 0$, there is a standard reduction sequence from $Rt_1 \ldots t_{m+3}$ to $Rt_1' \ldots t_{m+3}'$; $Rt_1't_2't_3' \ldots t_{m+3}' \geq_1 t_1't_4' \ldots t_{m+3}'$, and since, according to our hypotheses and the remark at the beginning of the proof, $\mathrm{Comp}''(t_1't_4't_5' \ldots t_{m+3}')$, also $\mathrm{Comp}''(Rt_1 \ldots t_{m+3})$.
If $t_3' \not\equiv St$ for any t, then $Rt_1't_2't_3' \ldots t_{m+3}'$ is already in normal form.

(b) Induction step. Assume (1) to be established for $n < k$.
If $\mathrm{Comp}''(t_i)$, $1 \leq i \leq m+3$, $\nu(t_3) = k$, there are standard reduction sequences from t_i to t_i', t_i' normal $(1 \leq i \leq m+3)$, $t_3' \equiv S^k t_3''$ for a suitable t_3''.
Then there is a standard reduction sequence from $Rt_1 \ldots t_{m+3}$ to

$Rt'_1 \ldots t'_{m+3}$; also $Rt'_1 t'_2 (S^k t''_3) t'_4 \ldots t'_{m+3} \geq_1 t'_2 (Rt'_1 t'_2 (S^{k-1} t''_3))(S^{k-1} t''_3) t'_4 \ldots t'_{m+3}$; and this term has a terminating standard reduction sequence by our hypotheses and the remark at the beginning of the proof.

Cases (vi) and (vii), where it is to be shown that $\mathrm{Comp}''(\Pi)$ and $\mathrm{Comp}''(\Sigma)$, are left to the reader.

2.2.7. Remarks.
(i) $\mathrm{Comp}'(t)$ and $\mathrm{Comp}(t)$ for all terms t of $\mathbf{N}$-$\mathbf{HA}^{\omega}$ follow directly from 2.2.6. However, $\mathrm{Comp}'(t)$ and $\mathrm{Comp}(t)$ can also be proved directly along the same lines as in 2.2.6.
(ii) In the definition of Comp', Comp" we might actually have left out the condition "and t reduces to normal form" in clause (ii).
To see this, note that one easily proves by induction that 0^{σ} (defined by $0^0 \equiv 0$, $0^{(\sigma)\tau} \equiv \Pi_{\sigma,\tau} 0^{\sigma}$) is normal and satisfies Comp''_{σ} (with the weakened definition of Comp"). Now if $\mathrm{Comp}''_{\sigma}(t)$, $\sigma \equiv (\sigma_1) \ldots (\sigma_m) 0$, then $\mathrm{Comp}''_0(t 0^{\sigma_1} \ldots 0^{\sigma_m})$. Hence there is a terminating standard reduction sequence of $t 0^{\sigma_1} \ldots 0^{\sigma_m}$, from which a terminating standard reduction sequence of t can be extracted. Similarly for Comp'.
(iii) If we are interested in the computability of closed terms only, a slight simplification might have been achieved by omitting clauses (iii) and (iv) in the proof of 2.2.6.

2.2.8. Lemma. If $t \in 0$, t a closed term of $\mathbf{N}$-$\mathbf{HA}^{\omega}$ in normal form, then t is a numeral.
Proof. We proceed by induction on the complexity of t. Suppose t is closed and normal. Then t has (possibly) one of the forms

$$0,\ S,\ St_1,\ R,\ Rt_1t_2 \ldots t_n,\ \Pi,\ \Pi t_1,\ \Sigma,\ \Sigma t_1,\ \Sigma t_1 t_2$$

(with $t_1, t_2, \ldots$ normal). But the only forms of this list which could have type 0, are

$$0, \quad St_1, \quad Rt_1t_2 \ldots t_n \quad (n \geq 3).$$

If $t \equiv 0$, we are done. If $t \equiv St_1$, then (since t_1 is closed, normal, of type 0) by induction hypothesis t_1 is a numeral, therefore so is t. Finally, if $t \equiv Rt_1t_2t_3 \ldots t_n$, then by induction hypothesis t_3 is a numeral, so t cannot be in normal form.

2.2.9. Theorem. (On assumption of consistency of $\mathbf{N}$-$\mathbf{HA}^{\omega}$.) Each closed term of type 0 reduces to a uniquely determined numeral.
Proof. Theorem 2.2.6 and lemma 2.2.8 imply that each closed term of type zero reduces to a numeral. The uniqueness of the numeral follows from the consistency of $\mathbf{N}$-$\mathbf{HA}^{\omega}$, since if $t \geq \bar{n}$, $t \geq \bar{m}$, $\bar{n} \neq \bar{m}$, it would follow that $\mathbf{N}$-$\mathbf{HA}^{\omega} \vdash \bar{n} = \bar{m}$.

2.2.10. Theorem. $\underset{\sim}{N}-\underset{\sim}{HA}^{\omega}$ is conservative w.r.t. closed prime formulae of type 0 over $\underset{\sim}{H}$, $\underset{\sim}{H}$ obtained by omitting induction from $\underset{\sim}{N}-\underset{\sim}{HA}^{\omega}$.
Proof. By refinement of the argument in 2.2.9, noting that $t_1 \geq t_2$ implies $\underset{\sim}{H} \vdash t_1 = t_2$. For let $\underset{\sim}{N}-\underset{\sim}{HA}^{\omega} \vdash t^o = s^o$. Then we can find $\bar{n}$, $\bar{m}$ such that $t^o \geq \bar{n}$, $s^o \geq \bar{m}$, hence $\underset{\sim}{H} \vdash t^o = \bar{n}$, $\underset{\sim}{H} \vdash s^o = \bar{m}$. By consistency of $\underset{\sim}{N}-\underset{\sim}{HA}^{\omega}$, $\bar{n} = \bar{m}$, hence $\underset{\sim}{H} \vdash t^o = s^o$.

2.2.11. Remark. In 2.5.6 it will be shown how to prove uniqueness of normal form for all types (i.e. every terminating reduction sequence of t terminates in the same term) by means of a model for $\underset{\sim}{N}-\underset{\sim}{HA}^{\omega}$.

2.2.12 - 2.2.19. Strong computability.

2.2.12. We shall now refine the preceding discussion, by proving a stronger theorem: each reduction sequence starting from a term t terminates. We shall call such a theorem a strong normalization theorem. A term t is said to be strongly normalizable, if all reduction sequences starting from t do terminate. In order to prove this theorem, we have to modify our definition of computability to a definition of strong computability.

2.2.13. Definition. Strong computability for terms of type σ, denoted by SC_{σ}, is defined for all $\sigma \in \underset{\approx}{T}$ as follows:
(i) $SC_0(t)$ iff $t \in 0$ and every reduction sequence starting from t terminates.
(ii) $SC_{(\sigma)\tau}$ iff $\forall t'(SC_{\sigma}(t') \Rightarrow SC_{\tau}(tt'))$.
We put $SC \equiv_{def} \cup \{SC_{\sigma} \mid \sigma \in \underset{\approx}{T}\}$.

2.2.14. Lemma. Let $x \in (\sigma_1)\dots(\sigma_n)0$, and let $t_i \in \sigma_i$, $1 \leq i \leq n$ be terms such that t_i is strongly normalizable; then $SC_0(xt_1\dots t_n)$.
Proof. Obvious.

2.2.15. Lemma. If $SC(t^{\sigma})$, then t^{σ} is strongly normalizable; also $SC(x^{\sigma})$.
Proof. We establish the assertion of the lemma by induction over the type structure.

Assume for al subtypes ρ of $(\sigma)\tau$ the assertion of the lemma to hold. Let $SC(t^{(\sigma)\tau})$. Then $SC(t^{(\sigma)\tau}x^{\sigma})$. Let $t, t^{(1)}, t^{(2)}, \dots$ be any reduction sequence starting from t. We now define a reduction sequence starting from $t^{(\sigma)\tau}x^{\sigma}$, say $tx, t_o^{(1)}, t_o^{(2)}, \dots$ as follows: $t_o^{(k)} \equiv t^{(k)}x$ as long as $t^{(k)}$ is defined; if $t, t^{(1)}, t^{(2)}, \dots$ breaks off at $t^{(p)}$, we take for $t_o^{(p+1)}, t_o^{(p+2)}, \dots$ a standard reduction sequence starting from $t^{(p)}x$. By induction hypothesis and $SC(tx)$, $tx, t_o^{(1)}, t_o^{(2)}, \dots$ terminates, and so has an initial segment of the form $tx, t^{(1)}x, \dots, t^{(k)}x$ such that

$t, t^{(1)}, t^{(2)}, \ldots, t^{(k)}$ is a terminating reduction sequence for t. Now assume $\tau \equiv (\sigma_1)\ldots(\sigma_m)0$, and let $SC_{\sigma_i}(t_i)$, $1 \leq i \leq m$, $SC_\sigma(t)$. Then t, t_i are strongly normalizable; hence by lemma 2.2.14, $SC_0(x^{(\sigma)\tau}tt_1\ldots t_m)$, and thus $SC_{(\sigma)\tau}(x^{(\sigma)\tau})$.

2.2.16. Remark. Instead of using $SC_\rho(x^\rho)$, we might also have established inductively $SC_\rho(0^\rho)$ (where $0^0 \equiv 0$, $0^{(\sigma)\tau} \equiv \Pi_{\sigma,\tau}0^\sigma$, as in 2.2.7) together with the induction hypothesis.

2.2.17. Definition. A reduction tree of a term t consists of a pair $\langle T, \varphi\rangle$, where T is a non-empty set of natural numbers representing finite sequences such that $n * \hat{x} \in T \Rightarrow n \in T$, and φ a function which assigns terms to the elements of T, such that

(a) $\varphi\langle\,\rangle = t$.

(b) If $n \in T$, $\varphi n = t'$, and $t'_1, \ldots, t'_n$ is a complete list of terms (without repetitions) which are obtained by a single contraction from t', then $n * \langle i\rangle \in T$ for $1 \leq i \leq n$, and $\varphi(n * \langle i\rangle) = t'_i$.

To make the description definite, we may assume that in a uniform way an ordering is prescribed among the possible contractions for all terms. The length of a reduction tree $\langle T, \varphi\rangle$ is the number of elements in T.

2.2.18. Definition (to be used in the proof of 2.2.19). Let $SC_0(t)$; then by 2.2.13 and the fan theorem (or classically, König's lemma) the reduction tree of t is finite, hence there are only finitely many terms in normal form, $t_1, \ldots, t_n$, such that $t \geq t_i$ for $1 \leq i \leq n$. Let $t_i = S^{p(i)}t'_i$, t'_i not of the form St^* for any t^*. Then $\nu(t) = \max\{p(i) \mid 1 \leq i \leq n\}$.
Remark. In giving this definition, we have made an appeal (on the meta-level) to the fan theorem, by assuming that a finitely branching tree with all its branches finite is itself finite. The implicit appeal to the fan theorem in the proof of 2.2.19 (via this definition of ν) can be avoided in two different ways:

(i) by giving a proof of the uniqueness of normal form which does not depend on the strong normalization theorem itself (2.2.23), and which enables us to define $\nu(t)$ as the p such that $t' \equiv S^p t''$, t'' not of the form St''', t' normal, $t \geq t'$; or

(ii) by strengthening $SC_0(t)$ to: the reduction tree of t is finite. This requires in the proof of 2.2.19 manipulation and recombination of reduction trees, which is notationally awkward.

2.2.19. Theorem. For all terms t of $\underset{\sim}{N}$-$\underset{\sim}{HA}^\omega$, $SC(t)$, and hence t is strongly normalizable.

Proof. We first note that

(I) If $SC(t_i)$, $1 \leq i \leq n$, then for each t formed by repeated application from $t_1, \ldots, t_n$, $SC(t)$.

(II) If $SC(t)$, $t \geq t'$, then $SC(t')$, as readily follows from the definition of SC.

(III) If $t \in \sigma = (\sigma_1) \ldots (\sigma_n)0$, then $SC(t) \Leftrightarrow (\forall t_1 \in SC_{\sigma_1}) \ldots (\forall t_n \in SC_{\sigma_n})$ ($tt_1 \ldots t_n$ is strongly normalizable).

By (I), it now suffices to prove $SC(\varphi)$ for φ a constant or variable of our theory (cf. lemma in 2.1.2).

(i) $SC(0)$ is immediate.

(ii) If $SC_0(t)$, then $SC_0(St)$, since any reduction sequence starting from St must necessarily be of the form $St, St_1, St_2, \ldots$, where $t, t_1, t_2, \ldots$ is a reduction sequence. Hence $SC_1(S)$.

(iii) $SC(\Pi_{\sigma,\tau})$ holds. For let $t_1, \ldots, t_n$ be terms such that $\Pi_{\sigma,\tau}t_1 \ldots t_n \in 0$, and suppose $SC(t_i)$ $(1 \leq i \leq n)$. Now consider an arbitrary reduction sequence starting from $\Pi_{\sigma,\tau}t_1 \ldots t_n$. This will be of the form

$$\Pi_{\sigma,\tau}t_1 \ldots t_n, \Pi_{\sigma,\tau}t_1^{(1)} \ldots t_n^{(1)}, \ldots, \Pi_{\sigma,\tau}t_1^{(k)} \ldots t_n^{(k)},$$
$$t_1^{(k)}t_3^{(k)} \ldots t_n^{(k)}, \ldots, t_1^{(k+1)}t_3^{(k+1)} \ldots t_n^{(k+1)}, \ldots$$

(Such a k must occur, otherwise one of the sequences $t_i, t_i^{(1)}, t_i^{(2)}, \ldots$ would, after omission of repetitions, become an infinite reduction sequence, contradicting $SC(t_i)$ and 2.2.15.)

Now $t_1t_3 \ldots t_n \geq t_1^{(k)}t_3^{(k)} \ldots t_n^{(k)}$, and by (I) $SC(t_1t_3 \ldots t_n)$, hence by (II) $SC(t_1^{(k)}t_3^{(k)} \ldots t_n^{(k)})$. Now $t_1^{(k)}t_3^{(k)} \ldots t_n^{(k)}, t_1^{(k+1)}t_3^{(k+1)} \ldots$ $\ldots t_n^{(k+1)}, \ldots$ is a reduction sequence starting from $t_1^{(k)}t_3^{(k)} \ldots t_n^{(k)}$, and therefore terminates.

(iv) $SC(\Sigma_{\rho,\sigma,\tau})$ is proved similarly.

(v) For $SC(x^\sigma)$, see 2.2.15.

(vi) Now we have to show that R_σ is strongly computable. Now if $R_\sigma \in (\sigma_1) \ldots (\sigma_n)0$, we have to show that $SC_0(Rt_1 \ldots t_n)$ for $t_1, \ldots, t_n$ such that $SC_{\sigma_i}(t_i)$, $1 \leq i \leq n$. We apply a sub-induction w.r.t. $\nu(t_3)$.

(vi)[a]. If $\nu(t_3) = 0$, a reduction sequence α starting from $Rt_1 \ldots t_n$ has one of the following forms ($\langle f_i \rangle_i$ indicating the sequence $f_0, f_1, f_2, \ldots$):

(1) $\langle Rt_1^{(i)} \ldots t_n^{(i)} \rangle_i$, with $t_1^{(o)} \ldots t_n^{(o)} \equiv t_1 \ldots t_n$,

(2) $Rt_1 \ldots t_n, Rt_1^{(1)} \ldots t_n^{(1)}, Rt_1^{(2)} \ldots t_n^{(2)}, \ldots, Rt_1^{(p-1)} \ldots t_n^{(p-1)},$
$Rt_1^{(p)}t_2^{(p)}0\ t_4^{(p)} \ldots t_n^{(p)}, t_1^{(p)}t_4^{(p)} \ldots t_n^{(p)}, t^{(p+2)}, t^{(p+3)}, \ldots$.

In case (1) α obviously terminates, in the second case α terminates since by (I), (II) $SC(t_1^{(p)}t_4^{(p)}\dots t_n^{(p)})$, and

$t_1^{(p)}t_4^{(p)}\dots t_n^{(p)},\ t^{(p+2)},\ t^{(p+3)},\ \dots$

is a reduction sequence starting from $t_1^{(p)}t_4^{(p)}\dots t_n^{(p)}$.

(vi)[b]. Assume $SC(Rt_1t_2t_3\dots t_n)$ to have been proved for all $t_1,t_2,\dots,t_n$ such that $SC(t_i)$, $1\le i\le n$, and $\nu(t_3)\le k$. Now consider $Rt_1t_2t_3\dots t_n$ with $SC(t_i)$, $1\le i\le n$, $\nu(t_3)=k+1$. Let α be any reduction sequence starting from $Rt_1t_2t_3\dots t_n$. α has one of the following forms:

(1) $\langle Rt_1^{(i)}\dots t_n^{(i)}\rangle_i$ where $Rt_1\dots t_n \equiv Rt_1^{(o)}\dots t_n^{(o)}$.
(Actually, as follows from 2.2.23, this case cannot occur.)

(2) $Rt_1\dots t_n,\ Rt_1^{(1)}\dots t_n^{(1)},\ \dots,\ Rt_1^{(p-1)}\dots t_n^{(p-1)},$
$Rt_1^{(p)}t_2^{(p)}0\ t_4^{(p)}\dots t_n^{(p)},\ t_1^{(p)}t_4^{(p)}\dots t_n^{(p)},\ t^{(p+2)},\ t^{(p+3)},\ \dots$.
This case may be dealt with as under (vi)[a]. (Actually, this case is also excluded, by 2.2.23.)

(3) $Rt_1\dots t_n,\ Rt_1^{(1)}\dots t_n^{(1)},\ \dots,\ Rt_1^{(p-1)}\dots t_n^{(p-1)},$
$Rt_1^{(p)}t_2^{(p)}(St_o)t_4^{(p)}\dots t_n^{(p)},\ t_2^{(p)}(Rt_1^{(p)}t_2^{(p)}t_o)t_ot_4^{(p)}\dots t_n^{(p)},$
$t^{(p+2)},\ t^{(p+3)},\ \dots$.

In case (1), it is obvious that α must terminate; (2) is referred to (vi)[a]; in case (3) we use (I), (II), and our induction hypothesis (since $\nu(t_o)\le k$).

Our next task will be to prove uniqueness of normal form. We first show how to do this by a method due to J.B. Rosser (Rosser 1935).

2.2.20. Definition (for use in 2.2.21). We define inductively a notion of "bounded" reducibility ($\ge^*$) by

(a) $t\ge^* t'$ and $t_1\ge^* t_1' \Rightarrow tt_1\ge^* tt'$.

(b) $\left\{\begin{array}{l}\Pi tt'\ge^* t,\ \Sigma tt't''\ge^* tt''(t't''),\ Rtt'0\ge^* t,\\ Rtt'(St'')\ge^* t'(Rtt't'')t'',\ t\ge^* t.\end{array}\right.$

In other words, if $t\ge^* t'$, there is a derivation sequence of assertions $t_o\ge^* t_o',\ t_1\ge^* t_1',\ t_2\ge^* t_2',\ \dots,\ t_n\ge^* t_n'$ where $t_n\equiv t$, $t_n'\equiv t'$, such that each $t_i\ge^* t_i'$ either holds by (b), or is obtained from $t_j\ge^* t_j'$, $t_k\ge^* t_k'$ ($j,k<i$) by rule (a).

2.2.21. Lemma. If $t\ge^* t'$, $t\ge^* t''$, then there is a t''' such that $t'\ge^* t'''$, $t''\ge^* t'''$.

Proof. Let $t\ge^* t'$, $t\ge^* t''$; we may assume $t'\not\equiv t''$. Let $t\ge^* t'$, $t\ge^* t''$ be established by derivation sequences of length n, m respectively;

we apply induction w.r.t. $n+m$.

(i) $t \geq^* t'$ holds since $t' \equiv t$. Then take $t''' \equiv t''$.

(ii) $t \equiv \Pi t_1 t_2$, $t' \equiv t_1$. Then $t'' \equiv \Pi t_1' t_2'$, $t_1 \geq^* t_1'$, $t_2 \geq^* t_2'$. Take $t''' \equiv t_1'$.

(iii) $t \equiv \Sigma t_1 t_2 t_3$, $t' \equiv t_1 t_3 (t_2 t_3)$. Then $t'' \equiv \Sigma t_1' t_2' t_3'$, $t_i \geq^* t_i'$ for $i = 1, 2, 3$. Take $t''' \equiv t_1' t_3' (t_2' t_3')$.

(iv) $t \equiv Rt_1 t_2 0$, $t' \equiv t_1$; then $t'' \equiv Rt_1' t_2' 0$. Take $t''' \equiv t_1'$. Similarly, if $t \equiv Rt_1 t_2 (St_3)$, $t' \equiv t_2 (Rt_1 t_2 t_3) t_3$.

(v) Let $t \equiv t_1 t_2$, $t' \equiv t_1' t_2'$, and let the final assertion $t \geq^* t'$ in the derivation sequence hold by application of (a) in 2.2.20 to $t_1 \geq^* t_1'$, $t_2 \geq^* t_2'$.
If $t \geq^* t''$ holds by (b), we may deal with this case as under (i), (ii), (iii), (iv).
Hence assume $t'' \equiv t_1'' t_2''$, and the derivation sequence of $t \geq^* t''$ ends with an application of rule (a) in 2.2.20, so $t_1 \geq^* t_1''$, $t_2 \geq^* t_2''$.
Then, by induction hypothesis there are t_1''', t_2''' such that $t_1' \geq^* t_1'''$, $t_1'' \geq^* t_1'''$, $t_2' \geq^* t_2'''$, $t_2'' \geq^* t_2'''$, hence $t_1' t_2' \geq^* t_1''' t_2'''$, $t_1'' t_2'' \geq^* t_1''' t_2'''$.

2.2.22. Lemma. If $t \geq t'$, $t \geq t''$, there is a t''' such that $t' \geq t'''$, $t'' \geq t'''$.

Proof. $\geq$ is the transitive closure of $\geq^*$, i.e. if $t \geq t'$, then there is a sequence $t_0, \ldots, t_n$ such that $t_0 \equiv t$, $t_n \equiv t'$, $t_i \geq^* t_{i+1}$ for $0 \leq i \leq n$. Let us write $t \geq^*_n t'$ if there is such a sequence consisting of $n+1$ terms. So $t \geq^* t' \Longleftrightarrow t \geq^*_1 t'$. Also $t \geq t' \Longleftrightarrow \exists n (t \geq^*_n t')$.
Now we show: If $t \geq^*_n t'$, $t \geq^*_m t''$, there is a t''' such that $t' \geq^*_m t'''$, $t'' \geq^*_n t'''$. Proof by induction on $n+m$. Assume the assertion to hold for $n+m < k$; let now $t \geq^*_n t'$, $t \geq^*_m t''$, $n+m = k$; let e.g. $n > 1$.
We can find t_0 such that $t \geq^*_{n-1} t_0$, $t_0 \geq^*_1 t'$. Construct (fig. 1) (induction hypothesis) t_1'', $t_0 \geq^*_m t_1$, $t'' \geq^*_{n-1} t_1$; and t_2 such that $t' \geq^*_m t_2$, $t_1 \geq^*_1 t_2$.
Then $t' \geq'_m t_2$, $t'' \geq^*_n t_2$.

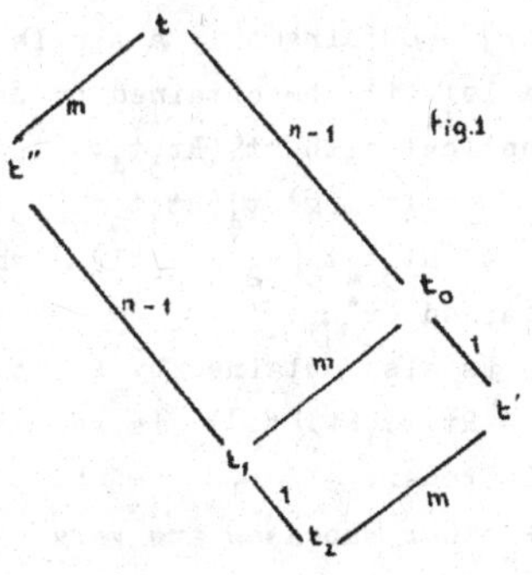

2.2.23. Theorem. The normal form of terms of $\mathbf{N}$-$\mathbf{HA}^\omega$ is uniquely determined.

Proof. Let $t \geq t'$, $t \geq t''$, t' and t'' normal; then by 2.2.22, there is a t''' such that $t' \geq t'''$, $t'' \geq t'''$. Then t''' is normal, and $t' \equiv t''' \equiv t''$.

2.2.24. Remarks. The essential idea of the preceding method is a clever method of "counting" contractions: simultaneous contractions of disjoint subterm occurrences count for a "single" step (expressed by $\underline{\succ}^*$) whereas $\succ_1$ refers to a single contraction under all circumstances. It is also worth noting that the method described is very "elementary"; the methods used are "quantifier-free", and explicit definitions of $\underline{\succ}^*$, $\underline{\succ}$ are Σ_1^0 in character.

Below we describe an alternative method (2.2.25 - 26), which is logically less elementary, but mathematically slightly simpler. On the other hand, the preceding method does not require a strong normalization theorem to be proved first, and also applies to the type-free systems of combinatory logic and the λ-calculus. In the case of the theories based on the λ-operator as a primitive, the second method is even simpler (less cases to check) whereas the first method becomes somewhat more complicated to apply.

A third method to obtain uniqueness of normal form is via the embedding in a model (HRO); see 2.5.5 - 2.5.6 . This method is only easily applicable to the combinatorial version of $\underset{\sim}{\mathrm{N}}$ - $\underset{\sim}{\mathrm{HA}}^\omega$.

2.2.25. Lemma. If $t \underline{\succ}_1 t'$, $t \underline{\succ}_1 t''$, then there is a t^* such that $t' \underline{\succ} t^*$, $t'' \underline{\succ} t^*$.

Proof. For the proof we must distinguish two cases.

(a) The redexes t_1, t_2 in t which are contracted in reducing t to t', t'' respectively, are disjoint. Then obviously t^* is obtained by contracting both t_1 and t_2 in t; the order in which the contractions are executed is irrelevant.

(b) Let again t_1, t_2 be the redexes in t which are contracted to obtain t', t'' respectively, and assume now t_2 to be a subterm of t_1 (the case where t_1 is a subterm of t_2 is obviously completely similar).

Now we have to distinguish various cases according to the form of t_1.

For example, let t_1 be a redex of the form $Rt_3t_4(St_5)$, and let t_2 occur as a subterm of t_5 (so that we may write t_5 as $t_6[t_2]$, where $t_6[x]$ contains only a single occurrence of x).

Now let t' be obtained by contracting t_1 to $t_4(Rt_3t_4t_5)t_5$, which is identical with $t_4(Rt_3t_4t_6[t_2])t_6[t_2]$. If we then apply two contractions to t', replacing $t_4(Rt_3t_4t_6[t_2])t_6[t_2]$ by $t_4(Rt_3t_4t_6[t_2'])t_6[t_2]$, and then by $t_4(Rt_3t_4t_6[t_2'])t_6[t_2']$, where t_2' is the contraction of t_2, we have obtained t^*.

t^* is also obtained by first replacing t_2 by t_2' in t'', so that $t_1 \equiv Rt_3t_4(St_6[t_2])$ is replaced by $Rt_3t_4(St_6[t_2'])$, and then contracting this redex.

The other subcases are very similar.

2.2.26. Second proof of 2.2.23. Let t be any term; by 2.2.19 $SC(t)$. Hence the reduction tree of t is finite (intuitive appeal to the fan theorem, or König's lemma). Let $\mu(t)$ denote the number of nodes in the reduction tree of t. We prove by induction on $\mu(t)$ that the normal form of t is uniquely determined.

The assertion is obvious for $\mu(t) = 1$. Assume the assertion to hold for all t' with $\mu(t') < k$, and let $\mu(t) = k$. Let $t, t_1, t_2, \ldots, t_l$ and $t, t_1', t_2', \ldots, t_m'$ be two reduction sequences starting from t, and ending in normal form. If $t_1 \equiv t'$, then $t_l \equiv t_m'$ by induction hypothesis.
If $t_1 \not\equiv t_1'$, we can find a t^* so that $t_1 \geq t^*$, $t_1' \geq t^*$ (by 2.2.25); let t^* reduce to a normal term t^{**}. Then, since $\mu(t_1) < k$, $\mu(t_1') < k$, $t^{**} \equiv t_l$, $t^{**} \equiv t_m'$; hence $t_l \equiv t_m'$.

2.2.27 - 2.2.34. Computability for theories based on λ-conversion.

Instead of dealing with the terms of $\underset{\sim}{N}$-$\underset{\sim}{HA}^{\omega}$, we define a set of terms with λ-abstraction as a primitive.

2.2.27. Definition of Tm'. $Tm' = \bigcup \{Tm_\sigma' \mid \sigma \in \underset{\approx}{T}\}$, where the Tm_σ' are defined by

Tm' (i). $0 \in Tm_0'$, $S \in Tm_{(o)o}'$, $R_\sigma \in Tm_\tau'$ with $\tau = (\sigma)((\sigma)(0)\sigma)(0)\sigma$, $\sigma \in \underset{\approx}{T}$.

Tm' (ii). If x^σ is a variable of type σ, then $x^\sigma \in Tm_\sigma'$.

Tm' (iii). If $t \in Tm_\tau'$, then $\lambda x^\sigma . t \in Tm_{(\sigma)\tau}'$.

Tm' (iv). If $t \in Tm_{(\sigma)\tau}'$, $t' \in Tm_\sigma'$, then $tt' \in Tm_\tau'$.

2.2.28. Contractions. "contr" is in general a relation between terms. "$[x^\sigma]$" in the notation "$t[x^\sigma]$" refers only to the occurrences of x^σ free in t. The contraction relation is given by

(c) $Rts0$ contr. t;

(d) $Rts(St')$ contr. $s(Rtst')t'$;

(h) $(\lambda x^\sigma . t[x^\sigma])t'$ contr $t[t']$ if no variable free in t' becomes bound in $t[t']$;

(j) $\lambda x^\sigma . tx^\sigma$ contr. t for t not containing x^σ free;

(k) $\lambda x^\sigma . t[x^\sigma]$ contr. $\lambda y^\sigma . t[y^\sigma]$.

2.2.29. Reductions $\geq$, $>_1$.

We define (strict, standard) reductions similar to before. We say that t' is obtained from t by an α-reduction (resp. β-, η-reduction) if t' is obtained from t by replacing a redex in t by its contractum according to contraction rule (k) (resp. (h), (j)).
We write $t >_1 t'$ if t' is obtained from t by some α-reduction

followed by a contraction according to (h), (j), (c), or (d). We write $t \geq t'$ if there is a sequence $t_1, \dots, t_n$ such that $t \equiv t_1$, $t_n \equiv t'$, $t_i >_1 t_{i+1}$ for $1 \leq i \leq n$.

A term t is normal if $t \geq t'$ implies that t' is obtained from t by α-reductions.

2.2.30. Computability and strong computability for λ-based theories.

Predicates "Comp" and "SC" may be defined as before (2.2.5, 2.2.13). In order to prove $SC(t)$ for each t, we in fact have to use induction over a stronger property, to be called strong computability under substitution (SC^*), which is defined as follows:

Definition. t is said to be strongly computable under substitution (notation $SC^*(t)$) if for each substitution of strongly computable terms of appropriate types (renaming bound variables if necessary to avoid clashes of variables) for some occurrences of variables free in t, the resulting term is strongly computable (cf. 2.1.4).

2.2.31. Theorem. $SC^*(t)$ for each $t \in Tm'$.

Proof. The proof is very similar to the argument in 2.2.19.

(i) $SC^*(0)$ is immediate.

(ii) $SC^*(S)$ is proved as in 2.2.19, (ii).

(iii) Let $\sigma = (\sigma_1) \dots (\sigma_n)0$. Then $SC^*(x^\sigma)$ means:

(a) for all $t_1, \dots, t_n$, $t_i \in \sigma_i$, $t_i \in SC$ $(1 \leq i \leq n)$, $SC(x^\sigma t_1 \dots t_n)$, and

(b) for all $t, t_1, \dots, t_n$, $t_i \in \sigma_i$, $t_i \in SC$ $(1 \leq i \leq n)$, $t \in \sigma$, $t \in SC$, $SC(tt_1 \dots t_n)$.

(b) is obvious; (a) is proved as in 2.2.19 (v).

(iv) $SC^*(R_\sigma)$: compare 2.2.19, $(vi)^a$, $(vi)^b$.

(v) Let $SC^*(\tilde{t}[x^\sigma])$, $\tilde{t}[x^\sigma] \in (\tau_1) \dots (\tau_n)0$, and let $\lambda x^\sigma . t[x^\sigma]$ be obtained from $\lambda x^\sigma . \tilde{t}[x^\sigma]$ by substitution of strongly computable terms for some occurrences of free variables, and possibly some renaming of bound variables.

We wish to show that for arbitrary $t_0 \in \sigma$, $t_i \in \tau_i$, $t_i \in SC$ for $0 \leq i \leq n$, $SC((\lambda x^\sigma . t[x^\sigma])t_0 t_1 \dots t_n)$.

A reduction sequence starting from $(\lambda x^\sigma . t[x^\sigma])t_0 t_1 \dots t_n$ is either of the form: $(\lambda x^\sigma . t[x^\sigma])t_0 \dots t_n$, $(\lambda x^\sigma . t^{(1)}[x^\sigma])t_0^{(1)} \dots t_n^{(1)}$, ..., $(\lambda x^\sigma . t^{(k)}[x^\sigma])t_0^{(k)} \dots t_n^{(k)}$, $(t^{*(k)}[t_0^{(k)}])t_1^{(k)} \dots t_n^{(k)}$, $\bar{t}^{(k+2)}$, $\bar{t}^{(k+3)}$, ... where $t^{*(k)}[x^\sigma]$ is obtained from $t^{(k)}[x^\sigma]$ by some α-reductions, and where

$(t^{*(k)}[t_o^{(k)}])t_1^{(k)} \ldots t_n^{(k)} \equiv \mathfrak{t}^{(k+1)}, \mathfrak{t}^{(k+2)}, \mathfrak{t}^{(k+3)}, \ldots$ is a reduction sequence starting from $(t^{*(k)}[t_o^{(k)}])t_1^{(k)} \ldots t_n^{(k)}$; or the reduction sequence is of the following form:

$(\lambda x^\sigma.t[x^\sigma])t_o \ldots t_n, \ldots, (\lambda x^\sigma.t^{(k)}[x^\sigma])t_o^{(k)} \ldots t_n^{(k)}, (s)t_o^{(k)} \ldots t_n^{(k)},$
$\mathfrak{t}^{(k+2)}, \mathfrak{t}^{(k+3)}, \ldots,$

where $t^{(k)}[x^\sigma]$ is of the form sx^σ, s not containing x^σ free, and where $(s)t_o^{(k)} \ldots t_n^{(k)}, \mathfrak{t}^{(k+2)}, \mathfrak{t}^{(k+3)}, \ldots$ is a reduction sequence starting from $st_o^{(k)} \ldots t_n^{(k)}$. The second form of reduction sequence is in fact a special case of the first form, so that we may further restrict our attention to the first form.

We see that the reduction sequence terminates, since $SC^*[\tilde{x}[x^\sigma]]$ implies $SC^*(t^{*(k)}[x^\sigma])$, hence also $SC(t^{*(k)}[t_o^{(k)}])$ since $SC(t_o)$ implies $SC(t_o^{(k)})$.

2.2.32. Lemma. If $t >_1 t'$, $t >_1 t''$, then there is a t^* such that $t' \geq t^*$, $t'' \geq t^*$.

Proof. Similar to 2.2.25.

2.2.33. Theorem. Each reduction sequence starting from $t \in Tm'$ terminates in a term in normal form, which is uniquely determined up to renaming bound variables.

Proof. Immediate by combining theorem 2.2.31 with the previous lemma.

2.2.34. Remark. An alternative method for proving 2.2.32 is obtained by adaptation of Rosser's method (cf. 2.2.20 - 2.2.23) to the λ-calculus (as in Martin-Löf 1971 C, Barendregt 1971; Hindley, Lercher and Seldin 1972; the adaptation is due to W.W. Tait and P. Martin-Löf).

2.2.35. Discussion and comparison of proofs of computability for terms of HA^ω in the literature.

Tait 1965 proves normalization by means of quantifier-free ϵ_o-induction, for a system of infinite terms, into which the λ-terms (as in 2.2.27) can be embedded. This procedure is very similar to the consistency proof for HA^c obtained by embedding HA^c into classical arithmetic with the ω-rule.

Hanatani 1966 shows by cut-elimination for sequent calculi that to each closed term t of type 0 we can find a numeral $\bar{n}$ such that in the system H obtained by omission of induction from $N-HA^\omega$, $H \vdash \mathfrak{t} = \bar{n}$. (His set of primitives is that of Spector 1962).

Tait 1967 introduces the computability predicate "Comp", but restricts attention to "strict reductions" (2.2.2). By means of the computability predicate Tait gives a proof of normalization for closed terms of $\underset{\sim}{N}-\underset{\sim}{HA}^{\omega}$ (the proof applies to open terms as well, as we have seen). By restricting his attention to strict reductions, it is easy to show the uniqueness of the resulting normal form. (Tait 1967 is an elaboration of Tait 1963.)

The papers Sanchis 1967, Diller 1968, Dragalin 1968 apply bar-induction arguments to suitably defined partial orderings.

Diller 1968 shows computability for closed terms of type 0, relative to a specified order of computation, by means of bar induction; it is shown that the argument may be replaced by a transfinite induction up till the first ω-critical number, assigning ordinals to terms.

Sanchis 1967 uses an argument very similar to Diller's bar induction, to obtain a strong normalization theorem for the terms of $\underset{\sim}{N}-\underset{\sim}{HA}^{\omega}$, but relative to a weaker set of contraction rules; instead of (d) in 2.2.2 he only permits

(d') $\quad Rt_1t_2(\overline{n+1})$ contr. $t_2(Rt_1t_2\bar{n})\bar{n}$

(disregarding the fact that his R_σ is in fact our $\lambda xyz.R_\sigma x(\lambda vu.yuv)z$). It is not obvious how to generalize his notion of successor of a term so as to be applicable to our stronger reductions.

The proof of Stenlund 1971 is based on Sanchis' proof, but Stenlund restricts his attention to reduction sequences where a definite order is prescribed. For yet another presentation, see Hindley, Lercher and Seldin 1972.

Hinata 1967 discusses the λ-version of $\underset{\sim}{N}-\underset{\sim}{HA}^{\omega}$, and assigns to each term a tree, expressing the construction of the term from simpler terms. Let us call such trees construction trees. The construction tree of a term t (which is not uniquely determined) has a term associated with each node; t is associated to the end node. Hinata then describes an ordinal assignment (of ordinals $< \varepsilon_0$) to construction trees, and a process for transforming trees into other trees such that the term associated to the end node of the transformed tree is a reduction of the term associated to the end node of the original tree. If a tree T' is obtained from T by this transformation process, then T' may be assigned an ordinal which is lower than the ordinal assigned to T. The proof can be given in quantifier-free arithmetic extended with ε_0-induction.

Hinata's contractions are almost the same as those described in 2.2.28. There are two differences: he does not introduce the recursor as a constant, but instead uses $\rho t_1t_2t_3$ for $\varphi(t_1)$, if φ has been defined primitive recursively by

$$\varphi(0) = t_2,$$
$$\varphi(Sz) = t_3(\varphi z)z.$$

Then $(\lambda x.t[x])(t')$ is contracted to $t[t']$, $\rho t_1 t_2(S^n t_3)$ contracts to $t_2 \ldots (t_2(t_2(\rho t_1 t_2 t_3)t_3)(St_3))(S^2 t_3) \ldots (S^n t_3)$ if $n > 0$, t_3 not of the form St'.

Obviously, Hinata's result implies a corresponding result for the combinatorial version. His rule of definition by recursion permits the definition of the recursor R_σ, as we have seen in 1.7.5.

Howard 1970 proves a normalization theorem for a λ-version of $\underset{\sim}{N}$-$\underset{\sim}{HA}^\omega$ by means of assignment of ordinals less than ϵ_0 to terms.

Howard's version of $\underset{\sim}{N}$-$\underset{\sim}{HA}^\omega$ contains as primitive constants all numerals, the abstraction operator λ with contraction (h), successor S with $S\bar{n}$ contr. $\overline{n+1}$, recursor R_σ and constants R_σ^n for each $n \geq 0$ with contractions $R_\sigma \bar{n}$ contr. R_σ^n, $R_\sigma^0 ts$ contr. s, $R_\sigma^{n+1} ts$ contr. $t\bar{n}(R^n ts)$.

Howard first considers "restricted reductions", where only contractions of closed subterms are permitted (i.e. contractions from the "outside", as opposed to strict reductions, which involve contractions from the "inside"). If $t \geq t'$ by a restricted reduction, and α, β are ordinals assigned to t, t' respectively, then $\beta < \alpha$.

Next, to allow the consideration of arbitrary reductions, Howard describes a non-unique assignment of ordinals to terms, such that if $t \geq_1 t'$ is an arbitrary reduction, and α is any ordinal assigned to t, then there is some ordinal β assigned to t' such that $\beta < \alpha$.

This method yields a direct proof of a strong normalization theorem by means of primitive recursive arithmetic extended by quantifier-free induction up to ϵ_0 (for any primitive recursive well-ordering of order type ϵ_0 satisfying certain adequacy conditions).

§ 3. More about computability.

2.3.1. Computability in $\underset{\sim}{I}-\underset{\sim}{HA}^{\omega}+IE_0$.

Let us add to $\underset{\sim}{I}-\underset{\sim}{HA}^{\omega}$ the following set of axioms

IE_0 $\begin{cases} Ets = 1 & \text{if } t,s \text{ are closed terms in normal form} \\ \text{and } t \not\equiv s. \end{cases}$

Now we extend the notions of contraction and (strict-, standard-)reduction from $\underset{\sim}{N}-\underset{\sim}{HA}^{\omega}$ to $\underset{\sim}{I}-\underset{\sim}{HA}^{\omega}+IE_0$, by adding to the contraction rules

(e) Ett contr. 0,

(f) Ets contr. 1 if t,s are distinct closed terms in normal form.

2.3.2. Theorem. All terms of $\underset{\sim}{I}-\underset{\sim}{HA}^{\omega}+IE_0$ are standard computable, hence reduce to a term in normal form by a standard reduction.

Proof. We only have to add to the proof in 2.2.6 an argument showing E_σ to be standard computable.

(viii) Assume t,t' to be terms reducing to normal forms t_1,t_1' respectively by a standard reduction.

Ett' reduces by a standard reduction to Et_1t_1'.

If t_1 or t_1' is not closed, $t_1 \not\equiv t_1'$, Et_1t_1' is normal. If $t_1 \equiv t_1'$, Et_1t_1' standard reduces to 0; and if t_1t_1' are closed, $t_1 \not\equiv t_1$, Et_1t_1' standard reduces to 1.

2.3.3. Corollary. In $\underset{\sim}{I}-\underset{\sim}{HA}^{\omega}+IE_0$ every closed term of type 0 reduces to a numeral.

2.3.4. Remarks. (i) The consistency of $\underset{\sim}{I}-\underset{\sim}{HA}^{\omega}+IE_0$ requires that the normal form is uniquely determined, i.e. $\underset{\sim}{I}-\underset{\sim}{HA}^{\omega}+IE_0 \vdash t=t'$, $\underset{\sim}{I}-\underset{\sim}{HA}^{\omega}+IE_0 \vdash t=t''$, t, t', t'' closed, t', t'' in normal form must imply $t' \equiv t''$. The uniqueness of normal form is also insured by the model (HRO - version) described in 2.5.5.

(ii) The proofs of the strong normalization theorem (2.2.19) and the proof in 2.2.23 may be extended also to $\underset{\sim}{I}-\underset{\sim}{HA}^{\omega}+IE_0$.

As a corollary to remark (i) and 2.3.2 we have

2.3.5. Corollary. In $\underset{\sim}{I}-\underset{\sim}{HA}^{\omega}+IE_0$, for closed terms t,s

$$\vdash t=s \quad \text{or} \quad \vdash t \neq s.$$

2.3.6. The equality axioms IE_1.

The axioms IE_0 are implied by the following stronger set of axioms IE_1:

$\underset{\sim}{\mathrm{IE}}_1$(a) Let t_1, t_2 be two distinct terms with the same type taken from the set of terms consisting of Π_{σ_1,σ_2}, $\Sigma_{\sigma_3,\sigma_4,\sigma_5}$, S, R_{σ_6}, E_{σ_7}, $\Pi_{\sigma_8,\sigma_9}x_1$, $\Sigma_{\sigma_{10},\sigma_{11},\sigma_{12}}x_2$, $R_{\sigma_{13}}x_3$, $E_{\sigma_{14}}x_4$, $\Sigma_{\sigma_{15},\sigma_{16},\sigma_{17}}x_5x_6$, $R_{\sigma_{18}}x_7x_8$, for all $\sigma_1, \ldots, \sigma_{18} \in \underset{\sim}{T}$. Then $t_1 \neq t_2$ is an axiom.

$\underset{\sim}{\mathrm{IE}}_1$(b)

$$\begin{cases} x \neq x' \vee y \neq y' \rightarrow Rxy \neq Rx'y' \\ x \neq x' \vee y \neq y' \rightarrow \Sigma xy \neq \Sigma x'y' \,. \end{cases}$$

$\underset{\sim}{\mathrm{IE}}_1$(c) If $\sigma \neq \sigma'$, then $\Sigma_{\rho,\sigma,\tau}xy \neq \Sigma_{\rho,\sigma',\tau}x'y'$ is an axiom.

It follows from results in 2.5.9 and 2.5.10 that

$$\underset{\sim}{\mathrm{I}}\text{-}\underset{\sim}{\mathrm{HA}}^\omega \subsetneqq \underset{\sim}{\mathrm{I}}\text{-}\underset{\sim}{\mathrm{HA}}^\omega + \mathrm{IE}_0 \subsetneqq \underset{\sim}{\mathrm{I}}\text{-}\underset{\sim}{\mathrm{HA}}^\omega + \mathrm{IE}_1 \,.$$

The HRO-version described in 2.5.5 is also a model for $\underset{\sim}{\mathrm{I}}\text{-}\underset{\sim}{\mathrm{HA}}^\omega + \mathrm{IE}_1$. The proof of computability in $\underset{\sim}{\mathrm{I}}\text{-}\underset{\sim}{\mathrm{HA}}^\omega + \mathrm{IE}_1$ is also easily given, extending the contractions in the obvious way.

It has sometimes been argued that the rule IE_0 is "unnatural", since it seems to point to a confusion between "use" and "mention". This impression is mistaken, and probably due to too much exclusive contemplation of the term model.

A glance at the HRO-version given in 2.5.5 may do something to dispel the doubts, since there not every object is denoted by a term; on the other hand, the even stronger set IE_1 shows that IE_0 is the consequence of axioms who do not at all have the look of a syntactic criterion smuggled into the semantics.

2.3.7. Standard computability of terms in languages with cartesian product type.

Let us consider $\underset{\sim}{\mathrm{N}}\text{-}\underset{\sim}{\mathrm{HA}}^\omega_p$, the conservative extension of $\underset{\sim}{\mathrm{N}}\text{-}\underset{\sim}{\mathrm{HA}}^\omega$ defined in 1.8.2.

We add to our contractions

(g) $D'(Dtt')$ contr. t, $D''(Dtt')$ contr. t', $D(D't)(D''t)$ contr. t

and our concept of standard reducibility is correspondingly enlarged.

We extend the notion of standard computability by a clause

(iii) $\mathrm{Comp}''_{\sigma\times\tau}(t)$ iff $\mathrm{Comp}''_\sigma(D't)$ and $\mathrm{Comp}''_\tau(D''t)$, and τ is normalizable by a standard reduction.

Note that the existence of a terminating standard reduction sequence for a term implies the existence of a terminating standard reduction sequence for all its subterms. Hence we may drop the condition " t is normalizable by a standard reduction" in clauses (ii), (iii). (This is seen by establishing simultaneously by induction w.r.t. σ that (a) $0^{\sigma} \in \mathrm{Comp}^*_{\sigma}$ and (b) $\mathrm{Comp}^*_{\sigma}(t) \Rightarrow t$ has a terminating standard reduction; here $0^0 \equiv 0$, $0^{(\sigma)\tau} \equiv \Pi_{\sigma,\tau}0^{\sigma}$, $0^{\sigma\times\tau} \equiv D0^{\sigma}0^{\tau}$, and Comp* indicates the weakened form of Comp" with "t is normalizable by a standard reduction" dropped. Cf. 2.2.7 (ii).)

Further we note that the following lemma holds for standard reducibility ($\geq'$):

Lemma. $t \geq' t' \Rightarrow (\mathrm{Comp}''(t) \Leftrightarrow \mathrm{Comp}''(t'))$.

The proof is entirely straightforward, by induction on the type of t, t', using $t \geq' t' \Leftrightarrow D't \geq' D't' \Leftrightarrow D''t \geq' D''t'$, and $t \geq' t' \Rightarrow (tt'' \geq' t't'')$, if t'' has a terminating standard reduction sequence.

We extend the proof that every term of $\underset{\sim}{N}-\underset{\sim}{HA}^{\omega}$ is standard computable to the terms of $\underset{\sim}{N}-\underset{\sim}{HA}^{\omega}_p$; this requires consideration of some additional cases in the proof of 2.2.6.

Case (ix) (extension of case (ii)).

Let Δ be the following inductively defined class:

(1) $\mathrm{Comp}''(t_i)$, $1 \leq i \leq n \Rightarrow x^{\sigma}t_1 \ldots t_n \in \Delta$; $x^{\sigma} \in \Delta$

(2) $t^{\sigma\times\tau} \in \Delta$, $\mathrm{Comp}''(t_i)$, $1 \leq i \leq n \Rightarrow (D't)t_1 \ldots t_n \in \Delta$, $(D''t)t_1 \ldots t_n \in \Delta$, $D't, D''t \in \Delta$.

Let Δ_0 denote the subset of Δ containing terms of type 0 only. We readily see that $\Delta_0 \subseteq \mathrm{Comp}'' \Rightarrow \mathrm{Comp}''(x^{\sigma})$. By induction over Δ we prove that if $t \in \Delta$, then t has a terminating standard reduction sequence not containing a term of the form $Dt't''$. For x^{σ} this is obviously true. Suppose the assertion to have been established for $t \in \Delta$, then it is obviously also true for $D't$ or $D''t$. If it holds for t, $D't$ or $D''t \in \Delta$, then also for $tt_1 \ldots t_n$ or $(D't)t_1 \ldots t_n$, or $(D''t)t_1 \ldots t_n$, $t_1 \ldots t_n \in \mathrm{Comp}''$. Therefore $\Delta_0 \subseteq \mathrm{Comp}''$, so $\mathrm{Comp}''(x^{\sigma})$.

Case (x). Let $s, t \in \mathrm{Comp}''$. Then $s \geq' s'$, $t \geq' t'$, $s', t' \in \mathrm{Comp}''$, s', t' normal (lemma). Now $\mathrm{Comp}''(Dst) \Leftrightarrow (\mathrm{Comp}''(D'(Dst))$ and $\mathrm{Comp}''(D''(Dst))) \Leftrightarrow (\mathrm{Comp}''(D'(Ds't'))$ and $\mathrm{Comp}''(D''(Ds't'))) \Leftrightarrow (\mathrm{Comp}''(s')$ and $\mathrm{Comp}''(t'))$, using $D'(Dst) \geq' s'$, $D''(Dst) \geq' t'$, and the lemma.

Case (xi). $\mathrm{Comp}''(D')$ is established as follows. Let $\mathrm{Comp}''(t)$, then we have to show $\mathrm{Comp}''(D't)$, $t \geq' t'$, $\mathrm{Comp}''(t')$, t' normal. Since t' must have a cartesian product as type, $\mathrm{Comp}''(t') \Leftrightarrow \mathrm{Comp}''(D't)$ and $\mathrm{Comp}''(D''t)$.

$\mathrm{Comp}''(D'')$ is established similarly.

2.3.8. Computability relative assignment of functions.

The concept is taken from Tait 1967. In the remainder of this section, we restrict our attention to standard reduction sequences. For simplicity, we restrict attention to computability relative an assignment of a function to a single, fixed type 1 variable, say x^1. If α is the function assigned to x^1, we add to our contractions:

$$x^1\bar{n} \text{ contr. } \overline{\alpha n} \text{ (i.e. the numeral representing } \alpha n).$$

2.3.9. Theorem. All terms of $\underset{\sim}{N}$ - $\underset{\sim}{HA}^\omega$ possess a standard reduction sequence relative α to a term in α-normal form.

Proof. As compared to ordinary standard reductions, there is only a single additional case to consider: x^1 must be discussed separately and distinguished from other type 1 variables. We have to show that x^1t_o possesses a standard reduction relative α for any t_o possessing a standard reduction relative α. Let $t_o, t_1, \ldots, t_n$ be a standard reduction relative α, t_n in α-normal form. If t_n is not a numeral, $x^1t_o, \ldots, x^1t_n$ is a standard reduction sequence for x^1t_o; if t_n is a numeral, say $\bar{m}$, then $x^1t_o, \ldots, x^1t_n = x^1\bar{m}$, $\overline{\alpha m}$ is a standard reduction sequence relative α.

2.3.10. Remark (i). The result of the reduction obviously depends on the extension of α only, i.e. if $\forall x(\alpha x = \beta x)$, the α- and β-normal forms of all terms will be equal. However, we can say more. For if t is a term of type 0, containing x^1 as a free variable, $\lambda x^1.t$ represents a type 2 functional in $\underset{\sim}{N}$ - $\underset{\sim}{HA}^\omega$. Now the type 2 functionals definable in this manner in $\underset{\sim}{N}$ - $\underset{\sim}{HA}^\omega$ are obviously continuous; for if we reduce t to α-normal form, there is a finite reduction sequence where we used finitely many instances of

$$x^1\bar{n}_i \text{ contr. } \overline{\alpha n}_i, \qquad 1 \leq i \leq k.$$

Let $n = \max\{n_i \mid 1 \leq i \leq k\} + 1$. Then obviously

$$\bar{\alpha}n = \bar{\beta}n \Rightarrow \text{the } \alpha\text{- and } \beta\text{-normal forms of } t \text{ are equal,}$$

which implies:

$$\underset{\sim}{N} - \underset{\sim}{HA}^\omega \vdash \bar{x}^1\bar{n} = \bar{y}^1\bar{n} = \bar{\alpha}n \rightarrow t[x^1] = t[y^1].$$

(ii). The preceding discussion may be readily extended to the assignment of functions to a finite set or to all type 1 variables; our restriction to a single type 1 variable was motivated by considerations of notational simplicity only, since all our applications concern this special case.

2.3.11- 2.3.13. Arithmetization of computability.

2.3.11. If we arithmetize by the device of gödelnumbering the concepts of standard reduction etc., we are led to consider the question of how much of our results on computability can be formalized in $\underset{\sim}{HA}$ or $\underset{\sim}{N}$ - $\underset{\sim}{HA}^\omega$.

Let "SRED" denote, as before, an arithmetical Σ_1^0-predicate, such that

$SRED(\ulcorner t\urcorner, \ulcorner s\urcorner)$ intuitively expresses: "there is a standard reduction sequence from t to s" (so "SRED" has the character of a restricted provability predicate for a fragment of $\underset{\sim}{N}$ - $\underset{\sim}{HA}^{\omega}$; "SRED" is a <u>restricted</u> provability predicate in the sense that it concerns proofs in a certain standard form only: a reduction sequence $t_1, t_2, \ldots, t_n$ representing a proof of $t_1 = t_2$, $t_2 = t_3$, ..., $t_{n-1} = t_n$ successively).

Now inspection of the proof of 2.2.6 shows that if an arithmetical predicate "Comp" " were definable such that (the arithmetized version of) clauses (i) and (ii) (cf. 2.2.5) would be provable in $\underset{\sim}{N}$ - $\underset{\sim}{HA}^{\omega}$, then, since the proof of $\underset{\sim}{\forall} t\, \mathrm{Comp}''(t)$ uses essentially only arithmetical principles, the arithmetized version of $\forall n(\mathrm{Term}(n) \rightarrow \mathrm{Comp}''(n))$ (where $\mathrm{Term}(n)$ is the primitive recursive predicate expressing that n is the gödelnumber of a term) would be provable in $\underset{\sim}{HA}$. Now let $t_0, t_1, t_2, t_3, t_4, \ldots$ be a primitive recursive enumeration of closed terms of type 1; let $f(x,y)$ be the primitive recursive function such that $f(x,y) = \ulcorner s(t_x \bar{y})\urcorner$.
It would follow from the assumption that Comp" was arithmetically definable that

$$\vdash \forall xy\, \exists ! z\, SRED(f(x,y), \ulcorner \bar{z}\urcorner)$$

hence especially

$$\vdash \forall x\, \exists ! z\ \ SRED(f(x,x), \ulcorner \bar{z}\urcorner) .$$

Therefore we can find a provably total recursive function φ, represented by a closed term t of type 1 (cf. 3.4.29)

$$\vdash \forall x\ SRED(f(x), \ulcorner \overline{tx}\urcorner) .$$

Since on the other hand it will follow from 2.3.13 below that

$$\vdash SRED(\ f(x), \ulcorner \overline{tx}\urcorner) \longleftrightarrow t_x \bar{x} + 1 = t\bar{x}$$

we obtain a contradiction (take $t \equiv t_x$).

Hence our assumption that Comp" was arithmetically definable must be false. The intuitive reason for this is rather clear: if we arithmetize Comp''_{σ} for σ of increasing level, the logical complexity of the formulae of $\underset{\sim}{HA}$ representing Comp''_{σ} increases indefinitely.

At the same time it is clear that if we restrict ourselves to the applicative set of all terms constructed from constants and variables of type level $\leq n$, this restricted predicate Comp^n is arithmetically definable.
As a consequence, for each given closed term t its computability is provable in arithmetic, or even more generally, if t contains $x_1, \ldots, x_n \in 0$ free, and does not contain other free variables, then

$$\underset{\sim}{HA} \vdash \forall x_1 \ldots x_n\, \exists y\, SRED(\ulcorner t(\bar{x}_1, \ldots, \bar{x}_n)\urcorner, \ulcorner \bar{y}\urcorner) .$$

Still more generally, if we arithmetize computability with bounded type level for α-reductions (w.r.t. a fixed type 1 variable x^1) we obtain

$$\underset{\sim}{N}-\underset{\sim}{HA}^{\omega} \vdash \exists y\, \mathrm{SRED}(\alpha, \ulcorner t(x^1, \bar{x}_1,\ldots,\bar{x}_n)\urcorner, \ulcorner \bar{y}\urcorner)$$

where $\mathrm{SRED}(\alpha, n, m)$ expresses in $\underset{\sim}{N}-\underset{\sim}{HA}^{\omega}$ arithmetically standard reducibility relative to α.

2.3.12. Standard gödelnumbering. We find it convenient to make some assumptions about the (standard) gödelnumbering to be used for terms of $\underset{\sim}{N}-\underset{\sim}{HA}^{\omega}$ and certain extensions. Let a code number $\tilde{c}$ be assigned to each constant (and variable) c. Then the code number of the terms is defined inductively as follows:

(i) If t is a variable or constant, then $\ulcorner t\urcorner \equiv \langle\tilde{t}\rangle$.

(ii) If $t \equiv t_1 t_2$, then $\ulcorner t\urcorner = \ulcorner t_1\urcorner * \langle\ulcorner t_2\urcorner\rangle$.

2.3.13. Theorem. Let t be any term of $\underset{\sim}{N}-\underset{\sim}{HA}^{\omega}$ constructed from constants, type 0 variables and (possibly) the type 1 variable x^1; let $x_1,\ldots,x_n$ be a list containing all the type 0 variables occurring free in t. Let α be another type 1 variable of $\underset{\sim}{N}-\underset{\sim}{HA}^{\omega}$. Then

(1) $$\underset{\sim}{N}-\underset{\sim}{HA}^{\omega} \vdash t(\alpha,x_1,\ldots,x_n) = y \leftrightarrow \mathrm{SRED}(\alpha, \ulcorner t(x^1,\bar{x}_1,\ldots,\bar{x}_n)\urcorner, \ulcorner \bar{y}\urcorner).$$

Proof (W.A. Howard *).

We define for each type $\sigma \in \underset{\approx}{T}$, a binary relation $\mathrm{VAL}_\sigma(x^0, y^\sigma)$ expressing: the term with gödelnumber x has the functional y^σ as value. If x is not the gödelnumber of a closed term of type σ, $\mathrm{VAL}_\sigma(x^0,y^\sigma)$ is false. The definition is as follows.

(i) $$\mathrm{VAL}_0(x^0,y^0) \equiv_{\mathrm{def}} \mathrm{SRED}(\alpha, x, \ulcorner \bar{y}\urcorner)$$

(ii) $$\mathrm{VAL}_{(\sigma)\tau}(x^0,y^{(\sigma)\tau}) \equiv_{\mathrm{def}} \forall z^0 u^\sigma(\mathrm{VAL}_\sigma(z,u) \rightarrow \mathrm{VAL}_\tau(x * \langle z\rangle, yu)).$$

Note that

(2) $$\mathrm{VAL}_{(\sigma)\tau}(x,y) \,\&\, \mathrm{VAL}_\sigma(x',y') \rightarrow \mathrm{VAL}_\tau(x * \langle x'\rangle, yy').$$

We also need

Lemma. (a) $\mathrm{VAL}_\tau(x,y^\tau) \rightarrow \exists x'[\mathrm{SRED}(\alpha,x,x') \,\&\, \mathrm{Normal}(x')]$

(b) $\mathrm{VAL}_\tau(\ulcorner 0^\tau\urcorner, 0^\tau)$

which is readily established simultaneously, by induction on τ. ("Normal" is the arithmetization of "... is in normal form".)

We now establish $\mathrm{VAL}(\langle\tilde{c}\rangle, c)$ for constants c, $\mathrm{VAL}(\ulcorner\bar{x}\urcorner, x)$ for numerical variables x, and $\mathrm{VAL}(\ulcorner x^1\urcorner, \alpha)$ if x^1 is the fixed type 1 variable to which the function α is assigned.

---

*) In a letter dated May 18, 1972.

(a) $\underset{\sim}{N}\text{-}\underset{\sim}{HA}^{\omega} \vdash \forall x(VAL_0(\ulcorner \bar{x} \urcorner, x))$ is immediate.

(b) Assume $VAL_0(y,z)$, i.e. $SRED(\alpha, y, \ulcorner \bar{z} \urcorner)$. Then $SRED(\alpha, \ulcorner x^1 \urcorner * \langle y \rangle, \ulcorner \overline{\alpha z} \urcorner)$, i.e. $VAL_0(\ulcorner x^1 \urcorner * \langle y \rangle, \alpha z)$, so $VAL_1(\ulcorner x^1 \urcorner, \alpha)$.

(c) $VAL_0(x_1,y_1) \to VAL_0(\ulcorner S \urcorner * \langle x_1 \rangle, Sy_1)$ is obvious by the properties of SRED, hence $VAL_1(\ulcorner S \urcorner, S)$.

(d) Let $\sigma \equiv (\sigma_1) \ldots (\sigma_m)0$, $VAL_\sigma(x_1,y_1)$, $VAL_\tau(x_2,y_2)$, $VAL_{\sigma_i}(x_{i+2},y_{i+2})$ for $1 \leq i \leq m$.

Then by our assumptions $VAL_0(x_1 * \langle x_3, \ldots, x_{m+2} \rangle, y_1 y_3 \ldots y_{m+2})$, hence since $SRED(\alpha, x_1 * \langle x_3, x_4, \ldots, x_{m+2} \rangle, y_1 y_3 y_4 \ldots y_{m+2})$, also $SRED(\alpha, \ulcorner \Pi_{\sigma,\tau} \urcorner * \langle x_1, x_2, \ldots, x_{m+2} \rangle, \Pi_{\sigma,\tau} y_1 y_2 y_3 \ldots y_{m+2})$. Here we have also used that $VAL_\tau(x_2,y_2)$ implies $\exists x(SRED(\alpha, x_2, x) \,\&\, \text{Normal}(x))$, by the lemma.

(e) $VAL(\ulcorner \Sigma_{\rho,\sigma,\tau} \urcorner, \Sigma_{\rho,\sigma,\tau})$ is proved similarly.

(f) $VAL(\ulcorner R_\sigma \urcorner, R_\sigma)$. For notational simplicity, let $\sigma \equiv (\tau)0$; we first establish, by induction on y

$$(3) \qquad \begin{cases} \forall x_0 y_0 x_1 y_1 (VAL_\sigma(x_0,y_0) \,\&\, VAL_{(\sigma)(0)\sigma}(x_1,y_1) \to \\ \quad \to VAL_\sigma(\ulcorner R \urcorner * \langle x_0, x_1, \ulcorner \bar{y} \urcorner \rangle, Ry_0 y_1 y)) \end{cases}$$

Basis. Let $VAL_\sigma(x_0,y_0)$, $VAL_{(\sigma)(0)\sigma}(x_1,y_1)$. We wish to show $VAL_0(\ulcorner R \urcorner * \langle x_0, x_1, \ulcorner 0 \urcorner, x_3 \rangle, Ry_0 y_1 0 y_3)$ for all x_3, y_3 such that $VAL_\tau(x_3,y_3)$. By our hypothesis, $VAL_0(x_0 * \langle x_3 \rangle, y_0 y_3)$.
By the properties of SRED,

$$SRED(\alpha, \ulcorner R \urcorner * \langle x_0, x_1, \ulcorner 0 \urcorner, x_3 \rangle, y_0 y_3)$$

i.e. $$SRED(\alpha, \ulcorner R \urcorner * \langle x_0, x_1, \ulcorner 0 \urcorner, x_3 \rangle, Ry_0 y_1 0 y_3).$$

Induction step. Assume (3); we wish to establish (3) with Sy instead of y. With the induction hypothesis, the argument is as straightforward as before.

Now we are ready to show

$$(4) \qquad \begin{cases} VAL_\sigma(x_0,y_0) \,\&\, VAL_{(\sigma)(0)\sigma}(x_1,y_1) \,\&\, VAL_0(x_2,y_2) \,\&\, VAL_\tau(x_3,y_3) \to \\ \quad \to VAL_0(\ulcorner R \urcorner * \langle x_0, x_1, x_2, x_3 \rangle, Ry_0 y_1 y_2 y_3). \end{cases}$$

Note that $VAL_0(x_2,y_2)$ implies $SRED(\alpha, x_2, \ulcorner \bar{y}_2 \urcorner)$.
Also, by (3) applied to y_2 for y

$$SRED(\alpha, \ulcorner R \urcorner * \langle x_0, x_1, \ulcorner \bar{y}_2 \urcorner, x_3 \rangle, Ry_0 y_1 y_2 y_3)$$

and therefore by the properties of SRED

$$SRED(\alpha, \ulcorner R \urcorner * \langle x_0, x_1, x_2, x_3 \rangle, Ry_0 y_1 y_2 y_3).$$

This establishes (4), and hence $VAL(\ulcorner R_\sigma \urcorner, R_\sigma)$.

By (2) and (a) - (f), it is obvious that (1) holds.

Remark. Since standard reduction sequences correspond to a (very restricted) class of proofs, (1) may be viewed as a (weak) uniform reflection principle.

§ 4. Models based on partial recursive function application: HRO, HEO.

2.4.1 - 2.4.5. General remarks on models of $\underset{\sim}{N}-\underset{\sim}{HA}^{\omega}$.

2.4.1. Models; normal, extensional models.

A model for $\underset{\sim}{N}-\underset{\sim}{HA}^{\omega}$ is given by specifying domains for the range of the variables, and interpreting the constants, including equality. Hence our models are models w.r.t. many-sorted predicate logic, not always w.r.t. predicate logic with equality. This is only natural, since the interpretation of equality varies in the different systems studied.

If the equality relation of $\underset{\sim}{N}\text{-}\underset{\sim}{HA}^{\omega}$ is interpreted by the identity in the domains of the model, the model may be called normal. The model is called extensional, if equality is interpreted in the model as the (definable) extensional equality between the elements of the model (cf. 1.6.12).

All models of $\underset{\sim}{N}-\underset{\sim}{HA}^{\omega}$ studied in this chapter are ω-models, i.e. the natural numbers (objects of type 0) receive their standard interpretation.

2.4.2. Notation. If N is any model of $\underset{\sim}{N}-\underset{\sim}{HA}^{\omega}$, let then N_σ denote the objects of type σ in the model, and let $Ap_N^{\sigma,\tau}$ denote a binary operation (of type $(N_{(\sigma)\tau}\times N_\sigma)N_\tau$) representing application (of type $(\sigma)\tau$ to type σ) in N. (So $Ap(x^{(\sigma)\tau}, y^\sigma)$ interprets $x^{(\sigma)\tau}y^\sigma$.) If c is a constant of $\underset{\sim}{N}-\underset{\sim}{HA}^{\omega}$, let c_N be its interpretation in N. We abbreviate Ap(x,y) also as xy.

2.4.3. Submodel, homomorphism, embedding.

Let N, M be models of $\underset{\sim}{N}-\underset{\sim}{HA}^{\omega}$.

φ is a homomorphism from N into M, if φ maps N_σ into M_σ for all $\sigma\in T$, and $\varphi(Ap_N(x,y)) = Ap_M(\varphi x,\varphi y)$, $\varphi(C_N) = C_M$ for all constants c of $\underset{\sim}{N}-\underset{\sim}{HA}^{\omega}$.

φ is an embedding if φ is a bi-unique homomorphism.

N is a submodel of M, if $N_\sigma\subseteq M_\sigma$ for all $\sigma\in\underset{\approx}{T}$, and $Ap_N^{\sigma,\tau}$ is the restriction of $Ap_M^{\sigma,\tau}$ to $N_{(\sigma)\tau}\times N_\sigma$, for all $\sigma,\tau\in\underset{\approx}{T}$, and $c_N = c_M$ for all constants c of $\underset{\sim}{N}-\underset{\sim}{HA}^{\omega}$.

2.4.4. Definition (of the extensional equivalence relation $\approx$).

For any model N of $\underset{\sim}{N}-\underset{\sim}{HA}^{\omega}$, we define by induction over the type structure:

(i) For $x,y\in N_0$: $x\approx y \equiv_{def} x=y$

(ii) For $x,y\in N_{(\sigma)\tau}$: $x\approx y \equiv_{def} \underset{\sim}{\forall} z\in N_\sigma(xz\approx yz)$.

If $=$ is interpreted by $\approx$ in the model, N is extensional, and is then a model of $\underset{\sim}{E}-\underset{\sim}{HA}^{\omega}$.

2.4.5. Theorem (Zucker 1971, § 8). If M is a model of $\underset{\sim}{N}-\underset{\sim}{HA}^{\omega}$, then there is a standard procedure for constructing an extensional model M^E of $\underset{\sim}{E}-\underset{\sim}{HA}^{\omega}$ which is almost a submodel of M, i.e. M^E satisfies $Ap_{M^E}^{\sigma,\tau} = Ap_M^{\sigma,\tau} |(M^E_{(\sigma)\tau} \times M^E_{\sigma})$ ($|$ expressing restriction) and $M^E_{\sigma} \subseteq M_{\sigma}$, $C_{M^E} = C_M$ for all constants c, except equality, whereas for equality $x =_M y \Rightarrow x =_{M^E} y$.

Proof. We define binary relations I'_{σ} on M_{σ}, by induction on σ, as follows:

$$I'_{\sigma}(x,y) \equiv_{def} x = y$$

$$I'_{(\sigma)\tau}(x,y) \equiv_{def} x \in M_{(\sigma)\tau} \ \&\ y \in M_{(\sigma)\tau} \ \&\ \forall uv(I'_{\sigma}(u,v) \to I'_{\tau}(xu, yv)) .$$

If we put $M^E_{\sigma}(x) \equiv_{def} I'_{\sigma}(x,x)$, and we define application for M^E as the restriction of application in M, and interpret constants in M^E as in M, then M^E becomes a submodel of M.

When restricted to M^E, $I'_{(\sigma)\tau}$ coincides with $\approx$ for M^E, by a straightforward induction on the type structure. We also have to verify that for constants c^{σ} of $\underset{\sim}{N}-\underset{\sim}{HA}^{\omega}$, c^{σ}_M belongs to M^E_{σ}, which is straightforward. For example, consider $(R_{\sigma})_M$ (abbreviated as R'_{σ}). We have to show that for $x \in M^E_{\sigma}$, $y \in M^E_{(\sigma)(0)\sigma}$, $z \in M^E_0$, it follows that $R'_{\sigma}xyz \in M^E_{\sigma}$. We prove this by induction on z.

2.4.6. The classical set-theoretical model of $\underset{\sim}{E}-\underset{\sim}{HA}^{\omega}$.

For completeness sake, we mention here the most obvious classical model of $\underset{\sim}{E}-\underset{\sim}{HA}^{\omega}$ (for which we have no interesting applications), the full set-theoretical model S.

If we consider say ZF-set theory, and identify the natural numbers with a standard set, say the ordinal ω, we may define the set-theoretical model S of $\underset{\sim}{E}-\underset{\sim}{HA}^{\omega}$ in ZF in an obvious way; objects of type $(\sigma)\tau$ are then all set-theoretical mappings from the set of objects of type σ into a set of objects of type τ; the objects of type 0 are the elements of ω.

2.4.7. Models based on partial recursive function application.

The models for $\underset{\sim}{N}-\underset{\sim}{HA}^{\omega}$ described in the remainder of this section are based on partial recursive function application between natural numbers (denoted by Kleene-brackets: $\{.\}.$). Our basic models are HRO, HEO, the Hereditarily Recursive Operations and the Hereditarily Effective Operations.

Later on, in section 6 we shall describe analogous models based on continuous function application (written as $\{\alpha\}[\beta]$ in Kleene and Vesley 1965, and Kleene 1969; we use $.|.$ instead of $\{.\}[.]$).

2.4.8. Description of HRO.

We first define, for each $\sigma \in \underset{\approx}{T}$, a set of natural numbers V_{σ}, as follows.

$$V_0(x) \equiv_{def} 0 = 0$$

$$V_{(\sigma)\tau}(x) \equiv_{def} \forall y \in V_\sigma \exists z \in V_\tau(\{x\}(y) \simeq z)\text{ , or equivalently}$$

$$\forall y \in V_\sigma \exists u(T(x,y,u) \,\&\, V_\tau(Uu))\,.$$

Now the hereditarily recursive operations of type σ consist of all pairs (x,σ) with $x \in V_\sigma$. Since we may assume σ to be represented by a natural number hereditarily recursive operations may be supposed to be represented by natural numbers.

Application is partial recursive function application:

$$(x, (\sigma)\tau)(y, \sigma) = (\{x\}(y), \tau)\,.$$

HRO becomes a model of $\underset{\sim}{I}$-$\underset{\sim}{HA}^\omega$, if we can find numbers $[\Pi]$, $[\Sigma]$, $[R]$, $[S]$, $[E]$ such that, if we abbreviate $\{\ldots\{\{\{t_0\}(t_1)\}(t_2)\}\ldots\}(t_n)$ by $\{t_0\}(t_1,\ldots,t_n)$,

$$\{[\Pi]\}(x,y) \simeq x$$

$$\{[\Sigma]\}(x,y,z) \simeq \{\{x\}(z)\}(\{y\}(z))$$

$$\{[S]\}(x) \simeq Sx$$

$$\begin{cases}\{[R]\}(x,y,0) \simeq x \\ \{[R]\}(x,y,Sz) \simeq \{y\}(\{[R]\}(x,y,z), z)\end{cases}$$

$$\{[E]\}(x,y) \simeq sg|x-y|\,.$$

Such numbers are constructed as follows.

We put

$$[\Pi] = \Lambda x \Lambda y\,.\,x$$

$$[\Sigma] = \Lambda x \Lambda y \Lambda z\,.\,\{\{x\}(z)\}(\{y\}(z))$$

$$[S] = \Lambda x\,.\,Sx$$

$$[E] = \Lambda x \Lambda y\,.\,sg(|x-y|)\,.$$

To construct a number $[R]$, we may either use the fact that in Kleene's formalization of recursion theory (Kleene 1969, § 1.1) definition by primitive recursion is included, and combine this with the fact that the rule of definition by recursion permits us to construct a uniform recursor (cf. 1.7.5), or, if we do not wish to use this fact, we may appeal to the recursion theorem, noting that there exists a partial recursive function $\psi(u,x,y,z)$ such that

$$\psi(u,x,y,0) \simeq x$$

$$\psi(u,x,y,Sz) \simeq \{y\}(\{u\}(x,y,z), z)\,,$$

hence by the recursion theorem we find a number $[R]$ such that

$$\psi([R],x,y,z) \simeq \{[R]\}(x,y,z)\,.$$

Then $\Pi_{\sigma,\tau}$, $\Sigma_{\rho,\sigma,\tau}$, S, E_σ, R_σ are represented by

$$([\Pi], (\sigma)(\tau),\sigma)$$

$([\Sigma], ((\rho)(\sigma)\tau)((\rho)\sigma)(\rho)\tau)$
$([S], (0)0)$
$([E], (\sigma)(\sigma)0)$
$([R], (\sigma)((\sigma)(0)\sigma)(0)\sigma)$.

Thus HRO is a model of $\underset{\sim}{I}-\underset{\sim}{HA}^{\omega}$, if we interprete application, Π, Σ, S, E_σ, R_σ as indicated above, 0 by $(0,0)$, and $=$ by identity.

To each closed term t^σ of $\underset{\sim}{I}-\underset{\sim}{HA}^{\omega}$ we can thus find a number $[t]$ such that $([t],\sigma)$ represents t in HRO, and

$$\underset{\sim}{HA} \vdash V_\sigma([t]).$$

2.4.9. Remark on terminology. HRO becomes a model if $\underset{\sim}{I}-\underset{\sim}{HA}^{\omega}$ only by specifying $[\Pi]$, $[\Sigma]$, $[S]$, $[R]$, $[E]$. However, to avoid ponderous circumlocutions, or the introduction of special designations for the variants, we shall talk somewhat loosely about HRO as "a model for $\underset{\sim}{I}-\underset{\sim}{HA}^{\omega}$" and if we wish to refer to a specific choice of $[\Pi]$, $[\Sigma]$, $[S]$, $[E]$, $[R]$, we shall speak of a "version of HRO".

2.4.10. The formal theories $\underset{\sim}{HRO}$, $\underset{\sim}{HRO}^-$.

$\underset{\sim}{HRO}$ is an extension of $\underset{\sim}{I}-\underset{\sim}{HA}^{\omega}$ in which it is asserted that the objects of type σ coincide with the hereditarily recursive operations of type σ.

The language of $\underset{\sim}{HRO}$ is obtained by adding constants $\Phi_\sigma \in (\sigma)0$, $\Phi'_{\sigma,\tau} \in ((\sigma)\tau)(\sigma)0$ for all $\sigma,\tau \in \underset{\sim}{T}$ to the language of $\underset{\sim}{I}-\underset{\sim}{HA}^{\omega}$. $\underset{\sim}{HRO}$ is axiomatized by addition of the following axioms to $\underset{\sim}{I}-\underset{\sim}{HA}^{\omega}$:

G1. $\Phi_0 x^0 = x^0$

G2. $\Phi_\sigma x^\sigma = \Phi_\sigma y^\sigma \leftrightarrow x^\sigma = y^\sigma$

G3. $T(\Phi_{(\sigma)\tau} x^{(\sigma)\tau}, \Phi_\sigma y^\sigma, \Phi'_{\sigma,\tau} x^{(\sigma)\tau} y^\sigma)$

G4. $\Phi_\tau x^{(\sigma)\tau} y^\sigma = U(\Phi'_{\sigma,\tau} x^{(\sigma)\tau} y^\sigma)$ (T, U as in 1.3.9 A)

G5. $\forall x \in V_\sigma \exists y^\sigma (\Phi_\sigma y = x)$.

G1 - 4 express that all objects of finite type are hereditarily recursive operations; G5 expresses that every hereditarily recursive operation is an object of finite type. $\underset{\sim}{HRO}^-$ is obtained by deleting G5 from $\underset{\sim}{HRO}$.

HRO is a model for $\underset{\sim}{HRO}$. To see this, we only have to interpret Φ_σ by $(\Lambda x.x, (\sigma)0)$ and $\Phi'_{\sigma,\tau}$ by $(\Lambda x \Lambda y . \min_z T(x,y,z), ((\sigma)\tau)(\sigma)0)$.

2.4.11. Description of HEO.

We define, for each $\sigma \in \underset{\sim}{T}$, a set of natural numbers W_σ and an equivalence relation I_σ as follows:

$$W_0(x) \equiv_{def} x = x \ , \ I_0(x,y) \equiv_{def} (x = y).$$

$$W_{(\sigma)\tau}(x) \equiv_{def} \forall y \in W_\sigma\ \exists u(T(x,y,u) \,\&\, W_\tau(Uu) \,\&$$
$$\& \forall y \in W_\sigma\ \forall y' \in W_\sigma\ \forall uu'(I_\sigma(y,y') \,\&\, T(x,y,u) \,\&$$
$$\& T(x,y',u') \to I_\tau(Uu,Uu')) .$$

$$I_{(\sigma)\tau}(x,y) \equiv_{def} W_{(\sigma)\tau}(x) \,\&\, W_{(\sigma)\tau}(y) \,\&\, \forall z \in W_\sigma\ \forall uu'(T(x,z,u) \,\&$$
$$\& T(y,z,u') \to I_\tau(Uu,Uu')) .$$

Now the hereditarily effective operations (HEO) of type σ consist of all pairs (x,σ) with $x \in W_\sigma$.
If we interpret application, $=_0$, 0, S, Π, Σ, R as in the case of HRO, we have obtained a model for $\underset{\sim}{E}$ - $\underset{\sim}{HA}^\omega$. I_σ corresponds to extensional equality between objects of type σ.
Remark (i). An extension $\underset{\sim}{HEO}$ or $\underset{\sim}{HEO}^-$, analogous to $\underset{\sim}{HRO}$ or $\underset{\sim}{HRO}^-$ does not exist. With HEO as a model, Φ_1 should then assign a gödelnumber to each object of type 1, such that

$$\forall x^1 y^1 [\forall z^0 (xz = yz) \to \Phi_1 x = \Phi_1 y]$$

and this would make equality between objects of type 1 recursively decidable, which is well known to be false.

(ii). For each closed term $t \in \sigma$, as in the case of HRO

$$\underset{\sim}{HA} \vdash W_\sigma([t]) .$$

Note that the [t] assigned for HRO, HEO are the same.

2.4.12. Theorem. HRO^E and HEO are distinct. In fact, if we write $W_\sigma^!$ for HRO_σ^E, we have

$$W_0 = W_0^!,\quad W_1 = W_1^!,\quad W_2 = W_2^!,\quad W_3 \supset W_3^!,\quad W_n \not\subseteq W_n^!,$$
$$W_n^! \not\subseteq W_n \quad \text{for } n > 3 .$$

Proof. $W_0^! = W_0$, $W_1^! = W_1$ is obvious.
$I^!_{(1)0}(x,y) \equiv_{def} x \in V_1 \,\&\, y \in V_1 \,\&\, \forall uv(I_1^!(u,v) \to \{x\}(u) \simeq \{y\}(v))$ which is equivalent to $x \in W_1^! \,\&\, y \in W_1^! \,\&\, \forall uv(I_1(u,v) \to \{x\}(u) \simeq \{y\}(v))$, which is in turn $I_{(1)0}(x,y)$.
Hence, since $x \in W_2^! \leftrightarrow I_2^!(x,x) \leftrightarrow I_2(x,x) \leftrightarrow x \in W_2$, it follows that $W_2 = W_2^!$.

Let $z \in W_3^!$, then by definition

$$z \in V_3 \,\&\, \forall uv(I_2(u,v) \to \{z\}(u) \simeq \{z\}(v)) .$$

Since $W_2 \subset V_2$, it follows that $\{z\}(w)$ is defined for all $w \in W_2$, hence $z \in W_3$.

We note that $\Lambda x.x \in V_2 - W_2$.
Now we construct x_0 such that

$$\{x_0\}(x) \simeq 0 \quad \text{for } x \neq \Lambda z.z ,$$
$$\{x_0\}(x) \quad \text{undefined for } x = \Lambda z.z .$$

Then $x_0 \in W_3$, but $x_0 \notin W_3'$, since this would imply $x_0 \in V_3$, whereas $\{x_0\}(\Lambda z.z)$ is undefined.

This construction may be generalized; let u_0, u_1 be two numbers $\in W_n$ such that $u_0 \neq u_1$, $\{u_0\}(x) = \{u_1\}(x) = 0$ for all x, then, since $I_n(u_0,u_1)$ for all $n \geq 1$, $\Lambda x.x \notin W_{n+1}$ for $n \geq 1$, since $\{\Lambda x.x\}(u_0) \neq \{\Lambda x.x\}(u_1)$. Therefore, as before, $x_0 \in W_{n+2} - W'_{n+2}$, $x_0 \notin V_{n+2}$.

Now we construct x_1 such that

$$\begin{cases} \{x_1\}(x) \simeq 0 & \text{for } x \neq x_0 \\ \{x_1\}(x_0) & \text{undefined.} \end{cases}$$

Then obviously $x_1 \in W'_n - W_n$ for $n > 3$, since x_1 is defined on V_{n-1}, but not on W_{n-1}.

Open problem. Are there mathematically interesting functionals which occur in HRO^E, but not in HEO, or in HEO, but not in HRO^E?

2.4.13. Definition of $[t]_{HRO}$, $[t]_{HEO}$ (for terms t of $\underset{\sim}{N}-\underset{\sim}{HA}^{\omega}$).
For applications in the future, it is simplest if we restrict attention to terms containing type 0 variables from a fixed recursive infinite set V with infinite complement. Let Γ denote some 1-1 mapping of higher type variables onto the type 0 variables not in V. We then define $[t]_{HRO}$ $(\equiv [t]_{HEO})$ by induction on the complexity of t, as follows:

$$[0]_{HRO} \equiv 0, \quad [x^0]_{HRO} \equiv x^0, \quad [x^\sigma]_{HRO} \equiv \Gamma x^\sigma (\sigma \neq 0)$$
$$[S]_{HRO} \equiv [S], \quad [R_\sigma]_{HRO} \equiv [R], \quad [\Pi]_{HRO} \equiv [\Pi], \quad [\Sigma]_{HRO} \equiv [\Sigma],$$

(where $[S], [R], [\Pi], [\Sigma], [E]$ are chosen as in 2.4.8) and

$$[tt']_{HRO} \equiv \{[t]_{HRO}\}[t']_{HRO}, \text{ unless } t \equiv S;$$
$$[St']_{HRO} \equiv S[t']_{HRO}.$$

2.4.14. Theorem (Provable faithfulness of HRO, uniformly in type 0 variables). Let t be any type 0 term containing only type 0 variables free. Then

$$\underset{\sim}{N}-\underset{\sim}{HA}^{\omega} \vdash [t]_{HEO} \simeq [t]_{HRO} \simeq t.$$

Proof. For closed t, $[t]_{HEO}$ is represented by a pseudo-term constructed from 0, S, and $\{.\}.$.
Let us assume a gödelnumbering for such pseudo-terms to be given, e.g. as follows:

$$\ulcorner 0 \urcorner = j(0,0)$$
$$\ulcorner St \urcorner = j(1, \ulcorner t \urcorner)$$
$$\ulcorner \{t\}(t') \urcorner = j(2, j(\ulcorner t \urcorner, \ulcorner t' \urcorner)).$$

gnpt(x) is the primitive recursive predicate which holds iff x is the gödelnumber of a pseudo-term of the described kind.
A binary predicate $A(x,y)$ can be explicitly defined (Ch. I, § 4) such that

$$A(x,y) \leftrightarrow [x = j(0,0) \;\&\; y = 0] \vee$$
$$\vee [j_1x = 1 \;\&\; \exists z(A(j_2x,z) \;\&\; y = Sz)] \vee$$
$$\vee [j_1x = 2 \;\&\; \exists uvw(A(j_1j_2x,u) \;\&\; A(j_2j_2x,v) \;\&\; Tuvw \;\&\; Uw=y)] .$$

Obviously $A(x,y) \rightarrow \mathrm{gnpt}(x)$.
Compl(x) is the primitive recursive function such that if gnpt(x), then Compl(x) is the complexity of the term represented by x.
One readily proves by induction on Compl(x)

$$\forall yy'[A(x,y) \;\&\; A(x,y') \rightarrow y = y'] .$$

Let t be a pseudo-term constructed from 0, S, type 0 variables $x_1,\ldots,x_n$ and $\{.\}$. Let us write $\bar{t}$ for $t(\bar{x}_1,\ldots,\bar{x}_n)$.
Then

(1) $\quad \underset{\sim}{HA} \vdash \forall y(t \simeq y \leftrightarrow A(\ulcorner \bar{t} \urcorner,y))$.

Proof by induction on the complexity of t.
$\forall y(0 \simeq y \leftrightarrow A(j(0,0),y))$ is obvious.
$\forall y(x_i \simeq y \leftrightarrow A(\ulcorner \bar{x}_i \urcorner,y))$ is readily proved by induction on x_i.
Let $t \equiv St'$, and assume $\forall y(t' \simeq y \leftrightarrow A(\ulcorner \bar{t}' \urcorner,y))$.
$A(\ulcorner S\bar{t}' \urcorner,Sy) \leftrightarrow A(\ulcorner \bar{t}' \urcorner,y)$, hence $\forall y(St' \simeq y \leftrightarrow A(\ulcorner S\bar{t}' \urcorner,y))$.
Finally, let $t \equiv \{t'\}(t'')$.
Assume $\forall y(t' \simeq y \leftrightarrow A(\ulcorner \bar{t}' \urcorner,y))$, $\forall y(t'' \simeq y \leftrightarrow A(\ulcorner t'' \urcorner,y))$.
Now

$$A(\ulcorner \{\bar{t}'\}(\bar{t}'') \urcorner,y) \leftrightarrow \exists uvw[A(\ulcorner \bar{t}' \urcorner,u) \;\&\; A(\ulcorner \bar{t}'' \urcorner,v) \;\&\; Tuvw \;\&\; Uw = y] \leftrightarrow$$
$$\exists uvw[t' \simeq u \;\&\; t'' \simeq v \;\&\; Tuvw \;\&\; Uw = y] \leftrightarrow \{t'\}(t'') \simeq y .$$

Thus (1) is proved.

Now we define by induction on the type structure $\mathrm{Int}_\sigma(x^0,y^\sigma)$ ("the pseudo-term with gödelnumber x is the HEO - interpretation of the functional y^σ"):

(i) $\quad \mathrm{Int}_0(x^0,y^0) \equiv A(x,y)$

(ii) $\quad \mathrm{Int}_{(\sigma)\tau}(x,y) \equiv \forall x'y'[\mathrm{Int}_\sigma(x',y') \rightarrow \mathrm{Int}_\tau(j(2,j(x,x')),yy')]$.

Now we prove, entirely parallel to 2.3.13, that for terms $t(x_1,\ldots,x_n)$ constructed by application from constants and type 0 variable $x_1,\ldots,x_n$, that

(2) $\quad \underset{\sim}{N}-\underset{\sim}{HA}^\omega \vdash \mathrm{Int}(\ulcorner[\bar{t}]_{HEO}\urcorner,t)$

where $\bar{t}$ is an abbreviation for $t(\bar{x}_1,\ldots,\bar{x}_n)$.
Combining (1) and (2), we find for type 0 terms $t(x_1,\ldots,x_n)$ of $\underset{\sim}{N}-\underset{\sim}{HA}^\omega$:

$$\underset{\sim}{N}-\underset{\approx}{HA}^{\omega} \vdash A(\ulcorner[\bar{t}]_{HEO}\urcorner, t)$$

$$\underset{\sim}{N}-\underset{\approx}{HA}^{\omega} \vdash A(\ulcorner[\bar{t}]_{HEO}\urcorner, t) \leftrightarrow [t]_{HEO} \simeq t ,$$

which together yield the assertion of the theorem.

2.4.15. Corollary. All closed type 1 terms of $\underset{\sim}{N}-\underset{\approx}{HA}^{\omega}$ represent provably recursive functions of $\underset{\approx}{HA}$.

Proof. Let t be a closed type 1 term; then there is a numeral [t] such that $\underset{\approx}{HA} \vdash [t] \in V_1$ (cf. end of 2.4.8). Also $\underset{\sim}{N}-\underset{\approx}{HA}^{\omega} \vdash [tx]_{HRO} \simeq tx$, where $[tx]_{HRO} \equiv \{[t]\}(x)$.

2.4.16. Generalization. In $\underset{\approx}{HA}^{c}$ we can easily define a version of HRO based on A - partial recursive functions instead of recursive functions, with $\{x\}^{A}(y)$ taking the place of $\{x\}(y)$.

2.4.17. Historical note. HEO, for pure types, is described in Kreisel 1959, p. 117.

A (form of) HRO first appears in Kreisel 1958 B, lecture 60. A variant formulated in the theory of combinators is briefly indicated in Tait 1968, p. 191, lines -10 to -2. Troelstra rediscovered HRO and made extensive use of it in Troelstra 1971.

2.4.18. Sketch of a variant of HRO satisfying $\beta\eta$ - conversion.

An intensional variant of $\underset{\sim}{N}-\underset{\approx}{HA}^{\omega}$ with the λ - operator as a primitive, is most easily formulated by introducing intensional equality as follows: we require reflexivity, symmetry, transitivity and monotonicity. Reduction is defined syntactically: $t \geq t$, $t \geq t'$ and $t' \geq t'' \Rightarrow t \geq t''$, $t \geq t' \Rightarrow tt'' \geq t't''$, $t \geq t' \Rightarrow t''t \geq t''t'$, $\lambda x.t \geq \lambda y[x/y]t$ (y not free in t), $(\lambda x.t)t' \geq [x/t']t$, $t \geq t' \Rightarrow \lambda x.t \geq \lambda x.t'$, $Rtt'0 \geq t$, $Rtt'(St'') \geq t'(Rtt't'')t''$, $E_{\sigma}tt' \geq 0$ if t, t' are distinct closed terms in normal form, $E_{\sigma}tt \geq 1$ if t is closed, normal. Finally we add a schema $t = t'$ if t reduces to t'. Otherwise $\lambda\underset{\sim}{I}-\underset{\approx}{HA}^{\omega}$ is similar to $\underset{\sim}{I}-\underset{\approx}{HA}^{\omega}$.

A model for $\lambda\underset{\sim}{I}-\underset{\approx}{HA}^{\omega}$, similar to HRO can be obtained as follows.

We consider the $\lambda K - \beta\eta$ calculus, with an additional constant $\underline{E}$, and introduce $\beta\eta\delta$ - conversion as $\beta\eta$ - conversion and in addition a rule of δ - conversion:

$$\underline{E}tt' \text{ conv } [I,K] \text{ if } t,t' \text{ are distinct closed terms in normal form}$$

$$\underline{E}tt \text{ conv } I \quad \text{ if } t \text{ is closed, normal,}$$

where $I \equiv_{def} \lambda x.x$, $U \equiv_{def} \lambda xy.x$, $[t_1, t_2] \equiv_{def} \lambda z.zt_1t_2$.

For such a system the Church - Rosser theorem of uniqueness of normal form is provable (cf. Curry - Feys 1958, § 3D.6, chapter 4).

We put

$$\underline{0} \equiv I, \quad \underline{n+1} = [\underline{n},K], \quad \underline{S} \equiv \lambda x[x,K].$$

Obviously, $\underline{\underline{S}}\underline{n} \geq \underline{n+1}$. H.P. Barendregt has shown (Barendregt C) that we can find a term $\underline{R}$ such that

$\underline{R}xy\underline{0} = x$

$\underline{R}xy(\underline{S}z) = y(\underline{R}xyz)z$

$\underline{R}$ is in normal form, and when t_1, t_2 are in normal form, then $\underline{R}t_1$, $\underline{R}t_1t_2$ have a normal form.

We now define our HRO - analogue λ - HRO as follows.

$$t \in V_0^\lambda \equiv_{def} \exists n(t \equiv \underline{n})$$

$$t \in V_{(\sigma)\tau}^\lambda \equiv_{def} t \text{ normal, closed, } \forall t' \in V_\sigma^\lambda \exists t'' \in V_\tau^\lambda(tt' = t'') .$$

The objects of type σ are now pairs (x,σ), x a (gödelnumber of a) term of V_σ^λ. Obviously, $(\ulcorner I \urcorner, 0)$ is going to represent 0, $(\ulcorner \underline{S} \urcorner, 1)$ represents successor, $(\ulcorner \underline{E} \urcorner, (\sigma)(\sigma)0)$ represents E_σ, $(\ulcorner \underline{R} \urcorner, (\sigma)((\sigma)(0)\sigma)(0)\sigma)$ represents R_σ.

Another possibility for constructing a HRO - analogue is the following: add to the language of the λ - calculus four additional constants $\underline{0}, \underline{S}, \underline{E}, \underline{R}$, satisfying the reduction rules $\underline{E}tt$ conv $\underline{0}$ if t is closed, normal, and $\underline{E}tt'$ conv $\underline{\underline{S}}\underline{0}$ if t,t' are distinct, closed terms in formal form, $\underline{R}tt'\underline{0}$ conv t, $\underline{R}tt'(\underline{S}t'')$ conv $t'(\underline{R}tt't'')t''$. Abbreviate $\underline{\underline{S}}\underline{0}$ as $\underline{1}$, $\underline{\underline{S}}\underline{n}$ as $\underline{n+1}$.

Extend now the Church - Rosser theorem to this extended λ - calculus, and then proceed as before.

2.4.19. Pairing in HRO, HEO.

It is easy to extend HRO, HEO to models for $\underset{\sim}{I}$ - $\underset{\sim}{HA}_p^\omega$, $\underset{\sim}{E}$ - $\underset{\sim}{HA}_p^\omega$ by adding to the definition

$$V_{\sigma \times \tau}(x) \equiv_{def} V_\sigma(j_1 x) \;\&\; V_\tau(j_2 x)$$

and similarly

$$W_{\sigma \times \tau}(x) \equiv_{def} W_\sigma(j_1 x) \;\&\; W_\tau(j_2 x)$$

and

$$I_{\sigma \times \tau}(x,y) \equiv_{def} I_\sigma(j_1 x, j_1 y) \;\&\; I_\tau(j_2 x, j_2 y) ,$$

and representing D, D', D" by $(\Lambda xy.\, j(x,y), (\sigma)(\tau)\sigma \times \tau)$, $(\Lambda x.j_1 x, (\sigma \times \tau)\sigma)$, $(\Lambda x.j_2 x, (\sigma \times \tau)\tau)$ respectively.

The models so extended we shall usually also denote by HRO, HEO.

§ 5. Term models of $\underset{\sim}{N}$ - $\underset{\sim}{HA}^{\omega}$.

2.5.1. Definitions. Let CTM_{σ} be the set of closed terms of type σ in $\underset{\sim}{N}$ - $\underset{\sim}{HA}^{\omega}$, and $CTNF_{\sigma}$ the set of closed terms of type σ in normal form in $\underset{\sim}{N}$ - $\underset{\sim}{HA}^{\omega}$. We put $CTM = \bigcup \{CTM_{\sigma} \mid \sigma \in \underset{\approx}{T}\}$, $CTNF = \bigcup \{CTNF_{\sigma} \mid \sigma \in \underset{\approx}{T}\}$.

CTM becomes a model of $\underset{\sim}{N}$ - $\underset{\sim}{HA}^{\omega}$, if we let the variables x^{σ} range over CTM_{σ}, we interpret application of t to s as juxtaposition ts, equality $=_{\sigma}$ as equality of normal form, and 0, S, Π, Σ, R as themselves. Let us denote, for simplicity, this model also by CTM.

CTNF becomes a model of $\underset{\sim}{N}$ - $\underset{\sim}{HA}^{\omega}$, if we let the variables x^{σ} range over $CTNF_{\sigma}$, application Ap assigns to t, s the term t' in normal form such that $ts \geq t'$, $=_{\sigma}$ is interpreted as proper equality (equality in CTNF), and the constants 0, S, Π, Σ, R are interpreted by themselves. Again we denote this model by "CTNF".

Note that for the proof that CTM, CTNF are models of $\underset{\sim}{N}$ - $\underset{\sim}{HA}^{\omega}$, we have to make use of the fact that every term of $\underset{\sim}{N}$ - $\underset{\sim}{HA}^{\omega}$ possesses a unique normal form.

2.5.2. Definitions. Let CTM'_{σ}, $CTNF'_{\sigma}$ be the closed terms of type σ of $\underset{\sim}{I}$ - $\underset{\sim}{HA}^{\omega}$ and the normal closed terms of type σ of $\underset{\sim}{I}$ - $\underset{\sim}{HA}^{\omega}$ respectively. $CTM' = \bigcup \{CTM'_{\sigma} \mid \sigma \in \underset{\approx}{T}\}$, $CTNF' = \bigcup \{CTNF'_{\sigma} \mid \sigma \in \underset{\approx}{T}\}$. CTM', CTNF' can be made into models of $\underset{\sim}{I}$ - $\underset{\sim}{HA}^{\omega}$, also denoted by CTM', CTNF', similar to the models CTM, CTNF.

2.5.3. Some properties of CTM, CTM', CTNF, CTNF'.

(i) In CTM application is primitive recursive, $=_{\sigma}$ is recursive, but not provably recursive in $\underset{\sim}{HA}$ (for standard gödelnumberings). The second assertion is established by a well-known type of diagonal argument: let h_x denote the xth closed type 1 term; the enumeration may be supposed to be primitive recursive in x. The (gödelnumber of) $h_{\bar{x}}\bar{x}$ is again a primitive recursive function of x. Suppose f(x,y) is a provably recursive function such that f(x,y) = 0 if x,y are gödelnumbers of closed terms with the same normal form, 1 elsewhere.

Then $1 \dot{-} f(\ulcorner h_{\bar{x}}\bar{x} \urcorner, \ulcorner 0 \urcorner)$ is a provably recursive function of x, denoted by a term t (cf. 3.4.29); say $t \equiv h_{\bar{y}_0}$. Now $h_{\bar{y}_0}\bar{y}_0 = 0 \leftrightarrow$ $1 \dot{-} f(\ulcorner h_{\bar{y}_0}(\bar{y}_0) \urcorner, \ulcorner 0 \urcorner) = 1 \leftrightarrow h_{\bar{y}_0}\bar{y}_0 \neq 0$; contradiction. Similarly for CTM'.

(ii) In CTNF, $=_{\sigma}$ is primitive recursive, application is recursive, but not provably recursive (for the standard gödelnumberings) in $\underset{\sim}{HA}$. Similarly for CTNF'.

In this case the non-provable recursiveness is established in an even

more straightforward way, utilizing the diagonal function $h_x x+1$.
(iii) The domains of the variables in CTM, CTNF, CTM', CTNF' are recursive (in contrast to HRO!).
(iv) $QF-AC_{00}$ does not hold in CTM, CTNF, CTM', CTNF'.
Proof. Let $\bar{n}$ be the gödelnumber of a recursive function which is not provably recursive in HA. Hence $\forall x \exists y\, T(\bar{n}, x, y)$ holds, and $\min_y T(\bar{n},x,y)$ is a recursive function of x, but not provably recursive in HA.
$QF-AC_{00}$ for CTM would require the existence of a $t^1 \in$ CTM such that $\forall x\, T(\bar{n}, x, t^1x)$; but since all $t^1 \in$ CTM are interpreted by provably recursive functions in HRO, it follows that $\neg \forall x T(\bar{n},x,t^1x)$. Similarly for CTNF, CTM', CTNF'.
2.5.4. Lemma. For standard gödelnumberings of partial recursive functions there exists a two-place primitive recursive function φ such that (cf. Rogers 1967, § 7.2, in proof of theorem IV).

$$\forall xyz(\{x\}(z) \simeq \{\varphi(x,y)\}(z))$$
$$\forall xx'yy'(x \neq x' \vee y \neq y' \leftrightarrow \varphi(x,y) \neq \varphi(x',y')) .$$

2.5.5. Theorem. There exists a version of the model HRO, such that the model CTNF' can be embedded in HRO (is isomorphic to a submodel of HRO).
Proof. Let $\langle x,y,z\rangle \equiv \nu_3(x,y,z)$ (1.3.9 (C)). We define the required version of HRO by re-defining the numbers $[c]$ representing the constants c (cf. 2.4.8) as follows.

Let $[0]=0$, and let $\bar{r}$ be any numeral such that

$$\{\bar{r}\}(x,y,0) \simeq x$$
$$\{\bar{r}\}(x,y,Sz) \simeq \{y\}(\{\bar{r}\}(x,y,z),z)$$

where $\{t\}(t_0,\dots,t_n)$ is an abbreviation $\{\dots\{\{\{t\}(t_0)\}(t_1)\}\dots\}(t_n)$.
We put

$$[S] = \varphi(\Lambda x.x+1, \langle 0,0,0\rangle),$$
$$[\Pi] = \varphi(\Lambda x.\varphi(\Lambda y.x, \langle x,x,2\rangle), \langle 1,1,1\rangle),$$
$$[\Sigma_{\rho,\sigma,\tau}] = \varphi(\Lambda x.\varphi(\Lambda y.\varphi(\Lambda z.\varphi(\{\{x\}(z)\}(\{y\}(z)),\ulcorner\sigma\urcorner), \langle x,y,5\rangle), \langle x,x,4\rangle), \langle 3,3,3\rangle),$$
$$[R] = \varphi(\Lambda x.\varphi(\Lambda y.\varphi(\Lambda z. \{\bar{r}\}(x,y,z), \langle x,y,8\rangle), \langle x,x,7\rangle), \langle 6,6,6\rangle).$$
$$[E] = \varphi(\Lambda x.\varphi(\Lambda y. sg|x-y|), \langle x,x,10\rangle), \langle 9,9,9\rangle).$$

If $t_0, t_1 \in$ CTNF, then each has one of the forms of the following list ($s,t \in$ CTNF):

$$\Pi,\ \Pi s,\ \Sigma,\ \Sigma s,\ \Sigma st,\ R,\ Rs,\ Rst,\ S,\ Ss,\ 0,\ E,\ Es .$$

a) If t_0, t_1 correspond to different forms in the list, then $[t_0] \neq [t_1]$; e.g. if $t \equiv \Sigma_{\rho,\sigma,\tau}st, t_1 \equiv Rs't'$, then

$$[t_0] = \varphi(\Lambda z.\varphi(\{\{[s]\}(z)\}(\{[t]\}(z)),\ulcorner\sigma\urcorner),\langle[s],[t],5\rangle)$$
$$[t_1] = \varphi(\Lambda z.\ \{\bar{r}\}([s'],[t'],z),\ \langle[s'],[t'],8\rangle)\ .$$

Now $[t_0]\neq[t_1]$, since $\varphi(x,\ \langle y,z,5\rangle)\neq\varphi(x',\ \langle y',z',8\rangle)$ for all x', y', z', x, y, z.

b) If $t_0 \not\equiv t_1$ and t_0, t_1 correspond to the same form on the list, then also $[t_0]\neq[t_1]$, or type $(t_0)\neq$ type (t_1). The proof now proceeds by induction on the sum of the complexities of t_0 and t_1. For example, let $t_0 \equiv \Sigma_{\rho,\sigma,\tau}st$, $t_1 \equiv \Sigma_{\rho',\sigma',\tau'}s't'$, then $[t_0]$ is as under (a), $[t_1] = \varphi(\Lambda z.\ \varphi(\{\{[s']\}(r)\}(\{[t']\}(z)),\ulcorner\sigma'\urcorner),\ \langle[s'],\ [t'],\ 5\rangle)$. $[t_0]\equiv[t_1]$ would imply $[s]=[s']$, $[t]=[t']$ so then either type $(s)\neq$ type (s'), or type $(t)\neq$ type (t'), by (a) and induction hypothesis.
If type $(s)\neq$ type (s'), it follows that $\rho\neq\rho'$ or $\sigma\neq\sigma'$ or $\tau\neq\tau'$.
If type $(t)\neq$ type (t'), it follows that $\rho\neq\rho'$ or $\sigma\neq\sigma'$.
If $\sigma\neq\sigma'$, then obviously $[t_0]\not\equiv[t_1]$. If $\rho\neq\rho'$ or $\tau\neq\tau'$, then type $(t_0)\neq$ type (t_1), since type $(t_0)\equiv(\rho)\tau$, type $(t_1)\equiv(\rho')\tau'$.

Alternative proof. In the preceding proof, we have kept the assignment as uniform in the types as possible; if we use a slightly different definition of $[\Pi_{\sigma,\tau}]$, $[\Sigma_{\rho,\sigma,\tau}]$, $[R_\sigma]$, $[E_\sigma]$, we need less verification (case (b) in the preceding proof is simpler), but the uniformity in the types is gone.
The new definitions are:

$$[\Pi_{\sigma,\tau}] = \varphi(\Lambda x.\varphi(\Lambda y.\varphi(x,\ j(\ulcorner\sigma\urcorner,\ulcorner\tau\urcorner)),\ \langle x,x,2\rangle),\ \langle 1,1,1\rangle)$$
$$[\Sigma_{\rho,\sigma,\tau}] = \varphi(\Lambda x.\varphi(\Lambda y.\varphi(\Lambda z.\varphi(\{\{x\}(z)\}(\{y\}(z)),\ \langle\ulcorner\rho\urcorner,\ulcorner\sigma\urcorner,\ulcorner\tau\urcorner\rangle),\ \langle x,y,5\rangle),\ \langle x,x,4\rangle),\ \langle 3,3,3\rangle)$$
$$[R_\sigma] = \varphi(\Lambda x.\varphi(\Lambda y.\varphi(\Lambda z.\varphi(\{\bar{r}\}(x,y,z),\ulcorner\sigma\urcorner),\ \langle x,y,8\rangle),\ \langle x,x,7\rangle),\ \langle 6,6,6\rangle)$$
$$[E_\sigma] = \varphi(\Lambda x.\varphi(\Lambda y.\varphi(sg|x-y|,\ulcorner\sigma\urcorner),\ \langle x,x,10\rangle),\ \langle 9,9,9\rangle)\ .$$

2.5.6. Alternative proof of the uniqueness of nornal form.

Since for $t,t'\in$ CTNF', $t \geq t'$ implies $[t]=[t']$, it follows from 2.3.2 and 2.5.5 that each closed term of $\underset{\sim}{I}$ - $\underset{\sim}{HA}^{\omega}$ + IE_0 (and hence each closed term of $\underset{\sim}{N}$ - $\underset{\sim}{HA}^{\omega}$) possesses a unique normal form.

2.5.7. Corollary to 2.5.5, 2.5.6. If t,t' are closed terms of $\underset{\sim}{N}$ - $\underset{\sim}{HA}^{\omega}$, then $\underset{\sim}{N}$ - $\underset{\sim}{HA}^{\omega} \vdash t = t'$ iff t,t' reduce to the same normal form; hence, if $\underset{\sim}{N}$ - $\underset{\sim}{HA}^{\omega} \vdash t = t'$ then $t = t'$ can be proved in qf - $\underset{\sim}{N}$ - $\underset{\sim}{HA}^{\omega}$ without the use of induction.

This may be rephrased as a conservative extension result: $\underset{\sim}{N}$ - $\underset{\sim}{HA}^{\omega}$ is conservative over qf - $\underset{\sim}{N}$ - $\underset{\sim}{HA}^{\omega}$ without induction, for closed prime formulae.
Similarly for $\underset{\sim}{I}$ - $\underset{\sim}{HA}^{\omega}$ + IE_0.

2.5.8. Theorem. For suitable versions of HRO (i.e. the ones defined in 2.5.5), HRO is a model for $\underset{\sim}{I}-\underset{\sim}{HA}^{\omega}+IE_1$; hence, as a corollary of 2.5.5, CTNF' is also a model of $\underset{\sim}{I}-\underset{\sim}{HA}^{\omega}+IE_1$.

Proof. Similar to the argument in 2.5.5, we can show that IE_1 is satisfied.

Remark. CTNF' is a minimal model of $\underset{\sim}{N}-\underset{\sim}{HA}^{\omega}$ w.r.t. equality, i.e. two closed terms t,t' have the same interpretation in the model iff they reduce to the same normal form, i.e. if $\underset{\sim}{N}-\underset{\sim}{HA}^{\omega} \vdash t = t'$.

Each model of $\underset{\sim}{N}-\underset{\sim}{HA}^{\omega}+IE_0$ must be minimal w.r.t. equality between closed terms, as will be obvious.

2.5.9. Examples of versions of HRO where distinct normal terms are represented by the same element in the version of HRO.

(i) The first example is suggested by the necessity of referring to σ in the definition of $[\Sigma_{\rho,\sigma,\tau}]$ in the proof of 2.2.5. If we can find normal closed terms $t \in (\rho)(\sigma)\tau$, $t' \in (\rho)(\sigma')\tau$, $s \in (\rho)\sigma$, $s' \in (\rho)\sigma'$ such that $\sigma \neq \sigma'$, $[t] = [t']$, $[s] = [s']$ (under the assignment described in 2.4.8), then $[\Sigma_{\rho,\sigma,\tau} ts] = [\Sigma_{\rho,\sigma',\tau} t's']$, type $(\Sigma_{\rho,\sigma,\tau} ts)$ = type $(\Sigma_{\rho,\sigma',\tau} t's')$ = $= (\rho)\tau$, $\Sigma_{\rho,\sigma,\tau} ts \not\equiv \Sigma_{\rho,\sigma',\tau} t's'$.

Take $\rho = \sigma_1$, $\tau = \sigma_1$, $\sigma = (\sigma_2)\sigma_1$, $\sigma' = (\sigma_3)\sigma_1$, $\sigma_2 \neq \sigma_3$; $t \equiv \Pi_{\sigma_1,\sigma}$, $t' \equiv \Pi_{\sigma_1,\sigma'}$; $s \equiv \Pi_{\sigma_1,\sigma_2}$, $s' \equiv \Pi_{\sigma_1,\sigma_3}$; then all our requirements are met.

(ii) $R(SO)(\Sigma(\Pi\Pi')S)$ (where Π, Π' denote $\Pi_{\sigma,\tau}$, $\Pi_{\sigma',\tau'}$ for appropriate $\sigma, \tau, \sigma', \tau'$) is extensionally equal to the successor function S.

Now we modify our description of [R] in 2.5.5 as follows. Let $\bar{r}$ as before denote a numeral, satisfying

$$\{\bar{r}\}(x,y,0) \simeq x, \quad \{\bar{r}\}(x,y,Sz) \simeq \{y\}(\{\bar{r}\}(x,y,z),z).$$

Then let $\psi(x,y) \equiv \Lambda z.\ \{\bar{r}\}(x,y,z)$; ψ is primitive recursive. We put

$$\psi'(x,y) = sg|1-x| \cdot |y-\bar{n}| \cdot \psi(x,y) + (1 \dot{-} (|y-\bar{n}| + |1-x|)) \cdot [S],$$

where $\bar{n} \equiv [\Sigma(\Pi\Pi')S]$.

Now we put $[R] = \Lambda x \Lambda y.\ \psi'(x,y)$. It is then obvious that

$$\{[R]\}(x,y,0) \simeq x$$
$$\{[R]\}(x,y,Sz) \simeq \{y\}(\{[R]\}(x,y,z),z)$$

(this is proved by distinguishing cases: $y = \bar{n}\ \&\ x = 1$, or $x \neq 1 \vee y \neq \bar{n}$).

Also

$$\{[R]\}(0,\bar{n}) = [S].$$

From the preceding examples it is obvious that we cannot assert $\underset{\sim}{I}-\underset{\sim}{HA}^{\omega} \vdash t \neq t'$ whenever t,t' are closed terms with different normal forms.

Remark. The second example is based on another idea than the one used in the first example. The first example is based on the "type-ambiguity": $\Sigma_{\rho,\sigma,\tau}tt'$ and $\Sigma_{\rho,\sigma',\tau}t't''$ are of the same type for $\sigma \neq \sigma'$.

The second example picks more or less arbitrarily two closed terms of type 1, with different normal forms, but representing extensionally equal functions, and identifies them in the model.

2.5.10. IE_0 is weaker than IE_1.

We wish to show that the axiom schema $E_\sigma ts = 1$ if t,s are distinct closed terms in normal form, $E_\sigma tt = 0$ (i.e. the schema IE_0) does not imply IE_1, for example, it does not follow that

(1) $y \neq y \rightarrow Rxy \neq Rxy'$.

More precisely, we can find versions of HRO for which IE_0 is obviously valid, but for which (1) fails.

To see this, we argue as follows. Take any closed term of $\underset{\sim}{N}$-$\underset{\sim}{HA}^\omega$ of type (0)(0)0 not containing R_0, say e.g. $\Pi_{0,0}$, and let $\bar{\bar{n}}$ be any numeral such

$$\{\{\bar{\bar{n}}\}(x)\}(y) \simeq x.$$

The function φ of lemma 2.5.4 may be chosen such that $\bar{\bar{n}}$ is not in the range of φ.

Now we define our version of HRO as in the first proof of 2.5.5, but with one exception: we define $[R_0]$ as

$$\varphi(\Lambda x.\ \varphi(\Lambda y.\ \psi(\Lambda z.\ \{\bar{r}\}(x,y,z))\langle x,y,8\rangle),\ \langle x,x,7\rangle),\ \langle 6,6,6\rangle),$$

where ψ is given by

(2a) $\psi(u,\ \langle x,y,8\rangle) = \varphi(u,\ \langle x,y,8\rangle)$ if $y \neq \bar{\bar{n}}$

(2b) $\psi(u,\ \langle x,\bar{\bar{n}},8\rangle) = \varphi(u,\ \langle x,[\Pi_{0,0}],8\rangle)$.

We note that $\bar{\bar{n}}$ is also outside the range of ψ.

Let us indicate the number assigned to a closed term t by the original assignment as $[t]'$, and by the new one as $[t]$; then $[t] = [t]'$, as we can show by an induction on the complexity of t. For a closed term t_0 in normal form, which is not a numeral, is of one of the forms

$$\Pi,\ \Pi t,\ \Sigma,\ \Sigma t,\ \Sigma ts,\ R,\ Rt,\ Rts,\ S,\ E,\ Et$$

where t, s themselves are in normal form. So the corresponding numbers $[t_0]$, $[t_0]'$ are in the range of φ, ψ; since $\bar{\bar{n}}$ was chosen outside that range, an easy induction on the complexity of t_0 yields that in evaluating $[t_0]$ we never have to use clause (2b), hence $[t_0] = [t_0]'$.

Therefore IE_0 holds. But, obviously, (1) is false:

$$\{[R_o]\}(x,\bar{n}) = \{[R_o]\}(x,[\Pi_{0,0}])$$

whereas $\bar{n} \neq [\Pi_{0,0}]$.

2.5.11. Remark. Presupposing the theory of combinators, Tait's version (Tait 1968, p. 191, lines -10 to -2) of HRO is a slightly more direct way of achieving the result of 2.5.5. However, for our purposes the present definition of HRO is more flexible. Similarly, the HRO - variant satisfying $\beta\eta\delta$ - conversion, described in 2.4.18, contains a λ - term model isomorphically embedded.

2.5.12. Remark on the properties of gödelnumberings used.

The construction in 2.5.5 made essential use of the lemma 2.5.4 on standard gödelnumberings. One might wonder to what extent the results depend on the gödelnumbering chosen. An answer is provided by Rogers 1958. The "fully effective" numberings there are precisely the numberings which can be brought into recursive one-to-one correspondence with a standard gödelnumbering. Therefore any fully effective numbering satisfies 2.5.4 and yields the result in 2.5.5.

2.5.13. Historical note.

Term models for $\underset{\sim}{N}$ - $\underset{\sim}{HA}^{\omega}$ first appeared in Tait 1963, Appendix B, which is a preliminary draft of Tait 1967.

A detailed comparison between term models and HRO is made in Kreisel 1971, Appendix I.

§ 6. Models based on continuous function application: ICF, ECF.

2.6.1. Contents of the section.

In the present section, we study models of $\underset{\sim}{N}$ - $\underset{\sim}{HA}^{\omega}$ similar to HRO, HEO, but based on continuous function application instead. The hereditarily continuous functionals (ECF) make their appearance in Kreisel 1959 and Kleene 1959A (as countable functionals); the intensional continuous functionals ICF are introduced in Kreisel 1962 (page 154).

2.6.2 - 2.6.10 describe ECF, ICF and discuss the existence of moduli of continuity and uniform continuity in these models.

2.6.11, 2.6.12 extend the faithfulness theorem from HEO to ECF; 2.6.13 - 2.6.21 are devoted to the recursive density theorem for ECF and the equivalence between ECF($\mathcal{R}$) (= ECF relativized to recursive functions) and HEO.

2.6.22 discusses the models ECF^r, ICF^r, obtained by taking the recursive elements of ECF, ICF relative to a universe of functions satisfying bar-induction.

2.6.23 describes variants ECF*, ICF* of ECF, ICF respectively, where application is defined in a more uniform way than for ECF, ICF.

2.6.25 describes the interpretation of pairing operators in ECF, ICF, ECF*, ICF*.

2.6.26 describes the analogues $\underset{\sim}{ICF}$, $\underset{\sim}{ICF}^-$ to $\underset{\sim}{HRO}$, $\underset{\sim}{HRO}^-$ imtroduced in 2.4.10.

Directions for use. For most applications in connection with modified realizability and the Dialectica interpretation (§ 3.4, § 3.5), it suffices to study 2.6.2 - 2.6.10; a few results in § 3.5 (obtained with the help of the Dialectica interpretation) require 2.6.20. 2.6.11 - 2.6.12 are used in § 7, in studying derivable instances of the rule of extensionality.

2.6.2. Below we shall assume $\mathcal{U}$ to be a universe of functions of type 1, closed under "recursive in", (in short, $\mathcal{U}$ is a model of $\underset{\sim}{EL}$). α, β, γ are variables ranging over $\mathcal{U}$. We introduce V^1_σ for each $\sigma \in \underset{\sim}{T}$, analogous to V_σ for HRO, as follows:

$$x \in V^1_0 \equiv x = x$$

$$\alpha \in V^1_{(0)0} \equiv \alpha = \alpha$$

$$\alpha \in V^1_{(\sigma)0} \equiv \forall\beta \in V^1_\sigma\ \exists x(\alpha(\beta) \simeq x)\,, \qquad \text{for } \sigma \neq 0$$

$$\alpha \in V^1_{(\sigma)\tau} \equiv \forall\beta \in V^1_\sigma\ \exists\gamma \in V^1_\tau(\alpha|\beta \simeq \gamma)\,, \quad \text{for } \sigma, \tau \neq 0$$

$$\alpha \in V^1_{(0)\tau} \equiv \forall x\ \exists\gamma \in V^1_\tau(\alpha|\lambda y.x \simeq \gamma)\,, \qquad \text{for } \tau \neq 0\ .$$

The objects of type σ of the model $ICF(\mathcal{U})$ (the intensional continuous functionals relative to the universe $\mathcal{U}$, in short: ICF) consist of the pairs $(x,0)$, $x \in V_0^1$ if $\sigma = 0$, and of (α,σ) with $\alpha \in V_\sigma^1$ if $\sigma \neq 0$.

Equality is defined as $(x,0) = (y,0) \equiv_{def} x = y$, $(\alpha,\sigma) = (\beta,\sigma) \equiv_{def} \forall x(\alpha x = \beta x)$ (short: $\alpha = \beta$) for $\sigma \neq 0$. The interpretation of application depends on the type. We put:

$$(\alpha,1)(x,0) = (\alpha x,0)$$
$$(\alpha,(0)\sigma)(x,0) = (\alpha|\lambda y.x,\sigma) \quad \text{for } \sigma \neq 0$$
$$(\alpha,(\sigma)0)(\beta,\sigma) = (\alpha(\beta),0) \quad \text{for } \sigma \neq 0$$
$$(\alpha,(\sigma)\tau)(\beta,\sigma) = (\alpha|\beta,\tau) \quad \text{for } \sigma,\tau \neq 0 .$$

Further we have to show how the constants may be interpreted.
Let us write $[c]^1$ (in short: $[c]$) for the function or number such that $([c],\sigma)$ represents the constant $c \in \sigma$ in our model.

(a) $[0] \equiv 0$

(b) $[S] \equiv \lambda x.Sx$

(c) $[\Pi_{0,0}] \equiv \Lambda^1 x\Lambda^0 y.x$, $[\Pi_{0,\sigma}] \equiv \Lambda^1 x\Lambda^0 \alpha.x$,

$[\Pi_{\sigma,0}] \equiv \Lambda^1\alpha\Lambda^1 y.\alpha$, $[\Pi_{\sigma,\tau}] \equiv \Lambda^1\alpha\Lambda^1\beta.\alpha$ for $\sigma,\tau \neq 0$.

(d) $[\Sigma_{0,\sigma,0}] \equiv \Lambda^1\alpha\Lambda^1\beta\Lambda^0 z(\alpha|\lambda x.z)(\beta z)$,

$[\Sigma_{0,\sigma,\tau}] \equiv \Lambda^1\alpha\Lambda^1\beta\Lambda^1 z(\alpha|\lambda x.z)|(\beta|\lambda x.z)$ $(\sigma,\tau \neq 0)$,

$[\Sigma_{\rho,\sigma,0}] \equiv \Lambda^1\alpha\Lambda^1\beta\Lambda^0\gamma(\alpha|\gamma)(\beta|\gamma)$ $(\rho \neq 0)$,

$[\Sigma_{\rho,0,\tau}] \equiv \Lambda^1\alpha\Lambda^1\beta\Lambda^0\gamma(\alpha|\gamma)(\lambda y.\beta(\gamma))$ $(\rho,\tau \neq 0)$,

$[\Sigma_{\rho,\sigma,\tau}] \equiv \Lambda^1\alpha\Lambda^1\beta\Lambda^1\gamma(\alpha|\gamma)|(\beta|\gamma)$ $(\rho,\sigma,\tau \neq 0)$,

$[\Sigma_{0,0,\tau}] \equiv \Lambda^1\alpha\Lambda^1\beta\Lambda^1 z(\alpha|\lambda x.z)|\lambda y.\beta z$ $(\tau \neq 0)$.

(e) Construction of $[R_\sigma]$. Subcase $\sigma = 0$.

$$\epsilon(\delta,\alpha,\beta,\gamma) = \begin{cases} \alpha 0 & \text{if } \gamma 0 = 0, \\ \beta(\delta(\alpha,\beta,\lambda z(\gamma 0 \dot{-} 1)),\lambda z.\gamma 0 \dot{-} 1) & \\ \text{if } \gamma 0 > 0 . & \end{cases}$$

Then, by the recursion theorem analogue 1.9.16, we find δ_0 such that

$$\delta_0(\alpha,\beta,\gamma) = \begin{cases} \alpha 0 & \text{if } \gamma 0 = 0 \\ \beta(\delta_0(\alpha,\beta,\lambda z(\gamma 0 \dot{-} 1)),\lambda z.\gamma 0 \dot{-} 1)) & \text{if } \gamma 0 \neq 0, \end{cases}$$

and therefore we may take

$$[R_0] \equiv \Lambda^1 x\Lambda^1\beta\Lambda^0 y.\ \delta_0(\lambda z.x,\beta,\lambda z.y) .$$

Subcase $\sigma \neq 0$. We can find an ϵ such that

$$\epsilon|(\delta,\alpha,\beta,\gamma) = \begin{cases} \alpha, & \text{if } \gamma 0 = 0 \\ \beta|(\delta|(\alpha,\beta,\lambda z.\gamma 0 \dot{-} 1),\lambda z.\gamma 0 \dot{-} 1) & \text{if } \gamma 0 > 0, \end{cases}$$

and then by 1.9.16 a δ_1 such that

$$\delta_1|(\alpha, \beta, \gamma) = \begin{cases} \alpha & \text{if } \gamma 0 = 0, \\ \beta|(\delta_1|(\alpha, \beta, \lambda z.\gamma 0 \dot{-} 1), \lambda z.\gamma 0 \dot{-} 1) & \text{if } \gamma 0 \neq 0, \end{cases}$$

and then we may take

$$[R_\sigma] \equiv \Lambda^1\alpha\Lambda^1\beta\Lambda^1 y.\ \delta_1|(\alpha, \beta, \lambda z.y).$$

Note further that ICF is actually a model of $\underset{\sim}{\text{Int}} - \underset{\sim}{\text{HA}}^\omega$ (i.e. the theory obtained by adding to $\underset{\sim}{\text{N}} - \underset{\sim}{\text{HA}}^\omega$: $x^1 = y^1 \leftrightarrow \forall z(x^1 z = y^1 z)$).

2.6.3. <u>Theorem.</u> A model ICF possesses a modulus-of-continuity functional, i.e. in the model there is an object $\varphi_{mc} \in (2)(1)0$ such that

MC $\quad \bar{y}(\varphi_{mc}xy) = \bar{z}(\varphi_{mc}xy) \rightarrow xy = xz$.

This is provable in $\underset{\sim}{\text{EL}}$.

<u>Proof.</u> Let $\alpha \in V_2^1$; we define $\varphi[\alpha]$ such that

$$\varphi[\alpha]n = 0 \text{ if } \forall m \leq n \ (\alpha m = 0)$$
$$\varphi[\alpha]n = \text{lth}(m) + 1 \text{ if } m \leq n,\ \alpha m \neq 0,\ \forall m' < m\ (\alpha m' = 0).$$

Then put $[\varphi_{mc}] = \Lambda^1\alpha.\ \varphi[\alpha]$.

We have to show

(1) $\quad \bar{\beta}(\varphi[\alpha](\beta)) = \bar{\gamma}(\varphi[\alpha](\beta)) \rightarrow \alpha(\beta) \simeq \alpha(\gamma)$.

If $\varphi[\alpha](\beta) = x$, $\alpha(\bar{\beta}x) \neq 0$, $\alpha(\bar{\beta}y) = 0$ for $y < x$; hence (1) is immediate.

2.6.4. <u>Theorem.</u> If $\mathcal{U}$ satisfies $\underset{\sim}{\text{EL}} + \text{FAN}$, then we have in ICF($\mathcal{U}$) an object $\varphi_{uc} \in (2)0$ ("uc" for uniform continuity) such that

MUC $\quad \forall z^2 \forall x^1 \in B \forall y^1 \in B(\bar{x}(\varphi_{uc}z) = \bar{y}(\varphi_{uc}z) \rightarrow zx = zy)$,

where, as before, $x^1 \in B$ abbreviates $\forall u^0(xu \leq 1)$.

<u>Proof.</u> Let $\alpha \in V_2^1$, then $\forall\beta\exists x\ (\alpha(\bar{\beta}x) \neq 0)$, hence by FAN

$$\exists z \forall \beta \in B \exists x \leq z(\alpha(\bar{\beta}x) \neq 0).$$

We define

$$\varphi_{uc}(\alpha) = \min_z \forall\beta \in B\ \exists x \leq z(\alpha(\bar{\beta}x) \neq 0).$$

It remains to be shown that φ_{uc} is represented by $([\varphi_{uc}], (2)0)$ in ICF, for suitable $[\varphi_{uc}]$, as follows

Let $B_z = \{n \mid \text{lth}(n) = z \ \&\ \forall i < z((n)_i \leq 1)\}$, and put

$$[\varphi_{uc}](m) = \begin{cases} \min_z[z \leq \text{lth}(m) \ \&\ \forall n \in B_z\ \exists n' \leq n((m)_{n'} \neq 0)] + 1 \\ \qquad \text{if there is such a } z, \\ 0 \qquad \text{otherwise.} \end{cases}$$

$[\varphi_{uc}]$ is obviously recursive.

2.6.5. <u>The extensional model</u> ECF <u>of the hereditarily continuous functionals</u>.

Now we describe, relative to a universe of functions $\mathcal{U}$ satisfying $\underset{\sim}{EL}$, the model ECF($\mathcal{U}$) (in short: ECF) which is similar to HEO, but based on continuous function application. We introduce simultaneously domains W^1_σ and equivalence relations I^1_σ for all $\sigma \in \underset{\sim}{T}$, as follows.

$$x \in W^1_0 \equiv x = x, \quad I^1_0(x,y) \equiv (x = y)$$

$$\alpha \in W^1_{(0)0} \equiv \alpha = \alpha, \quad I^1_1(\alpha,\beta) \equiv (\alpha = \beta)$$

$$\alpha \in W^1_{(\sigma)0} \equiv \forall \beta \in W^1_\sigma \, \exists x(\alpha(\beta) \simeq x) \ \& \ \forall \gamma\gamma'(I^1_\sigma(\gamma,\gamma') \to \alpha(\gamma) \simeq \alpha(\gamma')) \qquad \sigma \neq 0$$

$$I^1_{(\sigma)0}(\alpha,\beta) \equiv \forall \gamma \in I^1_\sigma(\alpha(\gamma) \simeq \beta(\gamma)) \ \& \ W^1_{(\sigma)0}(\alpha) \ \& \ W^1_{(\sigma)0}(\beta), \quad \sigma \neq 0$$

$$\alpha \in W^1_{(\sigma)\tau} \equiv \forall \beta \in W^1_\sigma \, \exists \gamma \in W^1_\tau(\alpha|\beta \simeq \gamma) \ \& \ \forall \beta\beta'\delta\delta'(I^1_\sigma(\beta,\beta') \ \& \ \alpha|\beta \simeq \delta \ \& \ \alpha|\beta' \simeq \delta' \to I^1_\tau(\delta,\delta'))$$

$$I^1_{(\sigma)\tau}(\alpha,\beta) \equiv \forall \gamma \in W^1_\sigma(\alpha|\gamma \simeq \beta|\gamma) \ \& \ W^1_{(\sigma)\tau}(\alpha) \ \& \ W^1_{(\sigma)\tau}(\beta) \qquad (\sigma,\tau \neq 0)$$

$$\alpha \in W^1_{(0)\tau} \equiv \forall x \, \exists \gamma \in W^1_\tau(\alpha|\lambda y.x \simeq \gamma)$$

$$I^1_{(0)\tau}(\beta,\gamma) \equiv \forall x(\beta|\lambda y.x \simeq \gamma|\lambda y.x) \ \& \ W^1_{(0)\tau}(\beta) \ \& \ W^1_{(0)\tau}(\gamma) \qquad (\tau \neq 0).$$

The objects of type σ of the model ECF($\mathcal{U}$) (the extensional continuous functionals relative to the universe $\mathcal{U}$) are the pairs $(x,0)$, $x \in W^1_0$ if $\sigma = 0$, and (α,σ), $\alpha \in W^1_\sigma$ if $\sigma \neq 0$.

Equality at type σ is interpreted as I^1_σ; application and the other constants are interpreted as in ICF($\mathcal{U}$).

ECF is (provably in $\underset{\sim}{EL}$) a model for $\underset{\sim}{E}$-$\underset{\sim}{HA}^\omega$.

2.6.6. <u>Theorem</u>. (<u>Kreisel</u> 1962, lemma 7) If $\mathcal{U}$ satisfies $\underset{\sim}{EL}$ + FAN, then there is a "fan-functional" $\varphi_{uc} \in (2)0$ in ECF (provably in $\underset{\sim}{EL}$) such that

$$\forall z^2 \, \forall x^1 \in B \, \forall y^1 \in B(\bar{x}(\varphi_{uc} z) = \bar{y}(\varphi_{uc} z) \to zx = zy)$$

where, as before $x^1 \in B$ abbreviates $\forall u^0(xu \leq 1)$.

<u>Proof</u>. In the proof of 2.6.4 we must replace $[\varphi_{uc}]$ by $[\varphi_{uc}]'$ defined by (B_z as in 2.6.4)

$$[\varphi_{uc}]'(m) = \begin{cases} \min_y[y \leq z \ \& \ \forall n \in B_z \, \forall n' \in B_z \, \forall m'(m' \leq n \ \& \\ \quad \& \ m' \leq n' \ \& \ \mathrm{lth}(m') = y \to (m)[n] = (m)[n'] \neq 0)] + 1, \\ \text{where} \quad z = [\varphi_{uc}](m) \dot{-} 1, \ \text{if} \ [\varphi_{uc}](m) \neq 0 \\ 0 \quad \text{otherwise} \end{cases}$$

and where

$$(m)[n] = \begin{cases} (m)_y & \text{if} \ y = \min_x[x \leq n \ \& \ (m)_y \neq 0] \\ 0 & \text{otherwise}. \end{cases}$$

Obviously, if $\alpha \in W_2^1$, $\exists x([\varphi_{uc}]'(\bar{\alpha}x) \neq 0)$.
Also $[\varphi_{uc}]'m \leq [\varphi_{uc}]m$.
We have to show $[\varphi_{uc}]' \in W_2^1$, i.e. $I_2^1(\alpha,\alpha') \to [\varphi_{uc}]'(\alpha) = [\varphi_{uc}]'(\alpha')$.
Suppose $I_2^1(\alpha,\alpha')$, and let $\alpha \in m$, $\alpha' \in m'$ such that $[\varphi_{uc}]'m = x_1 + 1$, $[\varphi_{uc}]'m' = x_2 + 1$.
By the definition of $[\varphi_{uc}]'$, also $[\varphi_{uc}]m \neq 0$, $[\varphi_{uc}]m' \neq 0$. Now it is readily seen that $[\varphi_{uc}]((m)$ is not changed if we replace in the definition of $[\varphi_{uc}]'(m)$ z by $\max([\varphi_{uc}](m) \dot{-} 1, [\varphi_{uc}](m') \dot{-} 1)$, and similarly for $[\varphi_{uc}]'(m')$. Thus $[\varphi_{uc}]'(m) = [\varphi_{uc}]'(m')$.

2.6.7. <u>Theorem</u>. ECF does not contain a modulus-of-continuity functional φ_{mc}. (<u>Kreisel</u> 1962, after remark 10).
<u>Proof</u>. We shall show, more particularly, that in the model there is no φ such that

$$\forall y < \varphi x^2(\beta y = 0) \to x^2\beta = x^2(\lambda z.0) .$$

In order to show this, we consider functionals ψ_o, $\psi_{m,1}$ of type 2 such that

$$\psi_o = \lambda\alpha.0 ;$$
$$\psi_{m,1}\alpha = \begin{cases} 1 & \text{if } \forall y \leq m_o(\alpha y = 0) \ \& \ \alpha(m_o+1) > m \\ \psi_o\alpha = 0 & \text{in all other cases} \end{cases}$$

where $m_o = \varphi\psi_o$ for given φ.
Now we choose representatives α_o, α_1 of ψ_o, $\psi_{m,1}$ as follows.

(i) $\alpha_o n = 1$ if $n = (\overline{\lambda y.0})z * \langle Su\rangle * n'$ for suitable $z \leq m_o$, u, n'
or $n = (\overline{\lambda y.0})(m_o+1) * n'$ for suitable n'
$\alpha_o n = 0$ in all other cases

(ii) $\alpha_1 n = 2$ if $n = (\overline{\lambda y.0})(m_o+1) * \langle Sm + y\rangle * n'$ for suitable y, n'
$\alpha_1 n = 1$ if $n = (\overline{\lambda y.0})(m_o+1) * \langle u\rangle * n'$ for some $u \leq m$, and suitable n'
$\alpha_1 n = \alpha_o n$ in all other cases.

Note that $\alpha_o(\beta)$ is defined for all β, and in fact equal to 0; $\alpha_o\bar{\beta}(m_o+1) = 1$ for all β.
$\alpha_1(\beta)$ is always determined from $\bar{\beta}(m_o+2)$.

Now assume φ to be represented by γ; then we can find a v, $\alpha_o \in v$ such that $\gamma v \neq 0$; if we choose v sufficiently large, $\alpha_o n = 1$ for some $n < \text{lth}\, v$, $\text{lth}\, n > m_o+1$.
We put

$$m = \max\{(n)_{m_o+1} \mid n < \text{lth}(v) \ \& \ \text{lth}\, n > m_o+1\} .$$

It follows that $\alpha_1 \in v$ for this choice of m.

For let $n < lth\, v$. If $lth(n) \leq m_o+1$, $\alpha_1 n = \alpha_o n$; if $lth(n) > m_o+1$, $\alpha_1 n = 2$ is excluded, since then $(n)_{m_o+1}$ would be less than m, contradicting our choice of m. Hence either $\alpha_1 n = \alpha_o n$, or we are in the case where $n = (\overline{\lambda y.0})(m_o+1) * \langle u \rangle * n'$, $u \leq m$. But in this case $\alpha_o n = \alpha_1 n = 1$. Thus we have

$$\gamma(\alpha_o) = \gamma(\alpha_1) = m_o+1 .$$

On the other hand, for

$$\beta = \lambda x.(m+1)(1 \dot{-} |x - Sm_o|)$$

it follows that $\alpha_1(\lambda x.0) = 0$, $\alpha_1(\beta) = 1$, whereas $\forall y < \gamma(\alpha_1)(\beta y = 0) \rightarrow \alpha_1(\beta) = \alpha_1(\lambda x.0)$ becomes false, since (by $\gamma(\alpha_1) = m_o+1$) $\forall y < \gamma(\alpha_1)(\beta y = 0)$ holds.

2.6.8. We shall now pay some special attention to $ICF(\mathcal{R})$ and $ECF(\mathcal{R})$, where $\mathcal{R}$ is the universe of (total) recursive functions. The fact that $\mathcal{R}$ is closed under "recursive in" can now be established in $\mathbf{HA}$, and the metamathematics below can be established in $\mathbf{HA}$.

2.6.9. <u>Theorem</u> (<u>Kleene</u> & <u>Vesley</u> 1965, lemma 9.8 in § 9.3).
Let us use B_r for: α is recursive and $\forall x(\alpha x \leq 1)$. Then we can find a primitive recursive predicate Rx such that

(1) $\forall \alpha \in \mathcal{R} \exists x R(\bar{\alpha} x)$

(2) $\forall z \exists \alpha \in B_r \forall x \leq z \neg R(\bar{\alpha} x)$

and therefore

(3) $\neg \exists z \forall \alpha \in B_r \exists x \leq z R(\bar{\alpha} x)$.

<u>Proof</u>. Briefly, we define R so that the tree of unsecured sequences w.r.t. R has infinite branches, but no infinite recursive branches. We put

$$W_0(x,y) \equiv T(j_2 x, x, y) \ \&\ \forall z \leq y \neg T(j_1 x, x, z)$$
$$W_1(x,y) \equiv T(j_1 x, x, y) \ \&\ \forall z \leq y \neg T(j_2 x, x, z) .$$

Note that

$$\exists y W_0(x,y) \rightarrow \neg \exists y W_1(x,y), \quad \exists y W_1(x,y) \rightarrow \neg \exists y W_o(x,y) .$$

We put

$$W(i,x,y) \equiv W_{sg(i)}(x,y)$$

and define

$$Rx \equiv \exists u < lth(x) \exists y < lth(x) \dot{-} u\, W((x)_u, u, y) .$$

Then for any α with $\forall x(\alpha x \leq 1)$

(4) $\quad R\bar{\alpha}x \equiv \exists u{<}x\,\exists y < x \dot{-} u\ W_{sg(\alpha u)}(u,y)$.

Now let $\alpha \in \mathcal{R}$. Then there are n, n_o, n_1, $n = j(n_o,n_1)$ such that

(5) $$\begin{cases} \alpha u \neq 0 \leftrightarrow \exists y T(n_o,u,y) \leftrightarrow \exists y T(j_1 n,u,y) \\ \alpha u = 0 \leftrightarrow \exists y T(n_1,u,y) \leftrightarrow \exists y T(j_2 n,u,y) . \end{cases}$$

Case 1. Let $\alpha n \neq 0$. Then $\exists y T(j_1 n, n, y) \;\&\; \neg \exists y T(j_2 n, n, y)$, hence $\exists y W_1(n,y)$, so $\exists y W_{sg(\alpha n)}(n,y)$.

Case 2. Let $\alpha n = 0$; then similarly $\exists y W_{sg(\alpha n)}(n,y)$.

We now take n for u, $n+y+1$ for x in (1), where $W_{sg(\alpha n)}(n,y)$. We see that $R\bar{\alpha}x$ holds; thus we have established (1).

Now take any z and define

$$\varphi u = \begin{cases} 1 & \text{if } u < z \;\&\; \exists y < z \dot{-} u\ W_0(u,y) \\ 0 & \text{otherwise.} \end{cases}$$

Then $\varphi \in B_r$. Let $x \leq z$, assume $R\bar{\varphi}x$. Then there should be u, y such that $u < x \leq z$, $y < x \dot{-} u \leq z \dot{-} u$ such that $W_{\varphi u}(u,y)$. This leads to a contradiction:

Case (a). $\varphi u = 1$; then $W_1(u,y)$, so $\neg \exists y\, W_0(u,y)$, conflicting with the definition of φ;

Case (b). $\varphi u = 0$; similarly.

This establishes (2).

2.6.10. Corollary. ECF($\mathcal{R}$) and ICF($\mathcal{R}$) contain a type 2 object φ which is continuous, but not uniformly continuous on B_r, i.e.

$$\forall \alpha \exists x \forall \beta (\bar{\alpha}x = \bar{\beta}x \rightarrow \varphi\alpha = \varphi\beta)$$

but

$$\neg \exists z \forall \alpha \in B_r\ \forall \beta \in B_r (\bar{\alpha}z = \bar{\beta}z \rightarrow \varphi\alpha = \varphi\beta) .$$

Proof. We take for φ:

$$\varphi\alpha = n \equiv Rn \;\&\; \forall m (m < n \rightarrow \neg Rm) \;\&\; \alpha \in n .$$

The representation is by γ such that $\gamma m = n+1$ if $Rn \;\&\; \forall m(m < n \rightarrow \neg Rm)$ $\&\; n \leq m$, $\gamma m = 0$ in all other cases.

2.6.11. Definition of $[t]_{ICF}$, $[t]_{ECF}$. We define pseudo-terms $[t]_{ICF}$ ($\equiv [t]_{ECF}$), for terms $t \in \underset{\sim}{N}\text{-}\underset{\sim}{HA}^{\omega}$, built from constants of $\underset{\sim}{N}\text{-}\underset{\sim}{HA}^{\omega}$, variables of type 0 and variables of type 1. Let $v_o^o, v_1^o, v_2^o, \ldots$ be the type 0 variables, and let $v_o^1, v_1^1, v_2^1, \ldots$ be the type 1 variables of our theory.

For any constant c of $\underset{\sim}{N}\text{-}\underset{\sim}{HA}^{\omega}$, we put $[c]_{ICF} \equiv [c]$, where $[c]$ is as defined in 2.6.2.

Furthermore we put

$[v_i^o]_{ICF} \equiv v_i^o$,

$[v_i^1]_{ICF} \equiv v_i^1$,

and inductively

$[tt']_{ICF} \equiv [t]_{ICF} \mid [t']_{ICF}$, if t, tt' are both types $\neq 0$;

$[tt']_{ICF} \equiv [t]_{ICF}([t']_{ICF})$, if t' is of type $\neq 0$, $tt' \in 0$;

$[tt']_{ICF} \equiv [t]_{ICF}[t']_{ICF}$, if $t \in 1$, $t' \in 0$.

2.6.12. Theorem. Let t be a term of type 0, constructed from type 0 and type 1 variables (say $x_1,\ldots,x_n$, $\alpha_1,\ldots,\alpha_m$) and constants. Then

$$\underset{\sim}{N}\text{-}\underset{\sim}{HA}^{\omega} + AC_{oo} \vdash t \simeq y \leftrightarrow [t]_{ICF} \simeq [t]_{ECF} \simeq y .$$

Proof. The proof is very similar to the proof of theorem 2.4.14. We construct a code number $\ulcorner t \urcorner$ for pseudo-terms t, containing at most $\alpha_1,\ldots,\alpha_m$ free, as follows

$$\ulcorner \varphi \mid \varphi' \urcorner = j(1, j(\ulcorner\varphi\urcorner, \ulcorner\varphi'\urcorner))$$
$$\ulcorner \varphi(\varphi') \urcorner = j(2, j(\ulcorner\varphi\urcorner, \ulcorner\varphi'\urcorner))$$
$$\ulcorner \varphi t \urcorner = j(3, j(\ulcorner\varphi\urcorner, \ulcorner t'\urcorner))$$
$$\ulcorner [c]_{ICF} \urcorner = j(0, \tilde{c})$$ where $\tilde{c}$ is some code number for the constant

$$\ulcorner \alpha_i \urcorner = j(4+i, 0) .$$

As a result, $\ulcorner [t]_{ICF} \urcorner$ can be computed effectively.

We now construct a predicate $A(x, \beta, \alpha_1,\ldots,\alpha_m)$ such that, for $t \in 0$,

(1a) $$\underset{\sim}{N}\text{-}\underset{\sim}{HA}^{\omega} + AC_{oo} \vdash A(\ulcorner \bar{t} \urcorner, \beta, \alpha_1,\ldots,\alpha_m) \leftrightarrow t(x_1,\ldots,x_n,\alpha_1,\ldots,\alpha_m) \simeq \beta 0$$

and for $t \notin 0$

(1b) $$\underset{\sim}{N}\text{-}\underset{\sim}{HA}^{\omega} + AC_{oo} \vdash A(\ulcorner \bar{t} \urcorner, \beta, \alpha_1,\ldots,\alpha_m) \leftrightarrow t(x_1,\ldots,x_n,\alpha_1,\ldots,\alpha_m) \simeq \beta .$$

Here $\bar{t}$ abbreviates $t(\bar{x}_1,\ldots,\bar{x}_n,\alpha_1,\ldots,\alpha_m)$. For convenience we abbreviate $t(x_1,\ldots,x_n,\alpha_1,\ldots,\alpha_m)$, $A(x, \beta, \alpha_1,\ldots,\alpha_m)$ by $t(x_1,\ldots,x_n)$, $A(x,\beta)$ respectively.

$A(x,\beta)$ is defined such that

$$\begin{aligned} A(x,\beta) \leftrightarrow {} & [x = j(0,\tilde{0}) \;\&\; \beta 0 = 0] \vee B(x,\beta) \vee \\ & \vee [j_1 x = 1 \;\&\; \exists\gamma_1\gamma_2[A(j_1 j_2 x, \gamma_1) \;\&\; A(j_2 j_2 x, \gamma_2) \;\&\; \beta \simeq \gamma_1 \mid \gamma_2]] \vee \\ & \vee [j_1 x = 2 \;\&\; \exists\gamma_1\gamma_2[A(j_1 j_2 x, \gamma_1) \;\&\; A(j_2 j_2 x, \gamma_2) \;\&\; \beta 0 \simeq \gamma_1(\gamma_2)]] \vee \\ & \vee [j_1 x = 3 \;\&\; \exists\gamma_1\gamma_2[A(j_1 j_2 x, \gamma_1) \;\&\; A(j_1 j_2 x, \gamma_2) \;\&\; \beta 0 \simeq \gamma_1(\gamma_2 0)]] \vee \\ & \vee \exists i[x = j(4+1,0) \;\&\; \beta = \alpha_i] , \end{aligned}$$

where $B(x,\beta)$ is a disjunction of all clauses of the form

$$x = j(0,\tilde{c}) \;\&\; \beta = [c]$$

for c a constant of $\underset{\sim}{N}\text{-}\underset{\sim}{HA}^{\omega}$, $c \neq 0$.

(1) is now proved by induction on the complexity of t.
The remainder of the proof is entirely parallel to the proof of 2.4.14. The addition of AC_{00} to $\underset{\sim}{N}-\underset{\sim}{HA}^{\omega}$ was to insure $\underset{\sim}{EL}$ to be available, so that we could rely on Kleene's formalization of recursive functionals (Kleene 1969). In fact, $QF-AC_{00} \cap \mathcal{L}(\underset{\sim}{HA})$ would have been enough.

2.6.13. The equivalence between $ECF(\mathcal{R})$ and HEO.

For Kreisel's concept of hereditarily continuous functional (let us say ECF_K), relativized to recursive neighbourhood functions (i.e. $ECF_K(\mathcal{R})$) one can show classically that $ECF_K(\mathcal{R})$ and HEO represent the same class of functionals. (Full details are not in the published literature; published is only Kreisel 1959, Kreisel - Lacombe - Shoenfield 1959; more details, and improvements, are in the privately circulated Stanford report (Tait 1963, Harrison 1963) and the unpublished course notes Kreisel 1958 B.) Via the equivalence between Kreisel's notion of hereditarily continuous functional, and Kleene's notion of countable functional (Kleene 1959A; see Hinata and Tugué 1969) the result then also holds for Kleene's notion. In fact, it is technically even simpler to formalize the result directly for Kleene's countable functionals (hereditarily restricted to functionals with recursive associates), or for our model $ECF(\mathcal{R})$.

In subsections 14 - 19, 21 below the materials needed for the equivalence proof are given; the proof can actually be carried out in $\underset{\sim}{HA}+M_{PR}$. We have refrained from paraphrasing Kleene's proof of the recursive density theorem for the countable functionals, for our model $ECF(\mathcal{R})$; the proof, as it stands, is unperspicuous; it is to be hoped that adapting the more informative and perspicuous arguments of Tait 1963, Kreisel 1958 B to $ECF(\mathcal{R})$, and extending the discussion to impure types also, will yield a more satisfactory exposition in the future.

Because of the coding, for extensional functionals, of objects of arbitrary types of our type structure by objects of pure type (1.8.5 - 1.8.8) we may restrict our proof of equivalence to the pure types.

Kleene's treatment of the countable functionals has, when compared with our introduction of $ECF(\mathcal{U})$, a different conceptual background: Kleene has really a classical hierarchy of functionals in mind, coded by "associates" ("neighbourhood functions" in Kreisel's terminology); $ECF(\mathcal{U})$ is introduced by talking exclusively about associates or neighbourhood functions, not about the functionals themselves. A somewhat artificial aspect of Kleene's and our notion, as compared to Kreisel's, is the fixed ordering of neighbourhoods of higher type, as a consequence of the fact that only initial segments of the neighbourhood functions (which correspond to Kreisel's neighbourhoods) play a rôle. From a topological point of view, this is indeed arbitrary.

However, there are definite technical and heuristic advantages in using ECF instead of Kreisel's hereditarily continuous functionals; 1°) the similarities and differences between ECF and ICF are readily described (and the topological point of view, which is natural for ECF, is much less relevant for ICF); 2°) formalizing is simpler for ECF, we can use the available apparatus of Kleene 1969; 3°) there is a heuristically useful analogy between HRO, HEO on the one hand, and ICF, ECF on the other hand.

2.6.14. Definition. Let V be a set of total recursive functions. We put

$$V^* \equiv_{def} \{x \mid \{x\} \in V\}.$$

E(V) is to be the set of (gödelnumbers of) effective operations defined on V, i.e.

$$z \in E(V) \equiv_{def} \forall x \in V^* \ \forall y \in V^*(\{x\} = \{y\} \rightarrow \{z\}(x) = \{z\}(y)).$$

V is said to have a recursively dense basis, enumerated by Θ, if Θ is recursive and $\forall n(\Theta n \in V^*)$ and

$$\forall \alpha \in V \ \forall x \ \exists n(\overline{\{\Theta n\}}(x) = \bar{\alpha}x).$$

2.6.15. Theorem (Kreisel - Lacombe - Shoenfield 1959). In $\underset{\sim}{HA} + M_{PR}$, if V has a recursively dense basis enumerated by Θ, then there is a partial recursive modulus of continuity for E(V), i.e.

$$\text{KLS} \quad \begin{cases} \exists m \ \forall z \in E(V) \ \forall y \in V^*(!\{m\}(z,y) \ \& \ \forall u \in V^*(\overline{\{u\}}(\{m\}(z,y)) = \\ = \overline{\{y\}}(\{m\}(z,y)) \rightarrow \{z\}(u) = \{z\}(y)). \end{cases}$$

Proof. We define a partial recursive function $\{\varphi(k,y,z)\}$, φ a primitive recursive function of k, y, z as follows:

$$\{\varphi(k,y,z)\}(x) \simeq \begin{cases} \{y\}(x) \quad \text{if} \quad (\forall n \leq x)\neg(Tkkn \ \& \ Un = 0); \\ \{\Theta m\}(x) \quad \text{otherwise, where} \\ m \simeq \min_n[\{\Theta n\} \in \overline{\{y\}}(u) \ \& \ \{z\}(\Theta n) \neq \{z\}(y)] \\ \text{where} \quad u \simeq \min_v[Tkkv \ \& \ Uv = 0]. \end{cases}$$

We abbreviate φ(k, y, z) as p_k, and define ψ(y,z), ψ primitive recursive by

$$\{\psi(y,z)\}(k) \simeq \begin{cases} 0 \ \text{if} \ !\{z\}(p_k) \ \& \ \{z\}(p_k) = \{z\}(y) \\ 1 \ \text{if} \ !\{z\}(p_k) \ \& \ \{z\}(p_k) \neq \{z\}(y) \\ \text{undefined if} \ \neg ! \{z\}(p_k). \end{cases}$$

We abbreviate ψ(y,z) as q.

Assume $y \in V^*$, $z \in E(V)$. We first establish, using M_{PR}, that $\{q\}(q) = 0$. Suppose $\neg\{q\}(q) = 0$, then $\forall n \neg(Tqqn \ \& \ Un = 0)$, hence $\{p_k\} = \{y\}$; but then

$!\{z\}(p_k)$, hence $\neg\neg\{q\}(q) = 0$ therefore (M_{PR}) $\{q\}(q) = 0$.
Therefore the partial recursive function with index

$$m \equiv \Lambda zy \,.\, \min_v T(\psi(y,z),\psi(y,z),v)$$

is defined for all $z \in E(V)$, $y \in V^*$. Put $Mzy \equiv \{m\}(z,y)$. Note that

$$z \in E(V) \;\&\; y \in V^* \rightarrow U(Mzy) = \{q\}(q) = 0 \,.$$

Now suppose $z \in E(V)$, $y,u \in V^*$, $\overline{\{u\}}(Mzy) = \overline{\{y\}}(Mzy)$.
We will show

(1) $\qquad \{z\}(u) = \{z\}(y)$.

Since Θ enumerates a recursively dense basis for V, we can find a least n such that

$$\overline{\{\Theta n\}}(Mzy) = \overline{\{u\}}(Mzy) = \overline{\{y\}}(Mzy) \,.$$

First we show that

(2) $\qquad \{z\}(\Theta n) = \{z\}(y)$.

Suppose $\neg(2)$. Then since

$$n = \min_m(\{\Theta m\} \in \overline{\{y\}}(Mzy) \;\&\; \{z\}(\Theta m) \neq \{z\}(y)) \,,$$

we have by the definition of φ $\{p_q\} = \{\Theta n\}$, and hence $\{z\}(p_q) = \{z\}(\Theta n) \neq \{z\}(y)$. But $\{q\}(q) = 0$, so by the definition of ψ, $!\{z\}(p_q) \;\&\; \{z\}(p_q) = \{z\}(y)$. This contradicts $\neg(2)$, so $\neg\neg(2)$, hence (2) holds. Similarly

(3) $\qquad \{z\}(\Theta n) = \{z\}(u)$.

Thus we obtain (1) from (2), (3).

2.6.16. Remark. The proof as presented here is close to the proof in the original paper and in Rogers 1967 (pp. 362 - 364) ; the modification is from Beeson 1972. Another proof is in Gandy 1962, which we find less intuitive however.

2.6.17. Theorem (Refinement of KLS ; Kreisel, Lacombe, Shoenfield 1959). Let us say that f is a normal associate for $z \in E(V)$ if f is total, $\forall y \in V^* \, \exists n(f(\overline{\{y\}}(n)) = \{z\}(y) + 1))$, $\forall nm(fn \neq 0 \rightarrow f(n * m) = fn)$. Then we may strengthen the preceding theorem as follows : for each $z \in E(V)$, V a set of total recursive functions with a recursively dense base, there exists a recursive associate for z (provable in $\mathbf{HA} + M_{PR}$).
Proof. We define a recursive h such that (m as in 2.6.15) :

$$ht = \begin{cases} 1 + \{z\}(y) & \text{if } y < \text{lth}(t) = n,\ \exists u < n\ Tzyu, \\ & \exists u' < n(T(\{m\}(z), y, u') \ \&\ Uu' \leq n), \\ & \forall j \leq \{m\}(z,y)\ \exists k < n[Tyjk \ \&\ Uk = (t)_j], \\ & \exists n' < n[\forall j < \{m\}(z,y)(\{\Theta n'\}(j) = \{y\}(j))], \\ 0 & \text{otherwise.} \end{cases}$$

Obviously $ht \neq 0 \rightarrow ht = h(t * n)$.
Let $\{y\} \in V^*$. Then, for suitable u, u', k_j, n

(1) $Tzyu$

(2) $T(\{m\}(z), y, u')$

(3) $j < \{m\}(z,y) \rightarrow Tyjk_j$

(4) $\overline{\{\Theta n\}}(\{m\}(z,y)) = \overline{\{y\}}(\{m\}(z,y))$.

Now we can always find a t, $\{y\} \in t$ such that $\text{lth}\, t > \max(u, u', Uu', k_j$ for all $j < \{m\}(z,y), n)$; then $Uk_j = (t)_j$. For this t, $ht \neq 0$; moreover, from the definition we see $ht = 1 + \{z\}(y)$.

2.6.18. Corollary. If V is a set of total recursive functions with a recursively dense basis, then to any $\{z\}, z \in E(V)$ there is a partial recursive functional coinciding with $\{z\}$ on V.

2.6.19. Theorem (Existence of a recursively dense basis for ECF, provably in EL.) For each $j \geq 1$ there are a primitive recursive predicate $\text{Cons}_j(x,y)$ and a primitive recursive function $\lambda x.\text{ext}_j(x,y,z)$ such that (provably in HA)

(i) $\text{Cons}_j(x,y) \rightarrow \lambda z.\text{ext}_j(z,x,y) \in W_j^1 \ \&\ \lambda z.\text{ext}_j(z,y,x) \in W_j^1$

(ii) $\text{Cons}_j(x,y) \rightarrow \overline{(\lambda z.\text{ext}_j(z,x,y))}\text{lth}(x) = x \ \&\ \overline{(\lambda z.\text{ext}_j(z,y,x))}\text{lth}(y) = y$.

Corollary. If we put $\text{Cons}_j(x) \equiv_{def} \text{Cons}_j(x,x)$, and $\text{ext}_j(y,x) \equiv_{def} \text{ext}_j(y,x,x)$ then

$$\text{Cons}_j(x) \rightarrow \lambda y.\text{ext}_j(y,x) \in W_j^1 \ \&\ \lambda y.\text{ext}_j(y,x) \in x.$$

Intuitively, $\text{Cons}_j(x,y)$ may be read as: x, y represent neighbourhoods with a non-empty intersection; $\lambda z.\text{ext}_j(x,y)$ and $\lambda z.\text{ext}_j(y,x)$ are elements belonging to this intersection.

Proof. The proof is given in detail in Kleene 1959A, pp. 86 - 89; for Kreisel's related (and in fact equivalent) notion, the proof is in Kreisel 1959. Since the proof is not very informative and rather long, we shall omit it and refer to Kleene 1959A.

Kleene's set of associates of type j correspond to our W_j^1, but does not coincide with it; let us call it $\bar{W}_j^1$ for the time being. The principal distinction between $\bar{W}^1_j$ and W^1_j is that the elements of $\bar{W}^1_j$ must satisfy the additional condition

$$\alpha \in \bar{W}^1_j \ \& \ \beta \in \bar{W}^1_{j-1} \ \& \ \alpha(\bar{\beta}x) \neq 0 \rightarrow \alpha(\bar{\beta}(x+y)) = \alpha(\bar{\beta}x) .$$

If we wish to use W^1_j instead, we have to make the appropriate changes in Kleene's proof.

2.6.20. <u>Theorem</u>. $QF\text{-}AC_{\sigma\tau}$ (relative to the language of $\underset{\sim}{E}\text{-}\underset{\sim}{HA}^{\omega}$) holds for ECF (provably in $\underset{\sim}{EL}$).

<u>Proof</u>. We have to make use in an essential way of the recursive density theorem. Suppose

(1) $\quad \forall x^{\sigma} \exists y^{\tau} A(x,y)$,

A quantifier-free. Let Γ, Γ', Ω be the mappings effecting the reduction to pure types, described in 1.8.5 - 8.

We can find a functional φ_A such that

(2) $\quad \varphi_A\, x^{\Omega\sigma} y^{\Omega\tau} = 0 \leftrightarrow A(\Gamma' x^{\Omega\sigma}, \Gamma' y^{\Omega\tau})$.

φ_A a term of $\underset{\sim}{E}\text{-}\underset{\sim}{HA}^{\omega}$.

By the recursive density theorem there exists an element $\gamma \in W^1_{(0)\Omega\tau}$ enumerating a recursively dense base for the objects of type $\Omega\tau$ in ECF. Let $\alpha \in W^1_{(\Omega\sigma)(\Omega\tau)0}$ represent φ_A.

Let us define

$$\psi(\beta) \equiv j_1 \min_u [(\overline{\alpha|\beta})((\overline{\gamma|\lambda z.j_1u})j_2u) = 1] .$$

Note that for $\beta \in W^1_{\Omega\sigma}$, it follows that there exists a $\delta \in W^1_{\Omega\tau}$ such that $(\alpha|\beta)(\delta) = 0$ (by (1), (2)). Now this implies $(\alpha|\beta)\bar{\delta}z = 1$ for suitable z. Hence there is a u such that $\gamma|\lambda w.u \in \bar{\delta}z$; and therefore $\psi(\beta)$ is defined.

If we put

$$\epsilon \equiv \Lambda^1\beta.(\gamma|\ \lambda z.\ \psi(\beta)),$$

then readily

$$\forall\beta \in W^1_{\Omega\sigma}(\alpha|\beta)(\epsilon(\beta)) = 0$$

and therefore also in ECF

$$\exists z^{(\sigma)\tau}\, \forall x^{\sigma} A(x, zx) .$$

2.6.21. <u>Theorem</u>. HEO and ECF($\mathcal{R}$) are isomorphic w.r.t. extensional equality; i.e. we can find, for the pure types, $n \geq 1$, partial recursive functions g_n, h_n such that (if $W^*_m \equiv_{def} \{x \mid \{x\} \in W^1_m\}$, and $[x](y)$ denotes application according to ECF($\mathcal{R}$), i.e. $[x](y) \equiv_{def} \{x\}(\{y\})$), then for $x_n, x'_n \in W_n$, $x_{n-1} \in W_{n-1}$, $y_n, y'_n \in W^*_n$, $y_{n-1} \in W^*_{n-1}$:

$g_n(x_n) \in W_n^*$, $h_n(y_n) \in W_n$, and

$(i)_n$ $[g_n(x_n)](g_{n-1}(x_{n-1})) = \{x_n\}(x_{n-1})$

$(ii)_n$ $\{h_n(y_n)\}(h_{n-1}(y_{n-1})) = [y_n](y_{n-1})$

$(iii)_n$ $I_n(h_n g_n(x_n), x_n)$

$(iv)_n$ $I'_n(\{g_n h_n y_n\}, \{y_n\})$

$(v)_n$ $I_n(x_n, x'_n) \to I'_n(\{g_n x_n\}, \{g_n x'_n\})$

$(vi)_n$ $I'_n(\{y_n\}, \{y'_n\}) \to I_n(h_n y_n, h_n y'_n)$.

Proof. By induction on n. For $n = 1$ immediate.
Assume g_k, h_k to have been defined for $1 \leq k \leq n$, and $(i)_k - (vi)_k$ to have been proved for $k \leq n$.

Let h be the gödelnumber of a partial recursive function such that $\{h\}(x,y) \simeq [x](g_n(y))$. By the $s-m-n$-theorem there is a primitive rec. φ such that $\{h\}(x,y) \simeq \{\varphi(h,x)\}(y)$; take $h_{n+1}(x) \equiv \varphi(h,x)$. It follows that $\{h_{n+1}(y_{n+1})\}(x_n) \simeq [y_{n+1}](g_n(x_n))$, hence for $y_{n+1} \in W^*_{n+1}$, $h_{n+1}(y_{n+1}) \in W_{n+1}$.

Now let F be the operation on elements of W_n^1 defined by $F(\{y_n\}) = \{x_{n+1}\}(h_n(y_n))$. By $(vi)_n$, F is really an operation on W_n^1, not just on W_n^*. As we have proved, W_n^1 possesses a recursively dense basis, hence by theorem 2.6.18 we can extend F to a partial recursive functional F' such that

$$F'(\alpha) \simeq U \min_y T(z_o, \bar{\alpha}y)$$

(Kleene 1969, *34.1 on page 69). Now define

$$\varphi n = \begin{cases} U\, m+1 & \text{if } T(z_o, m)\ \&\ m \leq n \text{ for some } m \\ 0 & \text{otherwise.} \end{cases}$$

Then $F'(\alpha) \simeq x \leftrightarrow \exists y(\varphi(\bar{\alpha}y) = x+1)$.
φ is uniformly recursive in x_{n+1}, so φ is a function with gödelnumber $g_{n+1}(x_{n+1})$; this is the required g_{n+1}.
Then $[g_{n+1}(x_{n+1})](y_n) = F(\{y_n\}) = \{x_{n+1}\}(h_n(y_n))$.
Now the verification of $(i)_{n+1} - (vi)_{n+1}$ is completely routine:
$(v)_{n+1}$, $(vi)_{n+1}$ follow from $(v)_n$, $(vi)_n$; $(i)_{n+1}$ by $(iii)_n$, $(v)_n$;
$(ii)_{n+1}$ by $(v)_n$, $(iii)_{n+1}$: $\{h_{n+1}g_{n+1}(x_{n+1})\}(x_n) = \{h_{n+1}g_{n+1}(x_{n+1})\}(h_n g_n x_n)$ (by $(iii)_n$) $= [g_{n+1}(x_{n+1})](g_n(x_n))$ (by $(ii)_n$) $= \{x_{n+1}\}(x_n)$ (by $(i)_n$);
so $I_n(h_{n+1}g_{n+1}(x_{n+1}), x_{n+1})$. $(iv)_{n+1}$ similarly.

Remark. For Kreisel's definition of ECF($\mathcal{R}$) the proof was given in detail in Harrison 1963; here the situation is even simpler.

2.6.22. The models $ECF^r(\mathcal{U})$, $ICF^r(\mathcal{U})$.

Let $\mathcal{U}$ be a universe of functions which is a model for EL. A kind of hybrid between $ICF(\mathcal{R})$ and $ICF(\mathcal{U})$, and similarly between $ECF(\mathcal{R})$ and $ECF(\mathcal{U})$ is obtained by defining the set of objects of type σ as the recursive elements of the set of objects of type σ in $ICF(\mathcal{U})$, $ECF(\mathcal{U})$ respectively. Let us call the resulting models $ICF^r(\mathcal{U})$, $ECF^r(\mathcal{U})$ respectively. More formally:

(a). $V^r_\sigma \equiv V^1_\sigma \cap \mathcal{R}$, $W^r_\sigma \equiv W^1_\sigma \cap \mathcal{R}$, I^r_σ is the restriction of I^1_σ to $W^r_\sigma \times W^r_\sigma$.

(b). The objects of type σ in $ICF^r(\mathcal{U})$, $ECF^r(\mathcal{U})$ are now the pairs (α,σ), $\alpha \in V^r_\sigma$, and (α,σ), $\alpha \in W^r_\sigma$ respectively (α to be replaced by a number for $\sigma = 0$).

It is easy to see that $ICF^r(\mathcal{U})$, $ECF^r(\mathcal{U})$ are again models of $N\text{-}HA^\omega$, $E\text{-}HA^\omega$ respectively. Especially interesting is the case where $\mathcal{U}$ is a universe satisfying bar induction; hybrid models for such universes we shall often simply denote by ICF^r, ECF^r.

The G-realizability of Moschovakis 1971 may be viewed as "abstract" modified realizability relative to ICF^r (cf. 3.4.2) by interpreting the objects of finite type as elements of ICF^r (cf. 3.4.15).

2.6.23. A variant of ICF and ECF.

Sometimes it is a disadvantage that the definition of the application operation is not uniform in all types for ICF, ECF. This disadvantage can be removed by considering the following variants ICF^*, ECF^* of ICF, ECF. We redefine the species V^1_σ by

$$\alpha \in V^1_0 \equiv (\alpha = \lambda x.\alpha 0)$$

$$\alpha \in V^1_1 \equiv \exists \beta(\alpha = \Phi\beta)\text{, where } \Phi \text{ is defined as follows:}$$

$$(\Phi\beta)0 = 0, \quad (\Phi\beta)(\hat{x}) = 0$$

$$(\Phi\beta)(\hat{x} * \hat{y} * n) = \begin{cases} \beta y & \text{if } lth(n) > y \\ 0 & \text{if } lth(n) \le y. \end{cases}$$

Note that $(\Phi\beta) \mid \lambda z.x = \lambda z.\beta x$.

$$\alpha \in V^1_{(\sigma)\tau} \equiv \forall \beta \in V^1_\sigma \exists \gamma \in V^1_\tau (\alpha|\beta \simeq \gamma)\text{, for } \sigma \neq 0 \text{ or } \tau \neq 0.$$

Application is now always interpreted as $\cdot|\cdot$.

The interpretation of the constants is then adapted as follows:

(a) $[0] \equiv \lambda x.0$

(b) $[S] \equiv \Phi S$, so that the numeral $\bar{n}$ is represented by $\lambda x.\bar{n}$.
(In general, a function β is represented by $\Phi\beta$.)

(c) $[\Pi_{\sigma,\tau}] = \Lambda^1\alpha\Lambda^1\beta.\alpha$
(d) $[\Sigma_{\rho,\sigma,\tau}] = \Lambda^1\alpha\Lambda^1\beta\Lambda^1\gamma.(\alpha|\gamma)|(\beta|\gamma)$
(e) $[R_\sigma]$ as for the subcase $\sigma \neq 0$ in 2.6.2 (but now for all σ).

Equality between terms of type σ is interpreted as

$$(\alpha,\sigma) = (\beta,\sigma) \longleftrightarrow \forall x(\alpha x = \beta x) .$$

As a pleasant corollary we have :

$$(\Phi\alpha,1) = (\Phi\beta,1) \longleftrightarrow \forall x(\alpha x = \beta x) .$$

2.6.24. Remark. The latter pleasant property would not hold, if we would have bluntly defined V_1^1 by

$$\alpha \in V_1^1 \equiv_{def} \forall\beta \in V_0^1 \exists\gamma \in V_0^1(\alpha|\beta \simeq \gamma)$$

since then each function may be represented by many different elements of V_1^1.

The redefined model, ICF^*, is not isomorphic to ICF, for types more complex than 0, 1 ; i.e. to a single element of V_σ^1 in ECF there usually corresponds an infinity of elements of V_σ^1 in ECF^*, since the predicate $\alpha|\beta \simeq \gamma$, for given β, γ does not determine α uniquely.
ECF^* is the analogous variant of ECF, with the obvious definitions.

2.6.25. Pairing operators in ICF, ECF, ICF^*, ECF^*.

In case we wish to extend our type structure with cartesian products the obvious interpretation of the pairing operators would be given by :

$$[D_{0,0}] \equiv \lambda xy.j(x,y), \quad [D'_{0,0}] \equiv j_1, \quad [D''_{0,0}] \equiv j_2$$

$$[D_{0,1}] \equiv \Lambda^0 x.\Lambda^1\alpha(\lambda z.j(x,\alpha z)), \quad [D'_{0,1}] \equiv \Lambda^0\alpha.j_1\alpha 0,$$
$$[D''_{0,1}] \equiv \Lambda^1\alpha.\lambda z.j_2\alpha z .$$

$$[D_{1,0}] \equiv \Lambda^1\alpha\Lambda^1 x(\lambda z.j(\alpha z,x)) ; \quad [D'_{1,0}] \equiv \Lambda^1\alpha.\lambda z.j_1\alpha z,$$
$$[D''_{1,0}] \equiv \Lambda^1\alpha.j_2\alpha 0 .$$

$$[D_{\sigma,\tau}] \equiv \Lambda^1\alpha\Lambda^1\beta.\lambda z.j(\alpha z,\beta z), \quad [D'_{\sigma,\tau}] \equiv \Lambda^1\alpha.\lambda z.j_1\alpha z,$$
$$[D''_{\sigma,\tau}] \equiv \Lambda^1\alpha.\lambda z.j_2\alpha z , \quad \text{for } \sigma,\tau \neq 0 .$$

For ICF^*, ECF^* we define $[D_{\sigma,\tau}]$, $[D'_{\sigma,\tau}]$, $[D''_{\sigma,\tau}]$ as for ICF, ECF in the case $\sigma,\tau \neq 0$, but now for all σ,τ.

2.6.26. The systems $\underset{\sim}{ICF}^-$, $\underset{\sim}{ICF}$.

Similar to $\underset{\sim}{HRO}^-$, $\underset{\sim}{HRO}$ we can describe two extensions of $\underset{\sim}{N}-\underset{\sim}{HA}^\omega$ which express that the objects of finite type are intensional continuous functionals, or precisely the intensional continuous functions respectively, by adding constants $\Phi^* \in (\sigma)1$, for $\sigma \in \underset{\sim}{T}$, $\Phi^{*\prime}_{\sigma,\tau} \in ((\sigma)\tau)(\sigma)1$ for all $\sigma,\tau \in \underset{\sim}{T}$

and axioms

G*1 $\Phi^*_0\ x^0 = \lambda y.x^0,\ \Phi^*_1 x^1 = \Phi x^1$ (Φ as in 2.6.23)

G*2 $\Phi^*_\sigma\ x^\sigma = \Phi^*_\sigma\ y^\sigma \leftrightarrow x^\sigma = y^\sigma$

G*3 $\Phi^*_{(\sigma)\tau} x^{(\sigma)\tau} \mid \Phi^*_\sigma\ y^\sigma = \Phi^*_\tau\ x^{(\sigma)\tau} y^\sigma$

G*4 $\Phi^{*\,\prime}_{\sigma,\tau} y^{(\sigma)\tau} z^\sigma =$

$$\lambda x.(\min_u[\Phi^*_{(\sigma)\tau} y^{(\sigma)\tau}(\hat{x} * (\overline{\Phi^*_\sigma z^\sigma})u) \neq 0])$$

G*5 $\forall \alpha \in V^1_\sigma\ \exists y^\sigma(\Phi^*_\sigma y^\sigma = \alpha)$

where V^1_σ is defined in 2.6.23. Using the original definition of V^1_σ, the definition becomes slightly more complicated, because we have to distinguish more cases for the axioms.

$\underset{\sim}{N}$ - $\underset{\sim}{HA}^\omega$ + G*1 - G*4 is $\underset{\sim}{ICF}^-$, $\underset{\sim}{N}$ - $\underset{\sim}{HA}^\omega$ + G*1 - G*5 is $\underset{\sim}{ICF}$.

It is also possible to define systems $\underset{\sim}{ECF}^-$, $\underset{\sim}{ECF}$, but they are without practical interest.

§ 7. Extensionality and continuity in $\underset{\sim}{N}-\underset{\sim}{HA}^{\omega}$.

2.7.1. In this section we bring together some results on continuity and extensionality rules and axioms in $\underset{\sim}{N}-\underset{\sim}{HA}^{\omega}$, as applications of our results on computability and the models of $\underset{\sim}{N}-\underset{\sim}{HA}^{\omega}$.

2.7.2. Extensionality and hereditary extensionality.

Extensional equality $=_e$ for type $\sigma \equiv (\sigma_1)\dots(\sigma_n)0$ is simply defined by

$$x^0 =_e y^0 \equiv_{def} x^0 = y^0$$
$$x^\sigma =_e y^\sigma \equiv_{def} \forall z_1^{\sigma_1} \dots z_n^{\sigma_n}(xz_1\dots z_n = yz_1\dots z_n)\ .$$

Hereditary extensional equality $\approx$ is defined over the type structure by

$$x^0 \approx y^0 \equiv_{def} x^0 = y^0$$
$$x^{(\sigma)\tau} \approx y^{(\sigma)\tau} \equiv_{def} \forall x_1^\sigma y_1^\sigma\ (x_1 \approx y_1 \rightarrow xx_1 \approx yy_1)\ .$$

Note that

$$x^0 \approx y^0 \leftrightarrow x^0 =_e y^0 \leftrightarrow x^0 = y^0$$

and

$$x^1 \approx y^1 \leftrightarrow \forall z^0 u^0(z^0 = u^0 \rightarrow x^1 z^0 = y^1 u^0)$$
$$\leftrightarrow \forall z^0(x^1 z^0 = y^1 z^0)$$
$$\leftrightarrow x^1 =_e y^1$$

(and similarly $x^\sigma \approx y^\sigma \leftrightarrow x^\sigma =_e y^\sigma$ if $\sigma \equiv (0)(0)\dots(0)0$).

The axiom of extensionality states

$$\mathrm{EXT}_{\sigma,\tau} \qquad x^\sigma =_e y^\sigma \rightarrow z^{(\sigma)\tau}x^\sigma =_e z^{(\sigma)\tau}y^\sigma .$$

The corresponding axiom of hereditary extensionality

(1) $$x^\sigma \approx y^\sigma \rightarrow\ z^{(\sigma)\tau}x^\sigma \approx z^{(\sigma)\tau}y^\sigma$$

for all σ, τ is equivalent to $\mathrm{EXT}_{\sigma,\tau}$ for all σ, τ.

Assume $\mathrm{EXT}_{\sigma,\tau}$ for all σ, τ, then we prove by induction over the type structure that $x^\sigma =_e y^\sigma \leftrightarrow x^\sigma \approx y^\sigma$.

Let $x^\sigma =_e y^\sigma \leftrightarrow x^\sigma \approx y^\sigma$, $x^\tau =_e y^\tau \leftrightarrow x^\tau \approx y^\tau$.

Then

$$z^{(\sigma)\tau} =_e u^{(\sigma)\tau} \leftrightarrow \forall x^\sigma(zx =_e ux) \leftrightarrow \forall x^\sigma(zx \approx ux)\ .$$

Also, $x^\sigma \approx y^\sigma \leftrightarrow x^\sigma =_e y^\sigma \rightarrow u^{(\sigma)\tau}x^\sigma =_e u^{(\sigma)\tau}y^\sigma \leftrightarrow ux \approx uy$,

hence $z^{(\sigma)\tau} =_e u^{(\sigma)\tau} \rightarrow \forall x^\sigma y^\sigma(x \approx y \rightarrow zx \approx uy) \leftrightarrow z \approx u$

and since $x =_e x$, so $x \approx x$, $z \approx u$ implies $\forall x(zx =_e ux)$.

Therefore (1) is implied by $EXT_{\sigma,\tau}$.
Conversely, assume (1) for all σ, τ. We note that $x^\sigma \approx x^\sigma$ for all σ; for it holds for $\sigma = 0$, and if $\sigma = (\tau)\rho$, then
$x^{(\tau)\rho} \approx x^{(\tau)\rho} \leftrightarrow \forall z^\tau z_1^\tau (z \approx z_1 \to x^{(\tau)\rho} z \approx x^{(\tau)\rho} z_1)$; and the right hand side of this equivalence holds because of (1).
Now assume $x^\sigma \approx y^\sigma \leftrightarrow x^\sigma =_e y^\sigma$, $x^\tau \approx y^\tau \leftrightarrow x^\tau =_e y^\tau$.
Then

$$x^{(\sigma)\tau} =_e y^{(\sigma)\tau} \to \forall z^\sigma (xz =_e yz) \to \forall z^\sigma (xz \approx yz) .$$

Also, if $z \approx u$, then $yz \approx yu$ (since $y \approx y$); therefore $\forall z^\sigma u^\sigma (z \approx u \to xz \approx yu)$, which is equivalent to $x \approx y$.
Conversely, $x^{(\sigma)\tau} \approx y^{(\sigma)\tau} \leftrightarrow \forall z_1 z_2 (z_1 \approx z_2 \to xz_1 \approx yz_2)$

$$\to \forall z_1 (xz_1 \approx yz_1) \to \forall z_1 (xz_1 =_e yz_1)$$

so $x^{(\sigma)\tau} =_e y^{(\sigma)\tau}$.
Therefore also (1) implies $EXT_{\sigma,\tau}$.

2.7.3. Theorem. In $\underset{\sim}{N}$ - $\underset{\sim}{HA}^\omega$, for any term t built from constants, type 1 and type 0 variables, $\underset{\sim}{N}$ - $\underset{\sim}{HA}^\omega \vdash t \approx t$.
Proof. (W.A. Howard)
We note that if $t_i \approx t_i'$ $(1 \leq i \leq n)$, and t is constructed by application from $t_1, \ldots, t_n$, and t' is constructed as t, but everywhere with t_i' instead of t_i, then $t \approx t'$.
One readily verifies, with the help of this remark, that $0 \approx 0$, $x^0 \approx x^0$, $x^1 \approx x^1$, $S \approx S$, $\Pi \approx \Pi$, $\Sigma \approx \Sigma$. $R \approx R$ is established proving by induction on z^0: $\forall x x_1 y y_1 z_1^0 (x \approx x_1 \ \& \ y \approx y_1 \ \& \ z^0 = z_1^0 \to Rxyz^0 \approx Rx_1 y_1 z_1^0)$.

2.7.4. Corollary.
(i) $\underset{\sim}{N}$ - $\underset{\sim}{HA}^\omega \vdash x^\sigma \approx y^\sigma \to t^{(\sigma)\tau}[x^\sigma] \approx t^{(\sigma)\tau}[y^\sigma]$
where $t[z^\sigma]$ is any term constructed from type 0 variables, type 1 variables, z^σ, and constants.
(ii) $\underset{\sim}{N}$ - $\underset{\sim}{HA}^\omega \vdash \forall x(\alpha x = \beta x) \to t[\beta] \approx t[\beta]$
and hence in particular, if t is of type 0 or 1:
$\underset{\sim}{N}$ - $\underset{\sim}{HA}^\omega \vdash \forall x(\alpha x = \beta x) \to t[\alpha] =_e t[\beta]$.
Proof. (i) is immediate; (ii) follows by our remark that $\alpha =_e \beta \leftrightarrow \alpha \approx \beta$.

2.7.5. Theorem. We have the following derived rules of extensionality

$$\underset{\sim}{N} - \underset{\sim}{HA}^\omega \vdash t =_e s \Rightarrow \underset{\sim}{N} - \underset{\sim}{HA}^\omega \vdash F[t] =_e F[s]$$

where t, s, $F[t]$, $F[s]$ are terms built from constants and variables of types 0 and 1, and where $F[t]$, $F[s]$ are of type 0, 1 or 2.

First proof. Note that the case where $F[t]$ is of type 0 includes the other cases; for if $F[t]$ is of type 2, then $F[t] =_e F[s]$ is equivalent to $F[t]\alpha = F[s]\alpha$, where α is a new type 1 variable not occurring free in $F[t]$, $F[s]$. Similarly if $F[t]$ is of type 1. So we may restrict our attention to the case where $F[t] \in 0$. Now we use theorem 2.6.12 on ECF, and a conservative extension result to be proved in 3.6.6(ii), implying AC to be conservative over $\underset{\sim}{N}-\underset{\sim}{HA}^{\omega}$ for universally quantified equations between terms of type 0.

If $\vdash t =_e s$, then t, s are represented in ECF by extensionally equal functionals; and since ECF is extensional, it follows that

$\underset{\sim}{N}-\underset{\sim}{HA}^{\omega} \vdash [F[t]]_{ECF} \simeq [F[s]]_{ECF}$.

Combining this with (2.6.12)

$$\underset{\sim}{N}-\underset{\sim}{HA}^{\omega} \vdash [F[t]]_{ECF} \simeq F[t]$$
$$\underset{\sim}{N}-\underset{\sim}{HA}^{\omega} \vdash [F[s]]_{ECF} \simeq F[s]$$

we obtain

$$\underset{\sim}{N}-\underset{\sim}{HA}^{\omega} \vdash F[t] = F[s].$$

Remark. For the case where $F[t]$ is of type 0 or 1, and $t, s, F[t]$ do not contain variables of type 1 , we may also use theorem 2.4.14, using HEO instead of ECF; we do not need the conservative extension result of 3.6.6(ii) in this case.

Kreisel's notes (Kreisel 1971A) contain a sketch for another proof, not appealing to the uniform faithfulness (2.4.14 or 2.6.12) but only to faithfulness of HEO w.r.t. numerals and using partial reflection principles instead. We failed to find a satisfactory reconstruction of this proof.

Second proof. (For reducing the case where F is of type 2 to the case where F is of type 1.) This argument may be combined with one of the other arguments for the case where F is of type 1.

Let us, for simplicity, once again restrict our attention to the case t, s, $F[t]$, $F[s]$ closed, $F[t]$ of type 2. $F[t]$, $F[s]$ represent constant functionals of type 2, which have a provable modulus of continuity by 2.7.8.

Hence there are terms t_1, s_1 of type 2 such that

$$(1) \quad \begin{cases} \underset{\sim}{N}-\underset{\sim}{HA}^{\omega} \vdash \bar{\alpha}(t_1\alpha) = \bar{\beta}(t_1\alpha) \to F[t]\alpha = F[t]\beta \\ \underset{\sim}{N}-\underset{\sim}{HA}^{\omega} \vdash \bar{\alpha}(s_1\alpha) = \bar{\beta}(s_1\alpha) \to F[s]\alpha = F[s]\beta . \end{cases}$$

Now let t_2 be a term of type $(0)1$, such that t_2n is defined by

$$\begin{cases} t_2nx = (n)_x & \text{for } x < lth(n) \\ t_2nx = 0 & \text{elsewhere.} \end{cases}$$

Then obviously, by (1)

$$\underset{\sim}{N}-\underset{\sim}{HA}^{\omega} \vdash \forall\alpha(F[t]\alpha = F[s]\alpha) \longleftrightarrow \forall x(F[t](t_2x) = F[s](t_2x)),$$

hence $\underset{\sim}{N}-\underset{\sim}{HA}^{\omega} \vdash F[t]\alpha = F[s]\alpha$ follows from

$$\underset{\sim}{N}-\underset{\sim}{HA}^{\omega} \vdash F'[t]x = F'[s]x$$

where $F'[z] = \lambda x.F[z](t_2x)$.

2.7.6. Counterexample (H.P. Barendregt). If in 2.7.5 we remove the restriction that $F[t]$, $F[s]$ contain variables of type 0 or 1 only, we can give a counterexample to

$$\underset{\sim}{N}-\underset{\sim}{HA}^{\omega} \vdash t =_e s \Rightarrow \underset{\sim}{N}-\underset{\sim}{HA}^{\omega} \vdash F[t] =_e F[s]$$

with $F[t] \in 0$, t, s closed.

Take e.g. $t \equiv \lambda x.0$, $s \equiv \lambda x.x \dot{-} x$ (where $x \dot{-} x$ may be supposed to be defined as $Rx(\lambda uv.prd(u))x \equiv Rx(\lambda uv.(R0(\lambda u'v'.v')u))x)$, and let $F \equiv \lambda y^1.x^2y^1$. Then $Ft = x^2t$, $Fs = x^2s$.
$\underset{\sim}{N}-\underset{\sim}{HA}^{\omega} \vdash t = s$ obviously holds; but $\underset{\sim}{N}-\underset{\sim}{HA}^{\omega} \not\vdash Ft = Fs$, for then $x^2t = x^2s$ would have to hold in all versions of HRO, i.e. if we take $(\Lambda x.x, 2)$ for x^2, then $[t]_{HRO^*} \simeq [s]_{HRO^*}$ in the HRO-version HRO* into which the term model CTNF' can be embedded (2.5.5). Since t, s have different normal forms, this is obviously false.

2.7.7. Counterexample (R. Statman). We can find a closed F of type 3, such that

$$\underset{\sim}{N}-\underset{\sim}{HA}^{\omega} \not\vdash x^2 =_e y^2 \rightarrow Fx^2 = Fy^2 .$$

Take F to be $\lambda x^2.x^2[\lambda z^0.(x^2(\lambda w^0.z^0))]$, and choose two elements $x_1^2 = (\bar{n}, 2)$, $x_2^2 = (\bar{m}, 2)$ of the version of HRO described in 2.5.5, such that $\bar{n}, \bar{m}$ are distinct indices of the same 1-1 total, recursive function. F is represented by $(\bar{p}, 3)$ for suitable $\bar{p}$. Since Fx^2 is in full

$$x^2[\Sigma(\Pi x^2)(\Sigma(\Pi\Pi)(\Pi z^0))]$$

it is obvious that since the version of HRO considered satisfies the equality axioms IE_1, that $\Sigma(\Pi x_1^2)(\Sigma(\Pi\Pi)(\Pi z^0))$ and $\Sigma(\Pi x_2^2)(\Sigma(\Pi\Pi)(\Pi z^0))$ in the model are different, hence also $Fx_1^2 \neq Fx_2^2$ in the model.

A similar counterexample has been given by H.P. Barendregt.

2.7.8. Theorem. Every closed term $t \in 2$ of $\underset{\sim}{N}-\underset{\sim}{HA}^{\omega}$ possesses a provable modulus of continuity in $\underset{\sim}{N}-\underset{\sim}{HA}^{\omega}$, i.e. a closed term $t' \in 2$ such that

$$N-HA^{\omega} \vdash \bar{\alpha}(t'\alpha) = \bar{\beta}(t'\alpha) \rightarrow t\alpha = t\beta$$

(α, β variables of type 1).
Proof. From 2.3.13 we know that

(1) $\underset{\sim}{N}-\underset{\sim}{HA}^{\omega} \vdash t\alpha = y \longleftrightarrow SRED(\alpha, \ulcorner tx^1 \urcorner, \ulcorner \bar{y} \urcorner)$

where x^1 is the type 1 variable to which α is assigned in the reduction process (cf. 2.3.8).

Also, because of the derivability of computability for terms of bounded type level (cf. 2.3.11)

(2) $\underset{\sim}{N}-\underset{\sim}{HA}^{\omega} \vdash \forall\alpha \exists! y\, SRED(\alpha, \ulcorner tx^1 \urcorner, \ulcorner \bar{y} \urcorner)$.

Now $SRED(\alpha, x, y)$ may be written as $\exists z\, SR(\alpha, x, y, z)$, where $SR(\alpha, x, y, z)$ expresses : z is the (number of a) standard reduction sequence relative to α of x to y .

Note that f , defined by

$$fn = \max\{n_i \mid 1 \leq i \leq k\} + 1$$

where $n_1, \ldots, n_k$ is a list of all numbers for which $\alpha\bar{n}_i$ contr $\overline{\alpha n_i}$ has been used in the reduction sequence n , may be taken to be a primitive recursive function of n .

Also

(3) $\underset{\sim}{N}-\underset{\sim}{HA}^{\omega} \vdash \bar{\alpha}(fn) = \bar{\beta}(fn) \rightarrow (SR(\alpha, \ulcorner t \urcorner, \ulcorner s \urcorner, n) \longleftrightarrow SR(\beta, \ulcorner t \urcorner, \ulcorner s \urcorner, n))$.

Combining (1), (2), (3) :

$$\underset{\sim}{N}-\underset{\sim}{HA}^{\omega} \vdash \forall\alpha \exists z \forall\beta(\bar{\alpha}z = \bar{\beta}z \rightarrow t\alpha = t\beta) .$$

Now, using a result from the next chapter (closure under a rule of choice, see 3.7.1(ii)) we find that there must be a $t' \in z$, t' a closed term of $\underset{\sim}{N}-\underset{\sim}{HA}^{\omega}$, such that

$$\underset{\sim}{N}-\underset{\sim}{HA}^{\omega} \vdash \forall\beta(\bar{\alpha}(t'\alpha) = \bar{\beta}(t'\alpha) \rightarrow t\alpha = t\beta) .$$

2.7.9. <u>Product topology</u>.

Already at type 3 , the functionals represented by closed terms of $\underset{\sim}{N}-\underset{\sim}{HA}^{\omega}$ are not necessarily continuous w.r.t. the product topology. Take for example

$$F \equiv \lambda z^2 . z^2[\lambda x^0 . z^2(\Pi_{0,0} x^0)] .$$

F is discontinuous w.r.t. the product topology at $\lambda\alpha.0$, since given $\alpha_1, \ldots, \alpha_k$ we can find a constant t^2 such that

$$t^2\alpha_i = 0 \quad (1 \leq i \leq k) , \quad Ft^2 \neq 0 ,$$

while $F(\lambda\alpha.0) = 0$. t^2 is defined as follows :

$$t^2\alpha \begin{cases} = 0 \text{ if } \exists i\, (1 \leq i \leq k \,\&\, \bar{\alpha}_i k = \bar{\alpha} k) \\ = m+1 \text{ otherwise, where } m = \max\{\alpha_i(y) \mid 1 \leq i \leq k,\ 0 \leq y < k\} . \end{cases}$$

Now $Ft^2 \neq 0$; for $(\overline{\Pi_{0,0}0})k, \ldots, (\overline{\Pi_{0,0}k})k$ are all distinct, hence one of them, say $(\overline{\Pi_{0,0}k_0})k$ $(0 \leq k_0 \leq k)$ is distinct from all $\bar{\alpha}_1 k, \ldots, \bar{\alpha}_k k$, and therefore $t^2(\Pi_{0,0}k_0) = m+1$; but then $(\lambda x^0.t^2(\Pi_{0,0}x))k$ differs from all of $\bar{\alpha}_1 k, \ldots, \bar{\alpha}_k k$, and thus Ft^2 takes the value $m+1$.

2.7.10. "Floating product topology". In Kreisel 1971A, Kreisel noted the following continuity property for functionals of $\underset{\sim}{N}-\underset{\sim}{HA}^{\omega}$, at arguments definable in $\underset{\sim}{N}-\underset{\sim}{HA}^{\omega}$, for types $((\sigma)0)0$. Let $F \in ((\sigma)0)0$, $t \in (\sigma)0$ be closed terms of $\underset{\sim}{N}-\underset{\sim}{HA}^{\omega}$, then we can find a finite number of terms $t_1[z^{(\sigma)0}], \ldots, t_n[z^{(\sigma)0}]$ of type σ such that

$$\forall z^{(\sigma)0}(\bigwedge_{1 \leq i \leq n} (tt_i[t] = z[t_i z] \rightarrow Ft = Fz).$$

The proof is an elaboration of Kreisel's sketch.

Proof. Let us call a term not containing t as a subterm t-free. Let $t \in (\sigma)0$ be a closed term in normal form, and let $F[z^{(\sigma)0}]$ be a t-free term of type 0, and let $F[t]$ be closed.

By the results on computability, $F[t]$ reduces to a numeral, say $\bar{x}$, by a standard reduction sequence $[z/t]F_m, \ldots, [z/t]F_0 = \bar{x}$, where $F_m, \ldots, F_0$ are t-free.

Let $t_1', \ldots, t_n'$ be the set of all terms of type σ occurring in this reduction sequence, and let $t_1[z], \ldots, t_n[z]$ be t-free terms such that $t_i[t] = t_i'$, $1 \leq i \leq n$. Given $F[z]$ and t, $t_1, \ldots, t_n$ are uniquely determined.

Now we shall prove, for fixed t, by induction on the length of the standard reduction sequence of $F[t]$, for all t-free $F[z]$ such that $F[t]$ is closed, that

(1) $$\forall z^{(\sigma)0} (\bigwedge_{1 \leq i \leq n} (tt_i[t] = zt_i[z]) \rightarrow F[t] = F[z]).$$

Basis. If the reduction sequence for $F[t]$ has length 1, $F[t]$ is a numeral and we are done.

Induction step. Assume (1) to have been proved for all t-free $F[z] \in 0$ with $F[t]$ closed, and standard reduction sequence of $F[t]$ of length $\leq k$. Now assume $F[z]$ to be a t-free term of type 0 with $F[t]$ closed, with a reduction sequence

(2) $$[z/t]F_k,\ [z/t]F_{k-1}, \ldots, [z/t]F_0 = \bar{x}$$

and let

(3) $$\bigwedge_{1 \leq i \leq n} (tt_i[t] = zt_i[z])$$

for the terms t_i obtained from this reduction sequence in the manner described above. By the induction hypothesis, $[z/t]F_{k-1} \equiv F_{k-1}$ ((3) implies the

hypothesis with respect to F_{k-1}).
$[z/t]F_{k-1}$ is obtained from $[z/t]F_k$ by application of a contraction to a subterm of one of the forms $\Sigma s_1 s_2 s_3$, $R s_1 s_2 s_3$ or $\Pi s_1 s_2$. We have to distinguish two cases.

(a) The leftmost occurrence of Σ, Π, R in the subterm contracted is not part of a subterm of the form t.

For example, let $\Sigma s_1 s_2 s_3$ be contracted into $s_1 s_3(s_2 s_3)$, and let $s_i \equiv [z/t]s_i'$, s_i' t-free. Then F_{k-1} is obtained from F_k by contracting a corresponding occurrence of $\Sigma s_1' s_2' s_3'$ into $s_1' s_3'(s_2' s_3')$. Now obviously $F_k = F_{k-1}$, $[z/t]F_k = [z/t]F_{k-1}$, and since by induction hypothesis $[z/t]F_{k-1} = F_{k-1}$, it follows that $F_k = [z/t]F_k$.

(b) The principal occurrence of Σ, Π, R in the subterm contracted does not belong to t, so $[z/t]F_{k-1}$ is obtained from $[z/t]F_k$ by contraction of a subterm of the form ts, s of type σ; so s must be of the form $t_i[t]$, and $t_i[z]$ is t-free. Let the result of the contraction be of the form $s_1[t]$, $s_1[z]$ being t-free.

The reduction sequence for $[z/t]F_{k-1}$ implicitly contains a reduction sequence of length $\leq k$ for $s_1[t]$, hence the hypothesis (3) for $s_1[t]$ is contained in the hypothesis (3) for $F[t]$; therefore, by our induction hypothesis, $s_1[t] = s_1[z]$. Also $tt_i[t] = zt_i[z]$, hence under our assumptions

$$s_1[z] = s_1[t] = tt_i[t] = zt_i[z].$$

Now $F_{k-1} = [z/t]F_{k-1}$; since $s_1[z] \in 0$, replacing the relevant occurrence $s_1[z]$ in F_{k-1} by $zt_i[z]$ yielding F_k, does not change the value, so $F_{k-1} = F_k$; since also $[z/t]F_k = [z/t]F_{k-1}$, we have $[z/t]F_k = F_k$.

§ 8. Other models of $\underset{\sim}{N} - \underset{\sim}{HA}^{\omega}$.

2.8.1. Contents. In this section we rather briefly comment on some other models of $\underset{\sim}{N} - \underset{\sim}{HA}^{\omega}$ occurring in the literature. The first of these models is given by a concept of importance in its own right: Kleene's notion of recursive functional of higher type, as described by his schemata S1 - 9 (see Kleene 1959, 1963A). Unfortunately, this concept has up till now only been investigated from a classical point of view, e.g. Kleene assumes his functionals of type $(\sigma)0$ to be defined on all classical functionals of type σ ("classical" in the sense of "existing in say the intended model of ZF-set theory"). It would be interesting to know which class of functionals is singled out from a given constructively meaningful model for the theory of finite types by S1 - 9. Because of this lacuna in the literature, we shall be rather brief and restrict ourselves to some remarks on the recursive functions of finite type.

Two other models for the theory of finite types were introduced in Scarpellini 1971A. They were introduced because they provided models for the theory of bar-recursive functionals (cf. 1.9.26). See below, in 2.8.5 and 2.9.9 - 2.9.12.

Two models introduced by Howard are briefly discussed in 2.8.6.

2.8.2. The schemata S1 - 9.

Kleene uses in his description (Kleene 1959) only variables of pure types; let $\underline{\underline{z}}, \underline{\underline{z}}', \ldots$ be used for sequences of variables of pure types. The first S1 - S8 schemata are as follows.

S1) $\varphi(x^0, \underline{\underline{z}}) = Sx^0$

S2) $\varphi(\underline{\underline{z}}) = \bar{n}$

S3) $\varphi(x^0, \underline{\underline{z}}) = x^0$

S4) $\varphi(\underline{\underline{z}}) = \psi(\chi(\underline{\underline{z}}), \underline{\underline{z}})$

S5) $\begin{cases} \varphi(0, \underline{\underline{z}}) = \psi(\underline{\underline{z}}) \\ \varphi(Sx^0, \underline{\underline{z}}) = \chi(x^0, \varphi(x^0, \underline{\underline{z}}), \underline{\underline{z}}) \end{cases}$

S6) $\varphi(\underline{\underline{z}}) = \psi(\underline{\underline{z}}_1)$ ($\underline{\underline{z}}$ not empty, consisting of $k+1$ variables and $\underline{\underline{z}}_1$ obtained by shifting the $k+1^{st}$ variable in $\underline{\underline{z}}$ to the front)

S7) $\varphi(y^1, x^0, \underline{\underline{z}}) = yx$

S8) $\varphi(y^{k+2}, \underline{\underline{z}}) = y^{k+2}(\lambda u^k . \chi(y^{k+2}, u^k, \underline{\underline{z}}))$

Permuting the order of variables of different type is regarded as immaterial, provided only the order of variables of the same type is retained.

S1 - S8 characterize Kleene's primitive recursive functionals of finite type.

In the schemata φ is always the functional to be defined, ψ, χ are supposed to have been defined before ; all functionals considered are assumed to be numerical valued. (The functionals are considered throughout from the extensional point of view.)

To each functional we can assign an index ; the index for S1, S2, S3, S7 uniquely determines the intended functional, the index for a functional introduced by S4 - S6, S8 can be computed primitive recursively from indices χ, ψ.

The primitive recursive functionals are generalized to partial recursive functionals, by reading $\simeq$ for $=$ in S1 - 8 (where $\simeq$ intuitively means the same as in elementary recursion theory) and adding a schema which permits "self-reference" by introducing the index as an argument:

S9) $\varphi(x^0, \underline{\underline{z}}, \underline{\underline{z}}') \simeq \{x^0\}(\underline{\underline{z}})$.

Of course, φ in S9 itself also obtains an index, primitive recursive in the numbers of arguments of each type in $\underline{\underline{z}}$ and $\underline{\underline{z}}'$.
S1 - S9 may be viewed as constituting a generalized inductive definition of the relation $\{x^0\}(\underline{\underline{z}}) \simeq y^0$.
Kleene shows that the class of functionals determined by S1 - 9 is closed under the minimum operator, and definition by cases (Kleene 1959, XVI, XVIII); but closure under the minimum operator cannot replace S9 (in contrast to the theory of recursive functions).

2.8.3. Recursive functionals as a model for $\underset{\sim}{\mathrm{E}}$ - $\underset{\sim}{\mathrm{HA}}^\omega$.

From inspection of S1 - S9 it will be clear that the only fact which needs to be verified and which is not immediate from the definitions, is closure under definition by recursion as given for numerical types 1.8.9 (iv). Via the methods of coding finite sequences of pure types into pure types, as in 1.8.7, this is equivalent to showing closure under the schema

$$\begin{cases} \varphi(0, y^j, \underline{\underline{z}}) = \psi(y^j, \underline{\underline{z}}) \\ \varphi(Sx^0, y^j, \underline{\underline{z}}) = \chi(x^0, y^j, \lambda y^j.\varphi(x^0, y^j, \underline{\underline{z}}), \underline{\underline{z}}) . \end{cases}$$

(φ the functional to be defined, χ, ψ given functions.)
But this is precisely Kleene's XXIV (section 4.5 in Kleene 1959).

2.8.4. Remark. That the functionals generated by S1 - S9 do not contain a function representing a modulus of uniform continuity has been proved by R.O. Gandy (unpublished).

2.8.5. Scarpellini's models. (Scarpellini 1971A, 1972A)

The starting point for the definition of these models is based on the following axiomatic characterization of convergence of sequences :

Let X be a set on which a relation $\rightarrow$ of convergence between elements of $(0)X$ and X is given (we write $\langle p_n \rangle_n \rightarrow p$ or for short $p_n \rightarrow p$; $(0)x$ is the class of infinite sequences of elements of X) such that

1) If $\langle p_n \rangle_n \rightarrow p$, and $k_1 < k_2 < \ldots$ then $\langle p_{k_i} \rangle_i \rightarrow p$,
2) If $p_n = p$ for almost all n, then $\langle p_n \rangle_n \rightarrow p$,
3) If not $\langle p_n \rangle_n \rightarrow p$, then there is a sequence $k_1 < k_2 < k_3 \ldots$ such that no sub-sequence of $\langle p_{k_i} \rangle_i$ converges to p,
4) If $\langle p_n \rangle_n \rightarrow p$, $\langle p_n \rangle_n \rightarrow q$, then $p = q$.

Then $(X, \rightarrow)$ is called an L - space.

Let $(X_i, \rightarrow_i)$, $i = 1,\ldots,s$ and $(Y, \rightarrow)$ be L - spaces. A mapping f from $(X_1 \times \ldots \times X_s)$ to Y is said to be continuous, if $f(x_{1,n}, \ldots, x_{s,n}) \rightarrow f(x_1,\ldots,x_s)$ whenever $\langle x_{i,n} \rangle_n \rightarrow_i x_i$, for $1 \leq i \leq s$. Let us denote the species of continuous mappings from $X_1 \times \ldots \times X_s$ to Y by $C(X_1,\ldots,X_s,Y)$. $C(X_1,\ldots,X_s,Y)$ is made into an L - space again by defining

$$\langle f_n \rangle_n \rightarrow f \equiv_{def} \forall \langle x_{1,n} \rangle_n \ldots \forall \langle x_{s,n} \rangle_n \forall x_1 \ldots x_s ((\langle x_{1,n} \rangle_n \rightarrow x_1 \& \ldots$$
$$\ldots \& \langle x_{s,n} \rangle_n \rightarrow x_s) \rightarrow (\langle f_n(x_{1,n},\ldots,x_{s,n}) \rangle_n \rightarrow f(x_1,\ldots,x_s))) .$$

Now Scarpellini's first model M_1 (called $\underset{\sim}{S}$ in <u>Scarpellini</u> 1971) is described as follows, for the type structure $\underset{\sim}{T}_0$ (cf. 1.8.9).

$S_0 = N$ are the objects of type 0, with the following notion of convergence:
$\langle x_n \rangle_n \rightarrow x \equiv_{def} \exists k \forall m (x_{k+m} = x)$.
$S_{\sigma_1 \times \ldots \times \sigma_p} = S_{\sigma_1} \times \ldots \times S_{\sigma_p}$, $S_{(\sigma_1 \times \ldots \times \sigma_p)\tau} \equiv_{def} C(S_{\sigma_1}, \ldots, S_{\sigma_p}, S_\tau)$ with the notion of convergence obtained from the notions in $S_{\sigma_1}, \ldots, S_{\sigma_p}, S_\tau$ as described above.

M_1 can be shown to be a model for the bar-recursive functionals.

The description of the second model (called $\underset{\sim}{K}$ in <u>Scarpellini</u> 1971 A) is too long to be reproduced here, but may be viewed as a refinement of the first model.

2.8.6. <u>Compact and hereditarily majorizable functionals</u>.

W.A. Howard has described two other models for $\underset{\sim}{N}$ - $\underset{\sim}{HA}^\omega$. The first concept, that of the <u>compact</u> functionals over the type structure $\underset{\sim}{T}$, is defined as follows:

(i) A species of natural numbers is <u>compact</u> iff it is finite.
(ii) A species X of functionals of type $(\sigma)\tau$ is <u>compact</u> iff, for each compact species Y of functionals of type σ, $\{x^{(\sigma)\tau} y^\sigma \mid x \in X \ \& \ y \in Y\}$ is compact.
(iii) A single functional t^σ is <u>compact</u> iff $\{t^\sigma\}$ is compact.

It can then be shown that the functionals of $\underset{\sim}{N}$ - $\underset{\sim}{HA}^\omega$ are compact (in fact, even the functionals of the theory obtained by adding bar recursion of type 0 (BR_0) are compact).

The second concept, that of hereditarily majorizable functionals is introduced as follows.
We introduce a concept $Maj(x_1,x_2)$ (x_1 "majorizes" x_2) by definition over the type structure $\underset{\sim}{T}$ as follows:

(i) $Maj_0(x_1,x_2) \equiv_{def} x_1 \geq x_2$

(ii) $Maj_{(\sigma)\tau}(x_1,x_2) \equiv \forall y_1^\sigma y_2^\sigma (Maj_\sigma(y_1,y_2) \rightarrow Maj_\tau(x_1y_1,x_2y_2))$.

$Maj = \bigcup \{Maj_\sigma \mid \sigma \in \underset{\sim}{T}\}$.

We now define: a class X of functionals is hereditarily majorized by a class Y of functionals of $\forall x \in X \exists y \in Y\, Maj(y,x)$. It is not hard to show that the class of functionals of $\underset{\sim}{N}-\underset{\sim}{HA}^\omega$ is hereditarily majorized by itself.

Howard B makes the following application of this concept: the Dialectica translation (§ 3.5) of the simplest non-trivial case of the extensionality axiom

$$\forall y^2\ \forall\alpha\beta[\forall x(\alpha x = \beta x) \rightarrow y^2\alpha = y^2\beta]$$

is of the form

(1) $$\exists X\ \forall y^2\ \forall\alpha\beta[\alpha(Xy^2\alpha\beta) = \beta(Xy^2\alpha\beta) \rightarrow y^2\alpha = y^2\beta]\,;$$

and it can be shown that X satisfying (1) cannot be hereditarily majorizable by a functional from $\underset{\sim}{N}-\underset{\sim}{HA}^\omega$, and therefore the simplest non-trivial instance of the extensionality axiom has no Dialectica interpretation by a functional from $\underset{\sim}{N}-\underset{\sim}{HA}^\omega$. For more information, see Howard B (Appendix of this volume).

§ 9. Computability and models for extensions of $\underset{\sim}{N}$ - $\underset{\sim}{HA}^{\omega}$.

2.9.1. Contents of the section. In 2.9.2 - 2.9.4, 2.9.6 we describe extensions of computability arguments; in 2.9.5, 2.9.7 - 2.9.11 various extensions of our models for $\underset{\sim}{N}$ - $\underset{\sim}{HA}^{\omega}$ are described; see especially 2.9.9, for a simple model for bar recursion of higher type.

2.9.2. Extension of computability to (the functionals of) $\underset{\sim}{N}$ - $\underset{\sim}{IDB}^{\omega}$ and related theories.

For the theories of first-and second-order trees, computability is discussed in chapter VI.

In Howard 1972, a first-order theory $\underset{\sim}{U}$ is considered, obtained by extending $\underset{\sim}{HA}$ with a species of "abstract constructive ordinals", introduced by a g.i.d.. $\underset{\sim}{U}$ is embedded in a theory $\underset{\sim}{V}$ analogous to qf - $\underset{\sim}{WE}$ - $\underset{\sim}{HA}^{\omega}$, for objects of finite type over natural numbers and ordinals. Howard shows computability of terms in $\underset{\sim}{V}$ by means of an ordinal assignment, thereby determining the "proof-theoretic ordinal" of $\underset{\sim}{U}$ as Bachmann's (Bachmann 1950) $\varphi_{\varepsilon_{\Omega+1}}(1)$. $\underset{\sim}{U}$ is proof-theoretically equivalent to $\underset{\sim}{IDB}$.

In Troelstra 1971A, there is a proof of computability for the closed terms of $\underset{\sim}{N}$ - $\underset{\sim}{IDB}^{\omega}$ which differs from the method used in chapter VI. Although the method in chapter VI is more elegant, we thought it not without interest to demonstrate the other method also, for the case of $\underset{\sim}{N}$ - $\underset{\sim}{IDB}^{\omega}$ (cf. 2.9.6).

The contraction rules are as for $\underset{\sim}{N}$ - $\underset{\sim}{HA}^{\omega}$ with product types and pairing, and in addition

$$I(\Phi_1 t)s \quad \text{contr } St$$
$$I(\Phi_2 ts)t_0 \quad \text{contr } It(\langle s\rangle * t_0)$$
$$I(\Phi_3 t)0 \quad \text{contr } 0, \quad I(\Phi_3 t)(St') \text{ contr } I(t(C_1 t'))(C_2 t');$$

(C_1, C_2 are constants of $\underset{\sim}{N}$ - $\underset{\sim}{HA}^{\omega}$ such that $\langle C_1\rangle * (C_2 t) = St$ for all $t \in O$), and furthermore

$$\text{If } It0 \geq' Ss, \text{ then } \Psi_\sigma tt't'' \text{ contr } t's$$
$$\text{If } It0 \geq' 0, \text{ then } \Psi_\sigma tt't'' \text{ contr } t''(\lambda^* v.\Psi_\sigma(\Phi_2 tv)t't'')t$$

where $\geq'$ means standard reduction, defined in terms of "contr" as in 2.2.2 and λ^*v indicates the combinatorially defined λ-operator.

2.9.3. Computability for bar-recursive functionals ($\underset{\sim}{N}$ - $\underset{\sim}{HA}^{\omega}$ + BR).

Proof of computability for bar-recursive functionals are to be found in Tait 1971, Luckhardt 1970, 1973 and Scarpellini 1971A; another exposition is in Girard 1972 (following Tait's proof).

The methods of Tait 1971 and Luckhardt 1973 are essentially the same. A straightforward extension of the computability proof as given for $\underset{\sim}{N}$ - $\underset{\sim}{HA}^{\omega}$ does not work, because the induction hypothesis is too weak. Specifically, suppose we wish to show B_σ to be computable; then we have to show that for computable t, t', t" $B_\sigma tt't"$ is computable. But if we would have defined the computability predicate w.r.t. closed terms only, the assumption that t, t', t" are computable is weaker (at least prima facie) than the assumption that they are computable w.r.t. larger classes of terms; e.g. a functional of type 2 which is computable for recursive arguments need not be computable for arbitrary arguments, but it is functionals of the latter kind we are intuitively thinking of.

Tait's solution is to throw in additional terms according to the clause:

> If α is a function from natural numbers to natural numbers, then the pair (τ,α) is a term of type $(0)\tau$.

Luckhardt's formulation (Luckhardt 1973) is similar.

The intended interpretation of these additional terms is as follows: We assume all terms to be coded by numerical functions such that it is decidable whether a function codes a term of a given type or not. Then, if α is a numerical function, and $\alpha_m = \lambda n.\alpha j(m,n)$, (τ,α) is interpreted as follows: if α_m codes a term of type τ, then $(\tau,\alpha)(m) = \alpha_m^{(\tau)}$ is the term coded by α_m; if α_m does not code a term of type τ, then $(\tau,\alpha)(m) = \alpha_m^{(\tau)}$ is a fixed constant of type τ, say 0^τ.

For the new terms we add contraction rules $(\sigma,\alpha)\bar{n}$ contr $\alpha_n^{(\sigma)}$.

Scarpellini's method is slightly different. Whereas the original computability argument, at least for closed terms, only referred, so to speak, to the model of the closed terms themselves, Tait's and Luckhardt's method for $\underset{\sim}{N}$ - $\underset{\sim}{HA}^{\omega}$ + BR uses an embedding of the closed terms in a "generalized term model". A difference is, that Tait uses (informally) DC_1 and classical logic; Luckhardt formalizes his treatment in a system with intuitionistic logic + EBI_D. Scarpellini on the other hand uses instead of this generalized term model his model M_1 (cf. 2.8.5); at a certain stage in the proof, the computability of the bar-recursive constants is reduced to establish, by means of bar induction, truth of certain assertions in the model M_1 (Scarpellini 1971A, page 135); the method is thereby closer in spirit to the idea used in Troelstra 1971A, § 4 for the computability of $\underset{\sim}{N}$ - $\underset{\sim}{IDB}^{\omega}$.

2.9.4. Computability for Girard's system of functionals.

A first proof is in Girard 1971; a more detailed exposition is to be found in Girard 1972. The contractions are as suggested by the equations in 1.9.27:

$0^{(\sigma)\tau}t$ contr 0^{τ}, $D'(0^{\sigma\times\tau})$ contr 0^{σ}, $D''(0^{\sigma\times\tau})$ contr 0^{τ},

$I_{\forall\alpha\sigma[\alpha],\tau}0^{\forall\alpha\sigma[\alpha]}t^{\tau}$ contr $0^{\sigma[\tau]}$, $St\alpha t(0^{\exists\alpha\sigma[\alpha]})$ contr $t(0^{\sigma[\alpha]})$,

$Rtt'0$ contr t, $Rtt'\,\overline{n+1}$ contr $t'(Rtt'\bar{n})\bar{n}$, $(\lambda x.t)t'$ contr $[x/t']t$,

$D'(Dt_1t_2)$ contr t_1, $D''(Dt_1t_2)$ contr t_2,

$I_{\forall\alpha\sigma[\alpha],\tau}\,DT\alpha t^{\sigma[\alpha]})$ contr $t^{\sigma[\tau]}$,

$ST\alpha t^{(\sigma[\alpha])\rho}(I_{\exists\alpha\sigma[\alpha],\tau}t_1^{\sigma[\tau]})$ contr $t^{(\sigma[\tau])\rho}t_1$.

2.9.5. Extensions of HRO, HEO to models for other systems.

Let us first consider HRO. HRO is easily extended to a model K-HRO for $\underset{\sim}{I}$-$\underset{\sim}{IDB}^{\omega}$, by addition of

$$V_K = \{x \mid \{x\} \in K\}$$

and representing I by

$$(\Lambda x.x,\ (K)(0)0),$$

Φ_1, Φ_2 by

$$(\Lambda x \Lambda y.Sx,\ (0)K),\ (\Lambda x \Lambda y \Lambda n.x(\hat{y} * n),\ (K)(0)K)$$

and Φ_3 by

$$(\Lambda x \Lambda y.(1 \dot{-} y)\{\{x\}(y)_0\}(tl(y)),\ ((0)K)K).$$

The representation of Ψ_σ is constructed by means of the recursion theorem. Let v_1 be the gödelnumber of a partial recursive function such that

$$\{w\}(0) \neq 0 \rightarrow \{\bar{v}_1\}(v_0,w,x,y) \simeq \{x\}(\{w\}(0) \dot{-} 1)$$
$$\{w\}(0) = 0 \rightarrow \{\bar{v}_1\}(v_0,w,x,y) \simeq \{\{y\}(\Psi(v_0,w,x,y))\}(w)$$

where

$$\Psi(v,w,x,y) \simeq \Lambda u.\{\{\{v\}(\{[\Phi_2]\}(w)\}(u)\}(x)\}(y),$$

and Λu is to be chosen so as to correspond to the syntactically defined λ-operator in the axioms for Ψ_σ. Now by the recursion theorem, there is a v such that

$$\{w\}(0) \neq 0 \rightarrow \{\{\{\bar{v}\}(w)\}(x)\}(y) \simeq \{x\}(\{w\}(0) \dot{-} 1),$$
$$\{w\}(0) = 0 \rightarrow \{\{\{\bar{v}\}(w)\}(x)\}(y) \simeq \{y\}(\Psi(\bar{v},w,x,y)\}(w).$$

By induction over K w.r.t. e, it can be shown that for all e, $w \in V_K$, $x \in V_{(0)\sigma}$, $y \in V_{((0)\sigma)(K)\sigma}$, we have

$$e = \{w\} \rightarrow\ !\ \{\{\{\bar{v}\}(w)\}(x)\}(y).$$

For if $\{w\} \in K$, $\{w\}(0) \neq 0$, then obviously $!\{\{\{\bar{v}\}(w)\}(x)\}(y)$. Assume

$\{w\}(0) = 0$, and suppose $\{\{\{\bar{v}\}(z)\}(x)\}(y)$ to be defined for all z such that $\{z\} = \lambda n.\{w\}(\hat{x} * n)$ for some x. Then $\Lambda u.\Psi(v_1,w,x,y) \in V_{(o)\sigma}$, and so $!\{\{\{\bar{v}\}(w)\}(x)\}(y)$. Hence we may take $[\Psi_\sigma] = \bar{v}$.

K-HEO is defined similarly as a model for $\underset{\sim}{E}$-$\underset{\sim}{IDB}^{\omega}$. We may represent the constants by the same numerals as in K-HRO; we define W_σ, I_σ as for HEO, extending the definitions by

$$W_K \equiv_{def} V_K,$$

$$I_K(x,y) \equiv_{def} I_{(o)o}(x,y) \ \& \ x \in V_K \ \& \ y \in V_K.$$

The other developments in § 2.4, § 2.5 also carry over without essential change.

2.9.6. Application of K-HRO: Computability of the closed terms of $\underset{\sim}{N}$-$\underset{\sim}{IDB}^{\omega}$.

The contraction rules have already been listed in 2.9.2.

We define a closed term to be in normal form if no contractions are possible and it is not of the form $\Psi_\sigma tt't''$, or $I_\sigma tt'$ (t, t', t'' in normal form).

We then define "Comp" exactly as in 2.2.5, 2.3.7 (standard computability) with a clause added:

(iv) $\quad \mathrm{Comp}''_K(t) \equiv_{def} \forall t'(\mathrm{Comp}''_o(t') \rightarrow \mathrm{Comp}''_o(Itt'))$.

Now, from the fact that K-HRO is a model for $\underset{\sim}{N}$-$\underset{\sim}{IDB}^{\omega}$, we can find, for each closed term $t \in K$ a number $n \in V_K$ such that if $\mathrm{Comp}''_K(t)$, then

$$It\bar{m} \geq' \bar{m}' \leftrightarrow \{n\}(m) = m'.$$

For in K-HRO, natural numbers are interpreted by themselves.

If $[t] = n$, i.e. (n, K) represents t in K-HRO, then obviously $It\bar{m} \geq' \bar{m}'$ implies $It\bar{m} = \bar{m}'$, hence in the model, $\{n\}(m) = m'$, and similarly in the other direction.

Now, if we wish to show that every closed term is standard computable, it is sufficient, as in 2.2.6, to verify that all constants are computable.

We first note that a computable closed term of type 0 reduces to a numeral.

(i) The constants of $\underset{\sim}{N}$-$\underset{\sim}{HA}^{\omega}$, D, D', D" have been treated before.

I is trivially computable.

(ii) Φ_3 is computable; for assume $\mathrm{Comp}''_{(o)K}(t)$, $\mathrm{Comp}''_o(s)$. Then $s \geq' 0$ or $s \geq' S\bar{n}$, and $t \geq' t_1$ for some normal t_1.

In the first case, $I(\Phi_3 t)s \geq' I(\Phi_3 t_1)0$ contr. 0; in the second case $I(\Phi_3 t)s \geq' I(\Phi_3 t_1)(S\bar{n})$ contr. $I(t_1(C_1\bar{n}))(C_2\bar{n})$. C_1, C_2 are closed terms of $\underset{\sim}{N}$-$\underset{\sim}{HA}^{\omega}$, hence have already been shown to be computable, i.e. $C_1\bar{n} \geq' \bar{n}_1$, $C_2\bar{n} \geq' \bar{n}_2$, so $I(\Phi_3 t_1)(S\bar{n}) \geq I(t_1\bar{n}_1)\bar{n}_2$, which is computable by our assumptions. The verification of the computability of Φ_1, Φ_2 is left to the reader.

(iii) Ψ_σ is computable. Assume $\mathrm{Comp}''_K(t)$, $\mathrm{Comp}''_{(o)\sigma}(t')$, $\mathrm{Comp}''_{((o)\sigma)(K)\sigma}(t'')$. We now define an operator $\Phi_2^{[n]} \in (K)K$ for each finite sequence of natural numbers n, as follows:

$$\Phi_2^{[o]}t \equiv_{def} t$$
$$\Phi_2^{[n*\langle x\rangle]}t \equiv_{def} \Phi_2(\Phi_2^{[n]}t)\bar{x}$$

Now let $t \geq' t_1$, $t' \geq' t_1'$, $t'' \geq' t_1''$ (t_1, t_1', t_1'' normal). We prove that $\Psi_\sigma tt't''$ is computable. This is proved, for fixed t', t'', by induction over the unsecured sequences of $\{[t]\}$ (i.e. by induction over $\{n \mid \{[t]\}(n) = 0\}$). (1.9.18, (1)), i.e. we show that

(1) $\{[t]\}(m) \neq 0 \Rightarrow \mathrm{Comp}''(\Psi_\sigma(\Phi_2^{[m]}tt't''))$.

(2) $\forall x\, \mathrm{Comp}''(\Psi_\sigma(\Phi_2^{[m*\langle x\rangle]}tt't'')) \Rightarrow \mathrm{Comp}''(\Psi_\sigma(\Phi_2^{[m]}tt't''))$.

We note that

(3) $\{[\Phi_2^{[m]}t]\}(x) = \{[t]\}(\bar{m} * \langle x\rangle)$.

Further we note that by a straightforward induction on $\mathrm{lth}(m)$

(4) $\forall m(\mathrm{Comp}''_K(\Phi_2^{[m]}t))$.

Assume now $\{[t]\}(\bar{m}) = S\bar{u}$, then by (3) $\{[\Phi_2^{[m]}t]\}(0) = S\bar{u}$, and therefore by 2.9.2 $\Psi_\sigma(\Phi_2^{[m]}t)t't'' \geq' t_1'\bar{u}$, which is computable by hypothesis.

Assume $\{[t]\}(\bar{m}) = 0$, i.e. $\{[\Phi_2^{[m]}t]\}(0) = 0$, and suppose $\forall x\, \mathrm{Comp}(\Psi_\sigma(\Phi_2^{[m*\langle x\rangle]}t))t't''$; since $I(\Phi_2^{[m]}t)0 \geq' 0$ by our assumption, $\Psi_\sigma(\Phi_2^{[m]}t)t't'' \geq' t_1''(\lambda^*x.\Psi_\sigma(\Phi_2((\Phi_2^{[m]}t_1)x)t_1t_1'')t_1'$, which is computable by our assumptions.

With an application of 1.9.18, (1), we conclude that $\Psi(\Phi_2^{[o]}t)t't''$, i.e $\Psi_\sigma tt't''$ is computable, hence Ψ_σ is computable.

2.9.7. Extension of HRO, HEO to Girard's system of functionals.

This extension of HRO appears first in Troelstra A, and is extended in Girard 1972 to the intuitionistic theory of types; there also the corresponding extension of HEO is described.

In describing the analogue of HRO for Girard's theory of functionals (let us denote this analogue by HRO^2), the problem is how to interpret the objects of variable type. Noting that each type σ is supposed to contain a constant 0_σ ($= 0^\sigma$) it is reasonable to interpret each variable type as a species of gödelnumbers of partial recursive functions, containing at least one element. It is quite convenient if we could achieve this element to be always 0.

In order to do this, we note that for our standard pairing $j(0,0) = 0$, and we select a special gödelnumbering for the partial recursive functions such that

(1) $\{0\}(x) \simeq 0$ for all x.

Such a gödelnumbering may be constructed as follows. In a given standard gödelnumbering, let x_0 be a gödelnumber of the function $\lambda x.0$, and let T be the T-predicate corresponding to this gödelnumbering, and let φ be defined by

$$\varphi(x) = \begin{cases} x_0 & \text{if } x = 0 \\ 0 & \text{if } x = x_0 \\ x & \text{otherwise.} \end{cases}$$

We obtain a new gödelnumbering by interchanging 0 and x_0, and its T-predicate T' is defined by

$$T'(x,y,z) \equiv_{def} T(\varphi x,y,z) .$$

For the new gödelnumbering we easily obtain, as before, the s-m-n-theorem and the recursion theorem.

Let us now introduce special variables for species of one argument, containing 0; we suppose to each type variable α such a new species variable V_α to be assigned.

We put further

$$x \in V_{(\sigma)\tau} \equiv_{def} \forall y \in V_\sigma \, \exists z \in V_\tau(\{x\}(y) \simeq z) ,$$

$$x \in V_{\sigma \times \tau} \equiv_{def} j_1 x \in V_\sigma \;\&\; j_2 x \in V_\tau$$

$$x \in V_{\forall\alpha\sigma[\alpha]} \equiv_{def} \forall V_\alpha(x \in V_{\sigma[\alpha]})$$

$$x \in V_{\exists\alpha\sigma[\alpha]} \equiv_{def} \exists V_\alpha(x \in V_{\sigma[\alpha]}) .$$

Since $0 \in V_\alpha$ by definition, we readily prove that $0 \in V_\tau$ for all $\tau \in \underset{\sim}{T}$. S, $\Pi_{\sigma,\tau}$, $\Sigma_{\rho,\sigma,\tau}$, R_σ, $D_{\sigma,\tau}$, $D'_{\sigma,\tau}$, $D''_{\sigma,\tau}$ and E_σ are interpreted as before, 0_σ is interpreted as $(0, \sigma)$ for all σ.

$I_{\forall\alpha\sigma[\alpha],\tau}$ is interpreted as $(\Lambda x.x, (\forall\alpha.\sigma[\alpha])(\sigma[\tau]))$, $I'_{\exists\alpha\sigma[\alpha],\tau}$ as $(\Lambda x.x, (\sigma[\tau])(\exists\alpha.\sigma[\alpha]))$.

If $t[x_1^{\sigma_1},\ldots] \in \sigma[\alpha]$ is a term, not containing free variables which contain α free in their type, then t is represented by a p-term $t'[x_1,\ldots]$ such that

$$x_1 \in V_{\sigma_1} \;\&\; \ldots \rightarrow \,!t'[x_1,\ldots] \;\&\; t'[x_1,\ldots] \in V_{\sigma[\alpha]} .$$

Then also since V_{σ_1}, V_{σ_2}, ... do not depend on V_α,

$$x_1 \in V_{\sigma_1} \;\&\; \ldots \rightarrow \,!t'[x_1,\ldots] \;\&\; t'[x_1,\ldots] \in V_{\forall\alpha\sigma[\alpha]} .$$

So t' is seen to represent $DT\alpha t$.

If $t[x_1^{\sigma_1},\ldots] \in (\sigma[\alpha])\tau$, α not occurring free in τ, and not occurring in a type of a variable free in t, then there is a p-term $t'[x_1,\ldots]$ such that

$$x_1 \in V_\sigma \ \& \ldots \rightarrow\ !t'[x_1,\ldots] \ \&\ t'[x_1,\ldots] \in V_{(\sigma[\alpha])\tau},$$

which represents $t[x_1^{\sigma_1},\ldots]$ in HRO^2.
It follows that

$$y \in V_{\sigma[\alpha]} \ \&\ x_1 \in V_{\sigma_1} \ \& \ldots \rightarrow\ !\{t'[x_1,\ldots]\}(y) \ \&\ \{t'[x_1,\ldots]\}(y) \in V_\tau .$$

Then also, since $V_{\sigma_1}, V_{\sigma_2}, \ldots, V_\tau$ do not depend on V_α,

$$\exists V_\alpha(y \in V_{\sigma[\alpha]}) \ \&\ x_1 \in V_{\sigma_1} \& \ldots \rightarrow\ !\{t'[x_1,\ldots]\}(y) \ \&\ \{t'[x_1,\ldots]\}(y) \in V_\tau .$$

So

$$x_1 \in V_{\sigma_1} \ \& \ldots \rightarrow\ !t'[x_1,\ldots] \ \&\ t'[x_1,\ldots] \in V_{(\exists\alpha\sigma[\alpha])\tau},$$

whence $t'[x_1,\ldots]$ is seen to represent $\mathbf{ST}\alpha t$.

The corresponding model HEO^2 is constructed by taking for I_α, the equivalence relation on W_α representing extensional equality between objects of type α, an arbitrary equivalence relation on W_α. Note that in the definition of HEO^2, the field of I_α is exactly W_α. Therefore we only need to consider variables I_α for equivalence relations with 0 in their field; W_α is then automatically defined as $W_\alpha(x) \equiv_{def} I_\alpha(x,x)$. $I_{(\sigma)\tau}$, $I_{\sigma\times\tau}$ are defined in terms of I_σ, I_τ as before in the case of HEO. $I_{\forall\alpha.\sigma[\alpha]}(x,y) \equiv_{def} \forall I_\alpha(I_{\sigma[\alpha]}(x,y))$, $I_{\exists\alpha.\sigma[\alpha]}(x,y) \equiv_{def} \exists I_\alpha(I_{\sigma[\alpha]}(x,y))$.

HRO^2 especially can be useful in connection with an extension of modified realizability to $\underset{\sim}{\mathbf{HAS}}$ (cf. 3.4.27)

2.9.8. In a very similar way it is possible to construct models $ICF^2(\mathcal{U})$, $ECF^2(\mathcal{U})$, which may be conceived as extensions of the models $ICF(\mathcal{U})$, $ECF(\mathcal{U})$. For the sake of "homogeneity" the second definition in 2.6.23 of the V_σ^1 is to be preferred for our analogy. More precisely, it is inconvenient that V_0^1 consists of natural numbers, and V_1^1 of sequences, if we wish to permit substitution of V_0^1 as well as V_1^1 for V_α^1. Hence we select for our definition of the analogues $ICF^2(\mathcal{U})$, $ECF^2(\mathcal{U})$ the second definition. The definition is in fact completely routine, once we have redefined $|$ as an operation $||$ such that

(1) $\lambda x.0 \,||\, \alpha \simeq \lambda x.0$

(similar to the replacement of $\{x\}(y)$ by $\{x\}'(y) \equiv_{def} \{\varphi x\}(y)$ in 2.9.7).

An operation $||$ satisfying (1) can be defined by $\alpha \,||\, \beta \simeq \gamma \equiv_{def} \Gamma\alpha|\beta \simeq \gamma$, where Γ is given by

$$(\Gamma\alpha)x = \begin{cases} 1 & \text{if } \alpha x = 0 \\ 0 & \text{if } \alpha x = 1 \\ \alpha x & \text{otherwise.} \end{cases}$$

2.9.9. Models for $\underset{\sim}{N}-\underset{\sim}{HA}^{\omega}+BR$.

The simplest model is presumably $ICF(\mathcal{U})$, if $\mathcal{U}$ is any class of functions satisfying EBI_D (cf. 1.9.21). For example, we may take $\mathcal{U}$ to be the classical universe of functions. The proof is given in the system $\underset{\sim}{EL}+EBI_D$, in 2.9.10.

Other models are the term models of Luckhardt 1970, 1973 and Tait 1971, and the models of Scarpellini 1971A,1972A (cf. 2.9.3).

Similarly, $ECF(\mathcal{U})$ is an extensional model for bar recursion. The proof for this is also given in $\underset{\sim}{EL}+EBI_D$ (2.9.10). The corresponding result for the term model of Luckhardt 1970 is proved in Luckhardt 1972 (1973). Scarpellini established the corresponding result for his second model (cf. Scarpellini 1971A) in Scarpellini 1972A.

2.9.10. Theorem. If $\mathcal{U}$ satisfies $\underset{\sim}{EL}+EBI_D$, then $ICF(\mathcal{U})$ can be shown to be a model for $\underset{\sim}{N}-\underset{\sim}{HA}^{\omega}+BR$, $ECF(\mathcal{U})$ for $\underset{\sim}{E}-\underset{\sim}{HA}^{\omega}+BR$, in $\underset{\sim}{EL}+EBI_D$.

Proof. Let $\alpha \in V^1_{((0)\sigma)0}$, $\gamma \in (V^1_\sigma)^\smile \equiv V^1_{\sigma^\smile}$ ($\sigma^\smile$ denoting the type of finite sequences of elements of type σ), $\beta \in V^1_{(\sigma^\smile)\tau}$, $\delta \in V^1_{((\sigma)\tau)(\sigma^\smile)\tau}$. We define $\tilde{\gamma}$ by

$$\begin{cases} (\tilde{\gamma})_x = (\gamma)_{Sx} & \text{for } x < \mathrm{lth}(\gamma) \\ (\tilde{\gamma})_{Sx+u} = \Theta, & \text{if } \Theta \text{ represents } 0^\sigma \text{ in } V^1_\sigma. \end{cases}$$

Then we can find an ε such that

$$\alpha(\tilde{\gamma}) < \mathrm{lth}\,\gamma \rightarrow \varepsilon\,|\,(\xi, \alpha, \beta, \delta, \gamma) \simeq \beta\,|\,\gamma$$
$$\alpha(\tilde{\gamma}) \geq \mathrm{lth}\,\gamma \rightarrow \varepsilon\,|\,(\xi, \alpha, \beta, \delta, \gamma) \simeq \delta\,|\,(\Lambda\eta.\,\xi\,|\,(\alpha, \beta, \delta, \gamma * \hat{\eta}), \gamma).$$

By the recursion theorem analogue (1.9.16) we can find ε_0 such that

$$\alpha(\tilde{\gamma}) < \mathrm{lth}\,\gamma \rightarrow \varepsilon_0\,|\,(\alpha, \beta, \delta, \gamma) \simeq \beta\,|\,\gamma$$
$$\alpha(\tilde{\gamma}) \geq \mathrm{lth}\,\gamma \rightarrow \varepsilon_0\,|\,(\alpha, \beta, \delta, \gamma) \simeq \delta\,|\,(\Lambda^1\eta.\,\varepsilon_0\,|\,(\alpha, \beta, \delta, \gamma * \hat{\eta}), \gamma).$$

One then proves, by an application of EBI_D, taking for R: V^1_σ, for $Q\gamma$: $!(\varepsilon_0\,|\,(\alpha, \beta, \delta, \gamma))$, and for $P\gamma$: $\alpha(\tilde{\gamma}) < \mathrm{lth}\,\gamma$, that $\varepsilon_0\,|\,(\alpha, \beta, \delta, \gamma)$ is always defined if $\alpha, \beta, \delta, \gamma$ satisfy the conditions listed in the beginning.

For the case of $ECF(\mathcal{U})$, we must also show extensionality conditions to be satisfied, but this can be proved in the same manner by an application of EBI_D.

2.9.11. Corollary to the proof. $ICF(\mathcal{U})$, $ECF(\mathcal{U})$ can be shown to be models for $\underset{\sim}{N}-\underset{\sim}{HA}^{\omega}+BR_0$, resp. $\underset{\sim}{E}-\underset{\sim}{HA}^{\omega}+BR_0$ if $\mathcal{U}$ satisfies $\underset{\sim}{EL}+BI_D$.

2.9.12. Remark. Kleene's recursive functionals, defined by the schemata S1 - S9, yield a model of ECF($\mathcal{U}$) if $\mathcal{U}$ is supposed to satisfy $\underset{\sim}{EL} + BI_D$; by a recursion theorem analogue (cf. 2.9.16) we can show the existence of a partial recursive functional satisfying the equations for BR_o (with $\simeq$ instead of $=$), and using BI_D we can prove the functional to be total, as before. The form of the recursion theorem to be applied in this case is found in Kleene 1959, XIV in subsection 3.12.

Chapter III

REALIZABILITY AND FUNCTIONAL INTERPRETATIONS

§ 1. A theme with variations: Kleene's $\Gamma | C$.

3.1.1. Introductory remarks. In Kleene 1952 (§ 82), S.C. Kleene introduced the notion of $\Gamma\vdash$-realizability to obtain certain proof-theoretic results for intuitionistic arithmetic, such as the well-known disjunction property $\vdash A \vee B \Rightarrow \vdash A$ or $\vdash B$ ($A \vee B$ closed). $\Gamma\vdash$-realizability was based on the idea of combining certain deducibility properties with realizability (realizability is discussed in extenso in the next section).
In Kleene 1962 and 1963, Kleene simplified his proof of the disjunction and existential definability property by introducing the $\Gamma | C$-relation, obtained by "omitting the realizability from $\Gamma\vdash$-realizability". Expressed otherwise, $\Gamma\vdash$-realizability may be viewed as the hybrid between realizability proper (in the sense of Kleene 1952, § 82) and the $\Gamma | C$-relation.

As an introduction to the various variants of realizability, we shall in this section study the $\Gamma | C$-relation, its variants and generalizations. For its model-theoretic equivalent, see chapter V.
Contents of the section. In subsections 1 - 10 the notion $\Gamma | C$ is defined, the soundness theorem proved, and some properties of $\Gamma | C$ and corollaries of the soundness theorem are given. In subsection 11 it is shown as an application that $\underset{\sim}{HA} \nvdash IP_0^c$. Subsection 12 discusses $\emptyset | C$ for $\underset{\sim}{HA} + M_{PR}$ ($\emptyset$ empty set).

In subsections 13 - 15 a variant of $\Gamma | C$ is discussed which yields a very simple proof of the rule IPR (with parameters).

In subsections 16 - 18 Kreisel's method for dealing with derived rules with parameters, using partial reflection principles, is described and applied to obtain closure under Church's rule.

Subsections 19 - 24 are devoted to the use of (variants of) $\Gamma | C$ for extensions of arithmetic.

3.1.2. Definition. We define $\Gamma | C$, for Γ a set of closed formulas, C a closed formula, by induction on the logical complexity of C, as follows. ("$\Gamma | \vdash A$" abbreviates "$\Gamma | A$ and $\Gamma \vdash A$", "$\Gamma \vdash A$" abbreviates "$\underset{\sim}{HA} + \Gamma \vdash A$").

(i) $\Gamma | P \equiv \Gamma \vdash P$ for prime P

(ii) $\Gamma | A \& B \equiv \Gamma | A$ and $\Gamma | B$

(iii) $\Gamma | A \vee B \equiv \Gamma | \vdash A$ or $\Gamma | \vdash B$

(iv) $\Gamma \mid A \to B \equiv \Gamma \| \!\vdash A \Rightarrow \Gamma \mid B$

(v) $\Gamma \mid \forall x\, Ax \equiv \Gamma \mid A\bar{n}$ for all numerals $\bar{n}$

(vi) $\Gamma \mid \exists x\, Ax \equiv \Gamma \| \!\vdash A\bar{n}$ for some numeral $\bar{n}$.

If $A(x_1,\ldots,x_n)$ is a formula containing at most $x_1,\ldots,x_n$ free, then we write $\Gamma \mid A(x_1,\ldots,x_n)$ iff $\Gamma \mid \forall x_1\ldots x_n A(x_1,\ldots,x_n)$.

3.1.3. Lemma. Let $A(x_1,\ldots,x_n)$ contain at most $x_1,\ldots,x_n$ free; let $t_1,\ldots,t_n$ be a set of closed terms, $\bar{t}_i$ the numeral corresponding to t_i under the standard interpretation. Then

(i) $\Gamma \vdash A(t_1,\ldots,t_n)$ iff $\Gamma \vdash A(\bar{t}_1,\ldots,\bar{t}_n)$

(ii) $\Gamma \mid A(t_1,\ldots,t_n)$ iff $\Gamma \mid A(\bar{t}_1,\ldots,\bar{t}_n)$.

Proof. (i) Straightforward, by induction on the logical complexity of A. For the basis we use the fact that all true closed prime formulae (hence in particular $t_i = \bar{t}_i$, $\bar{t}_i = t_i$) are derivable in HA.
(ii) is proved similarly. As an example of the induction step, let $A \equiv B \to C$, and assume

$$\Gamma \mid B(t_1,\ldots,t_n) \text{ iff } \Gamma \mid B(\bar{t}_1,\ldots,\bar{t}_n), \quad \Gamma \mid C(t_1,\ldots,t_n) \text{ iff } \Gamma \mid C(\bar{t}_1,\ldots,\bar{t}_n).$$

Let $\Gamma \mid B(t_1,\ldots,t_n) \to C(t_1,\ldots,t_n)$, and let $\Gamma \| \!\vdash B(\bar{t}_1,\ldots,\bar{t}_n)$. Then by induction hypothesis and (i), $\Gamma \| \!\vdash B(t_1,\ldots,t_n)$, hence $\Gamma \mid C(t_1,\ldots,t_n)$; by the induction hypothesis, $\Gamma \mid C(\bar{t}_1,\ldots,\bar{t}_n)$. Thus $\Gamma \mid B(\bar{t}_1,\ldots,\bar{t}_n) \to C(\bar{t}_1,\ldots,\bar{t}_n)$; $\Gamma \mid B(\bar{t}_1,\ldots) \to C(\bar{t}_1,\ldots) \Rightarrow \Gamma \mid B(t_1,\ldots) \to C(t_1,\ldots)$ is shown similarly.

3.1.4. Theorem (Soundness theorem). We show that, if $\Gamma \mid C$ for each $C \in \Gamma$ (all elements of Γ closed)

$$\Gamma \vdash A \Rightarrow \Gamma \mid A.$$

Proof. By induction on the length of a deduction of A from Γ in HA. We select again Gödel's formalization (1.1.4) as the basis for our verification.

Let $\Gamma \vdash A$, $A \in \Gamma$, then $\Gamma \mid C$ by the hypothesis of the theorem.

PL2). Let $\Gamma \vdash Ax$, $\Gamma \vdash Ax \to Bx$, $\Gamma \mid Ax$, $\Gamma \mid Ax \to Bx$.

Hence, for all $\bar{n}$, $\Gamma \| \!\vdash A\bar{n}$, therefore $\forall n(\Gamma \mid B\bar{n})$; therefore $\Gamma \mid Bx$.
For simplicity, we omit parameters in most other cases.

PL3). Let $\Gamma \vdash A \to B$, $\Gamma \vdash B \to C$, $\Gamma \mid A \to B$, $\Gamma \mid B \to C$.

Assume $\Gamma \| \!\vdash A$. Then $\Gamma \| \!\vdash B$, hence $\Gamma \mid C$; so $\Gamma \| \!\vdash A \to C$.

PL7). Let $\Gamma \| \!\vdash A \,\&\, B \to C$, and assume $\Gamma \| \!\vdash A$, $\Gamma \| \!\vdash B$.

It follows that $\Gamma \| \!\vdash A \,\&\, B$, hence $\Gamma \mid C$; so $\Gamma \mid A \to (B \to C)$.

PL8). Similarly, in the other direction.

PL9). Assume $\Gamma \Vdash 1=0$; then Γ is inconsistent, hence $\Gamma | P$ for all closed prime formulae P. Then one readily proves by induction on the logical complexity of A, that $\Gamma | A$ for all A, hence $\Gamma | 1=0 \rightarrow A$.

PL10), 11), 12). Immediate.

PL13). Assume $\Gamma \Vdash A \rightarrow B$, and let $\Gamma \Vdash C \vee A$.

Then $\Gamma \Vdash C$ or $\Gamma \Vdash A$; hence by our first assumption, $\Gamma \Vdash C$ or $\Gamma \Vdash B$. So $\Gamma | C \vee B$ under the assumption $\Gamma \Vdash C \vee A$, i.e. $\Gamma | C \vee A \rightarrow C \vee B$.

Q1). Assume $\Gamma \Vdash C \rightarrow Ax$, so for all $\bar{n}$, $\Gamma | A\bar{n}$, i.e. $\Gamma | \forall x Ax$. So $\Gamma | C \rightarrow \forall x Ax$.

Q2). Let $\Gamma \Vdash \forall x A(x,y)$, then $\Gamma \Vdash A(\bar{n},\bar{m})$ for all numerals $\bar{n}, \bar{m}$; hence if $t(y)$ is a term containing only y free, $t(\bar{m})$ is closed, and if $\bar{t}(\bar{m})$ is the corresponding numeral under the standard interpretation, $\Gamma | A(\bar{t}(\bar{m}), \bar{m})$, and with lemma 3.1.3, $\Gamma | A(t(\bar{m}), \bar{m})$. Thus $\Gamma \Vdash \forall x A(x,y) \rightarrow A(t(y), y)$.

Q3). Let $\Gamma \Vdash A(t(y), y)$, then $\Gamma \Vdash A(t(\bar{n}), \bar{n})$ for all numerals $\bar{n}$ (t containing at most y free). Then also by lemma 3.1.3, $\Gamma \Vdash A(\bar{t}(\bar{n}), \bar{n})$, hence $\Gamma | \exists x A(x,\bar{n})$. This holds for all $\bar{n}$, so $\Gamma | A(t(y),y) \rightarrow \exists x A(x,y)$.

Q4). Assume $\Gamma \Vdash Ax \rightarrow C$, i.e. $\Gamma \Vdash A\bar{n} \rightarrow C$ for all $\bar{n}$ (x being the only variable free in $Ax \rightarrow C$).
Let now $\Gamma \Vdash \exists x Ax$; then $\Gamma \Vdash A\bar{n}$ for some $\bar{n}$, hence by our assumption $\Gamma | C$. So $\Gamma | \exists x Ax \rightarrow C$.

The verification of the non-logical axioms is mostly trivial; consider e.g. $Sx \neq 0$, i.e. $Sx = 0 \rightarrow 1 = 0$. If $\Gamma \Vdash S\bar{n} = 0$, Γ is inconsistent, hence also $\Gamma \vdash 1 = 0$ which implies $\Gamma | 1=0$, so $\Gamma | Sx = 0 \rightarrow 1 = 0$.

The only non-trivial case which remains is the induction axiom. Assume $\Gamma \Vdash A0 \,\&\, \forall y(Ay \rightarrow A(Sy))$. Then $\Gamma \Vdash A0$, $\Gamma \Vdash A\bar{n} \rightarrow A(S\bar{n})$ for all $\bar{n}$; by induction one proves $\Gamma \Vdash \forall x Ax$, so $\Gamma | A0 \,\&\, \forall y(Ay \rightarrow A(Sy)) \rightarrow \forall x Ax$.

3.1.5. Corollaries. Assume $B, C, D, \exists x Ax$ to be closed. If $C | C$, then in **HA**:

(i) $\vdash C \rightarrow \exists x Ax$ iff $\vdash C \rightarrow A\bar{n}$ for some numeral $\bar{n}$

(ii) $\vdash C \rightarrow B \vee D$ iff $\vdash C \rightarrow B$ or $\vdash C \rightarrow D$

(iii) $\vdash C \rightarrow \exists x Ax \;\Rightarrow\; \vdash \exists x(C \rightarrow Ax)$.

Proof. (i) Assume $C | C$, $\vdash C \rightarrow \exists x Ax$. Then $C \vdash \exists x Ax$, so $C | \exists x Ax$, i.e. $C \Vdash A\bar{n}$ for some numeral $\bar{n}$; hence $\vdash C \rightarrow A\bar{n}$.

(ii) can be proved in the same way, but can also be obtained as an immediate consequence of (i) and $\vdash B \vee D \leftrightarrow \exists x[(x=0 \rightarrow B) \,\&\, (x \neq 0 \rightarrow D)]$.

(iii) is an immediate consequence of (i).

3.1.6. <u>Lemma</u>. $\neg C \mid \neg C$ for closed C.
<u>Proof</u>. Assume $\neg C \mid\!\vdash C$. Since $\neg C \vdash C$ implies $\neg C \vdash 1=0$, we have $\neg C \mid 1=0$. Therefore $\neg C \mid \neg C$.

3.1.7. <u>Corollary</u>. In $\underset{\sim}{HA}$, for closed C, $\exists x A$

IPR^c $\qquad \vdash (\neg C \to \exists x A) \Rightarrow \vdash \exists x(\neg C \to A)$.

<u>Proof</u>. 3.1.5 (i), 3.1.6.

3.1.8. <u>Theorem</u>. Let C be closed. If for all closed $\exists x Ax$

$$\underset{\sim}{HA} \vdash C \to \exists x\, Ax \Rightarrow \underset{\sim}{HA} \vdash C \to A\bar{n} \quad \text{for some } \bar{n},$$

then for all closed D such that $C \vdash D$, also $C \mid D$.
<u>Proof</u>. By induction on the logical complexity of D.
(i) For prime formulae D the assertion is obvious.
(ii) If $D \equiv D_1 \& D_2$, then $C \vdash D$ implies $C \vdash D_1$, $C \vdash D_2$, hence $C \mid D_1$ & $C \mid D_2$, so $C \mid D$.
(iii) If $D \equiv D_1 \vee D_2$, then $C \vdash D$ implies by hypothesis $C \vdash D_1$ or $C \vdash D_2$, hence $C \mid\!\vdash D_1$ or $C \mid\!\vdash D_2$, so $C \mid D_1 \vee D_2$.
(We use here the fact that the assumption of the theorem also implies $\underset{\sim}{HA} \vdash C \to C_1 \vee C_2 \Rightarrow \underset{\sim}{HA} \vdash C \to C_1$ or $\underset{\sim}{HA} \vdash C \to C_2$ for closed $C_1 \vee C_2$, again by $C_1 \vee C_2 \leftrightarrow \exists x[(x=0 \to C_1) \& (x \neq 0 \to C_2)]$.) etc. etc.
<u>Remark</u>. This theorem shows that $C \mid C$ is a necessary and sufficient condition for 3.1.5 (i).

3.1.9. <u>Corollaries</u>.
(i) If $C \mid C$ and $C \leftrightarrow C'$, then $C' \mid C'$.
(ii) If C is a Harrop formula, then $C \mid C$.
<u>Proof</u>. (i) By 3.1.5 and 3.1.8, taking C itself for D.
(ii) Harrop formulas satisfy $\neg\neg C \leftrightarrow C$ (1.10.8); then use (i) and 3.1.6.
<u>Remark</u>. We do not know of a simpler and more straightforward way to obtain invariance of $C \mid C$ under equivalence.

3.1.10. <u>Example</u>. We wish to show by an example that the class of formulae equivalent to a Harrop formula is properly included in the class of formulae C such that $C \mid C$. The example is taken from <u>T.T. Robinson</u> 1965.

By the Rosser version of Gödel's first incompleteness theorem, we can construct for any system $\underset{\sim}{H}$ containing a sufficient amount of arithmetic, a Π^0_1-sentence G such that $\underset{\sim}{H} \nvdash G$, $\underset{\sim}{H} \nvdash \neg G$ on assumption of the consistency of $\underset{\sim}{H}$.

Let G_1 be a rosser sentence of $\underset{\sim}{HA}^c$, then $\underset{\sim}{HA} + \neg G_1$ is consistent; let G_2 be the rosser sentence of $\underset{\sim}{HA} + \neg G_1$.
Let $A \equiv \neg G_1 \to G_2 \vee \neg G_2$.
Then $\underset{\sim}{HA} \nvdash A$, $\underset{\sim}{HA} \nvdash \neg A$. For assume $\underset{\sim}{HA} \vdash \neg G_1 \to G_2 \vee \neg G_2$, then by $\neg G_1 \mid \neg G_1$ it would follow that $\underset{\sim}{HA} \vdash \neg G_1 \to G_2$ or $\underset{\sim}{HA} \vdash \neg G_1 \to \neg G_2$, i.e. $\underset{\sim}{HA} + \neg G_1 \vdash G_2$ or $\underset{\sim}{HA} + \neg G_1 \vdash \neg G_2$, contrary to our assumptions.
Assume $\underset{\sim}{HA} \vdash \neg(\neg G_1 \to G_2 \vee \neg G_2)$, then $\underset{\sim}{HA}^c \vdash \curlywedge$, i.e. $\underset{\sim}{HA}^c \vdash \neg G_1$, contrary to our assumptions.

Now assume $\underset{\sim}{HA} \vdash A \leftrightarrow B$, B a Harrop formula; then $\neg\neg A \leftrightarrow A$, hence $(\neg G_1 \to G_2 \vee \neg G_2) \leftrightarrow (\neg G_1 \to \neg\neg(G_2 \vee \neg G_2))$. But then $\underset{\sim}{HA} \vdash \neg G_1 \to G_2 \vee \neg G_2$, contrary to what we just proved.

Finally, we wish to show $A \mid A$. Assume $\underset{\sim}{HA} + A \vdash \neg G_1$, then also $\underset{\sim}{HA}^c + A \vdash \neg G_1$, i.e. $\underset{\sim}{HA}^c \vdash \neg G_1$, which is impossible. Therefore the implication $A \mid \neg G_1$ and $\underset{\sim}{HA} + A \vdash \neg G_1 \Rightarrow A \vdash G_2 \vee \neg G_2$ is vacuously true.

3.1.11. Theorem. (Application of $\Gamma \mid C$). $\underset{\sim}{HA} \nvdash IP_0^c$. ($IP_0^c$: closed IP_0.)
Proof. Let G_1 be a rosser sentence for $\underset{\sim}{HA}^c$, G_2 a rosser sentence for $\underset{\sim}{HA} + G_1$. Then

(1) $\qquad \underset{\sim}{HA} \nvdash G_1 \to G_2 \vee \neg G_2 \qquad \underset{\sim}{HA}^c \vdash G_1 \to G_2 \vee \neg G_2$.

For assume $\underset{\sim}{HA} \vdash G_1 \to G_2 \vee \neg G_2$. Then, since $G_1 \mid G_1$ (G_1 being negative) $\underset{\sim}{HA} \vdash G_1 \to G_2$ or $\underset{\sim}{HA} \vdash G_1 \to \neg G_2$, i.e. $\underset{\sim}{HA} + G_1 \vdash G_2$ or $\underset{\sim}{HA} + G_1 \vdash \neg G_2$, which has been excluded by our assumptions.
Let $B \equiv G_1 \to G_2 \vee \neg G_2$. Also

(2) $\qquad B \mid B$.

For $B \mid G_1$ and $\underset{\sim}{HA} + B \vdash G_1 \Rightarrow (B \mid G_2$ and $\underset{\sim}{HA} + B \vdash G_2)$ or
$(B \mid \neg G_2$ and $\underset{\sim}{HA} + B \vdash \neg G_2)$
is true because the premiss is vacuous: assume $\underset{\sim}{HA} + G_1 \to (G_2 \vee \neg G_2) \vdash G_1$, then $\underset{\sim}{HA}^c \vdash G_1$ which contradicts our assumptions.
Now consider the following consequence of IP_{PR}^c (= IP_{PR} restricted to closed formulae):

(3) $\qquad B \to [(G_1 \to G_2) \vee (G_1 \to \neg G_2)]$.

Assume $\underset{\sim}{HA} \vdash$ (3). Then, since $B \mid B$

$$\underset{\sim}{HA} + B \vdash G_1 \to G_2 \quad \text{or} \quad \underset{\sim}{HA} + B \vdash G_1 \to \neg G_2$$

i.e.

$$\underset{\sim}{HA}^c \vdash G_1 \to G_2 \quad \text{or} \quad \underset{\sim}{HA}^c \vdash G_1 \to \neg G_2$$

which implies

$$\underset{\sim}{HA}^c + G_1 \vdash G_2 \quad \text{or} \quad \underset{\sim}{HA}^c + G_1 \vdash \neg G_2$$

which is false by assumption. Therefore $\mathbf{HA} \nvdash (3)$.

3.1.12. Addition of M_{PR}.

We can extend the soundness theorem for $|A$ from $\mathbf{HA}$ to $\mathbf{HA} + M_{PR}$, assuming ω-consistency of $\mathbf{HA} + M_{PR}$, and using M_{PR} on the meta-level.

For this extension we have to show that for any instance F of M_{PR}, $|$ holds. Let F be

$$\neg \forall x \neg A(x,y) \rightarrow \exists x A(x,y) .$$

$|F$ is equivalent to: for each numeral $\bar{n}$,

$$| \neg \forall x \neg A(x,\bar{n}) \rightarrow \exists x A(x,\bar{n})$$

which in turn is

$$| \vdash \neg \forall x \neg A(x,\bar{n}) \Rightarrow \exists m (\| \!\!\vdash A(\bar{m},\bar{n})) .$$

Now assume $\vdash \neg \forall x \neg A(x,\bar{n})$, then not for all $\bar{m}$, $\vdash \neg A(\bar{m},\bar{n})$ (because of the ω-consistency of $\mathbf{HA} + M_{PR}$) hence, using M_{PR} metamathematically, $\vdash A(\bar{m},\bar{n})$ for some numeral $\bar{m}$, hence also $\| \!\!\vdash A(\bar{m},\bar{n})$ (assuming A to be prime).

Thus we may establish DP, ED for $\mathbf{HA} + M_{PR}$.

3.1.13. Definition (of a variant of $\Gamma | C$).

This variant was introduced in de Jongh B , and enables us to obtain some results also for open formulas. Below, E is a single formula, not necessarily closed. The definition of $E \underset{\sim}{|} A$ is by induction on the complexity of A.

(i) $E \underset{\sim}{|} P \equiv E \rightarrow D$ for P prime

(ii) $E \underset{\sim}{|} A \& B \equiv E \underset{\sim}{|} A \& E \underset{\sim}{|} B$

(iii) $E \underset{\sim}{|} A \vee B \equiv [E \underset{\sim}{|} A \& (E \rightarrow A)] \vee [(E \underset{\sim}{|} B) \& (E \rightarrow B)]$

(iv) $E \underset{\sim}{|} A \rightarrow B \equiv (E \underset{\sim}{|} A) \& (E \rightarrow A) \rightarrow E \underset{\sim}{|} B$

(v) $E \underset{\sim}{|} \forall x A x \equiv \forall x (E \underset{\sim}{|} A x)$

(vi) $E \underset{\sim}{|} \exists x A x \equiv \exists x ((E \underset{\sim}{|} A x) \& (E \rightarrow A x))$.

Note that $E \underset{\sim}{|} A$ is represented by a formula of $\mathbf{HA}$, in contrast to $E | A$, which is a metamathematical property of E, A.

3.1.14. Theorem. $\mathbf{HA} \vdash E \rightarrow A \Rightarrow \mathbf{HA} \vdash E \underset{\sim}{|} E \rightarrow E \underset{\sim}{|} A$.

Proof. By induction on the length of derivations, one shows $\mathbf{HA} + E \vdash A \Rightarrow \mathbf{HA} + E \underset{\sim}{|} E \vdash E \underset{\sim}{|} A$.

The details are very similar to those in 3.1.4, and are therefore omitted; or the reader may consult de Jongh B

3.1.15. <u>Corollaries</u>. In $\underset{\sim}{\text{HA}}$:

(i) $\vdash A \to \exists x Bx \Rightarrow A \underset{\sim}{\downarrow} A \vdash \exists x(A \to Bx)$

$\vdash A \to B \vee C \Rightarrow A \underset{\sim}{\downarrow} A \vdash (A \to B) \vee (A \to C)$

(ii) $\vdash \neg A \to \exists x Bx \Rightarrow \vdash \exists x(\neg A \to Bx)$ (IPR)

$\vdash \neg A \to B \vee C \Rightarrow \vdash (\neg A \to B) \vee (\neg A \to C)$.

<u>Proof</u>. (i) Assume $\vdash A \to \exists x Bx$, then $\vdash A \underset{\sim}{\downarrow} A \to A \underset{\sim}{\downarrow} \exists x Bx$, hence $\vdash A \underset{\sim}{\downarrow} A \to \exists x((A \underset{\sim}{\downarrow} Bx) \,\&\, (A \to Bx))$, so $A \underset{\sim}{\downarrow} A \vdash \exists x(A \to Bx)$; the second assertion is a direct consequence of the first one.

(ii). Note that $\vdash \neg A \underset{\sim}{\downarrow} \neg A$, since $\neg A \underset{\sim}{\downarrow} A \,\&\, \neg A \to A$ implies $\neg A \to 1 = 0$; apply (i).

3.1.16 - 3.1.18. <u>A method of dealing with variables using partial reflection principles</u>.

3.1.16. We shall describe a simple instance of a method, first used by Kreisel in <u>Kreisel</u> 1959A, to obtain proof-theoretic closure conditions for instances where parameters are present, by means of partial reflection principles.

Let $Pr_n(y) \equiv_{def} \exists x\, Proof_n(x,y)$, where $Proof_n$ is defined as in 1.5.1.

If $A(x_1,\dots,x_n)$ is a formula, $\ulcorner A(\bar{x}_1,\dots,\bar{x}_n)\urcorner$ may be conceived as a function of $x_1,\dots,x_n$ (a suggestive notation is $\ulcorner A \urcorner(x_1,\dots,x_n)$; but then it should be tacitly understood that only closed formulae have a gödel<u>number</u>, formulae with free variables have a function assigned to them). We define in $\underset{\sim}{\text{HA}}$ the formalized $\Gamma \mid C$ - relation, for empty Γ, restricted to complexity $\leq n$ (notation $|_n$) as follows ($x_1,\dots,x_n$ containing all the variables free in P, A, B) :

(i) $|_n\, P(x_1,\dots,x_n) \equiv Pr_n(\ulcorner P(\bar{x}_1,\dots,\bar{x}_n)\urcorner)$

(ii) $|_n\, (A \,\&\, B) \equiv (|_n\, A) \,\&\, (|_n\, B)$

(iii) $|_n\, (A \vee B) \equiv (|_n\, A \,\&\, Pr_n(\ulcorner A(\bar{x}_1,\dots,\bar{x}_n)\urcorner)) \vee (|_n\, B \,\&\, Pr_n(\ulcorner B(\bar{x}_1,\dots,\bar{x}_n)\urcorner))$

(iv) $|_n\, (A \to B) \equiv (|_n\, A) \,\&\, Pr_n(\ulcorner A(\bar{x}_1,\dots,\bar{x}_n)\urcorner) \to |_n\, B$

(v) $|_n\, \forall x Ax \equiv \forall x(|_n\, Ax)$

(vi) $|_n\, \exists x Ax \equiv \exists x(|_n\, Ax \,\&\, Pr_n(\ulcorner A(\bar{x},\bar{x}_1,\dots,\bar{x}_n)\urcorner)$.

Then we prove

3.1.17. <u>Theorem</u>. If $\vdash_n$ denotes provability in $\underset{\sim}{\text{HA}}$, restricted to formulae of logical complexity $\leq n$, then

$$\vdash_n A \Rightarrow \underset{\sim}{\text{HA}} \vdash (|_n\, A) .$$

<u>Proof</u>. Completely parallel to the proof of 3.1.4 ; we have to use repeatedly the fact that whenever $\vdash_n A(x_1,\dots)$, we can also show $\vdash Pr_n(\ulcorner A(\bar{x}_1,\dots)\urcorner)$. More details in <u>de Jongh</u> B

3.1.18. Closure under Church's rule CR.

Now we are able to make applications. For example, assume $\mathbf{HA} \vdash \forall x \exists y A(x,y)$. Then also $\vdash_n \exists y A(x,y)$ for suitable n; hence $\vdash (|_n \exists y A(x,y))$, i.e. $\vdash \exists y (|_n A(x,y) \mathbin{\&} Pr_n(\ulcorner A(\bar{x},\bar{y})\urcorner))$. Hence we find $\vdash \forall x \exists y \exists z\, Proof_n(z, \ulcorner A(\bar{x},\bar{y})\urcorner)$, which implies, since $Proof_n$ is a recursive predicate, that for some numeral $\bar{n}$

$$\vdash \forall x\, \exists y[T(\bar{n},x,y) \mathbin{\&} Pr_n(\ulcorner A(\bar{x}, \overline{Uy})\urcorner)].$$

By the partial reflection principle,

$$\vdash_n \forall x\, \exists y[T(\bar{n},x,y) \mathbin{\&} A(x,Uy)].$$

Thus we have shown closure under Church's rule (CR).

For further applications of partial reflection principles of the same kind, see chapter IV, § 4.

3.1.19. Extension and generalization of $\Gamma | C$ to higher-order systems.

In Moschovakis 1967, $\Gamma | C$ has been extended to certain systems of intuitionistic analysis containing function variables. In order to prove a soundness theorem, one usually has to extend the systems considered with uncountably many additional function symbols (since in the usual systems constructed from a denumerable set of symbols, not every function has a name in the system).

Friedman A considers a generalization of $\Gamma | C$ applicable to various higher-order systems with species variables, such as the theory of finite types with impredicative comprehension. Here also one has to consider certain definitional extensions of the systems one is interested in, to provide enough "names" of objects for establishing a soundness theorem. In 3.1.21 - 3.1.23 we have described Friedman's method for the simplest interesting case: $\mathbf{HAS}$.

$\Gamma | C$ in its original form, without complicated tricks, quite easily extends to $\mathbf{HAS}_0 + PCA$. To see this, we add species constants C_A, A a formula of $\mathbf{HA}$, with axioms

$$C_{A(x_1,\ldots,x_n)}(x_1,\ldots,x_n) \leftrightarrow A(x_1,\ldots,x_n)$$

to $\mathbf{HAS}_0 + PCA$; the resulting system $\mathbf{H}$ is obviously a definitional extension of $\mathbf{HAS}_0 + PCA$. We then add to the clauses for $\Gamma | C$ in 3.1.2 ($\vdash$ referring now to $\mathbf{H}$):

(vii) $\Gamma | \forall X A(X) \equiv \Gamma | A(C_B)$ for all C_B (with the right number of arguments)

(viii) $\Gamma | \exists X A(X) \equiv \Gamma | A(C_B)$ and $\mathbf{H} \vdash A(C_B)$ for some C_B.

Clause (i) is extended by $\Gamma | C_B(\bar{x}_1,\ldots,\bar{x}_n) \equiv \Gamma | B(\bar{x}_1,\ldots,\bar{x}_n)$.

If $A(x_1,\ldots x_n, X_1,\ldots,X_m)$ is a formula of $\underset{\sim}{H}$ containing at most $x_1,\ldots,x_n, X_1,\ldots,X_m$ free, then $\Gamma \mid A(x_1,\ldots,x_n, X_1,\ldots,X_m)$ is defined as $\Gamma \mid A(\bar{x}_1,\ldots,\bar{x}_n, C_{B_1},\ldots,C_{B_m})$ for all numerals $\bar{x}_1,\ldots,\bar{x}_n$, and all constants $C_{B_1},\ldots,C_{B_m}$ with the appropriate number of arguments.

The idea of this extension is simply this: in order to prove a soundness theorem, we need at least a name for each definable species. Since the family of all arithmetical species is a model for $\underset{\sim}{HAS}_0 + PCA$, it is obvious that the arithmetical species are exactly the definable ones; hence we add constants for each arithmetical species. For an application of this extension, see 3.1.20.

Extremely similar is an extension to $\underset{\sim}{N}-\underset{\sim}{HA}^{\omega}$, $\underset{\sim}{I}-\underset{\sim}{HA}^{\omega}$, $\underset{\sim}{E}-\underset{\sim}{HA}^{\omega}$, where we put

$$\Gamma \mid \forall x^{\sigma} A x^{\sigma} \text{ iff } \Gamma \mid A t^{\sigma} \text{ for all closed } t^{\sigma}$$
$$\Gamma \mid \exists x^{\sigma} A x^{\sigma} \text{ iff } \Gamma \Vdash A t^{\sigma} \text{ for some closed } t^{\sigma}.$$

This yields DP, ED' for $\underset{\sim}{N}-\underset{\sim}{HA}^{\omega}$ etc.

The methods of Moschovakis 1967 can also be readily extended to the systems IDB and IDB_1 (1.9.18). The proof is for the greater part routine, see our remarks in 3.1.24.

3.1.20. Theorem. (i). The soundness theorem for $\Gamma \mid C$ extends to $\underset{\sim}{H}$ as defined in 3.1.19 (under the assumption that $\Gamma \mid E$ for $E \in \Gamma$).
(ii). For $\underset{\sim}{HAS}_0 + PCA$ ($\neg A$, $\exists x Bx$, $C \vee D$, $\exists X E(X)$ closed):

$\vdash \neg A \to \exists x Bx \Rightarrow \vdash \neg A \to B\bar{n}$ for some $\bar{n}$.
$\vdash \neg A \to C \vee D \Rightarrow \vdash (\neg A \to C)$ or $\vdash (\neg A \to D)$
$\vdash \neg A \to \exists X E(X) \Rightarrow \vdash \neg A \to \exists X(\forall x_1\ldots x_n(F(x_1\ldots x_n) \leftrightarrow X x_1\ldots x_n) \,\&\, E(X))$, for some F.

Proof. (i) We have to verify the quantifier rules and axioms and the comprehension schema.

Q1). Assume $\Gamma \vdash C \to A(X)$, $\Gamma \mid C \to A(X)$. Let $\Gamma \Vdash C$, then $\Gamma \mid A(C_B)$ for all C_B, so $\Gamma \mid \forall X A(X)$.
Hence $\Gamma \mid C \to \forall X A(X)$.

Q2). $\Gamma \mid \forall X A(X) \to A(T)$, where T is a species variable or constant, is verified as follows: let $\Gamma \Vdash \forall X A(X)$, then $\Gamma \mid A(Y)$ and $\Gamma \mid A(C_B)$ by definition.

Q3). $\Gamma \mid A(T) \to \exists X A(X)$, T a species variable or constant, is also immediate.

Q4). Let $\Gamma \Vdash A(X) \to C$, assume $\Gamma \Vdash \exists X A(X)$. Then $\Gamma \Vdash A(C_B)$ for a suitable constant C_B, and since also $\Gamma \Vdash A(C_B) \to C$, $\Gamma \mid C$.

P A). We restrict ourselves to the case where A contains a single (unary) species variable X free, and a single numerical variable x. Consider the instance

$$\forall X \exists Y \forall x[A(X,x) \leftrightarrow Yx] .$$

We have to show

$$\forall C_B \exists C_{B'} \forall x(\Gamma \Vdash A(C_B,\bar{x}) \Leftrightarrow \Gamma \Vdash C_{B'}\bar{x}) .$$

This is simple : replace in $A(X,x)$ every occurrence of Xt by Bt . The result is a formula $B'x$.
Now let $\Gamma \Vdash A(C_B,\bar{x})$; since $A(C_B,\bar{x}) \leftrightarrow B'\bar{x}$ is provable, we obtain $\Gamma \Vdash B'\bar{x}$, hence $\Gamma \Vdash C_{B'}\bar{x}$.
Conversely, if $\Gamma \Vdash C_{B'}\bar{x}$, then $\Gamma \Vdash B'\bar{x}$ etc.
(ii) The first two assertions are proved in the same manner as before.
The third assertion is established as follows :
Let $\vdash \neg A \rightarrow \exists X E(X)$, then also $H \vdash \neg A \rightarrow \exists X E(X)$, hence $H + \neg A \Vdash \exists X E(X)$ (using that $\neg A \mid \neg A$) ; therefore $H + \neg A \Vdash E(C_B)$, hence $H \vdash \neg A \rightarrow E(C_B)$.

3.1.21 - 3.1.23. Treatment of HAS .

3.1.21. Definition. We define a conservative extension H of HAS by adding constants $C_{B,V}$ for every formula B of H not containing species variables free, and for all sets V of n - tuples of closed terms, if B contains n numerical variables free, and adding axioms

$$C_{B,V}(x_1,\ldots,x_n) \leftrightarrow B(x_1,\ldots,x_n) .$$

H is obviously a definitional extension. Now we define $R(A)$ by induction on the complexity of A , as follows ($\vdash$ referring to deducibility in H) :

(i) $R(t=s) \equiv \vdash t = s$
$R(C_{B,V}(t_1,\ldots,t_n)) \equiv (t_1,\ldots,t_n) \in V$
(ii) $R(A \& B) \equiv R(A)$ and $R(B)$
(iii) $R(A \vee B) \equiv (R(A)$ and $\vdash A)$ or $(R(B)$ and $\vdash B)$
(iv) $R(A \rightarrow B) \equiv R(A)$ and $\vdash A \Rightarrow R(B)$
(v) $R(\forall x A x) \equiv$ For all $\bar{x}$, $R(A\bar{x})$
(vi) $R(\exists x A x) \equiv$ For some $\bar{x}$, $R(A\bar{x})$ and $\vdash A(\bar{x})$
(vii) $R(\forall X A(X) \equiv$ For all $C_{B,V}, R(A(C_{B,V}))$
(viii) $R(\exists X A(X) \equiv$ For some $C_{B,V}, R(A(C_{B,V}))$ and $\vdash A(C_{B,V})$.

We shall put $R(A(x_1,\ldots,x_n, X_1,\ldots,X_m))$ $(x_1,\ldots,x_n, X_1,\ldots,X_m$ containing all the variables free in A) if $R(A(\bar{x}_1,\ldots,\bar{x}_n, C_{B_1,V_1},\ldots)$ for all numerals $\bar{x}_1,\ldots,\bar{x}_n$, all constants $C_{B_1,V_1},\ldots,C_{B_m,V_m}$

3.1.22. Theorem (Soundness theorem ; adapted from Friedman A).
Let HAS $\vdash A$, then $R(A)$.

Proof. The verification for the axioms and rules of arithmetic and predicate logic is completely similar to the proof of 3.1.4, with Γ empty. It remains to verify the comprehension schema.

For simplicity, consider the instance

$$\forall Y \exists X \forall x[A(Y,x) \leftrightarrow Xx] ,$$

A not containing species variables free besides Y.

We have to show for all $C_{B,V}$

$$R(\exists X \forall x[A(C_{B,V},x) \leftrightarrow Xx]) .$$

Now take any $C_{B,V}$ and let $B'x \equiv A(C_{B,V},x)$; we wish to show

$$R(\forall x[A(C_{B,V},x) \leftrightarrow C_{B',W}\, x]) \;\&\; \vdash \forall x[A(C_{B,V},x) \leftrightarrow C_{B',W}\, x]$$

where

$$W = \{t \mid R(A(C_{B,V},t),\ t \text{ closed}\} .$$

Hence, since $\vdash \forall x[A(C_{B,V},x) \leftrightarrow C_{B',W}\, x]$ is obvious, it remains to be shown that

$$R(A(C_{B,V},\bar{x})) \;\&\; \vdash A(C_{B,V},\bar{x}) \Leftrightarrow R(C_{B',W}\,\bar{x}) \;\&\; \vdash C_{B',W}\,\bar{x} .$$

First, assume $\vdash A(C_{B,V},\bar{x})$, $R(A(C_{B,V},\bar{x}))$.
Then also $\vdash C_{B',W}\,\bar{x}$; and since $R(C_{B',W}\,\bar{x}) \Leftrightarrow \bar{x} \in W \Leftrightarrow R(A(C_{B,V},\bar{x}))$, $R(C_{B',W}\,\bar{x})$ too.
Conversely, assume $R(C_{B',W}\,\bar{x}) \;\&\; \vdash C_{B',W}\,\bar{x}$; then obviously $\bar{x} \in W$, hence $R(A(C_{B,V},\bar{x}))$; also $\vdash A(C_{B,V},\bar{x})$.

Remark on the proof. The method of defining R may be motivated as follows. If we attempt to extend the original definition of $|A$ to the second-order case, we encounter the following problem. The natural formulation for $|\ \forall X A(X)$ would be: $|\ \forall X A(X)$ iff for each substitution of a predicate B for X, $|A(B)$. But since the logical complexity of B can be arbitrarily large, the definition $|$ is not well founded, i.e. $|A$ is not defined in terms of $|B$ for formulae B with complexity less than A.

For this reason, the constants $C_{B,V}$ are introduced; they are treated as having logical complexity 0. Instead of asking

$$R(C_{B,V}(t_1,\ldots,t_n)) \quad \text{iff} \quad R(B(t_1,\ldots,t_n)) ,$$

which would be expected in view of the axioms for $C_{B,V}$, but which would make the definition of R not well founded, we use the (arbitrary) set V to determine $R(C_{B,V}(t_1,\ldots,t_n))$.

Using the freedom made possible by the set V, we defined a V by reference to the notion R itself (an example of an impredicative procedure), in order to establish $R(F)$ for instances F of the comprehension schema.

We cannot establish the soundness theorem in the form: $\underset{\sim}{H} \vdash A \Rightarrow R(A)$, since then we would be required to show $R(C_{B,V}(\bar{x}_1,\ldots,\bar{x}_n) \leftrightarrow B(\bar{x}_1,\ldots,\bar{x}_n))$ for arbitrary numerals $\bar{x}_1,\ldots,\bar{x}_n$, leading back to the problems we so carefully tried to avoid.

3.1.23. Corollary. In HAS ($\exists xAx$, $B \vee C$, $\exists XD(X)$ closed)

(i) $\vdash \exists xAx \Rightarrow \vdash A\bar{n}$ for a numeral $\bar{n}$

(ii) $\vdash B \vee C \Rightarrow \vdash B$ or $\vdash C$

(iii) $\vdash \exists XD(X) \Rightarrow \vdash \exists X[\forall x_1 \ldots x_n[Xx_1\ldots x_n = A(x_1,\ldots,x_n)] \& D(X)]$, for some A.

Proof. Completely similar to the first-order case for (i), (ii). In (iii), we first obtain $\underset{\sim}{H} \vdash D(T)$ for a closed species term T; then the assertion of (iii) readily follows.

3.1.24. Extension of Moschovakis 1967 to IDB, IDB_1.

The idea of Moschovakis 1967 is to construct suitably conservative extensions of the theories considered by addition of new (possibly uncountably many) function symbols so as to have "names" for all functions which can be shown to exist in the system. The clauses for $\Gamma | A$ are extended with:

$$\Gamma \mid \forall\alpha A\alpha \text{ if } \Gamma \mid A\varphi \text{ for each closed functor } \varphi$$

$$\Gamma \mid \exists\alpha A\alpha \text{ if } \Gamma \Vdash A\varphi \text{ for some closed functor } \varphi.$$

In the case of IDB, IDB_1 we have to add to the work done in Moschovakis 1967 a verification of $| C$ for instances C of K1, K2, K3, to obtain $\underset{\sim}{H} \vdash A \Rightarrow | A$ for $\underset{\sim}{H} \equiv IDB$, IDB_1. K1, K2 do not cause us trouble. The instances of K3 may be verified with the help of the lemma (for a proof see Kreisel and Troelstra 1970, 3.2.1):

$$Ka \;\&\; \forall n(an \neq 0 \rightarrow Pn) \;\&\; \forall n(\forall x(P(n * \hat{x}) \rightarrow Pn) \rightarrow P0$$

(induction over unsecured sequences). We use this lemma on the meta-level. For simplicity, let us assume Q to be a formula with a single free variable α; let $\varphi_1,\varphi_2,\ldots$ range over closed functors. Assume

(1) $\Vdash \forall\alpha(A_K(Q,\alpha) \rightarrow Q\alpha)$

(2) $\Vdash K\varphi_1$.

So

(3) $\underset{\sim}{\forall}\varphi_2 \Vdash A_K(Q,\varphi_2) \rightarrow Q\varphi_2$.

Let us put $\varphi_{1,n} = \lambda m.\varphi_1(\bar{n} * m)$.

(a) If $\varphi_1\bar{n} \neq 0$, say $\varphi_1\bar{n} = S\bar{p}$, then $\vdash \varphi_{1,\bar{n}}(x) = S\bar{p}$, so $\forall x(| \varphi_{1,\bar{n}}\bar{x} = S\bar{p})$, i.e. $\Vdash \forall x(\varphi_{1,\bar{n}}x = S\bar{p})$. Hence also $\Vdash \exists y \forall x(\varphi_{1,\bar{n}}x = Sy)$, i.e. $\Vdash A_K(Q, \varphi_{1,\bar{n}})$, hence by (3) $|Q(\varphi_{1,\bar{n}})$.

(b) Assume $\underset{\sim}{\forall} y \mid Q(\varphi_{1,\bar{n} * \langle\bar{y}\rangle})$.
If $\varphi_{1,\bar{n}}0 \neq 0$, then $\Vdash Q(\varphi_{1,\bar{n}})$ by (a). Let $\varphi_{1,\bar{n}}0 = 0$. By (1) and (2) $\vdash \forall y K\varphi_{1,\bar{n} * \langle y\rangle}$, $\vdash \forall y Q(\varphi_{1,\bar{n} * \langle y\rangle})$, and since also by hypothesis $\underset{\sim}{\forall} y \mid Q(\varphi_{1,\bar{n} * \langle\bar{y}\rangle})$, it follows that $\Vdash \forall y Q(\varphi_{1,\bar{n} * \langle y\rangle})$, hence also $\Vdash \varphi_{1,\bar{n}}0 = 0 \;\&\; \forall y Q(\varphi_{1,\bar{n} * \langle y\rangle})$. Now apply (3) and we find $\mid Q(\varphi_{1,\bar{n}})$.
Therefore

$$\underset{\sim}{\forall} y(\mid Q(\varphi_{1,\bar{n} * \langle\bar{y}\rangle})) \Rightarrow \mid Q(\varphi_{1,\bar{n}}) .$$

Apply now our lemma on the meta level, and we find $\mid Q(\varphi_1)$. Therefore

$$\mid \forall\alpha(A_K(Q,\alpha) \to Q\alpha) \to \forall\alpha(K\alpha \to Q\alpha) .$$

Q. e. d.

<u>Remark.</u> TI(<) can be dealt with in a very similar way.

3.1.25. <u>Concluding remarks.</u>

$\Gamma \mid C$ and its variants demonstrate various devices and phenomena in a simple context which one meets also in realizability notions and normalization theorems for natural deduction systems.

For example, the original non-formalized version of Kleene's $\Gamma \vdash$ realizability was similar to $\Gamma \mid C$, and could not deal directly with derived or admissible rules involving parameters. However, while the formalized variant $E \underset{\sim}{\mid} C$ yields certain results with parameters, the existential instantiation rule does not generalize thus; but in the case of its analogue $\underset{\approx}{q}$ - realizability, discussed in the next section, we obtain such a generalization: Church's rule. In this section, we used a formalization plus partial reflection principles to obtain this.

The introduction of constants $C_{A,V}$, in Friedman's treatment of $\underset{\sim}{\text{HAS}}$ by means of a concept generalized from $\Gamma \mid C$, is very similar to the use of Girard's "candidats de réductibilité" in <u>Girard</u> 1971, Prawitz's assignments η in Appendix B of <u>Prawitz</u> 1971, and the use of computability predicates in <u>Martin-Löf</u> 1971A.

§ 2. Realizability notions based on partial recursive function application.

3.2.1. Introduction. Realizability (by numbers) was first introduced by S.C. Kleene in Kleene 1945, and was intended as a kind of reinterpretation of intuitionistic arithmetic, so as to bring out more explicitly the intended constructive interpretation of the logical operators. As such, it may be viewed as a variant of the abstract interpretation schema first introduced by Heyting (see Heyting 1934, 1956 A), elaborated and made more precise in Kreisel 1962D, Kreisel 1965, 2.3 (for an informal description, see e.g. Troelstra 1969, § 2). As we shall see from the definition and results in the sequel, Kleene's notion is not just a variant of, but essentially differs from the interpretation intended by Heyting. Hence, it cannot be said to make the intended meaning of the logical operators more precise. As a "philosophical reduction" of the interpretation of the logical operators it is also only moderately successful ; e.g. negative formulae are essentially interpreted by themselves.

On the other hand, realizability possesses some nice formal properties, which provide it with some matematical interest of its own ; but more important, realizability and the many variants deriving from it turn out to be very convenient tools in the development of intuitionistic proof theory. In this section, we restrict attention to realizability and variants based on partial recursive function application ; in the next section we turn to realizability notions based on continuous function application. Many details might have been developed simultaneously for those two concepts of application, but, as long as an elegant axiomatic theory for partially defined application is lacking, only at the cost of considerable notational complexity, thereby obscuring the simplicity of the underlying ideas.

Contents of the section. Subsections 2 - 8 are devoted to the definition of realizability, the soundness theorem, and some direct consequences of these.

In subsections 9 - 19 $\underset{\sim}{r}$ - realizability is characterized, and ECT_0 introduced.

Subsections 21 - 22 extend this to $\underset{\sim}{HA} + M$, subsections 23 - 24 to $\underset{\sim}{HA} + TI(<)$, and $\underset{\sim}{HA} + M + TI(<)$. Subsections 25 - 26 describe realizability provable in $\underset{\sim}{HA}^c$, with an application. The non-realizability of IP is discussed in 27 - 28. Subsections 29 - 31 are devoted to extensions of the results to some other systems such as $\underset{\sim}{IDB}$, $\underset{\sim}{HAS}$.

Subsection 32 describes possible generalizations, with as an application the result that $\underset{\sim}{HA}^c$ is not finitely axiomatizable over $\underset{\sim}{HA}$.

Subsection 33 compares $\underset{\sim}{q}$ - and $\underset{\sim}{r}$ - realizability.

3.2.2. Definition. For each formula $A(x_1,\ldots,x_n)$ of HA containing at most $x_1,\ldots,x_n$ free, we shall construct another formula of HA, denoted by $x\,\underset{\approx}{r}_P A(x_1,\ldots,x_n)$ containing (at most) $x,x_1,\ldots,x_n$ free, $x \notin \{x_1,\ldots,x_n\}$; $x\,\underset{\approx}{r}_P A(x_1,\ldots,x_n)$ is to be called the (P-)realizability predicate of A. Here P(A) is assumed to be a property of A expressible by a formula of HA, containing at most $x_1,\ldots,x_n$ free. The definition is by induction on the logical complexity of A.

$\underset{\approx}{r}_P$(i) $\quad x\,\underset{\approx}{r}_P Q \equiv_{def} Q$ for Q prime

$\underset{\approx}{r}_P$(ii) $\quad x\,\underset{\approx}{r}_P (A\&B) \equiv_{def} (j_1x\,\underset{\approx}{r}_P A)\ \&\ (j_2x\,\underset{\approx}{r}_P B)$

$\underset{\approx}{r}_P$(iii) $\quad x\,\underset{\approx}{r}_P (A\vee B) \equiv_{def} [(j_1x=0 \rightarrow (j_2x\,\underset{\approx}{r}_P A) \& P(A))\ \&\ (j_1x\neq 0 \rightarrow (j_2x\,\underset{\approx}{r}_P B) \& P(B))]$

$\underset{\approx}{r}_P$(iv) $\quad x\,\underset{\approx}{r}_P (A\rightarrow B) \equiv_{def} \forall u((u\,\underset{\approx}{r}_P A) \& P(A) \rightarrow \exists v(T(x,u,v)\ \&\ Uv\,\underset{\approx}{r}_P B))$

$\underset{\approx}{r}_P$(v) $\quad x\,\underset{\approx}{r}_P (\exists yBy) \equiv_{def} (j_2x\,\underset{\approx}{r}_P B(j_1x)) \& P(A(j_1x))$

$\underset{\approx}{r}_P$(vi) $\quad x\,\underset{\approx}{r}_P (\forall yBy) \equiv_{def} \forall y\, \exists z(T(x,y,z)\ \&\ Uz\,\underset{\approx}{r}_P By)$.

3.2.3. Examples.

A). P(A) is universal, e.g. $P(A) \equiv (0=0)$. In this case, P may be omitted (modulo logical equivalence) in the definition of $\underset{\approx}{r}_P$. The result is a formalized version of Kleene's original realizability (a formalized version was first developed in Nelson 1947). In this case we write $x\,\underset{\approx}{r}A$ for $x\,\underset{\approx}{r}_P A$, and we speak about "$\underset{\approx}{r}$-realizability" instead of P-realizability.

B). $P(A) \equiv A$. We call the resulting notion $\underset{\approx}{q}$-realizability, and we write $x\,\underset{\approx}{q}A$ for $x\,\underset{\approx}{r}_P A$.

C). $P(A(x_1,\ldots,x_n)) \equiv \mathrm{Prov}(\ulcorner A(\bar{x}_1,\ldots,\bar{x}_n)\urcorner)$.

D). Let us call the realizability notion to be introduced in this example $\underset{\approx}{p}$-realizability, and let us write $x\,\underset{\approx}{p}A$ for $x\,\underset{\approx}{r}_P A$. Then we define

$$P(A(x_1,\ldots,x_n)) \equiv \mathrm{Prov}(\ulcorner \exists x(x\,\underset{\approx}{p}A(\bar{x}_1,\ldots,\bar{x}_n))\urcorner).$$

In other words, P(A) and $x\,\underset{\approx}{p}A$ have to be introduced simultaneously. This notion was introduced in Beeson 1972; for an application of a slightly modified notion see § 9 in this chapter.

E). Generalization of (B): $P(A) \equiv C \rightarrow A$, C fixed, closed, etc. etc.

3.2.4. Theorem (Soundness).

(i). Let P be any property defined relative to every formula A of HA, and expressed by a formula P(A) of HA, such that

(A) $\mathbf{HA} \vdash A \Rightarrow \mathbf{HA} \vdash P(A)$

(B) $\Gamma \vdash P(A),\ \Gamma \vdash P(A\rightarrow B) \Rightarrow \Gamma \vdash P(B)$,

or equivalently $\vdash P(A)\ \&\ P(A\rightarrow B) \rightarrow P(B)$.

Then, for closed A:

(1) $\mathbf{HA} \vdash A \Rightarrow \exists n(\mathbf{HA} \vdash \bar{n}\,\underset{\approx}{r}_P A)$.

In fact, the numeral $\bar{n}$ in (1) does not depend on P.

(ii). Let $\underset{\sim}{H} \equiv \underset{\sim}{HA} + \Gamma$ be obtained by adding a set of closed axioms Γ to $\underset{\sim}{HA}$, and suppose (A), (B) to be fulfilled w.r.t. $\underset{\sim}{H}$ instead of $\underset{\sim}{HA}$. Assume moreover

(C) $A \in \Gamma \Rightarrow \underset{\sim}{H} \vdash \exists x(x \underset{\approx}{r}_P A)$.

Then, for closed A

(2) $\underset{\sim}{H} \vdash A \Rightarrow \underset{\sim}{H} \vdash \exists x(x \underset{\approx}{r}_P A)$.

(iii). Let $\underset{\sim}{H}$ be as in (ii), and assume instead of (C):

(C') $A \in \Gamma \Rightarrow \exists n(\underset{\sim}{H} \vdash \bar{n} \underset{\approx}{r}_P A)$

(D) $\underset{\sim}{H}$ is conservative over $\underset{\sim}{HA}$ w.r.t. closed Σ_1^0-formulae.

Then, for closed A

(3) $\underset{\sim}{H} \vdash A \Rightarrow \exists n(\underset{\sim}{H} \vdash \bar{n} \underset{\approx}{r}_P A)$

($\bar{n}$ in fact not depending on P).

Proof. (i). By induction on the length of deductions. We select the system described in 1.1.4 for our verifications. The induction step (automatically also including the basis step in this case) splits into a number of cases corresponding to the axiom schema or rule applied to obtain the end formula of the deductions. For each instance F of an axiom schema, we establish $\exists n(\underset{\sim}{HA} \vdash \bar{n} \underset{\approx}{r}_P F^*)$ (where F^* denotes the universal closure of F), and for any application of a rule: $F_1, \ldots, F_n \Rightarrow F$ we show that assuming $F_1^*, \ldots, F_n^*$, $\bar{n}_1 \underset{\approx}{r}_P F_1^*, \ldots, \bar{n}_k \underset{\approx}{r}_P F_k^*$ (for some $\bar{n}_1, \ldots, \bar{n}_k$), and $P(F_1^*), \ldots, P(F_n^*)$ all to hold, we can establish in $\underset{\sim}{HA}$ $\bar{n} \underset{\approx}{r}_P F^*$ for some $\bar{n}$.

PL2). Assume $\forall x Ax$, $\forall x(Ax \to Bx)$, $\bar{n} \underset{\approx}{r}_P \forall x Ax$, $\bar{m} \underset{\approx}{r}_P \forall x(Ax \to Bx)$, $\forall x P(Ax)$, $\forall x(P(Ax \to Bx))$. Then $\{\bar{n}\}(x) \underset{\approx}{r}_P Ax$, $\{\bar{m}\}(x) \underset{\approx}{r}_P (Ax \to Bx)$, so $\{\bar{n}\}(x) \underset{\approx}{r}_P Ax \,\&\, P(Ax)$, therefore $\{\{\bar{m}\}(x)\}(\{\bar{n}\}(x)) \underset{\approx}{r}_P Bx$.
Thus $\Lambda x.\{\{\bar{m}\}(x)\}(\{\bar{n}\}(x)) \underset{\approx}{r}_P \forall x Bx$. Also $\forall x P(Bx)$.

Below we shall usually consider the cases without additional free parameters, for simplicity.

PL3). Assume $A \to B$, $B \to C$, $\bar{n} \underset{\approx}{r}_P A \to B$, $\bar{m} \underset{\approx}{r}_P B \to C$, $P(A \to B)$, $P(B \to C)$. Then, if we assume $(x \underset{\approx}{r}_P A) \,\&\, P(A)$, it follows that $P(B)$, (and hence $P(C)$); also $\{\bar{n}\}(x) \underset{\approx}{r}_P B$, $\{\bar{m}\}(\{\bar{n}\}(x)) \underset{\approx}{r}_P C$, so $\Lambda x.\{\bar{m}\}(\{\bar{n}\}(x)) \underset{\approx}{r}_P A \to C$.

PL7). Assume $A \,\&\, B \to C$, $\bar{n} \underset{\approx}{r}_P A \,\&\, B \to C$, $P(A \,\&\, B \to C)$.
Assume $x \underset{\approx}{r}_P A$, $P(A)$, $y \underset{\approx}{r}_P B$, $P(B)$.
It follows that $P(A \to (B \to C))$, hence $P(C)$. Also $j(x,y) \underset{\approx}{r}_P A \,\&\, B$; also by $P(A \to (B \to A \,\&\, B))$, $P(A)$, $P(B)$ we have $P(A \,\&\, B)$, hence $\{\bar{n}\}(j(x,y)) \underset{\approx}{r}_P C$.
Thus $\Lambda x \Lambda y.\{\bar{n}\}(j(x,y)) \underset{\approx}{r}_P A \to (B \to C)$.

PL8). Assume $A \to (B \to C)$, $\bar{n} \underset{\approx}{r}_P A \to (B \to C)$, $P(A \,\&\, B \to C)$.

Let $x\,\mathbf{r}_P A \& B$, $P(A \& B)$. Since $P(A \& B \rightarrow A)$, $P(A \& B \rightarrow B)$, it follows that $P(A)$, $P(B)$; thus $\{\{\bar{n}\}(j_1x\}(j_2x)\,\mathbf{r}_P C$. Therefore $\Lambda x.\{\{\bar{n}\}(j_1x)\}(j_2x)\,\mathbf{r}_P A\&B \rightarrow C$.

PL9). $0\,\mathbf{r}_P \curlywedge \rightarrow A$, and e.g. $\Lambda x.0\,\mathbf{r}_P \forall x(\curlywedge \rightarrow Ax)$.

PL10). $\Lambda x.j_2x\,\mathbf{r}_P A \vee A \rightarrow A$; $\Lambda x.j(x,x)\,\mathbf{r}_P A \rightarrow A \& A$.

PL11). $\Lambda x.j(0,x)\,\mathbf{r}_P A \rightarrow A \vee B$, $\Lambda x.j_1x\,\mathbf{r}_P A \& B \rightarrow A$.

PL12). $\Lambda x.j(1 \dot{-} j_1x, j_2x)\,\mathbf{r}_P A \vee B \rightarrow B \vee A$,

$\Lambda x.j(j_2x, j_1x)\,\mathbf{r}_P A \& B \rightarrow B \& A$.

PL13). Assume $A \rightarrow B$, $P(A \rightarrow B)$, $\bar{n}\,\mathbf{r}_P A \rightarrow B$.

Let $x\,\mathbf{r}_P C \vee A$. Then $j_1x = 0 \& (j_2x\,\mathbf{r}_P C) \& P(C)$, or $j_1x \neq 0 \& (j_2x\,\mathbf{r}_P A) \& P(A)$. If $j_1x = 0$, $x\,\mathbf{r}_P C \vee B$; if $j_1x \neq 0$, $\{\bar{n}\}(j_2x)\,\mathbf{r}_P B$ and $P(B)$, hence $j(j_1x, \{\bar{n}\}(j_2x))\,\mathbf{r}_P C \vee B$.

Thus

$$\Lambda x.[(1 \dot{-} j_1x)x + sgj_1x.j(j_1x, \{\bar{n}\}(j_2x))]\,\mathbf{r}_P C \vee B.$$

Q1). Assume $\forall y \forall x(Cy \rightarrow A(x,y))$, $\forall yx P(Cy \rightarrow A(x,y))$, $\bar{n}\,\mathbf{r}_P \forall yx(Cy \rightarrow A(x,y))$.

Let $z\,\mathbf{r}_P Cy$, $P(Cy)$.

Then $\{\bar{n}\}(y,x,z)\,\mathbf{r}_P A(x,y)$, so $\Lambda x.\{\bar{n}\}(y,x,z)\,\mathbf{r}_P \forall x A(x,y)$, hence $\Lambda y \Lambda z \Lambda x.\{\bar{n}\}(y,x,z)\,\mathbf{r}_P \forall y(Cy \rightarrow \forall x A(x,y))$.

Q2). $\Lambda y.\{y\}(t)\,\mathbf{r}_P \forall x Ax \rightarrow At$.

Q3). $\Lambda y.j(t,y)\,\mathbf{r}_P At \rightarrow \exists x Ax$.

Q4). Assume $\forall x(Ax \rightarrow C)$, $\bar{n}\,\mathbf{r}_P \forall x(Ax \rightarrow C)$, $\forall x(P(Ax \rightarrow C))$.

Then, if $u\,\mathbf{r}_P \exists x Ax$, $P(\exists x Ax)$, it follows that $j_2u\,\mathbf{r}_P A(j_1u)$, $P(A(j_1u))$; hence $\{\bar{n}\}(j_1u, j_2u)\,\mathbf{r}_P C$, so $\Lambda u.\{\bar{n}\}(j_1u, j_2u)\,\mathbf{r}_P(\exists x Ax \rightarrow C)$.

The equality axioms are realized as follows:

$\Lambda x.0\,\mathbf{r}_P \forall x(x=x)$, $\Lambda x \Lambda y \Lambda z \Lambda u.0\,\mathbf{r}_P \forall xyz(x=y \& z=y \rightarrow x=y)$,

$\Lambda x_1 \ldots \Lambda x_n \Lambda u.0\,\mathbf{r}_P \forall x_1 \ldots x_n(x_i = x_i' \rightarrow \varphi(\ldots, x_i, \ldots) = \varphi(\ldots, x_i', \ldots))$;

$\Lambda x \Lambda u.0\,\mathbf{r}_P \forall x(Sx \neq 0)$, $\Lambda x \Lambda y \Lambda z.0\,\mathbf{r}_P \forall xy(Sx = Sy \rightarrow x=y)$.

All defining axioms for primitive recursive functions are realized by 0, hence their universal closure by $\Lambda x_1 \Lambda x_2 \ldots .0$.

Assume $u\,\mathbf{r}_P A0 \& \forall x(Ax \rightarrow A(Sx))$, $P(A0 \& \forall x(Ax \rightarrow A(Sx)))$. We can find a partial recursive function φ such that (recursion theorem)

$$\varphi(u,0) \simeq j_1u,$$
$$\varphi(u,Sx) \simeq \{j_2u\}(x, \varphi(u,x)).$$

By induction we show $\varphi(u,x)$ to be defined for all x, hence $\Lambda u \Lambda x.\varphi(u,x)\,\mathbf{r}_P [A0 \& \forall x(Ax \rightarrow A(Sx)) \rightarrow \forall x Ax]$.

(ii). Assume $\underline{HA} + \Gamma \vdash A$, A closed; then $\underline{HA} \vdash F \rightarrow A$, F a conjunction of formulae from Γ. Hence by (C): $\underline{H} \vdash \exists x(x\,\mathbf{r}_P A)$, using $P(F)$ (obtained from (A) and (B)).

(iii). Similarly, if $\underline{HA} + \Gamma \vdash A$, we find $\underline{HA} \vdash F \rightarrow A$, F a conjunction of

formulae from Γ, hence $\underset{\sim}{H} \vdash ! \{\bar{n}\}(\bar{m}) \& \{\bar{n}\}(\bar{m}) \underset{\approx}{r}_P A$.
Now $! \{\bar{n}\}(\bar{m})$ in $\underset{\sim}{H}$, hence by (D) $\underset{\sim}{HA} \vdash ! \{\bar{n}\}(\bar{m})$, and therefore, since in $\underset{\sim}{HA}$ all provable closed Σ_1^0-formulae are true, $\underset{\sim}{HA} \vdash \bar{m}_0 \simeq \{\bar{n}\}(\bar{m})$ for suitable m_0, so $\underset{\sim}{H} \vdash \bar{m}_0 \underset{\approx}{r}_P A$

Remarks. (A). Under (ii), (iii) it is sufficient to assume (A), (B) for $\underset{\sim}{HA}$ and to add, besides (C), resp. (C'), (D),

$$A \in \Gamma \Rightarrow \underset{\sim}{H} \vdash P(A).$$

For assume $\underset{\sim}{H} + \Gamma_1 \vdash P(A)$, $\underset{\sim}{H} + \Gamma_1 \vdash P(A \to B)$ (Γ_1 finite); then $\underset{\sim}{HA} + \Gamma_0 + \Gamma_1 \vdash P(A)$, $\underset{\sim}{HA} + \Gamma_0 + \Gamma_1 \vdash P(A \to B)$ ($\Gamma_0 \subseteq \Gamma$, Γ_0 finite). Then $\underset{\sim}{HA} + \Gamma_0 + \Gamma_1 \vdash P(B)$. This establishes (B) for $\underset{\sim}{H}$.

Now let $\underset{\sim}{H} \vdash A$; then $\underset{\sim}{HA} \vdash (F_1 \to (F_2 \to \dots (F_n \to A) \dots))$, where $F_1, \dots, F_n \in \Gamma$. Hence with (A), (B) for $\underset{\sim}{HA}$ $\underset{\sim}{HA} + P(F_1) + \dots + P(F_n) \vdash P(A)$, hence $\underset{\sim}{H} \vdash P(A)$. This establishes (A) for $\underset{\sim}{H}$.

(B). The following variant of (iii) in the theorem also holds, as will be clear from the proof:

(iii)' Let $\underset{\sim}{H}$ be as in (ii), and assume instead of (C)

(C"). $\quad A \in \Gamma \Rightarrow \exists t(\underset{\sim}{H} \vdash t \underset{\approx}{r}_P A)$

where t is supposed to range over p-terms. Then for closed formulae A:

$$\underset{\sim}{H} \vdash A \Rightarrow \exists t(\underset{\sim}{H} \vdash t \underset{\approx}{r}_P A).$$

3.2.6. Remark. The proof of the soundness theorem by induction on the length of deductions, gives a quite elementary (primitive recursive) method for constructing, for each deduction of a closed formula A in $\underset{\sim}{HA}$, a p-term t such that $\underset{\sim}{HA} \vdash (t \underset{\approx}{r}_P A) \& !t$. Formalizing we find primitive recursive φ_1, φ_2 such that

$$\underset{\sim}{HA} \vdash \mathrm{Proof}_{HA}(x, \ulcorner A \urcorner) \to \mathrm{Proof}_{HA}(\varphi_1 x, \ulcorner t_{\varphi_2 x} \underset{\approx}{r}_P A \urcorner)$$

for closed A, where $t_{\varphi_2 x}$ denotes the p-term with number $\varphi_2 x$.
On the other hand, we do not have provably recursive functions ψ_1, ψ_2 such that

(1) $\quad \underset{\sim}{HA} \vdash \mathrm{Proof}_{HA}(x, \ulcorner A \urcorner) \to \mathrm{Proof}_{HA}(\psi_1 x, \ulcorner \overline{\psi_2 x} \underset{\approx}{r}_P A \urcorner)$.

This is seen as follows: take for A: $\forall x \exists y T(\bar{z}, x, y)$.
We now construct ψ as follows:

$$\exists z \, \mathrm{Proof}_{HA}(u, \ulcorner \forall x \exists y T \bar{z} x y \urcorner) \to \psi(u,x) \simeq j_1 \{\psi_2 u\}(x)$$

$$\neg \exists z \, \mathrm{Proof}_{HA}(u, \ulcorner \forall x \exists y T \bar{z} x y \urcorner) \to \psi(u,x) \simeq 0.$$

Under the assumption that (1) holds, ψ is provably total; hence also $\psi' x =_{def} U\psi(x,x) + 1$ is provably total, which yields a contradiction by a simple diagonal argument.

3.2.7. <u>Verification of the conditions of the soundness theorem for our examples.</u>

For examples (A), (B), (E) the verification of the conditions is immediate. For example (C) we use the well-known properties of the canonical proof predicates for $\underset{\sim}{\mathrm{HA}}$.

In the case of example (D), the conditions for the soundness theorem have to be verified simultaneously with the inductive proof of the soundness theorem itself. See § 3.9.

3.2.8. <u>Lemma.</u>

(i) $\forall u(u \underset{\approx}{r} \neg A) \leftrightarrow \forall v(\neg v \underset{\approx}{r} A) \leftrightarrow \exists u(u \underset{\approx}{r} \neg A)$,

so $u \underset{\approx}{r} \neg A \leftrightarrow \forall v(\neg v \underset{\approx}{r} A)$.

(ii) $u \underset{\approx}{r} \neg\neg A \leftrightarrow \neg\neg \exists w(w \underset{\approx}{r} A)$.

<u>Proof.</u> Straightforward by application of the definition of $\underset{\approx}{r}$ - realizability.

3.2.9 - 3.2.19. <u>Analysis of $\underset{\approx}{r}$ - realizability.</u>

3.2.9. <u>Definition.</u> A formula is said to be <u>almost negative</u> if it does not contain $\vee$, and $\exists$ only in front of an equation between terms (i.e. $\exists x(t=s)$ for $\underset{\sim}{\mathrm{HA}}$).

<u>Note</u> that, modulo logical equivalence, for $\underset{\sim}{\mathrm{HA}}$ the almost negative formulae are the formulae constructed from Σ_1^0 - formulae by means of $\forall$, &, $\rightarrow$.

3.2.10. <u>Lemma.</u> For all formulae A in the language of $\underset{\sim}{\mathrm{HA}}$, $x \underset{\approx}{r} A$ is logically equivalent to an almost negative formula.

<u>Proof.</u> By induction on the complexity of A.

For example, assume the lemma to be proved for A, B; then using $x \underset{\approx}{r}(A \rightarrow B) \leftrightarrow \forall u(u \underset{\approx}{r} A \rightarrow \exists v\, Txuv \,\&\, \forall w(Txuw \rightarrow Uw \underset{\approx}{r} B)$ we can rewrite $x \underset{\approx}{r} A \rightarrow B$ as an almost negative formula.

3.2.11. <u>Lemma.</u> Let $A(\underline{\underline{a}})$ be an almost negative formula of arithmetic, and let $\underline{\underline{a}}$ be a string of number variables, containing all the variables free in A. Then there is a partial recursive function ψ_A (expressed as a p - term of $\underset{\sim}{\mathrm{HA}}$) such that

(i) $\underset{\sim}{\mathrm{HA}} \vdash \exists u(u \underset{\approx}{r} A) \rightarrow A$

(ii) $\underset{\sim}{\mathrm{HA}} \vdash A(\underline{\underline{a}}) \rightarrow\, !\psi_A(\underline{\underline{a}}) \,\&\, \psi_A(\underline{\underline{a}}) \underset{\approx}{r} A(\underline{\underline{a}})$

(iii) $\underset{\sim}{\mathrm{HA}} \vdash u \underset{\approx}{r} A \leftrightarrow u \underset{\approx}{q} A$.

Note that (i) and (ii) together imply $\underset{\sim}{\mathrm{HA}} \vdash \exists u(u \underset{\approx}{r} A) \leftrightarrow A$ for almost negative A.

<u>Proof.</u> (i), (ii), (iii) are proved simultaneously by induction on the logical complexity of A; ψ_A is defined by induction on the logical complexity of A, as follows:

(a) $\psi_{t=s}(\underline{\underline{a}}) \simeq 0$.
(b) If $A \equiv \exists y(t=s)$, take $\psi_A(\underline{\underline{a}}) \equiv j(\min_y[t=s], 0)$.
(c) If $A \equiv B \& C$, take $\psi_A(\underline{\underline{a}}) \equiv j(\psi_B(\underline{\underline{a}}), \psi_C(\underline{\underline{a}}))$.
(d) If $A \equiv \forall x Bx$, take $\psi_A(\underline{\underline{a}}) \equiv \Lambda x . \psi_{B(x)}(\underline{\underline{a}}, x)$.
(e) If $A \equiv B \rightarrow C$, take $\psi_A(\underline{\underline{a}}) \equiv \Lambda u . \psi_C(\underline{\underline{a}})$.

Now we turn to the proof of (i), (ii), (iii).
For prime formulae, (i), (ii), (iii) are obvious.
(A). Let $A \equiv \exists x(t(x)=s(x))$. $\exists u(u \underset{\approx}{r} \exists x(t(x)=s(x)))$ implies $\exists u(j_2 u \underset{\approx}{r} (t(j_1 u) = s(j_1 u)))$ which is equivalent to $\exists v(t(v)=s(v))$.
This establishes (i).
Now assume $\exists u(t(u)=s(u))$. Then $\min_z[t(z)=s(z)]$ is defined, call it u'; then $t(u') = s(u')$, and $j(u',0) \equiv \psi_{\exists u(t=s)}(\underline{\underline{a}})$ realizes $A(\underline{\underline{a}})$. This proves (ii).
(iii) is obvious.
(B). Let $A \equiv B \rightarrow C$, and assume (i), (ii), (iii) for B, C.
If $u \underset{\approx}{r}(B \rightarrow C)$, and B, then $!\psi_B(\underline{\underline{a}})$, $!\psi_B(\underline{\underline{a}}) \underset{\approx}{r} B$, so $!\{u\}(\psi_B(\underline{\underline{a}}))$, $\{u\}(\psi_B(\underline{\underline{a}})) \underset{\approx}{r} C$. Therefore C holds, and thus $\exists u(u \underset{\approx}{r}(B \rightarrow C)) \rightarrow (B \rightarrow C)$; this establishes (i).
Conversely, assume $B \rightarrow C$, and let $u \underset{\approx}{r} B$. Then B holds, hence C holds, and thus $!\psi_C(\underline{\underline{a}})$, $\psi_C(\underline{\underline{a}}) \underset{\approx}{r} C$. Therefore $\psi_{B \rightarrow C}(\underline{\underline{a}}) \underset{\approx}{r} B \rightarrow C$. This proves (ii).
Finally, $u \underset{\approx}{r}(B \rightarrow C) \leftrightarrow \forall x(x \underset{\approx}{r} B \rightarrow\, !\{u\}(x) \& \{u\}(x) \underset{\approx}{r} C) \leftrightarrow$
$\leftrightarrow \forall x((x \underset{\approx}{q} B) \& B \rightarrow\, !\{u\}(x) \& \{u\}(x) \underset{\approx}{q} C) \leftrightarrow u \underset{\approx}{q} (B \rightarrow C)$.
(Induction hypothesis is used thrice in the second equivalence: $x \underset{\approx}{r} B \leftrightarrow x \underset{\approx}{q} B$, $\{u\}(x) \underset{\approx}{r} C \leftrightarrow \{u\}(x) \underset{\approx}{q} C$, and $x \underset{\approx}{r} B \leftrightarrow (x \underset{\approx}{r} B) \& B$.)
(C). The other cases: $A \equiv B \& C$, $A \equiv \forall x B$ are left to the reader.

3.2.12. Lemma. If A is an arithmetical formula, then A is provably equivalent (in $\underset{\sim}{HA}$) to an almost negative formula, iff we can find a partial recursive ψ for which (i) and (ii) of the previous lemma are provable, i.e.
(i) $\underset{\sim}{HA} \vdash \exists x(x \underset{\approx}{r} A) \rightarrow A$, (ii) $A\underline{\underline{a}} \rightarrow\, !\psi\underline{\underline{a}} \& \psi\underline{\underline{a}} \underset{\approx}{r} A$.

Proof. The "only if" part has already been established (3.2.11).
Now assume (i), (ii) to hold for A. $\psi\underline{\underline{a}}$ may be supposed to be represented by a p-term. If z does not occur in $\underline{\underline{a}}$, $\Lambda z . \psi\underline{\underline{a}}$ can be given as a primitive recursive function $\varphi\underline{\underline{a}}$.
Now by (i), (ii)

$$A\underline{\underline{a}} \leftrightarrow\, !\psi\underline{\underline{a}} \& \psi\underline{\underline{a}} \underset{\approx}{r} A\underline{\underline{a}}$$

hence

$$A\underline{\underline{a}} \leftrightarrow [\exists u T(\varphi\underline{\underline{a}}, 0, u) \& \forall v(T(\varphi\underline{\underline{a}}, 0, v) \rightarrow Uv \underset{\approx}{r} A\underline{\underline{a}})];$$

the right hand side of this expression is obviously almost negative.

3.2.13. Remark. Note that $\underset{\sim}{HA} \vdash A \leftrightarrow \exists x(x \underset{\sim}{r} A)$ for a formula A iff A is provably equivalent to an almost negative formula or an existentially quantified almost negative formula.

3.2.14. Definition. Let ECT_0 denote the following schema, for A almost negative :

ECT_0 $\quad \forall x[A \to \exists y By] \to \exists u \forall x[A \to \exists v(Tuxv \,\&\, B(Uv))]$

(y not occurring free in A).

Note that for $A \equiv 0 = 0$, we obtain CT_0 (Church's thesis)

CT_0 $\quad \forall x \exists y\, By \to \exists u\, \forall x(\exists v\, Tuxv \,\&\, B(Uv))$.

ECT_0 stands for "extended Church's thesis".

3.2.15. Lemma. For any universal closure A of an instance of ECT_0 there is a numeral $\bar{n}$ such that

$$\underset{\sim}{HA} \vdash \bar{n} \underset{\sim}{r} A, \qquad \underset{\sim}{HA} + ECT_0 \vdash \bar{n} \underset{\sim}{q} A .$$

Proof. Consider an instance of ECT_0 :

(1) $\quad \forall x[A \to \exists y By] \to \exists u \forall x[A \to \exists v(Tuxv \,\&\, B(Uv))]$.

For simplicity, we assume that there are no additional free variables in A, B besides x, y . Assume

$$u \underset{\sim}{r} \forall x[A \to \exists y By]$$

and abbreviate $t \equiv \{u\}(x, \psi_A)$; then

$$\forall x[A \to {!t} \,\&\, t \underset{\sim}{r} \exists y By]$$

or equivalently

$$\forall x[A \to {!t} \,\&\, j_2 t \underset{\sim}{r} B(j_1 t)] .$$

Put $\varphi_1 \equiv \Lambda x. j_1 t$, $\varphi_2 \simeq \min_u T(\varphi_1, x, u)$, $\varphi(u) = j(\varphi_1, \Lambda x \Lambda w. j(\varphi_2, j(0, j_2 t)))$. Then

$$\varphi(u) \underset{\sim}{r} \exists z \forall x[A \to \exists v(T(z, x, v) \,\&\, B(Uv))] .$$

Hence $\Lambda u. \varphi(u) \underset{\sim}{r}$ (1) .

Similarly in the presence of additional variables, or for $\underset{\sim}{q}$ - realizability.

3.2.16. Theorem (Idempotency of realizability ; Nelson 1947).

$$\exists x(x \underset{\sim}{r} \exists y(y \underset{\sim}{r} A)) \leftrightarrow \exists y(y \underset{\sim}{r} A) .$$

Proof. By lemma 3.2.10 and remark 3.2.13.

3.2.17. Notation. Let us abbreviate "$\underset{\sim}{r}$ - realizability which is provable in the formal system $\underset{\sim}{H}$" as "$\underset{\sim}{H}$ - $\underset{\sim}{r}$ - realizability". Similarly, we use the expression "$\underset{\sim}{H}$ - $\underset{\sim}{r}$ - realizable". Similar definitions with $\underset{\sim}{q}$ instead of $\underset{\sim}{r}$.

3.2.18. Theorem (Characterization of $\underset{\sim}{HA}$ - $\underset{\sim}{r}$ - realizability).

(i) $\underset{\sim}{HA} + ECT_0 \vdash A \leftrightarrow \exists x(x \underset{\sim}{r} A)$.

(ii) $\underset{\sim}{HA} + ECT_0 \vdash A \Leftrightarrow \underset{\sim}{HA} \vdash \exists x(x \underset{\sim}{r} A)$.

Proof. (i) is shown by induction on the complexity of A. Consider e.g. the case $A \equiv B \rightarrow C$; $(B \rightarrow C) \leftrightarrow (\exists x(x \underset{\sim}{r} B) \rightarrow \exists y(y \underset{\sim}{r} C)) \leftrightarrow \forall x(x \underset{\sim}{r} B \rightarrow \exists y(y \underset{\sim}{r} C)) \leftrightarrow$ $\leftrightarrow \exists z \forall x(x \underset{\sim}{r} B \rightarrow \exists v(Tzxv \,\&\, Uv \underset{\sim}{r} C)) \leftrightarrow \exists x(z \underset{\sim}{r} B \rightarrow C)$. The third equivalence required an appeal to ECT_0 and lemma 3.2.10.
(ii). The implication from right to left follows from (i); the implication from left to right is verified thus: let $\underset{\sim}{HA} + ECT_0 \vdash A$, then $\underset{\sim}{HA} \vdash F \rightarrow A$, F a conjunction of universal closures of instances of ECT_0, hence $\underset{\sim}{HA} \vdash \exists x(x \underset{\sim}{r} F)$; also $\underset{\sim}{HA} \vdash \exists y(y \underset{\sim}{r} F \rightarrow A)$ (by the soundness theorem), so $\underset{\sim}{HA} \vdash \exists z(z \underset{\sim}{r} A)$.

3.2.19. Corollary to the proof of 3.2.18. (Characterization of $\underset{\sim}{H}$ - $\underset{\sim}{r}$ - realizability for certain extensions $\underset{\sim}{H}$ of $\underset{\sim}{HA}$).
Let $\underset{\sim}{H} \equiv \underset{\sim}{HA} + \Gamma$, Γ a set of (closed) additional axioms, such that

(1) $\quad A \in \Gamma \Rightarrow \underset{\sim}{H} \vdash \exists x(x \underset{\sim}{r} A)$.

Then

(i) $\quad \underset{\sim}{H} + ECT_0 \vdash A \leftrightarrow \exists x(x \underset{\sim}{r} A)$

(ii) $\quad \underset{\sim}{H} + ECT_0 \vdash A \Leftrightarrow \underset{\sim}{H} \vdash \exists x(x \underset{\sim}{r} A)$.

Proof. (i) follows immediately from 3.2.18 (i).
One direction of (ii) follows from (i).
For the implication from left to right, note that by assumption (1), and the soundness theorem for $\underset{\sim}{HA}$: $\underset{\sim}{H} \vdash B \Rightarrow \underset{\sim}{H} \vdash \exists x(x \underset{\sim}{r} B)$ (using once again the deduction theorem for $\underset{\sim}{HA}$). Then argue as for 3.2.18 (ii).

3.2.20. Theorem.

(i) $\underset{\sim}{HA} + ECT_0$ is consistent relative to $\underset{\sim}{HA}$

(ii) $\underset{\sim}{HA} + ECT_0$ is ω-consistent on assumption of the truth of $\underset{\sim}{HA}$.

Proof. (i) is an immediate corollary of 3.2.18 (ii) or 3.2.15.
(ii). Assume $\underset{\sim}{HA} + ECT_0 \vdash A\bar{n}$ for each numeral $\bar{n}$, and also $\underset{\sim}{HA} + ECT_0 \vdash \neg \forall x Ax$. Then $\underset{\sim}{HA} \vdash \bar{m}_n \underset{\sim}{r} A\bar{n}$, for each numeral $\bar{n}$ and for suitable $\bar{m}_n$ depending on n; also $\underset{\sim}{HA} \vdash \forall u \neg (u \underset{\sim}{r} \forall x Ax)$. Hence $\forall u(\neg u \underset{\sim}{r} \forall x Ax)$ is true, by our assumptions. Now let

$$\varphi(y) \simeq j_1 \min_z [\mathrm{Proof}_{HA}(j_2 z, \ulcorner \overline{j_1 z} \ \underset{\sim}{r}\ A(\bar{y}) \urcorner)] .$$

Then $\Lambda y.\varphi(y)\ \underset{\sim}{r}\ \forall y A y$ (since the truth of HA implies the truth of a uniform reflection principle) contradicting $\forall u(\neg u\ \underset{\sim}{r}\ \forall x A x)$; therefore $HA + ECT_0 \not\vdash \neg \forall x A x$.

Remark. ECT_0 cannot be generalized to arbitrary formulas A , as is illustrated by the following counterexample. Obviously,

$$(1) \quad \begin{cases} \forall x[(\exists y Txxy \vee \neg \exists y Txxy) \rightarrow \\ \rightarrow \exists z((z > 0 \ \& \ T(x,x,z \dot{-} 1)) \vee (z=0 \ \& \neg \exists y Txxy))] . \end{cases}$$

ECT_0 generalized to arbitrary A would yield, when applied to (1), the existence of a partial recursive function with gödelnumber u_0 such that

$$(2) \quad \begin{cases} \forall x[(\exists y Txxy \vee \neg \exists y Txxy) \rightarrow \exists w(Tu_0 xw \ \& \\ \& \ \{(Uw > 0 \rightarrow T(x,x,Uw \dot{-} 1)) \ \& \ (Uw=0 \rightarrow \neg \exists y Txxy)\})] . \end{cases}$$

On the other hand, $\forall x \neg\neg(\exists y Txxy \vee \neg \exists y Txxy)$, therefore with (2)

$$\forall x \neg\neg \exists w(Tu_0 xw \ \& \ \{(Uw > 0 \ \& \ T(x,x,Uw \dot{-} 1)) \vee (Uw=0 \ \& \neg \exists y Txxy)\}) .$$

Now let v_0 be such that $\exists w T v_0 x w \leftrightarrow \{u_0\}(x) \simeq 0$; then $\exists w T v_0 v_0 w \leftrightarrow \{u_0\}(v_0) \simeq 0 \leftrightarrow \neg \exists y T v_0 v_0 y$, which is contradictory ; hence (2) is false.

In fact, this counterexample even refutes a schema

$$\forall x[A \rightarrow \exists ! y B y] \rightarrow \exists u \forall x[A \rightarrow !\{u\}(x) \ \& \ B(\{u\}(x))] .$$

Later, we shall prove that $HA + CT_0 \not\vdash ECT_0$ (3.1.14).

3.2.21. Lemma. Let F be the universal closure of an instance of Markov's schema

M $\quad \forall x(A \vee \neg A) \ \& \ \neg\neg \exists x A \rightarrow \exists x A$.

Then there exists a numeral $\bar{n}$ such that

$$HA + M \vdash \bar{n}\ \underset{\sim}{r}\ F , \qquad HA + M \vdash \bar{n}\ \underset{\sim}{q}\ F .$$

Proof. Let an instance F of M

$$\forall x(Ax \vee \neg Ax) \ \& \ \neg\neg \exists x Ax \rightarrow \exists x Ax$$

be given, and assume for simplicity that A does not contain variables free besides x .

Assume

$$u\ \underset{\sim}{r}\ \forall x(Ax \vee \neg Ax) \ \& \ \neg\neg \exists x Ax .$$

Then

$$\forall x(!\{j_1u\}(x) \& ([j_1(\{j_1u\}(x)) = 0 \& j_2(\{j_1u\}(x)) \underset{\sim}{r} Ax] \vee$$
$$\vee [j_1(\{j_1u\}(x)) \neq 0 \& j_2(\{j_1u\}(x)) \underset{\sim}{r} \neg Ax])) \& j_2u \underset{\sim}{r} \neg\neg \exists xAx .$$

Let $\varphi(u) \simeq \min_x[j_1(\{j_1u\}(x)) \simeq 0]$. $\forall x(j_1\{j_1u\}(x) \neq 0)$ would imply $\forall x\exists w(w \underset{\sim}{r} \neg Ax)$, equivalent to $\forall x\forall u\neg(u \underset{\sim}{r} Ax)$, i.e. $\forall x\neg\exists u(u \underset{\sim}{r} Ax)$. On the other hand, $j_2u \underset{\sim}{r} \neg\neg\exists xAx \leftrightarrow \neg\neg\exists x\exists w(w \underset{\sim}{r} Ax) \leftrightarrow \neg\forall x \neg\exists w(w \underset{\sim}{r} Ax)$ (3.2. 8 (ii)); hence contradiction.

Thus

$$\neg\forall x(j_1\{j_1u\}(x) \neq 0) ,$$

i.e.

$$\neg\neg\exists x\exists v(T(j_1u,x,v) \& j_1Uv = 0)$$

hence with M, $\exists x[j_1\{j_1u\}(x) = 0]$.

Thus $!\varphi(u)$, and

$$j(\varphi(u), j_2(\{j_1u\}(\varphi(u)))) \underset{\sim}{r} \exists xAx ,$$

and so $\Lambda u.j(\varphi(u), j_2(\{j_1u\}(\varphi(u)))) \underset{\sim}{r} F$.

Similarly for $\underset{\sim}{q}$-realizability.

Remarks. (i). As we shall see later, not all instances of M are $\underset{\sim}{HA}$-$\underset{\sim}{r}$-realizable.

(ii). In the presence of CT_0, M is equivalent to the weaker schema

M_{PR} $\qquad \neg\neg\exists xAx \to \exists xAx$

for A primitive recursive. For let $\forall x(Bx \vee \neg Bx)$. By Church's thesis, there is a u such that $\forall x\exists y[Tuxy \& (Uy=0 \to Bx) \& (Uy\neq 0 \to \neg Bx)]$. Hence

$$\neg\neg\exists xBx \to \exists xBx$$

is equivalent to

$$\neg\neg\exists x\exists y[Tuxy \& Uy=0] \to \exists x\exists y[Tuxy \& Uy=0]$$

which can be obtained as an instance of M_{PR}.

Note that in the presence of M_{PR}, every almost negative formula is equivalent to a negative formula, by $\exists xA \leftrightarrow \neg\forall x\neg A$.

3.2.22. Corollaries.

(i) $\underset{\sim}{HA} + ECT_0 + M \vdash A \Leftrightarrow \underset{\sim}{HA} + M \vdash \exists x(x \underset{\sim}{r} A)$

(ii) $\underset{\sim}{HA} + ECT_0 + M$ is consistent relative to $\underset{\sim}{HA}$.

Proof. (i) is immediate from 3.2.18 (ii) and 3.2.21.

(ii). (i) implies that $\underset{\sim}{HA} + ECT_0 + M$ is consistent relative to $\underset{\sim}{HA} + M$ (a proof of $1 = 0$ in $\underset{\sim}{HA} + ECT_0 + M$ gives rise to a proof of $1 = 0$ in $\underset{\sim}{HA} + M$; and $\underset{\sim}{HA} + M$ is consistent relative to $\underset{\sim}{HA}$ since $\underset{\sim}{HA}^c$ is consistent relative to $\underset{\sim}{HA}$ (§ 1.10).

3.2.23. Lemma. For each closure F of an instance of $TI(<)$, we have

$$\exists n(\underset{\sim}{HA} + TI(<) \vdash \bar{n} \underset{\sim}{r} F \& \bar{n} \underset{\sim}{q} F).$$

Proof. For simplicity we restrict attention to a closed instance

$$\forall u((\forall v < u)Av \to Au) \to \forall uAu$$

of $TI(<)$, and prove the lemma for $\underset{\sim}{r}$-realizability.
Assume

$$w \underset{\sim}{r} \forall u((\forall v < u)Av \to Au),$$

so

$$\forall u(\{w\}(u) \underset{\sim}{r} ((\forall v < u)Av \to Au))$$

i.e.

$$\forall uw'(w' \underset{\sim}{r} (\forall v < u)Av \to \{w\}(u,w') \underset{\sim}{r} Au).$$

$w' \underset{\sim}{r} (\forall v < u)Av$ implies $\forall v < u(\{w'\}(v,0) \underset{\sim}{r} Av)$, since $\forall v < uAv$ abbreviates $\forall v(v < u \to Av)$ with $v < u$ quantifier-free. Now if we put

$$\varphi_c(x,u,v) \simeq \begin{cases} 0 & \text{if } u \geq v \\ x & \text{if } u < v \end{cases}$$

we can easily find a partial recursive φ such that

$$\varphi(z,w,u) \simeq \{w\}(u, \Lambda v \Lambda x.\varphi_c(\{z\}(w,v),v,u)).$$

By the recursion theorem, there is an $\bar{n}$ such that

$$\{\bar{n}\}(w,u) \simeq \{w\}(u, \Lambda v \Lambda x.\varphi_c(\{\bar{n}\}(w,v),v,u)).$$

We now easily prove $\forall u(!\{\bar{n}\}(w,u))$, and $\{\bar{n}\}(w,u) \underset{\sim}{r} Au$, by $TI(<)$ w.r.t. u. E.g. for the latter assertion it suffices to show

$$\forall u(\forall v < u(\{\bar{n}\}(w,v) \underset{\sim}{r} Av) \to \{\bar{n}\}(w,u) \underset{\sim}{r} Au)$$

which is completely straightforward; etc. etc.

3.2.24. Theorem. Let $\underset{\sim}{H}$ be $\underset{\sim}{HA} + TI(<)$ or $\underset{\sim}{HA} + M + TI(<)$.
(i) For A closed, $\underset{\sim}{H} + ECT_0 \vdash A \Rightarrow \exists n(\underset{\sim}{H} \vdash \bar{n} \underset{\sim}{r} A$ and $\underset{\sim}{H} + ECT_0 \vdash \bar{n} \underset{\sim}{q} A)$
(ii) $\underset{\sim}{H} + ECT_0 \vdash A \Leftrightarrow \underset{\sim}{H} \vdash \exists x(x \underset{\sim}{r} A)$.

Proof. (i). Immediate from 3.2.15, 3.2.21, 3.2.23.
(ii). Immediate from 3.2.19 (ii).

3.2.25. Theorem (Characterization of $\underset{\sim}{HA}^c$ - $\underset{\sim}{r}$ - realizability).

$$\underset{\sim}{HA}^c \vdash \exists x(x \underset{\sim}{r} A) \Leftrightarrow \underset{\sim}{HA} + M + ECT_0 \vdash \neg\neg A.$$

Proof. Assume $\underset{\sim}{HA}^c \vdash \exists x(x \underset{\sim}{r} A)$. Also $\underset{\sim}{HA} + M \vdash x \underset{\sim}{r} A \leftrightarrow Bx$, where B is the negative formula obtained from $x \underset{\sim}{r} A$ by replacing every subformula of the form $\exists y(t=s)$ by $\neg\forall y\neg(t=s)$. Then $\mathbf{HA}^c \vdash \neg\forall x\neg Bx$, and since $\mathbf{HA}^c$ is is conservative over $\mathbf{HA}$ w.r.t. negative formulae (§ 1.10), $\mathbf{HA} \vdash \neg\forall x\neg Bx$. Hence $\mathbf{HA} + M \vdash \neg\neg\exists x(x\,\underset{\sim}{r}\,A)$. Since $\mathbf{HA} + ECT_0 \vdash A \leftrightarrow \exists x(x\,\underset{\sim}{r}\,A)$ (3.2.18 (i)), we find $\mathbf{HA} + ECT_0 + M \vdash \neg\neg A$.

Conversely, if $\mathbf{HA} + ECT_0 + M \vdash \neg\neg A$, then $\mathbf{HA}^c \vdash \exists x(x\,\underset{\sim}{r}\,\neg\neg A)$, so $\mathbf{HA}^c \vdash \neg\neg\exists x(x\,\underset{\sim}{r}\,A)$ (3.2.8 (ii)), hence $\mathbf{HA}^c \vdash \exists x(x\,\underset{\sim}{r}\,A)$.

As an application of this characterization we prove

3.2.26. Theorem. Every universal closure of an instance of

IP_0 $\qquad \forall x[A \vee \neg A] \;\&\; [\forall xA \to \exists yB] \to \exists y[\forall xA \to B]$

is $\mathbf{HA}^c$-$\underset{\sim}{r}$-realizable.

Proof. Consider an instance of IP_0 with one additional parameter z. To show that it is $\mathbf{HA}^c$-realizable, we have to show in $\mathbf{HA} + M + ECT_0$ (y not free in A)

$$\neg\neg\forall z[\forall x(A \vee \neg A) \wedge (\forall xA \to \exists yB) \to \exists y(\forall xA \to B)].$$

Let us abbreviate this as $\neg\neg\forall z\underline{\underline{A}}$. Assume

(1) $\qquad \neg\forall z\underline{\underline{A}}$.

Because of $\neg\forall z P \leftrightarrow \neg\neg\exists z\neg P$, this is equivalent to

(2) $\qquad \neg\neg\exists z\neg\underline{\underline{A}}$.

Now assume

(3) $\qquad \neg\underline{\underline{A}}$.

Because of $\neg(P\to Q) \leftrightarrow \neg\neg P \,\&\, \neg Q$, $\neg\neg(P\,\&\,Q) \leftrightarrow \neg\neg P \,\&\, \neg\neg Q$, this is equivalent to

(4) $\qquad \neg\neg\forall x(A \vee \neg A) \;\&\; \neg\neg(\forall xA \to \exists yB) \;\&\; \neg\exists y(\forall xA \to B)$.

Assume

(5) $\qquad \forall x(A \vee \neg A)$,

(6) $\qquad \forall xA \to \exists yB$,

(7) $\qquad \neg\exists y(\forall xA \to B)$.

From (7)

$$\forall y\neg(\forall xA \to B),$$

hence

$$\forall y(\neg\neg\forall xA \,\&\, \neg B),$$

so

$$\neg\neg\forall xA \;\&\; \forall y\neg B.$$

Because of (5), this yields

$$\forall x A \ \& \ \forall y \neg B$$

which contradicts (6), hence

$$\neg [(5) \ \& \ (6) \ \& \ (7)].$$

Therefore $\neg (4)$, i.e. $\neg (3)$, hence $\neg \exists z \neg \underline{\underline{A}}$, and thus $\neg (2)$.
This refutes (1), and so the desired conclusion has been established (in HA in fact !).

Remark. It can be shown that IP_0 is not in general HA - r - realizable. Namely, by a version of de Jongh's theorem (5.6.16), we can find Σ_1^0 - formulae A, B, C such that

$$HA \nvdash (\neg A \rightarrow (B \vee C)) \rightarrow ((\neg A \rightarrow B) \vee (\neg A \rightarrow C)).$$

This formula is logically equivalent to a formula of the class Γ_0 (see 3.6.3) for which $HA + ECT_0$ is conservative over HA, hence the formula is provable in $HA + ECT_0$ (i.e. HA - realizable) iff it is provable in HA.

3.2.27. Theorem.
(i) $HA + M_{PR} + CT_0 + IP$ is inconsistent.
(ii) Not all (closures of) instances of

IP $\quad (\neg A \rightarrow \exists y B) \rightarrow \exists y (\neg A \rightarrow B)$

are realizable (and certainly not HA^c - r - realizable).

Proof. (i). By M_{PR}

(1) $\quad \forall x [\neg\neg \exists y Txxy \rightarrow \exists y Txxy]$,

hence with IP

(2) $\quad \forall x \exists y [\neg\neg \exists y Txxy \rightarrow Txxy]$,

and by CT_0

$$\exists u \forall x [\, ! \{u\}(x) \ \& \ (\neg\neg \exists y Txxy \rightarrow T(x,x,\{u\}(x))].$$

This implies that $\exists y Txxy$ is recursive in x, which is contradictory.
(ii). Consider

(3) $\quad \forall x \{(\neg\neg \exists y Txxy \rightarrow \exists y Txxy) \rightarrow \exists z (\neg\neg \exists y Txxy \rightarrow Txxz)\}$.

This is $(1) \rightarrow (2)$. Now we know that

$$HA + M + CT_0 + (1) \rightarrow (2) \vdash \curlywedge ,$$

hence also

$$HA + M + ECT_0 \vdash \neg (3).$$

Hence $\underset{\sim}{HA}^c \vdash \exists x(x \underset{\approx}{r} \neg(3))$, so $\underset{\sim}{HA}^c \vdash \neg\exists x(x \underset{\approx}{r} (3))$, by 3.2.8 (i).

3.2.28. Remark. Kleene proves a slightly stronger theorem (Kleene 1965A): he shows that the universal closure of

$$(1) \qquad (\neg A \to B \vee C) \to (\neg A \to B) \vee (\neg B \to C)$$

is for certain A, B, C is not realizable. We can also obtain this result by slightly refining the argument in 3.2.27, sub (ii).
As our instance of (1) we take for $\neg A \to (B \vee C)$

$$\neg\neg\exists y Txxy \to T(x,x,0) \vee (\exists y > 0) Txxy$$

which again follows from M_{PR}.

Note that our disproof of realizability is classical (at least it uses M); but the disproof of provable realizability is intuitionistic (it uses: $\underset{\sim}{HA}^c$ is consistent).

3.2.29. Extensions to other systems.

An extension of realizability to the language of $\underset{\sim}{EL}$ with function variables is obtained by interpreting the function variables as ranging over recursive functions, so we put:

$\underset{\approx}{r}$(vii) $\quad x \underset{\approx}{r} \forall\alpha A\alpha \equiv \forall y \in V_1(!\{x\}(y) \;\&\; \{x\}(y) \underset{\approx}{r} A(\{y\}))$
$\underset{\approx}{r}$(viii) $\quad x \underset{\approx}{r} \exists\alpha A\alpha \equiv j_2 x \underset{\approx}{r} A(\{j_1 x\}) \;\&\; j_1 x \in V_1$.

Here $A(\{t\})$ is shorthand for a formula $A^*(t)$, obtained by systematically eliminating each occurrence of α in $A(\alpha)$ by application of

$$t'[\alpha t''] = t''' \leftrightarrow \exists u[\alpha t''=u \;\&\; t'[u]=t''']$$

and replacing $\alpha t'' = u$ (t' not containing α) by $\exists v(T(t,t'',v) \& Uv=u)$. In short, if we replace in $A\alpha$ α by $\{t\}$, $=$ by $\simeq$, then $A(\{t\})$ contains p-terms, and the "prime formulae" of the form $t \simeq t'$ must be interpreted as abbreviations as in Kleene 1969 (cf. 1.3.10).

If we wish to extend this further to $\underset{\sim}{IDB}$, we must put (cf. Kreisel - Troelstra 1970, 3.7.1)

$\underset{\approx}{r}$(i)' $\quad x \underset{\approx}{r} K\varphi \equiv K\varphi \;\&\; \varphi = \{x\}$,

where $\varphi = \{x\} \equiv_{def} \forall y(\exists z Txyz \;\&\; \forall x(Txyu \to Uu=\varphi y))$.

For higher types we may extend realizability similarly, interpreting the higher-order quantifiers as ranging over the V_σ of HRO or the W_σ of HEO.

Realizability can be extended to the language of $\underset{\sim}{HAS}$ as follows. We associate with each variable V_i^n of $\underset{\sim}{HAS}$ a variable V_i^{n+1}; below we shall write X^* for V_i^{n+1} if $X \equiv V_i^n$.

$\underset{\sim}{r}(i)''$ $\quad x \underset{\sim}{r} X(t_1,\ldots,t_n) \equiv X^*(x,t_1,\ldots,t_n)$
$\underset{\sim}{r}(ix)$ $\quad x \underset{\sim}{r} \forall X A(X) \equiv \forall X^*(x \underset{\sim}{r} A(X))$
$\underset{\sim}{r}(x)$ $\quad x \underset{\sim}{r} \exists X A(X) \equiv \exists X^*(x \underset{\sim}{r} A(X))$.

For an application of this extension see 3.2.31.

3.2.30. Realizability for IDB. The treatment is very similar to the treatment for HA, hence we give a sketch (following Troelstra 1971A, § 5). For any formula A of IDB, let A^r denote the formula obtained by relativizing function quantifiers to recursive functions, and let us define almost negative formulae as before (3.2.9). Note that if A is almost negative, then also A^r is (equivalent to) an almost negative formula, since

$$(\forall\alpha A\alpha)^r \leftrightarrow \forall\alpha\forall x(\alpha = \{x\} \rightarrow A^r\alpha)$$

where, as before

$$\alpha = \{x\} \equiv_{def} \forall y(\exists z Txyz \;\&\; \forall u(Txyu \rightarrow Uu = \alpha y)) ,$$

and

$$(\exists\alpha(t[\alpha] = s[\alpha]))^r \leftrightarrow \exists\alpha \in V_1(t[\alpha] = s[\alpha]) \leftrightarrow \exists n(t[f_n] = s[f_n])$$

where $f_n = \lambda x.(n)_x$ (1.3.9 C). (As a lemma we have to show for any term $t[\alpha]$ in IDB that it depends continuously on α (provably in IDB).)

One then proves that for almost negative $A(\underline{x},\underline{\alpha})$ ($\underline{x}$ a sequence of numerical variables, $\underline{\alpha} \equiv \alpha_1,\ldots,\alpha_n$ a sequence of function variables) the existence of a partial recursive $\psi_A(\underline{x},\underline{y})$ ($\underline{y} \equiv y_1,\ldots,y_n$ a sequence of numerical variables, $\underline{x} \cap \underline{y} = \emptyset$) such that

$$\exists u(u \underset{\sim}{r} A) \rightarrow A^r$$
$$A^r(\underline{x},\underline{\alpha}) \;\&\; \alpha_1 = \{y_1\} \;\&\ldots\&\; \alpha_n = \{y_n\} \rightarrow$$
$$\rightarrow\; !\psi_A(\underline{x},\underline{y}) \;\&\; \psi_A(\underline{x},\underline{y}) \underset{\sim}{r} A(\underline{x},\underline{\alpha}) .$$

(A direct corollary, by the preceding remark, of the analogues of 3.2.11.) Using this, we find also for each universal closure F of ECT_0 in $\mathcal{L}$(IDB), IDB $\vdash \bar{n} \underset{\sim}{r} F$, in view of which we then obtain, after extending the soundness theorem for IDB as in Kreisel - Troelstra 1970, 3.7.2

$$\text{IDB} + ECT_0 \vdash A \leftrightarrow \exists x(x \underset{\sim}{r} A)$$
$$\text{IDB} + ECT_0 \vdash A \Leftrightarrow \text{IDB} \vdash \exists x(x \underset{\sim}{r} A) .$$

We might also have considered $\underset{\sim}{q}$-realizability, and treated it similarly. For an extremely detailed treatment, see Celluci 1971.

3.2.31. Theorem. HAS + CT_0 + UP is consistent relative to HAS, where UP (the uniform principle) can be stated as

UP $\quad \forall X \exists x A(X,x) \rightarrow \exists x \forall X A(X,x)$.

Proof. Straightforward, by extending the soundness theorem for HA + CT_0 to HAS + CT_0 + UP; cf. e.g. Kreisel - Troelstra 1970, 3.7 , and Troelstra A, §3.

3.2.32. Some generalizations.

A generalization inspired by Laüchli 1970 (there in the context of a notion closer to modified realizability) consists in the introduction of a family of realizabilities, as follows. Let $\{U_i \mid i \in I\}$, (I an index set) be a collection of species of natural numbers such that

$$\bigcap \{U_i \mid i \in I\} = \emptyset .$$

We now define $\underset{\approx}{r}^{(i)}$ realizability by the clauses

$\underset{\approx}{r}(i)'''$ $\quad x \underset{\approx}{r}^{(i)} A \equiv \neg A \rightarrow x \in U_i$

and $\underset{\approx}{r}(ii)$ - $\underset{\approx}{r}(vi)$ ($\underset{\approx}{r}^{(i)}$ everywhere replacing $\underset{\approx}{r}$), and then define realizability with respect to the family $\{U_i \mid i \in I\}$ by

$$x \underset{\approx}{r}^{I} A \equiv_{def} \forall i \in I\, (x \underset{\approx}{r}^{(i)} A).$$

Technically, this concept is not so easy to characterize as ordinary $\underset{\approx}{r}$-realizability. We may expect, in view of the result of Laüchli 1970, something like (an approximation to) completeness with respect to the schemata of intuitionistic predicate logic (if the concept is treated classically). But CT_0 anyway remains valid. Yet it seems to me such a completeness result is of doubtful interest.

Another, rather obvious, possibility of generalization is of greater practical interest: the use of A - recursive functions $\{t\}^A$ everywhere replacing $\{t\}$ in the definition of $x \underset{\approx}{r} B$, for an arbitrary arithmetical unary predicate Ax. The rôle of HA is then taken over by HA + $\forall x(Ax \vee \neg Ax)$; the soundness theorem becomes provable for this system. This idea has been used in Smorynski B to show that HA^c is essentially unbounded over HA.

Sketch of the proof: Define Σ^0_{n+1} - realizability as realizability by functions $\{x\}^A$, A a complete Π^0_n - set. Then establish a soundness theorem for HA, and show that for any A with alternating-quantifier complexity $\leq n$, there is a primitive recursive φ_A such that $HA^c \vdash A(x_1,\ldots,x_n) \leftrightarrow$ $\leftrightarrow \varphi_A(x_1,\ldots,x_n) \underset{\approx}{r} A(x_1,\ldots,x_n)$. Finally, note that for a predicate $Ax \in \Sigma^0_{n+1} - \Delta^0_{n+1}$, $HA^c \nvdash \exists y[y \underset{\approx}{r} \neg \forall x(Ax \vee \neg Ax)]$. Then it readily follows that for each set of axioms Δ of bounded alternating-quantifier complexity, $HA^c + \Delta \neq HA + \Delta$, so HA^c is an essentially unbounded extension of HA.

For similar uses of A - recursive functions in the context of modified realizability, see 3.4.31.

3.2.33. Comparison of $\underset{\approx}{q}$ - realizability with $\underset{\approx}{r}$ - realizability.

$\underset{\approx}{q}$ - realizability does not admit a simple characterization such as given in 3.2.18 for $\underset{\approx}{r}$ - realizability. It is not even closed under deduction, for the $\underset{\approx}{q}$ - realizability of $A \vee A$ implies that A is true, and formally $\underset{\approx}{HA} \vdash \exists x(x \underset{\approx}{q} A \vee A) \Leftrightarrow \underset{\approx}{HA} \vdash A$. Hence, if A is $\underset{\approx}{HA}$ - $\underset{\approx}{q}$ - realizable, but not provable (e.g. a suitable instance of CT_0), then $\underset{\approx}{HA} \vdash \exists x(x \underset{\approx}{q} A)$, $\underset{\approx}{HA} \vdash \exists x(x \underset{\approx}{q} A \rightarrow (A \vee A))$, but not $\underset{\approx}{HA} \vdash \exists x(x \underset{\approx}{q}(A \vee A))$.

$\underset{\approx}{q}$ - realizability may be viewed as a "hybrid" of the "$\Gamma \mid C$" relation in § 3.1 and realizability.

§ 3. Realizability notions based on continuous function application.

3.3.1. Introduction. Realizability by functions (in the present sense) is first introduced in Kleene & Vesley 1965, not formalized. Formalized versions are discussed in Kleene 1965 , Kleene 1968, Kleene 1969, Part II. For many details we shall rely on these publications, especially Kleene 1969. The general development is similar to that of the preceding section.

3.3.2. Definition. We define the P-^{1}realizability predicates ($\underset{\approx}{r}_P^1$-realizability) for formulae in the language of $\underset{\sim}{EL}$. P(A) is again a formula of $\underset{\sim}{EL}$, with its set of free variables contained among the free variables of A. The realizability predicate $\alpha\,\underset{\approx}{r}_P^1 A$ contains, besides variables free in A, the new function variable α. As before, the definition is by induction on the logical complexity of A.

$\underset{\approx}{r}_P^1$(i) $\quad \alpha\,\underset{\approx}{r}_P^1 A \equiv A$ for A prime,

$\underset{\approx}{r}_P^1$(ii) $\quad \alpha\,\underset{\approx}{r}_P^1 (A \& B) \equiv j_1\alpha\,\underset{\approx}{r}_P^1 A \;\&\; j_2\alpha\,\underset{\approx}{r}_P^1 B$,

$\underset{\approx}{r}_P^1$(iii) $\quad \alpha\,\underset{\approx}{r}_P^1 (A \vee B) \equiv (j_1\alpha 0=0 \to j_2\alpha\,\underset{\approx}{r}_P^1 A \;\&\; P(A))$
$\qquad\qquad \& \; (j_1\alpha 0\neq 0 \to j_2\alpha\,\underset{\approx}{r}_P^1 B \;\&\; P(B))$,

$\underset{\approx}{r}_P^1$(iv) $\quad \alpha\,\underset{\approx}{r}_P^1 (A\to B) \equiv \forall\beta(\beta\,\underset{\approx}{r}_P^1 A \;\&\; P(A) \to {!}\alpha|\beta \;\&\; \alpha|\beta\,\underset{\approx}{r}_P^1 B)$,

$\underset{\approx}{r}_P^1$(v) $\quad \alpha\,\underset{\approx}{r}_P^1 \forall xAx \equiv \forall x({!}\alpha|\lambda y.x \;\&\; \alpha|\lambda y.x\,\underset{\approx}{r}_P^1 A(x))$,

$\underset{\approx}{r}_P^1$(vi) $\quad \alpha\,\underset{\approx}{r}_P^1 \exists xAx \equiv j_2\alpha\,\underset{\approx}{r}_P^1 A(j_1\alpha 0) \;\&\; P(A(j_1\alpha 0))$,

$\underset{\approx}{r}_P^1$(vii) $\quad \alpha\,\underset{\approx}{r}_P^1 \forall\beta A\beta \equiv \forall\beta({!}\alpha|\beta \;\&\; \alpha|\beta\,\underset{\approx}{r}_P^1 A(\beta))$,

$\underset{\approx}{r}_P^1$(viii) $\quad \alpha\,\underset{\approx}{r}_P^1 \exists\beta A\beta \equiv j_2\alpha\,\underset{\approx}{r}_P^1 A(j_1\alpha) \;\&\; P(A(j_1\alpha))$.

Note that this is Kleene's notion of $\underset{\approx}{r}$-realizability and $\underset{\approx}{q}$-realizability in Kleene 1969, Part II, if we take $P(A)\equiv 0=0$, and $P(A)\equiv A$ respectively, disregarding a difference in the choice of pairing functions and codings of sequences of natural numbers. Let us indicate $\underset{\approx}{r}_P^1$-realizability for $P(A) \equiv 0=0$ (so that we may omit P(A) altogether) as $\underset{\approx}{r}^1$-realizability, and $\underset{\approx}{r}_P^1$-realizability for $P(A) \equiv A$ as $\underset{\approx}{q}^1$-realizability. Only these two notions have practical interest for us. We write correspondingly $\varphi\,\underset{\approx}{r}^1 A$, $\varphi\,\underset{\approx}{q}^1 A$.

3.3.3. Theorem (Soundness).
(i) Let $\underset{\sim}{EL}$ be the system as described in 1.9.10. Then, for any A such that $\underset{\sim}{EL} \vdash A$, there is a p-functor φ containing free only variables free in A such that

$$\underset{\sim}{EL} \vdash {!}\varphi \;\&\; (\varphi\,\underset{\approx}{r}^1 A), \quad \underset{\sim}{EL} \vdash {!}\varphi \;\&\; (\varphi\,\underset{\approx}{q}^1 A).$$

(ii) Assume $A_1,\dots,A_s$ to be closed, and suppose in $\underset{\sim}{EL}$ $A_1,\dots,A_s \vdash A$. Then there is a p-functor φ, containing free only variables free in A or variables $\alpha_1,\dots,\alpha_s$ such that in $\underset{\sim}{EL}$

$$\alpha_1 \underset{\approx}{r}^1 A_1, \dots, \alpha_s' \underset{\approx}{r}^1 A_s \vdash !\varphi \ \& \ (\varphi \underset{\approx}{r}^1 A)$$

$$\alpha_1 \underset{\approx}{q}^1 A_1, \dots, \alpha_s \underset{\approx}{q}^1 A_s \vdash !\varphi \ \& \ (\varphi \underset{\approx}{q}^1 A).$$

Corollaries.

(iii) Let $\underset{\sim}{H} \equiv \underset{\sim}{EL} + \Gamma$, Γ a set of closed additional axioms. If there are (closed) p-functors φ such that

$$F \in \Gamma \Rightarrow \underset{\sim}{H} \vdash !\varphi \ \& \ (\varphi \underset{\approx}{r}^1 A)$$

then

$$\underset{\sim}{H} \vdash A \Rightarrow \underset{\sim}{H} \vdash !\psi \ \& \ (\psi \underset{\approx}{r}^1 A)$$

for some p-functor ψ containing free only variables free in A, and similarly with $\underset{\approx}{q}^1$ instead of $\underset{\approx}{r}^1$.

(iv) Let $\underset{\sim}{H} \equiv \underset{\sim}{EL} + \Gamma$, $\underset{\sim}{H}' \equiv \underset{\sim}{EL} + \Gamma'$, Γ, Γ' sets of closed axioms. Then, if $\underset{\sim}{H} \subseteq \underset{\sim}{H}'$, and

$$A \in \Gamma' \Rightarrow \exists\varphi(\underset{\sim}{H} \vdash !\varphi \ \& \ (\varphi \underset{\approx}{r}^1 A))$$

it follows that for $\underset{\sim}{H}'$

$$\underset{\sim}{H}' \vdash A \Rightarrow \exists\varphi(\underset{\sim}{H} \vdash !\varphi \ \& \ \varphi \underset{\approx}{r}^1 A).$$

Proof. (i), (ii): See Kleene 1969, Theorem 50 A, B (page 80).
In comparing the formulations it should be remembered that in contrast to Kleene's usage, $A_1,\dots,A_s \vdash A$ was interpreted as: A is deduced from assumptions $A_1,\dots,A_s$ with the variables occurring free in $A_1,\dots,A_s$ held constant (in Kleene's terminology), i.e. the variables free in $A_1,\dots,A_s$ do not act as proper parameters of the rules Q1, Q4 (or $\forall$I, $\exists$I in natural deduction systems).

3.3.4. Theorem (Special instances of soundness).
Let $\underset{\sim}{EL} + (\alpha, \beta, \gamma)$ be extensions of $\underset{\sim}{EL}$ according to the following code:
$\alpha = 0, 1, 2, 3$ corresponds to the addition of nothing, $AC_{00}!$, AC_{00}, AC_{01} respectively;
$\beta = 0, 1, 2$ corresponds to the addition of nothing, FAN!, BI! respectively, where FAN! is

FAN! $\quad \forall\alpha \exists! x A(\alpha, x) \to \exists z \forall\alpha \exists y \forall\beta (\bar{\alpha}z = \bar{\beta}z \to A(\beta, y))$

and BI! is obtained by replacing $\forall\alpha\exists x$ in BI_D by $\forall\alpha\exists! x$.
$\gamma = 0, 1, 2, 3$ corresponds to the addition of nothing, C-N!, C-N, C-C respectively (C-N! is as C-N but with $\forall\alpha\exists! x$ replacing $\forall\alpha\exists x$, C-C can be

formulated as $\forall\alpha\exists\beta(\alpha,\beta) \to \exists\gamma\forall\alpha(!\gamma|\alpha \,\&\, A(\alpha, \gamma|\alpha))$).

Then, if $\underset{\sim}{H} \equiv \underset{\sim}{EL} + (1, \beta, 0)$, $\underset{\sim}{H}' \equiv \underset{\sim}{EL} + (\alpha, \beta, \gamma)$ $(\alpha > 0)$ it follows that

$$\underset{\sim}{H}' \vdash A \Rightarrow \exists\varphi(\underset{\sim}{H} \vdash !\varphi \,\&\, \varphi \underset{\sim}{r}^1 A)$$

$$\underset{\sim}{H}' \vdash A \Rightarrow \exists\varphi(\underset{\sim}{H}' \vdash !\varphi \,\&\, \varphi \underset{\sim}{q}^1 A).$$

Proof. Kleene 1969, 5.10 (page 103 - 104).

Remark. Actually, a detailed inspection of the argument shows that $\underset{\sim}{H} \equiv \underset{\sim}{EL} + (1', \beta, 0)$, where 1' stands for $QF\text{-}AC_{00}$ (1.9.10), would suffice.

3.3.5. Definition. We extend the definition of $\varphi \underset{\sim}{r}^1 A$ and $\varphi \underset{\sim}{q}^1 A$ to the language of $\underset{\sim}{IDB}_o$ by insertion of clauses:

$\underset{\sim}{r}^1(i)'$ $\quad \alpha \underset{\sim}{r}^1 K\beta \equiv K\beta$

$\underset{\sim}{q}^1(i)'$ $\quad \alpha \underset{\sim}{q}^1 K\beta \equiv K\beta$.

So $\underset{\sim}{r}^1(i)$, $\underset{\sim}{q}^1(i)$ may be stated, as before as $\alpha \underset{\sim}{r}^1 A \equiv A$, $\alpha \underset{\sim}{q}^1 A \equiv A$ for A prime.

3.3.6. Theorem (Soundness for $\underset{\sim}{IDB}$).

$$\underset{\sim}{IDB} \vdash A \Rightarrow \exists\varphi(\underset{\sim}{IDB} \vdash !\varphi \,\&\, (\varphi \underset{\sim}{r}^1 A))$$

$$\underset{\sim}{IDB} \vdash A \Rightarrow \exists\varphi(\underset{\sim}{IDB} \vdash !\varphi \,\&\, (\varphi \underset{\sim}{q}^1 A))$$

Proof. By 3.3.3 (iii) and 3.3.4 it is sufficient to establish the $\underset{\sim}{r}^1$- and $\underset{\sim}{q}^1$-realizability for K1, K2 and each instance of K3 (cf. 1.9.18). We give the verification for $\underset{\sim}{r}^1$-realizability; the argument for $\underset{\sim}{q}^1$-realizability is obtained by slight additions.

(i) $\quad \Lambda\alpha\Lambda\beta.(\lambda x.0) \underset{\sim}{r}^1 \forall\alpha[\forall x(\alpha x = Sy) \to K\alpha]$.

(ii) Assume $\beta \underset{\sim}{r}^1 [\alpha 0=0 \,\&\, \forall x K(\lambda n.\alpha(\hat{x} * n))]$. Then

$\alpha 0=0 \,\&\, \forall x(!(j_2\beta)|\lambda y.x \,\&\, (j_2\beta)|\lambda y.x \underset{\sim}{r}^1 K(\lambda n.\alpha(\hat{x} * n))$.

Hence $\alpha 0=0 \,\&\, \forall x K(\lambda n.\alpha(\hat{x} * n)$, hence $K\alpha$, and therefore

$$\Lambda\alpha\Lambda\beta.\lambda x.0 \underset{\sim}{r}^1 \forall\alpha[\alpha 0=0 \,\&\, \forall x K(\lambda n.\alpha(\hat{x} * n)) \to K\alpha].$$

(iii) Assume

(1) $\quad \beta \underset{\sim}{r}^1 [\forall y Q(\lambda x.Sy) \,\&\, \forall\alpha(\alpha 0=0 \,\&\, \forall x Q(\lambda n.\alpha(\hat{x} * n)) \to Q\alpha)]$

(2) $\quad \gamma \underset{\sim}{r}^1 K\alpha$ $\quad$ (i.e. $K\alpha$ holds).

Then

(3) $\quad \forall y(!(j_1\beta)|\lambda z.y \,\&\, (j_1\beta)|\lambda z.y \underset{\sim}{r}^1 Q(\lambda z.Sy))$

(4) $\quad \forall\alpha(!(j_2\beta)|\alpha \,\&\, j_2\beta|\alpha \underset{\sim}{r}^1 (\alpha 0=0 \,\&\, \forall x Q(\lambda n.\alpha(\hat{x} * n)) \to Q\alpha))$.

(4) is equivalent to

(5) $\forall\alpha(!(j_2\beta)|\alpha \,\&\, [\forall\gamma(j_2\gamma\, \underset{\approx}{r}^1\, \forall x Q(\lambda n.\alpha(\hat{x} * n)) \,\&\, \alpha 0 = 0) \rightarrow$
$\rightarrow\, !((j_2\beta)|\alpha) \mid \gamma \,\&\, ((j_2\beta)|\alpha) \mid \gamma\, \underset{\approx}{r}^1\, Q\alpha])$.

There exists a functor φ such that

$$\alpha 0 \neq 0 \rightarrow \varphi|(\alpha,\beta,\gamma) \simeq j_1\beta \mid \lambda z.(\alpha 0 \dot{-} 1)$$
$$\alpha 0 = 0 \rightarrow \varphi|(\alpha,\beta,\gamma) \simeq ((j_2\beta)|\alpha) \mid j(\lambda x.1, \Lambda\delta.\gamma|(f|(\alpha,\delta)))$$

where $f|(\alpha,\delta) = \lambda n.\alpha(\langle\delta 0\rangle * n)$.

Using the recursion theorem analogue 1.9.16, we find a ψ such that

$$\alpha 0 \neq 0 \rightarrow \psi|(\alpha,\beta) \simeq j_1\beta|\lambda z.(\alpha 0 \dot{-} 1)$$
$$\alpha 0 = 0 \rightarrow \psi|(\alpha,\beta) \simeq (j_2\beta|\alpha) \mid j(\lambda x.1, \Lambda\delta.\psi|(f|(\alpha,\delta))) .$$

Then one proves by induction over K w.r.t. α: $K\alpha \rightarrow\, !\psi|(\alpha,\beta)$ & $\psi|(\alpha,\beta)\, \underset{\approx}{r}^1\, Q\alpha$ (cf. Kreisel - Troelstra 1970 3.7.2).

3.3.7 - 3.3.13. Characterization of $\underset{\approx}{r}^1$ - realizability.

3.3.7. Almost negative formulae are defined as in 3.2.9. Similar to 3.2.10 we have

Lemma. For all formulae A in the language of $\underset{\sim}{EL}$, $x\, \underset{\approx}{r}^1\, A$ is logically equivalent to an almost negative formula.

3.3.8. Lemma. Let $A(\underline{\underline{a}})$ be an almost negative formula of $\underset{\sim}{EL}$, and let $\underline{\underline{a}}$ be a string of number- and function variables, containing all the variables free in A. Then there is a p-functor ψ_A, such that

(i) $\underset{\sim}{EL} \vdash \exists\alpha(\alpha\, \underset{\approx}{r}^1 A) \rightarrow A$

(ii) $\underset{\sim}{EL} \vdash A(\underline{\underline{a}}) \rightarrow\, !\psi_A(\underline{\underline{a}})$ & $\psi_A(\underline{\underline{a}})\, \underset{\approx}{r}^1\, A(\underline{\underline{a}})$

(iii) $\underset{\sim}{EL} \vdash \alpha\, \underset{\approx}{r}^1\, A \longleftrightarrow \alpha\, \underset{\approx}{q}^1\, A$.

Proof. Quite similar to the proof of 3.2.11. We indicate the definitions of ψ_A :

(a) $\psi_{t=s}(\underline{\underline{a}}) \equiv_{def} \lambda x.0$

(b) $\psi_{\exists x[t(x)=s(x)]}(\underline{\underline{a}}) \equiv_{def} j(\lambda z.\min_y[t(y) = s(y)], \lambda z.0)$

(c) $\psi_{\exists\alpha[t(\alpha)=s(\alpha)]}(\underline{\underline{a}}) \equiv_{def} j(f_t, \lambda z.0)$,
where $f_n = \lambda x.(n)_x$ and where $t = \min_n[t(f_n) = s(f_n)]$

(d) $\psi_{A \,\&\, B}(\underline{\underline{a}}) \equiv j(\psi_A(\underline{\underline{a}}), \psi_B(\underline{\underline{a}}))$

(e) $\psi_{\forall x A x}(\underline{\underline{a}}) \equiv \Lambda^1 x.\ \psi_{Ax}(\underline{\underline{a}},x)$

(f) $\psi_{\forall\alpha A\alpha}(\underline{\underline{a}}) \equiv \Lambda^1\alpha.\ \psi_{A\alpha}(\underline{\underline{a}},\alpha)$

(g) $\psi_{A \rightarrow B}(\underline{\underline{a}}) \equiv \Lambda^1\alpha.\ \psi_B(\underline{\underline{a}})$

(cf. Kleene & Vesley 1965, Kleene 1965).

3.3.9. Definition. Let GC denote the following schema, with A almost negative, β not occurring free in α

$$\mathrm{GC} \quad \forall\alpha[A \to \exists\beta B\beta] \to \exists\gamma\forall\alpha[A \to {!}\gamma|\alpha \,\&\, B(\gamma|\alpha)] .$$

3.3.10. Lemma. For any universal closure A of an instance of GC there exists a closed p-functor φ such that

$$\underset{\sim}{\mathrm{EL}} \vdash {!}\varphi \,\&\, \varphi \underset{\approx}{r}^1 A, \quad \underset{\sim}{\mathrm{EL}} \vdash {!}\varphi \,\&\, \varphi \underset{\approx}{q}^1 A .$$

Proof. The proof is very similar to the proof of 3.2.15.
Consider an instance of GC not containing parameters, and assume

$$\delta \underset{\approx}{r}^1 \forall\alpha[A \to \exists\beta B\beta] .$$

Then

$$\forall\alpha[{!}\delta|\alpha \,\&\, \forall\epsilon(\epsilon \underset{\approx}{r}^1 A \to {!}(\delta|\alpha)|\epsilon \,\&\, (\delta|\alpha)|\epsilon \underset{\approx}{r}^1 \exists\beta B\beta)] .$$

By 3.3.7-8

$${!}\psi_A^1 \,|\, \alpha \,\&\, (\psi_A^1 \,|\, \alpha \underset{\approx}{r}^1 A) \longleftrightarrow A .$$

Now put

$$\varphi \equiv (\delta|\alpha)|(\psi_A^1|\alpha) , \quad \text{then}$$

$$\forall\alpha[A \to {!}\varphi \,\&\, j_2\varphi \underset{\approx}{r}^1 B\; j_1\varphi] .$$

We must construct a p-functor χ ($\equiv \chi[\delta]$) such that

$$\chi \underset{\approx}{r}^1 \exists\gamma\forall\alpha[A \to {!}\gamma|\alpha \,\&\, \forall\epsilon(\gamma|\alpha \simeq \epsilon \to B\epsilon)] .$$

Then χ must satisfy

$$A\alpha' \to C(j_1[(j_2\chi|\alpha)|\alpha'],\ j_1\chi, \alpha)$$
where $C(\delta_1, \gamma, \alpha) \equiv_{\mathrm{def}} \delta_1 \underset{\approx}{r}^1 {!}\gamma|\alpha$.
$$A\alpha' \,\&\, D(\epsilon', j_1\chi, \alpha, \epsilon) \to ((j_2(j_2\chi|\alpha)|\alpha')|\epsilon) \,|\, \epsilon' \underset{\approx}{r}^1 B\epsilon ,$$
where $D(\epsilon', \gamma, \alpha, \epsilon) \equiv_{\mathrm{def}} \epsilon' \underset{\approx}{r}^1 Bc$.

Note that ${!}\gamma|\alpha$ and $\gamma|\alpha \simeq \epsilon$ are almost negative. Hence (lemma 3.3.8) there is a function $\psi_{!\gamma|\alpha}$

$${!}\gamma|\alpha \longleftrightarrow {!}\psi_{!\gamma|\alpha}(\gamma,\alpha) \,\&\, \psi_{!\gamma|\alpha}(\gamma,\alpha) \underset{\approx}{r}^1 ({!}\gamma|\alpha)$$

$$\gamma|\alpha \simeq \epsilon \longleftrightarrow \exists\epsilon'(\epsilon' \underset{\approx}{r}^1 (\gamma|\alpha \simeq \epsilon)) .$$

Now take χ such that

$$j_1\chi \simeq \Lambda\alpha.j_1\varphi$$

$$j_1[(j_2\chi|\alpha)|\alpha'] \simeq \psi_{!\gamma|\alpha}(j_1\chi,\alpha) \equiv \psi_{!\gamma|\alpha}(\Lambda\alpha.j_1\varphi,\alpha)$$

$$((j_2[(j_2\chi|\alpha)|\alpha']) \,|\, \epsilon)\epsilon' \simeq j_2\varphi .$$

Hence we must take for χ:

$$\chi \equiv j[\Lambda\alpha.j_1\varphi,\ \Lambda\alpha\Lambda\alpha'.j\{\psi^1_{!\gamma|\alpha}(\Lambda\alpha.j_1\varphi,\alpha),\ \Lambda\varepsilon\Lambda\varepsilon'.j^2\varphi\}].$$

Then $\Lambda\delta.\chi$ $\underset{\approx}{r}^1$ - realizes our instance of GC.
Similarly for $\underset{\approx}{q}^1$ - realizability.

3.3.11. Theorem (Characterization of $\underset{\approx}{r}^1$ - realizability).
(i) $\underset{\sim}{EL} + GC \vdash A \leftrightarrow \exists\alpha(\alpha\,\underset{\approx}{r}^1\,A)$.
(ii) Let $\underset{\sim}{H}$ be any extension of $\underset{\sim}{EL}$ for which the soundness theorem (for $\underset{\approx}{r}^1$ - realizability) has been established, then

$$\underset{\sim}{H} \vdash \exists\alpha(\alpha\,\underset{\approx}{r}^1\,A) \Leftrightarrow \underset{\sim}{H} + GC \vdash A.$$

Proof. (i) is shown by induction on the complexity of A, completely similar to the argument in 3.2.18, sub (i).
(ii) The implication from left to right follows from (i). Now assume $\underset{\sim}{H} + GC \vdash A$. Then $\underset{\sim}{H} \vdash F \to A$, F a conjunction of universally closed instances of GC, so $\underset{\sim}{H} \vdash \exists\alpha(\alpha\,\underset{\approx}{r}^1\,F)$. Also $\underset{\sim}{H} \vdash \exists\beta(\beta\,\underset{\approx}{r}^1\,F \to A)$ by the soundness theorem, therefore $\underset{\sim}{H} \vdash \exists\alpha(\alpha\,\underset{\approx}{r}^1\,A)$.

3.3.12. Corollary. $\underset{\sim}{H} + GC$ is consistent relative to $\underset{\sim}{H}$ for $\underset{\sim}{H} \equiv \underset{\sim}{EL}$, $\underset{\sim}{EL} + BI!$

3.3.13. Remarks. (A). Under the assumption of CT, GC implies ECT_0.
To see this, we note that if

(1) $\forall x[Ax \to \exists y Bxy]$

where A is almost negative, then also

(2) $\forall\alpha[A(\alpha 0) \to \exists\beta B(\alpha 0, \beta 0)]$

and with GC

$$\exists\gamma\ \forall\alpha[A(\alpha 0) \to\ !\gamma|\alpha\ \&\ B(\alpha 0,\ (\gamma|\alpha)0)],$$

therefore, by CT there is a $z \in V_1$ such that

$$\forall x[A(x) \to\ !\{z\}|\lambda y.x\ \&\ B(x,\ (\{z\}|\lambda y,x)0)].$$

Now we can find a u such that

$$u \equiv \Lambda x.[\{z\}(\hat{0} * (\overline{\lambda y.x})(\min_w[\{z\}(\hat{0} * (\overline{\lambda y.x})w) \neq 0]) \dot{-} 1]$$

and hence

$$\exists u\ \forall x[Ax \to\ !\{u\}(x)\ \&\ B(x,\ \{u\}(x))].$$

It is open whether ECT_0 and CT imply GC.

(B). Just as for ECT_0, GC cannot be generalized so as to omit the restriction of the formula A being almost negative.
A counterexample may be constructed in a similar way as for ECT_0 (3.2.20). Take for $A(\alpha)$: $\exists x(\alpha x \neq 0) \vee \neg \exists x(\alpha x \neq 0)$, and for $B(\alpha,y) \equiv (y=0 \mathbin{\&} \neg \exists x(\alpha x \neq 0)) \vee (y \neq 0 \mathbin{\&} \alpha(y \dot{-} 1) \neq 0)$. Then

$$\forall\alpha[A\alpha \rightarrow \exists y B(\alpha,y)] \rightarrow \exists\gamma\ \forall\alpha[A\alpha \rightarrow !\gamma(\alpha) \mathbin{\&} B(\alpha,\gamma(\alpha))]$$

is a consequence of the generalized GC. However, the premiss is valid, but the conclusion would imply the existence of a γ such that

$$\forall\alpha[A\alpha \rightarrow !\gamma(\alpha) \mathbin{\&} B(\alpha,\gamma(\alpha))] .$$

Since $\neg\neg A\alpha$, also (using $\neg\neg\forall\alpha A \rightarrow \forall\alpha\neg\neg A$, $\neg\neg(P \rightarrow Q) \rightarrow (\neg\neg P \rightarrow \neg\neg Q)$)

$$\forall\alpha\ \neg\neg\exists z\{(\gamma(\bar{\alpha}z) \neq 0) \mathbin{\&} \forall y < z(\gamma(\bar{\alpha}y) = 0) \mathbin{\&} ((\gamma(\bar{\alpha}z) \dot{-} 1 = 0 \mathbin{\&} \neg\exists x(\alpha x \neq 0)) \vee$$
$$\vee\ (\gamma(\bar{\alpha}z) \dot{-} 1 \neq 0 \mathbin{\&} \alpha(\gamma(\bar{\alpha}z) \dot{-} 2) \neq 0))\} .$$

Now let δ be such that $!\delta(\alpha) \leftrightarrow \gamma(\alpha) \simeq 0$, and $!\delta(\alpha) \rightarrow \delta(\alpha) > 0$. Such a δ is easily defined: we put

$$\delta n = 0 \leftrightarrow \gamma n = 0$$
$$\delta n = 2 \text{ if } \exists m \leq n(\gamma m = 1 \mathbin{\&} \forall m' < m(\gamma m' = 0))$$
$$\delta n = 0 \text{ if } \exists m \leq n(\gamma m > 1 \mathbin{\&} \forall m' < m(\gamma m' = 0)) .$$

It follows that

$$!\delta(\delta) \mathbin{\&} \delta(\delta) > 0 \leftrightarrow \gamma(\delta) \simeq 0 \leftrightarrow \neg\exists x(\delta x \neq 0) \leftrightarrow \neg !\delta(\delta)$$

and since $!\delta(\delta) \rightarrow \delta(\delta) > 0$, we have a contradiction.

§ 4. Modified realizability.

3.4.1. Introduction. Modified realizability was first introduced and used in Kreisel 1959, 3.52, and later in Kreisel 1962 under the misleading name of generalized realizability (see section 10 of Kreisel 1962). Modified realizability in its abstract form provides interpretations of the various HA^{ω}-versions into themselves; the interpretation may be specialized (to an interpretation in (a subsystem of) a version of HA^{ω} or into another system) by specifying a model for the objects of finite type; thus Kleene's "special realizability" (Kleene & Vesley 1965, § 10) may be viewed as (a variant of) a specialization of modified realizability to ICF, the intensional continuous functions (cf. 2.6.2).

One of its most distinctive properties is that M_{PR} is not validated by modified realizability; this was already noted and used by Kreisel (Kreisel 1959, 3.52, Kreisel 1962, Thm 6) to show underivability of M_{PR} in systems of intuitionistic analysis. Kleene used his "special realizability" to the same purpose (Kleene & Vesley 1965, § 10). See below in 3.4.9.

On the other hand, modified realizability validates

$$\text{IP} \qquad (\neg A \rightarrow \exists y B) \rightarrow \exists y(\neg A \rightarrow B)$$

(y not free in A) (This fact is connected with its invalidating M_{PR}, cf. 3.4.12 (i), 3.2.27 (i)).

This property was used (although not yet in full generality) in Kreisel 1959 B, D for proof-theoretic applications ("derived rules").

Vesley 1970 also depends on a weakened version of this property.

Below we shall describe modified realizability (mr - realizability) and a variant (mq - realizability) which bears the same relationship to mr - realizability as q - realizability does to r - realizability.

Contents of the section. Subsections 2 - 6 are devoted to the definition of mr - and mq - realizability and to the soundness theorem.

In subsections 7, 8 mr - realizability has been axiomatized. Subsection 9 deals with variant formulations; subsection 11 compares r - realizability and mr - realizability where the objects of finite types are interpreted by HRO.

Subsections 12 - 25 discuss the realizability and non-realizability of various schemata such as M_{PR}, CT, CT_0, FAN, BI_0, TI(<), and contain a proof that $HA + CT_0 \nvdash ECT_0$. Subsections 27 - 28 are devoted to modified realizability for HAS (relative to HRO).

Subsection 29 applies modified realizability to obtain a characterization of the provably recursive functions of HA: they are exactly the recursive functions represented by closed terms of $N\text{-}HA^{\omega}$.

3.4.2. Definition. Let A be a formula in a language $\mathcal{L}$ obtained by extending the language of $\underset{\sim}{N}-\underset{\sim}{HA}^{\omega}$ by some (possibly none) constants for objects of finite type. Let P(A) be a property, definable in $\mathcal{L}$, such that the free variables of P(A) are contained among the free variables of A.

We define a predicate $\underline{\underline{x}}\ \mathrm{mr}_P\, A$ ($\underline{\underline{x}}$ P-modified realizes A), its free variables contained among the variables free in A, and variables of the string $\underline{\underline{x}}$ (the variables of $\underline{\underline{x}}$ not occurring free in A); the types of the variables in $\underline{\underline{x}}$ and the length of $\underline{\underline{x}}$ are determined by the logical construction of A only.

The definition of $\underline{\underline{x}}\ \mathrm{mr}_P\, A$ is by induction on the logical complexity of A.

mr_P(i) $\quad \underline{\underline{x}}\ \mathrm{mr}_P\, A \equiv A$; $\underline{\underline{x}}$ is the empty sequence, if A is prime.

In the other clauses, assume $\underline{\underline{x}}\ \mathrm{mr}_P\, A$, $\underline{\underline{y}}\ \mathrm{mr}_P\, B$ to be well-formed (i.e. the types in $\underline{\underline{x}}, \underline{\underline{y}}$ are correct)

mr_P(ii) $\quad \underline{\underline{x}},\underline{\underline{y}}\ \mathrm{mr}_P\,(A \& B) \equiv \underline{\underline{x}}\ \mathrm{mr}_P\, A \;\&\; \underline{\underline{y}}\ \mathrm{mr}_P\, B$

mr_P(iii) $\quad z,\underline{\underline{x}},\underline{\underline{y}}\ \mathrm{mr}_P\,(A \vee B) \equiv [(z=0 \rightarrow (\underline{\underline{x}}\ \mathrm{mr}_P\, A) \;\&\; P(A)) \;\&\; (z\neq 0 \rightarrow (\underline{\underline{y}}\ \mathrm{mr}_P\, B) \;\&\; P(B))]$

mr_P(iv) $\quad \underline{\underline{y}}\ \mathrm{mr}_P\,(A \rightarrow B) \equiv \forall \underline{\underline{x}}((\underline{\underline{x}}\ \mathrm{mr}_P\, A) \;\&\; P(A) \rightarrow \underline{\underline{y}}\underline{\underline{x}}\ \mathrm{mr}_P\, B)$

mr_P(v) $\quad \underline{\underline{x}}\ \mathrm{mr}_P\,(\forall y^{\sigma} A y^{\sigma}) \equiv \forall y^{\sigma}(\underline{\underline{x}} y^{\sigma}\ \mathrm{mr}_P\, A y^{\sigma})$

mr_P(vi) $\quad z^{\sigma},\underline{\underline{x}}\ \mathrm{mr}_P\,(\exists y^{\sigma} A y^{\sigma}) \equiv \underline{\underline{x}}\ \mathrm{mr}_P\, A z^{\sigma} \;\&\; P(A z^{\sigma})$.

3.4.3. Examples. Actually, there are only two examples which are of practical interest to us.

(A). $P(A) \equiv 0=0$. In this case we may omit P(A) altogether (modulo logical equivalence); we call the resulting notion mr-realizability, and write $\underline{\underline{x}}\ \mathrm{mr}\, A$ for $\underline{\underline{x}}\ \mathrm{mr}_P\, A$ in this case.

(B). $P(A) \equiv A$. We call the resulting notion mq-realizability, and we write $\underline{\underline{x}}\ \mathrm{mq}\, A$ for $\underline{\underline{x}}\ \mathrm{mr}_P\, A$ in this case.

Notational convention: It is sometimes more convenient to write A^0, $A_0\underline{\underline{x}}$, A^1, $A_1\underline{\underline{x}}$ for $\exists \underline{\underline{x}}(\underline{\underline{x}}\ \mathrm{mr}\, A)$, $\underline{\underline{x}}\ \mathrm{mr}\, A$, $\exists \underline{\underline{x}}(\underline{\underline{x}}\ \mathrm{mq}\, A)$, $\underline{\underline{x}}\ \mathrm{mq}\, A$ respectively. It is helpful to think of A^0 as the "modified realizability interpretation" or "mr-translation" of A, and of A^1 as the "mq-translation" of A.

3.4.4. Remarks.

(i). Note that the modified-realizability predicate $\underline{\underline{x}}\ \mathrm{mr}\, A$ for negative A is identical with A (not only logically equivalent) on the convention that $P(A) \equiv 0=0$ is omitted, i.e. $\underline{\underline{x}}$ is the empty sequence; and $A \leftrightarrow x\ \mathrm{mq}\, A$. This remark corresponds to 3.2.3 for r-realizability.

(ii). $x\ \mathrm{mr}\ \neg A \leftrightarrow \forall y(\neg y\ \mathrm{mr}\, A)$, and $\underline{\underline{x}}$ is again the empty sequence!

(iii). For all A, $\underline{x}\ \underset{\approx}{mr}\ A$ is a formula in the negative fragment. This is seen by induction on the logical complexity of A.

3.4.5. Theorem (Soundness theorem).
(i) Let $P(A)$ be a property as intended in the definition of $\underset{\approx}{mr}_P$ - realizability, satisfying

(A) $\quad \underline{H} \vdash A \Rightarrow \underline{H} \vdash P(A)$

(B) $\quad \underline{H} + \Gamma \vdash P(A \to B)\ \&\ P(A) \Rightarrow \underline{H} + \Gamma \vdash P(B)$, or equivalently
$\quad \underline{H} \vdash P(A)\ \&\ P(A \to B) \to P(B)$,

where $\underline{H} \equiv \underline{HA}^{\omega}$, $\underline{I}\text{-}\underline{HA}^{\omega}$, $\underline{E}\text{-}\underline{HA}^{\omega}$, $\underline{HRO}^{-}$, $\underline{WE}\text{-}\underline{HA}^{\omega}$. Then, for any closed A

$$\underline{H} \vdash A \Rightarrow \underline{H} \vdash \underline{t}\ \underset{\approx}{mr}_P\ A$$

for a suitable sequence t of closed terms of $\underline{H}$.
(ii) Let $\underline{H}'$ be $\underline{H} + \Gamma$, $\underline{H}$ one of the systems in (i), Γ a set of sentences, such that for $\underline{H}'$ (A), (B) and

(C) $\quad A \in \Gamma \Rightarrow \underline{H}' \vdash \underline{t}\ \underset{\approx}{mr}_P\ A\ \& P(A)$

for suitable sequences $\underline{t}$ of closed terms of $\underline{H}'$; then the assertion of (i) also holds for $\underline{H}'$.
Corollary: In particular,
(iii) The assertion of (i) also holds for $\underline{H}$, and for $\underline{H}'$ as in (ii), with respect to either $\underset{\approx}{mr}$ - realizability or $\underset{\approx}{mq}$ - realizability, if we replace (C) by

(C') $\quad A \in \Gamma \Rightarrow \underline{H} \vdash \underline{t}\ \underset{\approx}{mr}\ A$ (resp. $\underline{H} \vdash \underline{t}\ \underset{\approx}{mq}\ A$).

Remark. In (ii), for $\underset{\approx}{mr}$ -, $\underset{\approx}{mq}$ - realizability, (C) is automatically satisfied for Γ consisting of negative formulae (by 3.4.4 (i)).

Proof. Note that the assertion of the soundness theorems for $\underline{N}\text{-}\underline{HA}^{\omega}$ is equivalent to the following assertion: if $A(\underline{x})$ is of a formula of $\underline{N}\text{-}\underline{HA}^{\omega}$ containing $\underline{x}$ free, then $\underline{N}\text{-}\underline{HA}^{\omega} \vdash A(\underline{x}) \Rightarrow \underline{N}\text{-}\underline{HA}^{\omega} \vdash \underline{Tx}\ \underset{\approx}{mr}_P\ A(\underline{x})$ where $\underline{T}$ is a sequence of constant terms of $\underline{N}\text{-}\underline{HA}^{\omega}$. Now it is sufficient to prove this assertion by induction on the length of derivations, i.e. we show that the assertion holds for axioms, and secondly, if $F_1,\ldots,F_n \Rightarrow F$ is an instance of a rule, and the assertion holds for $F_1,\ldots,F_n$, and $P(F_1),\ldots,P(F_n)$, then the assertion holds for F (Again, we use Gödel's system (1.1.4) for our verification).

PL 2). Assume $\underline{Tx}\ \underset{\approx}{mr}_P\ A\underline{x}$, $\underline{T}'\underline{x}\ \underset{\approx}{mr}_P\ (A\underline{x} \to B\underline{x})$, $P(A\underline{x})$; then $\underline{T}'\underline{x}(\underline{Tx})\ \underset{\approx}{mr}_P\ B\underline{x}$; so with $\underline{T}'' \equiv \lambda\underline{x}.\underline{T}'\underline{x}(\underline{Tx})$, $\underline{T}''\underline{x}\ \underset{\approx}{mr}_P\ B\underline{x}$.

For simplicity in notation, we omit parameters in further cases.

PL 3). Assume $\underline{T}\ \underset{\approx}{mr}_P\ A \to B$, $\underline{T}'\ \underset{\approx}{mr}_P\ B \to C$, $P(A \to B)$, $P(B \to C)$.
If $\underline{x}\ \underset{\approx}{mr}_P\ A\ \& P(A)$, then $\underline{Tx}\ \underset{\approx}{mr}_P\ B\ \&\ P(B)$, hence $\underline{T}'(\underline{Tx})\ \underset{\approx}{mr}_P\ C$. Thus if

$\underline{T}'' \equiv \lambda \underline{x}.\underline{T}'(\underline{T}\underline{x})$, then $\underline{T}'' \;\mathrm{mr}_P\; (A \to C)$.

PL 7). Assume $\underline{T} \;\mathrm{mr}_P\; A \& B \to C$, $P(A \& B \to C)$. Suppose $(\underline{x} \;\mathrm{mr}_P\; A) \& P(A)$, $(\underline{y} \;\mathrm{mr}_P\; B) \& P(B)$, then $\underline{x}, \underline{y} \;\mathrm{mr}_P\; A \& B$ and $P(A \& B)$ (since $P(A \to (B \to (A \& B))$, $P(A)$, $P(B)$). Therefore $\underline{T}\underline{x}\underline{y} \;\mathrm{mr}_P\; C$, hence $\underline{T} \;\mathrm{mr}_P\; A \to (B \to C)$.

PL 8). Similarly.

PL 9). The empty sequence realizes $\wedge \to A$: $\langle\rangle \;\mathrm{mr}_P\; \wedge \to A$.

PL10). If $((z^0, \underline{x}, \underline{y}) \;\mathrm{mr}_P\; (A \vee A)) \& P(A \vee A)$, then $\underline{T}$ must be defined by cases

$$\underline{T} z^0 \underline{x}\underline{y} = \begin{cases} \underline{x} & \text{if } z^0 = 0 \\ \underline{y} & \text{if } z^0 \neq 0 \end{cases}$$

(one may take $\underline{T} = \lambda z \underline{x}\underline{y}.\underline{R}\underline{x}(\lambda \underline{u} v^0.\underline{y})z$), then $\underline{T} \;\mathrm{mr}_P\; A \vee A \to A$.

$\lambda \underline{x}.\underline{x}, \underline{x} \;\mathrm{mr}_P (A \to A \& A)$.

PL11). Let $\underline{x} \;\mathrm{mr}_P\; A \& P(A)$, then $(\underline{0}, \underline{x}, \underline{0}) \;\mathrm{mr}_P\; A \vee B$, hence $\lambda \underline{x}.[0, \underline{x}, \underline{0}] \;\mathrm{mr}_P\; A \to A \vee B$. (Here $\underline{0}$ is a sequence of $0^{\tau'}$s of the appropriate types.)

Let $(\underline{x}, \underline{y} \;\mathrm{mr}_P\; (A \& B)) \& P(A)$; then $\underline{x} \;\mathrm{mr}_P\; A$, hence $\lambda \underline{x}\underline{y}.\underline{x} \;\mathrm{mr}_P\; (A \& B \to A)$.

PL12). $\lambda z^0 \underline{y}\underline{x}.[(1 \dot{-} z^0), \underline{x}, \underline{y}] \;\mathrm{mr}_P (A \vee B \to B \vee A)$,

$\lambda \underline{y}\underline{x}.[\underline{x}, \underline{y}] \;\mathrm{mr}_P (A \& B \to B \& A)$.

PL13). Let $\underline{T} \;\mathrm{mr}_P (A \to B)$, $P(A \to B)$.

Assume $z^0, \underline{x}, \underline{y} \;\mathrm{mr}_P\; C \vee A$, $P(C \vee A)$. Then either $z^0 = 0$, $\underline{x} \;\mathrm{mr}_P\; C$, $P(C)$; or $z^0 \neq 0$, $\underline{y} \;\mathrm{mr}_P\; A$ and $P(A)$. In the second case, also $P(B)$, and $\underline{T}\underline{y} \;\mathrm{mr}_P\; B$. Hence, take $\underline{T}' \equiv \lambda z^0 \underline{x}\underline{y}.[z^0, \underline{x}, (\underline{T}\underline{y})]$.

Q 1). Let $\vdash \underline{t}y \;\mathrm{mr}_P\; C \to Ay^\sigma$, $\vdash P(C \to Ay)$.

Assume $\underline{z} \;\mathrm{mr}_P\; C$, $P(C)$. Then $\vdash P(Ay)$, hence $\vdash P(\forall y Ay)$.

Then $\underline{t}y^\sigma \underline{z} \;\mathrm{mr}_P\; Ay^\sigma$, so $\lambda y.\underline{t}y^\sigma \underline{z} \;\mathrm{mr}_P\; \forall y^\sigma Ay^\sigma$, and $\lambda \underline{z}y.\underline{t}y\underline{z} \;\mathrm{mr}_P\; C \to \forall y^\sigma Ay^\sigma$.

Q 2). Let $\underline{x} \;\mathrm{mr}_P\; \forall y^\sigma Ay$, $P(\forall y^\sigma Ay)$; then $\underline{x}t^\sigma \;\mathrm{mr}_P\; At^\sigma$, so $\lambda \underline{x}.\underline{x}t^\sigma \;\mathrm{mr}_P\; \forall y^\sigma Ay \to At^\sigma$.

Q 3). $\lambda \underline{x}.[t, \underline{x}] \;\mathrm{mr}_P\; (At \to \exists y^\sigma Ay)$.

Q 4). Let $\underline{t}y \;\mathrm{mr}_P\; (Ay^\sigma \to C)$, $P(Ay^\sigma \to C)$ and assume $y, \underline{x} \;\mathrm{mr}_P\; \exists y Ay^\sigma$, $P(\exists y Ay^\sigma)$. Then $\underline{x} \;\mathrm{mr}\; Ay \;\&\; P(Ay)$, and also $\underline{t}y\underline{x} \;\mathrm{mr}\; C$. Hence $\underline{t} \;\mathrm{mr}_P\; \exists y Ay^\sigma \to C$.

Non-logical axioms.

Equality axioms and defining axioms for the constants are trivial (because purely negative).

It remains to verify the induction schema.

Let

$$\underline{x} \;\mathrm{mr}_P\; A0,\quad \underline{y} \;\mathrm{mr}_P\; \forall z^0[Az \to A(Sz)],$$
$$P(A0 \;\&\; \forall z^0[Az \to A(Sz)]).$$

We put

$$\underline{T} \equiv \lambda \underline{x}\underline{y}z.\underline{R}\underline{x}(\lambda \underline{u} v^0.\underline{y}v\underline{u})z \;,$$

where $\underline{u}$, $\underline{x}$ are of the same type. Then $\underline{T}\underline{x}\underline{y}0 = \underline{x}$, $\underline{T}\underline{x}\underline{y}(Sz) = \underline{y}z(\underline{T}\underline{x}\underline{y}z)$.

By induction on z, we see that $\underline{Txyz}\ \mathbf{mr}_P\ Az$, hence
$T\ \mathbf{mr}_P\ A0\ \&\ \forall z^0[Az \to A(Sz)] \to \forall z^0 Az$.

The preceding verification of (i) holds alike for all systems $\mathbf{H}$. (ii) is an almost immediate corollary, in view of the deduction theorem; and (iii) is immediate.

3.4.6. Remark. If we are satisfied with the weaker statement in the soundness theorem

$$\mathbf{H}' \vdash A \Rightarrow \mathbf{H}' \vdash \exists\underline{x}(\underline{x}\ \mathbf{mr}\ A),$$

we may replace in (iii) of 3.4.5 (C') by

(C") $\quad A \in \Gamma \Rightarrow \mathbf{H} \vdash \exists\underline{x}(\underline{x}\ \mathbf{mr}\ A)$

3.4.7 - 3.4.8. Axiomatization of $\mathbf{mr}$ - realizability.

3.4.7. Lemma. (i) For instances $F\underline{x}$ (containing at most $\underline{x}$ free) of

$IP^\omega \qquad (\neg A \to \exists y^\sigma B) \to \exists y^\sigma(\neg A \to B) \qquad (y^\sigma$ not free in $A)$

or

$AC_{\sigma,\tau} \qquad \forall x^\sigma \exists y^\tau A(x,y) \to \exists z^{(\sigma)\tau}\forall x^\sigma A(x,zx)$

we can find sequences of closed terms $\underline{T}$ such that

$$\mathbf{H} \vdash \underline{Tx}\ \mathbf{mr}\ F\underline{x}, \quad \underline{Tx}\ \mathbf{mq}\ F\underline{x},$$

where $\mathbf{H} = \mathbf{HA}^\omega$, $\mathbf{I}$-$\mathbf{HA}^\omega$, $\mathbf{HRO}^-$, $\mathbf{N}$-$\mathbf{HA}^\omega$.

Proof. Trivial and straightforward, taking for IP^ω remark 3.4.4 (ii) into account. (Use the notation $\exists\underline{x}A_0(\underline{x})$ for the modified realizability interpretation here.)

3.4.8. Theorem (Characterization theorem for $\mathbf{mr}$ - realizability).
Let $\mathbf{H}$ be $\mathbf{HA}^\omega$, $\mathbf{N}$-$\mathbf{HA}^\omega$, $\mathbf{I}$-$\mathbf{HA}^\omega$, $\mathbf{HRO}^-$ or $\mathbf{E}$-$\mathbf{HA}^\omega$ or an extension in the same language for which the soundness theorem can be established.

(i). $\quad \mathbf{H} \vdash IP^\omega + AC \vdash A \leftrightarrow \exists\underline{x}(\underline{x}\ \mathbf{mr}\ A)$

(ii). $\quad \mathbf{H} + IP^\omega + AC \vdash A \Leftrightarrow \mathbf{H} \vdash \exists\underline{x}(\underline{x}\ \mathbf{mr}\ A)$.

Here $AC \equiv \bigcup_{\sigma,\tau \in \mathbf{T}} AC_{\sigma,\tau}$.

Proof. (i). By induction on the complexity of A; consider e.g. $A \equiv B \to C$; by the induction hypothesis, $(B \to C) \leftrightarrow (\exists\underline{x}(\underline{x}\ \mathbf{mr}\ B) \to \exists\underline{y}(\underline{y}\ \mathbf{mr}\ C)) \leftrightarrow$
$\forall\underline{x}(\underline{x}\ \mathbf{mr}\ B \to \exists\underline{y}(\underline{y}\ \mathbf{mr}\ C))$, and since $\underline{x}\ \mathbf{mr}\ B$ is in the negative fragment,
$\underline{x}\ \mathbf{mr}\ B \leftrightarrow \neg\neg\underline{x}\ \mathbf{mr}\ B$, hence by IP^ω, AC: $\forall\underline{x}(\underline{x}\ \mathbf{mr}\ B \to \exists\underline{y}(\underline{y}\ \mathbf{mr}\ C)) \leftrightarrow$
$\leftrightarrow \forall\underline{x}\exists\underline{y}(\underline{x}\ \mathbf{mr}\ B \to \underline{y}\ \mathbf{mr}\ C) \leftrightarrow \exists\underline{Y}\forall\underline{x}(\underline{x}\ \mathbf{mr}\ B \to \underline{Yx}\ \mathbf{mr}\ C) \leftrightarrow \exists\underline{Y}(\underline{Y}\ \mathbf{mr}\ (B \to C))$.
(ii). The implication $\Leftarrow$ follows by (i). Assume now $\mathbf{H} + IP^\omega + AC \vdash A$, then $\mathbf{H} \vdash F \to A$ for F a conjunction of closures of instances of IP^ω, AC; so

$\underset{\sim}{H} \vdash \underline{t}\,\mathbf{mr}\,F$, $\underset{\sim}{H} \vdash \underline{t}'\,\mathbf{mr}\,F \to A$, hence $\underset{\sim}{H} \vdash \underline{t}'\underline{t}\,\mathbf{mr}\,A$.

3.4.9. Inessential (but convenient) variants of **mr** - realizability.

First we note that we have rather arbitrarily fixed, in the definition of **mr** - realizability (3.4.2) that the empty sequence (with the "empty" sequence of types) should modified-realize true prime formulae.

Instead, we might have stipulated: $\underline{x}\,\mathbf{mr}\,A \equiv A$ for A prime, where $\underline{x}$ is a string of variables with an arbitrarily fixed sequence of types (but the same sequence of types for all prime formulae); this would not have made any difference in the proof of the soundness theorem and other results about modified realizability.

Secondly, if we had based ourselves on a theory with pairing operators and products of types, we might have redefined $x\,\mathbf{mr}\,A$ as follows:

Variant I.

(i) $x^{\sigma}\,\mathbf{mr}\,A \equiv A$ if A is prime (σ some fixed but otherwise arbitrary type).

(ii) $x^{\sigma_1 \times \sigma_2}\,\mathbf{mr}\,A \,\&\, B \equiv D'x^{\sigma_1 \times \sigma_2}\,\mathbf{mr}\,A \,\&\, D''x^{\sigma_1 \times \sigma_2}\,\mathbf{mr}\,B$.

(iii) $x^{\sigma \times (\sigma_1 \times \sigma_2)}\,\mathbf{mr}\,A \vee B \equiv ((D'x)\underline{0}=0 \to D'(D''x)\,\mathbf{mr}\,A) \,\&$
$\&((D'x)\underline{0} \neq 0 \to D''(D''x)\,\mathbf{mr}\,B)$, where $\underline{0}$ is a sequence $0^{\sigma_1}, \ldots, 0^{\tau_m}$ such that $t^{\sigma}0^{\tau_1} \ldots 0^{\tau_m} \in 0$, ($\sigma$ fixed, as in (i)).

(iv) $x^{(\sigma)\tau}\,\mathbf{mr}\,(A \to B) \equiv \forall y^{\sigma}(y\,\mathbf{mr}\,A \to xy\,\mathbf{mr}\,B)$.

(v) $x^{\sigma \times \tau}\,\mathbf{mr}\,\exists y^{\sigma} A y^{\sigma} \equiv (D''x)\,\mathbf{mr}\,A(D'x)$.

(vi) $x^{(\sigma)\tau}\,\mathbf{mr}\,\forall y^{\sigma} A y^{\sigma} \equiv \forall y^{\sigma}(xy\,\mathbf{mr}\,Ay^{\sigma})$.

Kleene's "special realizability" is based on yet another schema. He only needs types $\underset{\approx}{T}^{*}$ generated by:

(i) 1 is a type;

(ii) if σ is a type, then $(\sigma)0$ (written as $\sigma+1$) is a type,

(iii) if σ, τ are types, then so is $\sigma \times \tau$ (written by Kleene as (σ, τ)).

We then define, for all types σ, τ of $\underset{\approx}{T}^{*}$, $\sigma * \tau$ (representing $(\sigma)\tau$) as follows:

(a) $\sigma * 1 \equiv (\sigma \times 1)0$

(b) $\sigma * (\tau)0 \equiv (\sigma \times \tau)0$

(c) $\sigma * (\tau_1 \times \tau_2) \equiv (\sigma * \tau_1) \times (\sigma * \tau_2)$.

It is obvious that we can construct functionals $\Phi_{\sigma,\tau}$, $\Phi'_{\sigma,\tau}$ such that $(\Phi_{\sigma,\tau} x^{\sigma*\tau}) y^{\sigma} \in \tau$, $\Phi'_{\sigma,\tau} x^{(\sigma)\tau} \in \sigma*\tau$, $\Phi_{\sigma,\tau}$ and $\Phi'_{\sigma,\tau}$ are inverses, and

$\Phi_{\sigma,1} x^{\sigma*1} = \lambda y^{\sigma} \lambda z^{0}.x^{\sigma*1}(Dy^{\sigma}z^{0})$

$\Phi'_{\sigma,1} x^{(\sigma)1} = \lambda y^{\sigma \times 0}.\, x^{(\sigma)1}((D'y)(D''y))$

$\Phi_{\sigma,(\tau)0} x^{\sigma*(\tau)0} = \lambda y^{\sigma} z^{\tau}.x(Dyz)$

$$\Phi'_{\sigma,(\tau)0}x^{(\sigma\times\tau)0} = \lambda y^{\sigma\times\tau}.x((D'y),(D''y))$$
$$\Phi_{\sigma,(\tau_1\times\tau_2)}x^{\sigma*(\tau_1\times\tau_2)} = \lambda y^{\sigma}.D(\Phi_{\sigma,\tau_1}(D'x)y^{\sigma})(\Phi_{\sigma,\tau_2}(D''x)y^{\sigma})$$
$$\Phi'_{\sigma,(\tau_1\times\tau_2)}x^{\sigma(\tau_1\times\tau_2)} = D(\lambda y^{\sigma}D'(xy))(\lambda y^{\sigma}D''(xy)) .$$

Now we may construct

Variant II.

(i) $x^1 \text{ mr } A \equiv A$ (if A is prime) ;

(ii), (iii) (with $\sigma \equiv 1$), (v) as in variant I ;

(iv) $x^{\sigma*\tau} \text{ mr } (A \to B) \equiv \forall y^{\sigma}(y \text{ mr } A \to (\Phi_{\sigma,\tau}x)y \text{ mr } B)$;

(vi) $x^{\sigma*\tau} \text{ mr } \forall y^{\sigma}Ay^{\sigma} \equiv \forall y^{\sigma}((\Phi_{\sigma,\tau}x)y \text{ mr } Ay)$.

Kleene's "special realizability" may now be seen as based on variant II, with the ICF as model for the objects of finite type (modulo inessential coding differences).

The G-realizability of Moschovakis 1971 may be viewed as based on a slight modification of variant I (with $\sigma = 1$ in clauses (i), (iii)), with ICF^r as the model for the objects of finite type.

It is usually convenient to use variant I, when we wish to interpret the objects of finite type in the study of modified realizability. So, if we wish to interpret $N\text{-}HA_p^{\omega}$ by HRO or HEO, we take $\sigma = 0$ in clauses (i), (iii) of variant I ; if we wish to use ECF, ICF or ICF^r, we use $\sigma = 1$ in clauses (i), (iii).

3.4.10. Notational convention. If $\exists x(x \text{ mr } A)$ has been proved in H, we shall say that A is H-mr-realizable ; and similarly for H-mq-realizable.

If $\exists x(x \text{ mr } A)$ holds if we interpret the objects of finite type by a model M, we can say that A is M-mr-realizable ; and if this fact can be established in a theory H', we shall say that A is H',M-mr-realizable. Similarly with mq replacing mr.

Remark. The set of M-mr-realizable, H'-mr-realizable formulae is closed under deduction (provided H' satisfies some obvious requirements, which are fulfilled in all relevant examples of H').

3.4.11. Comparison of HRO-mr-realizability and r-realizability.

Let us use variant I of 3.4.9 (with $\sigma = 0$ in clauses (i), (iii)) for mr-realizability, and HRO as a model of $N\text{-}HA_p^{\omega}$. Thus we obtain an interpretation of HA in HA ; let us denote this interpretation also by $x \text{ mr } A$.

This can also be defined directly, as follows. We first associate with each formula A a domain of definition D_A, a unary predicate, by induction on the logical complexity of A. D_A depends <u>exclusively</u> on the logical structure of A.

(i) $D_A x \equiv [x = x]$ if A is prime (x not free in A).
(ii) $D_{A\&B}(x) \equiv D_A(j_1 x) \& D_B(j_2 x)$.
(iii) $D_{A\vee B}(x) \equiv (j_1 x=0 \rightarrow D_A(j_2 x)) \& (j_1 x \neq 0 \rightarrow D_B(j_2 x))$.
(iv) $D_{A\rightarrow B}(x) \equiv \forall y(D_A(y) \rightarrow !\{x\}(y) \& D_B(\{x\}(y)))$.
(v) $D_{\exists x A x}(y) \equiv D_{Ax}(j_2 y)$.
(vi) $D_{\forall x A x}(y) \equiv \forall x(!\{y\}(x) \& D_{Ax}(\{y\}(x)))$.

(A literal use of variant I would have required

(iii') $D_{A\vee B}(x) \equiv D_A(j_1 j_2 x) \& D_B(j_2 j_2 x)$

but our deviation is inessential, and gives a slight technical simplification.)

Now we define $x \, \mathbf{mr} \, A$ by

$\mathbf{mr}$(i) $x \, \mathbf{mr} \, A \equiv A$ for A prime (x not free in A).
$\mathbf{mr}$(ii) $x \, \mathbf{mr} \, A \& B \equiv (j_1 x \, \mathbf{mr} \, A \ \& \ j_2 x \, \mathbf{mr} \, B)$.
$\mathbf{mr}$(iii) $x \, \mathbf{mr} \, A \vee B \equiv (j_1 x=0 \rightarrow j_2 x \, \mathbf{mr} \, A) \& (j_1 x \neq 0 \rightarrow j_2 x \, \mathbf{mr} \, B)$.
$\mathbf{mr}$(iv) $x \, \mathbf{mr} \, A \rightarrow B \equiv D_{A\rightarrow B}(x) \ \& \ \forall y(y \, \mathbf{mr} \, A \rightarrow !\{x\}(y) \& \{x\}(y) \, \mathbf{mr} \, B)$.
$\mathbf{mr}$(v) $x \, \mathbf{mr} \, \exists y A y \equiv j_2 x \, \mathbf{mr} \, A(j_1 x)$.
$\mathbf{mr}$(vi) $x \, \mathbf{mr} \, \forall y A y \equiv \forall y(!\{x\}(y) \& \{x\}(y) \, \mathbf{mr} \, Ay)$.

<u>Note</u>: If (iii) had been replaced by (iii'), with a similar change in $\mathbf{mr}$(iii), we would have obtained exactly HRO - $\mathbf{mr}$ - realizability (say $x \, \mathbf{mr}' \, A$). We leave it to the reader to verify by a routine argument that $\mathbf{HA} \vdash \exists x(x \, \mathbf{mr} \, A) \leftrightarrow \exists y(y \, \mathbf{mr}' \, A)$, by induction on the logical complexity of A.

As induction hypothesis one should use a slightly stronger assertion, namely the existence of numerals $\bar{n}_A$, $\bar{m}_A$ such that

$$\vdash x \, \mathbf{mr} \, A \rightarrow !\{\bar{n}_A\}(x) \ \& \ \{\bar{n}_A\}(x) \, \mathbf{mr}' \, A$$

$$\vdash x \, \mathbf{mr}' \, A \rightarrow !\{\bar{m}_A\}(x) \ \& \ \{\bar{m}_A\}(x) \, \mathbf{mr} \, A$$

and

$$\vdash D_A(x) \rightarrow !\{\bar{n}_A\}(x),$$

$$\vdash D_A'(x) \rightarrow !\{\bar{m}_A\}(x)$$

(here D_A' is the predicate obtained by using clause (iii)' instead of (iii)).

It should also be noted that $x \, \mathbf{mr} \, A$ automatically implies $D_A(x)$ by our definitions.

It is now obvious that the essential difference between HRO - $\mathbf{mr}$ - realizability and $\mathbf{r}$ - realizability consists in the additional requirement in the case of implication: $D_{A\rightarrow B}(x)$, so that $\{x\}$ is not only a partial recursive function defined for each y such that $y \, \mathbf{mr} \, A$, but defined for all y such that $D_A(y)$. So the effect of using types in the definition of modified realizability is, when interpreted in this model, that a minimum domain of definition is prescribed for the realizing operations which depends only on the logical structure of the formula to be realized, not on its truth.

3.4.12 - 3.4.25. Realizability and non-realizability of various schemata.

3.4.12. Theorem.

(i) M_{PR} is not HRO - mr - realizable, nor EL, ICF - mr - realizable;

(ii) CT is HRO$^-$ - mr - realizable, but not HEO - mr - realizable.
CT_0 (w.r.t. HRO$^-$) is HRO$^-$ - mr - realizable, hence also HRO - mr - realizable.

Proof. (i) Either direct: if $F \equiv \forall x[\neg\neg\exists y\, Txxy \to \exists y\, Txxy]$ then the mr-translation F^o is logically equivalent to $\exists z^1 \forall x^o[\neg\forall y\neg Txxy \to T(x,x,zx)]$. HRO - mr - realizability would imply the existence of a recursive z such that $\forall x^o[\neg\forall y\neg Txxy \to T(x,x,zx)$, which would imply the recursive decidability of $\exists y\, Txxy$. Or indirect: IP and CT_0 are HRO - mr - realizable (3.4.7 and (ii) below), hence by 3.2.27 (i), M_{PR} cannot be HRO - mr - realizable.

Similarly, we cannot prove in EL that $\exists\alpha\forall x[\neg\forall y\neg Txxy \to T(x,x,\alpha x)]$, since this would imply a proof of the existence of a non-recursive function in EL (contradicting e.g. 3.2.30).

(ii) The HRO$^-$ - mr - realizability of CT is immediate from the axioms for HRO$^-$ (2.4.10). Then obviously CT is also HA, HRO - realizable. Since AC is HRO$^-$ - mr - realizable, CT_0 is HRO$^-$ - mr - realizable and HA, HRO - mr - realizable.

CT is not HEO - mr - realizable, since this would require $[CT]^o \equiv \exists z^2 u^{(1)(o)o} \forall x^1 \forall y^o(T(zx,y,uxy)\ \&\ xy = U(uxy))$ to hold in HEO; i.e. we would require an effective operation which would assign a gödelnumber to each recursive function, depending on the extension of the function only, which is obviously false (e.g. by the well-known example with $f_n x = 0$ if $\neg Tnnx$, $f_n x = 1$ if $Tnnx$; recursive decidability in n of $f_n = \lambda x.0$ would make $\exists y\, Tnny$ recursive. Another argument appeals to the Kreisel - Lacombe - Shoenfield theorem (2.6.15)).

3.4.13. Corollary.

(i) MP_{PR} is underivable in HA;

(ii) HA + IP + CT_0 is consistent relative to HA;

(iii) H + IP^ω + AC + CT, for $H \equiv HA^\omega$, N - HA^ω, I - HA^ω, HRO$^-$ is consistent relative to H.

Proof. Immediate by 3.4.12 and 3.4.8 (ii). For more refined consequences of 3.4.12, see § 3.6. An application of (ii), due to Beeson, is given in 3.4.4 below, settling the relationship between ECT_0 and CT_0: ECT_0 is not derivable from CT_0 in $\underset{\sim}{HA}$.

3.4.14. Theorem (Beeson 1972). $\underset{\sim}{HA} + CT_0 \not\vdash ECT_0$; in fact, $\underset{\sim}{HA} + ECT_0 + IP$ is inconsistent.

Proof. Put

$$Az \equiv \neg \forall n \neg T(z,z,n) \rightarrow \exists n\, T(z,z,n)\,,$$
$$B(z,u) \equiv \neg \forall n \neg T(z,z,n) \rightarrow T(z,z,u)\,.$$

A is almost negative. We shall derive a contradiction in $\underset{\sim}{HA} + ECT_0 + IP$; the first assertion of the theorem then follows from 3.4.13 (ii). In $\underset{\sim}{HA} + ECT_0$:

(1) $\forall z[Az \rightarrow \exists u B(z,y)] \rightarrow \exists v \forall z[Az \rightarrow \exists w(Tvzw \,\&\, B(z,Uw))]\,.$

With IP $Az \leftrightarrow \exists u B(z,u)$, hence by (1) there is a v such that

(2) $\forall z[Az \rightarrow \exists w(Tvzw \,\&\, B(z,Uw))]\,.$

Since $\forall z \neg\neg Az$, also

(3) $\forall z \neg\neg \exists w\, Tvzw$

and also

(4) $\forall z \forall w(Tvzw \rightarrow (\neg \exists u\, Tzzu \vee \exists u\, Tzzu))\,.$

(To see this, note that $\exists u\, Tzzu$ implies Az, hence $B(z,Uw)$; and since $\exists u\, Tzzu \rightarrow \neg \forall u \neg Tzzu$, it follows when combined with $B(z,Uw)$ that $T(z,z,Uw)$. Hence $\exists u\, Tzzu \leftrightarrow T(z,z,Uw)$.

Now apply ECT_0 to a rewriting of (4):

$$\forall z[\exists w\, Tvzw \rightarrow (\neg \exists u\, Tzzu \vee \exists u\, Tzzu)]$$

and we see that we may assume for some v_0

(5) $\forall z[\exists w\, Tvzw \rightarrow \exists w_0[Tv_0zw_0 \,\&\, (Uw_0=0 \rightarrow \neg \exists u\, Tzzu) \,\&\, (Uw_0 > 0 \rightarrow \exists u\, Tzzu)]\,.$

We define $v_1 = \Lambda z.\min_w(Tv_0zw \,\&\, Uw=0)$. Then

$$\exists w\, Tvv_1w \rightarrow \exists w_0\, Tv_0v_1w_0\,;$$

let $Tv_0v_1w_0$; if $Uw_0 = 0$, then $\exists u\, Tv_1v_1u$, but also by (5) $Uw_0 = 0 \rightarrow \neg \exists u\, Tv_1v_1u$; if $Uw_0 \neq 0$, then $\neg \exists u\, Tv_1v_1u$, but by (5) $\exists u\, Tv_1v_1u$. Hence $\neg \exists w\, Tvv_1w$; but this contradicts (3).

3.4.15. Theorem. CT is not ICF^r - $\underset{\approx}{mr}$ - realizable, but WCT:

WCT $\quad \forall\alpha \neg\neg \exists z \forall y \exists z[Txyz \,\&\, \alpha x = Uz]$

is $\underset{\sim}{EL}$, ICF^r - $\underset{\approx}{mr}$ - realizable, and $\underset{\sim}{EL}$, ICF^r - $\underset{\approx}{mq}$ - realizable.

Remark. This result is very similar to, and was suggested by, Moschovakis 1971.

Proof. ICF^r - mr - realizability of CT is refuted utilizing the fact that all objects of type 2 are continuous in ICF^r. For, ICF^r - mr - realizability of CT would require the existence of x^2, $z^{(1)(o)o}$ such that

$$\forall\alpha\forall y[T(x^2\alpha,y,z\alpha y)\ \&\ \alpha x = U(z\alpha y)],$$

which of course would imply $x^2\alpha$ to be a continuous in α, which is obviously false.

On the other hand, WCT^o is

(1) $$\forall x^1 \neg\forall y^o\forall v^1 \neg\forall z^o[T(y,z,vz)\ \&\ U(vz) = xz].$$

We also have, in ICF^r

$$\forall x^1\,\exists y^o\exists v^1\,\forall z^o[T(y,z,vz)\ \&\ U(vz) = xz],$$

which may be weakened to

$$\forall x^1\,\exists y^o\,\neg\neg\,\exists v^1\forall z^o[T(y,z,vz)\ \&\ U(vz) = xz],$$

i.e.

$$\forall x^1\,\exists y^o\,\neg\forall v^1\,\neg\forall z^o[T(y,z,vz)\ \&\ U(vz) = xz],$$

which in turn implies (1).

The ICF^r - mq - realizability follows by observing that in EL $\mathrm{WCT}^1 \leftrightarrow \mathrm{WCT}^o$.

3.4.16. Theorem. FAN is (EL + FAN), ECF - mr - realizable. (Kreisel 1962.)

Proof. FAN may be stated as follows: ($\Phi \equiv \lambda\alpha.\lambda x.\mathrm{sg}(\alpha x)$)

$$\forall\alpha\,\exists x\,A(\Phi\alpha,x) \rightarrow \exists z\,\forall\alpha\,\exists y\,\forall\beta((\overline{\Phi\alpha})z = (\overline{\Phi\beta})z \rightarrow A(\Phi\beta,y)).$$

We carry out a derivation in $\mathrm{E\text{-}HA}^\omega + \mathrm{IP}^\omega + \mathrm{AC} + \mathrm{MUC}$ (2.6.4).

Assume $\forall\alpha\,\exists x\,A(\Phi\alpha,x)$. By AC, $\exists z^2\,\forall\alpha\,A(\Phi\alpha,z^2\alpha)$. Hence by the axiom MUC:

$$\forall\alpha\forall\beta((\overline{\Phi\alpha})(\varphi_{uc}z^2) = (\overline{\Phi\beta})(\varphi_{uc}z^2) \rightarrow z^2(\Phi\alpha) = z^2(\Phi\beta))$$

and therefore $\exists z\,\forall\alpha\,\exists y\,\forall\beta\,((\overline{\Phi\alpha})z = (\overline{\Phi\beta})z \rightarrow A(\Phi\beta,y))$.

3.4.17. Theorem.

(i) WC - N is not ECF($\mathcal{U}$) - mr - realizable, for any universe $\mathcal{U}$ satisfying EL.

(ii) WC - N is EL, ICF - mr - realizable.

(iii) WC - N is IDB, ICF^r - mr - realizable.

Proof. (i) This can be shown by paraphrasing Kreisel's counterexample in 2.6.5. If we define

$$A(\alpha,x) \equiv_{\mathrm{def}} \forall z \leq j_1(x \dot{-} 1)(\alpha z = 0)\ \&\ \alpha(j_1(x \dot{-} 1)+1) > j_2(x \dot{-} 1)$$

$$B(x,\alpha,y) \equiv_{def} [(x=0 \rightarrow y=0) \;\&\; (x \neq 0 \rightarrow \{(A(\alpha,x) \rightarrow y=1) \;\&\; (\neg A(\alpha,x) \rightarrow y=0)\})] .$$

then obviously

$$\underset{\sim}{EL} \vdash \forall x \, \forall \alpha \, \exists ! y \, B(x,\alpha,y) .$$

For $x=0$, $B(0,\alpha,y)$ intuitively represents the ψ_0 of 2.6.7 i.e. $\forall \alpha B(0,\alpha,\psi_0\alpha)$; for $x \neq 0$, $B(x,\alpha,y)$ represents the $\psi_{j_2(x \dot{-} 1),1}$ of 2.6.7, with $m_0 = j_1(x \dot{-} 1)$, so if $x = j(m_0,m)+1$, then $B(x,\alpha,\psi_{m,1}\alpha)$ for all α. Application of WC - N would yield

$$\forall x \, \forall \alpha \, \exists z \, \exists ! y \, \forall \beta (\bar{\alpha}z = \bar{\beta}z \rightarrow B(x,\beta,y)) .$$

If WC - N were ECF - $\underset{\approx}{mr}$ - realizable, we would have to find $Z^{(o)(1)o}$, $Y^{(o)(1)o}$ such that

$$\bar{\alpha}(Zx\alpha) = \bar{\beta}(Zx\alpha) \rightarrow B(x,\beta,Yx\alpha) .$$

Now $Y0$ must be equal to ψ_o, $Y(j(m_o,m)+1)$ to $\psi_{m,1}$ (as is seen by taking $\alpha = \beta$). Then, by copying the remainder of the argument in 2.6.7, it follows that we cannot find a solution for Z.

(ii) We can show WC - N to be $\underset{\approx}{mr}$ - realizable in $\underset{\sim}{N}$ - $\underset{\sim}{HA}^{\omega}$ + MC (2.6.3), since

$$\underset{=}{x} \; \underset{\approx}{mr} \; \forall \alpha \, \exists x^o A(\alpha,x)$$

means

$$\underset{=}{x}\alpha \; \underset{\approx}{mr} \; \exists x^o A(\alpha,x)$$

so $\underset{=}{x}\alpha$ can be written as $(z^2\alpha, \underset{=}{y}\alpha)$, and

$$\underset{=}{y}\alpha \; \underset{\approx}{mr} \; A(\alpha, z^2\alpha) .$$

On the other hand, $\underset{\approx}{mr}$ - realizability of

$$\forall \alpha \, \exists x \, \exists y \, \forall \beta (\bar{\alpha}x = \bar{\beta}x \rightarrow A(\beta,y))$$

is equivalent to

(1) $$\exists X Y \underset{=}{U} \, \forall \alpha \, \forall \beta (\bar{\alpha}(X\alpha) = \bar{\beta}(X\alpha) \rightarrow \underset{=}{U}\alpha\beta \; \underset{\approx}{mr} \; A(\beta, Y\alpha)) .$$

If we take for X $\varphi_{mc}z^2$, for Y z^2, and for $\underset{=}{U}$ $\underset{=}{y}\beta$, then (1) is satisfied. The construction of the desired X, Y, U being uniform in $z^2, \underset{=}{y}$, WC - N is $\underset{\approx}{mr}$ - realizable. Since ICF can be shown to be a model for $\underset{\sim}{N}$ - $\underset{\sim}{HA}^{\omega}$ + MC in $\underset{\sim}{EL}$, the assertion of the theorem follows.

(iii) Similarly, using $\underset{\sim}{IDB}$ instead of $\underset{\sim}{EL}$.

3.4.18. Theorem.

(i) $\underset{\sim}{EL}$ + WC-N + IP^1 + WCT is consistent relative to $\underset{\sim}{IDB}$;

(ii) $\underset{\sim}{N}$-$\underset{\sim}{HA}^{\omega}$ + WC-N + IP^{ω} + AC + WCT is consistent relative to $\underset{\sim}{IDB}$;

(iii) $\underset{\sim}{\text{N}}\text{-}\underset{\sim}{\text{HA}}^{\omega} + \text{WC-N} + \text{IP}^{\omega} + \text{AC} + \text{WCT} \not\vdash \text{CT}$.

Here IP^1 is

IP^1 $\quad (\neg A \to \exists\alpha B) \to \exists\alpha(\neg A \to B) \quad$ (α not free in A).

<u>Proof</u>. (i), (ii) are immediate by 3.4.17 (iii), 3.4.15, 3.4.7 ; (iii) by (ii) and 3.4.15.

3.4.19. <u>Theorem</u>. FAN is not $\text{ECF}(\mathcal{R})$ - $\underset{\approx}{\text{mr}}$ - realizable.

<u>Proof</u>. We proceed similarly to the proof of 3.4.17. Now we use 2.6.10 and represent the φ defined there by

$$A(\alpha,n) \equiv Rn \,\&\, \forall m(m < n \to \neg Rm) \,\&\, \alpha \in n .$$

Obviously, if α ranges over $\mathcal{R}$, $\forall\alpha\exists! n\, A(\alpha,n)$; application of FAN then yields a statement which is false in $\text{ECF}(\mathcal{R})$ by 2.6.10, hence certainly not $\text{ECF}(\mathcal{R})$ - $\underset{\approx}{\text{mr}}$ - realizable (if we keep in mind that $A_0(\alpha,n) \leftrightarrow A(\alpha,n)$).

3.4.20. <u>Lemma</u>. BI_M is provably $\underset{\approx}{\text{mr}}$ - realizable in a theory $\underset{\sim}{\text{N}}\text{-}\underset{\sim}{\text{HA}}^{\omega} + \text{BR}_0 + \text{BI}_D$ + continuity axiom ; the continuity axiom is formulated as:

(1) $\quad \forall z^2 \forall x^1 \exists y^0 \forall u^1 (\bar{x}(y) = \bar{u}(y) \to zx = zu)$.

<u>Proof</u>. The modified realizability interpretation of the four premisses of BI_M takes the form

(1) $\quad \exists ZX\, \forall\alpha\, P_0(\bar{\alpha}(X\alpha), Z\alpha)$

(2) $\quad \exists Z_0\, \forall n^0 z(P_0(n,z) \to Q_0(n, Z_0 nz))$

(3) $\quad \exists Z_1\, \forall n^0 m^0 z\, (P_0(n,z) \,\&\, m \geq n \to P_0(m, Z_1 nmz))$

(4) $\quad \exists Z_2\, \forall un(\forall y\, Q_0(n * \hat{y}, uy) \to Q_0(n, Z_2 un))$

where $P^0(n) \equiv \exists z\, P_0(n,z)$, $Q^0(n) \equiv \exists u\, Q_0(n,u)$ (for simplicity in notation taking single variables z, u instead of $\underline{\underline{z}}$, $\underline{\underline{u}}$), and where we have modified the third hypothesis in BI_M to $\forall nm(Pn \,\&\, m \geq n \to Pm)$.

Let f_n denote the sequence $\lambda x(n)_x$.

The modified realizability interpretation of BI_M now requires that given X, Z, Z_0, Z_1, Z_2 as in (1) - (4), we can construct a U , uniformly in X, Z, Z_0, Z_1, Z_2 such that

$$Q_0(n, Un) .$$

Such a U is constructed by taking

$$X(f_n) < \text{lth}(n) \to Un = Z_0 n(Z_1(\bar{f}_n(Xf_n))n(Zf_n)) ,$$

$$X(f_n) \geq \text{lth}(n) \to Un = Z_2(\lambda y.U(n*\hat{y}))n .$$

By BR_0 we can find a U' such that $U'XZZ_0Z_1Z_2 = U$. To see that U satisfies our requirements, we note that, if we take for

$Pn \equiv X(f_n) < lth(n)$
$Qn \equiv Q_o(n, Un)$

then we find, since X is continuous: $\forall\alpha\,\exists x(X(f_{\bar{\alpha}x}) < x)$. Also, Pn implies $X(f_n) < lth(n)$, hence $\bar{f}_n(X(f_n)) < n$; also $P_o(\bar{f}_n(X(f_n)), Zf_n)$, hence $P_o(\bar{f}_n(Xf_n), Z_1(\bar{f}_n(Xf_n))n(Zf_n))$, and therefore $Q_o(n, Z_o n(Z_1(\bar{f}_n(Xf_n))n(Zf_n)))$, i.e. $Q_o(n, Un)$, so

$$Pn \to Qn.$$

Also

$$Pn \vee \neg Pn$$

and finally, if $\forall y Q(n * \hat{y})$, then this implies, assuming $X(f_n) \geq lth\, n$, that $\forall y Q_o(n * \hat{y}, U(n * \hat{y}))$, therefore by (4) $Q_o(n, Z_2(\lambda y. U(n * \hat{y}))n)$, i.e. $Q_o(n, Un)$; hence also

$$\forall y\, Q(n * \hat{y}) \to Qn.$$

Applying BI_D yields $\forall n Qn$, which we had to show.

Remark. For the Dialectica interpretation, Howard (in Howard 1968) manages to interpret BI_M by BR_o, without the additional help of BI_M itself. The treatment does not carry over automatically to modified realizability, however; presumably the lemma can be improved, and also extended to bar induction of higher types (under suitable additional assumptions) for lack of interesting applications, at present, we have refrained from carrying this out.

3.4.21. Corollaries. In $\underset{\sim}{EL} + BI_D$ it can be shown that in ICF and ICF^r BI_M is $\underset{\approx}{mr}$-realizable (cf. Kleene & Vesley 1965, §11, Moschovakis 1971).

Proof. Combine 3.4.20 with 2.9.12 (which also applies to ICF^r).

3.4.22. Modified realizability for $\underset{\sim}{HA} + TI(<)$.

Let $\underline{\underline{x}}$ be a sequence of length n, with types $\sigma_1, \ldots, \sigma_n$. For any given primitive recursive well-ordering $<$ of the natural numbers, we can define in $\underset{\sim}{N} - \underset{\sim}{HA}^{\omega}$ a sequence of constants $\underline{\underline{C}}$ such that

$$u \geq v \to \underline{\underline{Cx}}uv = 0^{\sigma *} \quad (\text{where } 0^{\sigma *} \equiv 0^{\sigma_1}, \ldots, 0^{\sigma_n})$$
$$u < v \to \underline{\underline{Cx}}uv = \underline{\underline{x}}.$$

Let us consider $\underset{\sim}{N} - \underset{\sim}{HA}^{\omega} + T_{<} + TI(<)$, where $T_{<}$ denotes a defining axiom for a new sequence $\underline{\underline{R}}^{<}$ such that:

$$T_{<} \qquad \underline{\underline{R}}^{<}\underline{\underline{x}}u = \underline{\underline{x}}u(\lambda v. \underline{\underline{C}}(\underline{\underline{R}}^{<}\underline{\underline{x}}v)vu) \qquad (u, v \in 0)$$

(λv the defined λ-operator). Then we have

3.4.23. Lemma. TI(<) is $\underset{\sim}{N}$-$\underset{\approx}{HA}^{\omega}+T_{<}+TI(<)$ - $\underset{\approx}{mr}$ - realizable.

Proof. Assume

$$\underline{\underline{w}}\ \underset{\approx}{mr}\ \forall u((\forall v<u)Av \to Au),$$

i.e.

$$\forall u(\underline{\underline{w}}u\ \underset{\approx}{mr}\ (\forall v<u)Av \to Au),$$

hence

$$\forall u\underline{\underline{w}}'(\underline{\underline{w}}'\ \underset{\approx}{mr}\ (\forall v<u)Av \to \underline{\underline{w}}u\underline{\underline{w}}'\ \underset{\approx}{mr}\ Au),$$

which is the same as

(1) $$\forall u\underline{\underline{w}}'((\forall v< u)(\underline{\underline{w}}'v\ \underset{\approx}{mr}\ Av) \to \underline{\underline{w}}u\underline{\underline{w}}'\ \underset{\approx}{mr}\ Au).$$

We wish to show

$$\underline{\underline{R}}^{<}\underline{\underline{w}}\ \underset{\approx}{mr}\ \forall yAy.$$

By TI(<), it is sufficient to show

$$\forall u(\forall v<u(\underline{\underline{R}}^{<}\underline{\underline{w}}v\ \underset{\approx}{mr}\ Av) \to \underline{\underline{R}}^{<}\underline{\underline{w}}u\ \underset{\approx}{mr}\ Au).$$

If $(\forall v<u)\underline{\underline{R}}^{<}\underline{\underline{w}}v\ \underset{\approx}{mr}\ Av$, then also $\forall v<u\ \underline{C}(\underline{\underline{R}}^{<}\underline{\underline{w}}v)vu\ \underset{\approx}{mr}\ Av$, since $v<u \to \underline{C}(\underline{\underline{R}}^{<}\underline{\underline{w}}v) = \underline{\underline{R}}^{<}\underline{\underline{w}}v$. Therefore, by (1)

$$\underline{\underline{w}}u(\lambda v.\underline{C}(\underline{\underline{R}}^{<}\underline{\underline{w}}v)vu)\ \underset{\approx}{mr}\ Au,$$

so $\underline{\underline{R}}^{<}\underline{\underline{w}}u\ \underset{\approx}{mr}\ Au$.

3.4.24. Lemma. In $\underset{\sim}{HA}+TI(<)$ HRO, HEO can be shown to be models for $\underset{\sim}{N}$-$\underset{\approx}{HA}^{\omega}+T_{<}+TI(<)$.

Similarly for ICF, ECF in $\underset{\sim}{EL}+TI(<)$.

Proof. We give the proof for HRO. HEO only requires an additional extensionality verification; for ICF, ECF the proofs are very similar.

Because of the presence of pairing in HRO we restrict our attention to the case where $\underline{\underline{x}} \equiv x$ (a single variable), $\underline{\underline{R}}^{<} \equiv R^{<}$, hence also $\underline{C} \equiv C$. Let [C] be the numeral such that ([C],σ) for suitable σ represents C in HRO. Then

$$u \geq v \to \{[C]\}(x,u,v) \simeq [0^{\sigma *}]$$

$$u < v \to \{[C]\}(x,u,v) \simeq x.$$

We wish to construct $\bar{n}$ such that

$$\{\bar{n}\}(x,u) \simeq \{x\}(u,\ \Lambda v.\{[C]\}(\{\bar{n}\}(x,v),v,u)$$

and $\bar{n}\in V_{(\tau)(0)\tau'}$ if $x\in V_{\tau}$, $R^{<}\in(\tau)(0)\tau'$.

We easily find an $\bar{m}$ such that

$$\{\bar{m}\}(n,x,u) \simeq \{x\}(u,\Lambda v.\{[C]\}(\{n\}(x,v),v,u);$$

the recursion theorem yields $\bar{n}$ as required. Then by an application of TI(<) one establishes $\bar{n} \in V_{(\tau)(o)\tau'}$.

3.4.25. Corollary. (i) TI(<) is HA + TI(<), HRO - mr - realizable.
(ii) M_{PR} is not derivable in HA + TI(<) + IP,
etc. etc.

Proof. Immediate, by 3.4.23, 3.4.24, 3.4.12 (i).

3.4.26. For modified realizability in a context of theories with generalized inductive definitions iterated once or twice, see § 6.7, § 6.8.

3.4.27. Modified realizability for HAS.

It is possible to extend the "abstract" (i.e. not relativized to a model) modified realizability to HAS by using Girard's system of functionals described in 1.9.27; cf. the analogous extension of the Dialectica interpretation in the next section (3.5.21), which is even more complicated.

It is simpler, and yields a more direct application, to describe HRO^2 - mr - realizability for HAS , as an extension of the definition in 3.4.11 above. We use HRO^2 as described in 2.9.7, with the gödelnumbering satisfying $\{0\}(x) \simeq 0$ for all x.

Then HRO^2 - mr - realizability is obtained by extending the clauses mr(i) - (vi) in 3.4.11 by adding to the clauses (i) - (vi) for D_A

(i)' $D_{V_i^n} x \equiv U^1_{j(n,i)}$, where $U^1_0, U^1_1, U^1_2, \ldots$ is a sequence of variables for unary species containing 0 (the addition of such variables is obviously a conservative extension of HAS). We shall write $D_{V_i^n}$ for $U^1_{j(n,i)}$.

(vii) $D_{\forall XA(X)}(x) \equiv \forall D_X(D_{A(X)}(x))$

(viii) $D_{\exists XA(X)}(x) \equiv \exists D_X(D_{A(X)}(x))$

and adding to the clauses mr(i) - (vi):

mr(i)' $x \text{ mr } V_i^n(t_1,\ldots,t_n) \equiv V_i^{n+1}(x,t_1,\ldots,t_n) \;\&\; D_{V_i^n}(x)$.

We shall write X^* for the V_j^{n+1} corresponding to $V_j^n \equiv X$.

mr(vii) $x \text{ mr } \forall XA(X) \equiv \forall X^*(x \text{ mr } A(X))$

mr(viii) $x \text{ mr } \exists XA(X) \equiv \exists X^*(x \text{ mr } A(X))$.

Similarly we may define ICF^2 - mr - realizability (cf. 2.9.7).

HRO^2 - mr - realizability as defined here is introduced in Troelstra A and 1971A.

3.4.28. Corollaries.

(i) HAS + IP + CT_0 is consistent relative to HAS

HAS + IP + CT_0 is conservative over HAS w.r.t. negative first-order formulae.

(ii) M_{PR} is not derivable in $\underset{\sim}{HAS} + CT_0 + IP$.

Proof. (i) The consistency follows by establishing the soundness theorem for $\underset{\sim}{HAS} + IP + CT_0$, and HRO^2 - mr - realizability; the conservative extension result follows by proving

$$0 \text{ mr } A \leftrightarrow A$$

for all negative formulae of $\mathcal{L}(\underset{\sim}{HA})$ (by induction on the logical complexity of A).

(ii) M_{PR} is refutable in $\underset{\sim}{HA} + CT_0 + IP$, and $\underset{\sim}{HAS} + IP + CT_0$ is consistent relative to $\underset{\sim}{HAS}$ (cf. Troelstra A, § 4).

3.4.29. Theorem (Characterization of provably recursive functions). Each provably recursive function of $\underset{\sim}{HA}$ is represented by a closed term of type 1 in $\underset{\sim}{N} - \underset{\sim}{HA}^{\omega}$, and conversely. I.e.

$\underset{\sim}{HA} \vdash \forall x \exists y\, T\bar{n}xy \Rightarrow \exists s^1 (\underset{\sim}{N} - \underset{\sim}{HA}^{\omega} \vdash \forall xy(T\bar{n}xy \rightarrow s^1x = Uy))$ and conversely

$$\forall s^1 \exists n(\underset{\sim}{N} - \underset{\sim}{HA}^{\omega} \vdash \forall x \exists y(T\bar{n}xy \,\&\, s^1x = Uy)),$$

where s^1 ranges over closed terms of type 1 of $\underset{\sim}{N} - \underset{\sim}{HA}^{\omega}$.

Proof. In one direction, we use modified realizability: if $\underset{\sim}{HA} \vdash \forall x \exists y\, T\bar{n}xy$, then by the soundness theorem for mr - realizability, for suitable t^1

$$\underset{\sim}{N} - \underset{\sim}{HA}^{\omega} \vdash T(\bar{n}, x, t^1x).$$

So we may take $s^1 \equiv \lambda x.U(t^1x)$.

Conversely, we may for example appeal to 2.4.14 and find that

$$\underset{\sim}{N} - \underset{\sim}{HA}^{\omega} \vdash \{[s^1]\}(x) \simeq s^1x,$$

where $([s^1], 1)$ is the standard representation of $[s^1]$ in HRO.

3.4.30. The theorem automatically extends to other theories such as $\underset{\sim}{HA} + TI(<)$; the provably recursive functions correspond exactly to the closed terms of type 1 of $\underset{\sim}{N} - \underset{\sim}{HA}^{\omega} + TI(<) + T_<$, in view of 3.4.23, 3.4.24.

Quite similar results may be extracted from the Dialectica interpretation, discussed in the next section.

3.4.31. The concept of HRO - mr - realizability, as described in 3.4.11, can also be generalized to HRO^A - mr - realizability. Here HRO^A is defined as HRO, but relative to A - partial recursive functions. Combining this with the ideas of 3.2.32 on A - realizability, and the argument in 3.4.12 (i), we can show that $\underset{\sim}{HA} + M$ is not finitely axiomatizable over $\underset{\sim}{HA}$.

§ 5. The Dialectica interpretation and translation.

3.5.1. Introduction. The Dialectica interpretation and translation were first introduced in Gödel 1958, for intuitionistic arithmetic. The purpose was to provide a consistency proof for intuitionistic arithmetic (and hence for classical arithmetic) by elementary "logic" (i.e. quantifiers especially) by an interpretation of an arithmetical statement by a quantifier-free formula in a theory of objects of finite type, where the concept of a constructive (computable, in Gödel's terminology : "berechenbare") object of finite type was to be regarded as primitive and intuitively evident.

Hence logic was to be eliminated in favour of a suitable basic concept of object of finite type. From footnote 3 in Gödel 1958, and in view of 3.5.6 below, it seems that a concept with decidable equality at all types as a primitive was intended *.

In Kreisel 1959, and in Spector 1962, which apply the interpretation to intuitionistic analysis, only equality between objects of type 0 is taken as a primitive ; equality between higher type objects is interpreted as extensional equality. Howard 1968 contains simplifications and refinements of Spector's work. (For a discussion of the rôle of extensionality, see 3.5.12 - 3.5.15 below.)

A characterization of Dialectica interpretable formulae of $\mathbf{WE}\text{-}\mathbf{HA}^{\omega}$ was first given explicitly in Yasugi 1963, after Kreisel already noted that (weakenings of) AC, IP_0, M implied the equivalence of a formula with its interpretation (cf. Kreisel 1959, 2.11, 3.5.1) and showed the interpretability of M (Kreisel 1959, footnote 1 on page 113). For a correction to Yasugi 1963, see the review Troelstra 1972.

Contents of the section. 3.5.2 - 3.5.3 are devoted to the definition of and the motivation behind the Dialectica interpretation.

In 3.5.4 the soundness theorem is proved, 3.5.6 gives a counterexample due to W.A. Howard showing the decidability of prime formulae to be essential for the Dialectica interpretation.

3.5.7 - 3.5.11 are devoted to the axiomatization of the class of Dialectica-interpretable formulae, 3.5.12 - 3.5.15 to the interpretability and non-interpretability of the extensionality axiom, with an application.
3.5.16 lists some miscellaneous properties regarding the Dialectica inter-

-----

* This, however, conflicts with the suggestion in the last line of Gödel 1958, since $\mathbf{I}\text{-}\mathbf{HA}^{\omega}$ requires for its Dialectica interpretation non-extensional functionals (e.g. to interpret E_σ) yielding functionals discontinuous on the binary tree, thus not satisfying the fan theorem (cf. 3.5.6 below).

pretability of CT, CT_0, C - N, FAN, IP relative to various models of $\mathbf{N}-\mathbf{HA}^\omega$.

3.5.17 describes the Diller - Nahm variant of the Dialectica interpretation, which yields an interpretation of $\mathbf{N}-\mathbf{HA}^\omega$ in an "almost quantifier-free" fragment of $\mathbf{N}-\mathbf{HA}^\omega$, and which does not require in the proof of the soundness theorem decidability of prime formulae.

3.5.18 describes the extension of the Dialectica interpretation to stronger systems. Since these extensions have already been discussed in some detail in published literature, we have restricted ourselves to a brief indication of the principles of the extensions, and a fairly detailed heuristic account of the motivation behind Girard's extension to theories with species variables.

We have not discussed Parsons 1970, 1972, since they deal exclusively with $\mathbf{HA}^c$ and subsystems of $\mathbf{HA}^c$ (but based on classical logic).

3.5.2. Definition. To each formula of $\mathcal{L}[\mathbf{HA}^\omega]$ or $\mathcal{L}[\mathbf{I}-\mathbf{HA}^\omega]$ or $\mathcal{L}[\mathbf{HRO}^-]$ we assign a translation $A^D \equiv \exists \underline{x}\, \forall \underline{y}\, A_D(\underline{x},\underline{y})$ in the same language. The types of $\underline{x}$, $\underline{y}$ depend on the logical structure of A only; the free variables of A^D are contained among the free variables of A, A_D is quantifier-free. The definition is by induction on the logical complexity of A.

d(i) If A is prime, then $A^D \equiv A_D \equiv A$.

For the other clauses, let $A^D \equiv \exists \underline{x}\, \forall \underline{y}\, A_D(\underline{x},\underline{y})$, $B^D \equiv \exists \underline{u}\, \forall \underline{v}\, B_D(\underline{u},\underline{v})$.

d(ii) $(A \& B)^D \equiv \exists \underline{x}\underline{u}\; \forall \underline{y}\underline{v}[A \& B]_D \equiv \exists \underline{x}\underline{u}\; \forall \underline{y}\underline{v}[A_D(\underline{x},\underline{y}) \& B_D(\underline{u},\underline{v})]$,

d(iii) $(A \vee B)^D \equiv \exists z^0 \underline{x}\underline{u}\; \forall \underline{y}\underline{v}[A \vee B]_D \equiv$
$\equiv \exists z^0 \underline{x}\underline{u}\; \forall \underline{y}\underline{v}[(z=0 \to A_D(\underline{x},\underline{y})) \& (z \neq 0 \to B_D(\underline{u},\underline{v}))]$.

d(iv) $(\exists z A z)^D \equiv \exists z\underline{x}\; \forall \underline{y}(\exists z A z)_D \equiv \exists z\underline{x}\; \forall \underline{y}\, A_D(\underline{x},\underline{y},z)$.

d(v) $(\forall z A z)^D \equiv \exists \underline{X}\; \forall z\underline{y}(\forall z A z)_D \equiv \exists \underline{X}\; \forall z\underline{y}\, A_D(\underline{X}z,\underline{y},z)$.

d(vi) For the sake of clarity, we describe the construction of $(A \to B)^D$ in a number of steps:

$$
\begin{aligned}
(A \to B)^D &\equiv (\exists \underline{x}\, \forall \underline{y}\, A_D \to \exists \underline{u}\, \forall \underline{v}\, B_D)^D \equiv && \text{(a)}\\
&\equiv [\forall \underline{x}(\forall \underline{y} A_D \to \exists \underline{u}\, \forall \underline{v}\, B_D)]^D \equiv && \text{(b)}\\
&\equiv [\forall \underline{x}\, \exists \underline{u}(\forall \underline{y}\, A_D \to \forall \underline{v}\, B_D)]^D \equiv && \text{(c)}\\
&\equiv [\forall \underline{x}\, \exists \underline{u}\, \forall \underline{v}(\forall \underline{y}\, A_D \to B_D)]^D \equiv && \text{(d)}\\
&\equiv [\forall \underline{x}\, \exists \underline{u}\, \forall \underline{v}\, \exists \underline{y}(A_D \to B_D)]^D \equiv && \text{(e)}\\
&\equiv [\exists \underline{U}\underline{Y}\, \forall \underline{x}\underline{v}(A_D(\underline{x},\underline{Y}\underline{x}\underline{v}) \to B_D(\underline{U}\underline{x},\underline{v}))] && \text{(f)}
\end{aligned}
$$

$$(A \to B)_D \equiv A_D(\underline{x},\underline{Y}\underline{x}\underline{v}) \to B_D(\underline{U}\underline{x},\underline{v}).$$

Note that with classical logic and AC, $A^D \equiv A$ for all A. In fact, for prime A, $A^D \equiv A$, and for conjunctions, disjunctions, extensional quantifications $(A \& B)^D \leftrightarrow A^D \& B^D$, $(A \vee B)^D \leftrightarrow A^D \vee B^D$, $(\exists z A)^D \leftrightarrow \exists z A^D$, even intuitionistically ; $(\forall zA)^D \leftrightarrow \forall zA^D$ with AC , and $(A \rightarrow B)^D \leftrightarrow (A^D \rightarrow B^D)$ by AC and classical logic as an inspection of (a) - (f) shows. For a more refined result see lemma 3.5.7 below.

Note also that (i) for formulae $A \equiv \exists x \forall y B$, B quantifier-free, $A^D \equiv A$, and (ii) for B quantifier-free, $(\neg\neg \exists xB)^D \equiv (\neg \forall x \neg B)^D \equiv \exists x \neg\neg B$ (which is equivalent to $\exists xB$ if prime formulae are stable, as in the above theories).

3.5.3. Motivation. The motivation for the particular choice of interpretation is twofold. If we first consider $\neg A \equiv A \rightarrow 1{=}0$, as a simplified instance of an implication, we find

$$(\neg A)^D \equiv \exists Y \forall x \neg A_D(x, Yx)$$
$$(\neg\neg A)^D \equiv \exists X \forall Y \neg\neg A_D(XY, Y(XY)) .$$

This should be compared with the so-called no-counterexample interpretation (Kreisel 1951) of a statement of arithmetic A.
Assume $\neg A$ to be brought, first into prenex normal form, then into $\exists\forall$-form by means of Skolem functions:

$$\exists f_1 f_2 \ldots \forall x_1 x_2 \ldots B(x_1, f_1(x), x_2, f_2(x_1, x_2), \ldots)$$

$\neg\neg A$ then becomes (classically)

$$\forall f_1 f_2 \ldots \exists x_1 x_2 \ldots \neg B(x_1, f_1(x_1), x_2, \ldots) .$$

So the impossibility of a counterexample (i.e. $\neg A$) is demonstrated by functionals $F_1, F_2, \ldots$ such that

$$\neg B(F_1(f_1, f_2, \ldots), f_1(F_1(f_1, f_2, \ldots)), F_2(f_1, f_2, \ldots), \ldots) ,$$

which corresponds to the Dialectica interpretation of $\neg\neg A$ (cf. also Gödel's explanation in Gödel 1958, end of page 285 and top of page 286).

Secondly, the purpose of the Dialectica translation is, to transform an arbitrary formula of HA^ω or $N\text{-}HA^\omega$ into $\exists\forall$-form. One may ask why the particular order of shifting quantifiers to the front in the case of implication (lines (a) - (e) in 3.5.2) has been chosen: any other order of shifting to the front, using the classical laws

(1) $\qquad (\forall xA \rightarrow B) \leftrightarrow \exists x(A \rightarrow B), \quad (A \rightarrow \exists xA) \leftrightarrow \exists x(A \rightarrow B)$

and the intuitionistic laws

(2) $\qquad (\exists xA \rightarrow B) \leftrightarrow \forall x(A \rightarrow B), \quad (A \rightarrow \forall xB) \leftrightarrow \forall x(A \rightarrow B)$

combined with AC would also have resulted in an $\exists\forall$-form. But the particular transformation chosen in the Dialectica translation is such that the applications of (1) needed are as weak as possible (i.e. A of minimal logical complexity) so that the transformation may be expected to remain as close as possible to the intended constructive interpretation.

This impression is confirmed by the fact that even $A \to A$ for suitable A asks for non-recursive realizations for its $\exists\forall$-transform in all cases except the Dialectica $\exists\forall$-transform.

The other possibilities of shifting quantifiers to the front are

$$(\exists \underline{x}\, \forall \underline{y}\, A \to \exists \underline{u}\, \forall \underline{v}\, B) \mapsto \forall \underline{x}\, \exists \underline{\underline{uy}}\, \forall \underline{v} [A \to B] \quad \text{(A)}$$

$$\mapsto \exists \underline{u}\, \forall \underline{\underline{xv}}\, \exists \underline{\underline{y}} [A \to B] \quad \text{(B)}$$

$$\mapsto \exists \underline{u}\, \forall \underline{x} \exists \underline{y}\, \forall \underline{v} [A \to B] \quad \text{(C)}$$

(A) yields with AC

(3) $$\exists \underline{\underline{UY}}\ \forall \underline{\underline{xv}} [A(\underline{x}, \underline{\underline{Yx}}) \to B(\underline{\underline{Ux}}, \underline{v})] .$$

For $A \equiv B \equiv \neg Tzxy$, a recursive realization of (3) asks for U_z, Y_z recursive in z and other arguments such that

$$\forall xv(\neg T(z,x,Y_z x) \to \neg T(z,U_z x,v)) .$$

This yields

$$\neg T(z,x,Y_z x) \to \exists u\, \forall v\, \neg T(z,u,v)$$

and this would make $\exists u\, \forall v\, \neg Tzuv$ recursively decidable, hence a contradiction follows.

(B) yields with AC

(4) $$\exists \underline{\underline{uY}}\ \forall \underline{\underline{xv}} [A(\underline{x}, \underline{\underline{\underline{Yxv}}}) \to B(\underline{u}, \underline{v})] .$$

We apply again to $A \equiv B \equiv \neg Tzxy$; for a recursive realization we are requested to find Y_z, U_z recursive in their arguments and z, such that

$$\forall xv(\neg T(z,x,Y_z xv) \to \neg T(z,u_z,v)) .$$

Hence

$$\forall xv(T(z,u_z,v) \to T(z,x,Y_z xv)) ,$$

i.e.

$$\forall v(T(z,u_z,v) \to \forall x\ T(z,x,Y_z xv))$$

$$\Rightarrow \quad \exists v\ T(z,u_z,v) \to \forall x\, \exists w\, T(z,x,w)$$

$$\Rightarrow \quad (\exists v T(z,u_z,v) \leftrightarrow \forall x\, \exists w\, T(z,x,w)) .$$

This would make a complete Π^0_z-predicate equivalent to a Σ^0_1-predicate, which is impossible.

(C) yields with AC

$$\exists \underline{\underline{uY}}\ \forall \underline{\underline{xv}} [A(\underline{x}, \underline{Yx}) \to B(\underline{u}, \underline{v})]$$

which is a special case of (4).

3.5.4 <u>Theorem</u> (<u>Soundness</u>). Let $\underset{\sim}{H} = \underset{\sim}{HA}^{\omega}$, $\underset{\sim}{I} - \underset{\sim}{HA}^{\omega}$, $\underset{\sim}{WE} - \underset{\sim}{HA}^{\omega}$, $\underset{\sim}{HRO}^{-}$. Then, for $A(\underline{\underline{z}})$ containing at most $\underline{\underline{z}}$ free

$$\underset{\sim}{H} \vdash A(\underline{\underline{z}}) \Rightarrow qf - \underset{\sim}{H} \vdash A_D(\underline{\underline{tzy}}, \underline{\underline{z}})$$

for a suitable sequence $\underline{\underline{t}}$ of closed terms of $\underset{\sim}{H}$.

<u>Proof</u>. We apply induction on the length of deductions in $\underset{\sim}{H}$; as our logical basis we take Gödel's system. For definiteness, we shall suppose the verification to be given first for $\underset{\sim}{H} \equiv \underset{\sim}{I} - \underset{\sim}{HA}^{\omega}$; afterwards we comment on the minor changes needed for the other systems considered.

<u>Logical axioms and rules.</u>

PL 2). Assume

(1) $\quad A_D(\underline{\underline{T}}_1\underline{\underline{z}}, \underline{\underline{y}}, \underline{\underline{z}})$

(2) $\quad A_D(\underline{\underline{x}}, \underline{\underline{T}}_2\underline{\underline{zxv}}, \underline{\underline{z}}) \to B_D(\underline{\underline{T}}_3\underline{\underline{zx}}, \underline{\underline{v}}, \underline{\underline{z}})$.

We have to construct T_4 such that

$$B_D(\underline{\underline{T}}_4\underline{\underline{z}}, \underline{\underline{v}}, \underline{\underline{z}}).$$

Apply (1) to $\underline{\underline{y}} = \underline{\underline{T}}_2\underline{\underline{zxy}}$, then $A_D(\underline{\underline{T}}_1\underline{\underline{z}}, \underline{\underline{T}}_2\underline{\underline{zxv}}, \underline{\underline{z}})$.

Take $\underline{\underline{x}} \equiv \underline{\underline{T}}_1\underline{\underline{z}}$, then

$$A_D(\underline{\underline{T}}_1\underline{\underline{z}}, \underline{\underline{T}}_2\underline{\underline{z}}(\underline{\underline{T}}_1\underline{\underline{z}})\underline{\underline{v}}, \underline{\underline{z}})$$

$$A_D(\underline{\underline{T}}_1\underline{\underline{z}}, \underline{\underline{T}}_2\underline{\underline{zxv}}, \underline{\underline{z}}) \to B_D(\underline{\underline{T}}_3\underline{\underline{z}}(\underline{\underline{T}}_1\underline{\underline{z}}), \underline{\underline{v}}, \underline{\underline{z}})$$

and by PL 2

$$B_D(\underline{\underline{T}}_3\underline{\underline{z}}(\underline{\underline{T}}_1\underline{\underline{z}}), \underline{\underline{v}}, \underline{\underline{z}}).$$

Take now for T_4: $\lambda\underline{\underline{z}}.\underline{\underline{T}}_3\underline{\underline{z}}(\underline{\underline{T}}_1\underline{\underline{z}})$.

In the remaining cases we shall not consider additional free parameters.

PL 3). Assume

(3) $\quad A_D(\underline{\underline{x}}, \underline{\underline{T}}_1\underline{\underline{xv}}) \to B_D(\underline{\underline{T}}_2\underline{\underline{x}}, \underline{\underline{v}})$

(4) $\quad B_D(\underline{\underline{u}}, \underline{\underline{T}}_3\underline{\underline{uw}}) \to C_D(\underline{\underline{T}}_4\underline{\underline{u}}, \underline{\underline{w}})$.

We wish to find T_5, T_6 such that

$$A_D(\underline{\underline{x}}, \underline{\underline{T}}_5\underline{\underline{xw}}) \to C_D(\underline{\underline{T}}_6\underline{\underline{x}}, \underline{\underline{w}}).$$

Take for $\underline{\underline{u}}$ in (4) $\underline{\underline{T}}_2\underline{\underline{x}}$, and take for $\underline{\underline{v}}$ in (3) $\underline{\underline{T}}_3\underline{\underline{uw}}$; then

$$A_D(\underline{\underline{x}}, \underline{\underline{T}}_1\underline{\underline{x}}(\underline{\underline{T}}_3(\underline{\underline{T}}_2\underline{\underline{x}})\underline{\underline{w}})) \to B_D(\underline{\underline{T}}_2\underline{\underline{x}}, \underline{\underline{T}}_3(\underline{\underline{T}}_2\underline{\underline{x}})\underline{\underline{w}}),$$

$$B_D(\underline{\underline{T}}_2\underline{\underline{x}}, \underline{\underline{T}}_3(\underline{\underline{T}}_2\underline{\underline{x}})\underline{\underline{w}}) \to C_D(T_4(\underline{\underline{T}}_2\underline{\underline{x}}), \underline{\underline{w}}).$$

Then

$$A_D(\underline{\underline{x}}, \underline{\underline{T}}_1\underline{\underline{x}}(\underline{\underline{T}}_3(\underline{\underline{T}}_2\underline{\underline{x}})\underline{\underline{w}})) \to C_D(T_4(\underline{\underline{T}}_2\underline{\underline{x}}), \underline{\underline{w}}).$$

So we may take $T_5 \equiv \lambda \underline{x}\underline{w}.(\underline{T}_1\underline{x}(\underline{T}_3(\underline{T}_2\underline{x})\underline{w}), \; T_6 \equiv \lambda\underline{x}.T_4(T_2\underline{x})$.

PL 7), PL 8). Let $C^D \equiv \exists \underline{p}\, \forall \underline{q}\, C_D(\underline{p},\underline{q})$.

Then $(A \& B \to C)^D \equiv (\exists \underline{x}\underline{u}\, \forall \underline{y}\underline{v}(A_D \& B_D) \to \exists \underline{p}\, \forall \underline{q}\, C_D)^D \equiv$

$\exists \underline{P}\underline{Y}\underline{V}\, \forall \underline{x}\underline{u}\underline{q}(A_D(\underline{x}, \underline{Y}\underline{x}\underline{u}\underline{q}) \& B_D(\underline{u}, \underline{V}\underline{x}\underline{u}\underline{q}) \to C_D(\underline{P}\underline{x}\underline{u}, \underline{q}))$,

and

$(A \to (B \to C))^D \equiv (\exists \underline{x}\, \forall \underline{y}\, A_D \to \exists \underline{P}\underline{V}\, \forall \underline{u}\underline{q}(B_D(\underline{u}, \underline{V}\underline{u}\underline{q}) \to C_D(\underline{P}\underline{u}, \underline{q})))^D \equiv$

$\exists \underline{Y}\underline{P}\underline{V}\, \forall \underline{x}\underline{u}\underline{q}(A_D(\underline{x}, \underline{Y}\underline{x}\underline{u}\underline{q}) \to (B_D(\underline{u}, \underline{V}\underline{x}\underline{u}\underline{q}) \to C_D(\underline{P}\underline{x}\underline{u}, \underline{q})))$

which is equivalent to $(A \& B \to C)^D$. Hence the two induction steps are obvious.

PL 9). $(1=0 \to A)^D \equiv (1=0 \to \exists \underline{x}\, \forall \underline{y}\, A_D(\underline{x}\underline{y}))^D \equiv \exists \underline{x}\, \forall \underline{y}(1=0 \to A_D(\underline{x},\underline{y}))$ can be trivially satisfied, by any $\underline{t}$ of suitable type.

PL10). $(A \vee A \to A)^D \equiv (\exists z^0 \underline{x}\underline{x}'\, \forall \underline{y}\underline{y}'[(z^0=0 \to A_D(\underline{x},\underline{y})) \& (z^0 \neq 0 \to A_D(\underline{x}',\underline{y}'))] \to$
$\to \exists \underline{x}''\, \forall \underline{y}''\, A_D(\underline{x}'',\underline{y}''))^D \equiv \exists \underline{Y}\underline{Y}'\underline{X}''\, \forall z^0 \underline{x}\underline{x}'\underline{y}''\{[(z^0=0 \to A_D(\underline{x}, \underline{Y}z^0\underline{x}\underline{x}'\underline{y}'')) \&$
$\& (z^0 \neq 0 \to A_D(\underline{x}', \underline{Y}'z^0\underline{x}\underline{x}'\underline{y}''))] \to A_D(\underline{X}''z^0\underline{x}\underline{x}', \underline{y}'')\}$.

Take for $\underline{Y}$, $\underline{Y}'$ term sequences $\underline{T}$, $\underline{T}'$ such that $\underline{T} \equiv \underline{T}' \equiv \lambda z\underline{x}\underline{x}'\underline{y}''.\underline{y}''$, and take for $\underline{X}''$ a $\underline{T}''$ such that

$$\underline{T}''z^0\underline{x}\underline{x}' = \begin{cases} \underline{x} & \text{if } z^0 = 0 \\ \underline{x}' & \text{if } z_0 \neq 0 . \end{cases}$$

$[A \to A \& A]^D \equiv [\exists \underline{x}\, \forall \underline{y}\, A_D \to \exists \underline{x}'\underline{x}''\, \forall \underline{y}'\underline{y}''(A_D(\underline{x}'\underline{y}') \& A_D(\underline{x}'',\underline{y}''))]^D \equiv$
$\equiv \exists \underline{Y}\underline{X}'\underline{X}''\, \forall \underline{x}\underline{y}'\underline{y}''[A_D(\underline{x}, \underline{Y}\underline{x}\underline{y}'\underline{y}'') \to A_D(\underline{X}'\underline{x},\underline{y}') \& A_D(\underline{X}''\underline{x},\underline{y}'')]$.

Now let T_{A_D} be as in 1.6.14, and take for $\underline{X}'$, $\underline{X}''$, $\underline{Y}$ term sequences $\underline{T}'$, $\underline{T}''$, $\underline{T}$ such that

$$\underline{T}' \equiv \lambda\underline{x}.\underline{x}, \quad \underline{T}'' \equiv \lambda\underline{x}.\underline{x}, \quad \underline{T}\underline{x}\underline{y}'\underline{y}'' = \begin{cases} \underline{y}' & \text{if } T_{A_D}\underline{x}\underline{y}' \neq 0 \\ \underline{y}'' & \text{if } T_{A_D}\underline{x}\underline{y}' = 0. \end{cases}$$

It is in the verification of this axiom schema only, that the decidability of prime formulae plays an essential rôle.

PL11). $[A \to A \vee B]^D \equiv [\exists \underline{x}\, \forall \underline{y}\, A_D \to \exists z^0\underline{x}'\underline{u}\, \forall \underline{y}'\underline{v}((z=0 \to A_D(\underline{x}',\underline{y}')) \&$
$\& (z \neq 0 \to B_D(\underline{u},\underline{v}))]^D \equiv \exists \underline{Y}Z\underline{X}'\underline{U}\, \forall \underline{x}\underline{y}'\underline{v}[A_D(\underline{x}, \underline{Y}\underline{x}\underline{y}'\underline{v}) \to \{(Z\underline{x}=0 \to A_D(\underline{X}'\underline{x}, \underline{y}')) \&$
$\& (Z\underline{x} \neq 0 \to B_D(\underline{U}\underline{x}, \underline{v}))\}]$.

Take for $\underline{Y}$, Z, $\underline{X}'$, $\underline{U}$: $\lambda\underline{x}\underline{y}'\underline{v}.\underline{y}'$, $\lambda\underline{x}.0$, $\lambda\underline{x}.\underline{x}$, $\lambda\underline{x}.\underline{x}$ respectively ($\underline{U}$ is in fact arbitrary).

Similarly for the other half of P11).

P12) is also routine.

P13). Assume

$$A_D(\underline{x}, \underline{T}\underline{x}\underline{v}) \to B_D(\underline{T}'\underline{x},\underline{v})$$

$$[C \vee A \to C \vee B]^D \equiv [\exists z_1^o \underline{px}\, \forall \underline{qy}\, [(z_1 = 0 \to C_D) \,\&\, (z_1 \neq 0 \to A_D)] \to$$
$$\to \exists z_2^o \underline{p}'\underline{u}\, \forall \underline{q}'\underline{v}\, [(z_2 = 0 \to C_D(\underline{p}'\underline{q}')) \,\&\, (z_2 \neq 0 \to B_D(\underline{u},\underline{v}))]] \equiv$$
$$\exists \underline{QY}Z_2\underline{P}'\underline{U}\, \forall z_1^o \underline{pxq}'\underline{v}\, \{[(z_1 = 0 \to C_D(\underline{p},\underline{Q}z_1\underline{pxq}'\underline{v})) \,\&\, (z_1 \neq 0 \to A_D(\underline{x},\underline{Y}z_1\underline{pxq}'\underline{v}))] \to$$
$$\to [(Z_2 z_1 \underline{px} = 0 \to C_D(\underline{P}'z_1\underline{px},\underline{q}')) \,\&\, (z_2 z_1 \underline{px} \neq 0 \to B_D(\underline{U}z_1\underline{px},\underline{v}))]\}.$$

Now take for $\underline{Q}$, $\underline{Y}$, Z_2, $\underline{P}'$, $\underline{U}$: $\lambda z_1\underline{pxq}'\underline{v}.\underline{q}'$, $\lambda z_1\underline{pxq}'\underline{v}.\underline{Txv}$, $\lambda z_1\underline{px}.\underline{z}_1$, $\lambda z_1\underline{px}.\underline{p}$, $\lambda z_1\underline{px}.\underline{T}'\underline{x}$.

Q 1). Let

$$B_D(\underline{u},\underline{Tzuy}) \to A_D(\underline{T}'\underline{z}\underline{u}, \underline{y}, z).$$

To interpret $B \to \forall zAz$ we have to find $\underline{T}''$, $\underline{T}'''$ such that

$$B_D(\underline{u}, T''\underline{uyz}) \to A_D(\underline{T}'''\underline{u}z, \underline{y}, z).$$

Take $T'' \equiv \lambda\underline{uyz}.\underline{Tzyu}$, $T''' \equiv \lambda\underline{u}z.\underline{T}'z\underline{u}$.

Q 2). $(\forall zAz \to At)^D \equiv (\exists \underline{X}\, \forall z\underline{y}\, A_D(\underline{X}z,\underline{y},z) \to \exists \underline{x}'\, \forall \underline{y}'\, A_D(\underline{x}',\underline{y}',t))^D \equiv$

$$\equiv \exists \underline{ZYX}'\, \forall \underline{Xy}'(A_D(\underline{X}(Z\underline{Xy}'), \underline{YXy}', Z\underline{Xy}') \to A_D(\underline{X}'\underline{X}, \underline{y}', t)).$$

Take for Z, $\underline{Y}$, $\underline{X}'$: $\lambda\underline{Xy}'.t$, $\lambda\underline{Xy}'.\underline{y}'$, $\lambda\underline{X}.\underline{X}t$.

Similarly for Q3, Q4.

<u>Non-logical axioms.</u> The "defining axioms" for the various functional constants are quantifier-free, and therefore unproblematic. The equality axioms are also quantifier-free (or purely universal, if one considers their universal closure); so we only have to verify the induction axiom, or equivalently, the rule $B0$, $\forall y(By \to B(Sy)) \Rightarrow \forall xBx$.

Let $(By)^D \equiv \exists \underline{u}\, \forall \underline{v}\, B_D(\underline{u},\underline{v},y,\underline{z})$ and suppose:

$$(5) \quad \begin{cases} B_D(\underline{T}_o\underline{z},\underline{v}, 0, \underline{z}) \\ B_D(\underline{u}, \underline{T}_1 y\underline{zuv}', y,\underline{z}) \to B_D(\underline{T}_2 y\underline{zu}, \underline{v}', Sy, \underline{z}). \end{cases}$$

Now we define by recursion $\underline{t}$ such that

$$\underline{t}0 = \underline{T}_o\underline{z}$$
$$\underline{t}(Sy) = \underline{T}_2 y\underline{z}(\underline{t}y).$$

Then

$$B_D(\underline{t}0, \underline{v}, 0, \underline{z})$$
$$B_D(\underline{t}y, \underline{T}_1 y\underline{z}(\underline{t}y)\underline{v}', y, \underline{z}) \to B_D(\underline{t}(Sy), \underline{v}', Sy, \underline{z}).$$

Now apply the induction lemma 1.7.10 with $Q(y,\underline{v}) \equiv_{def} B(\underline{t}y, \underline{v}, y, \underline{z})$, and with $\underline{T} \equiv_{def} \lambda y\underline{v}.\underline{T}_1 y\underline{z}(\underline{t}y)\underline{v}$. Then $B_D(\underline{t}y, \underline{v}, y, \underline{z})$ follows, and with $\underline{T}' \equiv \lambda\underline{z}.\underline{t}$, $B_D(\underline{T}'\underline{z}y, \underline{v}, y, \underline{z})$.

The preceding verification of the soundness theorem is directly applicable to $\underset{\sim}{I}-\underset{\sim}{HA}^{\omega}$. For $\underset{\sim}{HA}^{\omega}$, we must note that the verifications only require that $(\lambda \underline{x}.t[\underline{x}])\underline{t}' = t[\underline{t}']$ for certain terms $t, \underline{t}'$, where $=$ is interpreted in the metamathematical sense indicated in 1.6.15, remark (ii).

A fortiori, the verification is valid for $\underset{\sim}{WE}-\underset{\sim}{HA}^{\omega}$, noting that premiss and conclusion in an application of the extensionality rule are both purely universal.

For $\underset{\sim}{HRO}^{-}$, the soundness theorem is also immediate.

3.5.5. Remarks. (i). Similarly, for the slightly strengthened version of the extensionality rule EXT - R' in 1.6.12, the soundness theorem applies.
(ii). If the deduction theorem holds for the system considered, the soundness theorem also extends to deductions under hypothesis: if

$$A_1, \ldots, A_n \vdash B,$$

then

$$A_{1D}(\underline{x}_1, \underline{T}_1\underline{x}_1 \ldots \underline{x}_n\underline{v}), \ldots, A_{nD}(\underline{x}_n, \underline{T}_n\underline{x}_1 \ldots \underline{x}_n\underline{v}) \vdash B_D(\underline{U}\underline{x}_1 \ldots \underline{x}_n, \underline{v})$$

for some term sequence $\underline{T}$.
(iii). If we would have been satisfied with the weaker soundness theorem

$$\underset{\sim}{H} \vdash A(\underline{z}) \Rightarrow \underset{\sim}{H} \vdash A_D(\underline{t}\underline{z}, \underline{y}, \underline{z})$$

we could simplify the treatment of the induction schema, inasmuch we do not need the induction lemma and its tedious proof, since instead of using the induction lemma at the final step of the verification, we note that

$$\forall \underline{v}\, B_D(\underline{t}0, \underline{v}, 0, \underline{z})$$

$$\forall y\, (\forall \underline{v}\, B_D(\underline{t}y, \underline{v}, y, \underline{z}) \rightarrow \forall \underline{v}\, B_D(\underline{t}(Sy), \underline{v}, Sy, \underline{z})),$$

and therefore, applying the induction schema

$$\forall y\, \forall \underline{v}\, B_D(\underline{t}y, \underline{v}, y, \underline{z}).$$

(iv). If A^D holds for a certain model M of $\underset{\sim}{HA}^{\omega}$, we may call A M-Dialectica interpretable, and if this can be proved in a formal theory $\underset{\sim}{H}$, $\underset{\sim}{H}$-M-Dialectica interpretable. The M-Dialectica interpretable and $\underset{\sim}{H}$-M-Dialectica interpretable formulae are closed under deduction (provided $\underset{\sim}{H}$ is), by a reasoning similar to that needed for (ii) above.

3.5.6. Remark. $\underset{\sim}{N}-\underset{\sim}{HA}^{\omega}$ does not have a Dialectica interpretation into itself, as is shown by the following counterexample, due to W.A. Howard (in correspondence). We can derive by predicate logic (and the decidability of type 0 equality) in $\underset{\sim}{N}-\underset{\sim}{HA}^{\omega}$

$$\forall y^1 \neg \forall u^0 \neg (u = 0 \leftrightarrow y^1 = z^1).$$

If we treat $y^1 = z^1$ as a prime formula, the Dialectica translation becomes

$$[\forall y^1 \neg \forall u^0 \neg (u = 0 \leftrightarrow y = z)]^D \equiv \text{ (by note (ii) in 3.5.2)}$$
$$\equiv [\forall y^1 \exists u^0 \neg\neg (u = 0 \leftrightarrow y = z)]^D \equiv$$
$$\equiv \exists U^2 \forall y^1 \neg\neg (Uy = 0 \leftrightarrow y = z)]^D .$$

On the other hand, for no continuous U^2

$$\forall y^1 \neg\neg (Uy = 0 \leftrightarrow y = z)]^D ;$$

one easily shows $\exists y^1 \neg (Uy = 0 \leftrightarrow y = \lambda x^0.0)]^D$ for any given continuous U^2. Since all the closed type 2 terms of $\underset{\sim}{N}$ - $\underset{\sim}{HA}^\omega$ represent continuous functionals (2.3.10(i)), it follows that $\underset{\sim}{N}$ - $\underset{\sim}{HA}^\omega$ does not permit a Dialectica interpretation into itself.

We also see from this counterexample that the assumption that the Dialectica translation of any formula of some extension of $\underset{\sim}{N}$ - $\underset{\sim}{HA}^\omega$ should be provable in the same theory implies that equality between higher types is decidable in a weak sense, for it follows that

$$\neg\neg y = z \vee \neg y = z .$$

We do not know whether the assumption about the Dialectica interpretability perhaps even <u>implies</u> decidability of higher type equality (but conjecture that the assumption is logically weaker).

In <u>Luckhardt</u> 1973, page 55, the example of $\neg\neg(\exists x Ax \vee \neg \exists x Ax)$ (Ax quantifier-free) is used to illustrate the need for decidability for prime formulae.

3.5.7 - 3.5.11. <u>Axiomatization of Dialectica interpretability</u>.

3.5.7. <u>Lemma</u>. Let IP_0', M' denote the schemata

IP_0' $\quad (\forall \underline{\underline{x}} A \underline{\underline{x}} \to \exists y By) \to \exists y (\forall \underline{\underline{x}} A \underline{\underline{x}} \to By)$
for A quantifier-free, $\underline{\underline{x}}, y$ with arbitrary types,

M' $\quad \neg\neg \exists \underline{\underline{x}} A \underline{\underline{x}} \to \exists \underline{\underline{x}} A \underline{\underline{x}}$
for A quantifier-free, $\underline{\underline{x}}$ with arbitrary types.

Then for $\underset{\sim}{H} \equiv \underset{\sim}{HA}^\omega$, $\underset{\sim}{I}$ - $\underset{\sim}{HA}^\omega$, $\underset{\sim}{HRO}^-$

$$\underset{\sim}{H} + IP_0' + M' + AC \vdash A \leftrightarrow A^D .$$

<u>Proof</u>. We establish the lemma by induction on the logical complexity of A. Assuming $\underset{\sim}{H}' \vdash A \leftrightarrow A^D$, $\underset{\sim}{H}' \vdash B \leftrightarrow B^D$ for $\underset{\sim}{H}' \equiv \underset{\sim}{H} + IP_0' + M' + AC$, we readily see that in $\underset{\sim}{H}'$, $(A \& B)^D \leftrightarrow A \& B$, $(A \vee B)^D \leftrightarrow A \vee B$, $(\forall x^\sigma A)^D \equiv \forall x^\sigma A$, $(\exists x^\sigma A)^D \equiv \exists x^\sigma A$; so it remains to consider implication. Using the notation in 3.5.2 we see that the transition from $(A \to B)^D$ to (a) is justified by the induction hypothesis, the transition to (b) by intuitionistic logic, the transition to (c) by IP_0', the transition to (d) again by

intuitionistic logic, and the transition from (d) to (e) by M', since $\forall\underline{y}\, A_D \to B_D$ is equivalent to $B_D \vee (\neg B_D \,\&\, \neg \forall\underline{y}\, A_D)$, i.e. $B_D \vee (\neg B_D \,\&\, \neg\neg \exists\underline{y}\, \neg A_D)$, and thus by M' $B_D \vee (\neg B_D \,\&\, \exists\underline{y}\, \neg A_D)$; hence $\exists\underline{y}[B_D \vee (\neg B_D \,\&\, \neg A_D)]$, and hence $\exists\underline{y}(A_D \to B_D)$.
The transition from (e) to (f) is justified by AC.

3.5.8. Lemma. The axiom of choice AC relative to $\mathcal{L}[\underline{H}]$, for $\underline{H} \equiv \underline{HA}^\omega$, $\underline{I}-\underline{HA}^\omega$, $\underline{HRO}^-$,

$$\mathrm{AC} \qquad \forall x^\sigma \exists y^\tau A(x,y) \to \exists z^{(\sigma)\tau} \forall x^\sigma A(x, zx)$$

is Dialectica interpretable in $\mathrm{qf}-\underline{H}$.
Proof. $[\forall x^\sigma \exists y^\tau A(x,y)]^D$ and $[\exists z^{(\sigma)\tau} \forall x^\sigma A(x, zx)]^D$ are identical (modulo renaming bound variables), hence interpreting an instance of AC reduces to interpreting an instance $B \to B$ of PL1.

3.5.9. Lemma. Each instance of the schemata M' and IP_0' in $\mathcal{L}[\underline{H}]$ (for $\underline{H} \equiv \underline{I}-\underline{HA}^\omega$, $\underline{HA}^\omega$, $\underline{HRO}^-$) is Dialectica interpretable in $\mathrm{qf}-\underline{H}$.
Proof. (i). $(\neg\neg \exists\underline{x} A\underline{x})^D \equiv \exists\underline{x} A\underline{x}$ (by note (ii) in 3.5.2); so the interpretation of an instance of M' reduces to interpreting an instance $C \to C$ of PL1.
(ii). Similarly for IP_0'.

3.5.10. Theorem. For $\underline{H} \equiv \underline{HA}^\omega$, $\underline{I}-\underline{HA}^\omega$, $\underline{HRO}^-$, $\underline{WE}-\underline{HA}^\omega$.
(i). $\underline{H} + M^\omega + IP_0^\omega + AC \vdash A \leftrightarrow A^D$
(ii). $\underline{H} + M^\omega + IP_0^\omega + AC \vdash A\underline{z} \Rightarrow \mathrm{qf}-\underline{H} \vdash A_D(\underline{tzy}, \underline{y})$.
Here M^ω, IP_0^ω are the schemata M, IP_0 extended to all finite types:

M^ω $\qquad \forall\underline{x}(A \vee \neg A) \,\&\, \neg\neg \exists\underline{x} A \to \exists\underline{x} A$

IP_0^ω $\qquad \forall\underline{x}(A \vee \neg A) \,\&\, (\forall\underline{x} A \to \exists\underline{y} B) \to \exists\underline{y}(\forall\underline{x} A \to B)$.

Proof. (i) and (ii) hold for M' and IP_0' in place of M^ω, IP_0^ω resp., by 3.5.7 - 3.5.9 and 3.5.5 (ii). To establish the theorem, it is therefore sufficient to show M^ω and IP_0^ω to be derivable in $\underline{H} + M' + IP_0' + AC$.
Consider any instance of M^ω

$$\forall\underline{x}(A \vee \neg A) \,\&\, \neg\neg \exists\underline{x} A \to \exists\underline{x} A .$$

By lemma 3.5.7, $\forall\underline{x}(A \vee \neg A)$ is equivalent in $\underline{H} + M' + IP_0' + AC$ to its own Dialectica interpretation, i.e. to $[\forall\underline{x}(A \vee \neg A)]^D$. Let $\exists\underline{y} \forall\underline{z} A_D(\underline{x},\underline{y},\underline{z}) \equiv A^D$, then $[\forall\underline{x}(A \vee \neg A)]^D \equiv (\forall\underline{x}[\exists\underline{y} \forall\underline{z} A_D(\underline{x},\underline{y},\underline{z}) \vee \exists\underline{Z} \forall\underline{v} \neg A_D(\underline{x},\underline{vZ},\underline{v})])^D \equiv$
$\equiv \exists U\underline{YZ}\, \forall\underline{xzv}([U\underline{x}=0 \to A_D(\underline{x},\underline{Yx},\underline{z})] \,\&\, [U\underline{x}\neq 0 \to \neg A_D(\underline{x},\underline{v},\underline{Zxv})])$.
From this we derive the existence of a U such that $U\underline{x} = 0 \leftrightarrow A\underline{x}$.
For let U, $\underline{Y}$, $\underline{Z}$ satisfy $(\forall\underline{x}(A \vee \neg A))^D$, i.e.

(1) $\underline{U}\underline{x} = 0 \to A_D(\underline{x}, \underline{Y}\underline{x}, \underline{z})$

(2) $\underline{U}\underline{x} \neq 0 \to \neg A_D(\underline{x}, \underline{v}, \underline{Z}\underline{x}\underline{v})$.

If $\underline{U}\underline{x} = 0$, then $\exists \underline{y}\, \forall \underline{z}\, A_D(\underline{x}, \underline{Y}\underline{x}, \underline{z})$, i.e. $A\underline{x}$.
Conversely, if $\underline{U}\underline{x} \neq 0$, then $\forall \underline{z}\, A_D(\underline{x}, \underline{y}, \underline{z})$ would imply $A_D(\underline{x}, \underline{y}, \underline{Z}\underline{x}\underline{y})$, which contradicts (2). Hence $\neg \exists \underline{y}\, \forall \underline{z}\, A_D(\underline{x}, \underline{Y}\underline{x}, \underline{z})$, so $\neg A$.
Now relative to $\mathbf{H} + M' + IP_0' + AC$ the instance of M^ω becomes equivalent to

$$\neg\neg \exists \underline{x}(\underline{U}\underline{x} = 0) \to \exists \underline{x}(\underline{U}\underline{x} = 0)$$

which is an instance of M', and therefore derivable.
Similarly for IP_0^ω.

3.5.11. Corollary. If $\mathbf{H} + \Gamma$ is an extension of $\mathbf{H}$ by a set of axioms Γ in $\mathcal{L}[\mathbf{H}]$ ($\mathbf{H} \equiv \mathbf{I\text{-}HA}^\omega$, $\mathbf{HRO}^-$, $\mathbf{HA}^\omega$, $\mathbf{WE\text{-}HA}^\omega$) so that the soundness theorem for $\mathbf{H} + \Gamma$ holds in the form

$$\mathbf{H} + \Gamma \vdash A \Rightarrow \mathbf{H} + \Gamma \vdash A^D$$

then

$$\mathbf{H} + \Gamma + M^\omega + IP_0^\omega + AC \vdash A \longleftrightarrow A^D$$
$$\mathbf{H} + \Gamma + M^\omega + IP_0^\omega + AC \vdash A \Leftrightarrow \mathbf{H} + \Gamma \vdash A^D .$$

Remark. In systems $\mathbf{H}$ extending $\mathbf{HA}^\omega$ ($\mathbf{HRO}^-$, $\mathbf{I\text{-}HA}^\omega$) but with the same language, each assertion of $\mathbf{H}$ can be brought into a normal form $\exists \underline{x}\, \forall \underline{y}\, A_D(\underline{x}, \underline{y})$, A_D quantifier-free. If the theorems of $\mathbf{H}$ can be shown to be $\mathbf{H}'$, M-Dialectica interpretable, $\mathbf{H}' \subseteq \mathbf{H}$, the result is a kind of normal form for assertions of $\mathbf{H}'$, obtained by interpreting the quantifiers in A^D for $A \in \mathcal{L}[\mathbf{H}']$ in M. Taking $M \equiv ICF$, $\mathbf{H}' \equiv \mathbf{EL}$, $\mathbf{H} \equiv \mathbf{HA}^\omega$ this yields (practically) the result of Vesley 1972.

3.5.12 - 3.5.15. The interpretability of the extensionality axiom.

3.5.12. Theorem. In $ECF(\mathcal{U})$, for any universe $\mathcal{U}$ satisfying $\mathbf{EL}$, the extensionality axiom (2.7.2) is Dialectica interpretable.
Proof. We have to show that

(1) $$[\forall z^{(\sigma)\tau} x^\sigma y^\sigma (x =_e y \to zx =_e zy)]^D$$

is valid in ECF.
Let $\underline{u}, \underline{v}$ be sequences of variables such that $x\underline{u}$, $z\underline{v}$ become terms of type 0. Then (1) can be stated as

$$[\forall zxy(\forall \underline{u}(x\underline{u} = y\underline{u}) \to \forall \underline{v}(zx\underline{v} = zy\underline{v}))]^D \equiv$$
(2) $$\equiv \exists \underline{U}\, \forall zxy\underline{v}(x(\underline{U}zxy\underline{v}) = y(\underline{U}zxy\underline{v}) \to zx\underline{v} = zy\underline{v}) .$$

Since we are working in a model with extensionality, we may make use of the reductions of the type structure (1.8.5 - 1.8.8) to reduce (2) to

(3) $\exists U\, \forall zxyv(x(Uzxyv) = y(Uzxyv) \to zxv = zyv)$

where $z \in (j+1)(\sigma)j$, $x,y \in j+1$, $v \in \sigma$, $U \in ((j+1)(\sigma)0)(j+1)(j+1)(\sigma)j$.
We construct the desired U in our model as follows.
Let $\gamma \in W^1_{(j+1)(\sigma)o}$, $\alpha,\beta \in W^1_{j+1}$, $\delta \in W^1_\sigma$, and let $\epsilon \in W^1_{(o)j}$ be an enumerating function of the recursively dense basis for W^1_j (cf. 2.6.19).
Let $\xi \in W^1_\sigma$ represent 0^σ say. Now we construct a φ (given by some p-functor), depending continuously on $\alpha, \beta, \gamma, \delta$, as follows:
1°) If $(\gamma \mid \alpha)(\delta) = (\gamma \mid \beta)(\delta)$, we put $\varphi = \xi$;
2°) If $(\gamma \mid \alpha)(\delta) \neq (\gamma \mid \beta)(\delta)$, there are initial segments $\bar{\gamma}x$, $\bar{\alpha}x$, $\bar{\beta}x$, $\bar{\delta}x$ such that $(\gamma|\alpha)(\delta)$, $(\gamma|\beta)(\delta)$ can be computed from them; so $\bar{\alpha}x$, $\bar{\beta}x$ must represent initial segments of different functionals, hence there are elements of the recursively dense basis of W^1_j to which α, β assign different values. Let φ be $(\epsilon)_{\min_z[\alpha((\epsilon)_z) \neq \beta((\epsilon)_z)]}$. φ is obviously extensional in $\alpha,\beta,\gamma,\delta$, so $\Lambda^1\gamma\Lambda^1\alpha\Lambda^1\beta\Lambda^o\delta.\varphi$ gives us the required U.

3.5.13. <u>Lemma</u>. Let F be any formula in the negative fragment of $\mathcal{L}[HA^\omega]$.
Then

$$(\underline{HA}^\omega)^c + QF\text{-}AC \vdash F \leftrightarrow F^D$$

where QF-AC is

QF-AC $\quad \forall \underline{x}\, \exists \underline{y}\, A(\underline{x},\underline{y}) \to \exists \underline{z}\, \forall \underline{x}\, A(\underline{x},\underline{\underline{zx}})$ (A quantifier-free).

<u>Proof</u>. We show by induction on the complexity of F, that F^D takes the form:

$$\exists \underline{x}\, \forall \underline{y}\, F^*(\underline{\underline{xy}}, \underline{y}), \qquad F^*(\underline{\underline{xy}}, \underline{y}) \equiv F_D(\underline{x}, \underline{y}),$$

and simultaneously that

$$(\underset{\sim}{HA}^\omega)^c + QF\text{-}AC \vdash F \leftrightarrow F^D.$$

The assertions are both obvious for F prime. Assume the assertions to have been established for F, G. Then

(a) $(\forall zF)^D \equiv \exists \underline{X}\, \forall z\underline{y}\, F^*(\underline{X}z\underline{y}, z, \underline{y}) \leftrightarrow \forall z\underline{y}\, \exists \underline{x}\, F^*(\underline{x}, z, \underline{y}) \leftrightarrow$
$\leftrightarrow \forall z\, \exists \underline{X}\, \forall \underline{y}\, F^*(\underline{\underline{Xy}}, z, \underline{y}) \leftrightarrow \forall z\, F^D \leftrightarrow \forall z\, F$
(repeated use of QF-AC, induction hypothesis).

Let $F^D \equiv \exists \underline{x}\, \forall \underline{y}\, F^*(\underline{\underline{xy}}, \underline{y})$, $G^D \equiv \exists \underline{u}\, \forall \underline{v}\, G^*(\underline{\underline{uv}}, \underline{v})$.

(b) $(F \,\&\, G)^D \equiv \exists \underline{\underline{xu}}\, \forall \underline{\underline{yv}}(F^*(\underline{\underline{xy}}, \underline{y}) \,\&\, G^*(\underline{\underline{uv}}, \underline{v})) \leftrightarrow \exists \underline{x}\, \forall \underline{y}\, F^*(\underline{\underline{xy}}, \underline{y}) \,\&\, \exists \underline{u}\, \forall \underline{v}\, G^*(\underline{\underline{uv}}, \underline{v})$
$\leftrightarrow F^D \,\&\, G^D$.

(c) $(F \to G)^D \equiv \exists \underline{\underline{YU}}\, \forall \underline{\underline{xv}}[F^*(\underline{x}(\underline{\underline{\underline{Yxv}}}), \underline{\underline{\underline{Yxv}}}) \to G^*(\underline{\underline{\underline{Uxv}}}, \underline{v})] \leftrightarrow$
$\leftrightarrow \forall \underline{\underline{xv}}\, \exists \underline{\underline{yu}}[F^*(\underline{\underline{xy}}, \underline{y}) \to G^*(\underline{u}, \underline{v})]$ (by QF-AC) $\leftrightarrow$
$\leftrightarrow \forall \underline{\underline{xv}}[\forall \underline{y}\, F^*(\underline{\underline{xy}}, \underline{y}) \to \exists \underline{u} G^*(\underline{u}, \underline{v})]$ (classical logic) $\leftrightarrow$

$\leftrightarrow [\exists \underline{x}\, \forall \underline{y}\, F^*(\underline{x}\underline{y},\underline{y}) \rightarrow \forall \underline{v}\, \exists \underline{u}\, G^*(\underline{u},\underline{v})]$ (classical logic) $\leftrightarrow$

$\leftrightarrow [\exists \underline{x}\, \forall \underline{y}\, F^*(\underline{x}\underline{y},\underline{y}) \rightarrow \exists \underline{u}\, \forall \underline{v}\, G^*(\underline{u}\underline{v},\underline{v})]$ (QF - AC) $\leftrightarrow$

$\leftrightarrow (F^D \rightarrow G^D) \leftrightarrow (F \rightarrow G)$.

3.5.14. Corollary. $\underset{\sim}{E} - \underset{\sim}{HA}^\omega + M^\omega + IP_0^\omega + AC$ is conservative over $\underset{\sim}{HA}$ w.r.t. negative formulae. (Kreisel.)

Proof. Assume

$$\underset{\sim}{E} - \underset{\sim}{HA}^\omega + M^\omega + IP_0^\omega + AC \vdash F ,$$

F a negative arithmetical formula. Then there are finitely many instances of the extensionality axiom, say $F_1, \ldots, F_n$ such that

$$\underset{\sim}{HA}^\omega + M^\omega + IP_0^\omega + AC \vdash F_1 \rightarrow (F_2 \rightarrow \ldots (F_n \rightarrow F)\ldots)$$

and therefore

$$\underset{\sim}{HA}^\omega + F_1^D + \ldots + F_n^D + F^D .$$

Note also

$$(\underset{\sim}{HA}^\omega)^c + QF - AC \vdash F \leftrightarrow F^D$$

hence

$$(\underset{\sim}{HA}^\omega)^c + QF - AC + F_1^D + \ldots + F_n^D \vdash F .$$

If we not interpret $\underset{\sim}{HA}^\omega$ by ECF($\mathcal{R}$) , then by 3.5.12, 2.6.20 and 1.10.12 it follows that $\underset{\sim}{HA} \vdash F$.

3.5.15. The non-interpretability of the extensionality axiom.

By means of the model of the hereditarily majorizable functionals (2.8.6) it is shown in Howard B that the extensionality axiom is not Dialectica interpretable by a functional of $\underset{\sim}{E} - \underset{\sim}{HA}^\omega$. As a consequence, for the variant of $\underset{\sim}{WE} - \underset{\sim}{HA}^\omega$ where the rule of extensionality is formulated as :

(1) $$ty_1 \ldots y_n = sy_1 \ldots y_n \Rightarrow F[t] = F[s]$$

(F[t] of type 0 , $ty_1 \ldots y_n$, $sy_1 \ldots y_n$ of type 0, $y_1, \ldots, y_n$ a sequence of variables not occurring free in any assumption on which the deduction of $ty_1 \ldots y_n = sy_1 \ldots y_n$ depends), the deduction theorem does not hold. For then (1) would imply

$$\forall y_1 \ldots y_n (ty_1 \ldots y_n = sy_1 \ldots y_n) \Rightarrow F[t] = F[s]$$

and by the deduction theorem, taking x, y for t, s , and zx for F[x] , it follows that

$$x =_e y \rightarrow zx = zy ,$$

which is the extensionality axiom.

3.5.16. Theorem

(i). CT, CT_0 are $\underset{\sim}{HRO}^-$ - Dialectica interpretable, hence also HRO - Dialectica interpretable.

(ii). C - N - continuity is $\underset{\sim}{EL}$ - ICF - Dialectica interpretable (w.r.t. $\mathcal{L}[\underset{\sim}{HA}^\omega]$). FAN is $\underset{\sim}{EL}$ - ICF - and $\underset{\sim}{EL}$ - ECF - Dialectica interpretable (w.r.t. $\mathcal{L}[\underset{\sim}{HA}^\omega]$).

(iii). The negation of an instance of IP is $\underset{\sim}{HRO}^-$ - Dialectica interpretable, so IP is not $\underset{\sim}{I}$ - $\underset{\sim}{HA}^\omega$ - Dialectica interpretable.

Proof. (i). $[CT]^D$ is obviously $\underset{\sim}{HRO}^-$ - Dialectica interpretable, and AC is Dialectica interpretable, hence also CT_0.

(ii). The axiom MC for the modulus-of-continuity functional (2.6.3) is quantifier-free:

(1). $\bar{y}(\varphi_{mc}xy) = \bar{z}(\varphi_{mc}xy) \rightarrow xy = xz \qquad (y, z \in 1,\ x \in 2)$.

Therefore (1) is identical with its own Dialectica translation. Since the existence of φ_{mc} in ICF (2.6.3) can be established in $\underset{\sim}{EL}$, it is $\underset{\sim}{EL}$, ICF - Dialectica interpretable. Since AC + (1) implies C - N, C - N is $\underset{\sim}{EL}$ - ICF - Dialectica interpretable.

The existence of the fan-functional MUC (2.6.4) can also be expressed in a quantifier-free form

$$(\overline{\Phi x})(\varphi_{uc}z) = (\overline{\Phi y})(\varphi_{uc}z) \rightarrow z(\Phi x) = z(\Phi y)$$

where $\Phi \equiv \lambda x^1 \lambda z. sg(x^1 z)$. Then the $\underset{\sim}{EL}$, ICF - and $\underset{\sim}{EL}$, ECF - Dialectica interpretability of FAN is obtained similarly.

(iii). $M + IP + CT_0$ is inconsistent (3.2.27); this directly yields the desired conclusion, in combination with (i).

3.5.17. The Diller - Nahm variant of the Dialectica interpretation.

In Diller - Nahm A, a variant of the Dialectica translation is described which interprets $\underset{\sim}{N}$ - $\underset{\sim}{HA}^\omega$ into $\underset{\sim}{N}$ - $\underset{\sim}{HA}^\omega$, as follows. To each formula A of $\underset{\sim}{N}$ - $\underset{\sim}{HA}^\omega$ an interpretation $A^\wedge$ of the form $\exists \underline{x} \forall \underline{y} A_\wedge(\underline{x}, \underline{y})$ is assigned, where $A_\wedge$ may contain bounded universal quantifiers, but no unbounded quantifiers $\forall$ or $\exists$. In the definition of the translation by induction on the logical complexity, bounded universal quantification $\forall x^0 < t$ is counted as a separate logical operator.

The inductive clauses for prime formulae, &, v, $\forall$, $\exists$ are as for the Dialectica translation. Let $B^\wedge \equiv \exists \underline{\underline{v}} \forall \underline{\underline{w}} B_\wedge(\underline{\underline{v}}, \underline{\underline{w}})$, $C^\wedge \equiv \exists \underline{\underline{y}} \forall \underline{\underline{z}} C_\wedge(\underline{\underline{y}}, \underline{\underline{z}})$. Then

$$(B \rightarrow C)^\wedge \equiv \exists \underline{\underline{X}}\underline{\underline{W}}\underline{\underline{Y}} \forall \underline{\underline{v}}\underline{\underline{z}}((\forall x < \underline{\underline{X}}\underline{\underline{v}}\underline{\underline{z}}) B_\wedge(\underline{\underline{v}}, \underline{\underline{W}}x\underline{\underline{v}}\underline{\underline{z}}) \rightarrow C_\wedge(\underline{\underline{Y}}\underline{\underline{v}}, \underline{\underline{z}})).$$

For $\underline{\underline{w}}$ empty it follows that

$$(B \rightarrow C)^\wedge \leftrightarrow \exists \underline{\underline{Y}} \forall \underline{\underline{v}}\underline{\underline{z}}[B_\wedge(\underline{\underline{v}}) \rightarrow C_\wedge(\underline{\underline{Y}}\underline{\underline{v}}, \underline{\underline{z}})].$$

Further we put

$$((\forall x < t)C)^{\wedge} \equiv \exists \underline{Y}\, \forall \underline{z}(\forall x < t)\, C_{\wedge}(\underline{Y}x, \underline{z}) .$$

One easily verifies

$$((\forall x < t)C)^{\wedge} \leftrightarrow (\forall x(x < t \to C))^{\wedge} .$$

Note that if we require $\underline{X}\underline{v}\underline{z} = 1$, the resulting translation is equivalent to the Dialectica interpretation.

The clause for implication may be conceived as being obtained by replacing in

$$\exists \underline{v}\, \forall \underline{w}\, B_{\wedge}(\underline{v},\underline{w}) \to \exists \underline{y}\, \forall \underline{z}\, C_{\wedge}(\underline{y},\underline{z})$$

$\forall \underline{w}\, B_{\wedge}(\underline{v},\underline{w})$ by the equivalent assertion

$$\forall \underline{W}\, \forall Z\, \forall z < Z\; B_{\wedge}(\underline{v},\underline{W}z)$$

and then use the same transformations as for the implication in the case of the Dialectica translation (3.5.2).

Extending $\mathrm{qf}\text{-}\mathbf{N}\text{-}\mathbf{HA}^{\omega}$ by the addition of bounded universal quantification as a new "propositional" operator, with axioms

$$(\forall x < 0)A$$
$$(\forall x < St)A \to (\forall x < t)A$$
$$\left.\begin{array}{l}(\forall x < St)A \to [x/t]A \\ (\forall x < t)A \;\&\; [x/t]A \to (\forall x < St)A\end{array}\right\} \quad t \text{ free for } x \text{ in } A,$$

(let us denote this system for the time being as $\mathbf{H}$) Diller and Nahm obtain

$$\mathbf{N}\text{-}\mathbf{HA}^{\omega} \vdash A \;\Rightarrow\; \mathbf{H} \vdash A_{\wedge}(\underline{t},\underline{x})$$

for a suitable sequence of terms $\underline{t}$ of $\mathbf{N}\text{-}\mathbf{HA}^{\omega}$.

The crucial point in the proof is to show how to interpret $A \to A \& A$. Let us suppose $A^{\wedge} \equiv \exists \underline{x}\, \forall \underline{y}\, A_{\wedge}(\underline{x},\underline{y})$. Then

$$(A \to A \& A)^{\wedge} \equiv [\exists \underline{x}_1 \forall \underline{y}_1 A_{\wedge}(\underline{x}_1,\underline{y}_1) \to \exists \underline{x}_2\underline{x}_3 \forall \underline{y}_2\underline{y}_3 (A_{\wedge}(\underline{x}_2,\underline{y}_2) \& A_{\wedge}(\underline{x}_3,\underline{y}_3))]^{\wedge} \equiv$$
$$\equiv \exists X\underline{Y}_1\underline{X}_2\underline{X}_3 \forall \underline{x}_1\underline{y}_2\underline{y}_3 [(\forall x < X\underline{x}_1\underline{y}_2\underline{y}_3) A_{\wedge}(\underline{x}_1, \underline{Y}_1 x \underline{x}_1\underline{y}_2\underline{y}_3) \to$$
$$\to A_{\wedge}(\underline{X}_2\underline{x}_1,\underline{y}_2) \& A_{\wedge}(\underline{X}_3\underline{x}_1,\underline{y}_3)] .$$

Now take $X \equiv \lambda \underline{x}_1\underline{y}_2\underline{y}_3 . 2$, $\underline{X}_2 \equiv \underline{X}_3 \equiv \lambda \underline{x}_1 . \underline{x}_1$, and by cases $\underline{Y}_1 0 \underline{x}_1\underline{y}_2\underline{y}_3 = \underline{y}_2$, $\underline{Y}_2(Sz)\underline{x}_1\underline{y}_2\underline{y}_3 = \underline{y}_3$.
In all other cases X can be taken to be identical 1, and the treatment is then as for the Dialectica interpretation.

It is also easy to verify that for theories $\mathbf{H}$ with decidable prime formulae, $\mathbf{H} \vdash A^{D} \Leftrightarrow \mathbf{H} \vdash A^{\wedge}$.

An advantage of the present variant is, that in the construction of the interpretations for the provable formulae, the prime formulae do not play a special rôle. In principle, any set of formulae closed under propositional operations and bounded universal quantification, and containing the prime formulae, can replace the prime formulae in the definition of the translation, and the soundness theorem might be established in the same manner.

3.5.18. Shoenfield's variant. By combining the translation ' into the negative fragment (§ 1.10) with the Dialectica interpretation (or: equivalently, restricting one's attention to the negative fragment) one obtains a Dialectica interpretation for classical arithmetic and systems $(\underset{\sim}{HA}^{\omega})^c$ etc.

Shoenfield described a variant, assigning to any formula A in the $\neg, \vee, \forall$ fragment a formula $A^S \equiv \forall \underline{x} \exists \underline{y} A_S(\underline{x}, \underline{y})$, A_S quantifier free, as follows (Shoenfield 1967, § 8.3):

Let $A^S \equiv \forall \underline{x} \exists \underline{y} A_S(\underline{x}, \underline{y})$, $B^S \equiv \forall \underline{u} \exists \underline{v} B_S(\underline{u}, \underline{v})$.

(i) $A^S \equiv A \equiv A_S$ for A quantifier free

(ii) $(\neg B)^S \equiv \forall \underline{Y} \exists \underline{x} \neg B_S(\underline{x}, \underline{Y}\underline{x})$

(iii) $(A \vee B)^S \equiv \forall \underline{x}\underline{u} \exists \underline{y}\underline{v}(A_S \vee B_S)$

(iv) $(\forall w B)^S \equiv \forall w \underline{x} \exists \underline{y} B_S(w, \underline{x}, \underline{y})$.

As compared to the translation ^{D}, for a formula in the fragment considered, $A^S \leftrightarrow A$ requires, besides intuitionistic logic, only QF-AC together with M' (M' as defined in 3.5.7), namely in case (ii). (Cf. lemma 3.5.13, and theorem 3.5.10(i).)

3.5.19 - 3.5.21. Extending the Dialectica interpretation to stronger systems.

3.5.19. For the system $\underset{\sim}{EL}$, with the schema DNS

DNS $\quad \forall x^{\sigma} \neg\neg A \rightarrow \neg\neg \forall x^{\sigma} A$

added, which is equivalent to full classical analysis (cf. 1.10.9) a Dialectica interpretation by means of bar-recursive functionals (closed under the defining schemata of $\underset{\sim}{N}$-$\underset{\sim}{HA}^{\omega}$ and BR_{σ} for all σ) was first given in Spector 1962. Spector gave the interpretation in an extensional quantifier-free system of bar-recursive functionals, containing the rule EXT-R' (1.6.12).

A more elegant presentation is given in Howard 1968, also for a system with a strong rule of extensionality (contrary to what is said there, the treatment given there does not automatically apply to an intensional version). Howard obtains other results besides: The general schema of bar induction

$$BI_{\sigma} \begin{cases} [\forall x^{(0)\sigma} \exists y P(\bar{x}y) \; \& \\ \forall \xi \eta (P\xi \rightarrow P(\xi * \eta)) \; \& \\ \forall \xi (P\xi \rightarrow Q\xi) \; \& \\ \forall \xi (\forall y^{\sigma} Q(\xi * \langle y^{\sigma} \rangle) \rightarrow Q\xi)] \rightarrow Q\langle\, \rangle \end{cases}$$

(ξ, η variables for finite sequences of objects of type σ, $*$ denoting concatenation, $\bar{\ }$ course-of-values etc.), is interpreted in $\underset{\sim}{BR}_\sigma$ ($\equiv \underset{\sim}{WE}-\underset{\sim}{HA}^\omega + BR_\sigma$) then it is established that the ' - translation (translation into the negative fragment, 1.10.2) of

$$\forall n \forall y^\sigma \exists z^\sigma A(n,y,z) \rightarrow \exists u^{(o)\sigma} \forall n A(n, un, u(Sn))$$

can be derived in $\underset{\sim}{WE}-\underset{\sim}{HA}^\omega + BI_\sigma + EXT-R'$ for suitable σ ($\underset{\sim}{WE}-\underset{\sim}{HA}^\omega$ also with types for finite sequences).

Let Rule - BI_ν be the rule corresponding to BI_ν, and Rule - BR_ν the rule corresponding to definition by bar recursion:
Let $T_1 \in ((0)\sigma)0$, T_2, T_3 be given closed terms, let x be a variable for sequences of type σ, u a variable of type σ. Then there is a constant t such that

$$\text{Rule} - BR_\sigma \quad \begin{cases} T_1[x] < lth(x) \rightarrow tx = T_1 x \\ T_1[x] \geq lth(x) \rightarrow tx = T_2(\lambda u.t(x * \langle u \rangle))x. \end{cases}$$

Howard shows

a) BI_σ is derivable in $\underset{\sim}{WE}-\underset{\sim}{HA}^\omega + \text{Rule} - BI_{((o)\sigma)\sigma} + AC + EXT - R'$;
b) BR_σ is derived from Rule - BR_ν for suitable ν;
c) A direct functional interpretation of systems with Rule - BI_σ in quantifier-free systems with Rule - BR_σ is given.

Detailed expositions are also given in Girard 1972 and Luckhardt 1970, 1971, 1973. Luckhardt pursues in detail the approach of Spector 1962. His principal aim is to give a consistency proof; in this context, he handles the axiom of extensionality by a process of relativization to extensional functionals (similar to the interpretation of $\underset{\sim}{HAS}$ in $\underset{\sim}{HAS}$ without EXT, cf. 1.9.6).

It is also possible to give a Dialectica interpretation for systems $\underset{\sim}{H}$, in qf - $\underset{\sim}{H}$, where $\underset{\sim}{H}$ is a theory similar to $\underset{\sim}{IDB}^\omega$, i.e. a theory based on arithmetic together with a generalized inductive definition (classically: a definition of a complete Π^1_1 - set) extended to all finite types, and qf - $\underset{\sim}{H}$ a corresponding quantifier-free fragment.

In Howard 1972 the Dialectica interpretation for a theory V^* of this type is given in detail. See also Zucker's discussion in § 6.8.

3.5.20. Church's thesis and bar recursion.

In Kreisel 1971 (pp. 126 - 127) a proof of Gödel is cited showing that the bar-recursive functionals satisfy the Dialectica interpretation of the negation of Church's thesis. (A more roundabout proof is already implicit in Spector 1962; see also Kreisel 1971, page 126). $\exists y \forall z[Txxy \vee \neg Txxz]$ holds in $\underset{\sim}{HA}^c$, hence by § 1.10, in $\underset{\sim}{HA}$

$$\neg\neg \exists y \forall z[Txxy \vee \neg Txxz]$$

is provable, since T is decidable. But then, applying $\forall x \neg\neg A \rightarrow \neg\neg \forall x A$ (i.e. DNS) with $A \equiv \exists y \forall z[Txxy \vee \neg Txxz]$ we find

$$\neg\neg \forall x \exists y \forall z[Txxy \vee \neg Txxz] ,$$

but this contradicts the non-recursiveness of $\exists y Txxy$. Since, by Spector 1962, DNS is Dialectica interpretable by bar-recursive functionals, it following that the interpretation of the negation of Church's thesis is satisfied by the bar-recursive functionals.

A proof is also found in Luckhardt 1970, 1973 (chapter IX).

3.5.21. In Girard 1971, an extension of the Dialectica interpretation is described for HAS, using Girard's system of functionals described in 1.9.27. In Girard 1972, this definition is still further extended to cover the theory of finite types (of species).

For technical details we refer the reader to Girard's papers: here we restrict ourselves to a heuristic motivation for the extension of the Dialectica interpretation. In Girard's presentation no sequences of adjacent like quantifiers are used; adjacent quantifiers are automatically contracted. Also for equations between terms:

$$[t_1 = t_2]^D \equiv_{def} \exists x^o \forall y^o[t_1 = t_2] ,$$

x, y not occurring free in t_1, t_2. Therefore, for any formula A of HAS, A^D is of the form $\exists x \forall y A_D$, A_D quantifier free.

As regards species variables, let us for simplicity restrict attention to unary species variables. To each species variable Z we suppose to be assigned in a one-to-one manner two variables α_Z, β_Z (α, β for short) and a variable x_Z of type $(\alpha)(\beta)(0)0$. Then

$$[Zu^o]^D \equiv_{def} \exists x^\alpha \forall y^\beta (x_Z xyu^o = 0) .$$

The interpretation for conjunctions, disjunctions and implications is adapted in an obvious way (contracting adjacent like quantifiers), e.g. if $A^D \equiv \exists x \forall y A_D$, $B^D \equiv \exists u \forall v B_D$, then

$$[A \& B]^D \equiv \exists x' \forall y'[A_D(D'x',D'y') \& B_D(D''x',D''y')] ,$$

etc. etc.

Now we shall discuss the choice of definition for $[\exists Z A(Z)]^D$ and $[\forall Z A(Z)]^D$. Let

$$[A(Z)]^D \equiv \exists x^\sigma \forall y^\tau A_D(x, y, x_Z, \underline{u}) .$$

The types σ, τ will in general contain α, β; $x_Z \in (\alpha)(\beta)(0)0$. Intuitively, $[\exists Z A(Z)]^D$ corresponds to

$$\exists\alpha\,\exists\beta\,\exists x_Z\,\exists x\,\forall y\,A_D(x, y, x_Z, \underline{\underline{u}}) .$$

(This is of course not a formula of our language, since we did not have quantifications over types.) Equivalently, for a variable $x' \in \sigma \times (\alpha)(\beta)(0)0 \equiv \rho[\alpha, \beta]$

$$\exists\alpha\,\exists\beta\,\exists x'\,\forall y\,A_D(D'x', y, D''x', \underline{\underline{u}})$$

which is replaced by

(1) $$\exists\alpha\,\exists\beta\,\exists x'\,\forall Y\,A_D(D'x', \mathrm{EXT}(\mathrm{EXT}\,Y\alpha)\beta)x', D''x', \underline{\underline{u}})$$

with $Y \in \forall\alpha\,\forall\beta\,((\rho[\alpha,\beta])\tau)$; $\mathrm{EXT}(\mathrm{EXT}\,Y\alpha)\beta)$ if of type $(\rho[\alpha,\beta])\tau$. Now to give $\alpha, \beta, x' \in \rho[\alpha,\beta]$ amounts to specification of an $X' \in \exists\alpha\,\exists\beta\,\rho[\alpha,\beta]$.

Using the rules for Girard's constants, we see that

(2) $$I_{\exists\alpha\exists\beta\rho[\alpha,\beta],\alpha}(I_{\exists\beta\rho[\alpha,\beta],\beta}x') \in \exists\alpha\,\exists\beta\,\rho[\alpha,\beta]$$

and

$$\mathrm{ST}\,\alpha(\mathrm{ST}\,\beta v) \in (\exists\alpha\,\exists\beta\,\rho[\alpha,\beta])0 \quad \text{for} \quad v \in (\rho[\alpha,\beta])0$$

and

$$\mathrm{ST}\,\alpha(\mathrm{ST}\,\beta\;\lambda x't[x'])(I_{\exists\alpha\exists\beta\rho[\alpha,\beta],\alpha}(I_{\exists\beta\rho[\alpha,\beta],\beta}x')) = t[x'] .$$

If we use $\lambda w.A_D$ as an abbreviation for (the characteristic function of) A_D as function of w, we find that (1) implies

(3) $$\exists X\,\forall Y\,\{\mathrm{ST}\,\alpha(\mathrm{ST}\beta(\lambda w.A_D(D'w, \mathrm{EXT}(\mathrm{EXT}\,Y\alpha)\beta)w, D''w, \underline{\underline{u}}))(X) = 0\} .$$

(To see that (1) implies (3), substitute the left hand side of (2) for X in (3).) We take (3) as $[\exists Z\,A(Z)]^D$.

Similarly, $[\forall Z\,A(Z)]^D$ corresponds heuristically to

$$\forall\alpha\,\forall\beta\,\forall x_Z\,\exists x\,\forall y\,A_D(x, y, x_Z, \underline{\underline{u}})$$

or equivalently

$$\forall\alpha\,\forall\beta\,\exists X\,\forall y'\,A_D(X(D'y'), D''y', D'y', \underline{\underline{u}})$$

where $x_Z \in (\alpha)(\beta)(0)0$, $y' \in (\alpha)(\beta)(0)0\times\tau$, $X \in ((\alpha)(\beta)(0)0)\sigma \equiv \rho[\alpha,\beta]$.

If we wish to give an X, uniformly in α, β we can do so by finding an $X' \in \forall\alpha\,\forall\beta\,\rho[\alpha,\beta]$. Then

$$\mathrm{EXT}\,\alpha(\mathrm{EXT}\,\beta x') \in \rho[\alpha,\beta] ,$$

and therefore we can put

(4) $$\exists X\,\forall\alpha\,\forall\beta\,\forall y'\,A_D(\mathrm{EXT}\,\alpha(\mathrm{EXT}\,\beta x)(D'y'), D''y', D'y', \underline{\underline{u}})$$

which in turn may be replaced by the stronger

(5) $$\exists X\,\forall Y\{\mathrm{ST}\,\alpha(\mathrm{ST}\,\beta(\lambda w.\,A_D(\mathrm{EXT}\alpha(\mathrm{EXT}\beta\,X)(D'w), D''w', D'w, \underline{\underline{u}}))(Y) = 0\} .$$

(To see that (5) implies (4), substitute $I_{\exists\alpha\exists\beta\,\rho[\alpha,\beta]\,\alpha}(I_{\exists\beta\,\rho[\alpha,\beta],\,\beta}y')$ for Y in (5).) We take (5) as our definition of $[\forall Z\, A(Z)]^{D}$.

§ 6. Applications : consistency and conservative extension results.

3.6.1. Contents of the section.

The present section utilizes $\underset{\sim}{r}$-, $\underset{\approx}{mr}$-realizability and the Dialectica interpretation to obtain conservative extension results. For the case of arithmetic, the results are given in extenso ; for analysis, only some of the more typical ones have been lifted out, since the treatment is very similar to the case for arithmetic, so it may be left to the reader to formulate further applications when he needs them. As one of the more interesting applications we point to the consistency of AC for HRO , and of AC! for HEO .

3.6.2. Theorem (summary). $\underset{\sim}{I}$-$\underset{\sim}{HA}^{\omega}$ + CT, $\underset{\sim}{E}$-$\underset{\sim}{HA}^{\omega}$ + CT, $\underset{\sim}{HRO}$ are conservative extensions of $\underset{\sim}{HA}$.

Proof. Immediate, by the fact that HRO can be shown (in $\underset{\sim}{HA}$) to be a model for $\underset{\sim}{I}$-$\underset{\sim}{HA}^{\omega}$, $\underset{\sim}{HRO}$, and HEO for $\underset{\sim}{E}$-$\underset{\sim}{HA}^{\omega}$.

3.6.3. Definition. $\Gamma_o(\Gamma_1,\Gamma_2)$ is the class of formulae (in the language of $\underset{\sim}{I}$-$\underset{\sim}{HA}^{\omega}$ or $\underset{\sim}{E}$-$\underset{\sim}{HA}^{\omega}$ or $\underset{\sim}{HA}$) such that in all their subformulae of the form $A \to B$ A is an almost negative formula (negative formula, purely universal formula) preceded by existential quantifiers.

Let Γ_n, Γ_{an}, Γ_{pr} stand for the classes of negative, almost negative and prenex formulae respectively.

Remarks. (i). $\Gamma_2 \subseteq \Gamma_1 \subseteq \Gamma_o$.

(ii). Since intuitionistically $(\exists x_1 \ldots x_n A \to B) \leftrightarrow \forall x_1 \ldots x_n (A \to B)$, we might have omitted (modulo logical equivalence) "preceded by existential quantifiers" in the definition of Γ_o, Γ_1, Γ_2.

(iii). Alternatively, we might have defined Γ_o, Γ_1, Γ_2 inductively :

(a) Prime formulae are in Γ_o (Γ_1, Γ_2)

(b) $A, B \in \Gamma_o \Rightarrow A \& B,\ A \vee B,\ \forall x A,\ \exists x A \in \Gamma_o$ (Γ_1, Γ_2)

(c) If A is almost negative (negative, purely universal), $B \in \Gamma_o$ (Γ_1, Γ_2) then $\exists x_1 \ldots x_n A \to B \in \Gamma_o$ (Γ_1, Γ_2).

3.6.4. Convention. We use the expression "$\underset{\sim}{H}$ is conservative over $\underset{\sim}{H}' \cap \Gamma$", where Γ is a class of formulae, as an abbreviation for : "$\underset{\sim}{H}$ is an extension of $\underset{\sim}{H}'$ which is conservative w.r.t. formulae of Γ".

3.6.5. Lemma.

(i) $A \in \Gamma_o \Rightarrow \underset{\sim}{HA} \vdash \exists x (x \underset{\sim}{r} A) \to A$

(ii) $A \in \Gamma_1 \Rightarrow \underset{\sim}{N}$-$\underset{\sim}{HA}^{\omega} \vdash \exists \underline{x} (\underline{x} \underset{\approx}{mr} A) \to A$

(iii) $A \in \Gamma_2 \Rightarrow \underset{\sim}{H} \vdash \exists \underline{x} \forall \underline{y} A_D(\underline{x}, \underline{y}) \to A$

where $\mathbf{H} \equiv \mathbf{HA}^{\omega}$, $\mathbf{N}\text{-}\mathbf{HA}^{\omega}$, $\mathbf{I}\text{-}\mathbf{HA}^{\omega}$, $\mathbf{HRO}^{-}$.

Proof. (i). We use the inductive definition of Γ_o (3.6.3, Remark (iii)) and establish (i) by induction over the definition of Γ_o.

(a) For prime formulae (i) is immediate.

(b) Assume (induction hypothesis) $\mathbf{HA} \vdash \exists x(x \,\underline{r}\, Ay) \rightarrow Ay$. Let $\exists x(x \,\underline{r}\, \forall y Ay)$, then $\exists x \forall y(!\{x\}(y) \;\&\; \{x\}(y) \,\underline{r}\, Ay)$, hence $\forall y \exists x(x \,\underline{r}\, Ay)$, therefore $\forall y Ay$, by our induction hypothesis. Similarly for $A \& B$.

Utilizing the same induction hypothesis, and assuming $\exists x(x \,\underline{r}\, \exists y Ay)$, we have $\exists x(j_2 x \,\underline{r}\, A(j_1 x))$, so $\exists x \exists y(x \,\underline{r}\, Ay)$, and thus by the induction hypothesis $\exists y Ay$. Similarly for $A \vee B$.

(c) Assume A to be almost negative; then $\exists x_1 \ldots x_n A \leftrightarrow \exists y(y \,\underline{r}\, \exists x_1 \ldots x_n A)$, since for almost negative B (by 3.2.11) $\exists x Bx \leftrightarrow \exists x \exists y(y \,\underline{r}\, Bx) \leftrightarrow$ $\leftrightarrow \exists z(j_2 z \,\underline{r}\, B(j_1 z)) \leftrightarrow \exists y(y \,\underline{r}\, \exists x Bx)$. Therefore, if $\exists y(y \,\underline{r}\, (\exists x A \rightarrow B))$, it follows that $\exists x A \rightarrow \exists z(z \,\underline{r}\, B)$. Using $\mathbf{HA} \vdash \exists z(z \,\underline{r}\, B) \rightarrow B$ as our induction hypothesis, we find $\exists x A \rightarrow B$.

(ii). The proof runs parallel to the proof of (i), now using $x \,\underline{mr}\, A \equiv A$ for A negative (3.4.4 (i)).

(iii). The proof is again more or less similar to the proof of (i), (ii):

(a) For prime formulae (iii) is obvious.

(b) Assume $\exists \underline{x} \forall \underline{y}\, A_D(\underline{x}, \underline{y}, z) \rightarrow Az$ (induction hypothesis). Suppose $(\forall z Az)^D$ be given, i.e. $\exists \underline{X} \forall z \underline{y}\, A_D(\underline{X}z, \underline{y}, z)$; then $\forall z \exists \underline{x} \forall \underline{y}\, A_D(\underline{x}, \underline{y}, z)$ hence $\forall z Az$, etc. etc.

(c) Assume $\exists \underline{x} \forall \underline{y}\, B_D(\underline{x}, \underline{y}) \rightarrow B$ (induction hypothesis).

Note that $(\exists \underline{z} A\underline{z})^D \equiv \exists \underline{z} A\underline{z}$ for purely universal A (3.5.2). Now let $(\exists \underline{z} A\underline{z} \rightarrow B)^D$, where $A\underline{z} \equiv \forall \underline{w} C(\underline{z}, \underline{w})$. Then $\exists \underline{X}\underline{W} \forall \underline{y}\underline{z}(C(\underline{z}, \underline{W}\underline{y}\underline{z}) \rightarrow B_D(\underline{X}\underline{z}, \underline{y}))$. Assume also $A\underline{z}$, i.e. $\forall \underline{w} C(\underline{z}, \underline{w})$; then $\forall \underline{y} C(\underline{z}, \underline{W}\underline{y}\underline{z})$, hence $\forall \underline{y} B_D(\underline{X}\underline{z}, \underline{y})$, i.e. $\exists \underline{x} \forall \underline{y} B_D(\underline{x}, \underline{y})$ and therefore by the induction hypothesis B; hence, eliminating our third hypothesis, $\exists \underline{z} A\underline{z} \rightarrow B$, and eliminating our second hypothesis, (iii) follows for $\exists \underline{z} A\underline{z} \rightarrow B$ as A.

Remark. By the proof (ii) and (iii) now hold for $\mathbf{I}\text{-}\mathbf{HA}^{\omega}$, $\mathbf{E}\text{-}\mathbf{HA}^{\omega}$, $\mathbf{HRO}^{-}$, $\mathbf{WE}\text{-}\mathbf{HA}^{\omega}$, $\mathbf{N}\text{-}\mathbf{HA}_p^{\omega}$ etc.

3.6.6. Theorem (Conservative extensions). AC is as in 3.5.8.

(i) $(\mathbf{HA} + \mathrm{ECT}_o) \cap \Gamma_o = \mathbf{HA} \cap \Gamma_o$

$(\mathbf{HA} + \mathrm{M} + \mathrm{ECT}_o) \cap \Gamma_o = (\mathbf{HA} + \mathrm{M}) \cap \Gamma_o$

(ii) $(\mathbf{H} + \mathrm{IP}^{\omega} + \mathrm{AC}) \cap \Gamma_1 = \mathbf{H} \cap \Gamma_1$, for $\mathbf{H} \equiv \mathbf{N}\text{-}\mathbf{HA}^{\omega}$, $\mathbf{HA}^{\omega}$

$\mathbf{I}\text{-}\mathbf{HA}^{\omega}$, $\mathbf{E}\text{-}\mathbf{HA}^{\omega}$, $\mathbf{HRO}^{-}$

(iii) $\mathbf{I}\text{-}\mathbf{HA}^{\omega} + \mathrm{IP}^{\omega} + \mathrm{AC} + \mathrm{CT}$ is conservative over $\mathbf{HA} \cap \Gamma_1$

(iv) $\mathbf{H} + \mathrm{IP}_o^{\omega} + \mathrm{AC} + \mathrm{M}^{\omega}$ is conservative over $\mathbf{H} \cap \Gamma_2$,

for $\mathbf{H} \equiv \mathbf{I}\text{-}\mathbf{HA}^{\omega}$, $\mathbf{WE}\text{-}\mathbf{HA}^{\omega}$, $\mathbf{HRO}^{-}$, $\mathbf{HA}^{\omega}$, hence conservative over $\mathbf{HA} \cap \Gamma_2$.

(v) $\underset{\sim}{I}$ - $\underset{\sim}{HA}^{\omega}$ + IP_0^{ω} + AC + M^{ω} + CT is conservative over $\underset{\sim}{HA} \cap \Gamma_2$.

Proof. (i). By the characterization theorem for realizability (3.2.18),

$$\underset{\sim}{HA} + ECT_0 \vdash A \Rightarrow \underset{\sim}{HA} \vdash \exists x(x \;\underline{r}\; A).$$

By lemma 3.6.5 (i)

$$\underset{\sim}{HA} \vdash \exists x(x \;\underline{r}\; A) \to A \quad \text{for } A \in \Gamma_0.$$

Therefore $(\underset{\sim}{HA} + ECT_0) \cap \Gamma_0 = \underset{\sim}{HA} \cap \Gamma_0$.

The second assertion is proved in the same manner (using 3.2.22 (i)).

(ii). Similarly, using the characterization theorem for modified realizability (3.4.8), and lemma 3.6.5 (ii).

(iii). Combine the characterization theorem for modified realizability with the fact that CT is HRO - $\underline{mr}$ - realizable in $\underset{\sim}{HA}$ (3.4.12 (ii)), and that HRO is a model for $\underset{\sim}{I}$ - $\underset{\sim}{HA}^{\omega}$, and use lemma 3.6.5 (ii).

(iv). Similar to (i), (ii) and (iii), using the characterization theorem for the Dialectica interpretation (3.5.10) and lemma 3.6.5 (iii).

(v). Using the reasoning of (iv) together with the fact that CT is HRO - Dialectica interpretable in $\underset{\sim}{HA}$.

Remark. The statements of the theorem in an obvious manner extend to extensions of the systems mentioned ($\underset{\sim}{HA}$, $\underset{\sim}{HA}$ + M in (i), $\underset{\sim}{H}$ in (ii), $\underset{\sim}{I}$ - $\underset{\sim}{HA}^{\omega}$ + IP^{ω} + AC + CT in (iii), $\underset{\sim}{H}$ in (iv), $\underset{\sim}{I}$ - $\underset{\sim}{HA}^{\omega}$ + IP_0^{ω} + AC + M^{ω} + CT in (v)) for which the appropriate soundness theorem is provable (3.2.19, 3.4.8, 3.5.11).

3.6.7. Corollaries (for $\underset{\sim}{HA}$).

(i) $\underset{\sim}{HA}$ + ECT_0, $\underset{\sim}{HA}$ + IP + CT_0 are conservative over $\underset{\sim}{HA} \cap \Gamma_n$.

(ii) $\underset{\sim}{HA}$ + ECT_0, $\underset{\sim}{HA}$ + IP + CT_0, $\underset{\sim}{HA}$ + IP_0 + M + CT_0 are conservative over $\underset{\sim}{HA} \cap \Gamma_{pr}$.

(iii) $\underset{\sim}{HA}$ + ECT_0 + M is conservative over $(\underset{\sim}{HA} + M) \cap \Gamma_{an}$, $(\underset{\sim}{HA} + M) \cap \Gamma_{pr}$.

3.6.8. Corollary of 3.6.5 (i) or 3.6.6 (i). If KLS_1 (= KLS relative to V ≡ set of all total recursive functions, see 2.6.15) is not derivable in HA, then KLS_1 is also not derivable in $\underset{\sim}{HA}$ + ECT_0.

Proof. KLS_1 is expressible as a formula of Γ_0.

3.6.9. Theorem (repeated from 3.5.14). $\underset{\sim}{E}$ - $\underset{\sim}{HA}^{\omega}$ + M^{ω} + IP_0^{ω} + AC is conservative over $\underset{\sim}{HA} \cap \Gamma_n$.

3.6.10 - 3.6.16. Axioms of choice for HRO, HEO.

3.6.10. Axioms of choice. For definiteness, we list the principal forms which concern us here:

$$AC_{\sigma,\tau} \qquad \forall x^{\sigma} \exists y^{\tau} A(x,y) \to \exists z^{(\sigma)\tau} \forall x^{\sigma} A(x,zx)$$

$AC_{\sigma,\tau}!$ is similar to $AC_{\sigma,\tau}$ but with $\exists!y^\tau$ instead of $\exists y^\tau$.

$AC = \bigcup\{AC_{\sigma,\tau} \mid \sigma,\tau \in \underline{T}\}$; similarly $AC!$.

$QF\text{-}AC_{\sigma,\tau_1,\dots,\tau_m}$ $\forall x^\sigma \exists y_1^{\tau_1} \dots \exists y_m^{\tau_m} A(x,y_1,\dots,y_m) \rightarrow$

$$\rightarrow \exists z_1^{(\sigma)\tau_1} \dots \exists z_m^{(\sigma)\tau_m} \forall x^\sigma A(x,z_1x,\dots,z_mx)$$

(A quantifier free).

$QF\text{-}AC = \bigcup\{QF\text{-}AC_{\sigma,\tau_1,\dots,\tau_m} \mid \sigma,\tau_1,\dots,\tau_m \in \underline{T},\ m \text{ arbitrary}\}$.

3.6.11. Lemma. $x \in V_\sigma$, $x \in W_\sigma$, $I_\sigma(x,y)$ (defined in 2.4.8, 2.4.11) are equivalent to almost negative formulae.

Proof. By induction over the type structure. For $\sigma = 0$ the truth of the assertion is immediate. Assume the lemma to hold for σ, τ. Then $x \in V_{(\sigma)\tau}$, $x \in W_{(\sigma)\tau}$, $I_{(\sigma)\tau}(x,y)$ may be re-written as

$$\forall y \in V_\sigma[\exists u Txyu \ \&\ \forall v(Txyv \rightarrow Uv \in V_\tau)],$$

$$\forall y \in W_\sigma[\exists u Txyu \ \&\ \forall v(Txyv \rightarrow Uv \in W_\tau)] \ \&$$
$$\&\ \forall yzuw[I_\sigma(y,z) \ \&\ Txyv \ \&\ Txzw \rightarrow I_\tau(Uv,Uw)],$$

$$x \in W_{(\sigma)\tau} \ \&\ y \in W_{(\sigma)\tau} \ \&\ \forall z \in W_\sigma \forall uw[Txzv \ \&\ Tyzw \rightarrow I_\tau(Uv,Uw)],$$

respectively. With the induction hypothesis for σ, τ the lemma for $(\sigma)\tau$ follows.

3.6.12. Theorem. $QF\text{-}AC_{\sigma,0}$ for formulae of $\underline{I}\text{-}\underline{HA}^\omega$ holds for HRO (provably in $\underline{HA}$).

Proof. For simplicity we restrict ourselves to an instance of $QF\text{-}AC_{\sigma,0}$ without parameters. Assume $[\forall x^\sigma \exists y^0 A(x,y)]_{HRO}$, A a quantifier-free formula of $\underline{I}\text{-}\underline{HA}^\omega$. ($[B]_{HRO}$ is the interpretation of B in HRO, defined in the obvious way.) Let $A^*(u,v) \equiv [A(x,y)]_{HRO}$, where u, v are the variables corresponding to x, y under the interpretation, then

$$\forall u \in V_\sigma \exists v A^*(u,v) \qquad \text{(by our hypothesis)}$$

and

$$\underline{HA} \vdash u \in V_\sigma \rightarrow A^*(u,v) \vee \neg A^*(u,v)$$

or equivalently

$$\underline{HA} \vdash \forall uv(u \in V_\sigma \rightarrow \exists z[(z=0 \rightarrow A^*(u,v)) \ \&\ (z \neq 0 \rightarrow \neg A^*(u,v))]).$$

Now apply the preceding lemma, and the closure of $\underline{HA}$ under ECR_0 (to be proved in 3.7.2 (i)), then for a certain numeral $\bar{n}$

$$\underline{HA} \vdash u \in V_\sigma \rightarrow \exists w(T(\bar{n},j(u,v),w) \ \&\ (Uw = 0 \leftrightarrow A^*(u,v))),$$

therefore

$$\underline{HA} \vdash \forall u \in V_\sigma \exists vw(T(\bar{n},j(u,v),w) \ \&\ Uw = 0).$$

Let $z = \Lambda u.j_1 \min_w [T(\bar{n}, j(u, j_1 w), j_2 w) \ \&\ U(j_2 w) = 0]$. Then

$$\forall u \in V_\sigma (!\{z\}(u) \ \&\ A^*(u, \{z\}(u))).$$

So $z \in V_{(\sigma)o}$, and thus $\exists z \in V_{(\sigma)o} \forall u \in V_\sigma A^*(u, \{z\}(u))$.

Remark. It is open whether QF-AC holds generally for HRO. By the results 2.6.20, 2.6.21, QF-AC holds (classically) for HEO.

3.6.13. Theorem.

(i) $\underset{\sim}{I}$-$\underset{\sim}{HA}^\omega$ + ECT_o + AC, and $\underset{\sim}{I}$-$\underset{\sim}{HA}^\omega$ + AC + CT are conservative over $\underset{\sim}{HA} \cap \Gamma_o$ (ECT_o w.r.t. the language of $\underset{\sim}{HA}$ only).

(ii) $[AC]_{HRO}$ (i.e. the class of $[A]_{HRO}$, $A \in \mathcal{L}(\underset{\sim}{I}$-$\underset{\sim}{HA}^\omega)$) is derivable in $\underset{\sim}{HA}$ + ECT_o; so it is consistent (relative to $\underset{\sim}{HA}$) to assume AC for HRO.

Proof. The first assertion is a consequence of the second statement; for interpreting $\underset{\sim}{I}$-$\underset{\sim}{HA}^\omega$ in HRO in $\underset{\sim}{HA}$ + ECT_o gives, because $[AC]_{HRO}$ is provable in $\underset{\sim}{HA}$ + ECT_o by assertion (ii), that $\underset{\sim}{I}$-$\underset{\sim}{HA}^\omega$ + ECT_o + AC is conservative over $\underset{\sim}{HA}$ + ECT_o, hence over $\underset{\sim}{HA} \cap \Gamma_o$ (3.6.6 (i)). (Similarly for $\underset{\sim}{I}$-$\underset{\sim}{HA}^\omega$ + AC + CT, since CT holds in HRO.)
So it remains to prove (ii). We shall derive $[AC]_{HRO}$ in $\underset{\sim}{HA}$ + ECT_o. Assume

$$\forall x \in V_\sigma \exists y \in V_\tau A(x,y).$$

$x \in V_\sigma$ is equivalent to an almost negative formula by 3.6.11, therefore with ECT_o

$$\exists u \forall x \in V_\sigma \exists v [Tuxv \ \&\ A(x,Uv) \ \&\ Uv \in V_\tau],$$

hence

$$\exists u \in V_{(\sigma)\tau} \forall x \in V_\sigma \exists v (Tuxv \ \&\ A(x,Uv)).$$

3.6.14. Theorem.

(i) $[AC!]_{HEO}$ (the set of interpretations in HEO of instances of AC! in the language of $\underset{\sim}{E}$-$\underset{\sim}{HA}^\omega$) is derivable in $\underset{\sim}{HA}$ + ECT_o, so it is consistent (relative to $\underset{\sim}{HA}$) to assume AC! for HEO.

(ii) $\underset{\sim}{E}$-$\underset{\sim}{HA}^\omega$ + AC! + ECT_o, and $\underset{\sim}{E}$-$\underset{\sim}{HA}^\omega$ + AC! + CT are conservative over $\underset{\sim}{HA} \cap \Gamma_o$ (ECT_o with respect to the language of $\underset{\sim}{HA}$ only).

Proof. (ii) is obtained from (i) in the same manner as in 3.6.13.
So it remains to prove (i). Assume

(1) $\forall x \in W_\sigma \exists y \in W_\tau A(x,y)$

(2) $I_\sigma(x,x') \ \&\ A(x,y) \ \&\ A(x',y') \to I_\tau(y,y')$.

$x \in W_\sigma$ is equivalent to an almost negative formula (3.6.11), so by ECT_o

there is a u such that

$$\forall x \in W_\sigma\ \exists v(Tuxv \,\&\, A(x,Uv) \,\&\, Uv \in W_\tau) .$$

Assume $I_\sigma(x,x')$, then there are v, v' such that

$$Tuxv,\quad Tux'v',\quad A(x,Uv),\quad A(x',Uv'),\quad Uv \in W_\tau,\quad Uv' \in W_\tau .$$

By (2) $I_\tau(Uv,Uv')$. Therefore $u \in W_{(\sigma)\tau}$, and thus

$$\exists u \in W_{(\sigma)\tau}\ \forall x \in W_\sigma\ \exists v(Tuxv \,\&\, A(x,Uv)) .$$

3.6.15. Remarks.

(i) $AC_{1,0}$ is false for HEO; for consider

$$\forall x^1 \exists y^0 \forall z^0 \exists v^0(Tyzv \,\&\, Uv = xz)$$

which holds for HEO; $AC_{1,0}$ would imply $\exists w^2 \forall x^1 z^0 \exists v^0(T(wx,z,v) \,\&\, Uv = xz)$, which obviously does not hold for HEO (cf. remark (i) under 2.4.11).

(ii) By the characterization theorem for modified realizability, for the system $\underset{\sim}{HRO}^-$ (3.4.8), it is consistent relative to $\underset{\sim}{HA}$ to assume that all objects of finite type are hereditarily recursive operations and satisfy AC, IP^ω (in the language of $\underset{\sim}{I}$ - $\underset{\sim}{HA}^\omega$). Note that this does not imply the consistency of AC for HRO, only the consistency of AC for some submodel of HRO.

(iii) It is open whether $[AC]_{HRO}$ implies ECT_0; it is obvious that it implies CT_0. CT does not imply $AC_{0,0}$: CTM satisfies CT, but not $AC_{0,0}$ (2.5.3 (iv)).

3.6.16. Theorem. The following schema

$$RDC_\sigma\quad \forall x^\sigma[Ax \to \exists y^\sigma(B(x,y) \,\&\, Ay^\sigma)] \to \forall x^\sigma[Ax^\sigma \to \exists z^{(0)\sigma}(z0 = x^\sigma \,\& \\ \&\, \forall y^0\ B(zy,\, z(y+1)))]$$

can also be shown to be consistent for HRO relative to $\underset{\sim}{HA}$.

Proof. The argument is more complicated than for AC. For simplicity we consider an instance of RDC_σ with parameters. Assume

(1) $\qquad [\forall x^\sigma[Ax \to \exists y^\sigma Bxy]]_{HRO}$,

i.e.

(2) $\qquad \forall x \in V^\sigma[A^*x \to \exists y \in V^\sigma(B^*(x,y) \,\&\, A^*y)$

where A^*x, $B^*(x,y)$ are obtained by interpreting A, B in HRO. In $\underset{\sim}{HA} + ECT_0$, A^*x is equivalent to $\exists z A'(x,z)$, A' almost negative (3.2.18 (i)), hence (2) is equivalent to

$$\forall x \forall z(x \in V_\sigma \,\&\, A'(x,z) \to \exists yz'(y \in V^\sigma \,\&\, B^*(x,y) \,\&\, A'(y,z')) .$$

By ECT_0

$$\exists uv \forall xz[x \in V_\sigma \ \& \ A'(x,z) \to \ !\{u\}(x,z) \ \& \ !\{v\}(x,z) \ \& \\ \& \ (\{u\}(x,z) \in V^\sigma \ \& \ B^*(x, \{u\}(x,z)) \ \& \ A'(\{u\}(x,z), \{v\}(x,z)))] .$$

Now define $\varphi(x,z,y)$, $\varphi'(x,z,y)$ by simultaneous recursion on y as follows (recursion theorem):

$$\varphi(x,z,0) \simeq x, \quad \varphi'(x,z,0) \simeq z$$
$$\varphi(x,z,Sy) \simeq \{u\}(\varphi(x,z,y), \varphi'(x,z,y))$$
$$\varphi'(x,z,Sy) \simeq \{v\}(\varphi(x,z,y), \varphi'(x,z,y)) .$$

By induction over y one then proves $\varphi(x,z,y)$, $\varphi'(x,z,y)$ to be defined for all y whenever $x \in V_\sigma \ \& \ A'(x,z)$, and moreover

$$\forall x \forall z[x \in V_\sigma \ \& \ A'(x,z) \to \forall y(\varphi(x,z,y) \in V_\sigma \ \& \ B(\varphi(x,z,y), \varphi(x,z,y+1))]$$

which implies

$$\forall x[A^*x \to \exists z \in V_{(0)\sigma}[\{z\}(0) = x \ \& \ \forall y B^*(\{z\}(y), \{z\}(y+1))]] .$$

Q. e. d.

3.6.17 - 3.6.20. <u>Extensions to analysis</u>.

3.6.17. <u>Exploiting the analogy</u>. We shall not attempt to give an exhaustive list of results for analysis which can be obtained analogously to the results in 3.6.2 - 3.6.16 for arithmetic, but restrict ourselves to some of the more interesting and striking applications. Once the analogy is clear, and the proof ideas of the preceding subsections are understood, the reader will have no difficulty in formulating and proving other applications to analysis for himself.

Roughly, among the formal systems $\underset{\sim}{EL}$ takes the place of $\underset{\sim}{HA}$ in the analogy; $\underset{\sim}{E} - \underset{\sim}{HA}^\omega$ remains the same; there is no direct analogue to $\underset{\wedge}{I} - \underset{\sim}{HA}^\omega$, we usually have to be satisfied in proving results for $\underset{\sim}{N} - \underset{\sim}{HA}^\omega$ instead.

As regards the models, ECF, ICF correspond to HEO, HRO respectively. The analogues of ECT_0, CT_0, CT are respectively GC (the "generalized continuity" in 3.3.9), C - N (1.9.19) and the assertion that type-two objects are continuous:

$$\forall z^2 \forall x^1 \exists y^0 \forall u^1(\bar{x}^1 y^0 = \bar{u}^1 y^0 \to z^2 x^1 = z^2 u^1) .$$

3.6.18. <u>Theorem</u> (Examples).

(i) $(\underset{\sim}{EL} + GC) \cap \Gamma_0 = \underset{\sim}{EL} \cap \Gamma_0$

$(\underset{\sim}{EL} + GC + M^1) \cap \Gamma_0 = (\underset{\sim}{EL} + M^1) \cap \Gamma_0$. Here M^1 is

M^1 $\quad \forall \alpha[A \vee \neg A] \ \& \ \neg\neg \exists \alpha A \to \exists \alpha A$.

(ii) $\underset{\sim}{H} + IP^1 + AC_{0,1}$ is conservative over $\underset{\sim}{H} \cap \Gamma_1$, for $\underset{\sim}{H} \equiv \underset{\sim}{EL}$, $\underset{\sim}{EL} + FAN$, $\underset{\sim}{EL} + WC\text{-}N$, where

IP^1 $\quad (\neg A \to \exists \alpha B) \to \exists \alpha(\neg A \to B)$.

(iii) $\underset{\sim}{EL} + IP_0^1 + AC_{0,1} + M$ is conservative over $\underset{\sim}{EL} \cap \Gamma_2$ where

IP_0^1 $\quad \forall\alpha[A \vee \neg A] \;\&\; [\forall\alpha A \rightarrow \exists\beta B] \rightarrow \exists\beta[\forall\alpha A \rightarrow B]$.

Proof. (i). Completely similar to 3.6.6 (i), using $\underset{=}{r}^1$ - realizability.
(ii). First note that $\underset{\sim}{HA}^\omega$ is conservative over $\underset{\sim}{EL}$ (by use of the model ECF or ICF in $\underset{\sim}{EL}$). Now let $F \in \Gamma_1 \cap \mathcal{L}(\underset{\sim}{EL})$. Then

$$\begin{aligned} \underset{\sim}{EL} + IP + AC_{0,1} \vdash F \;&\Rightarrow \underset{\sim}{HA}^\omega + IP^\omega + AC \vdash F \\ &\Rightarrow \underset{\sim}{HA}^\omega \vdash F \quad \text{(by 3.6.6 (ii))} \\ &\Rightarrow \underset{\sim}{EL} \vdash F . \end{aligned}$$

Similarly for $\underset{\sim}{EL} + FAN$, using 3.4.16 and taking $\underset{\sim}{E}-\underset{\sim}{HA}^\omega + FAN + MUC$ in the previous argument, instead of $\underset{\sim}{HA}^\omega$.
For $\underset{\sim}{EL} + WC - N$ we must use 3.4.17 (ii).
(iii). Let $F \in \Gamma_2$. Then (3.6.6 (iv))

$$\begin{aligned} \underset{\sim}{EL} + IP_0^1 + AC_{0,1} + M^1 \vdash F \;&\Rightarrow \underset{\sim}{HA}^\omega + IP_0^\omega + AC + M^\omega \vdash F \\ &\Rightarrow \underset{\sim}{HA}^\omega \vdash F \\ &\Rightarrow \underset{\sim}{EL} \vdash F , \end{aligned}$$

etc. etc.

3.6.19. Theorem.
(i) $QF-AC_{\sigma,0}$ holds for ICF (provable in $\underset{\sim}{EL}$).
(ii) $[AC]_{ICF}$ (AC w.r.t. $\mathcal{L}(\underset{\sim}{N}-\underset{\sim}{HA}^\omega)$) is derivable in $\underset{\sim}{EL} + GC$.
(iii) $[AC!]_{ECF}$ (AC! w.r.t. $\mathcal{L}(\underset{\sim}{N}-\underset{\sim}{HA}^\omega)$) is derivable in $\underset{\sim}{EL} + GC$.
Proof. Entirely similar to the proof of 3.6.12, 3.6.13 (ii), 3.6.14 (i).

3.6.20. Remarks.
(i). A consistency proof for $[AC]_{ICF}$ is also contained in Vesley 1972 (implicitly).
(ii). The results of Kleene 1965 are special cases of results of the type of 3.6.18 (i).

§ 7. Applications : proof-theoretic closure properties.

3.7.1. Contents of the section.

This section is devoted to establishing proof-theoretic closure properties by means of functional and realizability interpretations. We briefly discuss the principal closure properties considered. The first is

ED' $\quad \vdash \exists x A x \Rightarrow \exists t(\vdash At) \quad (\exists x Ax$ closed).

If x is a numerical variable, and the system is complete w.r.t. closed equations between terms of type 0 (which implies that each closed term of type 0 can be shown to be equal to a numeral), then ED' implies ED :

ED $\quad \vdash \exists x^0 A x \Rightarrow \exists n(\vdash A\bar{n}) \quad (A\bar{n}$ closed).

If the system considered contains enough arithmetic to prove $\exists x((x=0 \to A) \,\&\, (x \neq 0 \to B)) \leftrightarrow A \vee B$, ED in turn implies DP :

DP $\quad \vdash A \vee B \Rightarrow \vdash A$ or $\vdash B \quad (A \vee B$ closed).

Among the rules corresponding to the schema IP we consider

IPR' $\quad \vdash A \to \exists x^0 Bx \Rightarrow \vdash \exists x^0(A \to Bx) \quad$ (x not free in A , A negative)

and

IPR'$^{\omega}$ $\quad \vdash A \to \exists x^{\sigma} Bx \Rightarrow \vdash \exists x^{\sigma}(A \to Bx) \quad$ (x^{σ} not free in A , A negative).

Earlier, in § 3.1, we showed how to establish for certain systems the stronger rules :

IPR $\quad \vdash \neg A \to \exists x^0 Bx \Rightarrow \vdash \exists x^0(\neg A \to Bx) \quad$ (x not free in A)

and

IPR$^{\omega}$ $\quad \vdash \neg A \to \exists x^{\sigma} Bx \Rightarrow \vdash \exists x^{\sigma}(\neg A \to Bx) \quad$ (x not free in A)

For IPR , see also 4.2.13, 4.4.4.

Corresponding to AC we consider the rule

ACR $\quad \vdash \forall x^{\sigma} \exists y^{\tau} A(x,y) \Rightarrow \vdash \exists z^{(\sigma)\tau} \forall x^{\sigma} A(x,zx)$.

Of course it is possible to formulate other rules corresponding to DC and RDC .

Corresponding to CT, CT_0, ECT_0 we consider the rules :

CR $\quad \forall t^1 \exists n(\vdash \forall y^0(\{\bar{n}\}(y) \simeq t^1 y)$

CR_0 $\quad \vdash \forall x \exists y A(x,y) \Rightarrow \vdash \exists z \forall x \exists v(Tzxv \,\&\, A(x,Uv))$

and

ECR$_0$ $\quad \vdash \forall x(Ax \to \exists yBxy) \Rightarrow \vdash \exists u \forall x(Ax \to \exists v(Tuxv \,\&\, B(x,Uv)))$
(A almost negative).

In the case of applications to analysis, CR$_0$, ECR$_0$ are replaced by certain continuity rules; see 3.7.9.

Subsections 2 - 8 are devoted to applications to arithmetic and related systems. In 3.7.9 we discuss, briefly, analogous applications to systems stronger than arithmetic.

3.7.2. Theorem.

(i) Assume $\underset{\sim}{HA} + \Gamma$ to be conservative over $\underset{\sim}{HA}$ with respect to closed Σ_1^0 - formulae, and let Γ be a set of closed formulae such that

(1) $$F \in \Gamma \Rightarrow \exists n(\underset{\sim}{HA} + \Gamma \vdash \bar{n}\ \underset{\approx}{q}\ F).$$

Then $\underset{\sim}{HA} + \Gamma$ satisfies ED, DP, CR$_0$, ECR$_0$.

(ii) Let $\underset{\sim}{H}$ be $\underset{\sim}{HA}^\omega$, $\underset{\sim}{N} - \underset{\sim}{HA}^\omega$, $\underset{\sim}{I} - \underset{\sim}{HA}^\omega$, $\underset{\sim}{WE} - \underset{\sim}{HA}^\omega$, $\underset{\sim}{E} - \underset{\sim}{HA}^\omega$ or $\underset{\sim}{HRO}^-$, and let Γ be a collection of closed formulae such that

(2) $$F \in \Gamma \Rightarrow \exists s(\underset{\sim}{H} + \Gamma \vdash s\ \underset{\approx}{mq}\ F).$$

Then $\underset{\sim}{H} + \Gamma$ satisfies ED', ACR, IPR'$^\omega$.

Proof. Ad (i). By 3.2.4 (iii)

$$\underset{\sim}{HA} + \Gamma \vdash A \Rightarrow \exists n(\underset{\sim}{HA} + \Gamma \vdash \bar{n}\ \underset{\approx}{q}\ A).$$

Assume $\underset{\sim}{HA} + \Gamma \vdash \exists xAx$, $\exists xAx$ closed. Then for some $\bar{n}$, $\underset{\sim}{HA} + \Gamma \vdash \bar{n}\ \underset{\approx}{q}\ \exists xAx$, i.e. $\underset{\sim}{HA} + \Gamma \vdash j_2\bar{n}\ \underset{\approx}{q}\ A(j_1\bar{n}) \,\&\, A(j_1\bar{n})$.
Since we can find a numeral $\bar{n}'$ such that $\underset{\sim}{HA} \vdash j_1\bar{n} = \bar{n}'$, it follows that $\underset{\sim}{HA} + \Gamma \vdash A\bar{n}'$.

ECR$_0$ is obtained as follows. Assume

$$\underset{\sim}{HA} + \Gamma \vdash \forall x(Ax \to \exists xBxy) \qquad (A \text{ almost negative})$$

and assume that A, B do not contain free variables besides x, y. (The more general case with additional parameters can be reduced to this case by contraction of variables.) Then for some numeral $\bar{m}$:

$$\underset{\sim}{HA} + \Gamma \vdash \bar{m}\ \underset{\approx}{q}\ \forall x(Ax \to \exists yB(x,y)),$$

so

$$\underset{\sim}{HA} + \Gamma \vdash \forall x \forall u[u\ \underset{\approx}{q}\ Ax \,\&\, Ax \to\ !t \,\&\, (j_2t\ \underset{\approx}{q}\ B(x,j_1t)) \,\&\, B(x,j_1t)]$$

where $t \equiv \{\{\bar{m}\}(x)\}(u)$. By 3.2.11 (i) $u\ \underset{\approx}{q}\ Ax \to Ax$ for A almost negative, and also $u\ \underset{\approx}{q}\ Ax \to\ !\psi_A(x) \,\&\, \psi_A(x)\ \underset{\approx}{q}\ Ax$, hence

$$\underset{\sim}{HA} + \Gamma \vdash \forall x[Ax \to\ !t' \,\&\, B(x,j_1t')]$$

where $t' \equiv \{\{\bar{m}\}(x)\}(\psi_A(x))$. We can find a numeral $\bar{n}$ such that $\underset{\sim}{HA} \vdash \bar{n} = \Lambda x.j_1t'$ hence

$$\underset{\sim}{HA} + \Gamma \vdash \exists u \forall x[Ax \to !\{u\}(x) \,\&\, B(x, \{u\}(x))) .$$

The proof of (ii) is similar to the proof of (i), $\underset{\approx}{mq}$ - realizability taking the place of $\underset{\approx}{q}$ - realizability; since in this case we deal with arbitrary terms, a condition such as conservativeness w.r.t. closed Σ^0_1 - formulae is not necessary.

3.7.3. Remark. For closure under ECR_0, ACR the conditions (1), (2) on Γ in 3.7.2 respectively may be weakened to

$$F \in \Gamma \Rightarrow \underset{\sim}{HA} + \Gamma \vdash \exists y(y \,\underset{\approx}{q}\, F)$$

and

$$F \in \Gamma \Rightarrow \underset{\sim}{H} + \Gamma \vdash \exists \underline{\underline{x}}(\underline{\underline{x}} \,\underset{\approx}{mr}\, F) .$$

3.7.4. Theorem (Corollaries of 3.7.2).

(i) $\underset{\sim}{HA} + \Gamma$, $\Gamma \subseteq ECT_0$ or $M \subseteq \Gamma \subseteq ECT_0 \cup M$ satisfies ED, DP, CR_0 and ECR_0.

(ii) $\underset{\sim}{H} + \Gamma$, $\Gamma \subseteq IP^\omega \cup AC$ where $\underset{\sim}{H}$ may be $\underset{\sim}{HA}^\omega$, $\underset{\sim}{I} - \underset{\sim}{HA}^\omega$, $\underset{\sim}{WE} - \underset{\sim}{HA}^\omega$, $\underset{\sim}{E} - \underset{\sim}{HA}^\omega$, $\underset{\sim}{N} - \underset{\sim}{HA}^\omega$ or $\underset{\sim}{HRO}^-$, satisfies ED', ACR, IPR'^ω.

(iii) $\underset{\sim}{HA}$ is closed under IPR'.

Proof. (i). We use 3.7.2 (i), 3.2.4, 3.2.15, 3.2.21.
The conservativeness w.r.t. closed Σ^0_1 - formulae follows, for the addition of instances of ECT_0, from 3.6.6 (i) and for the addition of instances of M by 3.6.6 (iv).
(ii). Similarly, from 3.7.2 (ii) and 3.4.7.
(iii) From 3.7.2 (ii), using $\underset{\sim}{HRO}^-$.

3.7.5. Theorem. For $\underset{\sim}{H} \equiv \underset{\sim}{HA}^\omega$, $\underset{\sim}{I} - \underset{\sim}{HA}^\omega$, $\underset{\sim}{WE} - \underset{\sim}{HA}^\omega$, $\underset{\sim}{I} - \underset{\sim}{HA}^\omega + IE_0$, $\underset{\sim}{HRO}^-$: $\underset{\sim}{H} + AC + IP^\omega_0 + M^\omega$ satisfies ED', ACR; for $\underset{\sim}{H} \equiv \underset{\sim}{HA}^\omega$, $\underset{\sim}{I} - \underset{\sim}{HA}^\omega + IE_0$, $\underset{\sim}{WE} - \underset{\sim}{HA}^\omega$: $\underset{\sim}{H} + AC + IP^\omega_0 + M^\omega$ satisfies DP.

Proof. For $\underset{\sim}{H}' \equiv \underset{\sim}{H} + AC + IP^\omega_0 + M^\omega$ we have (3.5.10)

(1) $\qquad \underset{\sim}{H}' \vdash A \Rightarrow \exists \underline{\underline{t}}(\underset{\sim}{H}' \vdash \forall \underline{\underline{y}}\, A_D(\underline{\underline{t}}, \underline{\underline{y}}))$

and also

(2) $\qquad \underset{\sim}{H}' \vdash A \leftrightarrow \exists \underline{\underline{x}}\, \forall \underline{\underline{y}}\, A_D(\underline{\underline{x}}, \underline{\underline{y}})$.

The soundness theorem for $\underset{\sim}{I} - \underset{\sim}{HA}^\omega$ automatically extends to $\underset{\sim}{I} - \underset{\sim}{HA}^\omega + IE_0$. Therefore if

$$\underset{\sim}{H}' \vdash \exists x^\sigma Ax^\sigma \qquad\qquad (\exists x^\sigma Ax^\sigma \text{ closed})$$

then

$$\underset{\sim}{H}' \vdash \forall \underline{\underline{y}}\, A_D(t^\sigma, \underline{\underline{t}}, \underline{\underline{y}})$$

for suitable closed t^σ, $\underline{\underline{t}}$. Since $Ax^\sigma \leftrightarrow \exists \underline{\underline{z}}\, \forall \underline{\underline{y}}\, A_D(x^\sigma, \underline{\underline{z}}, \underline{\underline{y}})$, it follows that

$$\underset{\sim}{H}' \vdash A(t^\sigma) .$$

If any closed type 0 term can be shown to be equal to a numeral, as is possible in $\underset{\sim}{HA}^{\omega}$, $\underset{\sim}{I}-\underset{\sim}{HA}^{\omega}+IE_0$, $\underset{\sim}{WE}-\underset{\sim}{HA}^{\omega}$, then DP holds. For then if $\underset{\sim}{H}' \vdash A \vee B$, $A \vee B$ closed, we find $\bar{n}$, $\underline{\underline{s}}$, $\underline{\underline{t}}$ such that

$$\underset{\sim}{H}' \vdash \forall \underline{\underline{y}}\underline{\underline{v}}[(\bar{n}=0 \rightarrow A_D(\underline{\underline{t}},\underline{\underline{y}})) \;\&\; (\bar{n}\neq 0 \rightarrow B_D(\underline{\underline{s}},\underline{\underline{v}}))],$$

which in view of (1), (2) implies

$$\underset{\sim}{H}' \vdash A \quad \text{or} \quad \underset{\sim}{H}' \vdash B.$$

3.7.6. Lemma. Let $\underset{\sim}{N}-\underset{\sim}{HA}^{\omega} \subseteq \underset{\sim}{H}$, $\mathcal{L}[\underset{\sim}{H}] = \mathcal{L}[\underset{\sim}{N}-\underset{\sim}{HA}^{\omega}]$.
Then $\underset{\sim}{H}$ is closed under CR.

Proof. Consider any closed term t^1 of $\underset{\sim}{N}-\underset{\sim}{HA}^{\omega}$. Then there is a numeral $\bar{n} = [t^1]$ such that $([t^1], 1)$ represents t^1 in HRO; by 2.4.13

$$[t^1 y]_{HRO} \simeq \{[t']\}(y),$$

and by 2.4.14

$$\underset{\sim}{N}-\underset{\sim}{HA}^{\omega} \vdash t^1 y \simeq \{[t^1]\}(y).$$

3.7.7. Theorem.

$\underset{\sim}{H}+\Gamma$, $\Gamma \subseteq IP^{\omega} \cup AC$ for $\underset{\sim}{H} \equiv \underset{\sim}{N}-\underset{\sim}{HA}^{\omega}$, $\underset{\sim}{E}-\underset{\sim}{HA}^{\omega}$, is closed under CR_0, and under the rule

ECR_1 $\quad \vdash \forall x^0(Ax \rightarrow \exists y^0 B(x,y)) \Rightarrow \vdash \exists u^0 \forall x^0 \exists v^0[Tuxv \;\&\; (Ax \rightarrow B(x,Uv))]$
(A negative).

Proof. (a) By the preceding lemma, $\underset{\sim}{H}+\Gamma$ is closed under CR; furthermore, by 3.7.4 (ii), $\underset{\sim}{H}+\Gamma$ is closed under ACR and ED' tells us that ED holds for $\underset{\sim}{H}+\Gamma$, hence closure under CR_0 follows.
(b) $\underset{\sim}{H}+\Gamma$ is closed under IPR', and CR_0, and as a result ECR_1 follows.

3.7.8. Corollary. $\underset{\sim}{HA}$ is closed under ECR_1.
Proof. Immediate, since $\underset{\sim}{N}-\underset{\sim}{HA}^{\omega}$ is a conservative extension of $\underset{\sim}{HA}$.

3.7.9. Extension of the preceding methods to systems of analysis.

The use of $\underset{\approx}{r}^1$- and $\underset{\approx}{q}^1$-realizability to obtain properties such as DP, ED, and the following forms of Church's rule

$$\vdash \exists\alpha A\alpha \Rightarrow \exists n(\vdash \exists\alpha(\forall x \exists y\, T\bar{n}xy \;\&\; Uy = \alpha x) \;\&\; A\alpha))$$

($\exists\alpha A\alpha$ closed) and

$$\vdash \forall x \exists y\, A(x,y) \Rightarrow \exists n(\vdash \forall x \exists y\, T\bar{n}xy \;\&\; A(x,Uy))$$

for certain extensions of $\underset{\sim}{EL}$ are discussed in detail in Kleene 1969, 5.9. The extensions of $\underset{\sim}{EL}$ considered are the same as in 3.3.4.

In addition to the applications given by Kleene, we can also use $\underset{\approx}{q}^1$-

realizability to show closure under the following generalized continuity rule GCR:

GCR $\vdash \forall\alpha[A\alpha \rightarrow \exists\beta B(\alpha,\beta)] \Rightarrow \vdash \exists\gamma\forall\alpha[A\alpha \rightarrow !\gamma|\alpha \,\&\, B(\alpha,\gamma|\alpha)]$,

where A is almost negative. For example, we can obtain closure under GCR for the extensions of EL mentioned in 3.3.4.

A variant of mq - realizability specialized for ICF (by analogy this might be called mq^1 - realizability), or in other words, a q - variant of Kleene's special realizability, has not been investigated in the literature, but may be expected to yield similar results for extensions of EL which include IP (not only w.r.t. numerical, but also w.r.t. function variables).

Vesley 1972, after formalization, may be expected to yield results similar to those listed in 3.7.5 for extensions of EL containing forms of IP_0 and M, since Vesley 1972 amounts to carrying through a Dialectica translation within the language of analysis by interpreting the objects of finite type in ICF.

For systems in the language of HA, but proof-theoretically stronger, such as HA + TI(<) for example, the results obtained by q - realizability readily extend (in view of 3.2.23), so in 3.7.4 (i), $TI(<) \subseteq \Gamma \subseteq ECT_0 \cup TI(<)$ or $TI(<) \cup M \subseteq \Gamma \subseteq ECT_0 \cup TI(<) \cup M$ is also permissible, etc. etc.

§ 8. Markov's schema and Markov's rule.

3.8.1. Contents of the section.

In this section some miscellaneous information concerning Markov's rule and Markov's schema is brought together. M, M_{PR}, M^c_{PR} and the corresponding rules MR, MR_{PR}, MR^c_{PR} have already been introduced in 1.11.5.

In theories of finite type, such as $\underset{\sim}{N}$ - $\underset{\sim}{HA}^{\omega}$, the generalization M^{ω} (defined in 3.5.10) and the corresponding rule

$$MR^{\omega} \qquad \vdash \forall x^{\sigma}(A \vee \neg A),\ \vdash \neg\neg \exists x^{\sigma} A \Rightarrow \vdash \exists x^{\sigma} A$$

play a rôle.

It is useful to note that M, M^{ω} can be formulated as rules (a remark due to H. Luckhardt) : M, M^{ω} are (relative to $\underset{\sim}{HA}$, $\underset{\sim}{N}$ - $\underset{\sim}{HA}^{\omega}$ respectively) equivalent to

$$\forall x^{o}(A \vee \neg A) \to \neg\neg \exists x^{o} A \Rightarrow \forall x^{o}(A \vee \neg A) \to \exists x^{o} A$$

and

$$\forall x^{\sigma}(A \vee \neg A) \to \neg\neg \exists x^{\sigma} A \Rightarrow \forall x^{\sigma}(A \vee \neg A) \to \exists x^{\sigma} A$$

respectively. This can be seen by taking for Ax in the rule $\neg\neg \exists y By \to Bx$. Then certainly

$$\vdash \forall x(Ax \vee \neg Ax) \to \neg\neg \exists x Ax ,$$

since $\vdash \neg\neg \exists x Ax$. On the other hand, $\forall x(A \vee \neg A) \to \exists x A$ is $\forall x(\neg\neg \exists y By \to Bx) \vee \neg(\neg\neg \exists y By \to Bx)) \to \exists x(\neg\neg \exists y By \to Bx)$, which implies $\forall x[(\neg\neg \exists y By \to Bx) \vee \neg(\neg\neg \exists y By \to Bx)]\ \&\ \neg\neg \exists y By \to \exists x Bx$, hence $\forall y(By \vee \neg By)\ \&\ \neg\neg \exists y By \to \exists y By$.

As regards M_{PR} , we note that this is equivalent to the specific single instance

$$[\neg\neg \exists z Txyz \to \exists z Txyz] .$$

If we make use of the fact that in $\underset{\sim}{HA}$ it is provable that $\{x \mid \exists z T(j_1 x, j_2 x, z)\}$ and $\{x \mid \exists z T(x,x,z)\}$ are both complete Σ^0_1 - sets, and therefore recursively isomorphic (cf. Rogers 1967, § 7.2), then we see that M_{PR} is also equivalent to

$$[\neg\neg \exists z Txxz \to \exists z Txxz] .$$

We interpret $\neg M_{PR}$ as the negation of the universal closure of this formula.

3.8.2. Theorem. There exists a formula $\neg A$, A almost negative, such that for no negative formula B $\underset{\sim}{HA} \vdash \neg A \leftrightarrow B$. In other words, negations of almost negative formulae are not always equivalent to negative formulae.

Proof. Take $A \equiv \forall x[\neg\neg \exists y\, Txxy \to \exists y\, Txxy]$. Assume $\mathbf{HA} \vdash \neg A \leftrightarrow B$, B negative, then a fortiori : $\mathbf{HA}^c \vdash \neg\neg A \leftrightarrow \neg B$. Since $\mathbf{HA}^c \vdash A$, it follows that $\mathbf{HA}^c \vdash \neg B$, and therefore, since B is negative, $\mathbf{HA} \vdash \neg B$. Thus

(1) $\qquad \mathbf{HA} \vdash \neg\neg A$.

But $\neg \exists z^1 (z \,\mathbf{mr}\, A)$ relative to HRO, hence $\langle\,\rangle \,\mathbf{mr}\, \neg A$ (by 3.4.4 (ii)), so $\mathbf{HA} + \neg A$ is consistent, contradicting (1).

3.8.3. Theorem (Kreisel). The system $\mathbf{HA} + \neg M_{PR}$ is consistent, and closed under MR_{PR}.

Proof. We establish first

(1) $\qquad (\mathbf{HA} + \neg M_{PR} + CT_0) \cap \Gamma_1 = \mathbf{HA} \cap \Gamma_1$.

The system $\mathbf{HRO}^- + IP^\omega + AC$ is conservative over $\mathbf{HRO}^-$ w.r.t. Γ_1 (cf. 3.6.6 (ii)). CT_0 holds in $\mathbf{HRO}^- + AC$, and $\neg M_{PR}$ holds in $\mathbf{HA} + IP + CT_0$ (3.2.27), therefore $\mathbf{HA} + \neg M_{PR} + CT_0 \subseteq \mathbf{HRO}^- + IP^\omega + AC$. On the other hand, $\mathbf{HRO}^-$ is conservative over $\mathbf{HA}$, and thus (1) follows.

Now assume $\mathbf{HA} + \neg M_{PR} \vdash \neg\neg \exists y Axy$, A primitive recursive. Then by (1) $\mathbf{HA} \vdash \neg\neg \exists y Axy$, and by the closure of $\mathbf{HA}$ under MP_{PR}, $\mathbf{HA} \vdash \exists x Axy$, so $\mathbf{HA} + \neg M_{PR} \vdash \exists y Axy$.

3.8.4. Theorem. Let $\mathbf{HA} \subseteq \mathbf{H} \subseteq \mathbf{HA}^c$. Then the following two assertions are equivalent :

(i) In $\mathbf{H}$ every almost negative formula is provably equivalent to a negative formula

(ii) M_{PR} is derivable in $\mathbf{H}$.

Proof. (ii) $\Rightarrow$ (i) is immediate. Assume (i) and consider

$$F \equiv [\neg\neg \exists z Txyz \to \exists z Txyz]$$

and assume $\mathbf{H} \vdash F \leftrightarrow B$, $B \in \Gamma_n$. Then $\mathbf{HA}^c \vdash F \leftrightarrow B$, and so $\mathbf{HA}^c \vdash B$; therefore $\mathbf{HA} \vdash B$, and thus $\mathbf{H} \vdash F$. This proves (ii).

3.8.5. Theorem.

(i) The following rule

$$\mathbf{H} \vdash Ax^\sigma \vee \neg Ax^\sigma,\ \mathbf{H} \vdash \neg\neg \exists x^\sigma Ax \to \neg \exists x^\sigma A \ \Rightarrow$$
$$\Rightarrow \mathbf{H} \vdash (\forall x^\sigma \neg Ax) \vee (\exists x^\sigma Ax)$$

holds for $\mathbf{H} \equiv \mathbf{H}' + \Gamma$, $\mathbf{H}' \equiv \mathbf{HA}^\omega$, $\mathbf{N}\text{-}\mathbf{HA}^\omega$, $\mathbf{I}\text{-}\mathbf{HA}^\omega$, $\mathbf{E}\text{-}\mathbf{HA}^\omega$, $\mathbf{HRO}^-$, $\mathbf{WE}\text{-}\mathbf{HA}^\omega$, $\Gamma \subseteq IP^\omega + AC$. The rule also holds for $\mathbf{HA}$ (with $\sigma = 0$).

(ii) $\mathbf{HA} \nvdash M_{PR}$.

(iii) MR^ω holds in $\mathbf{I}\text{-}\mathbf{HA}^\omega$, $\mathbf{HRO}^-$, $\mathbf{WE}\text{-}\mathbf{HA}^\omega$, $\mathbf{HA}^\omega$.

MR holds in $\mathbf{HA}$.

Proof. (i) Assume A to contain x^σ, y^τ free, then

$$\underset{\sim}{H} \vdash \forall x^{\sigma} y^{\tau} \exists z^{0}[(z=0 \to A) \,\&\, (z \neq 0 \to A)] .$$

Then by ACR, ED' for $\underset{\sim}{H}$ (3.7.4 (ii)), if $\underset{\sim}{H} \vdash A \vee \neg A$ we find a closed term $t_A \in (\sigma)(\tau)0$ such that

$$\underset{\sim}{H} \vdash \forall x^{\sigma} y^{\tau}[(t_A xy=0 \to A) \,\&\, (t_A xy \neq 0 \to \neg A)] .$$

Then $\underset{\sim}{H} \vdash A \leftrightarrow t_A xy = 0$, and assuming $\underset{\sim}{H} \vdash \neg\neg \exists x^{\sigma} A \to \exists x^{\sigma} A$, we have

$$\underset{\sim}{H} \vdash \neg\neg \exists x^{\sigma}(t_A xy = 0) \to \exists x^{\sigma}(t_A xy = 0) .$$

By closure under IPR'^{ω}; ACR (or more directly, soundness for mq - realizability; cf. 3.7.4 (ii))

$\underset{\sim}{H} \vdash \exists z^{(\tau)\sigma} \forall y^{\tau}(\neg\neg \exists x^{\sigma}(t_A xy = 0) \to t_A(zy)y = 0)$; since

$$\underset{\sim}{H} \vdash t_A(zy)y = 0 \vee t_A(zy)y \neq 0 ,$$

we have

$$\underset{\sim}{H} \vdash \forall x^{\sigma}(\neg t_A xy = 0) \vee \exists x^{\sigma}(t_A xy = 0) ,$$

hence

$$\vdash \forall x^{\sigma} \neg A(x,y) \vee \exists x^{\sigma} A(x,y) .$$

To obtain the result for HA, use the result just obtained together with the fact that $I\text{-}HA^{\omega}$ is a conservative extension of HA.

(ii) See 1.11.5.

(iii) Let $\vdash \forall x^{\sigma}(A \vee \neg A)$, $\vdash \neg\neg \exists x^{\sigma} A$. Assume A to contain x^{σ}, y^{τ} free only. As under (i) we may construct a term t_A such that

$$\vdash t_A xy = 0 \leftrightarrow A(x,y) .$$

Then $\vdash \neg\neg \exists x^{\sigma} Axy$ is equivalent to $\vdash \forall y^{\tau} \neg\neg \exists x^{\sigma}(t_A xy = 0)$. The Dialectica translation of this is (equivalent to) $\exists z^{(\sigma)\tau} \forall y^{\tau}\, t_A(zy)y = 0$ (by 3.5.2, Note (ii)). Therefore we can find a closed term $s \in (\tau)\sigma$ such that $\vdash \forall y^{\tau}(t_A(sy)y = 0)$ (3.5.4), and therefore $\vdash \exists x^{\sigma} A(x,y)$.

For arithmetic we must again use the fact that $I\text{-}HA^{\omega}$ is a conservative extension of HA.

3.8.6. Corollary. In HA and HA^{c} the same gödelnumbers can be proved to represent total recursive functions (Kreisel 1958).

Proof. Assume $HA^{c} \vdash \exists u T(\bar{z},x,u)$; then $HA \vdash \neg \forall u \neg T(\bar{z},x,u)$, hence $HA \vdash \exists u\, T(\bar{z},x,u)$.

3.8.7. M, MR for systems stronger than arithmetic.

The underivability of M in systems of intuitionistic analysis with function variables was shown in Kleene and Vesley 1965, § 10, 11, by means of "special realizability" (i.e. mr - realizability relative to ICF).

Otherwise there are few systematic investigations in the literature ; see e.g. Scarpellini 1971, chapter IX. Another line of approach for the extension of results on M, MR and its variants is given by the present section: scan around for systems to which the methods of proof of the present section apply.

For theories with species variables, see for closure under MR_{PR}, chapter IV, 4.2.14 , and for MR, Girard 1972, VI, §2.

Further remarks on M, MR and its variants are found in this volume, in 1.11.5.

§ 9. Applications of $\underset{\approx}{p}$ - realizability.

3.9.1. Contents of the section.

In this section we consider a notion of $\underset{\approx}{p}$ - realizability (due to Beeson 1972), which does not quite fall under the schema of $\underset{\approx}{r}_p$ - realizability of § 3.2, but the treatment of which is rather similar (cf. remarks in 3.2.4). Our principal interest in $\underset{\approx}{p}$ - realizability is in an application, found in Beeson 1972; namely $\mathbf{HA} \nvdash$ KLS. The result has a certain intrinsic interest, settling the old problem about the intuitionistic provability of KLS_1 See the definition in 3.9.9), but our principal reason for including it here is that it provides an example of an application of realizability techniques up till now not accessible by other methods.

At the end of the section we have summarized (without proofs) some other related results from Beeson 1972. (Postscript: a new simplified proof is in Beeson B; see 3.9.13.)

Beeson 1972 discusses matters in somewhat ~~greater generality~~, but in our presentation we have restricted ourselves to the essentials needed for the application.

3.9.2. Definition of $\underset{\approx}{p}$ - realizability.

A definition is obtained by replacing the clauses (iii), (iv), (v) in 3.2.2 by

(iii) $x \underset{\approx}{p} A(x_1,\ldots) \vee B(x_1,\ldots) \equiv_{def} [j_1x{=}0 \rightarrow j_2x\underset{\approx}{p}A(x_1,\ldots) \,\&\, Pr(\ulcorner \overline{j_2x}\underset{\approx}{p}A(\bar{x}_1,\ldots)\urcorner)$
$\&\ (j_1x{\neq}0 \rightarrow j_2x\underset{\approx}{p}B(x_1,\ldots) \,\&\, Pr(\ulcorner \overline{j_2x}\underset{\approx}{p}B(\bar{x}_1,\ldots)\urcorner)$

(iv) $x \underset{\approx}{p} A(x_1,\ldots) \rightarrow B(x_1,\ldots) \equiv_{def} \forall u(u \underset{\approx}{p} A(x_1,\ldots) \,\&\, Pr(\ulcorner \bar{u} \underset{\approx}{p} A(\bar{x}_1,\ldots)\urcorner) \rightarrow$
$\rightarrow\ !\{x\}(u) \,\&\, \{x\}(u) \underset{\approx}{p} B(x_1,\ldots))$

(v) $x \underset{\approx}{p} \exists y A(y,x_1,\ldots) \equiv_{def} j_2x \underset{\approx}{p} A(j_1x,x_1,\ldots) \,\&\, Pr(\ulcorner \overline{j_2x} \underset{\approx}{p} A(\overline{j_1x},x_1,\ldots)\urcorner)$

and replacing in the other clauses $\underset{\approx}{r}_p$ by $\underset{\approx}{p}$.

Here $Pr(\ulcorner A\urcorner)$ abbreviates $\exists y\, Proof_{HA}(y,\ulcorner A\urcorner)$, and in $A(x_1,\ldots)$ " $x_1,\ldots$ " is supposed to be a list containing all the variables free in A.

Note that the definition is more general * than $\underset{\approx}{r}_p$ - realizability as used in § 3.2, since to the predicate $P(A)$ there now corresponds a predicate $P(A,y)$, y a free variable not in A.

Let us introduce as an abbreviation:

$$t \underset{\approx}{pp} A(x_1,\ldots) \equiv_{def} t \underset{\approx}{p} A(x_1,\ldots) \,\&\, Pr(\ulcorner \bar{t} \underset{\approx}{p} A(\bar{x}_1,\ldots)\urcorner),$$

where if $t \equiv t[x_1,\ldots,x_n]$, $\bar{t}$ indicates the numeral giving the value of

-----

* Perhaps $P(A(x_1,\ldots)) \equiv Pr(\ulcorner \exists(y \underset{\approx}{p} A(\bar{x}_1,\ldots))\urcorner)$ might also have worked, but it certainly would have complicated the treatment.

$t[\bar{x}_1,\ldots,\bar{x}_n]$.

We collect in a lemma some properties needed in the proof of the soundness theorem.

3.9.3. Lemma. $\vdash$ indicates provability in **HA**.

(i) $\vdash \mathrm{Pr}(\ulcorner A \urcorner) \rightarrow \mathrm{Pr}(\ulcorner \mathrm{Pr}\ulcorner A \urcorner \urcorner)$ (A closed)

(ii) $\vdash \mathrm{Pr}(\ulcorner A \rightarrow B \urcorner) \ \&\ \mathrm{Pr}(\ulcorner A \urcorner) \rightarrow \mathrm{Pr}(\ulcorner B \urcorner)$ (A, B closed)

(iii) $\vdash x \underset{\approx}{pp} A(y_1,\ldots,y_n) \rightarrow \mathrm{Pr}(\ulcorner \bar{x} \underset{\approx}{pp} A(\bar{y}_1,\ldots,\bar{y}_n)\urcorner)$

(iv) $\vdash x \underset{\sim}{p} A(t,y_2,\ldots,y_n) \leftrightarrow [y_1/t](x \underset{\sim}{p} A(y_1,y_2,\ldots,y_n))$

(v) $\vdash x \underset{\approx}{pp} A(z_1,\ldots) \ \&\ y \underset{\approx}{pp} (A(z_1,\ldots) \rightarrow B(z_1,\ldots)) \rightarrow$
$\rightarrow\ !\{y\}(x) \ \&\ \{y\}(x) \underset{\approx}{pp} B(z_1,\ldots)$

(vi) $\vdash x \underset{\approx}{pp} A \ \&\ y \underset{\approx}{pp} B \rightarrow t[x,y] \underset{\sim}{p} C \Rightarrow$
$\vdash x \underset{\approx}{pp} A \ \&\ y \underset{\approx}{pp} B \rightarrow t[x,y] \underset{\approx}{pp} C$ (t primitive recursive)
$\vdash x \underset{\approx}{pp} A \ \&\ y \underset{\approx}{pp} B \rightarrow j(x,y) \underset{\approx}{pp} A \ \&\ B$

(vii) $\vdash w \underset{\approx}{pp} \forall x Bx \rightarrow\ !\{w\}(x) \ \&\ !\{w\}(x) \underset{\approx}{pp} Bx$ ($\forall x Bx$ closed)

(viii) $\vdash t \underset{\sim}{p} A \Rightarrow\ \vdash t \underset{\approx}{pp} B$.

Proof. (i). Immediate, since Pr is a Σ^0_1-predicate. Cf. § 1.5.10.

(ii). Immediate.

(iii). Assume $x \underset{\sim}{p} A(y_1,\ldots) \ \&\ \mathrm{Pr}(\ulcorner \bar{x} \underset{\sim}{p} A(\bar{y}_1,\ldots)\urcorner)$. Then by (i) $\mathrm{Pr}(\ulcorner \mathrm{Pr}(\ulcorner \bar{x}\, p\, A(\bar{y}_1,\ldots)\urcorner)\urcorner)$, hence: $\mathrm{Pr}(\ulcorner \bar{x} \underset{\sim}{p} A(\bar{y}_1,\ldots)\urcorner) \ \&\ \mathrm{Pr}(\ulcorner \mathrm{Pr}(\ulcorner \bar{x}\, p\, A(\bar{y}_1,\ldots)\urcorner)\urcorner)$, i.e. $\mathrm{Pr}(\ulcorner \bar{x} \underset{\approx}{pp} A(\bar{y}_1,\ldots)\urcorner)$.

(iv). Proof by induction on the complexity of A.

(v). Let $x \underset{\approx}{pp} A(z_1,\ldots)$, $y \underset{\approx}{pp} A(z_1,\ldots) \rightarrow B(z_1,\ldots)$.
Then $!\{y\}(x) \ \&\ \{y\}(x) \underset{\sim}{p} A(z_1,\ldots) \rightarrow B(z_1,\ldots)$. Also $\mathrm{Pr}(\ulcorner \bar{x} \underset{\sim}{p} A(\bar{z}_1,\ldots)\urcorner)$, $\mathrm{Pr}(\ulcorner \bar{y} \underset{\sim}{p} A(\bar{z}_1,\ldots) \rightarrow B(\bar{z}_1,\ldots)\urcorner)$, hence $\mathrm{Pr}(\ulcorner \overline{\{y\}(x)} \underset{\sim}{p} B(\bar{z}_1,\ldots)\urcorner)$, and thus $\{y\}(x) \underset{\approx}{pp} B(z_1,\ldots)$.

(vi). Let $\vdash x \underset{\approx}{pp} A(z,\ldots) \ \&\ y \underset{\approx}{pp} B(z,\ldots) \rightarrow t[x,y] \underset{\sim}{p} C(z,\ldots)$, and assume $x \underset{\approx}{pp} A(z,\ldots) \ \&\ y \underset{\approx}{pp} B(z,\ldots)$. We have $\vdash \mathrm{Pr}(\ulcorner \bar{x} \underset{\approx}{pp} A(\bar{z},\ldots) \ \&\ \bar{y} \underset{\approx}{pp} B(\bar{z},\ldots) \rightarrow$ $\rightarrow t[\bar{x},\bar{y}] \underset{\sim}{p} C(\bar{z},\ldots)\urcorner)$ with (iii), $\mathrm{Pr}(\ulcorner \bar{x} \underset{\approx}{pp} A(\bar{z},\ldots)\urcorner)$, $\mathrm{Pr}(\ulcorner \bar{y} \underset{\approx}{pp} B(\bar{z},\ldots)\urcorner)$, therefore $\mathrm{Pr}(\ulcorner t[\bar{x},\bar{y}] \underset{\approx}{pp} C\urcorner)$. Also $\mathrm{Pr}(\ulcorner t[\bar{x},\bar{y}] = \overline{t[x,y]}\urcorner)$, therefore $\mathrm{Pr}(\ulcorner \overline{t[x,y]} \underset{\sim}{p} C(\bar{z},\ldots)\urcorner)$, $t[x,y] \underset{\approx}{pp} C$.

(vii). Let $w \underset{\approx}{pp} \forall x Bx$. Then $\forall x(!\{w\}(x) \ \&\ \{w\}(x) \underset{\sim}{p} Bx) \ \&\ \mathrm{Pr}(\ulcorner \bar{w} \underset{\sim}{p} \forall x Bx\urcorner)$, hence $\mathrm{Pr}(\ulcorner \overline{\{w\}(x)} \underset{\sim}{p} B\bar{x}\urcorner)$, and thus $\{w\}(x) \underset{\approx}{pp} Bx$.

(viii). Suppose $\vdash t \underset{\approx}{pp} A$. Then $\vdash \mathrm{Pr}(\ulcorner t \underset{\sim}{p} A\urcorner)$, hence $\vdash t \underset{\approx}{pp} A$.

3.9.4. Theorem (Soundness theorem). For closed A

$$\mathbf{HA} \vdash A \ \Rightarrow\ \exists n(\mathbf{HA} \vdash \bar{n} \underset{\sim}{p} A).$$

Proof. The proof is rather along the lines of the proof of 3.2.4, but we have to appeal repeatedly to 3.9.3; we establish, by induction on the length of a deduction of A,

$$\underset{\sim}{HA} \vdash A \;\Rightarrow\; \exists n(\underset{\sim}{HA} \vdash \bar{n}\, p\, A^*),$$

where A^* is the universal closure of A.
We may take Gödel's system for the verification. We select four typical examples of basis and induction step.
PL 2). Assume (induction hypothesis)

$$\vdash \bar{n}\, \underset{\approx}{p}\, \forall x A x, \quad \vdash \bar{m}\, \underset{\approx}{p}\, \forall x(Ax \to Bx).$$

Then $\vdash ! \{\bar{m}\}(x)\ \&\ \{\bar{m}\}(x)\, \underset{\approx}{p}\, Ax \to Bx$. Also $\vdash \bar{n}\, \underset{\approx}{pp}\, \forall x A$ (3.9.3 (viii)) hence $\vdash ! \{\bar{n}\}(x)\ \&\ \{\bar{n}\}(x)\, \underset{\approx}{pp}\, Ax$, and therefore $\vdash \Lambda x. \{\{\bar{m}\}(x)\}\{\bar{n}\}(x)\, \underset{\approx}{p}\, \forall x Bx$.
PL 7). Suppose $\vdash \bar{n}\, \underset{\approx}{p}\, \forall x(A\ \&\ B \to C)$. Then
$\Lambda x \Lambda v \Lambda w. \{\bar{n}\}(x, j(v,w))\, \underset{\approx}{p}\, \forall x(A \to (B \to C))$ (3.9.3 (vi)).
Q 3). Consider $\forall y(A(t[y]) \to \exists x A x)$. Let $\bar{n} = \Lambda y \Lambda u. j(t[y],u)$.
Assume $u\, \underset{\approx}{pp}\, A(t[y])$, then $! \{\bar{n}\}(u)$, and $\{\{\bar{n}\}(y)\}(u)\, \underset{\approx}{p}\, \exists x A x$ (with a tacit use of 3.9.3 (iv)).

Induction. Take $\bar{n}$ such that $\{\bar{n}\}(u,0) \simeq j_1 u$, $\{\bar{n}\}(u, Sx) \simeq \{j_2 u\}(x, \{\bar{n}\}(x))$.
We then easily show $\bar{n}\, \underset{\approx}{p}\, A(0)\ \&\ \forall x(Ax \to A(Sx)) \to \forall y A y$, similarly to 3.2.4.

3.9.5. Lemma. For prenex A, $\underset{\sim}{HA} \vdash x\, \underset{\approx}{p}\, A \to A$.
Proof. By induction on the construction of A. If A is quantifier free, the result is immediate.
Assume $A \equiv \forall y B y$. Then

$$\begin{aligned} x\, \underset{\approx}{p}\, A &\to \forall y(! \{x\}(y)\ \&\ \{x\}(y)\, \underset{\approx}{p}\, By) \\ &\to \forall y B y \qquad \text{(by induction hypothesis).} \end{aligned}$$

Assume $A \equiv \exists y B y$. Then

$$\begin{aligned} x\, p\, A &\to j_2 x\, \underset{\approx}{p}\, B(j_1 x)\ \&\ Pr(\ulcorner \overline{j_2 x}\, \underset{\approx}{p}\, B(\overline{j_1 x})\urcorner) \\ &\to B(j_1 x) \qquad \text{(by induction hypothesis)} \\ &\to \exists y B y. \end{aligned}$$

3.9.6. Lemma. For Π^0_2 formulae A

$$\underset{\sim}{HA} \vdash A \to \exists x(x\, \underset{\approx}{p}\, A).$$

Proof. Let A be of the form $\forall y \exists z P(y,z,u)$ (P prime). Assume $\forall y \exists z P(y,z,u)$. Obviously,

$$w = \min_z P(y,z,u) \to 0\, \underset{\approx}{p}\, P(y,w,u).$$

Note also

$$\begin{aligned} P(y,w,u) &\to Pr(\ulcorner P(\bar{y}, \bar{w}, \bar{u})\urcorner), \text{ so} \\ P(y,w,u) &\to Pr(\ulcorner 0\, \underset{\approx}{p}\, P(\bar{y}, \bar{w}, \bar{u})\urcorner). \end{aligned}$$

Let $t \equiv \Lambda y. j(\min_z P(y,z,u), 0)$. Then $! \{t\}(y)$, $j_1 \{t\}(y) = \min_z P(y,z,u)$, $j_2 \{t\}(y) = 0$, $\{t\}(y)\, \underset{\approx}{p}\, \exists z P(y,z,u)$.

Refinement. If A is a Π^0_2 formula, then there is a $\underset{\sim}{p}$-term t_A containing (only) the free variables of A, such that

$$\underset{\sim}{HA} \vdash A \rightarrow\, !t_A \;\&\; t_A \underset{\sim}{p} A\,.$$

3.9.7. Corollary. For Π^0_2 formulae A, there is a p-term t_A containing (only) the free variables of A, such that

$$\begin{aligned}\underset{\sim}{HA} \vdash A &\leftrightarrow\, !t_A \;\&\; t_A \underset{\sim}{p} A\\ &\leftrightarrow \exists x(x \underset{\sim}{p} A)\,.\end{aligned}$$

3.9.8. Lemma. For Σ^0_1 formulae A: $\underset{\sim}{HA} \vdash A \rightarrow \exists x(x \underset{\approx}{pp} A)$.

Proof. Let $A \equiv \exists y P y$, Py prime (and possibly containing other variables besides y).

First note that $0 \underset{\sim}{p} Py \equiv Py$, hence (§ 1.5)

$$\underset{\sim}{HA} \vdash 0 \underset{\sim}{p} Py \rightarrow Pr(\ulcorner 0 \underset{\sim}{p} Py \urcorner)\,, \text{ i.e.}$$

(1) $\quad \underset{\sim}{HA} \vdash 0 \underset{\sim}{p} Py \rightarrow 0 \underset{\approx}{pp} Py\,.$

Now suppose (in $\underset{\sim}{HA}$) that for some y, Py. Then $0 \underset{\sim}{p} Py$, so by (1)

(2) $\quad 0 \underset{\approx}{pp} Py\,.$

Hence

(3) $\quad j(0,y) \underset{\sim}{p} A\,.$

Also by (2) and 3.9.3 (iii) $Pr(\ulcorner \bar{0} \underset{\approx}{pp} P\bar{y} \urcorner)$, and hence

(4) $\quad Pr(\ulcorner j(0,\bar{y}) \underset{\sim}{p} A \urcorner)\,.$

Thus, from (3) and (4) $\exists x(x \underset{\approx}{pp} A)$.

3.9.9. Some definitions. Let us use KLS_n to denote the assertion of the Kreisel - Lacombe - Shoenfield theorem for $V \equiv$ objects of type n in $ECF(\mathcal{R})$.

$Tpt(y) \equiv$ "y is total and provably total", i.e.
$\quad \forall x \exists v\, Tyxv \;\&\; Pr(\ulcorner \forall x \exists w\, T(\bar{y},x,v) \urcorner)$

$Def(z) \equiv$ "z is defined on Tpt-indices", i.e.
$\quad \forall y(Tpt(y) \rightarrow \exists u Tzyu)$

$Ext(z) \equiv$ "z is extensional on Tpt-indices", i.e.
$\quad \forall yy'vv'(Tpt(y) \;\&\; Tpt(y') \;\&\; y \backsim y' \;\&\; Tzyv \;\&\; Tzyv' \rightarrow Uv = Uv')$

where

$$y \backsim y' \equiv \forall zz'vv'(Tyzu \;\&\; Tyz'v' \rightarrow Uv = Uv')\,.$$

$Mod(n,z) \equiv$ "n is a modulus of continuity for the functional z with respect to Tpt-indices", i.e.
$\quad \forall yy'vv'(Tpt(y) \;\&\; Tpt(y') \;\&\; \overline{\{y\}}(n) = \overline{\{y'\}}(n) \;\&\; Tzyv \;\&\; Tzy'v' \rightarrow Uv = Uv')$.

Cont(z) $\equiv$ "z is continuous w.r.t. Tpt - indices", i.e.

$\exists n\,\mathrm{Mod}(n,z)$.

3.9.10. Lemma. $\underset{\sim}{HA} \vdash \exists x(x \underset{\approx}{pp}\, y \in W_1) \leftrightarrow Tpt(y)$.

Proof. Suppose (in HA) $x \underset{\approx}{pp}\, y \in W_1$. Then by 3.9.7 and 3.9.3 (ii), $y \in W_1$ & $Pr(\ulcorner \bar{y} \in W_1 \urcorner)$, i.e. $Tpt(y)$.

For the converse implication, assume $Tpt(y)$, i.e. $y \in W_1$ & $Pr(\ulcorner \bar{y} \in W_1 \urcorner)$.

By 3.9.7, for some t

$$\vdash y \in W_1 \rightarrow\, !t[y] \,\&\, t[y] \underset{\approx}{p} (y \in W_1) .$$

Hence $Pr(\ulcorner \bar{y} \in W_1 \rightarrow\, !t[\bar{y}] \,\&\, t[\bar{y}] \underset{\approx}{p} (\bar{y} \in W_1)\urcorner)$, and thus

$$!t[y] \,\&\, t[y] \underset{\approx}{p} (y \in W_1) \,\&\, Pr(\ulcorner !t[\bar{y}] \,\&\, t[\bar{y}] \underset{\approx}{p} (\bar{y} \in W_1)\urcorner)$$

Therefore $\exists x(x \underset{\approx}{pp}\, y \in W_1))$.

3.9.11. Lemma. Assume $\bar{z}$ is such that

$$\underset{\sim}{HA} \vdash Def(\bar{z}) \,\&\, Ext(\bar{z})$$

but $\neg Cont(\bar{z})$. Then KLS_1 is not provable in HA .

Proof. Our first aim is to show that there is an $\bar{n}$ such that

(1) $\underset{\sim}{HA} \vdash \bar{n} \underset{\approx}{p}\, \bar{z} \in W_2$.

We may write

$$u \in W_2 \equiv Pu \,\&\, Qu \text{ , where}$$
$$Pu \equiv \forall y(y \in W_1 \rightarrow \exists v Tuyv) ,$$
$$Qu \equiv \forall yy'vv'[Q_1(y,y') \,\&\, Tuyv \,\&\, Tuy'v' \rightarrow Uv = Uv']$$

where

$$Q_1(y,y') \equiv y \in W_1 \,\&\, y' \in W \,\&\, y \backsim y' .$$

Let $\bar{n}_1 = \Lambda y \Lambda v.j(\min_v T\bar{z}yv, 0)$. Then

(2) $\vdash \bar{n}_1 \underset{\approx}{p}\, P\bar{z}$,

by lemma 3.9.10, since $Def(\bar{z})$.

Also, if $\bar{n}_2 \equiv \Lambda y \Lambda y' \Lambda w' \Lambda w' \Lambda x.0$, then

(3) $\vdash \bar{n}_2 \underset{\approx}{p}\, Q\bar{z}$

(by 3.9.10, and 3.9.5 applied to $y \backsim y'$, and $Ext(\bar{z})$).

Put $\bar{n} \equiv j(\bar{n}_1, \bar{n}_2)$, then (1) follows from (2) and (3).

Next we show that KLS_1 is not provable in HA as follows.

Assume $\underset{\sim}{HA} \vdash KLS_1$, then $\underset{\sim}{HA} \vdash \exists x(x \underset{\approx}{p}\, KLS_1)$, and then by 3.9.3

$\vdash \exists x(x \underset{\approx}{p}\, \forall z \in W_2\, Cont(z))$, and so by (1)

(4) $\vdash \exists x(x \underset{\approx}{p}\, Cont(\bar{z}))$, which implies for certain n

(5) $\exists x(x \underset{\approx}{p} \mathrm{Mod}(n,\bar{z}))$.

We will show that (5) implies in fact $\mathrm{Mod}(n,\bar{z})$. So suppose, for certain y, y', v, v'

(6) $\mathrm{Tpt}(y) \,\&\, \mathrm{Tpt}(y') \,\&\, \overline{\{y\}(n)} = \overline{\{y'\}(n)} \,\&\, T\bar{z}yv \,\&\, T\bar{z}y'v'$.

By lemma 3.9.10, applied to the first two conjuncts of (6)

$$\exists x(x \underset{\approx}{pp} (y \in W_1)) \,\&\, \exists x(x \underset{\approx}{pp} (y' \in W_1)) .$$

Also, by lemma 3.9.8. to the last three conjuncts of (6) (since $\overline{\{y\}(n)} = \overline{\{y'\}(n)}$ is expressible by a Σ^0_1-formula)

$$\exists x(x \underset{\approx}{pp} \overline{\{y\}(n)} = \overline{\{y'\}(n)}) \,\&\, \exists x(x \underset{\approx}{pp} T\bar{z}yv) \,\&\, \exists x(x \underset{\approx}{pp} T\bar{z}y'v') .$$

Applying all this to (5), we find $Uv = Uv'$ to be $\underset{\approx}{p}$-realizable, hence true, i.e. $\{\bar{z}\}(y) = \{\bar{z}\}(y')$. Hence $\mathrm{Cont}(\bar{z})$, giving us a contradiction.

3.9.12. <u>Theorem.</u> $\underset{\approx}{HA} \nvdash \mathrm{KLS}_1$.

<u>Proof.</u> We may construct a numeral $\bar{z}$ such that

$$\{\bar{z}\}(y) \simeq \begin{cases} 0 \text{ if } \exists_j(\mathrm{Proof}(j,\ulcorner \bar{y} \in W_1 \urcorner) \,\&\, \forall k \leq j(\{y\}(k) = 0)) ; \\ 0 \text{ if } \exists n(Q(y,n) \,\&\, \neg Q'(y,n)) ; \\ 1 \text{ if } \exists n(Q(y,n) \,\&\, Q'(y,n)) , \\ \quad \text{where } Q(y,n) \equiv \forall j < n[\neg \mathrm{Proof}(j,\ulcorner \bar{y} \in W_1 \urcorner) \,\& \\ \quad\quad \& \, \{y\}(n) \neq 0 \,\&\, \{y\}(j) = 0] , \text{ and where} \\ \quad Q'(y,n) \equiv \forall j < n \; \forall w < n(\mathrm{Proof}(j,\ulcorner \bar{w} \in W_1 \urcorner) \to \exists k < \{y\}(n) Twnk) ; \\ \text{undefined otherwise.} \end{cases}$$

We now proceed to show $\underset{\approx}{HA} \vdash \mathrm{Def}(\bar{z})$.

Assume $\mathrm{Tpt}(y)$. Then $\mathrm{Proof}(j,\ulcorner \bar{y} \in W_1 \urcorner)$ for some such j ; let this j be chosen minimal. Then

$$\forall k \leq j(\{y\}(k) = 0) \vee \neg \forall k \leq j(\{y\}(k) = 0).$$

In the first case, $\{z\}(y) = 0$. In the second case, let $n = \min_u(\{y\}(u) \neq 0)$. Since there is no $j' < n$ such that $\mathrm{Proof}(j',\ulcorner \bar{y} \in W_1 \urcorner)$, it follows that the second or third alternative in the definition of $\{\bar{z}\}(y)$ applies with the n as indicated.

Next we show $\underset{\approx}{HA} \vdash \mathrm{Ext}(\bar{z})$.

Let $\mathrm{Tpt}(y) \,\&\, \mathrm{Tpt}(y') \,\&\, y \backsim y'$. Then $\{\bar{z}\}(y)$ and $\{\bar{z}\}(y')$ are defined. Assume e.g. $\{\bar{z}\}(y) = 0$ and $\{\bar{z}\}(y') = 1$. Let again $n = \min_u \{y'\}(u) \neq 0$. Since $y \backsim y'$, also $n = \min_u \{y\}(u) \neq 0$. We have $\{\bar{z}\}(y') = 1$, therefore $Q'(y',n)$, hence also $Q'(y,n)$. But then, since $\{\bar{z}\}(y) = 0$, it cannot be computed according to the second or third clause in the definition of $\{\bar{z}\}(y)$, so it must be computed according to the first clause, hence there is a j such that

$$\mathrm{Proof}(j,\ulcorner\bar{y}\in W_1\urcorner)\ \&\ \forall k\leq j(\{y\}(k)=0).$$

Then $j<n$; also $y<n$ (since it occurs in a proof with gödelnumber j, $y<j$, on the usual definitions for "Proof"). Therefore, since $Q'(y,n)$, we have $\exists k<\{y\}(n)(Tynk)$. Since, for a standard gödelnumbering, $\neg Uk>k$, we have obtained a contradiction. Therefore $\{\bar{z}\}(y)=\{\bar{z}\}(y')$.

Finally we proceed to show that $\neg\mathrm{Cont}(\bar{z})$. For any given n_o, define

$$m_o = 1+\sup\{k \mid \exists w<n_o\ \exists j<n_o(\mathrm{Proof}(j,\ulcorner\bar{\bar{w}}\in W_1\urcorner)\ \&\ Twn_o k)\}.$$

Take a canonical index $y>n_o$ such that

$$\{y\}(x)=0 \text{ if } x\neq n_o,\ \{y\}(n_o)=m_o.$$

Then $\mathrm{Tpt}(y)$. We note that (a) the first clause cannot have been applied to compute $\{\bar{z}\}(y)$, since $\forall k\leq j(\{y\}(k)=0)\rightarrow j<n_o$, but $\mathrm{Proof}(j,\ulcorner\bar{\bar{y}}\in W_1\urcorner)\rightarrow$ $\rightarrow j>y>n_o$, and (b) if $j<n_o$, $w<n_o$, $\mathrm{Proof}(j,\ulcorner\bar{\bar{w}}\in W_1\urcorner)$, then by the definition of m_o there is a $k<m_o$, such that $Twn_o k$, so the third clause in the definition of $\{\bar{z}\}(y)$ must have been used, and therefore $\{\bar{z}\}(y)=1$.

On the other hand, $\{\bar{z}\}(\Lambda x.0)=0$. Hence n_o cannot be a modulus of continuity for $\bar{z}$ at $\Lambda x.0$ w.r.t. total and provably total indices.

Now the assertion of the theorem follows with 3.9.11.

3.9.11. Definitions. Let KLS_n denote KLS with $E(V)\equiv W_n$; and let $KLSR_n$ denote the rule corresponding to KLS_n:

$KLSR_n$
$$\vdash \bar{z}\in W_{n+1} \Rightarrow$$
$$\vdash \exists m\ \forall u\in W_n\ \forall v\in W_n(\overline{\{u\}}(\{m\}(u)) = \overline{\{v\}}(\{m\}(u))\rightarrow\{\bar{z}\}(u)=\{\bar{z}\}(v)).$$

Let

$$x\subseteq y \equiv_{def} \forall z(!\{x\}(z)\rightarrow !\{y\}(z)\ \&\ \{x\}(z)=\{y\}(z))$$

$$x\backsim y \equiv_{def} x\subseteq y\ \&\ y\subseteq x$$

$\mathrm{Ext}'(z) \equiv_{def}$ " z is extensional on partial indices", i.e.

$$\forall uvw[Tzuw\ \&\ u\backsim v\rightarrow\exists y(Tzvy\ \&\ Uy=Uw)]$$

$$\mathrm{Consis}(z)\equiv_{def}\forall uvw[Tzuw\ \&\ u\subseteq v\rightarrow\exists y(Tzvy\ \&\ Uy=Uw)]$$

$\mathrm{Pras}(x,z)\equiv_{def}$ " x is a partial recursive associate for z ", i.e.

$$\forall nm(f_n\subseteq f_m\ \&\ !\{x\}(n)\rightarrow\{x\}(n)=\{x\}(m))\ \&$$
$$\&\ \forall y(!\{z\}(y)\rightarrow\exists n(f_n\subseteq y\ \&\ \{x\}(n)=\{z\}(y)))$$

where f_n is a uniformly constructed gödelnumber for the finite partial function whose graph is coded by the number n.

Now the Myhill - Shepherdson theorem can be stated as:

MS $\quad(\mathrm{Ext}'(z)\rightarrow\mathrm{Consis}(z))\ \&\ ((\mathrm{Consis}(z)\rightarrow\exists x\,\mathrm{Pras}(x,z))$.

3.9.12. Some other results from Beeson 1972.

The following results, among others, are established in Beeson 1972:

(i) $\mathbf{HA} + M_{PR} \vdash MS$

(ii) MS, KLS are HRO-$\mathbf{mr}$-realizable, hence $\mathbf{HA} + MS + KLS_1 \nvdash M_{PR}$

(iii) $\mathbf{HA} \nvdash MS$ (by $\mathbf{p}$-realizability)

(iv) $\mathbf{HA}$ is closed under $KLSR_1$, $KLSR_2$.

Similar results hold if $\mathbf{HA}$ is replaced by $\mathbf{HA}$ + axioms of the form $TI(\prec)$, $\prec$ a primitive recursive well ordering.

3.9.13. Postscript added in proof. In Beeson B, a simplified treatment of $\mathbf{HA} \nvdash MS$, $\mathbf{HA} \nvdash KLS_1$ is given, omitting "realizing numbers" from $\mathbf{p}$-realizability. The resulting "realizability" can be seen as a variant of Kleene's $\Gamma \mid C$-relation (§ 3.1).

Chapter IV

NORMALIZATION THEOREMS FOR SYSTEMS OF NATURAL DEDUCTION

§ 1. The strong normalization theorem for HA

4.1.1. In this section we shall discuss normalization and normal form theorems for the natural deduction system for HA, as described in 1.3.6. There are different treatments available in the literature: Jervell 1971, Prawitz A and (implicitly, as a special case) Martin-Löf 1971; Prawitz 1971 contains a method for predicate logic which is easily adapted to arithmetic.

Jervell 1971 has been inspired by the method of Sanchis 1967 for establishing normal form theorems for the terms of theories of objects of finite type (such as $N\text{-}HA^{\omega}$; cf. our remarks in 2.2.35).

Technically the methods of Prawitz 1971 and Martin-Löf 1971 are similar, Martin-Löf 1971 has been inspired directly by Tait's introduction of computability predicates for terms in $N\text{-}HA^{\omega}$ (cf. 2.2.5); Prawitz's definition of validity and strong validity (Prawitz 1971, Appendix A) is similar to the definition of computability used by Martin-Löf 1971, but seems to have originated in the attempt to clarify Gentzen's ideas about an operational interpretation of the logical constants (cf. Prawitz 1971, II 2.2.2, A 1.3.1, Gentzen 1935, § 5).

Cut-elimination theorems for calculi of sequents for first-order arithmetic originate with Gentzen, especially Gentzen 1938 (for classical arithmetic). The methods of Gentzen 1938 have been further explored and extended. to systems of intuitionistic arithmetic in Scarpellini 1969, A and to intuitionistic analysis and other extensions of HA in Scarpellini A, 1970, 1971, 1972.

For some further general comments see also Prawitz 1971, III.

Here we restrict ourselves to systems of natural deduction. We rather closely follow the presentation in Prawitz 1971, Appendix A, but adapted and extended to the case of first-order arithmetic. (This adaptation yields a stronger result than stated in Martin-Löf 1971 in two respects: Prawitz also deals with permutative reductions, and proves a strong normalization theorem. It is the first addition (convenient, although not absolutely necessary for some of our applications) which is the cause of additional complication in the treatment as compared to Martin-Löf 1971.

For sections 1 - 3 the prerequisites are to be found in § 1.1 - 1.3 ; for sections 4 - 5, also § 1.4, 1.5 are needed.

Special notational convention for this chapter : successor is indicated by a lower case s .

4.1.2. Notational conventions about proof-trees

We use Π, Π', Π'', Π_1, ... for arbitrary deductions, and Σ, Σ', Σ'', ..., Σ_0, Σ_1, ... for arbitrary finite sequences of deductions. If we wish to indicate that Π or Σ possess a set of open assumptions of the form A , we write

$$\genfrac{}{}{0pt}{}{[A]}{\Pi} \quad , \quad \genfrac{}{}{0pt}{}{[A]}{\Sigma} .$$

[A] does not necessarily refer to all assumptions of the form A ; if we wish to refer to different sets of assumptions of the same form, we may use indexing (e.g. $[A]_\alpha$, $[A]_\beta$) to indicate the difference and use notations such as

$$\genfrac{}{}{0pt}{}{[A]_\alpha\,[A]_\beta}{\Pi} \ , \ \genfrac{}{}{0pt}{}{[A]_\alpha\,[A]_\beta}{\Sigma}$$

etc. In case of a single occurrence we may omit the square brackets around A .

If we wish to indicate that a deduction Π has a conclusion A , we write $\genfrac{}{}{0pt}{}{\Pi}{A}$. We write $\frac{\Sigma}{A}$ to indicate a derivation obtained by an application of a rule to the conclusions of Σ as premisses, with conclusion A .

If Π or $\genfrac{}{}{0pt}{}{\Pi}{A}$ is a derivation with conclusion A , we write

$\genfrac{}{}{0pt}{}{\Pi}{\genfrac{}{}{0pt}{}{[A]}{\Pi'}}$ or $\genfrac{}{}{0pt}{}{\Pi}{\genfrac{}{}{0pt}{}{[A]}{\Sigma}}$ for the derivation obtained by inserting Π in $\genfrac{}{}{0pt}{}{[A]}{\Pi'}$ or $\genfrac{}{}{0pt}{}{[A]}{\Sigma}$ for every open assumption of the set indicated by [A] . Similarly for $\genfrac{}{}{0pt}{}{\frac{\Sigma}{[A]}}{\Pi}$, $\genfrac{}{}{0pt}{}{\frac{\Sigma}{[A]}}{\Sigma'}$.

This concept of substitution of deductions for open assumptions must be understood as follows.

If B is an open assumption (occurrence) in $\genfrac{}{}{0pt}{}{\Pi'}{A}$, and [A] in $\genfrac{}{}{0pt}{}{[A]}{\Pi}$ indicates a set of n occurrences of open assumptions in Π , then all n occurrences of B in $\genfrac{}{}{0pt}{}{\Pi'}{\genfrac{}{}{0pt}{}{[A]}{\Pi}}$ deriving from the same occurrence of B in Π' under the substitution are in the same assumption-class (i.e. must be discharged or remain open simultaneously).

More precisely, assume the nodes (positions) in a deduction tree to be coded by finite sequences of natural numbers, such that 0 corresponds to the conclusion, and if n is a node with m nodes immediately above it, they are labelled from left to right as $n * \hat{0}, \ldots, n * \langle m-1\rangle$. Assumption classes may be specified as sets of top nodes. Let in $\begin{array}{c}[A]\\ \Pi\end{array}$, $[A]$ stand for an assumption class of top nodes α; the nodes in $\begin{array}{c}\Pi'\\ [A]\\ \Pi\end{array}$ then consist of $\{n * m \mid m \text{ a node of } \Pi', n \in \alpha\} \cup \{m \mid m \text{ a node of } \Pi\}$. The assumption class β in $\begin{array}{c}\Pi'\\ A\end{array}$ gives rise to a corresponding assumption class $\{n * m \mid n \in \alpha, m \in \beta\}$, if β has not been discharged in Π'; if β has been discharged in Π', there is a set of distinct assumption classes $\beta_n = \{n * m \mid m \in \beta\}$, one for each $n \in \alpha$.

Similarly for the notations $\begin{array}{c}\Sigma\\ [A]\\ \Pi\end{array}$, $\begin{array}{c}\Sigma\\ [A]\\ \Sigma'\end{array}$.

$\Pi(a)$, $\Sigma(a)$ may be used to denote a derivation or finite sequence of derivations with an improper parameter a; if the notation $\Pi(a)$, $\Sigma(a)$ has been introduced in the course of an argument, $\Pi(t)$, $\Sigma(t)$ refer to the derivation, resp. finite sequence of derivations, obtained by substituting t for a. As an alternative notation, we can use $[a/t]\Pi$, $[a/t]\Sigma$.

4.1.3. Description of the reduction processes

Below we shall assume all deductions to satisfy the general conditions on parameters indicated in 1.1.7, 1.3.6; if a reduction process as described below would violate this condition we shall assume proper parameters to be renamed automatically so as to satisfy the condition on parameters. The reduction processes are of two kinds: removal of redundant parameters, and application of reduction steps.

(A) Removal of redundant parameters. A parameter a in a deduction Π is said to be *redundant*, if a is an improper parameter of Π which does not occur in the open assumptions nor in the conclusion of Π.

A redundant parameter can be removed by substituting a constant term for it; for simplicity we shall always take 0 for this term, so the redundant parameter a is removed from Π by replacing Π by $[a/0]\Pi$.

It is worth noting that the reduction steps described sub (B) - (F) below do not introduce new redundant parameters.

A *reduction step* applied to a deduction Π consists in replacing a subdeduction Π' of Π by a "simpler" deduction Π'' with the same conclusion; we write Π' contr. Π'' (Π' *contracts to* Π''). The possible contractions are of four types: proper contractions, permutative contractions, induction contractions, λ-contractions. Let us call a reduction step obtained by application of a - contraction a - reduction.

(B) <u>Proper reductions</u>. If in a deduction an application of an I-rule, with conclusion A, is immediately followed by an application of an E-rule with major premiss A, the deduction may be simplified by cancelling both applications of rules. This gives rise to proper reductions, according to the following possible contractions:

1) $\&_i$ - contraction ($i = r, l$; r stands for "right", l for "left").

$$\frac{\dfrac{\begin{matrix}\Pi_l & \Pi_r\\ A_l & A_r\end{matrix}}{A_l \& A_r}}{A_i} \qquad \text{contr.} \qquad \begin{matrix}\Pi_i\\ A_i\end{matrix}$$

A &-contraction is a $\&_r$- or a $\&_l$-contraction.

2) →-contraction

$$\frac{\dfrac{\begin{matrix}[A]\\ \Pi_1\\ B\end{matrix}}{A \rightarrow B} \qquad \begin{matrix}\Pi_2\\ A\end{matrix}}{B} \qquad \text{contr.} \qquad \begin{matrix}\Pi_2\\ [A]\\ \Pi_1\\ B\end{matrix}$$

3) ∀-contraction

$$\frac{\dfrac{\begin{matrix}\Pi(a)\\ Aa\end{matrix}}{\forall x\, Ax}}{At} \qquad \text{contr.} \qquad \begin{matrix}\Pi(t)\\ At\end{matrix}$$

4) $\vee_i$ - contraction ($i = r, l$)

$$\frac{\dfrac{\begin{matrix}\Pi\\ A_i\end{matrix}}{A_l \vee A_r} \qquad \begin{matrix}[A_l]\\ \Pi_l\\ C\end{matrix} \qquad \begin{matrix}[A_r]\\ \Pi_r\\ C\end{matrix}}{C} \qquad \text{contr.} \qquad \begin{matrix}\Pi_i\\ [A_i]\\ \Pi_i\\ C\end{matrix}$$

A ∨-contraction is a $\vee_r$- or a $\vee_l$-contraction.

5) ∃-contraction

$$\frac{\dfrac{\begin{matrix}\Pi\\ At\end{matrix}}{\exists x\, Ax} \qquad \begin{matrix}[Aa]\\ \Pi'(a)\\ C\end{matrix}}{C} \qquad \text{contr.} \qquad \begin{matrix}\Pi\\ [At]\\ \Pi'(t)\\ C\end{matrix} \; .$$

(C) Permutative reductions. Permutative reductions make further applications of proper contractions possible by changing the order of application of certain rules.

6) $\vee E$-contraction

$$\dfrac{\dfrac{\begin{matrix}\Pi\\ A_1 \vee A_2\end{matrix} \quad \begin{matrix}[A_1]\\ \Pi_1\\ C\end{matrix} \quad \begin{matrix}[A_2]\\ \Pi_2\\ C\end{matrix}}{C} \qquad \Sigma}{A} \quad \text{contr.} \quad \dfrac{\begin{matrix}\Pi\\ A_1 \vee A_2\end{matrix} \quad \begin{matrix}[A_1]\\ \Pi_1\\ \dfrac{C \quad \Sigma}{A}\end{matrix} \quad \begin{matrix}[A_2]\\ \Pi_2\\ \dfrac{C \quad \Sigma}{A}\end{matrix}}{A}$$

(the lowest occurrence of C is major premiss of an elimination rule).

7) $\exists E$-contraction

$$\dfrac{\dfrac{\begin{matrix}\Pi\\ \exists x\, Ax\end{matrix} \quad \begin{matrix}[Aa]\\ \Pi'\\ C\end{matrix}}{C} \qquad \Sigma}{A} \quad \text{contr.} \quad \dfrac{\begin{matrix}\Pi\\ \exists x\, Ax\end{matrix} \quad \begin{matrix}[Aa]\\ \Pi'\\ \dfrac{C \quad \Sigma}{A}\end{matrix}}{A}$$

(the lowest occurrence of C acts as major premiss in an elimination rule).

(D) Induction reductions. Induction reductions simplify the induction term in applications of IND. The corresponding constructions are

8)

$$\dfrac{\begin{matrix}\Pi_0\\ A0\end{matrix} \quad \begin{matrix}[Aa]\\ \Pi(a)\\ A(sa)\end{matrix}}{A0} \quad \text{contr.} \quad \begin{matrix}\Pi_0\\ A0\end{matrix}$$

9)

$$\dfrac{\begin{matrix}\Pi_0\\ A0\end{matrix} \quad \begin{matrix}[Aa]\\ \Pi(a)\\ A(sa)\end{matrix}}{A(st)} \quad \text{contr.} \quad \begin{matrix}\dfrac{\begin{matrix}\Pi_0\\ A0\end{matrix} \quad \begin{matrix}[Aa]\\ \Pi(a)\\ A(sa)\end{matrix}}{[At]}\\ \Pi(t)\\ A(st)\end{matrix} .$$

(E) $\wedge$-reductions. The effect of $\wedge$-reductions is to lower the logical complexity of the conclusions in applications of $\wedge_I$. The corresponding contractions are

10) $\curlywedge\&$- contraction

$$\dfrac{\dfrac{\Sigma}{\curlywedge}}{A_1 \& A_2} \qquad \text{contr.} \qquad \dfrac{\dfrac{\dfrac{\Sigma}{\curlywedge}}{A_1} \qquad \dfrac{\dfrac{\Sigma}{\curlywedge}}{A_2}}{A_1 \& A_2}$$

11) $\curlywedge\vee$- contraction

$$\dfrac{\dfrac{\Sigma}{\curlywedge}}{A_1 \vee A_2} \qquad \text{contr.} \qquad \dfrac{\dfrac{\dfrac{\Sigma}{\curlywedge}}{A_1}}{A_1 \vee A_2}$$

12) $\curlywedge\to$- contraction

$$\dfrac{\begin{array}{c}\Pi\\ \curlywedge\end{array}}{A_1 \to A_2} \qquad \text{contr.} \qquad \dfrac{\dfrac{\begin{array}{c}\Pi\\ \curlywedge\end{array}}{A_2}}{A_1 \to A_2}$$

13) $\curlywedge\forall$- contraction

$$\dfrac{\begin{array}{c}\Pi\\ \curlywedge\end{array}}{\forall x Ax} \qquad \text{contr.} \qquad \dfrac{\dfrac{\begin{array}{c}\Pi\\ \curlywedge\end{array}}{A0}}{\forall x Ax}$$

14) $\curlywedge\exists$- contraction

$$\dfrac{\begin{array}{c}\Pi\\ \curlywedge\end{array}}{\exists x Ax} \qquad \text{contr.} \qquad \dfrac{\dfrac{\begin{array}{c}\Pi\\ \curlywedge\end{array}}{A0}}{\exists x Ax} \; .$$

(F) Immediate simplifications. Immediate simplifications consist in the removal of redundant applications of $\vee E$, $\exists E$. The contractions are

15) $\vee Es$ -contraction

$$\dfrac{\begin{array}{c}\Pi\\ A_1 \vee A_2\end{array} \qquad \begin{array}{c}\Pi_1\\ C\end{array} \qquad \begin{array}{c}\Pi_2\\ C\end{array}}{C} \qquad \text{contr.} \qquad \begin{array}{c}\Pi_i\\ C\end{array}$$

if in the derivation on the left side no assumptions are closed in

Π_i by the application of $\vee E$ shown.

16) $\exists$Es-contraction

$$\frac{\begin{matrix}\Pi & \Pi' \\ \exists x Ax & C\end{matrix}}{C} \quad \text{contr.} \quad \begin{matrix}\Pi' \\ C\end{matrix}$$

if in the derivation on the left hand no assumptions are closed in Π' by the application of $\exists E$ shown.

4.1.4. Definitions

A thread in a proof tree is a sequence of formula occurrences $A_1, \ldots, A_n$, such that A_1 is a top formula, A_n the end formula, and such that for each i, $i < n$, A_{i+1} is immediately below A_i.

A segment in a deduction Π is a sequence $A_1, \ldots, A_n$ of consecutive (i.e. A_{i+1} immediately below A_i) formula occurrences in a thread of Π such that

1) A_1 is not the conclusion of an application of $\vee E$, $\exists E$;
2) A_i, for each $i < n$, is a minor premiss of an application of $\vee E$, $\exists E$.
3) A_n is not a minor premiss of an application of $\vee E$, $\exists E$.

From the definition it is obvious that a segment consists of occurrences of the same formula. This permits us to transfer some terminology for individual formula occurrences to segments:
a segment $\sigma \equiv A_1, \ldots, A_n$ is said to be conclusion of an application α of a rule, if A_1 is conclusion of α;
σ is said to be (minor, major) premiss of an application α of a rule, if A_n is (minor, major) premiss of α.

A segment is said to be maximal, if it is conclusion of an introduction and major premiss of an elimination; a maximal formula (occurrence) is defined similarly.

A deduction Π is said to be in normal form (w.r.t. a set of contractions $\mathcal{R}$) if no contraction of $\mathcal{R}$ is applicable to Π. Π is said to be in strictly normal form, if it is in normal form and does not contain redundant parameters.

We write $\Pi \succ_1 \Pi'$ if Π' is obtained from Π by application of a single contraction to a subderivation of Π; $\Pi' \prec_1 \Pi \equiv_{def} \Pi \succ_1 \Pi'$. $\succ$ is the transitive relation generated by $\succ_1$; $\Pi \succeq \Pi' \equiv_{def} \Pi \succ \Pi'$ or $\Pi = \Pi'$, $\Pi \preceq \Pi' \equiv_{def} \Pi' \succeq \Pi$.

Let us call $\Pi_1, \Pi_2, \Pi_3, \ldots$ a reduction sequence (starting from Π_1) if for all i, $\Pi_{i+1} \prec_1 \Pi_i$. If the reduction sequence is finite, it is said to terminate (in the last deduction of the sequence) if the last term of the sequence is normal.

Similarly to 2.2.17, we define a <u>reduction tree</u> of a deduction Π as a pair $\langle T,\varphi\rangle$, where T is a non-empty set of natural numbers representing finite sequences such that $n * \hat{x} \in T \Rightarrow n \in T$, and φ a function which assigns deductions to the elements of T such that

(a) $\varphi\langle\ \rangle = \Pi$

(b) if $n \in T$, $\varphi n = \Pi'$, and $\Pi'_1, \ldots, \Pi'_n$ is a complete listing of the Π'' such that $\Pi' \;_1 \Pi''$, (without repetitions, some standard ordering of deductions being imposed), then $n * \langle i\rangle \in T$ for $1 \leq i \leq n$, and $\varphi(n * \langle i\rangle) = \Pi'_i$.

The <u>length</u> of a reduction tree $\langle T,\varphi\rangle$ is the number of elements in T.

4.1.5. <u>Remarks on reductions, normal form and normalization.</u>

A <u>normal form</u> theorem is a theorem of the type: If A is derivable from Γ then there is a (strictly) normal deduction of A from Γ.

A <u>normalization</u> theorem is of the form: For every deduction Π, there is a reduction sequence starting from Π which terminates.

A <u>strong normalization</u> theorem is of the form: All reduction sequences are finite.

Let $\mathcal{R}_p$, $\mathcal{R}_c$, $\mathcal{R}_\wedge$, $\mathcal{R}_s$ denote the sets of contractions according to 1 - 5 + 8 + 9, 1 - 9, 10 - 14, 15 - 16 respectively; we abbreviate unions by $\mathcal{R}_{c\wedge} \equiv \mathcal{R}_c \cup \mathcal{R}_\wedge$, $\mathcal{R}_{ps} \equiv \mathcal{R}_p \cup \mathcal{R}_s$ etc.

<u>Remark</u> I. A deduction which is normal w.r.t. $\mathcal{R}_p(\mathcal{R}_c)$ does not contain maximal formulas (segments). A deduction which is normal w.r.t. $\mathcal{R}_\wedge$ contains atomic applications (i.e. applications with atomic conclusion) only.

<u>Remark</u> II. There exists a primitive recursive procedure for transforming an arbitrary deduction into a normal deduction (w.r.t. $\mathcal{R}_{c\wedge}$) with the same conclusion, by the introduction of redundant parameters (elaborating a remark of <u>Jervell</u> 1971, page 106). For let in

$$\begin{array}{c}\Pi\\ A\\ \Pi'\end{array}$$

A be a maximal formula. We then transform this derivation as follows:

$$\dfrac{\dfrac{\dfrac{\begin{array}{c}\Pi\\ A\end{array}\quad 0=0}{A\ \&\ (0=0)}\qquad \dfrac{sa=sa}{A\ \&\ (sa=sa)}}{A\ \&\ (b=b)}\,\text{IND}}{\begin{array}{c}A\\ \Pi'\end{array}}\ .$$

The two occurrences of A are not maximal anymore.

We may deal similarly with maximal segments: if A in $\begin{matrix}\Pi\\ A\\ \Pi'\end{matrix}$ occurs as the last formula of a maximal segment, then the same transformation makes the segment non-maximal. Normalizing with respect to $\mathcal{R}_\wedge$ is itself a primitive recursive process: one needs n $\wedge$- contractions if n is the total number of occurrences of logical symbols in conclusions of $\wedge_I$- applications in the deduction considered.

Hence the normal form theorem w.r.t. $\mathcal{R}_{c\wedge}$, when only normal form, not strictly $\vee$ normal form is required, becomes trivial.

<u>Remark</u> III. If we first normalize a derivation w.r.t. $\mathcal{R}_\wedge$, then it remains normal w.r.t. $\mathcal{R}_\wedge$ when contractions from $\mathcal{R}_{cs}$ are applied.

<u>Remark</u> IV. Contractions from $\mathcal{R}_p$ may introduce new redundant applications of $\vee$E, $\exists$E. For example, consider

$$\dfrac{C \vee (B \to C) \qquad \dfrac{\dfrac{(2)\ C}{C \vee C}}{C \& C \to C \vee C} \qquad \dfrac{\dfrac{\dfrac{\dfrac{\dfrac{(1)\ C \& C}{C}}{C \vee C}}{C \& C \to C \vee C}{-(1)} \qquad \dfrac{\dfrac{\dfrac{B \quad B \to C\ (3)}{C}}{C \vee C}}{C \& C \to C \vee C}}{(C \& C \to C \vee C) \& (C \& C \to C \vee C)}{\&\,I}}{C \& C \to C \vee C}{\&\,E}}{C \& C \to C \vee C}{-(2),-(3)}\ .$$

Contracting the &- introduction and &- elimination marked yields a deduction with a redundant $\vee$E-application. On the other hand, contractions of $\mathcal{R}_s$ do not introduce maximal formulas or segments in $\mathcal{R}_c$- normal derivations (but they may do so in $\mathcal{R}_p$- normal derivations: a maximal segment may reduce to a maximal formula by applications of contractions from $\mathcal{R}_s$).

<u>Remark</u> V. Certain contraction steps may be conceived as preserving the intuitive proof idea corresponding to the formal deduction. This can be defended for proper contractions, and perhaps also for permutative reductions. Hence equality of normal forms with respect to $\mathcal{R}_c$ would be sufficient to ensure identity of the underlying proof idea. The converse is a more dubious hypothesis. For a discussion see <u>Kreisel</u> 1971, 1(c), pp. 114 - 117, <u>Prawitz</u> 1971, 3.5.6.

More generally, we might try to establish the conjecture: Π, Π' represent the same proof idea, if there is a sequence $\Pi \equiv \Pi_0, \Pi_1, \ldots, \Pi_n \equiv \Pi'$, such that for all i $(0 \leq i < n)$ $\Pi_{i+1} <_1 \Pi_i$ or $\Pi_i <_1 \Pi_{i+1}$. This conjecture is then obviously false, however, if we permit immediate simplifications. For example, the following two derivations represent intuitively distinct proofs, and are both in normal (from Prawitz 1971, 3.5.6)

$$\Pi = \left\{ \dfrac{\dfrac{\dfrac{A \quad (1)}{A \to A}\ -(1)}{B \to (A \to A)}}{(A \to A) \to (B \to (A \to A))} \qquad \dfrac{\dfrac{A \to A \quad (1)}{B \to (A \to A)}}{(A \to A) \to (B \to (A \to A))}\ -(1) \right\} = \Pi' .$$

Now consider the following derivation Π'' :

$$\dfrac{C \vee D \qquad \dfrac{\dfrac{\dfrac{A}{A \to A}}{B \to (A \to A)}}{(A \to A) \to (B \to (A \to A))} \qquad \dfrac{\dfrac{A \to A}{B \to (A \to A)}}{(A \to A) \to (B \to (A \to A))}}{(A \to A) \to (B \to (A \to A))} .$$

Now we may apply in two different ways a $\vee$Es -contraction, thereby obtaining either Π or Π'. Our conjecture would then lead to regarding Π, Π' as intuitively the same proof, as is illustrated by the sequence Π, Π'', Π'.

Π'' may be said to combine two different proof ideas (and hence represents itself a proof different from each of these). Since immediate simplifications make normal forms non-unique, the most plausible alternative seems to be, in the presence of a strong normalization theorem, to reformulate the conjecture as follows :

Conjecture : Two deductions Π, Π' represent the same intuitive proof idea, if the sets of normal deductions to which they reduce are identical.

Remark VI. If we would have failed to distinguish between $\&_r E$ and $\&_l E$, and correspondingly between $\&_r$ - and $\&_l$ - contractions, another source of non-uniqueness of normal forms would have resulted, as will become clear from the following example :

$$\dfrac{\dfrac{\dfrac{A \qquad A \to B}{B} \qquad \dfrac{C \qquad C \to B}{B}}{B \,\&\, B}}{B}$$

could then be reduced by a & - contractions to either $\dfrac{A \quad A \to B}{B}$ or $\dfrac{C \quad C \to B}{B}$.

4.1.6. Isomorphism and homomorphism between terms and deductions

References : Prawitz 1971, IV, especially IV 2.5 ; Howard A ; Girard 1972, Martin-Löf 1972.

We consider a very special case to illustrate the idea, namely pure intuitionistic implication logic. Let us assign a type to each formula A; for simplicity we denote this type also by A. We have "deduction variables" $\Pi_A, \Pi'_A, \Pi''_A, \ldots$ for each type A, and abstraction operators $\lambda\Pi_A$.

To any deduction Π (in the system for natural deduction with $\to$I, $\to$E as the only rules) we associate a "deduction term" $\Psi(\Pi)$ as follows.

To a derivation consisting of an assumption A only we assign a variable Π_A. Assumptions in the same class will have assigned the same variable to them. We define further inductively:

$$\Psi\begin{pmatrix} [A] \\ \Pi \\ \dfrac{B}{A \to B} \end{pmatrix} = \lambda\Pi_A . \Psi\begin{bmatrix} [A] \\ \Pi \\ B \end{bmatrix} ,$$

$$\Psi\begin{pmatrix} \Pi \quad\quad \Pi' \\ \dfrac{A \quad A \to B}{B} \end{pmatrix} = \Psi\begin{bmatrix} \Pi' \\ \overline{A \to B} \end{bmatrix} \Psi\begin{bmatrix} \Pi \\ A \end{bmatrix} .$$

Now we see that $\to$-contraction corresponds to λ-conversion:

$$\dfrac{\begin{matrix} \Pi' \\ A \end{matrix} \quad \dfrac{\begin{matrix} [A] \\ \Pi \\ B \end{matrix}}{A \to B}}{B} \text{ is mapped onto } (\lambda\Pi_A . \Psi(\Pi))(\Psi(\Pi'))$$

and
$$\begin{matrix} \Pi' \\ [A] \\ \Pi \\ B \end{matrix} \text{ onto } [\Pi_A / \Psi(\Pi')]\Psi(\Pi) .$$

This makes it understandable that the same techniques which have been used to prove reduction to normal form for terms of $\underset{\sim}{N}$-$\underset{\sim}{HA}^{\omega}$, also apply to deductions. In fact, for a suitably defined theory of terms with reduction relations between them, a complete isomorphism, with respect to the reduction relation can be obtained. This idea is exploited in Girard 1972.

If we introduce a type for individuals, $\forall$I and $\forall$E may be interpreted as λ-abstraction and substitution of individual terms respectively, and $\forall$-contractions as λ-conversion.

To $\forall$I, $\forall$E, $\exists$I, $\exists$E we have no direct analogue in the λ-version of $\underset{\sim}{N}$-$\underset{\sim}{HA}^{\omega}$; in this sense we have to extend the results from § 2.2.

"IND" behaves similarly to a special instance of the recursion operator: if we assign to a derivation

$$\dfrac{\begin{matrix} \Pi \\ A0 \end{matrix} \quad \begin{matrix} [Aa] \\ \Pi' \\ Asa \end{matrix}}{At}$$

the deduction term $R_{IND}(\Psi\Pi)(\Psi\Pi')t$, then the IND-reductions correspond to

$$R_{IND}(\Psi\Pi)(\Psi\Pi')0 \text{ conv. } \Psi\Pi$$
$$R_{IND}(\Psi\Pi)(\Psi\Pi')(st) \text{ conv. } (\Psi\Pi')(R_{IND}(\Psi\Pi)(\Psi\Pi')t) .$$

It should be noted that distinct assumption classes correspond to distinct deduction variables(an open assumption may be viewed as an "unspecified" deduction of the assumption). Having the same deduction variable Π_A assigned to all assumptions of the form A, corresponds in terms of a λ-calculus of having only one variable of type A available. Hence, for a natural correlation with a theory of terms (a fortiori for an isomorphism) we must distinguish assumption classes. The method of assumption classes corresponds to the second method for defining deductions in Prawitz 1965 (I, § 4, pages 29-31).

Historical note. The earliest publication where the analogy between types and terms on one hand, and formulas and deductions on the other hand, was explicitly noted, seems to be Curry-Feys 1958, 9E (for implicational calculus only). Their remarks concern not the equivalence between λ-calculus and natural deduction, but between the pure theory of combinators and a Hilbert type system for intuitionistic implicational logic.

4.1.7. Strong normalization for HA

Below we shall establish a strong normalization theorem relative $\mathcal{R}_{c\wedge}$. Carrying out immediate simplifications afterwards, we also obtain a normalization theorem relative $\mathcal{R}_{c\wedge s}$.

We might also have proved the strong normalization relative $\mathcal{R}_c$ only; first normalizing w.r.t. $\mathcal{R}_{\wedge}$, then using the strong normalization theorem relative $\mathcal{R}_c$ yields, by 4.1.5 Remark III a normalization theorem relative $\mathcal{R}_{c\wedge}$.

For applications, we only need a normalization theorem relative $\mathcal{R}_{c\wedge}$; so if the reader wishes, he may delete everything in the proof below referring to $\wedge$-contractions.

4.1.8. Notational conventions

a) "Rule" denotes a function which assigns to a deduction Π the final rule of Π. So Rule(Π) takes as value one of the basic rules, an I- or E-rule, $\wedge_I$ or IND.

b) Con(Π) is the conclusion of Π (the formula derived by Π). For implications $C \to D$, Premiss$(C \to D) = C$.

c) Rapp(Π) is the final application (instance) of Rule(Π) in Π. If Rapp(Π) has a proper parameter a, we put Param$(\Pi) = a$; Param(Π) may be fixed arbitrarily elsewhere.

d) If $\mathrm{Rule}(\Pi) \in \{\rightarrow I, \vee E, \exists E, IND\}$, $\mathrm{Ass}(\Pi)$ denotes the (index of the) class of assumptions discharged by $\mathrm{Rapp}(\Pi)$.

e) If $\mathrm{Rule}(\Pi) = \exists I$ then $\mathrm{Term}(\Pi)$ denotes the term t such that Π ends with At, $\exists x Ax$; or if $\mathrm{Rule}(\Pi) = IND$, then $\mathrm{Term}(\Pi)$ is the induction term.

f) "Subst" maps a triple (a, t, Π), consisting of a parameter a, a term t, and a deduction Π containing a as an improper parameter, on the proof $[a/t]\Pi$ obtained by substituting t for a throughout the proof; for other triples the value of $\mathrm{Subst}(a, t, \Pi)$ may be fixed arbitrarily.

g) "Sub" maps a quadruple (A, Π, Π', α), where $\mathrm{Con}(\Pi) = A$, onto a deduction Π'', which is obtained by substituting Π in Π' for every open assumption of the form A belonging to the assumption class (indexed by) α. The value of $\mathrm{Sub}(A, \Pi, \Pi', \alpha)$ may be fixed arbitrarily on other arguments.

h) Let $\mathrm{PRD}(\Pi)$ denote the set of deductions of the premisses of $\mathrm{Rapp}(\Pi)$. $\mathrm{Prd}_1(\Pi)$, $\mathrm{Prd}_1(\Pi)$, ... are the deductions of the first, second, ... premiss of $\mathrm{Rapp}(\Pi)$ respectively.

If $\mathrm{Rule}(\Pi)$ is an E-rule, $\mathrm{Prd}_1(\Pi)$ will be assumed to refer to the deduction of the major premiss.

i) $\mathrm{Norm}(\Pi)$: Π is strictly normal.

If we assume the discussion to be formalized, we may introduce certain code numbers for the various rules, and identifying the other syntactical objects with their gödel numbers, on a standard gödel numbering, all functions and predicates described in (a) - (i) then become primitive recursive.

4.1.9. Définition of strong validity.

Let us write SV as an abbreviation for the predicate of strong validity to be defined below. The definition of $SV(\Pi)$ ("Π is strongly valid") is primarily given by induction on the complexity of $\mathrm{Con}(\Pi)$; for derivations Π with $\mathrm{Con}(\Pi)$ of fixed complexity, the definition of $SV(\Pi)$ takes the form of a (non-iterated) generalized inductive definition.

A deduction Π is said to be strongly valid ($SV(\Pi)$ holds) if one of the following clauses applies:

(i) $\mathrm{Rule}(\Pi) \in \{\&I, \vee I, \exists I\}$, and $\forall \Pi' \in \mathrm{PRD}(\Pi)[SV(\Pi')]$.

(ii) $\mathrm{Rule}(\Pi) = \rightarrow I$, $\mathrm{Premiss}(\mathrm{Con}(\Pi)) = A$, then
$\forall \Pi'(SV(\Pi') \text{ and } \mathrm{Con}(\Pi') = A \Rightarrow SV(\mathrm{Sub}(A, \Pi', \Pi, \mathrm{Ass}(\Pi)))$.

(iii) $\mathrm{Rule}(\Pi) = \forall I$, $\mathrm{Param}(\Pi) = a$, $\forall t(SV(\mathrm{Subst}(a, t, \mathrm{Prd}_1(\Pi))))$.

(iv) $\mathrm{Rule}(\Pi)$ is not an I-rule, and

(a) $\forall \Pi'(\Pi' <_1 \Pi \Rightarrow SV(\Pi'))$ or Π_1 is normal.

(b) If $\mathrm{Rule}(\Pi) = \vee E$, then the reduction tree of $\mathrm{Prd}_1(\Pi)$ is finite, $SV(Prd_2(\Pi))$, $SV(Prd_3(\Pi))$, and if $Con\ Prd_1(\Pi) = A_1 \vee A_2$, then for each $\Pi' \leq Prd_1(\Pi)$: if Π'' is the sub-derivation above an endsegment of Π', with $Con(\Pi'') = A_i$, then $SV(Sub(A_i, \Pi'', Prd_{i+1}(\Pi), Ass(\Pi)))$.

(c) If $Rule(\Pi) = \exists E$, then the reduction tree of $Prd_1(\Pi)$ is finite *, $SV(Prd_2(\Pi))$, and if $Con\ Prd_1(\Pi) = \exists x Ax$, then for each Π'' which is a sub-derivation with $Con(\Pi'') = At$, immediately above an endsegment of a $\Pi' \leq Prd_1\Pi$, $SV(Sub(At, \Pi'', Subst(Param(\Pi), t, Prd_2(\Pi))))$.

More pictorially, clause (iv)(c) requires: If Π is of the form

$$\frac{\begin{matrix}\Pi_1 \\ \exists x Ax\end{matrix} \qquad \begin{matrix}[Aa] \\ \Pi_2(a) \\ C\end{matrix}}{C}$$

then $\begin{matrix}[Aa] \\ \Pi_2\end{matrix}$ strongly valid, the reduction tree of Π_1 is finite, and if Π_1 reduces to a Π' containing $\begin{matrix}\Pi'' \\ At\end{matrix}$ as a sub-derivation immediately above an endsegment of Π', then

$$\begin{matrix}\Pi'' \\ [At] \\ \Pi_2(t) \\ C\end{matrix}$$

should be strongly valid.

Similarly for clause (iv)(b).

A variant of the definition is obtained by reading clause (iv)(a) as follows:

(a)' Each reduction sequence from Π either terminates or passes through a deduction which is strongly valid by clause (i), (ii) or (iii).

4.1.10. <u>Lemma</u>. Assume $Rule(\Pi)$ to be an I-rule, so Π is of the form

$$\frac{\Pi'}{A} \quad \text{or} \quad \frac{\Pi' \quad \Pi''}{A},$$

and let $\Pi, \Pi_1, \Pi_2, \ldots$ be a reduction sequence starting from Π; and so $Rule(\Pi_i) = Rule(\Pi)$ for all i, Π_i being of the form

$$\frac{\Pi_i}{A} \quad \text{or} \quad \frac{\Pi_i' \quad \Pi_i''}{A}$$

---

* the requirement that the reduction tree of $Prd_1(\Pi)$ is finite is an additional requirement as compared to <u>Prawitz</u> 1971, and serves to facilitate formalization in § 4.

respectively. Then Π', Π'_1, Π'_2, ... and Π'', Π''_1, Π''_2, ... are reduction sequences after omission of repititions.

Proof. Trivial, by inspection of the various cases.

4.1.11. Lemma. Let $\Pi_1(a) \succeq \Pi_2(a)$, a not a proper parameter in $\Pi_1(a)$, and let $[A(a)]$ indicate a set of open hypotheses of the form $A(a)$ in $\Pi_1(a)$, then

$$\begin{matrix}\Pi \\ [A(t)] \\ \Pi_1(t)\end{matrix} \succeq \begin{matrix}\Pi \\ [A(t)] \\ \Pi_2(t)\end{matrix}$$

for each derivation $\begin{matrix}\Pi \\ A(t)\end{matrix}$.

Proof. Trivial, by inspection of cases.

4.1.12. Lemma. If $\Pi_1 \succeq \Pi_2$, $SV(\Pi_1)$, then $SV(\Pi_2)$.

Proof. We show by induction over SV w.r.t. Π_1 that $SV(\Pi_1) \Rightarrow \forall \Pi_2 (\Pi_1 \succeq \Pi_2 \Rightarrow SV(\Pi_2))$.

(a) If $\mathrm{Rule}(\Pi_1) = \&I$, then Π_1, Π_2 are of the form (lemma 4.1.10)

$$\frac{\Pi'_1 \quad \Pi''_1}{A}, \quad \frac{\Pi'_2 \quad \Pi''_2}{A}$$

respectively, and $\Pi'_1 \succeq \Pi'_2$, $\Pi''_1 \succeq \Pi''_2$; $SV(\Pi'_1)$, $SV(\Pi''_1)$ imply by the induction hypothesis $SV(\Pi'_2)$, $SV(\Pi''_2)$ hence $SV(\Pi_2)$.

(b) $\mathrm{Rule}(\Pi_1) = \rightarrow I$. Similarly, using 4.1.10 and clause (ii) for SV.

(c) $\mathrm{Rule}(\Pi_1) = \forall I$. Use lemma 4.1.10 and clause (i) for SV.

(d) $\mathrm{Rule}(\Pi_1) = \vee I, \exists I$. Similarly to case (a), (b).

(e) $\mathrm{Rule}(\Pi_1)$ is not an I-rule. Either Π_1 is normal, and then $\Pi_1 = \Pi_2$, or $SV(\Pi_1)$ holds because Π_1 is not normal and (iv)(a), (b), (c) are fulfilled. Then there is a Π', $SV(\Pi')$ such that $\Pi_2 \preceq \Pi' <_1 \Pi_1$, and the assertion for Π_1 follows from the assertion for Π'.

4.1.13. Theorem. Each strongly valid deduction has a finite reduction tree.

Proof. We prove $R(\Pi) \equiv [\Pi$ has a finite reduction tree$]$ by induction over the SV deductions.

(a) If $\mathrm{Rule}(\Pi)$ is an I-rule, it follows by lemma 4.1.10 that the induction hypotheses for the deductions from $PRD(\Pi)$ imply $R(\Pi)$.

(b) If $\mathrm{Rule}(\Pi)$ is not an I-rule, then either Π is normal, or $\forall \Pi' (\Pi' <_1 \Pi \Rightarrow R(\Pi'))$ is our induction hypothesis, which immediately implies $R(\Pi)$.

4.1.14. Remark. The preceding theorem is classically equivalent to: each reduction sequence terminates (by an appeal to König's lemma). Intuitionistically, the stronger conclusion follows by an appeal to the fan theorem from the weaker assertion. The weaker assertion might have been proved directly, by applying induction to $R'(\Pi) \equiv$ [all reduction sequences starting from Π terminate].

The advantage in proving the stronger assertion is in fact that it is directly expressible in HA, whereas the weaker assertion must be expressed in a conservative extension of HA containing function variables.

4.1.15. Strong validity under substitution.

We define : a deduction Π is strongly valid under substitution, if for each substitution of terms for parameters, and each replacement of open hypotheses in Π by SV deductions of these hypotheses, the resulting deduction is SV. (It is not assumed that all open hypotheses are replaced : an open hypothesis is a strongly valid deduction of itself.)

Our next aim is to prove each deduction to be SV under substitution. To do this, we need the following lemma.

4.1.16. Lemma. A deduction Π for which $\mathrm{Rule}(\Pi)$ is not an I-rule, is strongly valid if the following conditions are satisfied :

(I) The reduction tree of any $\Pi \in \mathrm{PRD}(\Pi)$ is finite.

(II) If $\mathrm{Rule}(\Pi) \in \{\&E, \to E, \forall E, \curlywedge_I\}$, then the elements of $\mathrm{PRD}(\Pi)$ are strongly valid ;

(III) If $\mathrm{Rule}(\Pi) = \mathrm{IND}$, i.e. Π is of the form

$$\dfrac{\begin{matrix}\Pi' \\ A0\end{matrix} \qquad \begin{matrix}[Aa] \\ \Pi''(a) \\ Asa\end{matrix}}{At}$$

then $\begin{matrix}\Pi' \\ A0\end{matrix}$ is SV, and for each $\mathrm{SV}\begin{pmatrix}\Pi''' \\ At\end{pmatrix}$,

$$\begin{matrix}\Pi''' \\ [At] \\ \Pi''(t) \\ Ast\end{matrix}$$ is strongly valid.

(IV) If $\mathrm{Rule}(\Pi) \in \{\vee\exists, \exists E\}$, then conditions (b), (c) in clause (iv) of the definition of strong validity are fulfilled.

Proof. To a deduction Π satisfying condition (I) we assign an induction value $(\alpha, \beta, \gamma, \delta, \epsilon)$, where

α = logical complexity of $\mathrm{Con}(\Pi)$,

β = length of reduction tree of $\mathrm{Prd}_1(\Pi)$, if $\mathrm{Rule}(\Pi)$ is an E-rule, 0 otherwise ;

γ = length of $\mathrm{Prd}_1(\Pi)$, if $\mathrm{Rule}(\Pi)$ is an E-rule, 0 otherwise ;

δ = sum of length of reduction trees of $\mathrm{PRD}(\Pi)$;

ϵ = complexity of induction term of $\mathrm{Rapp}(\Pi)$ if $\mathrm{Rule}(\Pi) = \mathrm{IND}$, 0 otherwise.

We assume the induction values to be ordered lexicographically, i.e.
$(\alpha_0, \alpha_1, \alpha_2, \alpha_3, \alpha_4) < (\alpha'_0, \alpha'_1, \alpha'_2, \alpha'_3, \alpha'_4) \equiv$
$\equiv \exists i < 5[\forall j < i\ (\alpha_j = \alpha'_j)$ and $\alpha_i < \alpha'_i]$.

Now the proof proceeds by induction on the induction value of deductions satisfying the conditions of the lemma. We have for a deduction which satisfies I - IV, only to verify (iv)(a) in the definition of strong validity in order to ensure that the deduction is strongly valid, since (iv)(b), (iv)(c) hold by IV.

If Π is normal, we are finished. If Π is not normal, we must show that each Π' such that $\Pi' <_1 \Pi$ is SV.

Case (a). Π' is obtained from Π by replacing a proper subtree of Π by a contraction. So if Π is

$$\frac{\Pi_1 \quad \cdots \quad \Pi_n}{A}$$

then Π' is

$$\frac{\Pi'_n \quad \cdots \quad \Pi'_n}{A}$$

where for some i $(1 \leq i \leq n)$ $\Pi_i \succ_1 \Pi'_i$, and for $j \neq i$ $\Pi_j = \Pi'_j$. Let the induction value of Π' be $(\alpha', \beta', \gamma', \delta', \epsilon')$. Obviously, $(\alpha', \beta', \gamma', \delta', \epsilon') < (\alpha, \beta, \gamma, \delta, \epsilon)$ since either Π_i is the major premiss, and then $\alpha = \alpha'$, $\beta' < \beta$, or $\alpha = \alpha'$, $\beta = \beta'$, $\gamma = \gamma'$ and $\delta' < \delta$. Π' obviously satisfies condition I of the lemma; II and III are satisfied because of lemma 4.1.12.

For example, if Rule(Π) = IND, then Rule(Π') = IND, and if

$$\Pi = \left\{ \frac{\begin{matrix} & [Aa] \\ \Pi_1 & \Pi_2 \\ A0 & Asa \end{matrix}}{At} \right. , \qquad \Pi' = \left\{ \frac{\begin{matrix} & [Aa] \\ \Pi'_1 & \Pi'_2 \\ A0 & Asa \end{matrix}}{At} \right. , \text{ where}$$

$\begin{matrix}[Aa] \\ \Pi_2(a) \\ Asa\end{matrix} \succ_1 \begin{matrix}[Aa] \\ \Pi'_2(a) \\ Asa\end{matrix}$, then by condition III for Π, and lemma 4.1.12 $\begin{matrix}\Pi_3 \\ [At] \\ \Pi'_2(t) \\ Ast\end{matrix}$

is SV for each SV $\begin{matrix}\Pi_3 \\ At\end{matrix}$.

(IV) is also satisfied for Π'. For assume Π to be

$$\frac{\dfrac{\Sigma}{B_1 \vee B_2} \quad \dfrac{\begin{matrix}[B_1] \\ \Sigma_1\end{matrix}}{A} \quad \dfrac{\begin{matrix}[B_2] \\ \Sigma_2\end{matrix}}{A}}{A} , \text{ and let } \Pi'_i = \left\{ \begin{matrix}[B_j] \\ \Sigma'_j \\ A\end{matrix} <_1 \begin{matrix}[B_j] \\ \Sigma_j \\ A\end{matrix} \right. ,$$

then Π_i' is strongly valid (condition II for Π, lemma 4.1.12), and if $\frac{\Sigma}{B_1 \vee B_2} \geq \Pi_o$, Π_o containing $\begin{matrix}\Pi'' \\ B_j\end{matrix}$ as subderivation immediately above an endsegment of Π_o, then

$$\begin{matrix}\Pi'' \\ [B_j] \\ \underline{\Sigma_j'} \\ A\end{matrix} \preceq \begin{matrix}\Pi'' \\ [B_j] \\ \underline{\Sigma_j} \\ A\end{matrix}, \text{ hence } \begin{matrix}\Pi'' \\ [B_j] \\ \underline{\Sigma_j'} \\ A\end{matrix} \text{ is SV, by lemma 4.1.12.}$$

If $\Pi_i' \equiv \frac{\Sigma'}{B_1 \vee B_2} \prec_1 \frac{\Sigma}{B_1 \vee B_2}$, and Π_i' reduces to Π_o with $\begin{matrix}\Pi'' \\ B_j\end{matrix}$ as a subderivation immediately above an endsegment of Π_o, then $\Pi_o \leq \Pi_i'$, and IV for Π' follows from IV for Π.

Similarly if Rule $(\Pi) = \exists E$.

Case (b). Π' is a proper reduction of Π, and case (a) does not apply. Then the major premiss B of the last inference of Π is the conclusion of an introduction.

When B is a conjunction, implication or universal quantification, we may apply, by condition II of the lemma, clauses (i) - (iii) in the definition of strong validity, and conclude that Π' is strongly valid. For example,

$$\text{let} \quad \Pi = \left\{ \frac{\begin{matrix}\Pi'' \\ A_1\end{matrix} \qquad \dfrac{\begin{matrix}[A_1] \\ \Pi''' \\ A_2\end{matrix}}{A_1 \rightarrow A_2}}{A_2} \right. , \quad \text{then} \quad \Pi' = \left\{ \begin{matrix}\Pi'' \\ [A_1] \\ \Pi''' \\ A_2\end{matrix} \right. .$$

Since $\begin{matrix}\Pi'' \\ A_1\end{matrix}$, $\dfrac{\begin{matrix}[A_1] \\ \Pi''' \\ A_2\end{matrix}}{A_1 \rightarrow A_2}$ are strongly valid by (II), the strong validity of Π' is immediate by clause (ii) in the definition of strong validity.

If B is a disjunction or existential formula, the strong validity of Π' is immediate by (IV).

Case (c). Rule (Π) = IND, and case (a) does not apply. Suppose

$$\Pi = \left\{ \frac{\begin{matrix}\Pi_1 \\ A0\end{matrix} \qquad \begin{matrix}[Aa] \\ \Pi_2(a) \\ Asa\end{matrix}}{A0} \right. , \qquad \Pi' = \left\{ \begin{matrix}\Pi_1 \\ A0\end{matrix} \right. ,$$

then the strong validity is immediate by (III).

Suppose $\Pi = \left\{ \begin{array}{c} \quad [Aa] \\ \Pi_1 \quad \Pi_2(a) \\ \underline{A0 \quad Asa} \\ Ast \end{array} \right.$, $\Pi' = \left\{ \begin{array}{c} \quad [Aa] \\ \Pi_1 \quad \Pi_2(a) \\ \underline{A0 \quad Asa} \\ [At] \\ \Pi_2(t) \\ A(st) \end{array} \right.$.

Now $\Pi'' = \left\{ \begin{array}{c} \quad [Aa] \\ \Pi_1 \quad \Pi_2(a) \\ \underline{A0 \quad Asa} \\ At \end{array} \right.$ is strongly valid, since it has a lower induction value $(\alpha'', \beta'', \gamma'', \delta'', \varepsilon'')$: $\alpha = \alpha''$, $\beta = \beta''$, $\gamma = \gamma''$, $\delta = \delta''$, $\varepsilon = \varepsilon'' + 1$. Then by condition III Π' is strongly valid.

Case (d). Rule $(\Pi) = \curlywedge_I$, and case (a) does not apply. A single example illustrates this case.

Let $\Pi = \left\{ \begin{array}{c} \Pi'' \\ \curlywedge \\ \hline A_1 \& A_2 \end{array} \right.$, $\Pi' = \left\{ \begin{array}{c} \begin{array}{cc} \Pi'' & \Pi'' \\ \curlywedge & \curlywedge \\ \hline A_1 & A_2 \end{array} \\ \hline A_1 \& A_2 \end{array} \right.$,

$\begin{array}{c} \Pi'' \\ \curlywedge \\ \hline A_1 \end{array}$, $\begin{array}{c} \Pi'' \\ \curlywedge \\ \hline A_2 \end{array}$ satisfy the conditions of the lemma and have a lower induction value, hence are SV ; therefore Π' is SV .

Case (e). Π' is obtained by a permutative reduction from Π, and case (a) does not apply. We discuss the case of an $\exists E$ - reduction, the case of $\vee E$ - reductions being similar.

Let $\Pi = \left\{ \begin{array}{c} \quad [Ba] \\ \Pi_0 \quad \Pi_1(a) \\ \underline{\exists x\, Bx \quad C} \\ \underline{C \quad \Sigma} \\ A \end{array} \right.$, $\Pi' = \left\{ \begin{array}{c} \quad [Ba] \\ \Pi_0 \quad \Pi_1(a) \\ \exists x\, Bx \quad \dfrac{C \quad \Sigma}{A} \\ \hline A \end{array} \right.$.

The lowest occurrence of C shown is major premiss in an elimination. We wish to verify that Π' satisfies conditions I - IV of the lemma.

Condition IV for Π' implies condition I for Π', since the reduction tree of $\dfrac{\Pi_0}{\exists x\, Bx}$ is finite because of condition I for $\exists$, and condition IV for Π' implies that $\begin{array}{c} [Ba] \\ \Pi_1(a) \\ \underline{C \quad \Sigma} \\ A \end{array}$ is strongly valid, hence by 4.1.3 the reduction tree of this deduction is finite.

Conditions II, III are vacuously fulfilled for Π', so it remains to be shown that condition IV is fulfilled. For this we have to show that

$$\Pi_2 = \left\{\begin{array}{c} [Ba] \\ \Pi_1(a) \\ \dfrac{C \qquad \Sigma}{A} \end{array}\right. \qquad \text{and} \qquad \Pi_3 = \left\{\begin{array}{c} \Pi_4 \\ [Bt] \\ \Pi_1(t) \\ \dfrac{C \qquad \Sigma}{A} \end{array}\right.$$

are strongly valid, under the assumption that $\begin{array}{c}\Pi_0\\ \exists x\, Bx\end{array}$ reduces to a Π_5 such that $\begin{array}{c}\Pi_4\\ Bt\end{array}$ is the sub-deduction of Π_5 immediately above an endsegment.

We note that the induction value $(\alpha_2, \beta_2, \gamma_2, \delta_2, \epsilon_2)$ of Π_2 is lower than $(\alpha, \beta, \gamma, \delta, \epsilon)$, since $\alpha_2 = \alpha$, $\beta_2 \leq \beta$, $\gamma_2 < \gamma$. The induction value $(\alpha_3, \beta_3, \gamma_3, \delta_3, \epsilon_3)$ of Π_3 is also lower than $(\alpha, \beta, \gamma, \delta, \epsilon)$, since $\alpha_2 = \alpha$, $\beta_3 < \beta$, as is seen by the following reasoning.

Assume e.g.

$$\dfrac{\begin{array}{c}\Pi_0\\ \exists x\, Bx\end{array} \qquad \begin{array}{c}[Ba]\\ \Pi_1(a)\\ C\end{array}}{C} \;\geq\; \left\{ \dfrac{\dfrac{\begin{array}{c}\Pi_n^*\\ \exists x\, D_n\end{array} \qquad \begin{array}{c}\dfrac{\begin{array}{c}\Pi_1^*\\ \exists x\, D_1\end{array} \qquad \dfrac{\begin{array}{c}\Pi_3\\ Bt\end{array}}{\exists x\, Bx}}{\exists x\, Bx}\\ \vdots \\ \exists x\, Bx\end{array}}{\exists x\, Bx} \qquad \begin{array}{c}[Ba]\\ \Pi_1(a)\\ C\end{array}}{C} \right.$$

(for simplicity we assume Π_4 to have a single endsegment) ; then by a number of permutative reductions, we obtain as a result a deduction Π_4' containing a sub-deduction

$$\dfrac{\dfrac{\begin{array}{c}\Pi_3\\ Bt\end{array}}{\exists x\, Bx} \qquad \begin{array}{c}[Ba]\\ \Pi_1(a)\\ C\end{array}}{C}$$

which by a proper $\exists$ - reduction reduces to Π_3. Therefore $\beta_3 < \beta$.

It is also obvious that Π_2, Π_3 satisfy (I) of the lemma, since Π satisfies this condition (cf. our argument for $\beta_3 < \beta$). Condition II is satisfied for Π_2, since if $\text{Rule}(\Pi_2) \in \{\vee E, \rightarrow E, \& E\}$, then

$$\Pi_6 = \left\{ \dfrac{\begin{array}{c}\Pi_0\\ \exists x\, Bx\end{array} \qquad \begin{array}{c}[Ba]\\ \Pi_1(a)\\ C\end{array}}{C} \right.$$

is strongly valid, hence $\begin{array}{c}[Ba]\\ \Pi_1(a)\\ C\end{array}$ is SV ; the derivations of Σ are also SV because of (II) for Π.

Condition II is satisfied for Π_3, since Σ is SV, and

$$\begin{matrix}\Pi_4\\ [Bt]\\ \Pi_1(t)\\ C\end{matrix}$$

is SV by clause (iv)(c) in the definition of SV, applied to Π_3.
Condition III is vacuously satisfied for Π_2, Π_3.
Condition IV is satisfied for Π_2, Π_3 because it is satisfied for Π.
For example, let Π be of the form

$$\dfrac{\dfrac{\begin{matrix}\Pi_0\\ \exists x\, Bx\end{matrix}\qquad \begin{matrix}[Ba]\\ \Pi_1(a)\\ \exists x\, Dx\end{matrix}}{\exists x\, Dx}\qquad \begin{matrix}[Db]\\ \Pi_7(b)\\ A\end{matrix}}{A}$$

and assume

$$\dfrac{\begin{matrix}\Pi_0\\ \exists x\, Bx\end{matrix}\qquad \begin{matrix}[Ba]\\ \Pi_1(a)\\ \exists x\, Dx\end{matrix}}{\exists x\, Dx} \;\geq\; \dfrac{\exists x\, Bx\qquad \begin{matrix}[Ba]\\ \Pi'_1(a)\\ \dfrac{Dt}{\exists x\, Dx}\end{matrix}}{\exists x\, Dx}\ .$$

Then by condition IV on Π, $\begin{matrix}[Ba]\\ \Pi'_1(a)\\ [Dt]\\ \Pi_7(t)\\ A\end{matrix}$ is SV, etc. etc.

Therefore we may apply the lemma to Π_2, Π_3 (induction hypothesis); this in turn shows IV to hold for Π'.
Therefore Π' satisfies condition I, and thus has an induction value $(\alpha', \beta', \gamma', \delta', \epsilon')$ for which obviously $\beta' \leq \beta$, $\gamma' < \gamma$. So Π' is strongly valid by the induction hypothesis.

4.1.17. <u>Theorem</u>. All deductions are strongly valid under substitution.

<u>Proof</u>. The proof is by induction on the length of a deduction.

1°) The basis step for deductions of length 1 is trivial.

2°) If Rule(Π) is an I-rule, the induction step is also trivial: clauses (i)-(iii) in the definition of strong validity.

3°) If Rule(Π) is not an I-rule, the induction step is also trivial: we have to apply the preceding lemma.

For this, it is sufficient to show that lemma 4.1.16 implies that a derivation Π:

$$\frac{\Pi_1 \ldots \Pi_n}{A},$$

Rule (Π) not an I-rule, is strongly valid under substitution if $\Pi_1, \ldots, \Pi_n$ are strongly valid under substitution. Let Π^* be any deduction obtained from Π by substituting terms for parameters which are not proper parameters in Π, and substituting strongly valid deduction of open hypotheses of Π for these hypotheses. Then Π^* is of the form

$$\frac{\Pi_1^*, \ldots, \Pi_n^*}{A^*}.$$

$\Pi_1^*, \ldots, \Pi_n^*$ are SV, if we assume $\Pi_1, \ldots, \Pi_n$ to be SV under substitution; so condition II holds for Π^*, condition I holds because of theorem 4.1.13; condition III holds also because of the strong validity under substitution of Π_2, which yields that the substitution of an SV deduction $\begin{matrix}\Pi' \\ A^*t\end{matrix}$, if Rule (Π) = IND and

$$\Pi_2^* = \begin{cases} [A^*a] \\ \Pi_2^*(a) \\ A^*sa \end{cases}, \quad \text{in} \quad \begin{matrix} [A^*t] \\ \Pi_2^*(t) \\ A^*st \end{matrix} \quad \text{is SV}.$$

Condition IV holds also for Π^*. For assume Rule (Π) to be $\exists E$, i.e.

$$\Pi_1 = \begin{cases} \Pi_1 \\ \exists x\, Bx \end{cases}, \quad \Pi_2 = \begin{cases} [Ba] \\ \Pi_2(a) \\ A \end{cases}.$$

Then $$\Pi_1^* = \begin{cases} \Pi_1^* \\ \exists x\, B^*x \end{cases}, \quad \Pi_2^* = \begin{cases} [B^*a] \\ \Pi_2^*(a) \\ A^* \end{cases}.$$

Assume Π_1^* to reduce to e.g.

$$\dfrac{\begin{matrix}\Pi_n'' \\ \exists x\, D_n x\end{matrix} \qquad \begin{matrix} \dfrac{\begin{matrix}\Pi_1'' \\ \exists x\, D_1 x\end{matrix} \qquad \dfrac{\begin{matrix}\Pi'' \\ B^*t\end{matrix}}{\exists x\, B^*x}}{\exists x\, B^*x} \\ \vdots \\ \exists x\, B^*x \end{matrix}}{\exists x\, B^*x}$$

(For simplicity we have assumed a single endsegment).

Since Π_1^* is SV, this is also SV, hence by clause (iv)(c) in the definition of SV, applied n times,

$$\dfrac{\begin{matrix}\Pi' \\ B^*t\end{matrix}}{\exists x\, B^*x} \text{ is SV; then by clause (i), } \begin{matrix}\Pi' \\ B^*t\end{matrix} \text{ is SV}.$$

The strong validity under substitution of Π_2 yields that

$$\begin{matrix} \Pi' \\ [B^*t] \\ \Pi_2^*t \\ A^* \end{matrix} \quad \text{is strongly valid.}$$

Similarly, if Rule$(\Pi) = \vee E$. This completes the proof of (IV) for Π^*, and thereby the proof of the theorem.

4.1.18. Remark. The proof can be greatly simplified if we disregard permutative reductions. Then in the proof of the lemma we may use as induction value $(\alpha, \delta, \epsilon)$, where α, δ and ϵ have the same meaning as before, and we can leave out the discussion of the troublesome case (e).

4.1.19. Uniqueness of normal form of deductions.

To establish uniqueness of normal form relative $R_{c\wedge}$, we have essentially the same methods available as in 2.2.27 - 34 for normal forms of terms in $\underset{\sim}{N} - \underset{\sim}{HA}^{\omega}$. Technically the easiest (but less elementary) method runs parallel to 2.2.32 - 33.

4.1.20. Lemma (Analogue of 2.2.31). If $\Pi \succ_1 \Pi'$, $\Pi \succ_1 \Pi''$, then there is a Π^* with $\Pi' \succeq \Pi^*$, $\Pi'' \succeq \Pi^*$.

Proof. We distinguish a number of cases and subcases.

1°). The contractions used to obtain Π', Π'' respectively are disjoint, i.e. were applied to disjoint subtrees of Π. Both contractions may be applied one after another, and their order is interchangeable ; the result is in both cases Π^*.

2°). The contraction which transforms Π into Π'' is applied to a subtree of the tree involved in the contraction from Π to Π'. We have to distinguish a number of subcases.

a) The reduction from Π to Π' is a &-reduction. Then we have a situation as follows:

$$\Pi = \left\{ \dfrac{\dfrac{\dfrac{\begin{matrix}\Pi_1\\A\end{matrix} \qquad \begin{matrix}\Pi_2\\B\end{matrix}}{A \,\&\, B}}{A} \qquad \Sigma}{\Pi_3} \right. , \qquad \Pi' = \left\{ \dfrac{\begin{matrix}\Pi_1\\A\end{matrix} \qquad \Sigma}{\Pi_3} \right. .$$

The contraction used for the transition from Π to Π'' applies to (a subtree of) $\begin{matrix}\Pi_1\\A\end{matrix}$ or $\begin{matrix}\Pi_2\\A\end{matrix}$. In both cases the contractions may be applied one after another, and are interchangeable ; the result is Π^*.

b) The reduction from Π to Π' is a permutative reduction. Then e.g.

$$\Pi = \left\{ \begin{array}{c} \dfrac{\dfrac{\dfrac{\begin{array}{ccc}\Pi_1 & \Pi_2 & \Pi_3\\ A\vee B & C & C\end{array}}{C} \quad \Sigma}{D}}{\Pi_1} \end{array}\right. \qquad \Pi' = \left\{ \begin{array}{c} \dfrac{\dfrac{\begin{array}{ccc}\Pi_1 & \dfrac{\begin{array}{cc}\Pi_2 & \\ C & \Sigma\end{array}}{D} & \dfrac{\begin{array}{cc}\Pi_3 & \\ C & \Sigma\end{array}}{D}\\ A\vee B & & \end{array}}{D}}{\Pi_1} \end{array}\right. .$$

In the reduction from Π to Π'' there are two possible cases :

b1) $\begin{array}{c}\Pi_1\\ A\vee B\end{array}$, $\begin{array}{c}\Pi_2\\ C\end{array}$, $\begin{array}{c}\Pi_3\\ C\end{array}$ are reduced ; then the contractions are again interchangeable, as under (a).

b2) The reduction from Π to Π'' reduces Σ to Σ''.

First applying the contraction from Σ to Σ'', then the $\vee$E-contraction has the same effect as first applying the $\vee$E-contraction, and then contraction of Σ to Σ'' twice. The result is in both cases Π^*.

c) All other cases can be treated in a very similar manner, see 2.2.25 (cf. also our remarks under 4.1.6).

4.1.21. Theorem. The normal form of a deduction in HA relative $R_{C\wedge}$ is uniquely determined.

Proof. Completely parallel to the proof in 2.2.26.

Remark. Of course, the other method mentioned in 2.2.34 can be applied also ; its transcription to the present context is completely routine.

§ 2. Applications of the normalization theorem.

4.2.1. Contents. In this section we investigate the structure of normal deductions in HA, and give some applications of the normalization theorem. Strictly positive parts (s.p.p.'s) are supposed to be defined as in 1.10.5.

4.2.2. Definition. A path in a deduction Π is a sequence of formula occurrences $A_1, \ldots, A_n$ such that

(i) A_1 is an assumption not to be discharged by $\vee E$, $\exists E$, or A_1 is the conclusion of an application of IND.

(ii) If A_i is not major premiss of an $\vee E$-, $\exists E$-application, then A_{i+1} is the formula occurrence immediately below A_i in Π; A_i is not minor premiss of an $\rightarrow E$ or IND if $i < n$.

(iii) If A_i is major premiss of a (non-redundant) application of $\vee E$, $\exists E$, then A_{i+1} is one of the assumptions discharged by the application;

(iv) A_n is the end formula (= conclusion) of Π, or a minor premiss of an application of $\rightarrow E$, or a premiss of an IND-application, or a premiss of a redundant $\vee E$, $\exists E$-application.

Remarks. (i) Every formula occurrence in Π belongs to a path. If we omit "or a premiss of a redundant $\vee E$, $\exists E$-application" in (iv), this is only true for deductions without redundant applications of $\vee E$, $\exists E$. The formula occurrences not belonging to a path on this alternative definition disappear if we apply immediate simplifications.

(ii) For practical purposes, the concept of spine defined below turns out to be more convenient and useful.

4.2.3. Definition. A spine $A_1, \ldots, A_n$ of a deduction Π is a sequence of formula occurrences in Π such that

(i) A_n is conclusion of Π, A_{i+1} occurs immediately below A_i for $1 \leq i < n$;

(ii) For $1 \leq i < n$ A_i is either premiss of an introduction or λ_I-application, or a basic rule, or A_i is major premiss of an elimination with conclusion A_{i+1};

(iii) A_1 is a top formula or the conclusion of an application of IND.

Remarks. (i) A deduction may have more than one spine, due to &I-applications.

(ii) If a spine does not pass through $\vee E$, $\exists E$-applications, it is a path.

(iii) The concept of spine was suggested by Martin-Löf's use of "main branch" (Martin-Löf 1971), but since it does not coincide with that concept, we have introduced a new term for it.

4.2.4. Lemma. In a normal deduction, a path can be divided into segments $\sigma_1, \dots, \sigma_n$; the segments may be divided into
(a) an elimination part (E-part) $\sigma_1, \dots, \sigma_{m-1}$, where each segment σ_i $(1 \leq i < m-1)$ is major premiss of an elimination, and contains σ_{i+1} as a subformula ;
(b) a minimum part $\sigma_m, \dots, \sigma_{m+k-1}$, in which each segment except the last one is premiss of $\wedge_I$ or a basic rule ;
(c) an introduction part (I-part) $\sigma_{m+k}, \dots, \sigma_n$, where each segment is the conclusion of an introduction and a subformula of the immediately preceding one.

Proof. Entirely straightforward, deriving a contradiction from the assumption that an introduction would precede a $\wedge_I$-application or a basic inference or an elimination, or that a $\wedge_I$-application with atomic conclusion or a basic inference would precede an elimination.

4.2.5. Lemma. In a path in a normal deduction, each formula in the E-part of the path is a s.p.p. of the first formula in the path.

Proof. Straightforward, noting that the conclusion of an application of the E-rule is s.p.p. of the major premiss of the application, and that the relation " A is s.p.p. of B " is transitive.

4.2.6. Lemma. In a normal deduction, a spine $A_1, \dots, A_n$ can be divided into three parts.
(a) an elimination part (E-part) $A_1, \dots, A_{m-1}$, where each A_j $(1 \leq j < m-1)$ is major premiss of an elimination with conclusion A_{j+1},
(b) a minimum part $A_m, \dots, A_{m+k-1}$, where each formula except the last one is premiss of $\wedge_I$ or a basic rule,
(c) an introduction part (I-part) $A_{m+k}, \dots, A_n$, where each A_j $(m+k \leq j < n)$ is premiss of an introduction with conclusion A_{j+1}.

Proof. Similar to the proof of 4.2.4.

Remark. If in a spine $A_1, \dots, A_n$ A_1 is a Harrop formula (1.10.5), i.e. does not contain disjunctions or existential formulae as s.p.p., the spine does not pass through $\vee$E- or $\exists$E-applications, and therefore coincides with a path.

4.2.7. Lemma. (i) In a normal deduction the top formula of a spine cannot be an assumption discharged by a $\vee$E- or $\exists$E-application.
(ii) The top formula of a spine of a strictly normal derivation with closed conclusion not ending with an introduction is an open assumption of the deduction, or a basic axiom.

Proof. (i) Let $\sigma \equiv A_1, \ldots, A_n$ be a spine of a normal deduction Π; if A_1 would be an assumption discharged by an $\vee$E - or $\exists$E - application, then A_1 would occur above a minor premiss of a $\vee$E - or $\exists$E - application, hence σ would have to pass through such a minor premiss which is excluded by the definition of spine.

(ii) Let $\sigma \equiv A_1, \ldots, A_n$ be once again a spine of a strictly normal deduction Π, not ending with an introduction. A_1 cannot be an assumption discharged by a $\vee$E - or $\exists$E - application, nor an assumption discharged by $\rightarrow$ - introduction, since no $\rightarrow$ - introduction occurs below A_1 (lemma 4.2.6, and our hypothesis that σ does not end with an introduction). It is also excluded that A_1 would be the conclusion of an IND - application, for then A_1 would be of the form Bb, b the induction term of the IND - application; but b would then be a redundant parameter, since no IND - application nor $\forall$I - application occurs below A_1, and also A_1 does not occur above a minor premiss of an $\exists$E - application, so b cannot be a proper parameter of $\forall$I -, $\exists$E - or IND - applications. Therefore A_1 is an open assumption of the deduction, or a basic axiom.

4.2.8. Theorem. A strictly normal deduction without open assumptions, of a closed formula, ends with an application of a basic inference, an atomic application of $\wedge_I$ or an application of an I - rule. If the conclusion of the deduction is not atomic, the final rule applied is an I - rule.

Proof. Let Π be a deduction satisfying the assumptions of the theorem. Since there are no open assumptions, a spine not ending with an introduction can only begin with a basic axiom belonging to the minimal part of the spine, hence the spine ends with an atomic formula also.

4.2.9. Corollary.

(i) If Π is a normal deduction of $A \vee B$, $A \vee B$ closed, without redundant parameters and without open assumptions, then Π contains a subdeduction of A or a subdeduction of B.

(ii) If Π is a normal deduction of $\exists x Ax$, $\exists x Ax$ closed, without redundant parameters and without open assumptions, then Π contains a subdeduction of $A\bar{n}$ for a suitable numeral $\bar{n}$.

Proof. Almost immediate by 2.8, since a normal deduction Π without redundant variables of $A \vee B$ must end with an introduction, i.e. $\vee_l$I or $\vee_r$I; hence Π is of the form

$$\dfrac{\begin{matrix}\Pi \\ A\end{matrix}}{A \vee B} \quad \text{or} \quad \dfrac{\begin{matrix}\Pi \\ B\end{matrix}}{A \vee B} \quad \text{etc. etc.}$$

4.2.10. Remark. Inspection shows that the proof of lemma's 4.2.6, 4.2.7 and theorem 4.2.8, as well as 4.2.11 below, do not require normalization with respect to permutative reduction rules, but only with respect to $R_{p\wedge}$.

4.2.11. Theorem. Let Γ be a finite set of closed Harrop formulae.

(i) In a strictly formal deduction of a closed formula A from assumptions Γ, the final rule applied is not $\vee E$ or $\exists E$; i.e. the endsegment of any main path has length 1.

(ii) In a strictly normal deduction of a closed disjunctive or existential formula from assumptions Γ the final rule applied is $\vee I$, $\exists I$ respectively.

Proof. (i) Let Π be a strictly normal deduction of a closed formula A from Γ, and assume the final rule applied in Π to be $\vee E$ or $\exists E$, say e.g. $\vee E$. Then a spine $\sigma \equiv A_1, \ldots, A_n$ in Π passes through the major premiss of this $\vee E$-application, i.e. A_{n-1} is of the form $C \vee D$, and coincides with its E-part.
By lemma 4.2.7 (ii), A_1 is a basic axiom or $A_1 \in \Gamma$. The first possibility is ruled out since A_{n-1} is not atomic. Now let A_i be the first formula occurrence in σ which is major premiss of a $\vee E$- or $\exists E$-application; $A_1, \ldots, A_i$ then coincides with an initial piece of a path, and hence by lemma 4.2.5 A_i is a s.p.p. of A_1. But since the elements of Γ are Harrop formulae, which do not contain disjunctive or existential formulae as s.p.p.'s, a contradiction follows. Hence there is no such A_i, and therefore Π cannot end with a $\vee E$- or $\exists E$-application.
(ii) Let Π be as before, and let $A \equiv B \vee C$ (for $A \equiv \exists x B$ the argument is similar). Π cannot end with an atomic application of $\wedge_I$ or a basic rule, since A is not atomic. If Π would end with an elimination, a spine σ would coincide with its E-part, and as under (i) it would follow that σ cannot pass through an application of $\vee E$ or $\exists E$; but then $A \equiv B \vee C$ would be a s.p.p. of an element of Γ, which is excluded. Therefore Π must end with a $\vee$-introduction.

4.2.12. Corollary. Let Γ be a finite set of closed Harrop formulae.

(i) A strictly normal deduction of $A \vee B$, $A \vee B$ closed, from assumptions Γ contains a subdeduction of A from Γ or a subdeduction of B from Γ.

(ii) A strictly normal deduction of $\exists x Ax$, $\exists x Ax$ closed, from assumptions Γ contains a subdeduction of $A\bar{n}$ from Γ, for a suitable numeral $\bar{n}$.

4.2.13. Corollary. (The IP-rule, without parameters)
$\vdash \neg A \rightarrow \exists x Bx \Rightarrow \vdash \exists x(\neg A \rightarrow Bx)$ (x not free in A, $\neg A \rightarrow \exists x Bx$ closed).

<u>Proof</u>. Let Π be a strictly normal deduction of $\neg A \to \exists x\, Bx$, then Π must end with an $\to$I-application (4.2.8), so Π is of the form

$$\dfrac{\Pi'' = \left\{ \begin{array}{c} [\neg A] \\ \Sigma \\ \hline \exists x\, Bx \end{array} \right.}{\neg A \to \exists x\, Bx} \quad .$$

By 4.2.11 Π'' must also end with an introduction, i.e. Π'' is of the form

$$\begin{array}{c} [\neg A] \\ \Sigma' \\ \hline Bt \\ \hline \exists x\, Bx \end{array} \quad .$$

We then obtain a derivation of $\exists x\,(\neg A \to Bx)$ as follows:

$$\begin{array}{c} [\neg A] \\ \Sigma' \\ \hline Bt \\ \hline \neg A \to Bt \\ \hline \exists x\,(\neg A \to Bx) \end{array} \quad .$$

4.2.14. <u>Theorem</u>. (Markov's rule MR_{PR}, cf. 1.11.5, § 3.5.)
Let P be any basic constant of $n+1$ arguments of our language, and let P^*y denote any formula with a single free variable y obtained by substituting numerals for all arguments of P except one.
Then $\vdash \neg \forall y\, P^*y \Rightarrow \vdash \exists y\, \neg P^*y$.

Note that this implies the validity of MR^c_{PR}, Markov's rule for primitive recursive predicates applied to closed formulae, since for any primitive recursive predicate $A(x)$, we can find a constant of our language P such that $P(x,0) \leftrightarrow Ax$; and $\neg \forall y\, Ay \leftrightarrow \neg \forall y\, P(y,0) \leftrightarrow \neg \forall y\, P^*y$ etc.

<u>Proof</u>. A closed strictly normal deduction of $\neg \forall y\, P^*y$ takes the following form

$$\dfrac{\Pi = \left\{ \begin{array}{c} [\forall y\, P^*y] \\ \Sigma \\ \hline \curlywedge \end{array} \right.}{\forall y\, P^*y \to \curlywedge} \quad .$$

We wish to show that the tree Π^*, obtained by deleting all assumption formulae of the form $\forall y\, P^*y$ from Π is again a proof tree, deriving $\curlywedge$ from assumptions of the form $P^*\bar{n}$.
To see this, we show that every path σ in Π begins with $\forall y\, P^*y$, $P^*\bar{n}$, followed by atomic formulae only, or with a basic axiom, followed by atomic formulae only.

Since Π ends with a minimal formula, the I-part of a spine of Π is empty. Hence, as before, the top formula of the spine, if not an axiom, must be an assumption formula $\forall y P^*y$; and since $\forall y P^*y$ does not contain a disjunction or existential quantifier, a spine does not pass through a $\vee$E- or $\exists$E-application. Hence a spine coincides with a path ending in Con(Π) (= main path) and vice versa.

Let σ be a main path (= spine) of Π beginning in an open assumption. Since σ only consists of an E-part and a minimal part, the first rule applied must be $\forall$E, so the path starts with $\forall y P^*y$, P^*t. t is a numeral, since if t would be of the form $s^n b$, b a parameter, b would be redundant, no $\forall$I-application occurring below P^*t, and P^*t not occurring above a minor premiss of an $\exists$-elimination. Therefore P^*t is a closed formula belonging to the minimum part of the path. Obviously, the path does not pass through an $\rightarrow$-elimination or IND-application, hence all paths are main paths. As a consequence, Π^* provides a deduction of $\wedge$ from a set of assumptions $P^*\bar{n}_1, \ldots, P^*\bar{n}_u$. Since P^* is decidable, it follows that we can construct a proof of $\exists x \neg P^*x$ from Π^*.

4.2.15. The result below is a modification of Theorem 6 in Scarpellini A, but Scarpellini's result is obtained using a calculus of sequents. We give a proof in the context of natural deduction systems.

Definition. Let Φ denote a class of formulae defined inductively by:

(i) Prime formulae belong to Φ, and formulae $\exists x P(x)$, $P(a)$ prime belong to Φ;

(ii) $A, B \in \Phi \Rightarrow A \& B \in \Phi$;

(iii) $B \in \Phi \Rightarrow A \rightarrow B \in \Phi$;

(iv) $A(a) \in \Phi \Rightarrow \forall x A(x) \in \Phi$.

4.2.16. Theorem. Let Ψ be a set of closed formulae, $\Psi \subseteq \Phi$, such that $\underset{\sim}{\mathrm{HA}} + \Psi$ is conservative over $\underset{\sim}{\mathrm{HA}}$ w.r.t. closed Σ^0_1-formulae. Then $\underset{\sim}{\mathrm{HA}} + \Psi$ satisfies ED and DP.

Proof. Note that a formula from Φ does not contain disjunctions as s.p.p.'s, and existential s.p.p.'s are of the form $\exists x P(x)$, $P(a)$ prime. Assume Π to be a strictly normal deduction of $\exists x Ax$, $\exists x Ax$ closed, from assumptions belonging to Ψ. If Π ends with an introduction, it must be an $\exists$-introduction, and then Π contains as a subdeduction Π', deriving $A\bar{n}$ from assumptions belonging to Ψ, for a suitable numeral $\bar{n}$.

So in the remainder of the argument, let σ be a spine of Π, and assume Π to end with an application of an E-rule (atomic $\wedge_I$- or basic rules are excluded, since the conclusion of Π is not atomic). Hence $\sigma \equiv A_1, \ldots, A_n$ coincides with its E-part. By lemma 4.2.7 (ii),

$A_1 \in \Psi \subseteq \Phi$, or A_1 is a basic axiom.

1°). Let A_i be the first major premiss of a $\vee E$- or $\exists E$-application (assuming there is such an application occurring in σ). Then A_i is an s.p.p. of A_1, of the form $B \vee C$ or $\exists x B$. The first is excluded, the second case is only possible if $A_1 \in \Psi$.

2°). A_i cannot contain a parameter free, since such a parameter would be redundant: A_i does not occur above a minor premiss of a $\vee E$- or $\exists E$-application, nor above an application of $\forall I$ or IND. Therefore Π is of the following form

$$\dfrac{\dfrac{\begin{matrix}\Pi' \\ A_i \equiv \exists x Bx\end{matrix} \quad \begin{matrix}[Ba] \\ \Pi''(a) \\ C\end{matrix}}{A_{i+1} \equiv C}}{\begin{matrix}\Pi''' \\ \hline \exists x Ax\end{matrix}} \quad .$$

3°). The only assumptions open in $\dfrac{\Pi'}{\exists x Bx}$ are assumptions from Ψ, since no $\to I$- or IND-application occurs below A_i. Hence $\underset{\sim}{HA} + \Psi \vdash \exists x Bx$, and thus $\underset{\sim}{HA} \vdash \exists x Bx$; therefore we can find a derivation Π^* in $\underset{\sim}{HA}$ of $B\bar{n}$ for some numeral $\bar{n}$, using basic rules and axioms only.

Now change Π into

$$\Pi_1 = \left\{ \begin{matrix} \Pi^* \\ [B\bar{n}] \\ \Pi''(\bar{n}) \\ A_{i+1} \\ \Pi''' \\ \hline \exists x\ Ax \end{matrix} \right. \quad .$$

4°). Π_1 is again strictly normal; to see this, we must realize that the minor premiss C in Π could not have been introduced by an I-rule, since below A_{i+1} only E-rules are applied.

So Π_1 satisfies the same general assumptions as Π. Π_1 contains fewer applications of E-rules than Π. Hence if we repeat the process, it must finally terminate in a derivation Π^* in which no spine passes through an $\exists$-elimination or $\vee$-elimination. But then a spine σ of Π^* must end either with an introduction, or $\exists x Ax$ is a s.p.p. of the top formula of σ. In the latter case, $\exists x Ax$ is a closed Σ_1^0-formula derivable in $\underset{\sim}{HA} + \Psi$, hence in $\underset{\sim}{HA}$. In both cases, we can find a deduction in $\underset{\sim}{HA} + \Psi$ of $A\bar{n}$ for a suitable numeral $\bar{n}$.

4.2.17. Corollary. $\underset{\sim}{HA} + M_{PR}$ satisfies ED, DP.

Proof. $\underset{\sim}{HA} + M_{PR}$ is conservative over closed Σ_1^0-formulae (see 3.6.7 (ii)); the universal closures of instances of M_{PR} all belong to Φ, hence the conditions of 4.2.16 are satisfied.

4.2.18. Theorem. Let $\underset{\sim}{H}$ be intuitionistic arithmetic with induction restricted to quantifier-free formulae, and let qf - $\underset{\sim}{HA}$ be the quantifier-free part of (the natural deduction system of) $\underset{\sim}{HA}$. Then $\underset{\sim}{H}$ is conservative over qf - $\underset{\sim}{HA}$.

Proof. Let Π' be a closed deduction in $\underset{\sim}{H}$ of a quantifier-free formula A; then we can find B such that qf - $\underset{\sim}{HA} \vdash A \leftrightarrow B$, B a prime formula. By the normalization theorem we can find a strictly normal, closed deduction Π of B.

Now consider any spine $A_1, \ldots, A_n$ of Π; A_1 is either a basic axiom, or an atomic conclusion of IND; in both cases the spine consists exclusively of atomic formulae, and especially, does not pass through an $\exists$- or $\vee$-elimination.

Let us define a spine of order 0 as a spine; a spine of order $n+1$ is defined as a spine, but its lowest formula is premiss of an application of IND, whose conclusion is top formula of a spine of order n.

We can easily establish, by induction on n, all spines of order n, for all n to consist of atomic formulae, and hence especially not passing through a $\vee$- or $\exists$-elimination. As a consequence, Π is a deduction in qf - $\underset{\sim}{HA}$.

4.2.19. Theorem. $\underset{\sim}{HA}$ is conservative w.r.t. closed formulae over the logic-free fragment without induction.

Proof. This theorem was established before by computability methods in combination with models for $\underset{\sim}{N}$ - $\underset{\sim}{HA}^{\omega}$ (2.5.6).
A proof via the normalization theorem is very similar to the argument under 4.2.14, but simpler: all spines in a closed deduction with closed conclusion must begin in a basic axiom, hence no $\vee$E- or $\exists$E-applications occur, and all formulae in the deduction are atomic.

4.2.20. Theorem (Reflection principle for closed Σ_1^0-formulae).
Let Proof_N denote the (standard) proof predicate for normal deductions in $\underset{\sim}{HA}$, i.e. $\mathrm{Proof}_N(x,y) \equiv_{def} \mathrm{Proof}(x,y) \,\&\, \mathrm{Norm}(x)$.
Then, for closed $\exists y Ay$

$$\underset{\sim}{HA} \vdash \mathrm{Proof}_N(x, \ulcorner \exists y Ay \urcorner) \rightarrow \exists y Ay .$$

Proof. By formalizing our argument in 4.2.11, 4.2.12.
$\underset{\sim}{HA} \vdash \mathrm{Proof}_N(x, \ulcorner \exists y Ay \urcorner) \rightarrow \exists y z \, \mathrm{Proof}_N(y, \ulcorner A\bar{z} \urcorner)$; formalizing 4.2.19 yields

$$\underset{\sim}{HA} \vdash \mathrm{Proof}_N(y, \ulcorner A\bar{z} \urcorner) \rightarrow \mathrm{Proof}_0(y, \ulcorner A\bar{z} \urcorner)$$

where Proof_0 is the proof predicate for the system without logic and induction; finally, using the reflection principle for this system, $\underset{\sim}{HA} \vdash \mathrm{Proof}_0(y, \ulcorner A\bar{z} \urcorner) \rightarrow A\bar{z}$ (cf. §1.5).

§ 3. Normalization for HA + IP, with applications.

4.3.1. In this section we study a natural deduction system for HA + IP, which is obtained by adding to the natural deduction system for HA the rule:

$$\text{IP}\ \frac{\neg A \rightarrow \exists x B}{\exists x(\neg A \rightarrow B)} .$$

As follows from the results in 4.2.13, 4.4.5 , this rule is derivable from null assumptions.

We add a new contraction (IP-contraction):

$$\begin{array}{c} [\neg A] \\ \Pi \\ \exists I\ \dfrac{Bt}{\exists x Bx} \\ \rightarrow I\ \overline{\neg A \rightarrow \exists x Bx} \\ \text{IP}\ \overline{\exists x(\neg A \rightarrow Bx)} \end{array} \quad \text{contr.} \quad \begin{array}{c} [\neg A] \\ \Pi \\ \rightarrow I\ \dfrac{Bt}{\neg A \rightarrow Bt} \\ \exists I\ \overline{\exists x(\neg A \rightarrow Bx)} \end{array} .$$

In Troelstra A, normalization for intuitionistic second order logic + IP, relative the reductions of R_C (extended to second order logic) + IP-reductions is discussed; the proof, an adaptation of the proof of strong normalization for intuitionistic second order logic in Prawitz 1971, App. B was only indicated.

Prawitz A contains a proof of a normalization theorem (including IP-reductions, but not including permutative reductions).

Here we prove strong normalization by extension of the method of § 1 (so ultimately by an extension of the method of Prawitz 1971, Appendix A).

4.3.2. Theorem. In HA + IP, every deduction has a finite reduction tree.

Proof. We only have to make some additions to the argument in § 1. The definition of strong validity remains unchanged; but we classify IP among the non-introduction rules.

Lemma's 4.1.10, 4.1.11, 4.1.12 carry over unchanged, just as theorem 4.1.13.

In lemma 4.1.16, we must read for II:

(II) If Rule(Π) $\in$ {& E, $\rightarrow$E, $\forall$E, $\bot_I$, IP}, then the elements of PRD(Π) are strongly valid.

The proof of 4.1.15 has to be extended with an additional case:

Case (f). Π' is obtained by an IP-contraction from Π.

Let $\Pi'' = \left\{ \dfrac{\dfrac{\begin{matrix}[\neg A]\\ \Pi\\ Bt\end{matrix}}{\exists x\, Bx}}{\neg A \to Bx} \right.$, $\Pi = \left\{ \dfrac{\Pi''}{\exists x\,(\neg A \to Bx)} \right.$, $\Pi' = \left\{ \dfrac{\dfrac{\begin{matrix}[\neg A]\\ \Pi\\ Bt\end{matrix}}{\neg A \to Bt}}{\exists x\,(\neg A \to Bx)} \right.$

Π'' is SV (by (II)), hence by SV (ii), for each strongly valid $\begin{matrix}\Pi'''\\ \neg A\end{matrix}$,

$$\dfrac{\begin{matrix}\Pi'''\\ [\neg A]\\ \Pi\\ Bt\end{matrix}}{\exists x\, Bx}$$

is SV. Therefore, by SV (i), for each SV deduction $\begin{matrix}\Pi'''\\ \neg A\end{matrix}$,

$$\begin{matrix}\Pi'''\\ [\neg A]\\ \Pi\\ Bt\end{matrix}$$

is SV. Therefore, by SV (ii), SV (i) again, Π' is strongly valid. Now theorem 4.1.17 carries over unchanged.

4.3.3. Definition. We define spine as in 4.2.3, but clause (ii) extended with: "or A_i is premiss of an application of IP".
We note that in a spine an introduction is never followed by an elimination, although an $\to$I - application may be followed by an IP - application.

4.3.4. Lemma. We consider deductions in HA + IP.

(i) In a normal deduction the top formula of a spine cannot be an assumption discharged by a $\vee$E - or $\exists$E - application.

(ii) A spine of a normal deduction, $A_1, \ldots, A_n$, may be divided into three parts:

(a) An E - IP - part, $A_1, \ldots, A_{i-1}$, where each A_k $(k < i-1)$ is either major premiss of an elimination, or premiss of an $\to$I -application discharging an assumption of the form $\neg B$, and then A_{k+1} is premiss of an IP - application and $k < i-2$, or A_k is premiss of an IP - application;

(b) A minimum part $A_i, \ldots, A_{i+j-1}$, where only basic rules and atomic instances of $\wedge_I$ are used;

(c) An I - part $A_{i+j}, \ldots, A_n$, where each A_k $(i+j \leq k < n)$ is premiss of an introduction.

(iii) The top formula of a spine of a strictly normal deduction not ending with an introduction with closed conclusion is a basic axiom, an open assumption of the deduction, or an assumption of the form $\neg B$ to be discharged by an $\to$ -introduction followed by IP.

Proof. (i) As for lemma 4.2.7 (i).
(ii) Let $A_1, \ldots, A_n$ be a spine; let A_i be the first formula not major premiss of an elimination.

(a) If A_i is atomic, let $i+j$ be the maximal index such that $A_i, \ldots, A_{i+j}$ are all atomic.
A_{i+j+1} (if existing) must be obtained by an introduction which cannot be followed by an IP-application nor an elimination. If two successive introductions occur in the spine, they can nevermore be followed by an IP-application or an elimination. So in this case the spine has the structure described in the lemma.

(b) If A_i is not atomic, there are the following possibilities:

(b') A_i is followed by a sequence of introductions ending in A_n, or $i = n$;

(b") A_i is followed by an $\rightarrow$-introduction and IP,

(b"') A_i is followed by an IP-application.

If (b") (resp. (b"')) is the case, we can consider $A_{i+2}, \ldots, A_n$ (resp. $A_{i+1}, \ldots, A_n$) and iterate our argument for $A_{i+2}, \ldots, A_n$ (resp. $A_{i+1}, \ldots, A_n$), etc. till we arrive at case (a) or (b'). Then we have shown that $A_1, \ldots, A_n$ can be split up into parts $\sigma_0 \equiv A_1, \ldots, A_{i_0}$, $\sigma_1 \equiv A_{i_0}, \ldots, A_{i_1}$, $\sigma_2 \equiv A_{i_1}, \ldots, A_{i_2}, \ldots, \sigma_m$, such that σ_k $(k < m)$ passes through a (possibly empty) sequence of eliminations followed by $\rightarrow$I, IP or by IP only, and σ_m passes through a sequence of eliminations, followed by a sequence of basic rules and atomic $\curlywedge_I$-applications, followed by a sequence of introductions. Hence the spine is of the form as described.

(iii) Similar to the argument for 4.2.7 (ii), using (i) and (ii) proved above.

4.3.5. Lemma. Let Π be a strictly normal deduction in HA + IP without open assumptions, with Con(Π) closed. Then a spine of Π does not contain IP-applications, and ends with an application of a basic rule, an atomic instance of $\curlywedge_I$ or an application of an I-rule. If the conclusion of the deduction is not atomic, the final rule applied is an I-rule.

Proof. Let $A_1, \ldots, A_n$ be a spine of Π, and let i be the maximal index such that $A_1, \ldots, A_i$ passes through eliminations only. Then A_i is a s.p.p. of A_1, since A_1 does not contain disjunctions or existential formulae as a s.p.p. By 4.3.6 (iii) A_i can be atomic only; so no IP-application or $\rightarrow$I-application followed by an IP-application is possible. Therefore a spine in this case has the same structure as in a normal deduction in HA, and for the remainder of the lemma we may argue as in 4.2.8.

4.3.6. **Theorem.** In $\underset{\sim}{HA} + IP$ every strictly normal deduction without open assumptions of $\exists x Ax$ contains a subdeduction of $A\bar{n}$, if $\exists x Ax$ is closed; similarly, strictly normal deductions without open assumptions of $A \vee B$, $A \vee B$ closed, contain either a subdeduction of A or a subdeduction of B. Hence also ($A \vee B$ closed, $\exists x Cx$ closed)

$$\underset{\sim}{HA} + IP \vdash A \vee B \Rightarrow \underset{\sim}{HA} + IP \vdash A \text{ or } \underset{\sim}{HA} + IP \vdash B$$
$$\underset{\sim}{HA} + IP \vdash \exists x Cx \Rightarrow \underset{\sim}{\exists} n (\underset{\sim}{HA} + IP \vdash C\bar{n}) .$$

Proof. Immediate by 4.3.5.

§ 4. Formalization of the normalization theorem, with applications.

4.4.1. Conventions. Below we shall describe the formalization of the normalization theorem (for subsystems of bounded logical complexity) in HA itself. In describing the formalization we tacitly identify the various syntactical categories and objects with arithmetic predicates and natural numbers ; $\mathrm{Rule}(\Pi)$, $\mathrm{Prd}_1(\Pi)$, $\mathrm{Con}(\Pi)$ etc. are now assumed to be primitive recursive numbers, $\mathrm{Norm}(\Pi)$ (Π is normal) a recursive predicate of Π etc. (This is automatically ensured for anyone of the standard gödel-numberings.)

We need some additional conventions and abbreviations.

(a) $\mathrm{Rule}(\Pi)$ is a number, such that $j_1 \mathrm{Rule}(\Pi) = 0$ for the I - rules
$j_1 \mathrm{Rule}(\Pi) \neq 0$ for the other rules ;

(b) $\mathrm{Red}_1(\Pi', \Pi)$ is a primitive recursive relation, expressing that Π' is obtained by a single contraction (applied to a subdeduction of Π) from Π ;

(c) $\mathrm{Red}(n', n) \equiv_{def} \exists m(\mathrm{lth}(m) > 0 \,\&\, (m)_o = n \,\&\, (m)_{\mathrm{lth}(m) \dot{-} 1} = n' \,\&$
$\& \forall i < \mathrm{lth}(m) \dot{-} 1[\mathrm{Red}((m)_i, (m)_{i+1})]$;

(d) $V_1(n,m)$ is a primitive recursive predicate expressing :
m is a subdeduction occurring immediately above an endsegment of n, and n, m are (code numbers of) deductions ;

(e) If Π' is a proper subdeduction from Π, $\Pi' < \Pi$.

4.4.2. Formal definition of strong validity for deductions with conclusion of bounded complexity.

Let us assume SV_{d-1}, the predicate of strong validity for deductions of formulae of logical complexity $< d$ to be given, and consider the following closure conditions on a set of deductions X :

(i) $\mathrm{Rule}(\Pi) = \& I \,\&\, SV_{d-1}(\mathrm{Prd}_1\Pi) \,\&\, SV_{d-1}(\mathrm{Prd}_2\Pi)$,

(ii) $\mathrm{Rule}(\Pi) = \vee I_l \,\&\, SV_{d-1}(\mathrm{Prd}_1\Pi) \vee$
$\vee\, \mathrm{Rule}(\Pi) = \vee I_r \,\&\, SV_{d-1}(\mathrm{Prd}_2\Pi)$,

(iii) $\mathrm{Rule}(\Pi) = \exists I \,\&\, SV_{d-1}(\mathrm{Prd}_1\Pi)$,

(iv) $\mathrm{Rule}(\Pi) = \rightarrow I \,\&\, \forall \Pi'[\mathrm{Con}\,\Pi' = \mathrm{Prem}(\mathrm{Con}\,\Pi) \,\&$
$\& SV_{d-1}(\Pi') \rightarrow SV_{d-1}\{\mathrm{Sub}(\mathrm{Prem}(\mathrm{Con}\,\Pi), \Pi', \mathrm{Prd}_1(\Pi), \mathrm{Ass}(\Pi))\}]$,

(v) $\mathrm{Rule}(\Pi) = \forall I \,\&\, \forall x[\mathrm{Term}(x) \rightarrow \mathrm{Subst}(\mathrm{Param}(\Pi), x, \mathrm{Prd}_1(\Pi))]$

(vi) $j_1 \mathrm{Rule}(\Pi) \neq 0 \,\&\, \mathrm{Norm}(\Pi)$,

(vii) $j_1 \mathrm{Rule}(\Pi) \neq 0 \,\&\, \forall \Pi'(\mathrm{Red}_1(\Pi', \Pi) \rightarrow X\Pi')) \,\&$
$\&[\mathrm{Rule}(\Pi) = \exists E \rightarrow \exists n\{n \neq 0 \,\&\, \forall \Pi'(\mathrm{Red}(\Pi', \mathrm{Prd}_1\Pi) \rightarrow$
$\rightarrow \exists i < \mathrm{lth}(n)(\Pi' = (n)_i) \,\&\, \forall i < \mathrm{lth}(n)(\mathrm{Red}((n)_i, \mathrm{Prd}_1(\Pi))) \,\&$
$\& X(\mathrm{Prd}_2\Pi) \,\&\, \forall i < \mathrm{lth}(n)\, \forall m \leq (n)_i\{V_1((n)_i, m) \,\&$
$\& \mathrm{Rule}(m) = \exists I \rightarrow X(\mathrm{Subst}(\mathrm{Param}(a), \mathrm{Term}(m), \mathrm{Prd}_2(\Pi)))\}\}] \,\&$

$\& [\mathrm{Rule}(\Pi) = \vee E \rightarrow \exists n\{n \neq 0 \;\&\; \forall \Pi' (\mathrm{Red}(\Pi', \mathrm{Prd}_1\, \Pi) \rightarrow$
$\rightarrow \exists i < \mathrm{lth}(n)(\Pi' = (n)_i) \;\&\; \forall i < \mathrm{lth}(n)(\mathrm{Red}((n)_i, \mathrm{Prd}_1(\Pi)) \;\&$
$\& X(\mathrm{Prd}_2\Pi) \;\&\; X(\mathrm{Prd}_3\Pi) \;\&$
$\& \forall i < \mathrm{lth}(n)\; \forall m \leq (n)_i \{V_1((n)_i, m) \;\&\; \mathrm{Rule}(m) = \vee I_r \rightarrow$
$\rightarrow X(\mathrm{Sub}(\mathrm{Con}(\mathrm{Prd}_1(m)), \mathrm{Prd}_1 n, \mathrm{Prd}_2\Pi, \mathrm{Ass}(\Pi)))\} \;\&$
$\& \forall i < \mathrm{lth}(n)\; \forall m \leq (n)_i \{V_1((n)_i, m) \;\&\; \mathrm{Rule}(m) = \vee I_l \rightarrow$
$\rightarrow X(\mathrm{Sub}(\mathrm{Con}(\mathrm{Prd}_1(m)), \mathrm{Prd}_1 n, \mathrm{Prd}_3\Pi, \mathrm{Ass}\,\Pi))\}]$.

Now let $A(X,\Pi)$ (Π a number variable ranging over proofs) be given as

$$(i) \vee (ii) \vee (iii) \vee \ldots \vee (vii) .$$

Now SV_d must satisfy

(1) $A(SV_d, \Pi) \rightarrow SV_d(\Pi)$

(2) $\forall \Pi\, (A(R,\Pi) \rightarrow R\Pi) \rightarrow \forall \Pi\, (SV_d(\Pi) \rightarrow R\Pi)$

for all arithmetical R.

Inspection of (i) - (vii) shows that $A(X,\Pi)$ can be slightly rewritten so as to become an element of the class Γ introduced in § 1.4. Hence, by 1.4.5, SV_d is arithmetically definable, and can be proved in HA to satisfy (1), (2).

Since the proof of the normalization theorem for deductions of formulae of bounded complexity uses only the methods of intuitionistic arithmetic, the arithmetic definability of SV_d implies:

4.4.3. Theorem. In HA we can find a primitive recursive function $\varphi(z,y)$ such that for each n

(i) $HA \vdash \mathrm{Proof}_n(z, \ulcorner A \urcorner) \rightarrow \exists y\, (\mathrm{Proof}_n(\varphi(z,y), \ulcorner A \urcorner) \;\&\; \mathrm{Norm}\, \varphi(z,y))$,

(ii) $HA \vdash \forall z\, [\exists x\, \mathrm{Proof}_n(z,x) \rightarrow \exists y\, \mathrm{Norm}\, \varphi(z,y)]$,

(iii) $HA \vdash \varphi(z,0) = z$,

(iv) $HA \vdash \forall y (\neg \mathrm{Norm}\, \varphi(z,y) \rightarrow \mathrm{Red}_1(\varphi(z, Sy), \varphi(z,y)))$,

(v) $HA \vdash \mathrm{Norm}\, \varphi(z,y) \rightarrow \forall y' > y (\varphi(z,y) = \varphi(z,y'))$.

Proof. The proof is for the greater part already contained in the discussion of the preceding subsection.
Let us suppose a standard order of carrying out reductions to be prescribed, and let $\varphi(z,y)$ be the primitive recursive function , such that $\varphi(z,y)$ denotes the (gödelnumber of the) result of the yth reduction step in the prescribed order, applied to the deduction (with gödelnumber) z ; if there is no such reduction step, $\varphi(z,y) = \varphi(z,y')$ for the first y' for which $\varphi(z,y')$ is normal.
On this definition, (iii), (iv), (v) obviously hold.
We note that for deductions of formulae of bounded complexity, we can prove in HA that every reduction sequence terminates (and a fortiori our

standard reduction sequence according to the prescribed order); hence (iii) holds.
(i) is then satisfied in view of the fact that if $\mathrm{Red}(\Pi', \Pi)$, and Π does not contain formulae of logical complexity $> n$, then Π' satisfies the same restriction.

4.4.4. Remark. Now it will be clear why there was a slight advantage in strengthening Prawitz's definition of strong validity by requiring in clauses (iv) (b), (c) (4.1.9) the reduction tree of $\mathrm{Prd}_1(\Pi)$ to be finite instead of only requiring the termination of all reduction sequences, i.e. the well foundedness of the reduction tree. For the well foundedness requires for its expression function symbols, the finiteness not; and thereby it is possible to define SV_d in the language of HA, so that it becomes possible to apply 1.4.5 directly, without any intermediate steps.

4.4.5. Theorem. (Cf. 3.1.15.)
IPR $\mathrm{HA} \vdash \neg A \to \exists y\, By \Rightarrow \mathrm{HA} \vdash \exists y (\neg A \to By)$, for A, B arbitrary, y not free in A.

Proof. Let us assume for simplicity A, $\exists y\, B$ to contain a single parameter a free, and let

(1) $\qquad \mathrm{HA} \vdash \neg Aa \to \exists y\, B(a,y)$.

Then obviously, since (1) implies the existence of a definite proof from which, primitive recursively in x, proofs of $A\bar{x} \to \exists y\, B(\bar{x}, y)$ may be obtained,

$$\mathrm{HA} \vdash \forall x \exists z\, \mathrm{Proof}_n(z, \ulcorner \Rightarrow \neg A\bar{x} \to \exists y\, B(\bar{x},y) \urcorner).$$

Formalizing the reasoning in 4.2.13, and combining this with 4.4.2, we obtain:

$$\mathrm{HA} \vdash \forall x \exists u\, \mathrm{Proof}_n(u, \ulcorner \Rightarrow \exists y\, (\neg A\bar{x} \to B(\bar{x}, y)) \urcorner),$$

and with the help of the reflection principle:

$$\mathrm{HA} \vdash \exists y\, (\neg Aa \to Bay).$$

The case of more parameters is readily reduced to the case of a single free parameter.

4.4.6. Theorem. HA is closed under Church's rule:

CR $\qquad \mathrm{HA} \vdash \forall x \exists y\, A(x,y) \Rightarrow \mathrm{HA} \vdash \exists z \forall x \exists u\, [Tzxu \,\&\, A(x, Uu)]$.

Proof. Assume, for simplicity, $\forall x \exists y\, A(x, y)$ to be closed, and assume

(1) $\qquad \mathrm{HA} \vdash \forall x \exists y\, A(x, y)$.

Then there is a numeral $\bar{z}$

$$\mathrm{HA} \vdash \mathrm{Proof}_n(\bar{z}, \ulcorner \exists y\, A(x, y) \urcorner),$$

and therefore a primitive recursive function φ' of x, such that

(2) $\quad \underset{\sim}{\mathrm{HA}} \vdash \mathrm{Proof}_n(\varphi'(x), \ulcorner \exists y\, A(\overline{x}, y)\urcorner)$.

Let Ψ be a primitive recursive function such that for every closed $\exists x\, Bx$:

(3) $\quad \underset{\sim}{\mathrm{HA}} \vdash \mathrm{Proof}\,(x, \ulcorner \exists y\, By\urcorner)\ \&\ \mathrm{Norm}\,(x) \rightarrow \mathrm{Proof}\,(j_1 \Psi x, \ulcorner B(\overline{j_2 \Psi x})\urcorner)$.

Then it follows that

$$\underset{\sim}{\mathrm{HA}} \vdash \forall x\, \exists yz\, \mathrm{Proof}_n(y, \ulcorner A(\overline{x}, \overline{z})\urcorner) ,$$

combining (2), 4.4.2, (3).

Now using the partial reflection principle :

$$\underset{\sim}{\mathrm{HA}} \vdash \forall x\, \exists yz\, \mathrm{Proof}_n(y, \ulcorner A(\overline{x}, \overline{z})\urcorner)\ \&\ A(x, z)$$

and therefore, since $\min_u \mathrm{Proof}_n(j_1 u, \ulcorner A(\overline{x}, \overline{j_2 u})\urcorner)$ is a recursive function,

$$\underset{\sim}{\mathrm{HA}} \vdash \exists z\, \forall x\, \exists u(Tzxu\ \&\ A(x, Uu)) .$$

§ 5. Normalization for second order logic and arithmetic.

4.5.1. Introductory remarks. Basic sources are Girard 1971, Martin-Löf 1971 A, Prawitz 1971 (Appendix B), Girard 1972.

In Prawitz 1970, Prawitz established a normal form theorem for intuitionistic second order logic utilizing a Beth-type semantics (inessentially different from Kripke semantics in this case). In Girard 1971, a system of terms with variable types (cf. also 1.9.27 in this volume) was described which made the extension of the Dialectica interpretation to second order logic with full impredicative comprehension possible. In establishing normalizability by defining a computability predicate, analogous to the treatment for the first order case (cf. § 4.1), we encounter the following difficulty: the induction on the type structure in the definition does not work, because what complexity is to be assigned to a variable type? If a complicated type is substituted for a type variable, suddenly a high complexity is introduced. The difficulty was overcome by Girard by the introduction of to some extent arbitrary predicates as computability predicates ("candidats de réductibilité") to be assigned to certain occurrences of terms; computability of other terms was then defined relative such assignments. It turns out that computability predicates relative such assignments are themselves examples of such "candidats de réductibilité"; thus, by an essential application of impredicative comprehension on the meta level, it becomes possible to prove normalizability for all terms in Girard's system.

The ideas of Girard 1971 were applied in Martin-Löf 1971 A and Prawitz 1971 to second order intuitionistic arithmetic. Martin-Löf has also extended the method to simple type theory, as did Girard in Girard 1972. In Girard 1972 the isomorphism between terms and deductions (cf. our remarks in 4.1.6) has been exploited to the full.

A simple treatment, using Girard's idea, in the context of Schütte's system for intuitionistic type theory is in Osswald 1972, Pohlers 1973. Troelstra A discusses various extensions of these results to other systems (also with applications; special attention has been given to the addition of IP, and the so called "uniformity principle"

$$\text{UP} \qquad \frac{\forall X\, \exists x\, A(X,x)}{\exists x\, \forall X\, A(X,x)}\ .$$

Intuitionistic analysis formalized with sequence variables only, formalized as a calculus of sequents, has been extensively investigated by the methods of "pure proof theory" in Scarpellini 1971, 1972; they will not be discussed in this section.

The present section is not intended as a self-contained treatment, but contains some supplementary discussions.

4.5.2. The system $M_2(S)$.

As the basis for our discussion we take the system of minimal (= intuitionistic) second order logic $M_2(S)$ over the basic system S ; S contains a constant 0 (zero), = (equality) and a function constant s (successor).

The individual variables are $v_0, v_1, v_2, \ldots$ (usually indicated by means of meta variables $x, y, z, \ldots$) and variables for p-ary relations $V_0^p, V_1^p, V_2^p, \ldots$ $p = 0, 1, 2, \ldots$ (usually indicated by meta variables $X^p, Y^p, Z^p, \ldots$).

First and second order quantifiers are distinguished by indices: $\exists_1, \exists_2, \forall_1, \forall_2$; the other logical operators are $\&, \to, \vee, \lambda$. For simplicity, we shall restrict our attention to the language based on $\to, \forall_1, \forall_2$, regarding the other operators as defined symbols by the second order definitions (cf. Prawitz 1965 V, § 1).

$$A \,\&\, B \equiv_{def} \forall X\,[(A \to ((B \to X) \to X)]$$
$$A \vee B \equiv_{def} \forall X\,[(A \to X) \to ((B \to X) \to X)]$$
$$\exists x\, A \equiv_{def} \forall X\,[\forall x(A \to X) \to X]$$
$$\exists Y\, A \equiv_{def} \forall X\,[\forall Y(A \to X) \to X]$$
$$\lambda \equiv_{def} \forall X\, X$$

(X a zero-place predicate variable).

There are two variants of the system, with the λ-operator as a primitive, and without.

In the system with λ as a primitive, second order terms are defined thus:

1°) A p-place relation constant or parameter is a λ-place second order term;

2°) If A is a formula, $\lambda x_1 \ldots x_p A$ is a p-place second order term.

In this case, besides the first order rules we add

$$\forall_2 I)\ \frac{A(P^n)}{\forall X^n A(X^n)} \qquad \forall_2 E)\ \frac{\forall X^n A(X^n)}{A(T^n)}$$

$$\lambda I)\ \frac{At_1 \ldots t_n}{\{\lambda x_1 \ldots x_n\, Ax_1 \ldots x_n\} t_1 \ldots t_n} \qquad \lambda E)\ \frac{\{\lambda x_1 \ldots x_n\, Ax_1 \ldots x_n\} t_1 \ldots t_n}{At_1 \ldots t_n}$$

Here in $\forall_2 I$ P^n must not be proper parameter of another preceding inference, and must not occur free in assumptions on which $A(P^n)$ depends. In λE, $t_1, \ldots, t_n$ must be free for $x_1, \ldots, x_n$ in A.

If λ is absent, we use the notation $\lambda x_1 \ldots x_n A$ and the concept of second order term as purely metamathematical, and in $\forall_2 E$, $A(T^n)$ should be understood as an abbreviation for the formula obtained from $A(X^n)$ by replacing each occurrence of $X^n t_1 \ldots t_n$ by $B^* t_1 \ldots t_n$, if T^n is the metamathematical expression $\lambda x_1 \ldots x_n . Bx_1 \ldots x_n$, and where B^* is obtained from B by renaming bound individual variables so as to make free $t_1, \ldots, t_n$ for substitution in $B^* x_1 \ldots x_n$.

The corresponding additional reduction rules are given by $\forall_2$-contractions and (for the first version) λ-contractions:

$\forall_2$-contraction

$$\frac{\dfrac{\begin{array}{c}\Pi(P^n)\\ A(P^n)\end{array}}{\forall X^n A(X^n)}}{A(T^n)} \quad \text{contr.} \quad \begin{array}{c}\Pi(T^n)\\ A(T^n)\end{array}$$

λ-contraction

$$\frac{\dfrac{\begin{array}{c}\Pi\\ At_1 \ldots t_n\end{array}}{\{\lambda x_1 \ldots x_n \; Ax_1 \ldots x_n\} t_1 \ldots t_n}}{At_1 \ldots t_n} \quad \text{contr.} \quad \begin{array}{c}\Pi\\ At_1 \ldots t_n\end{array} .$$

4.5.3. Formalizing the proof of the normalization theorem.

The proofs of normalizability of Prawitz 1971 and Martin-Löf 1971 A can both be formalized in HAS. We verify this here for Prawitz's proof, since this is somewhat easier to describe, especially since we already have followed Prawitz's treatment for the first order case; but for Martin-Löf's proof the details are similar.

The crucial point in the whole formalization is to show definability in HAS of the predicate "strong validity relative η" for all deductions with a conclusion with a bound on the complexity relative η. Hence we restrict our attention to this point.

An assignment η in the sense of Prawitz 1971 may be conceived as an assignment of regular sets to occurrences of second order terms (not necessarily to all occurrences, and not necessarily the same regular set to occurrences of the same term). If η is given for Con(Π), and Rule(Π) is an I-rule, then η is automatically extended to the conclusions of the premisses of Rapp(Π).

Below we are going to define $SV_d(\eta, \Pi)$, strong validity relative η, for derivations with Con(Π) of complexity $\leq d$ relative η. (Complexity relative η is counted as logical complexity, but where the term occurrences in the domain of η are counted as atomic, i.e. of complexity 0.)

Some conventions:

(a) Param(Π), Subst are extended to the second order case in the obvious way.

(b) Regular (N) : N is a regular set. Dom (η) : domain of η. degree (T) : number of arguments of T.

(c) In clause (v) below, $\eta \cup \binom{T}{N}$ is shorthand for the assignment η, extended by assigning N to all occurrences of T obtained by substituting T for the occurrences of Param (Π) in Con $\mathrm{Prd}_1(\Pi)$.

Now we consider the following closure conditions:

(i) $\quad$ $\mathrm{Rule}(\Pi) = \to I \;\&\; \forall \Pi'(\mathrm{Con}\,\Pi' = \mathrm{Prem}(\mathrm{Con}\,\Pi) \;\&\; SV_{d-1}(\eta,\Pi') \to$
$\to SV_{d-1}(\eta, \mathrm{Sub}(\mathrm{Prem}(\mathrm{Con}\,\Pi), \Pi', \mathrm{Prd}_1\Pi, \mathrm{Ass}(\Pi))))$.

(ii) $\quad$ $\mathrm{Rule}(\Pi) = \forall_1 I \;\&\; \forall t\,(SV_{d-1}(\eta, \mathrm{Subst}(\mathrm{Param}\,\Pi, t, \mathrm{Prd}_1\Pi))$.

(iii) $\quad$ $\mathrm{Rule}(\Pi) = \lambda I \;\&\; \mathrm{Con}\,\Pi \in \mathrm{Dom}(\eta) \;\&\; \Pi \in \eta(\mathrm{Con}\,\Pi)$.

(iv) $\quad$ $\mathrm{Rule}(\Pi) = \lambda I \;\&\; \mathrm{Con}\,\Pi \notin \mathrm{Dom}(\eta) \;\&\; SV_{d-1}(\eta, \mathrm{Prd}_1\Pi)$.

(v) $\quad$ $\mathrm{Rule}(\Pi) = \forall_2 I \;\&\; \forall T\,[\mathrm{degree}(T) = \mathrm{degree}(\Pi) \;\&\; \forall N\,(\mathrm{Regular}(N) \to$
$\to SV_{d-1}(\eta \cup \binom{T}{N}, \mathrm{Subst}(T, \mathrm{Param}(\Pi), \mathrm{Prd}_1\Pi)))]$.

(vi) $\quad$ $j_1\,\mathrm{Rule}(\Pi) = 0 \;\&\; [\mathrm{Norm}(\Pi) \vee \forall \Pi'(\mathrm{Red}_1(\Pi',\Pi) \to X(\eta,\Pi'))]$.

Now $SV_d(\eta,\Pi)$ should satisfy:

$$(1)\quad \begin{array}{l} A(SV_d, \Pi, \eta) \to SV_d(\eta, \Pi) \\ \forall \Pi\, \forall \eta(A(R, \Pi, \eta) \to R(\eta, \Pi)) \to \forall \Pi\ \eta(SV_d(\eta, \Pi) \to R(\eta, \Pi)) \end{array}$$

for R in the language of HAS, where $A(X, \Pi)$ is

$$(i) \vee (ii) \vee \ldots \vee (vi).$$

Now we put

$$SV_d(\eta, \Pi) \equiv_{\mathrm{def}} \forall X\,[\forall \Pi' \forall \eta'(A(X, \Pi', \eta') \to X(\eta', \Pi')]$$

and $SV_d(\eta, \Pi)$ satisfies (1).

4.5.4. Construction of a satisfaction relation.

Actually, what we shall define below combines characteristics of a satisfaction relation and a truth definition as used for the first order case in § 4.4. (See remark 4.5.5 below.)

We first define, in HAS, a satisfaction relation Sat_n for formulae of $M_2(S)$ of logical complexity bounded by n, for all n. S is the system with constants 0, =, s (zero, equality, successor).

$\mathrm{Sat}_o(X,x)$, the satisfaction relation for prime formulae, x the gödel-number of a prime formula, is defined by cases such that:

$$\mathrm{Sat}_o(X, \ulcorner \bar t_1 = \bar t_2 \urcorner) \leftrightarrow \mathrm{Val}(\ulcorner \bar t_1 \urcorner) = \mathrm{Val}(\ulcorner \bar t_2 \urcorner),$$

where $\bar t_i$ denotes the term $t_i[\bar x]$ if $t_i \equiv t_i[x]$.
$\mathrm{Val}(\ulcorner \bar x \urcorner) = x$, by definition, and

$$\mathrm{Sat}_o(X, \ulcorner V_i^p\, \bar t_1 \ldots \bar t_p \urcorner) \leftrightarrow X_{(p,i)}(\mathrm{Val}(\ulcorner \bar t_1 \urcorner), \ldots, \mathrm{Val}(\ulcorner \bar t_p \urcorner)).$$

Here $X_{(p,i)}(t_1', \ldots, t_p')$ abbreviates $X(j(j(p,i), \langle t_1', \ldots, t_p' \rangle))$.

Finally we put

$$Sat_o(X, \curlywedge) \leftrightarrow \curlywedge .$$

Now Sat_{n+1} is defined by cases from Sat_n such that

$$Sat_{n+1}(X, \ulcorner A \circ B \urcorner) \leftrightarrow Sat_n(X, \ulcorner A \urcorner) \circ Sat_n(X, \ulcorner B \urcorner)$$

for $\circ \equiv \rightarrow, \&, \vee$, and

$$Sat_{n+1}(X, \ulcorner (Qv_i)A(v_i)\urcorner) \leftrightarrow (Qv_i)Sat_n(X, \ulcorner A(\bar{v}_i)\urcorner) ,$$

for $Q \equiv \forall_1, \exists_1$, and

$$Sat_{n+1}(X, \ulcorner (QV_i^p)A(V_i^p)\urcorner) \leftrightarrow (QY)\forall Z^1(\forall y_1 y_2(j(y_1,y_2) \neq$$
$$\neq j(p,i) \rightarrow Z^1_{(y_1,y_2)} = X_{(y_1,y_2)}) \& Z_{(p,i)} = Y \rightarrow Sat_n(Z, \ulcorner A(V_i^p)\urcorner) ,$$

for $Q = \forall_2, \exists_2$; $T'_o = T'_1$ abbreviates $\forall x(T'_o x \leftrightarrow T'_1 x)$.
In the presence of the λ-operator we should also add

$$Sat_{n+1}(X, \ulcorner \{\lambda x_1 \ldots x_p . Ax_1 \ldots x_p\}t_1 \ldots t_p \urcorner) \leftrightarrow Sat_n(X, \ulcorner At_1 \ldots t_p \urcorner) .$$

Below we shall leave out the λ-operator, and also disregard a possible distinction between (bound) variables and parameters.
To complete our specifications, we add

$$\neg Sat_n(X,m)$$

if m is not a gödelnumber of a formula of complexity $\leq n$ without free numerical variables.

4.5.5. Remark. In defining a satisfaction relation Sat_n, it would have seemed more natural, if we would have defined Sat_n in a conservative extension of HAS with function variables as a relation $Sat_n(\alpha, X, n)$ such that if we put

$$Val(\alpha, \ulcorner t \urcorner) = \begin{cases} x & \text{if } \ulcorner t \urcorner = \ulcorner s^x 0 \urcorner = \ulcorner \bar{x} \urcorner , \\ x + \alpha y & \text{if } \ulcorner t \urcorner = \ulcorner s^x v_y \urcorner , \end{cases}$$

$$Sat_o(\alpha, X, \ulcorner t_1 = t_2 \urcorner) \leftrightarrow Val(\alpha, \ulcorner t_1 \urcorner) = Val(\alpha, \ulcorner t_2 \urcorner)$$
$$Sat_o(\alpha, X, \ulcorner V_i^p t_1 \ldots t_p \urcorner) \leftrightarrow X_{(p,i)}(Val(\alpha, \ulcorner t_1 \urcorner), \ldots, Val(\alpha, \ulcorner t_p \urcorner)) ,$$

and

$$Sat_{n+1}(\alpha, X, \ulcorner (Qv_i)A(v_i)\urcorner) \leftrightarrow$$
$$\leftrightarrow (Qy)\ \forall\beta(\forall x(x \neq i \rightarrow \beta x = \alpha x) \& \beta i = y \rightarrow Sat_n(\beta, X, \ulcorner A(v_i)\urcorner)) ,$$

for $Q \equiv \forall_1, \exists_1$, and otherwise as before.

Actually, for any given n, the two definitions of satisfaction relation coincide, as can be shown in HAS itself; and for our applications it is more convenient to deal with numerical quantifications in exactly the same manner as we did in the first-order case.

4.5.6. Construction of a satisfaction relation, continued.

Let us denote the class of all formulae of $\underset{\sim}{M}_2(S)$ obtained by repeated mutual substitution from the set of formulae of complexity $\leq n$, by $Fm^{(n)}$. We wish to extend the definition of our satisfaction relation to $Fm^{(n)}$; we achieve this by first defining $Sat_{n,p}$, a satisfaction relation obtained from formulae of complexity $\leq n$ by iterating substitution at most p times. We put

$$Sat_{n,o}(X, \ulcorner A \urcorner) \equiv_{def} Sat_n(X, \ulcorner A \urcorner),$$

and we define $Sat_{n,p+1}$ from $Sat_{n,p}$, such that

$$(1)\quad \begin{cases} Sat_{n,p+1}(X, \ulcorner A \urcorner) \leftrightarrow Sat_{n,p}(X, \ulcorner A \urcorner) \vee \\ \vee \exists B \exists T_o \ldots T_m \exists Y [Sat_n(Y, \ulcorner B \urcorner) \,\& \\ \& \exists i_o, \ldots, i_m, p_o, \ldots, p_m(\ulcorner A \urcorner = \ulcorner [V^{p_o}_{i_o} / T_o, \ldots, V^{p_m}_{i_m} / T_m] B \urcorner) \,\& \\ \& \forall k \leq m \, \forall w (Sat_{n,p}(X, \ulcorner T_k(\overline{(w)}_o, \ldots, \overline{(w)}_m) \urcorner) \leftrightarrow \\ \leftrightarrow Y_{(i_k, p_k)}((w)_o, \ldots, (w)_m)))] . \end{cases}$$

Here $V^{p_o}_{i_o}, \ldots, V^{p_m}_{i_m}$ is supposed to be a complete list of the second-order variables free in B. As it stands, (1) is not formally correct, because it is not an expression in the language of $\underset{\sim}{HAS}$. "B", "$T_o, \ldots, T_m$" in "$\exists B$", "$\exists T_o \ldots T_m$" should be understood as variables for gödelnumbers of formulae and finite sequences of second-order terms respectively.

The right hand side of (1) may be abbreviated as

$$C(\lambda x . Sat_{n,p}(X, x), X, \ulcorner A \urcorner) .$$

From this we see that we can define $Sat^{(n)}(X, m) = \bigcup_p Sat_{n,p}(X, m)$ by

$$\begin{aligned} Sat^{(n)}(X, m) \equiv_{def} \exists Z \exists y \, (Z_o = \lambda x . Sat_n(X, x) \,\& \\ \& \forall i < y \, (Z_{i+1} = \lambda x . C(Z_i, X, x)) \,\& \, Z_y m) . \end{aligned}$$

Here Z_i abbreviates $\lambda x . Z(j(i, x))$.

4.5.7. Lemma. $Sat^{(n)}(X, \ulcorner A(\bar{v}_{i_1}, \ldots, \bar{v}_{i_k}, V^{p_1}_{j_1}, \ldots, V^{p_m}_{j_m}) \urcorner) \leftrightarrow$ $\leftrightarrow A(v_{i_1}, \ldots, v_{i_k}, X_{(j_1, p_1)}, \ldots, X_{(j_m, p_m)}))$ is provable in $\underset{\sim}{HAS}$ for $A \in Fm^{(n)}$.

Proof. We first note that

$$\exists Z \, (Z_o = \lambda x . Sat_n(X, x) \,\&\, \forall i < y \, (Z_{i+1} = \lambda x . C(Z_i, X, x)) \,\&\, Z_y m)$$

is in fact equivalent to $Sat_{n,y}(X, m)$ as defined before.

Now it is sufficient to show that if A is obtained by iterating substitution of second-order terms p times, then the assertion of the lemma holds with $Sat_{n,p}$ replacing $Sat^{(n)}$. This is proved by induction on p. The basis is provided by the corresponding property for $Sat_n(X, \ulcorner A \urcorner)$, which is (similar to § 4.4) proved by a straightforward induction on the logical complexity of A.

Consider as an example $Sat_{n,p+1}(X, \ulcorner \forall v_i\ B(v_i) \urcorner)$, and assume $\forall v_i\ B(v_i)$ to be obtained by p+1-fold iterated substitution. Then either already $Sat_{n,p}(X, \ulcorner \forall v_i\ B(v_i) \urcorner)$; or there is a formula $C(V^{p_o}_{j_o}, \ldots, v_i)$ such that $A \equiv \forall v_i B v_i \equiv [V^{p_o}_{j_o}/T_o, \ldots]\ \forall v_i\ C(V^{p_o}_{j_o}, \ldots, v_i) \equiv$
$\equiv \forall v_i [V^{p_o}_{j_o}/T_o, \ldots]\ C(V^{p_o}_{j_o}, \ldots, v_i)$, (C a formula of logical complexity $\leq n$, $T_o, \ldots,$ obtained by substitution iterated at most p times, and not containing v_i free).
In this case, we use the induction hypothesis for $Sat_{n,p}$ relative $T_o, \ldots$, together with $Sat_n(Y, \ulcorner \forall v_i\ C(\ldots, v_i) \urcorner) \leftrightarrow \forall v_i\ Sat_n(Y, \ulcorner C(\ldots, \bar{v}_i) \urcorner)$, etc. etc.

4.5.8. <u>Lemma</u>. In HAS we can prove

(i) $Sat^{(n)}(X, \ulcorner A \circ B \urcorner) \leftrightarrow Sat^{(n)}(X, \ulcorner A \urcorner) \circ Sat^{(n)}(\ulcorner B \urcorner)$
for $\circ \equiv \to, \&, \vee$.

(ii) $Sat^{(n)}(X, \ulcorner (Qv_i)A(v_i) \urcorner) \leftrightarrow (Qv_i)\ Sat^{(n)}(X, \ulcorner A(\bar{v}_i) \urcorner)$
for $Q \equiv \forall_1, \exists_1$,

(iii) $Sat^{(n)}(X, \ulcorner (QV^p_i)A(V^p_i) \urcorner) \leftrightarrow$
$\leftrightarrow (QY)\ \forall Z^1(\forall y_1 y_2(j(y_1,y_2) \neq j(p,i) \to Z^1_{(y_1,y_2)} = X_{y_1,y_2})\ \&$
$\&\ Z^1_{(p,i)} = Y \to Sat^{(n)}(Z, \ulcorner A(V^p_i) \urcorner))$,
for $Q \equiv \forall_2, \exists_2$. (Abbreviations as before.)

<u>Proof</u>. Immediate, as a corollary to the previous lemma.

4.5.9. <u>Lemma</u>. Let $\Gamma = \{C_1, \ldots, C_m\}$ and let $Proof^{(n)}$ denote the proof predicate of $\underset{\sim}{M}_2(S)$, restricted to deductions involving only formulae of $Fm^{(n)}$. Then

$$\text{HAS} \vdash Proof^{(n)}(x, \ulcorner C_1(\bar{v}_1, \ldots), \ldots, C_m(\bar{v}_1, \ldots) \Rightarrow A(\bar{v}_1, \ldots) \urcorner) \to$$
$$\to [Sat^{(n)}(X, \ulcorner C_1(\bar{v}_1, \ldots) \urcorner)\ \&\ \ldots\ \&\ Sat^{(n)}(X, \ulcorner C_m(\bar{v}_1, \ldots) \urcorner) \to$$
$$\to Sat^{(n)}(X, \ulcorner A(\bar{v}_1, \ldots) \urcorner)].$$

<u>Proof</u>. By induction on the length of x, cf. § 1.5.

4.5.10. <u>Corollary</u>. (Partial reflection principle for $Fm^{(n)}$).

$$\text{HAS} \vdash Proof^n(x, \ulcorner \Rightarrow A(\bar{v}_1, \ldots, \bar{v}_n) \urcorner) \to A(v_1, \ldots, v_n).$$

<u>Proof</u>. Combine lemma 4.5.7 and 4.5.9.

4.5.11. Applications of the preceding results.

We note that now we can obtain various closure conditions on the set of theorems, also for cases with free parameters, exactly as in § 4.4 for first-order arithmetic.

To see this, we note that for any deduction Π, in which all formulae are of complexity $\leq n$, the formulae occurring in any derivation in the reduction tree of a derivation $[a_1, a_2, \ldots / t_1, t_2, \ldots]\Pi$ ($a_1, a_2, \ldots$ being improper parameters of Π) all belong to $Fm^{(n)}$.

For example, we can now obtain closure under Church's rule CR, and under IPR (with parameters) for HAS.

As an example of an application which does not only use the partial reflection principle, but also the satisfaction relation itself in its rôle as a truth definition, we discuss closure under the rule of choice in the next theorem.

4.5.12. Theorem. HAS is closed under the following rule of choice

$$\vdash \forall x\, \exists X^p A(x, X^p) \Rightarrow \vdash \exists Y^{p+1}\, \forall x\, A(x, \lambda y_1 \ldots y_p . Y^{p+1}(x, y_1, \ldots, y_p)) .$$

Proof. We sketch the argument: full details are long and tedious, and hardly instructive.

Assume $\vdash \forall x\, \exists X^n A(x, X^n)$. For simplicity, we assume $\forall x\, \exists X^n A(x, X^n)$ to be closed. Suppose all formulae in the given derivation of $\forall x\, \exists X^n A(x, X^n)$ to have logical complexity $\leq n$. Then, using the formalization of the normalization theorem in HAS (similar to § 4.4):

(1) $\quad \mathbf{HAS} \vdash \forall x\, \exists y\, \exists z \in Tm_{n,p}\ \ Proof^n(Y, \varphi(\ulcorner A(\bar{x}, X^p)\urcorner, X^p, z))$

where $Tm_{n,p}$ is class of gödelnumbers of second-order terms in $Fm^{(n)}$ with p arguments, and $\varphi(\ulcorner B(X^p)\urcorner, X^p, z)$ indicates the gödelnumber resulting by substitution of the second-order term with gödelnumber z for X^p in $B(X^p)$.

Using Church's rule and combining this with lemma 4.5.9, we find a HAS-provably-recursive f such that

(2) $\quad \mathbf{HAS} \vdash \forall x(Tm_{n,p}(fx)\ \&\ Sat^{(n)}(V, \varphi(\ulcorner A(\bar{x}, X^p)\urcorner, X^p, fx)) .$

Now we make use of the following facts.

(a) For all $A \in Fm^{(n)}$, and fixed n, the assertion of lemma 4.5.7 can itself be shown to be m-satisfiable in HAS, for a suitable $m > n$ (a formula B is said to be m-satisfiable if $Sat^{(m)}(X, \ulcorner B(\bar{v}_1, \ldots)\urcorner)$ holds for all $X, v_1, \ldots$).

(b) The replacement of a certain second-order term of $Fm^{(n)}$ by a provably equal one of $Fm^{(n)}$ (where equality of second-order terms is defined as

being coextensive), yields a provably equivalent formula ; this fact can be
shown to be m' - satisfiable for suitable m' .

fx in (2) may be supposed to represent a closed second-order term for
all x , since $\forall x \, \exists X^n A(x, X^n)$ was assumed to be closed. Intuitively

$$(3) \qquad \lambda v_1 \ldots v_p \cdot \mathrm{Sat}^{(n)}(Z, \varphi(\ulcorner X^p \bar{v}_1 \ldots \bar{v}_p \urcorner, X^p, fx)$$

for any arbitrary Z , represents the required $\lambda v_1 \ldots v_p Y^{p+1}(x, v_1 \ldots v_p)$.
By fact (a), the equality of (3) and the term with gödelnumber fx is
provably m - satisfiable for some m . Combining this with (2) and fact (b),
we find that, for suitable m' , in $\underline{\mathrm{HAS}}$:

$$\mathrm{Sat}^{(m')}(V, \varphi(\ulcorner A(\bar{x}, X^p)\urcorner, X^p, \lambda v_1 \ldots v_p \cdot \mathrm{Sat}^{(n)}(Z, \varphi(\ulcorner X^p_1 \bar{v}_1 \ldots \bar{v}_p \urcorner, X^p_1, fx)))) \, .$$

Thus we finally obtain, using 4.5.7 repeatedly :

$$\forall x \, A(x, \lambda v_1 \ldots v_p \, \mathrm{Sat}^{(n)}(\lambda z . 0 = 0, \varphi(\ulcorner X^p_1 \bar{v}_1 \ldots \bar{v}_p \urcorner, X^p_1, fx))).$$

For A containing species variables besides X^p , the argument becomes
only slightly more complicated.

<u>Remark</u>. Note that, if for the provably recursive function no symbol is
available, we must work in a definitional extension of $\underline{\mathrm{HAS}}$. For the
argument it is irrelevant whether $\underline{\mathrm{HAS}}$ is formulated as an extension of
$\underline{M}_2(S)$, with N defined and induction provable, or as a system with an
induction axiom.

One can similarly show closure under a rule of dependent choices :

$$\vdash \forall X^n \, \exists Y^n \, A(X, Y) \Rightarrow$$
$$\Rightarrow \vdash \forall X^n \, \exists Z^{n+1} (\lambda v_1 \ldots v_n . Z(0, v_1 \ldots v_n) = X \, \& $$
$$\& \, \forall y \, A(\lambda v_1 \ldots v_n . Z(y, v_1 \ldots v_n), \lambda v_1 \ldots v_n . Z(sy, v_1 \ldots v_n))) \, .$$

Chapter V

APPLICATIONS OF KRIPKE MODELS

C. A. Smorynski

§ 1. Kripke models.

5.1.1. Discussion. In Kripke 1965, S. Kripke introduced a set-theoretic semantics for the intuitionistic predicate calculus. In this Chapter, we study this set-theoretic machinery and apply it to the investigation of Heyting's Arithmetic. Since the set-theoretic approach may seem out of place in a study of intuitionistic systems, we remark in Section 5.1.26 on how intuitionistic proofs of some of the results can be recovered.

Kripke's model theory bears no resemblance to intuitionistic reasoning despite various attempts to make it a plausible interpretation of intuitionistic reasoning. (The reader who disagrees will certainly change his mind by the time he finishes this chapter.) Formally, however, the same logical laws are valid in the Kripke models and in the intuitionistic predicate calculus. This fact, combined with the ease in handling the Kripke models, makes them an extremely useful tool in the metamathematical investigation of Heyting's Arithmetic.

Before defining the Kripke models, let us consider one of these interpretations in order to motivate somewhat the formal definition of a Kripke model. The interpretation we consider is that of intuitionistic logic as a logic of "positivistic research". We have various "states of knowledge", which form themselves into a partial order. At each state of knowledge there is a collection of objects we have mentally constructed. A larger state of knowledge may require us to mentally construct new objects. Also, an atomic relation, e.g. an equation, may or may not be seen to hold on the basis of a given state of knowledge. Obviously, if it is seen to be true on the basis of a given state of knowledge, it must be seen to be true on the basis of any extension of the given state of knowledge. Further, this should hold for more complicated properties than atomic relations. The problem, then, is to find an interpretation of the logical connectives and quantifiers which preserve this property. Conjunction, disjunction, and existential quantification are straightforward - e.g. we see $\exists x Ax$ to be true on the basis of some state of knowledge iff we have some mentally constructed object a such that Aa is seen to be true on the basis of this state of knowledge.

The other connectives and quantifier are problematical and it is here that the interpretation loses its plausibility. Consider, e.g., the implication $A \rightarrow B$. If $A \rightarrow B$ is adjudged true on the basis of a state of knowledge, then $A \to B$ is also true in any extension of this state of knowledge and, if A is true in such an extension, so is B. The converse, that, if, for every extension of our knowledge, once we know A to be true we also know B to be true, then we know $A \to B$ to be true, is not at all obvious; but a usable condition to define the connective $\to$ is needed and we accept it. Negation and the universal quantifier are treated similarly.

The interpretations of &, $\vee$, and $\exists$ seem natural enough, but those of the more negative connectives and quantifiers are a little forced. The net result is that, to show that we cannot assert the truth of a statement on the basis of a given state of knowledge, we appeal not to the lack of positive knowledge - but to the fact that some extension of our knowledge contains false assertions. Modifying the treatment of the negative connectives might make the interpretation more palatable. Such a task, however, lies beyond the scope of this Chapter and we turn now to the formal definition of Kripke's models.

5.1.2. Definition. By a Kripke model (Kripke 1965) we shall mean a quadruple $\underline{K} = (K, \leq, D, \Vdash)$, where $(K, \leq)$ is a non-empty partially ordered set, D is a non-descreasing function associating elements of K with non-empty sets, and $\Vdash$ is a relation between elements of K and formulae with no free variables (but which may possess constants denoting elements of the $D\alpha$'s) which satisfies the following (where small greek letters denote elements of K):

i) for $A(x_1,\ldots,x_n)$ atomic, $\beta \geq \alpha$, $a_1,\ldots,a_n \in D\alpha$, if $\alpha \Vdash A(a_1,\ldots,a_n)$, then $\beta \Vdash A(a_1,\ldots,a_n)$;

ii) $\alpha \Vdash A \& B$ iff $\alpha \Vdash A$ and $\alpha \Vdash B$;

iii) $\alpha \Vdash A \vee B$ iff $\alpha \Vdash A$ or $\alpha \Vdash B$;

iv) $\alpha \Vdash A \to B$ iff $\forall \beta \geq \alpha(\beta \Vdash A \Rightarrow \beta \Vdash B)$;

v) $\alpha \Vdash \neg A$ iff $\forall \beta \geq \alpha(\beta \nVdash A)$;

vi) $\alpha \Vdash \exists x Ax$ iff $\exists a \in D\alpha(\alpha \Vdash Aa)$;

vii) $\alpha \Vdash \forall x Ax$ iff $\forall \beta \geq \alpha\ \forall b \in D\beta(\beta \Vdash Ab)$.

The relation " $\alpha \Vdash A$ " may be read " A is true at α " or, for those familiar with set theory, " α forces A ". The elements of K will be denoted by small greek letters and will be called nodes in order to avoid confusion with the elements of the domains of the nodes - i.e. elements of the sets $D\alpha$. The triple $(K, \leq, D)$ is often called a quantificational model structure (or qms). If we restrict our attention to the propositional calculus, a propositional model structure (pms) is just a partially ordered set $(K, \leq)$ and a propositional model is a triple $(K, \leq, \Vdash)$, where $\Vdash$ satisfies (i) - (v).

As in classical model theory, one may define a notion of validity: A will be called valid in the model $\underline{K}$ iff $\alpha \Vdash A$ for all $\alpha \in K$. A will be called valid (universally valid) if A is valid in every model $\underline{K}$. More generally, if Γ is a set of formulae, we say Γ entails A, written $\Gamma \models A$, iff A is valid in every model in which every formula of Γ is valid. We shall prove later on:

$$\Gamma \models A \quad \text{iff} \quad \Gamma \vdash A .$$

5.1.3. Some basic properties of Kripke models.

Before giving some examples of Kripke models, let us remark on some of their basic properties. The first is that conditions (ii) - (vii) on the forcing relation $\Vdash$ constitute the recursion clauses for an inductive definition of a forcing relation on a qms. In particular, if we specify which atomic formulae are forced at which nodes of the qms (in such a manner that (i) holds), then the relation extends uniquely (by using clauses (ii) - (vii)) to a forcing relation on that qms.

A second remark is that the first condition on atomic formulae specifies a property that holds for all formulae. I.e. if $\alpha \Vdash A$ and $\alpha \leq \beta$, then $\beta \Vdash A$. The proof of this is by induction on the length of a formula. For atomic formulae, the result is immediate. Let A be a conjunction, say $A = B \& C$. Then

$$\begin{aligned} \alpha \Vdash A &\Rightarrow \alpha \Vdash B \text{ and } \alpha \Vdash C \\ &\Rightarrow \beta \Vdash B \text{ and } \beta \Vdash C, \text{ by induction hypothesis,} \\ &\Rightarrow \beta \Vdash B \& C . \end{aligned}$$

Disjunction and existential quantification are handled similarly. For implication, negation, and universal quantification, we use the fact that we have required our condition defining $\alpha \Vdash A$ to hold for all $\beta \geq \alpha$. For example, let $A = B \rightarrow C$ and $\beta \geq \alpha$.

$$\begin{aligned} \alpha \Vdash B \rightarrow C &\Rightarrow \underset{\sim}{\forall} \gamma \geq \alpha (\gamma \Vdash B \Rightarrow \gamma \Vdash C) \\ &\Rightarrow \underset{\sim}{\forall} \gamma \geq \beta (\gamma \Vdash B \Rightarrow \gamma \Vdash C), \text{ since } \beta \geq \alpha \\ &\Rightarrow \beta \Vdash B \rightarrow C . \end{aligned}$$

Negation and universal quantification are treated similarly.

A final remark is that the truth of $\alpha \Vdash A$ depends only on those β which are $\geq \alpha$ - each clause in the definition of the forcing relation refers only to those $\beta \geq \alpha$. Let $\underline{K}$ be a model and define $\underline{K}_\alpha = (K_\alpha, \leq_\alpha, D_\alpha, \Vdash_\alpha)$ for $\alpha \in K$ by:

$$K_\alpha = \{\beta \in K : \beta \geq \alpha\},$$

$\leq_\alpha$ and D_α are the restrictions of $\leq$ and D to K_α, and $\Vdash_\alpha$ is defined by letting $\beta \Vdash_\alpha A$ iff $\beta \Vdash A$, for A atomic and $\beta \in K_\alpha$. We should expect that, for any A, $\alpha \Vdash A$ in $\underline{K}$ iff $\alpha \Vdash_\alpha A$ in $\underline{K}_\alpha$. Indeed, a simple induction on the length of A shows that $\beta \Vdash A$ in $\underline{K}$ iff $\beta \Vdash_\alpha A$ in $\underline{K}_\alpha$ for all $\beta \in K_\alpha$. Thus, to verify that $\alpha \Vdash A$, we need only look at those $\beta \geq \alpha$ (i.e. we may restrict ourselves to the model $\underline{K}_\alpha$).

5.1.4. Examples. Let us first consider examples of models for the propositional calculus. We indicate the model by drawing a graph, the vertices of which determine nodes of the model. A node α precedes a node β in the ordering if the vertex corresponding to α is connected by a series of ascending lines to the vertex corresponding to β. E.g. $\alpha \leq \beta$ in the pms:

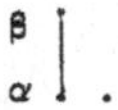

We indicate the forcing relation by writing atomic formulae next to the nodes forcing them. E.g. using the pms just given, we obtain a model by letting $\alpha \Vdash A$, $\beta \Vdash A, B$:

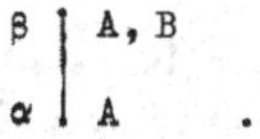

Observe that, in the model just given, (i) $\alpha \nVdash B \vee \neg B$; (ii) $\alpha \Vdash \neg\neg B$, but $\alpha \nVdash B$, whence $\alpha \nVdash \neg\neg B \rightarrow B$; (iii) β forces any tautology; and (iv) $\alpha \Vdash (C \rightarrow D) \vee (D \rightarrow C)$ for any formulae C, D.

One can get more complicated models by allowing the graphs to branch:

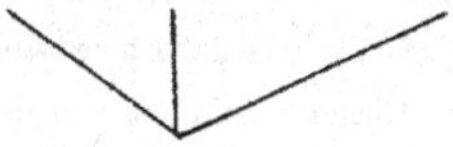

For the quantificational theory, we must add domains. Just as it is hard to draw models for classical theories, it will be hard to do this for intuitionistic theories. For simple cases, however, we may indicate the domains by listing their elements at each vertex of the graph. E.g.:

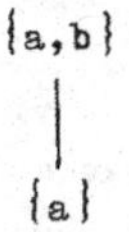

We may use this qms to construct a model:

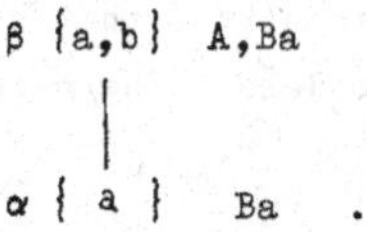

Here A is a propositional sentence. Observe that $\alpha \Vdash \forall x(A \vee Bx)$, but $\alpha \nVdash A \vee \forall x Bx$.

As one may easily verify, the formula $\forall x(A \vee Bx) \rightarrow A \vee \forall x Bx$ is valid in all models with constant domains (i.e. models in which D is a constant function), where we again assume that x does not occur free in A. (It is known that this class of models is <u>complete</u> for intuitionistic logic with this scheme added. Cf. <u>Gabbay</u> 1969 A or <u>Görnemann</u> 1971.)

Another interesting classically valid sentence which is not intuitionistically valid is $\neg\neg\forall x(Ax \vee \neg Ax)$. Consider the model :

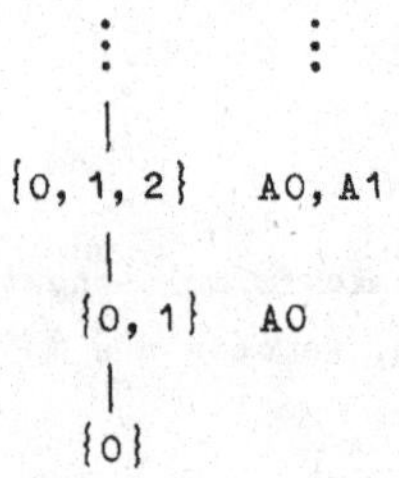

I.e. we have a sequence $\alpha_0 < \alpha_1 < \ldots$ of nodes with $D\alpha_n = \{0,\ldots,n\}$ and $\alpha_m \Vdash An$ iff $m > n$. Suppose $\alpha_0 \Vdash \neg\neg\forall x(Ax \vee \neg Ax)$. Then $\forall \beta \geq \alpha_0$ $\beta \nVdash \neg\forall x(Ax \vee \neg Ax)$. In particular, $\alpha_0 \nVdash \neg\forall x(Ax \vee \neg Ax)$. But then $\exists \beta \geq \alpha_0$ $\beta \Vdash \forall x(Ax \vee \neg Ax)$. Let $\beta = \alpha_n$. Letting $x = n$, $\alpha_n \Vdash An \vee \neg An$, i.e. $\alpha_n \Vdash An$ or $\alpha_n \Vdash \neg An$. But $\alpha_n \nVdash An$ by definition and $\alpha_n \nVdash \neg An$ since $\alpha_{n+1} \Vdash An$. It not only follows that $\alpha_0 \nVdash \neg\neg\forall x(Ax \vee \neg Ax)$, but, in fact, that $\alpha_0 \Vdash \neg\forall x(Ax \vee \neg Ax)$.

When we have a classical model, e.g. the standard model, ω, of arithmetic, instead of listing the domain and the atomic formulae to be forced, if we wish to force those atomic formulae true in the model, we simply place an ω at the vertex. E.g. if ω^+ and ω' are non-standard models of arithmetic, we will write

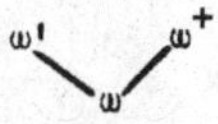

for the intended Kripke model.

We could continue to give several further examples of Kripke models, but feel it would be more instructive for the reader to construct some of his own. E.g. he may wish to construct countermodels to $\neg A \vee \neg\neg A$, $((A \rightarrow B) \rightarrow A) \rightarrow A$, $(A \rightarrow B) \vee (B \rightarrow A)$. We should like to stress that he should pay close attention to the geometry of his countermodels. The geometry of the Kripke models is the basic tool used in this Chapter.

5.1.5 - 5.1.11. The completeness theorem.

5.1.5. So far we have constructed a model theory for the intuitionistic predicate calculus and used this model theory to demonstrate the failure of certain basic laws of classical logic which are not intuitionistically valid. It is now our job to demonstrate how closely the model theory fits intuitionistic reasoning. Formally, the fit is exact:

5.1.6. Theorem. (The completeness theorem.) $\Gamma \vdash A$ iff $\Gamma \models A$.

The proof of soundness, $\Gamma \vdash A$ implies $\Gamma \models A$, is long but easy. One merely has to show that each axiom is valid and that the rules of inference preserve truth. E.g. consider the rule PL2: $A, A \to B \vdash B$. If $\underline{K} = (K, \leq, D, \Vdash)$ is given and $\alpha \in K$ is such that $\alpha \Vdash A$, $\alpha \Vdash A \to B$, then, by the definition of $\alpha \Vdash A \to B$, it follows that $\alpha \Vdash B$. Hence, this rule is sound.

The more ambitious reader may prove the soundness theorem for any of the formulations of the intuitionistic predicate calculus given in Chapter I.

We now turn to proving the completeness theorem. The weak form, $\vdash A$ iff $\models A$, is due to Kripke 1965. The form we shall prove, often called a strong completeness theorem, is due independently to Aczel 1968, Fitting 1969, and Thomason 1968. For the sake of subsection 5.1.26, we shall follow Thomason's treatment. These proofs are modelled on Henkin's proof for classical logic.

Let M be a first-order language containing

i) a denumerable set V_M of individual variables;

ii) a denumerable set C_M of individual constants, and

iii) for each $j \geq 0$, a denumerable set F^j_M of j-ary predicate letters.

Formulae are to be built up from atomic formulae by using $\&$, $\vee$, $\to$, $\neg$, $\exists$, and $\forall$. Fm_M will denote the set of such formulae. Note that Fm_M is denumerable. Sn_M will denote the set of sentences - i.e. the formulae with no free variables.

5.1.7. Definition. A set $\Gamma \subseteq Sn_M$ is called M-saturated if

i) Γ is consistent;

ii) $A \in Sn_M$ and $\Gamma \vdash A \Rightarrow A \in \Gamma$;

iii) $A, B \in Sn_M$ & $A \vee B \in \Gamma \Rightarrow A \in \Gamma$ or $B \in \Gamma$; and

iv) if $Ax \in Fm_M$, x is the only free variable in A and $\exists x Ax \in \Gamma$, then, for some $c \in C_M$, $Ac \in \Gamma$.

Those familiar with the algebraic representation theorems may consider a saturated set Γ to be a sort of counterpart to a prime filter in a distributive lattice. Basically, these prime filters will yield nodes of a model and their inclusion relations will yield an ordering. Matters are slightly complicated by the necessity of introducing new constants to successively enlarge the domains.

5.1.8. Lemma. Let $\Gamma \cup \{A\} \subseteq Sn_M$ and suppose $\Gamma \not\vdash A$. Let $\{c_1, c_2, \ldots\}$ be a denumerably infinite set of symbols disjoint from C_M and let M' be obtained by adding $\{c_1, c_2, \ldots\}$ to the constants C_M of M. Then there is an M'-saturated superset Γ_ω of Γ such that $A \notin \Gamma_\omega$.

Proof. Set $\Gamma_o = \Gamma$ and define Γ_{k+1} inductively as follows:

Case 1. k is even. Let $\exists xB$ be the first existential sentence of M' not already treated such that $\Gamma_k \vdash \exists xB$ and let c be the first constant in $\{c_1, c_2, \ldots\}$ not occurring in Γ_k. Then set $\Gamma_{k+1} = \Gamma_k \cup \{Bc\}$.

Case 2. k is odd. Let $B \vee B'$ be the first disjunctive formula of M' not already treated such that $\Gamma_k \vdash B \vee B'$. If $\Gamma_k \cup \{B\} \not\vdash A$, put $\Gamma_{k+1} = \Gamma_k \cup \{B\}$. Otherwise $\Gamma_{k+1} = \Gamma_k \cup \{B'\}$.

Finally, set $\Gamma_\omega = \bigcup_{k=o}^{\infty} \Gamma_k$. We must show that Γ_ω satisfies conditions (i) - (iv) of 5.1.7 above.

(i). We show by induction that $\Gamma_k \not\vdash A$. Let $\Gamma_{2n+1} \vdash A$. Then $\Gamma_{2n+1} = \Gamma_{2n} \cup \{Bc\}$ for some B, c where c does not occur in any formula of Γ_{2n}. Thus $\Gamma_{2n}, Bc \vdash A$, whence $\Gamma_{2n} \vdash Bc \to A$ and, by Q4, $\Gamma_{2n} \vdash \exists xB \to A$. But $\Gamma_{2n} \vdash \exists xB$, whence $\Gamma_{2n} \vdash A$, a contradiction.

Similarly, PL5 allows us to conclude that, if $\Gamma_{2n+2} \vdash A$, then $\Gamma_{2n+1} \vdash A$.

Hence, for all k $\Gamma_k \not\vdash A$. But $\Gamma_\omega \vdash A$ iff $\Gamma_k \vdash A$ for some k, from which it follows that $\Gamma_\omega \not\vdash A$.

(iii), (iv). If $B \vee C \in \Gamma_\omega$, then $\Gamma_i \vdash B \vee C$ for some i. Hence, for some odd $k \geq i$, $B \vee C$ is the first disjunction not treated. Thus $\Gamma_{k+1} = \Gamma_k \cup \{B\}$ or $\Gamma_k \cup \{C\}$, i.e. $B \in \Gamma_\omega$ or $C \in \Gamma_\omega$. Similarly, if $\exists xB \in \Gamma_\omega$, then $Bc \in \Gamma_\omega$ for some c.

(ii). If $\Gamma_\omega \vdash A$, then $\Gamma_\omega \vdash A \vee A$ and, by (iii), $A \in \Gamma_\omega$. Q.E.D.

5.1.9. Theorem. If $\underline{\Gamma}$ is M-saturated, then for some Kripke model $\underline{K} = (K, \leq, D, \Vdash)$, and for some $\alpha \in K$,

$$\underline{\Gamma} = \{A : \alpha \Vdash A\}.$$

In fact, α may be assumed to be a minimum element of $\underline{K}$.

Proof. Let $M_o = M$ and let M_{i+1} be obtained from M_i by adding the set $S_i = \{c_1^{i+1}, \ldots, c_n^{i+n}, \ldots\}$ to C_{M_i}, where $S_i \cap C_{M_i} = \emptyset$. Set $K = \{\Delta : \underline{\Gamma} \subseteq \Delta$ and Δ is M_i-saturated for some $i\}$. We define $\Delta \leq \Delta'$ iff $\Delta \subseteq \Delta'$, $D\Delta = C_{M_i}$, where Δ is M_i-saturated. Finally, for atomic formulae $A(c_1, \ldots, c_n)$ with $c_1, \ldots, c_n \in C_{M_i}$, let

$$\Delta \Vdash A(c_1, \ldots, c_n) \text{ iff } A(c_1, \ldots, c_n) \in \Delta.$$

We wish to show that this last equivalence holds for all applicable formulae (i.e. formulae with no free variables and whose parameters are from the proper language). For this, we need the following

5.1.10. Lemma. Let $\Delta \in K$.

(a) $B \to C \in \Delta$ iff $\forall \Delta' \supseteq \Delta \ (B \in \Delta' \Rightarrow C \in \Delta')$;

(b) $\neg B \in \Delta$ iff $\forall \Delta' \supseteq \Delta \ \ B \notin \Delta'$;

(c) $\forall x Bx \in \Delta$ iff $\forall \Delta' \supseteq \Delta \ \forall c \in D\Delta' \ \ Bc \in \Delta'$.

Proof. (a) If $B \to C \in \Delta$, $B \in \Delta'$, and $\Delta' \supseteq \Delta$, then $\Delta' \vdash C$ and, by saturation, $C \in \Delta'$. Conversely, suppose $B \to C \notin \Delta$. Then $\Delta \cup \{B\} \nvdash C$ and, for Δ M_i - saturated, there is, by lemma 5.1.8, an M_{i+1} - saturated $\Delta' \supseteq \Delta \cup \{B\}$ such that $C \notin \Delta'$. This contradicts the assumption $B \in \Delta' \Rightarrow C \in \Delta'$.

(b) Similar to (a).

(c) Again, one direction is trivial. Suppose, conversely, that $\forall \Delta' \supseteq \Delta \ \forall c \in D\Delta' \ Bc \in \Delta'$ and $\forall x Bx \notin \Delta$. Let Δ be M_i - saturated. Since $\Delta \nvdash \forall x Bx$, we conclude, by Q 1, that $\Delta \nvdash Bc$ for $c \in C_{M_{i+1}} - C_{M_i}$. Hence, there is an M_{i+1} - saturated $\Delta' \supseteq \Delta$ such that $\Delta' \nvdash Bc$, i.e. $Bc \notin \Delta'$.
Q. E. D.

We may now complete the proof of theorem 5.1.9 by proving by induction on the length of A that

$$\Delta \Vdash A(c_1, \ldots, c_n) \text{ iff } A(c_1, \ldots, c_n) \in \Delta ,$$

for $c_1, \ldots, c_n \in C_{M_i}$, where Δ is M_i - saturated. The case A is atomic follows by definition. The case $A = B \& C$ is trivial. Let $A = B \vee C$:

$$\begin{aligned} \Delta \Vdash B \vee C \ \ & \text{iff} \ \ \Delta \Vdash B \ \text{ or } \ \Delta \Vdash C \\ & \text{iff} \ \ B \in \Delta \ \text{ or } \ C \in \Delta, \ \text{ by induction hypothesis} \\ & \text{iff} \ \ B \vee C \in \Delta, \ \ \text{ by saturation.} \end{aligned}$$

Let $A = B \to C$:

$$\begin{aligned} \Delta \Vdash B \to C \ \ & \text{iff} \ \ \forall \Delta' \supseteq \Delta (\Delta' \Vdash B \to \Delta' \Vdash C) \\ & \text{iff} \ \ \forall \Delta' \supseteq \Delta (B \in \Delta' \to C \in \Delta'), \ \text{ by induction hypothesis} \\ & \text{iff} \ \ B \to C \in \Delta, \ \text{ by lemma 5.1.10.} \end{aligned}$$

The cases $A = \neg B$ and $\forall x B$ are similar.

Let $A = \exists x Bx$:

$$\begin{aligned} \Delta \Vdash \exists x Bx \ \ & \text{iff} \ \ \exists c \in D\Delta \ \Delta \Vdash Bc \\ & \text{iff} \ \ \exists c \in D\Delta \ Bc \in \Delta, \ \text{ by induction hypothesis} \\ & \text{iff} \ \ \exists x Bx \in \Delta, \ \ \text{ by saturation.} \end{aligned}$$

This completes the proof. Q. E. D.

We may now complete the proof of the completeness theorem.

5.1.11. Proof of theorem 5.1.6. We have yet to prove $\Gamma \models A$ implies $\Gamma \vdash A$. Let $\Gamma \nvdash A$ and find a saturated $\underline{\Gamma} \supseteq \Gamma$ such that $\underline{\Gamma} \nvdash A$. By theorem 5.1.9, there is a model $\underline{K} = (K, \leq, D, \Vdash)$ and $\alpha \in K$ such that for all B,

$\alpha \Vdash B$ iff $B \in \underline{\Gamma}$.

In particular, $\alpha \Vdash B$ for $B \in \Gamma$ and $\alpha \not\Vdash A$. Hence $\Gamma \not\models A$. Q.E.D.

5.1.12 - 5.1.18. <u>The Aczel slash</u>.

5.1.12. By theorem 5.1.9, for any M-saturated set Γ, there is a Kripke model $\underline{K}$ and a node α such that

$$\Gamma = \{A : \alpha \Vdash A\}.$$

The converse, that every such set is M'-saturated, where M' is obtained from M by extending C_M to include names for all elements of $D\alpha$, is an easy verification which we leave to the reader. As observed in <u>Aczel</u> 1968, we can obtain more information on M-saturation from the proof of theorem 5.1.9 than just this.

Observe that $\Gamma = \{A : \alpha \Vdash A\}$ for some α implies that α is a minimum element in the pms constructed. Thus, let us start with the model $\underline{K}$ constructed and add a new node α_0 such that $\alpha_0 \leq \alpha$ for all $\alpha \in K$, let $D\alpha_0 = C_M$, and extend the forcing relation by defining, for A atomic,

$$\alpha_0 \Vdash A \text{ iff } \Gamma \vdash A.$$

The <u>Aczel slash</u> is defined by

$$\Gamma \mid A \quad \text{iff} \quad \alpha_0 \Vdash A.$$

Also, define $|(\Gamma) = \{A : \Gamma \mid A\}$.

5.1.13. <u>Lemma</u>. $|(\Gamma)$ is M-saturated and $|(\Gamma) \subseteq \{A : \Gamma \vdash A\}$.
<u>Proof</u>. Clear.

5.1.14. <u>Theorem</u>. $|(\Gamma)$ is a maximal M-saturated subtheory of Γ.
<u>Proof</u>. Let $|(\Gamma) \subseteq \Delta \subseteq \{A : \Gamma \vdash A\}$, Δ M-saturated. We show that $\Delta \vdash A$ implies $A \in |(\Gamma)$.
(i). If A is atomic,

$$\Delta \vdash A \Rightarrow \Gamma \vdash A \Rightarrow A \in |(\Gamma).$$

(ii), (iii). $A = B \& C$, $B \vee C$. These cases are trivial.
(iv). Let $A = B \rightarrow C$:

$$\Delta \vdash B \rightarrow C \Rightarrow \Gamma \vdash B \rightarrow C.$$

a) $|(\Gamma) \vdash B$. Then $\Delta \vdash B$ and so $\Delta \vdash C$. Thus $|(\Gamma) \vdash C$.
b) $|(\Gamma) \nvdash B$. $B \rightarrow C \notin |(\Gamma)$ implies $\exists \beta \geq \alpha(\beta \Vdash B$ and $\not\Vdash C)$.

But $\beta = \Delta' \supset \Gamma$ and so $B \in \Delta' \Rightarrow C \in \Delta'$, a contradiction.
(v) $A = \neg B$. Similar to (iv).
(vi). Let $A = \exists x Bx$:

$$\Delta \vdash \exists x Bx \Rightarrow \exists a \in C_M \; \Delta \vdash Ba, \text{ by } M\text{-saturation}$$

$$\Delta \vdash \exists x Bx \Rightarrow Ba \in |(\Gamma)$$
$$\Rightarrow \exists x Bx \in |(\Gamma)\ .$$

(vii) $A = \forall x Bx$. Similar to (iv). Q. E. D.

5.1.15. Corollary. Let Γ be closed under deducibility. Then

$$\Gamma \text{ is } M\text{-saturated iff } \Gamma = |(\Gamma)$$
$$\text{iff } \Gamma \subseteq |(\Gamma)\ .$$

5.1.16. Corollary. The intuitionistic predicate calculus is saturated.

5.1.17. Corollary. $A|A$ in the sense of Aczel iff $A|A$ in the sense of Kleene (cf. § 3.1).

5.1.18. Theorem (Characterization of the Aczel slash by an inductive definition). The relation $\Gamma|A$ is inductively defined by the following:
(i) For atomic A

$$\Gamma|A \text{ iff } \Gamma \vdash A\ ;$$

(ii) $\Gamma | B \& C$ iff $\Gamma|B$ and $\Gamma|C$;
(iii) $\Gamma | B \vee C$ iff $\Gamma|B$ or $\Gamma|C$;
(iv) $\Gamma | B \to C$ iff $\Gamma \vdash B \to C$ and $(\Gamma|B \Rightarrow \Gamma|C)$;
(v) $\Gamma | \neg B$ iff $\Gamma \vdash \neg B$ and $\Gamma \not| B$;
(vi) $\Gamma | \exists x Ax$ iff $\Gamma | Aa$ for some $a \in C_M$;
(vii) $\Gamma | \forall x Ax$ iff $\Gamma \vdash \forall x Ax$ and $\Gamma|Aa$ for all $a \in C_M$.

Proof. (i) by definition ; (ii), (iii), and (vi) are obvious.
(iv) Let $\Gamma|B \to C$, i.e. $\alpha_0 \Vdash B \to C$. Then $\alpha_0 \Vdash B \Rightarrow \alpha_0 \Vdash C$, i.e. $\Gamma|B \Rightarrow \Gamma|C$. Since $|(\Gamma) \subseteq \Gamma$, $\Gamma \vdash B \to C$.
Conversely, if $\Gamma \not| B \to C$, i.e. $\alpha_0 \not\Vdash B \to C$, then either $\alpha_0 \Vdash B$ and $\alpha_0 \not\Vdash C$ or $\Delta \vdash B$, $\Delta \nvdash C$ for some saturated $\Delta \supseteq \Gamma$. The latter can only be true if $\Gamma \nvdash B \to C$; the former if $\Gamma|B$, $\Gamma \not| C$.
(v) and (vii) are similar. Q. E. D.

5.1.19 - 5.1.21. The operation $(\ ) \to (\Sigma\ )'$.

5.1.19. The Aczel slash, like the Kleene slash, may be used to prove saturation results (often called explicit definability results). In Aczel 1968, Aczel used the inductive characterization (theorem 5.1.18) to give a version of Kleene's slash - theoretic proof of the ED - property for HA. However, our interest in this chapter is primarily in the model theory and in model - theoretic proofs. Thus, let us ignore theorem 5.1.18 and reconsider what we did in proving theorem 5.1.14.

The proof of the completeness theorem involved our constructing a model of a theory Γ. We observed (i) that Γ is saturated iff it is the set of formulae forced by a minimum node of that model, and (ii) that, if we added a minimum node, we got a maximal saturated subtheory of Γ. We shall generalize the model-theoretic construction of (ii).

Let $\underline{K}$ be a Kripke model and let a language M with a non-empty set C_M of constants be given. We will let $\underline{K}'$ denote any model $(K', \leq', D', \Vdash)$ obtained by adding a new node α_0 to K such that

(i) $\alpha_0 \leq' \alpha$ for all $\alpha \in K$;

(ii) $D'\alpha_0 = C_M$;

(iii) if A is atomic, $\alpha_0 \Vdash' A \Rightarrow \alpha \Vdash A$ for all $\alpha \in K$.

Then, for $\alpha \in K$ and any formula A, $\alpha \Vdash' A$ iff $\alpha \Vdash A$. Of special interest is the case in which the implication in (iii) is replaced by an equivalence. This is the case we most often encounter.

By theorem 5.1.14, if the class of models of a theory Γ is closed under the operation $\underline{K} \to \underline{K}'$, then Γ is M-saturated. We shall give another proof of this shortly. First we must introduce another operation on models.

Let $\underline{F} = \{\underline{K}_\mu : \mu \in N\}$ be a family of Kripke models. The disjoint sum, $\Sigma\underline{F}$, of the model $\underline{K} = (K, \leq, D, \Vdash)$ defined by

(i) $K = \bigcup_{\mu \in N} K_\mu \times \{\mu\}$;

(ii) $(\alpha,\mu) \leq (\beta,\nu)$ iff $\mu = \nu$ and $\alpha \leq_\mu \beta$;

(iii) $D(\alpha,\mu) = D_\mu \alpha$;

(iv) for atomic A, $(\alpha,\mu) \Vdash A$ iff $\alpha \Vdash_\mu A$.

E.g. suppose $\underline{F}$ is the family consisting of the following models (where $C_M = \{a\}$):

```
         β {a,b}  Pa                      γ {a,b}  Pb
K_1 :      |                     K_2 :      |
         α {a}                            α {a,b}        .
```

Then $\Sigma\underline{F}$ is the model:

```
               (β,1)  {a,b}  Pa     (γ,2)  {a,b}  Pb
K_1 + K_2 :             |                    |
               (α,1)  {a}           (α,2)  {a,b}         .
```

The relation (iv) in the definition of $\Sigma\underline{F}$ may be shown by induction to hold for all A:

$$(\alpha,\mu) \Vdash A \quad \text{iff} \quad \alpha \Vdash_\mu A .$$

Remark. Alternatively, we may use the final remark of subsection 5.1.3 to prove this without another induction.

If $\underline{F}$ is a family of Kripke models, we can apply the two operations successively : $F \to \Sigma\underline{F} \to (\Sigma\underline{F})'$. E.g. for the family $\underline{F}$ given, $(\Sigma\underline{F})'$ is :

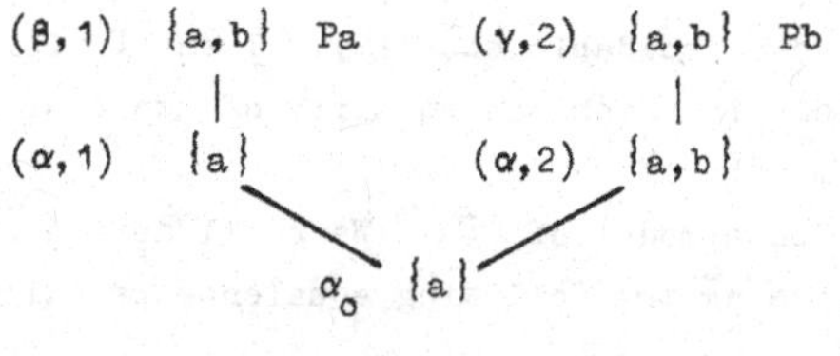

5.1.20. Theorem. Let the class of models of the theory Γ be closed under the operation $\underline{F} \to (\Sigma\underline{F})'$. Then Γ is M-saturated.

Proof. Let Ax contain only x free and let, for each $a \in C_M$, $\Gamma \not\vdash Aa$. Then, for each $a \in C_M$, we can find a model $\underline{K}_a$ such that $\underline{K}_a$ has a least node, say α_a, and $\alpha_a \not\Vdash Aa$.

Let $\underline{F} = \{\underline{K}_a : a \in C_M\}$ and let α_o be the least node of $(\Sigma F)'$. Suppose $\alpha_o \Vdash \exists x Ax$. Then, for some $a \in C_M = D\alpha_o$, $\alpha_o \Vdash Aa$. But $\alpha_a \geq \alpha_o$ and so $\alpha_a \Vdash Aa$, a contradiction.

(Recall that forcing at α_a in $\underline{K}_a$ is the same as that in $(\Sigma\underline{F})'$.)

Disjunction being handled similarly, we have the required result. Q. E. D.

Observing that the class of models of Γ is closed under the operation $\underline{F} \to \Sigma\underline{F}$, we have the immediate

5.1.21. Corollary. Let the class of models of the theory Γ be closed under the operation $\underline{K} \to \underline{K}'$. Then Γ is M-saturated.

Remark. The difference between using theorem 5.1.14 and theorem 5.1.20 to prove that Γ is saturated is that, to apply theorem 5.1.14, one has to show that a particular model $\underline{K}$ of Γ yields a model $\underline{K}'$ of Γ, while theorem 5.1.20 requires one to show that, for any model $\underline{K}$ of Γ, $\underline{K}'$ is a model of Γ. Model-theoretically, both tasks should be equally difficult. Theorem 5.1.18 makes the first task easier - but, it is the second approach that we will find more useful.

5.1.22 - 5.1.23. Models with equality.

5.1.22. In working with Kripke models, one may treat equality as a binary relation satisfying certain axioms. One doesn't always have the option one had in classical model theory to assume that equality is interpreted by actual identity - if equality is interpreted by identity, then $\forall xy(x = y \vee \neg x = y)$ is forced - but there are intuitionistic equality relations which are not decidable (e.g. the equality of the reals).

When equality is decidable, however, it suffices to consider the class of normal models - i.e. models in which the equality predicate is interpreted as actual identity.

5.1.23. Theorem. Let Γ have a decidable equality. Then Γ is strongly complete for the class of models in which the equality of two constants is forced iff they denote the same object.

Proof. Let $\underline{K} = (K, \leq, D, \Vdash)$ be a model of Γ. We shall define a corresponding normal model $\underline{K}^n$ by using the following equalence relation $\bigcup_{\alpha \in K} D\alpha$:

$$x \approx y \text{ iff } \exists\alpha(\alpha \Vdash x = y).$$

Let $[x] = \{y : x \approx y\}$ be the equivalence class of x under $\approx$. Define $\underline{K}^n$ by

(i) $K^n = K$;

(ii) $\leq^n = \leq$;

(iii) $D^n\alpha = \{[x] : x \in D\alpha\}$; and

(iv) for atomic A, $\alpha \Vdash^n A([a_1], \ldots, [a_n])$ if $\alpha \Vdash A(a_1', \ldots, a_n')$, where $a_i' \approx a_i$ and $a_i' \in D\alpha$.

By the standard induction on the length of A, the equivalence (iv) is seen to hold for all A. Q. E. D.

Since Heyting's arithmetic has a decidable equality, we shall, in the sequel, only consider normal models.

5.1.24. Function symbols. Another device we could use is function symbols. While we can show proof-theoretically that function symbols are eliminable, we cannot conclude from this that the theories determined by the classes of models with and without functions coincide. (To do this, we would have to prove completeness of the theories possessing function symbols with respect to their models possessing functions.) We shall, therefore, indicate the model-theoretic proof of the eliminability of function symbols for the special case of a theory with decidable equality.

Let Γ be a theory with the language M and let M possess function symbols. An interpretation of the symbol f in a model $\underline{K}$ is given by choosing a family of functions $\{f_\alpha : \alpha \in K\}$ such that (if f is n-ary) $f_\alpha : (D\alpha)^n \to D\alpha$ and, if $\alpha \leq \beta$, $f_\beta \restriction D\alpha = f_\alpha$. The interpretation of atomic formulae involving terms constructed by the use of such function symbols is handled as in classical model theory.

Suppose we now replace M by a language M' in which every n-ary function symbol is replaced by an $n+1$-ary relation symbol, as discussed in § 1.2. If Γ' is obtained by translating the axioms of Γ into M' and adding the function axioms, then, just as in classical model theory, there is a natural correspondence between models of Γ and models of Γ'.

This is proven by mimicking the classical proof. Thus we may restrict our attention to models with functions replacing certain relations (namely their graphs). The details are left to the reader.

5.1.25. Conventions. Let us finally make, in addition to our convention concerning models of HA that they be normal, a convention that they do not possess functions and the simplifying convention that they all possess minimum (or least) nodes, which we shall call origins. An origin of K will usually be denoted by α_0 and has the defining property that $\alpha_0 \leq \alpha$ for all $\alpha \in K$. (Observe that such models are not closed under $F \to \Sigma F$ and, hence, we shall have to apply theorem 5.1.20 rather than its corollary.)

5.1.26. Intuitionism?

What, one may ask, does all of this set-theoretic machinery have to do with intuitionism? We shall not attempt to answer this question - instead we merely outline how certain proofs obtained by the use of this machinery can be transformed into intuitionistically meaningful proofs. (See e.g. Mints 1969.)

The key to this transformation lies in the Hilbert - Bernays completeness theorem (cf. e.g. Kleene 1952), by which certain outwardly set-theoretic constructions may be replaced by arithmetical ones. Specifically, by arithmetizing the completeness theorem for classical logic, one can show that, for any r.e. theory T, if Con(T) is added to classical arithmetic, then a provably arithmetical model exists - i.e. there is a model with an arithmetically definable domain and arithmetically definable relations such that the translations of the axioms of T are all provable.

The same is true of the completeness theorem given above (especially in the treatment by Thomason). Thus, if we use the completeness theorem to prove (say) an independence result, we can prove the result in classical arithmetic augmented by some consistency statements. This is true of all the results of this chapter. If, in addition, the result is Π_2^0 (e.g. as in the case of an independence result), we know from a previous chapter that the proof in the classical system can be transformed into a proof in the corresponding intuitionistic system.

We shall not prove this result here, however, since most of the results we give can be obtained constructively by less devious means and since the only results which we need for our classical proofs are (i) the existence of arithmetically definable models for any r.e. theory (intuitionistic or classical) and (ii) the fact that the models are provably arithmetical if we add the statement of consistency of the theory to classical arithmetic. For classical theories, this is the Hilbert - Bernays completeness theorem.

For intuitionistic theories this almost reduces to the Hilbert - Bernays completeness theorem as follows: Observe that a Kripke model is a classical model when viewed as a structure in its own right. That is, given $\underline{K}$, Γ, and M, let M' be obtained by replacing each atomic formula $A(x_1,\ldots,x_n)$ by a new formula $A(\alpha, x_1,\ldots,x_n)$ and adding new atomic formulae $D(\alpha,x)$, $K(\alpha)$, and $\alpha \leq \beta$. (Let us assume for simplicity that there are no function symbols.) The relation $A(\alpha,x_1,\ldots,x_n)$ is to be interpreted by $\alpha \Vdash A(x_1,\ldots,x_n)$. We then translate all statements about $\underline{K}$ into M' as follows:

(i) for A atomic, $(\alpha \Vdash A(x_1,\ldots,x_n))^T = A(\alpha,x_1,\ldots,x_n)$;

(ii) $(\alpha \Vdash A \& B)^T = (\alpha \Vdash A)^T \& (\alpha \Vdash B)^T$;

(iii) $(\alpha \Vdash A \vee B)^T = (\alpha \Vdash A)^T \vee (\alpha \Vdash B)^T$;

(iv) $(\alpha \Vdash A \rightarrow B)^T = \forall \beta \geq \alpha((\beta \Vdash A)^T \rightarrow (\beta \Vdash B)^T)$;

(v) $(\alpha \Vdash \neg A)^T = \forall \beta \geq \alpha \neg(\beta \Vdash A)^T$;

(vi) $(\alpha \Vdash \exists x A x)^T = \exists x(D(\alpha,x) \& (\alpha \Vdash Ax)^T)$;

(vii) $(\alpha \Vdash \forall x A x)^T = \forall \beta \geq \alpha\ \forall x[D(\beta,x) \rightarrow (\beta \Vdash Ax)^T]$.

We define Γ' by taking, in addition to axioms asserting that we have a Kripke model (e.g. $(\alpha \Vdash A)^T \& \alpha \leq \beta \rightarrow (\beta \Vdash A)^T$), the axioms $(\alpha \Vdash A)^T$ for axioms A of Γ. Then Γ is r.e. iff Γ' is r.e. and we obtain an arithmetical model of Γ from one of Γ'. The only problem at this stage is that the provable arithmeticity of the models depends here on the consistency statement for Γ' rather than for Γ. However, this loss of precision will cause us no trouble.

§ 2. The treatment of Heyting's arithmetic

5.2.1 - 5.2.4. The operation $(\ ) \to (\Sigma\ )'$.

5.2.1. So far, aside from specializations of the form of the models used (to being normal, to not having functions, and to having origins), the only results which we have proven concern saturation or explicit definability. The result we wish to apply first to Heyting's arithmetic is theorem 5.1.20 which implies that, if we show the class of models of HA to be closed under the operation $(\ ) \to (\Sigma\ )'$, then we may conclude the following

5.2.2. Theorem (Explicit definability). If Ax has only x free and $HA \vdash \exists x Ax$, then $HA \vdash An$ for some n.

5.2.3. Theorem (Disjunction property). Let A, B be closed. If $HA \vdash A \vee B$, then $HA \vdash A$ or $HA \vdash B$.

To prove this, we shall have to choose a formulation of HA. The simplest one for our purposes is the one with constants $0, 1, \ldots$ for each natural number, relations $S(x,y)$, $A(x,y,z)$, and $M(x,y,z)$ defining the functions of successor, addition, and multiplication.

Typographically, we find it convenient to reserve in this chapter the letters n, m (possibly indexed) to denote numerals (in contrast to the other chapters, where n, m usually stood for numerical variables, and numerals were written with a bar: $\bar{n}, \bar{m}, \bar{x}, \bar{y}, \ldots$ etc.).

The axioms of HA are, in addition to the axioms of the predicate calculus with equality:

(i) $\neg S(x,0)$,
$\neg x = 0 \to \exists y\, S(y,x)$,
$S(x,y) \,\&\, S(x,z) \to y = z$,
$S(y,x) \,\&\, S(z,x) \to y = z$,
$\exists y\, S(x,y)$;

(ii) $A(x,y,z) \,\&\, A(x,y,w) \to z = w$,
$\exists z\, A(x,y,z)$,
$A(x,0,x)$,
$A(x,y,z) \,\&\, S(y,w) \,\&\, S(z,v) \to A(x,w,v)$;

(iii) $M(x,y,z) \,\&\, M(x,y,w) \to z = w$,
$\exists z\, M(x,y,z)$,
$M(x,0,0)$,
$M(x,y,z) \,\&\, S(y,w) \,\&\, A(z,x,v) \to M(x,w,v)$;

(iv) $S(n, n+1)$, for each constant n;

and the scheme, for any formula A whose free variables include x and do not include y :

  (v)   AO & ∀xy(Ax & S(x,y) → Ay) → ∀xAx .

Aesthetically, it is more pleasing to use a formulation with function symbols and, as shown in 5.1.24, we may do so. However, that would require a little more care in defining various structures and a little more work in proving results about them. We shall, occasionally, however, freely use the fact that there is a natural correspondence between models of our official system above and the system with function symbols (or, if one prefers, we shall abuse notation by using function symbols).

Our first step is to prove the following

5.2.4. **Theorem.** The class of models of $\underline{HA}$ is closed under the operation $(\ ) \to (\Sigma\ )'$.

Recall that, in the definition of $\underline{K} \to \underline{K}'$, we left open the problem of deciding which atomic formulae to force at α_0 , stating that we usually have $\alpha_0 \Vdash A$ iff $\alpha \Vdash A$ for all $\alpha \in K$. (Recall also that the proof of theorem 5.1.20 merely required us to have <u>some</u> model of the form $\underline{K}'$.) For $\underline{HA}$, there is no ambiguity - closed atomic formulae are decided by the theory and, if $\underline{K}'$ is to be a model of $\underline{HA}$, we must have $\alpha_0 \Vdash A$ iff A is true in the standard model.

Thus our operation $\underline{F} \to (\Sigma\underline{F})'$ is given by tacking on a new node α_0 below all nodes of $\Sigma\underline{F}$, setting $D\alpha_0 = \{0, 1, \ldots\}$, and letting $\alpha_0 \Vdash' A$ iff A is true, for any atomic A . E.g. if ω^+ and ω^* are non-standard models of classical arithmetic, then (using the graphic representation of subsection 5.1.4) $(\omega^+ + \omega^*)'$ is

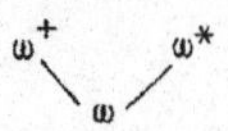

Proof of theorem 5.2.4. The assertion that $\underline{F}$ is a model of $\underline{HA}$ means that every axiom of $\underline{HA}$ is valid in every member of $\underline{F}$ (i.e. forced at each node of each model in $\underline{F}$). For $(\Sigma\underline{F})'$ not to be a model of $\underline{HA}$, some node α of $(\Sigma\underline{F})'$ must fail to force some axiom of $\underline{HA}$. Obviously, we cannot have $\alpha > \alpha_0$, since then $\alpha \in K$ for some $\underline{K} \in \underline{F}$ (making the obvious identification - i.e. ignoring the operation used to make members of $\underline{F}$ disjoint). Thus, to prove that $(\Sigma\underline{F})'$ is a model of $\underline{HA}$, it suffices to show that $\alpha_0 \Vdash A$ for each axiom A of $\underline{HA}$.

The only non-trivial case to consider is the induction axiom. For simplicity, we assume that Ax has only the variable x free. The general case is left to the reader (i.e. we let the reader verify the validity of the universal closure of the scheme with free variables).

Let $\alpha_0 \not\Vdash A0 \,\&\, \forall xy(Ax \,\&\, S(x,y) \rightarrow Ay) \rightarrow \forall xAx$. Then, for some $\beta \geq \alpha_0$, $\beta \Vdash A0 \,\&\, \forall xy(Ax \,\&\, S(x,y) \rightarrow Ay)$, but $\beta \not\Vdash \forall xAx$. Now we cannot have $\beta > \alpha_0$, since then $\beta \in K$ for some $\underline{K} \in \underline{F}$ and β forces all axioms of $\underset{\sim}{HA}$. Hence $\alpha_0 \Vdash A0 \,\&\, \forall xy(Ax \,\&\, S(x,y) \rightarrow Ay)$, but $\alpha_0 \not\Vdash \forall xAx$. Since $\alpha_0 \not\Vdash \forall xAx$, there is some $\beta \geq \alpha_0$ and some $b \in D\beta$ such that $\beta \not\Vdash Ab$. Again $\beta = \alpha_0$ and b is some natural number. Let m be the smallest such number. Since $\alpha_0 \Vdash A0$, m is a successor, say $n+1$, and, since m is the smallest number b such that $\alpha_0 \not\Vdash Ab$, $\alpha_0 \Vdash An$. But $\alpha_0 \Vdash \forall xy(Ax \,\&\, S(x,y) \rightarrow Ay)$, whence $\alpha_0 \Vdash An \,\&\, S(n, n+1) \rightarrow A(n+1)$. Thus $\alpha_0 \Vdash A(n+1)$, i.e. $\alpha_0 \Vdash Am$, a contradiction. Q. E. D.

Theorems 5.2.4 and 5.1.20 immediately yield theorems 5.2.2 and 5.2.3 as corollaries.

5.2.5 - 5.2.7. Applications of the operation $(\ ) \rightarrow (\Sigma\ )'$.

5.2.5. The closure of the class of models of $\underset{\sim}{HA}$ under $(\ ) \rightarrow (\Sigma\ )'$ is one of the basic tools of the Kripke model approach to studying $\underset{\sim}{HA}$. E.g. we have already used this to prove ED, the explicit definability property. Its use here is simply that it allows us to take countermodels to $A0, A1, \ldots$ and put them together to construct a countermodel to $\exists xAx$. It is in this construction of models that this operation is so useful. Consider, e.g., the old result of Kreisel's (Kreisel 1958):

5.2.6. Theorem. Let Ax have only x free and suppose $\vdash \forall x(Ax \vee \neg Ax)$, $\vdash$ denoting derivability in $\underset{\sim}{HA}$. Then

$$\begin{aligned} \vdash \forall xAx \vee \exists x \neg Ax \quad &\text{iff} \quad \vdash \neg \forall xAx \rightarrow \exists x \neg Ax \\ &\text{iff} \quad \vdash \exists y[\neg \forall xAx \rightarrow \neg Ay]. \end{aligned}$$

Proof. (Cf. also 3.8.5). We shall show that $\vdash \neg \forall xAx \rightarrow \exists x \neg Ax$ implies $\vdash \exists y[\neg \forall xAx \rightarrow \neg Ay]$ and leave the rest to the reader. Suppose $\vdash \neg \forall xAx \rightarrow \exists x \neg Ax$ and $\not\vdash \exists y[\neg \forall xAx \rightarrow \neg Ay]$. Then, $\not\vdash \neg \forall xAx \rightarrow An$ for each n. But $\vdash \neg An \rightarrow (\neg \forall xAx \rightarrow \neg An)$, whence $\not\vdash \neg An$. By the decidability of A and the DP, $\vdash An$.

On the other hand, $\vdash \forall xAx \rightarrow [\neg \forall xAx \rightarrow \neg A0]$, and so $\not\vdash \forall xAx$.

Let $\underline{K}$ be a model of $\underset{\sim}{HA}$ with $\alpha \in K$ such that $\alpha \not\Vdash \forall xAx$. Then $\exists \beta \geq \alpha \exists b \in D\beta\ \beta \not\Vdash Ab$. By decidability , $\beta \Vdash \neg Ab$ and hence $\beta \Vdash \neg \forall xAx$.

Now consider $\underline{K}_\beta$ (recall the definition from subsection 5.1.3) and especially $(\underline{K}_\beta)'$:

$$\begin{array}{c} \underline{K}_\beta \\ | \\ \omega \end{array} .$$

Observe that $\gamma \geq \alpha_0$ implies $\gamma = \alpha_0$ or $\gamma \geq \beta$ and that $\gamma \geq \beta$ implies $\gamma \Vdash \neg \forall x A x$. Also, $\alpha_0 \nVdash \forall x A x$ since, if $\alpha_0 \Vdash \forall x A x$, then $\beta \Vdash \forall x A x$ and one has a contradiction. Thus $\gamma \nVdash \forall x A x$ for all $\gamma \geq \alpha_0$ and $\alpha_0 \Vdash \neg \forall x A x$. We now use the fact that $\vdash \neg \forall x A x \rightarrow \exists x \neg A x$ to conclude $\alpha_0 \Vdash \exists x \neg A x$. But $D\alpha_0 = \{0, 1, \ldots\}$ and, for some n, $\alpha_0 \Vdash \neg A n$, a contradiction. Q. E. D.

5.2.7. We have not really used the basic operation $(\ ) \rightarrow (\Sigma\ )'$ in the direct construction of models. We turn our attention now to this task.

Let $\underline{T} = (T, \leq)$ be a finite tree. By a <u>terminal node</u> of the tree we shall mean a maximal node of the tree - i.e. a node with no successors. We shall let Ter denote the set of terminal nodes of T. For any node $\alpha \in T - \text{Ter}$, $S(\alpha)$ will denote the set of successors of α.

Let us assume that we have assigned models of classical arithmetic to each of the terminal nodes - say ω_α is assigned to $\alpha \in \text{Ter}$. We now associate with each $\alpha \in T$ a Kripke model $\underline{K}(\alpha)$ as follows:

(i) if $\alpha \in \text{Ter}$, $\underline{K}(\alpha) = \omega_\alpha$ (viewed as a one-node Kripke model);

(ii) if $\alpha \notin \text{Ter}$, $\underline{K}(\alpha) = (\sum_{\beta \in S(\alpha)} \underline{K}(\beta))'$.

Finally, define $\underline{K}_{\underline{T}} = \underline{K}(\alpha_0)$, where α_0 is the origin of $\underline{T}$.

5.2.8. <u>Theorem</u>. $\underline{K}_{\underline{T}}$ is a model of $\underset{\sim}{HA}$.

<u>Proof</u>. We show by bar induction that $\underline{K}(\alpha)$ is a model of $\underset{\sim}{HA}$. The theorem is trivial for terminal nodes. If α is not terminal, apply theorem 5.2.4. It follows that $\underline{K}(\alpha)$ is a model of $\underset{\sim}{HA}$ for all $\alpha \in K$. Letting $\alpha = \alpha_0$ completes the proof. Q. E. D.

<u>Note</u>. Obviously, we may replace the finiteness restriction on $\underline{T}$ by the well-foundedness restriction.

As an example, we know by Gödel's theorem that there is an independent sentence A of classical arithmetic. Thus there are models ω_1 and ω_2 of A and $\neg A$, respectively. Associating these models with the terminal nodes of the tree,

$\underline{T}$:

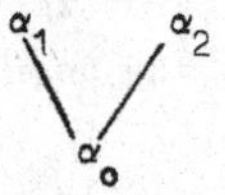

,

we have the model:

$\underline{K}_{\underline{T}}$:

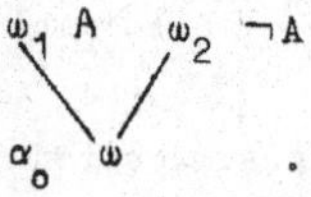

.

Observe e.g. that $\alpha_0 \nVdash A \vee \neg A$.

A stronger version of Gödel's theorem allows us, for any n, to find Σ_1^0 sentences $A_1,\ldots,A_n$ which are mutually independent over classical arithmetic. In particular, we can find models $\omega_1,\ldots,\omega_n$ such that A_j is true in ω_i iff $i=j$. (We shall discuss this further in section 3.) Letting $n=3$ and relabelling A_1, A_2, A_3 as A, B, and C, let ω_1, ω_2, and ω_3 be associated with the terminal nodes of the tree

$\underline{T}$:

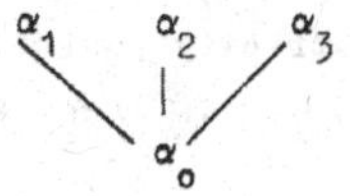

Then $\underline{K_T}$ is

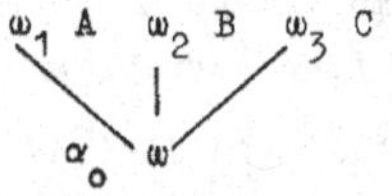

Observe that $\alpha_0 \Vdash\!\!\!\!/ (\neg A \to B \vee C) \to ((\neg A \to B) \vee (\neg A \to C))$. (See chapter III, section 2.2.6 for an application.)

Let ω_1, ω_2, and ω_3 be as in the preceding example and let $\underline{T}$ be :

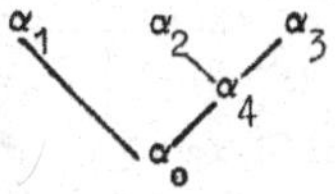

Associating ω_1, ω_2, and ω_3 with α_1, α_2, and α_3, we have

$\underline{K_T}$:

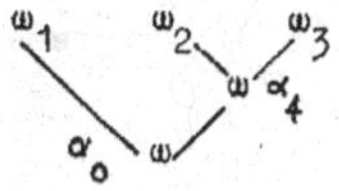

Observe that, although α_0 and α_4 both have copies of ω associated with them, they do not behave alike, e.g. $\alpha_4 \Vdash \neg A, \neg\neg(B \vee C)$, but $\alpha_0 \Vdash\!\!\!\!/ \neg A, \neg\neg(B \vee C)$.

5.2.9 - 5.2.12. <u>Formulae preserved under</u> $(\ )\to(\Sigma\ )'$.

5.2.9. If Γ is a set of sentences, we may ask whether or not various metamathematical properties of HA also hold for HA + Γ. For instance, one may ask whether or not the explicit definability theorem holds for HA + Γ or whether or not HA + Γ is closed under the derived rules given by theorem 5.2.6. Since the only property used in deriving these properties of HA is the closure of the class of models of HA under the operation $(\ )\to(\Sigma\ )'$, to prove these results for HA + Γ, we need only show that the class of models of HA + Γ is closed under this basic operation

Of course, to prove explicit definability, one could use the Aczel slash - its inductive definition makes it fairly usable. The operation $(\ )\to(\Sigma\ )'$

has the advantage that, if Γ and Δ are preserved by it, then $\Gamma+\Delta$ is preserved - i.e. if the validity of $\underset{\sim}{HA}+\Gamma$ is preserved by the operation $\underline{F}\to(\Sigma\underline{F})'$ and if the same holds of $\underset{\sim}{HA}+\Delta$, then $\underset{\sim}{HA}+\Gamma+\Delta$ is also preserved by this operation. Thus, the class of sets of formulae preserved by this operation exhibit better closure properties than the class of sets, Γ, of formulae which yield saturated extensions, $\underset{\sim}{HA}+\Gamma$, of $\underset{\sim}{HA}$.

5.2.10. Lemma. Let the sentence A have no strictly positive $\vee$ or $\exists$ (i.e. A is a Harrop sentence, see 1.10.5). Then A is preserved under the operation $(\ )\to(\Sigma\ )'$.

Proof. We shall prove this by induction on the length of A. To carry out the induction step corresponding to (v), we must make a convention involving free variables. Let A have $x_1,\ldots,x_n$ as free variables - we shall prove that $A(m_1,\ldots,m_n)$ is preserved for all numbers $m_1,\ldots,m_n$. The result then follows trivially for sentences.

(i) The preservation of atomic formulae follows by the decidability of atomic formulae in $\underset{\sim}{HA}$.

(ii) Let $A(m_1,\ldots,m_n)\,\&\,B(m_1,\ldots,m_n)$ be valid in $\underline{F}$. Then $A(m_1,\ldots,m_n)$ and $B(m_1,\ldots,m_n)$ are valid in $\underline{F}$. But each of these is preserved under $\underline{F}\to(\Sigma\underline{F})'$, whence $A\,\&\,B$ is valid in $(\Sigma\underline{F})'$.

(iii) Let $A(m_1,\ldots,m_n)\to B(m_1,\ldots,m_n)$ be valid in $\underline{F}$. For this implication to fail to be valid in $(\Sigma\underline{F})'$, we must have $\alpha_0\Vdash A(m_1,\ldots,m_n)$, $\alpha_0\not\Vdash B(m_1,\ldots,m_n)$. But then $A(m_1,\ldots,m_n)$ is valid in $\underline{F}$, whence $B(m_1,\ldots,m_n)$ is valid in $\underline{F}$. Again $B(m_1,\ldots,m_n)$ is preserved, whence $\alpha_0\Vdash B(m_1,\ldots,m_n)$, a contradiction.

(iv) Similar to (iii).

(v) Let $A(x,m_1,\ldots,m_n)$ be given, $\forall xA(x,m_1,\ldots,m_n)$ valid in $\underline{F}$. For $\forall xA(x,m_1,\ldots,m_n)$ to fail to be valid in $(\Sigma\underline{F})'$, we must have $\alpha_0\not\Vdash A(m,m_1,\ldots,m_n)$ for some $m\in D\alpha_0=\{0,1,\ldots\}$. But $A(m,m_1,\ldots,m_n)$ is valid in $\underline{F}$ and is preserved, leading to a contradiction. Q. E. D.

5.2.11. Theorem. The class $\mathfrak{P}$ of sets, Γ, such that the validity of $\underset{\sim}{HA}+\Gamma$ is preserved by the operation $(\ )\to(\Sigma\ )'$ has the following closure properties:

(i) $\mathfrak{P}$ is closed under arbitrary union;

(ii) if $\Gamma\in\mathfrak{P}$ and A is a Harrop-sentence, then $\Gamma\cup\{A\}\in\mathfrak{P}$;

(iii) if $\Gamma\in\mathfrak{P}$, A has only the variable x free, and $\underset{\sim}{HA}+\Gamma\vdash An$ for each numeral n, then $\Gamma\cup\{\forall xAx\}\in\mathfrak{P}$.

Proof. The only case we haven't proven already is (iii). The proof of this is basically the same as that of case (v) in the preceding proof.

5.2.12. <u>Corollary</u>. (<u>Friedman</u> A) Let $\Gamma \in \mathfrak{P}$. Then ED and DP hold for $\underset{\sim}{HA} + \Gamma$.

5.2.13 - 5.2.23. <u>Examples. Reflection principles and transfinite induction.</u>

5.2.13. Condition (iii) in the definition of $\mathfrak{P}$ was introduced in <u>Friedman</u> A for the purpose of proving results like corollary 5.2.12. By it, if we have an axiom scheme for which we wish to prove a preservation theorem, we need only prove the theorem for the scheme without free variables. For induction,

$$A0 \,\&\, \forall xy(Ax \,\&\, S(x,y) \to Ay) \to Ax\,,$$

we need only prove the preservation result for each instance,

$$A0 \,\&\, \forall xy(Ax \,\&\, S(x,y) \to Ay) \to An\,.$$

If we examine the proof we gave, we notice that we reduced the problem to proving the preservation of this last sentence. We shall now consider some further schemata and apply condition (iii) to prove preservation theorems for them.

Let $<$ be a primitive recursive (or even provably decidable - i.e. $\Vdash\!\!\!\!/ \vdash x<y \vee \neg x<y$) well-ordering of the natural numbers. By the scheme, $TI(<)$, of transfinite induction on $<$ is meant the following:

$$A0 \,\&\, \forall x[\forall y < xAy \to Ax] \to \forall xAx\,,$$

where, for convenience, 0 is taken to be the first element of the ordering.

5.2.14. <u>Lemma</u>. Let Γ be the subscheme of $TI(<)$ determined by the restriction that Ax have only x free. Then the preservation theorem holds for $\underset{\sim}{HA} + \Gamma$.

<u>Proof</u>. Let <u>F</u> be a family of models of $\underset{\sim}{HA} + \Gamma$ and observe that $<$ is a genuine well-ordering on ω. Thus, if Γ is not valid in $(\Sigma \underline{F})'$, we have

$$\alpha_0 \Vdash A0 \,\&\, \forall x[\forall y < xAy \to Ax]\,,$$
$$\alpha_0 \not\Vdash An\,.$$

Letting n_0 be the <u>least</u> such n, $\alpha_0 \Vdash Am$ for all $m < n_0$ and $\beta \Vdash \forall xAx$ for all $\beta > \alpha_0$, whence $\alpha_0 \Vdash \forall y < n_0 Ay$, whence $\alpha_0 \Vdash An_0$, a contradiction.

Q. E. D.

5.2.15. <u>Theorem</u>. The scheme $TI(<)$ is preserved.

<u>Proof</u>. Let Γ be as in lemma 5.2.14 and let $B(x_1,\ldots,x_n)$ denote the instance,

$$A(0,x_1,\ldots,x_n) \,\&\, \forall x[\forall y < xA(y,x_1,\ldots,x_n) \to A(x,x_1,\ldots,x_n)] \to \forall xA(x,x_1,\ldots,x_n)\,,$$

if $TI(<)$. For each choice $m_1,\ldots,m_n$ of numerals to replace

$x_1,\dots,x_n$, $B(m_1,\dots,m_n) \in \Gamma$, whence $\underset{\sim}{HA} + \Gamma \vdash B(m_1,\dots,m_n)$. It follows by condition (iii) that $\underset{\sim}{HA} + \Gamma + B \in \mathfrak{P}$. Thus $\underset{\sim}{HA} + TI(<) =$ $\bigcup \{\underset{\sim}{HA} + \Gamma + B \mid B \in TI(<)\} \in \mathfrak{P}$. Q. E. D.

5.2.16. Corollary. Let $\underset{\sim}{T}$ extend $\underset{\sim}{HA}$ by the addition of some schemata of transfinite induction on primitive recursive well-orderings. Then $\underset{\sim}{T}$ satisfies DP and ED.

5.2.17. To discuss the next set of schemata, let, for an r.e. extension $\underset{\sim}{T}$ of $\underset{\sim}{HA}$, $\mathrm{Proof}_T(x,y)$ be the canonical proof predicate. The properties of $\mathrm{Proof}_T(x,y)$ which we use are that

(i) Proof_T is decidable, and

(ii) $\exists n\ \underset{\sim}{HA} \vdash \mathrm{Proof}_T(n, \ulcorner A \urcorner)$ iff $\underset{\sim}{T} \vdash A$,

where $\ulcorner A \urcorner$ is the gödel number of A. If A contains the free variable y, we let $\ulcorner A\bar{y} \urcorner$ denote $s(y, \ulcorner A \urcorner)$, where s is a primitive recursive function such that

$$s(n, \ulcorner A \urcorner) = \ulcorner [y/n]A \urcorner \text{ is the gödel number of the sentence}$$

obtained by replacing the variable y in A by the numeral n.

We may use this notation to list the following schemata:

Local reflection for $\underset{\sim}{T}$, $RF(\underset{\sim}{T})$:

$RF(T)$ $\exists x\ \mathrm{Proof}_T(x, \ulcorner A \urcorner) \rightarrow A$, for sentences A.

Uniform reflection for $\underset{\sim}{T}$, $RFN(\underset{\sim}{T})$:

$RFN(T)$ $\forall y[\exists x\ \mathrm{Proof}_T(x, \ulcorner A\bar{y} \urcorner) \rightarrow Ay]$, for A containing only y free.

Uniform' reflection for $\underset{\sim}{T}$, $RFN'(\underset{\sim}{T})$:

$RFN'(T)$ $\forall y\, \exists x\ \mathrm{Proof}_T(x, \ulcorner A\bar{y} \urcorner) \rightarrow \forall y Ay$, for A containing only y free.

Consistency of $\underset{\sim}{T}$, $CON(T)$:

$CON(T)$ $\neg \exists x\ \mathrm{Proof}_T(x, \ulcorner 0=1 \urcorner)$.

ω-Consistency of $\underset{\sim}{T}$, ω-C $(\underset{\sim}{T})$:

ω-$CON(T)$ $\exists x\ \mathrm{Proof}_T(x, \ulcorner \neg \forall y Ay \urcorner) \rightarrow \neg \forall y\, \exists x\ \mathrm{Proof}_T(x, \ulcorner A\bar{y} \urcorner)$, for A containing only y free.

Feferman 1962, theorem 2.19 gives an intuitionistic proof of the following:

5.2.18. Lemma. The schemata $RFN(\underset{\sim}{T})$ and $RFN'(\underset{\sim}{T})$ are equivalent.

Thus, we need not consider $RFN'(\underset{\sim}{T})$. For the relative strengths of these reflection principles, see Feferman 1962 and Kreisel - Levy 1968.

5.2.19. Theorem. $CON(\underset{\sim}{T})$ and ω-$CON(\underset{\sim}{T})$ are preserved by the operation $(\ ) \rightarrow (\Sigma\ )'$.

Proof. $CON(\underset{\sim}{T})$ and ω-$CON(\underset{\sim}{T})$ have no strictly positive $\vee$ or $\exists$. Q.E.D.

5.2.20. Lemma. Let A be a sentence. If $\underset{\sim}{T} \vdash A$, then $\underset{\sim}{HA} + RF(\underset{\sim}{T}) \vdash A$.
Proof. Observe $\underset{\sim}{T} \vdash A$ implies $\underset{\sim}{HA} \vdash \exists x\, Proof_T(x, \ulcorner A \urcorner)$. $RF(\underset{\sim}{T})$ yields $\underset{\sim}{HA} + RF(\underset{\sim}{T}) \vdash A$. Q. E. D.

5.2.21. Theorem. If T is preserved by the operation $(\ ) \to (\Sigma\ )'$, then so is $\underset{\sim}{HA} + RF(\underset{\sim}{T})$.
Proof. Let $\underline{F}$ be a family of models of $\underset{\sim}{HA} + RF(\underset{\sim}{T})$ and let $(\Sigma\underline{F})'$ fail to be a model of $\underset{\sim}{HA} + RF(\underset{\sim}{T})$. Then

$$\alpha_0 \Vdash \exists x\, Proof_T(x, \ulcorner A \urcorner), \quad \alpha_0 \nVdash A,$$

for some sentence A. Thus, for some n, $\alpha_0 \Vdash Proof_T(n, \ulcorner A \urcorner)$.
But $Proof_T$ is decidable, whence $\underset{\sim}{HA} \vdash Proof_T(n, \ulcorner A \urcorner)$ and $\underset{\sim}{T} \vdash A$.
By lemma 2.4.6, $\underset{\sim}{T}$ is valid in $\underline{F}$, whence, by hypothesis, $\underset{\sim}{T}$ is valid in $(\Sigma\underline{F})'$. Thus $\alpha_0 \Vdash A$, a contradiction. Q. E. D.

5.2.22. Corollary. If $\underset{\sim}{T}$ is preserved by the operation $(\ ) \to (\Sigma\ )'$, then so is $\underset{\sim}{HA} + RFN(\underset{\sim}{T})$.

5.2.23. Corollary. If T is preserved by the operation $(\ ) \to (\Sigma\ )'$, then so are $\underset{\sim}{T} + RF(\underset{\sim}{T})$, $\underset{\sim}{T} + RFN(\underset{\sim}{T})$, $\underset{\sim}{T} + RF(\underset{\sim}{T} + RF(\underset{\sim}{T}))$, etc.

§ 3. Additional information from $(\ ) \to (\Sigma\ )'$: de Jongh's theorem.

5.3.1. Statement of de Jongh's theorem.

In addition to its use in proving explicit definability results and the validity of an occasional derived rule, we observed in 5.2.4 that we could use the operation $(\ ) \to (\Sigma\ )'$ to construct Kripke models of HA out of models of classical arithmetic. This last application has, as a corollary, a simple proof of the propositional case of an interesting theorem of de Jongh. In the sequel, Pp denotes intuitionistic propositional logic.

Let $A(p_1,\dots,p_n)$ be a propositional formula constructed from the propositional variables $p_1,\dots,p_n$. In an as yet unpublished paper (de Jongh A, see de Jongh 1970), D.H.J. de Jongh proved the following

5.3.2. Theorem. If $\mathrm{Pp} \nvdash A(p_1,\dots,p_n)$, then $\mathrm{HA} \nvdash A(B_1,\dots,B_n)$, for some sentences $B_1,\dots,B_n$ of arithmetic.

According to this theorem, if $A(p_1,\dots,p_n)$ is not an intuitionistic tautology, there are arithmetical substitution instances resulting in a sentence underivable in HA. Alternatively, we can view this as a completeness result if we define the validity of a formula $A(p_1,\dots,p_n)$ in HA to be the validity of the scheme $A(B_1,\dots,B_n)$ determined by $A(p_1,\dots p_n)$.

Actually, de Jongh proved a stronger result: The choice of substitution instances $B_1,\dots,B_n$ of $p_1,\dots,p_n$ can be made uniformly in all $A(p_1,\dots,p_n)$. A proof of this by means of Kripke models is more difficult and will be given in section 6.

Another result of de Jongh's is a completeness theorem for the predicate calculus. To date, the only proof of this result is de Jongh's original proof, which combines the use of Kripke models and realizability.

5.3.3 - 5.3.8. Preliminaries on the propositional calculus.

5.3.3. The proof of the completeness theorem given in 5.1.6 - 5.1.11 specializes easily to the propositional calculus. Kripke's original proof (Kripke 1965) also yields the completeness (but not strong completeness) of the intuitionistic propositional calculus, Pp, for the class of models whose underlying pms is a finite tree. Our first task is to retrieve this result. We do this by starting with a countermodel to a formula A and pluck out finitely many nodes needed to falsify A, splitting and ordering them into a tree in the process.

We will let σ, τ denote finite sequences. $\langle\ \rangle$ denotes the empty sequence. $\langle a \rangle$ denote the sequence whose only element is a. $\sigma * \tau$ will denote the concatenation of σ, τ - i.e. if $\sigma = \langle s_1,\dots,s_m\rangle$, $\tau = \langle t_1,\dots,t_n\rangle$, then $\sigma * \tau = \langle s_1,\ldots,s_m,t_1,\ldots,t_n\rangle$. In particular, $\sigma * \langle a\rangle = \langle s_1,\ldots,s_m,a\rangle$.

5.3.4. Theorem. (Finite tree theorem.) Let $(K, \leq, \Vdash)$ be a model with origin α_0 such that $\alpha_0 \nVdash A$. Then there is a finite tree model $(K^*, \leq^*, \Vdash^*)$ such that $\langle\ \rangle \nVdash^* A$.

Proof. Let S be the set of subformulae of A, and, for $\beta \in K$, let $S(\beta) = \{B \in S : \beta \Vdash B\}$.

Set $\beta_{\langle\ \rangle} = \alpha_0$.

Given β_σ, let $\beta_{\sigma*\langle 1\rangle},\ldots,\beta_{\sigma*\langle k\rangle}$ be a maximal set of $\gamma_1,\ldots,\gamma_k$ such that

(i) $\beta_\sigma \leq \gamma_i$ for all i,

(ii) $S(\beta_\sigma) \neq S(\gamma_i)$ for all i,

(iii) if $\beta_\sigma \leq \gamma \leq \gamma_i$, then $S(\gamma) = S(\beta_\sigma)$ or $S(\gamma) = S(\gamma_i)$, and

(iv) $S(\gamma_i) \neq S(\gamma_j)$ for $i \neq j$.

Now let $K^* = \{\sigma : \beta_\sigma$ has been defined$\}$, and let $\leq^*$ be the usual tree ordering. For atomic B, define $\sigma \Vdash^* B$ iff $\beta_\sigma \Vdash B$.

We prove by induction on the length of B, for $B \in S$, that $\sigma \Vdash^* B$ iff $\beta_\sigma \Vdash B$.

(i) The atomic case follows by definition.

(ii)-(iii). Let B be $C \& D$ or $C \vee D$. The proofs are trivial.

(iv) Let B be $C \to D$. Let $\sigma \nVdash^* C \to D$.

Then there is a $\tau \geq \sigma$ such that $\tau \Vdash^* C$, $\tau \nVdash^* D$. But then $\beta_\tau \Vdash C$, $\beta_\tau \nVdash D$ and, since $\beta_\sigma \leq \beta_\tau$, $\beta_\sigma \nVdash C \to D$.

Conversely, let $\beta_\sigma \nVdash C \to D$.

Case 1. $\beta_\sigma \Vdash C$. Then $\beta_\sigma \nVdash D$ and $\sigma \Vdash^* C$, $\sigma \nVdash^* D$. Then $\sigma \nVdash^* C \to D$.

Case 2. $\beta_\sigma \nVdash C$. Then there is a γ such that $\beta_\sigma \leq \gamma$, $\gamma \Vdash C$, $\gamma \nVdash D$. But, by construction, there is a $\tau \geq^* \sigma$ such that $S(\beta_\tau) = S(\gamma)$ and so $C \in S(\beta_\tau)$, $D \notin S(\beta_\tau)$. Thus $\tau \Vdash^* C$, $\tau \nVdash^* D$ and $\sigma \nVdash^* C \to D$.

(v) Let B be $\neg C$. The proof is similar to (iv). Q.E.D.

5.3.5. Corollary (Kripke). Pp is complete for the class of finite tree models, i.e. $\mathrm{Pp} \nvdash A$ iff A has a countermodel in a finite tree.

We shall find it convenient to work with a special class of trees. To prove completeness for them, we prove the following result (which generalizes a result of Gabbay 1969 B).

5.3.6. Theorem (Extension theorem). Let $(K_0, \leq_0)$ be a finite subtree of the finite tree $(K_1, \leq_1)$. Let $\Vdash_0$ be a forcing relation defined on $(K_0, \leq_0)$. Then there is a forcing relation $\Vdash_1$ on $(K_1, \leq_1)$ such that, for all $\alpha \in K$ and all formulae A,

$$\alpha \Vdash_0 A \ \text{ iff } \ \alpha \Vdash_1 A.$$

<u>Note</u>. By "subtree" we do not merely mean "tree which is a subordering of" - the result is false in this case. The successors of a node $\alpha \in K_o$ must be successors in the tree $(K_1, \leq_1)$. For convenience, we also require the origins of the two trees to coincide.

<u>Proof</u>. For $\alpha \in K_o$ and atomic A, define $\alpha \Vdash_1 A$ iff $\alpha \Vdash_o A$. For each $\beta \in K_o$, choose a terminal node $t_\beta \geq \beta$ in the tree $(K_o, \leq_o)$. Let $\alpha \in K_1 - K_o$. Then there is a maximum $\beta \in K_o$ such that $\alpha \geq_1 \beta$. Define, for atomic A, $\alpha \Vdash_1 A$ iff $t_\beta \Vdash_o A$.

We now show, for all A,

(i) if $\alpha \in K_o$, , $\alpha \Vdash_1 A$ iff $\alpha \Vdash_o A$,

(ii) if $\alpha \in K_1 - K_o$, $\alpha \Vdash_1 A$ iff $t_\beta \Vdash_o A$,

where t_β is defined as above.

(i) For atomic A, the result follows by definition.

(ii) - (iii) The cases $A = B \& C$, $B \vee C$ are trivial

(iv) Let $A = B \to C$.

(a) Let $\alpha \Vdash_o B \to C$, $\beta \geq_1 \alpha$ such that $\beta \Vdash_1 B$. If $\beta \in K_o$, $\beta \Vdash_o B$ by induction hypothesis and so $\beta \Vdash_o C$. Thus $\beta \Vdash_1 C$. If $\beta \in K_1 - K_o$, we have, for some $\gamma \leq_1 \beta$, $\beta \Vdash_1 B$ iff $t_\gamma \Vdash_o B$. Now $t_\gamma \geq \gamma \geq \alpha$ (since the predecessors of β are linearly ordered) and so $t_\gamma \Vdash_o C$, whence the induction hypothesis yields $\beta \Vdash_1 C$. Hence $\beta \geq_1 \alpha$ implies that, if $\beta \Vdash_1 B$, then $\beta \Vdash_1 C$ and we have $\alpha \Vdash_1 B \to C$.

(b) Let $\alpha \not\Vdash_o B \to C$. Then there is a $\beta \in K_o$, $\beta \geq_o \alpha$ such that $\beta \Vdash_o B$, $\beta \not\Vdash_o C$. Then $\beta \Vdash_1 B$, $\beta \not\Vdash_1 C$ and $\alpha \not\Vdash_1 B \to C$.

(c) Let $\alpha \in K_1 - K_o$. Let $\beta \geq_1 \alpha$ and let $\gamma \in K_o$ be maximal such that $\gamma \leq_1 \alpha$. Then γ is maximal in K_o such that $\gamma \leq_1 \beta$. By induction hypothesis, $\alpha, \beta \Vdash_1 B$ iff $t_\gamma \Vdash_o B$ and $\alpha, \beta \Vdash_1 C$ iff $t_\gamma \Vdash_o C$.

$$\begin{aligned} \alpha \Vdash_1 B \to C \quad & \text{iff} \quad \forall \beta \geq_1 \alpha (\beta \Vdash_1 B \Rightarrow \beta \Vdash_1 C) \\ & \text{iff} \quad \alpha \Vdash_1 B \Rightarrow \alpha \Vdash_1 C \\ & \text{iff} \quad t_\gamma \Vdash_o B \Rightarrow t_\gamma \Vdash_o C \\ & \text{iff} \quad t_\gamma \Vdash_o B \to C, \end{aligned}$$

since, for terminal t_γ, the forcing semantics is the same as in classical logic.

(v) The proof for negation is similar to that for case (iv). Q. E. D.

E.g. Consider the trees $(K_o, \leq_o)$ and $(K_1, \leq_1)$:

$(K_o, \leq_o)$: 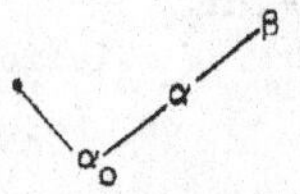

$(K_1, \leq_1)$: 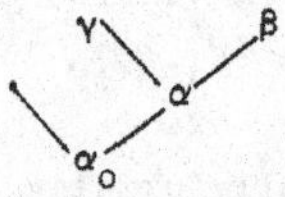

If we embed $(K_o, \leq_o)$ in $(K_1, \leq_1)$ in the obvious manner, and if we have a forcing relation $\Vdash_o$ on $(K_o, \leq_o)$, in order to extend to a relation $\Vdash_1$, we must decide how γ is to behave. We cannot necessarily let γ behave like α, because α has an extra node beyond it which may affect α's behavior. However, we can make γ behave like any terminal node beyond α - which in this case is β. Now, α cannot distinguish γ from β and hence α behaves the same in $(K_o, \leq_o, \Vdash_o)$ and $(K_1, \leq_1, \Vdash_1)$.

Note. In this proof, we need only assume $(K_o, \leq_o)$ is finite. The theorem holds when both $(K_o, \leq_o)$ and $(K_1, \leq_1)$ are infinite. In this case, the terminal nodes are replaced by complete sequences (in the sense of Cohen 1966).

Theorem 5.3.6 has the immediate corollary:

5.3.7. Corollary. Let $\{(K_n, \leq_n)\}_n$ be a sequence of finite trees with the property that every finite tree $(K, \leq)$ can be embedded as a subtree of some $(K_n, \leq_n)$. Then $P\rho$ is complete for the sequence $(K_n, \leq_n)$.

5.3.8. Examples.

A) The diagonal sequence. This is the sequence whose n-th element $(n \geq 1)$ is n-ary and of height n:

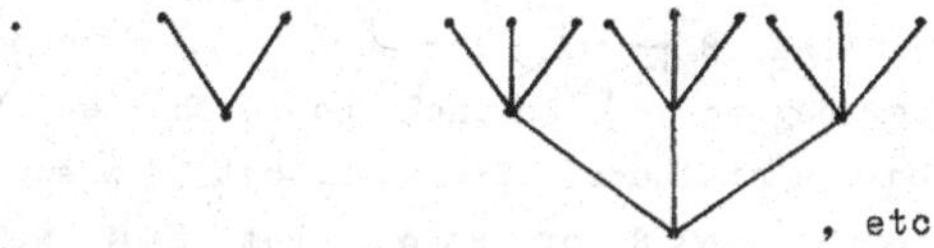

, etc.

B) The Jaskowski sequence. This economical sequence, due to Jaskowski (Jaskowski 1936) (see also Rose 1953, Gal, Rosser, and Scott 1958, Scott 1957, Gabbay 1969, and Mostowski 1966. Gabbay 1969 gives a treatment similar to ours for this sequence.), is obtained by letting the $n+1$-st tree be the result of taking n copies of the n-th tree and dropping a node below them:

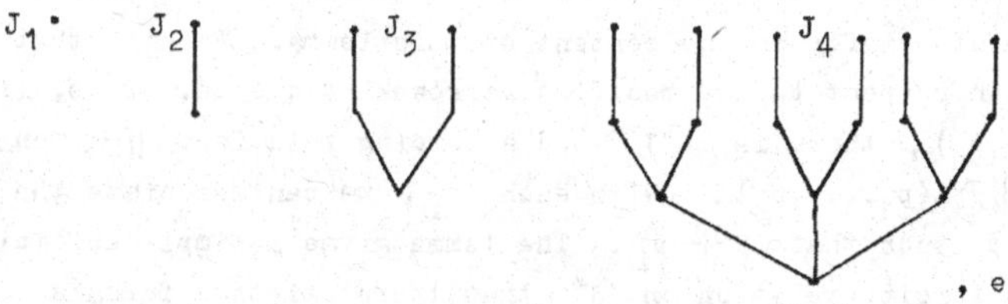

, etc.

C) The modified Jaskowski sequence.

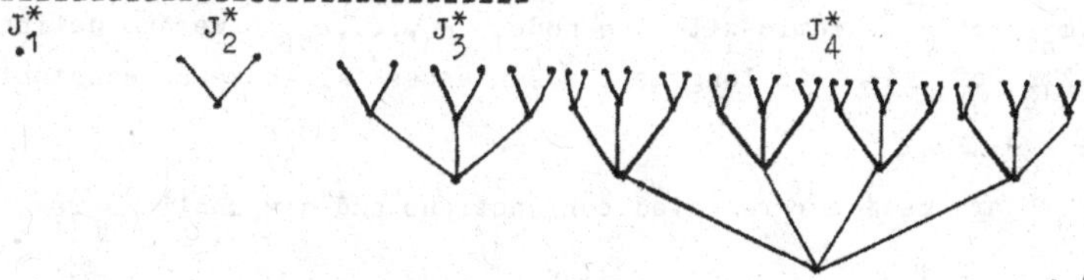

, etc.

Let us finish this subsection by remarking on a useful property of the modified Jaskowski trees (a property shared, incidentally by the diagonal trees): Every node of J_n^* is determined by the terminal nodes lying beyond it. From this, we can prove the following lemma:

5.3.9. Lemma. Let $\alpha_1, \ldots, \alpha_{n!}$ be the terminal nodes of J_n^*. Suppose we have a forcing relation, $\Vdash$, defined on J_n^* such that, for each i, there is a formula A_i such that

$$\alpha_j \Vdash A_i \quad \text{iff} \quad j = i .$$

Then, if S is a set of nodes of J_n^* such that $\alpha \in S$ and $\alpha \leq \beta$ imply $\beta \in S$, there is a formula A constructed from the A_i's, for which $S = \{\alpha : \alpha \Vdash A\}$. In particular, for any α, there is an A_α such that $\{\beta : \beta \geq \alpha\} = \{\beta : \beta \Vdash A_\alpha\}$.

Proof. Since $\alpha \in J_n^*$ is determined by those $\alpha_i \geq \alpha$, it is also determined by those $\alpha_i \not\geq \alpha$. Let $A_\alpha = \bigwedge_{\alpha_i \not\geq \alpha} \neg A_i$ for *) $\alpha \neq \alpha_0$ and let $A_{\alpha_0} = A_1 \rightarrow A_1$.

We must show that $\beta \Vdash A_\alpha$ iff $\beta \geq \alpha$. For α_0 this is trivial. Let $\alpha \neq \alpha_0$. If $\beta \Vdash A_\alpha$, the set of terminal nodes not beyond β includes those not beyond α (otherwise $\beta \not\Vdash \neg A_i$ for some i). Hence the set of terminal nodes of β (i.e. beyond β) is included in the set of terminal nodes of α. Let α_i be one of these. Since the set of predecessors of α_i is linearly ordered, either $\alpha \leq \beta$ or $\beta < \alpha$. But $\beta < \alpha$ would imply that α and β have the same terminal nodes beyond them, a contradiction. Thus $\beta \geq \alpha$. The converse is easy: $\alpha_i \not\geq \alpha$ implies $\alpha_i \not\geq \beta$ and $\beta \Vdash \neg A_i$ (since no extension of β forces A_i). Hence $\beta \Vdash A_\alpha$.

Finally, let S have the property stated in the lemma and let $A = \bigvee_{\alpha \in S} A_\alpha$. Q.E.D.

Let us comment briefly on the content of this lemma. We know that $P\rho$ is complete with respect to the modified Jaskowski sequence. Thus, if $P\rho \not\vdash A(p_1, \ldots, p_n)$, there is a J_n^* and a forcing relation, $\Vdash$, on J_n^* such that $\alpha_0 \not\Vdash A(p_1, \ldots, p_n)$. With each p_i, we can associate the set S_i of nodes β such that $\beta \Vdash p_i$. The lemma gives a simple sufficient condition on a forcing relation on J_n^* that there exists a formula behaving like p_i. Recall that, if we have associated non-standard models $\omega_1, \ldots, \omega_{n!}$ of arithmetic with the nodes $\alpha_1, \ldots, \alpha_{n!}$, we can define a model $K_{J_n^*}$ of HA. As long as the sentences A_i have no constants

------

*) $\bigwedge$, $\bigvee$ are used for repeated conjunctions and disjunctions respectively.

denoting non-standard numbers, the proof of the lemma carries through for $\underset{\sim}{K}_{J_n^*}$. This is the key to our simple proof of de Jongh's theorem.

5.3.10-12 The Gödel - Rosser - Mostowski - Kripke - Myhill theorem.

A straightforward iteration of the Gödel - Rosser theorem will give us independent sentences $A_1,\dots,A_m$ for any m. For the sake of obtaining the simplest possible substitution instances in theorem 5.3.2, we want the independent sentences $A_1,\dots,A_m$ to be Σ_1^0. This result has been proven by Mostowski 1961, Kripke 1963, and Myhill 1972.

We shall present Myhill's proof of the following

5.3.11. Theorem. Let $\underset{\sim}{T}_0, \underset{\sim}{T}_1, \dots$ be an r.e. sequence of consistent r.e. extensions of classical (Peano) arithmetic, $\underset{\sim}{HA}^c$. Then we can find a Σ_1^0 sentence A such that A is independent over each theory $\underset{\sim}{T}_i$.

Proof. Let X,Y be recursively inseparable sets and let X,Y be represented by formulae $\exists yR(x,y)$, $\exists yS(x,y)$, respectively, in such a way that:

$$n \in X \text{ iff } \exists yR(n,y) \text{ iff } \underset{\sim}{HA} \vdash \exists yR(n,y),$$
$$n \in Y \text{ iff } \exists yS(n,y) \text{ iff } \underset{\sim}{HA} \vdash \exists yS(n,y),$$

and $\underset{\sim}{HA} \vdash \neg(\exists yR(n,y) \;\&\; \exists yS(n,y))$.

Let $X_i = \{x : \underset{\sim}{T}_i \vdash \exists yR(x,y)\}$,
$X_i' = \{x : \underset{\sim}{T}_i \vdash \neg\exists yR(x,y)\}$,

and consider $X^* = \bigcup_i X_i$, $X' = \bigcup_i X_i'$.

By the reduction theorem in recursion theory (see e.g. Rogers 1967, p. 72), there are r.e. sets U, V such that

$$U \cup V = X^* \cup X',$$
$$U \cap V = \emptyset,$$
$$U \subseteq X^*, \text{ and } V \subseteq X'.$$

Clearly $X \subseteq U$. For, if $n \in X \cap V$, $n \in X'$ and, for some $\underset{\sim}{T}_i$, $\underset{\sim}{T}_i \vdash \exists yR(n,y)$ and $\underset{\sim}{T}_i \vdash \neg\exists yR(n,y)$, contradicting the consistency of $\underset{\sim}{T}_i$.

Also, $Y \subseteq V$, since $n \in Y$ implies $\underset{\sim}{T}_i \vdash \exists yS(n,y)$, whence $\underset{\sim}{T}_i \vdash \neg\exists yR(n,y)$. But $n \in U$ implies $\underset{\sim}{T}_i \vdash \exists yR(n,y)$ for some i, again contradicting consistency.

Now, U and V separate X and Y, whence there is an $n_0 \notin U \cup V$. Then, if we let $A = \exists yR(n_0,y)$, we see that A is independent over each $\underset{\sim}{T}_i$.

Q. E. D.

5.3.12. Corollary. For any m, we can find m Σ_1^0 sentences independent over $\underset{\sim}{HA}^c$.

<u>Proof</u>. Let A_1 be independent over $\underset{\sim}{HA}^c$; A_2 independent over $\underset{\sim}{HA}^c + A_1$, $\underset{\sim}{HA}^c + \neg A_1$; A_3 independent over $\underset{\sim}{HA}^c + A_1 + A_2$, $\underset{\sim}{HA}^c + A_1 + \neg A_2$, $\underset{\sim}{HA}^c + \neg A_1 + A_2$, $\underset{\sim}{HA}^c + \neg A_1 + \neg A_2$; etc. Q. E. D.

5.3.13 - 5.3.15. <u>de Jongh's theorem</u>. Let us now combine the results of 5.3.3 - 5.3.12 to prove theorem 5.3.2, which we restate here:

5.3.13. <u>Theorem</u>. If $\mathbf{P_P} \nvdash A(p_1,\dots,p_n)$, then $\underset{\sim}{HA} \nvdash A(B_1,\dots,B_n)$, for some sentence $B_1,\dots,B_n$ of arithmetic.

<u>Proof</u>. Let $\alpha_0 \Vdash\!\!\!/\, A(p_1,\dots,p_n)$ for some forcing relation on J_k^* and let $A_1,\dots,A_{n!}$ be independent over $\underset{\sim}{HA}^c$ and find, for each i, a model ω_i of $A_i + \bigwedge_{j\neq i} \neg A_j$. Associate these models with the terminal nodes of J_k^* and look at $\underline{K}_{J_k^*}$.

E.g. $\underline{K}_{J_3^*}$ is

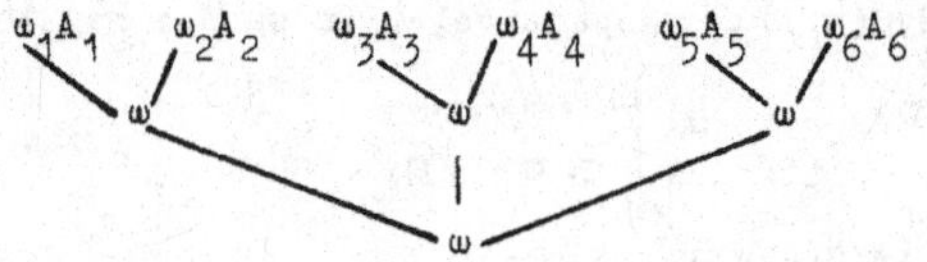

Let, for each p_i, S_i be the set of nodes forcing p_i. By lemma 5.3.9, we can find a sentence B_i of arithmetic built up from the A_j's such that $S_i = \{\beta : \beta \Vdash B_i\}$.

Now, a simple induction can be used to show that, for any formula $C(p_1,\dots,p_n)$ and any node β,

$$\beta \Vdash C(p_1,\dots,p_n) \quad \text{iff} \quad \beta \Vdash C(B_1,\dots,B_n),$$

under the two forcing relations. In particular,

$$\alpha_0 \Vdash\!\!\!/\, A(B_1,\dots,B_n). \qquad \text{Q. E. D.}$$

The sentence corresponding to S is (except in the trivial case $\alpha_0 \in S$) of the form,

$$\bigvee_i \bigwedge_j \neg A_{ij},$$

whence we have the following corollary due to Myhill:

5.3.14. <u>Corollary</u>. The substitution instances $B_1,\dots,B_n$ in theorem 5.3.13 may be taken to be disjunctions of Π_1^0 sentences.

Observe that one cannot use Π_1^0 sentences, because, if B is Π_1^0, $\underset{\sim}{HA} \vdash \neg\neg B \to B$.

Starting with $A_1,\dots,A_m$ Π_1^0 and independent, we have the following

5.3.15. Corollary. The substitution instances $B_1,\ldots,B_n$ in theorem 5.3.13 may be taken to be disjunctions of double negations of Σ_1^0 sentences.

Since the only properties of HA used in proving de Jongh's theorem were the closure of the class of models of HA under the operation $(\ )\to(\Sigma\ )'$, the consistency of HA with classical logic (so that $\omega_1,\ldots,\omega_m$ could be chosen), and the incompleteness of HA^c, we can conclude that de Jongh's theorem also holds for $HA+\Gamma$ for any r.e. $\Gamma\in\mathfrak{P}$ (as in section 5.2.11) which is consistent with classical logic. In particular, de Jongh's theorem holds for $HA+TI(<)$, $HA+RF(HA)$, $HA+RFN(HA)$, etc. If $HA+\Gamma$ is not consistent with classical logic, the independence of A is replaced by the independence of $\neg A$, so that models of A and $\neg A$ exist. Then the models $\omega_1,\ldots,\omega_m$ are replaced by Kripke models.

5.3.16. de Jongh's theorem for one propositional variable.

In Nishimura 1960, Iwao Nishimura characterized the lattice of formulae in one propositional variable in the intuitionistic propositional calculus. It happens that there are close relations between these lattices and pms's. From Nishimura's work, it is not hard to prove the following

5.3.17. Theorem. Let $(K,\leq,\Vdash)$ be the model shown below and let $A(p)$ be a formula in the variable p such that $\mathrm{Pp}\nVdash A(p)$. Then for some $\alpha\in K$, $\alpha\nVdash A(p)$.

$(K,\leq,\Vdash)$:

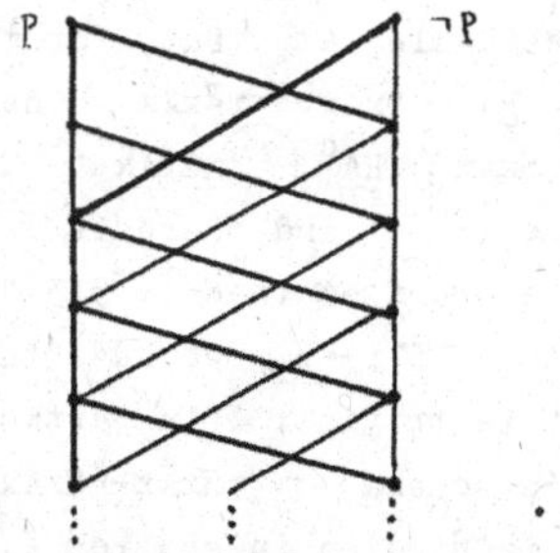

The proof of this lies beyond the scope of these notes. A proof avoiding the use of lattices may be found in de Jongh B.

Let B be Σ_1^0, independent over HA^c. Then there is a model ω^+ in which B is true. Consider the model

K :

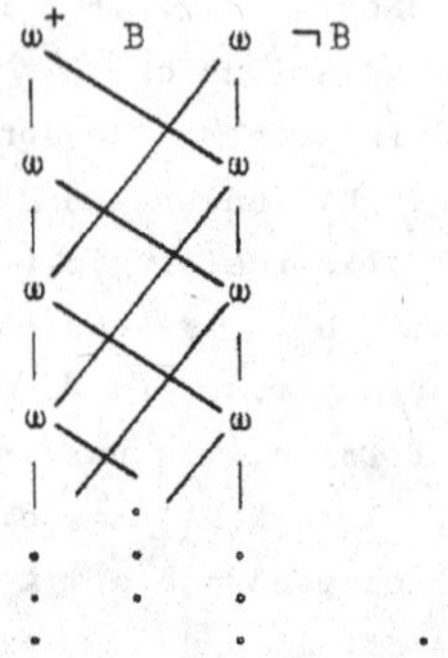

This allows us to prove the following result, due independently to de Jongh and ourselves.

5.3.18. Theorem. Let B be Σ^0_1, independent over $\underset{\sim}{\mathrm{HA}}^c$, and let $\mathcal{P}_p \nvdash A(p)$. Then $\underset{\sim}{\mathrm{HA}} \nvdash A(B)$.

Proof. Prove by induction on the length of $C(p)$ that, for $\alpha \in K$, $\alpha \Vdash C(p)$ in the first model iff $\alpha \Vdash C(B)$ in the second model. Then apply theorem 5.3.17. Q. E. D.

Recall theorem 5.2.6, by which e.g. we showed $\underset{\sim}{\mathrm{HA}} \vdash \forall x Ax \vee \exists x \neg Ax$ iff $\underset{\sim}{\mathrm{HA}} \vdash \neg \forall x Ax \to \exists x \neg Ax$, when $\underset{\sim}{\mathrm{HA}} \vdash \forall x(Ax \vee \neg Ax)$.

We may restate this equivalence as $\underset{\sim}{\mathrm{HA}} \vdash \exists x Ax \vee \neg \exists x Ax$ iff $\underset{\sim}{\mathrm{HA}} \vdash \neg\neg \exists x Ax \to \exists x Ax$, for decidable A. But, for this formulation, theorem 5.3.18 readily applies. In $\underset{\sim}{\mathrm{HA}} \nvdash \exists x Ax \vee \neg \exists x Ax$, then $\underset{\sim}{\mathrm{HA}} \nvdash \exists x Ax$ and $\underset{\sim}{\mathrm{HA}} \nvdash \neg \exists x Ax$, whence $\underset{\sim}{\mathrm{HA}}^c \nvdash \exists x Ax$, $\underset{\sim}{\mathrm{HA}}^c \nvdash \neg \exists x Ax$. The decidability of A implies that $\neg \exists x Ax$ is true in ω and there is a non-standard model ω^+ of $\exists x Ax$. Now, applying the proof of theorem 5.3.18 to the sentence $\exists x Ax$ and the propositional formula $\neg\neg p \to p$, we have the result.
The case $\underset{\sim}{\mathrm{HA}} \vdash \exists x Ax \vee \neg \exists x Ax$ is trivial. (In particular, we have an independence proof for Markov's schema $\neg\neg \exists x Ax \to \exists x Ax$, Ax primitive recursive. - Markov's schema is studied in section 4, below.)

Another point worth stressing is that, for one propositional variable, we have a best possible result: uniform Σ^0_1 substitution instances.

Finally, observe that we do not have the result for all $\underset{\sim}{\mathrm{HA}} + \Gamma$, $\Gamma \in \mathfrak{P}$, since we must have Γ valid in ω. One can get around this slightly by considering models:

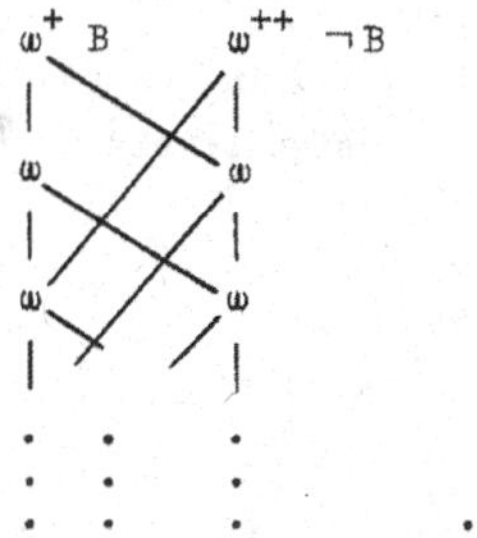

We leave the investigation of such results to the reader, taking time only to mention the following result of de Jongh's :

5.3.19. Theorem. Let B be a sentence such that $HA \nvdash \neg\neg B$, $HA \nvdash \neg\neg B \neg B$. Then, if $P_P \nvdash A(p)$, we have $HA \nvdash A(B)$.

Remark. We may use theorem 5.3.6 (the extension theorem) to prove the following : If $P_P \nvdash A(p)$, then $\alpha_0 \Vdash\!\!\!\!/\; A(p)$ for some finite tree model $\underline{K}$ in which p is forced only at terminal nodes (if at all). This will give a simple proof of theorem 5.3.18 without using the Nishimura pms.

5.3.20. Another theorem of de Jongh (digression).

The Nishimura pms was used by de Jongh to solve a problem of Kreisel. In the last paragraph of Kreisel - Levy 1968, Kreisel and Levy mention that, when one wants a truth definition for formulae of bounded complexity, one must include the number of nested implications and negations occurring in a formula as well as the number of nested alternating quantifiers in one's measure of complexity. The infinitude of the Nishimura lattice tells us that there are infinitely many propositional formulae in one variable which are non-equivalent over P_P. Kreisel asked for a proof that there is no truth definition within HA for the substitution instances - i.e. for any formula Tx and some sentence B, not all of the following equivalences are derivable:

$$T(\ulcorner A(B)\urcorner) \leftrightarrow A(B) ,$$

where A(p) ranges over all propositional formulae in one variable and $\ulcorner A(B)\urcorner$ denotes the gödel number of A(B).

We present de Jongh's proof of this result here :

5.3.21. Theorem. Let B be Σ_1^0, independent of HA^c, and let Tx have only x free. Then, for some propositional formula A(p),

$$HA \nvdash T(\ulcorner A(B)\urcorner) \leftrightarrow A(B) .$$

In other words, for any independent Σ_1^0 sentence B, there is no truth definition for the set of propositional formulae in B.

Proof. Since B is Σ_1^0 and independent, B is false in the standard model. Let ω^+ be a non-standard model of $\underset{\sim}{HA}^c + B$, and assign levels to the nodes of the Nishimura model as follows:

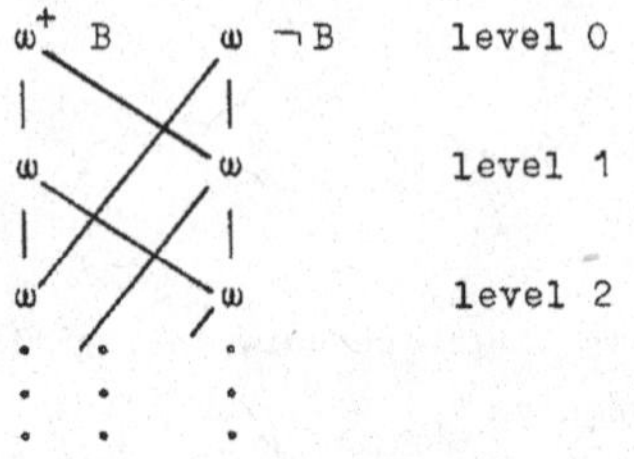

Let $C(x_1,\ldots x_n)$ be a formula with free variables as indicated. We shall prove by induction on the length of C that there is a level $n_C \geq 1$ such that, for any $m_1,\ldots,m_n$, if $C(m_1,\ldots,m_n)$ is forced by a node of level n_C, then $C(m_1,\ldots,m_n)$ is forced by all nodes.

(i) Let C be atomic. Then $n_C = 1$.

(ii)-(iii) If C is $D \& E$ or $D \vee E$, $n_C = \max(n_D, n_E)$ will do the trick.

(iv) Let C be $D \to E$. Then take $n_C = \max(n_D, n_E) + 1$. To see this, label the nodes as follows:

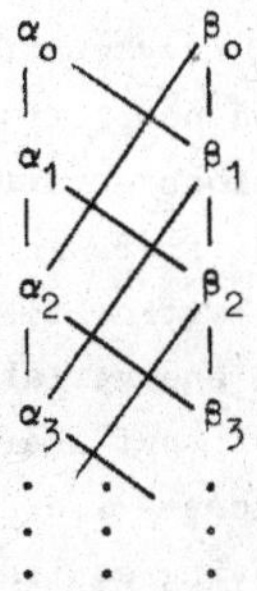

Let $n = \max(n_D, n_E)$ and let $\alpha_{n+1} \Vdash D \to E$. First, observe that $\alpha_n \Vdash D \to E$ and, if $\alpha_n \Vdash D$, $\alpha_n \Vdash E$ and all nodes force D and E, whence they force $D \to E$. If $\beta_n \Vdash D$, all nodes force D, whence $\alpha_n \Vdash D$, whence all nodes force $D \to E$.

If, for some γ, $\gamma \nVdash D \to E$, then there is a $\delta \geq \gamma$ such that $\delta \Vdash D$, $\delta \nVdash E$. If the level of δ is $\geq n$, then α_n or β_n is $\geq \delta$. But, if this is the case, α_n or β_n forces D and all nodes force the implication. On the other hand, there is no node of level $< n$ which is not $\geq \alpha_{n+1}$. Hence all nodes force $D \to E$.

The case in which $\beta_{n+1} \Vdash D \to E$ is similar.

(v) C is $\neg D$. Similar to (iv).

(vi) Let $C(x_1,\dots,x_n)$ be $\exists x\, D(x,x_1,\dots,x_n)$. Then $n_C = n_D$. Since $n_D \geq 1$, the domain at level n_D is $\{0, 1, \dots\}$. Let e.g. $\alpha_{n_D} \Vdash \exists x\, D(x,m_1,\dots,m_n)$. Then, for some m, $\alpha_{n_D} \Vdash D(m,m_1,\dots,m_n)$ and all nodes force $D(m,m_1,\dots,m_n)$, whence they force $\exists x\, D(x,m_1,\dots,m_n)$. β_{n_D} is handled similarly.

(vii) $C(x_1,\dots,x_n)$ is $\forall x\, D(x,x_1,\dots,x_n)$. Then $n_C = n_D$ and the proof is similar to that in case (vi).

Thus, any formula Tx will have a level n_T associated with it in such a way that, for all $\ulcorner A(B)\urcorner$, if $\alpha_{n_T} \Vdash T(\ulcorner A(B)\urcorner)$ or $\beta_{n_T} \Vdash T(\ulcorner A(B)\urcorner)$, then $\alpha \Vdash T(\ulcorner A(B)\urcorner)$ for all nodes α in the model. The proof is completed by the following lemma.

5.3.22. <u>Lemma</u>. For each level n, there is a sentence $A(B)$ which is forced at a node of level n, but at no nodes of level $n+1$ or higher.

The proof of this is not difficult, but is rather long and we omit it. The reader is referred to <u>de Jongh</u> B, for the proof. Alternatively, if he is willing to accept theorem 5.3.17 and the infinitude of the set of inequivalent formulae in one propositional variable, lemma 5.3.22 for arbitrarily large n follows by a simple cardinality argument - with only finitely many nodes of level $\leq n$ to distinguish these formulae, we can only find finitely many inequivalent formulae. By either approach, the proof of theorem 5.3.21 is completed. Q. E. D.

5.3.23. <u>Further results on de Jongh's theorem</u>.

In theorem 5.3.13 (theorem 5.3.2), we proved that, for any underivable $A(p_1,\dots,p_n)$, arithmetical $B_1,\dots,B_n$ can be found such that $A(B_1,\dots,B_n)$ is underivable in $\underset{\sim}{HA}$. de Jongh's original proof (<u>de Jongh</u> A) gave $B_1,\dots,B_n$ uniformly in all $A(p_1,\dots,p_n)$. <u>Friedman</u> A improved this by showing that, where uniformity is desired, any collection $B_1,\dots,B_n$ of Π^0_2 sentences independent over $\underset{\sim}{HA}^c$ augmented by all true Π^0_1 sentences will work. Friedman's proof made use of his generalization of the Kleene slash. We shall present a (Kripke) model-theoretic proof of his result in section 6, below.

In terms of the simplicity of the substitution instances, corollary 5.3.15 shows that $B_1,\dots,B_n$ (in the non-uniform version) may be taken to be disjunctions of double negations of Σ^0_1 sentences, which, classically, would be Σ^0_1. We obtain Σ^0_1 substitution instances in section 6.

When restricting one's attention to a particular number of variables, we outlined a proof of the existence of uniform Σ^0_1 counterexamples in 5.3.16 - 5.3.19 above. Using an alternate proof involving the arithmetization of the Kleene slash, <u>de Jongh</u> 1971 and B reproved his theorem 5.3.19 (cf. also 3.1.16).

§ 4. Markov's schema

5.4.1 - 5.4.3 . Markov's schema.

5.4.1. We have already encountered Markov's schema in our discussion of the Kripke models. In section 2, we presented a model-theoretic proof of Kreisel's version of the independence of Markov's schema (theorem 5.2.6) and in section 3, we observed that this result also came as a corollary to a special version of de Jongh's theorem for the special case of one propositional variable. Both proofs are off-shoots of the simple fact that Markov's schema is <u>not</u> preserved by the operation $(\ ) \to (\Sigma\ )'$.

Before discussing this last fact, let us consider several formulations of Markov's schema :

(i) $\forall x(Ax \vee \neg Ax) \;\&\; \neg\neg \exists x Ax \to \exists x Ax$,

(ii) $\forall xy(Axy \vee \neg Axy) \;\&\; \forall x\, \neg\neg \exists y Axy \to \forall x \exists y Axy$,

(iii) $\forall xy(Axy \vee \neg Axy) \to \forall x[\neg\neg \exists y Axy \to \exists y Axy]$,

(iv) $\forall x[\forall y(Axy \vee \neg Axy) \;\&\; \neg\neg \exists y Axy \to \exists y Axy]$,

(v) $\forall z[\forall xy(Axyz \vee \neg Axyz) \;\&\; \forall x\, \neg\neg \exists y Axyz \to \forall x \exists y Axyz]$,

where A contains only the free variables shown. Observe that the schemata obtained by replacing $x, y,$ or z by finite sequence of variables reduce to the present schemata via a pairing function. Thus, there is no loss of generality in considering only (i) - (v). For the treatment of Markov's schema by other methods, see also 1.11.5 and § 3.8.

5.4.2. Lemma. (v) $\longleftrightarrow$ (iv) $\to$ (iii) $\to$ (ii) $\to$ (i).

Proof. (v) $\to$ (iv). Trivial.

(iv) $\to$ (v). Let $\underline{K}$ be a model of (iv) and let $\alpha \in K$ be such that $\alpha \not\Vdash \forall z[\forall xy(Axyz \vee \neg Axyz) \;\&\; \forall x \neg\neg \exists y Axyz \to \forall x \exists y Axyz]$. Then, for some $\beta \geq \alpha$ and $b \in D\beta$, we have $\beta \Vdash \forall xy(Axyb \vee \neg Axyb)$, $\beta \Vdash \forall x \neg\neg \exists y Axyb$, and $\beta \not\Vdash \forall x \exists y Axyb$.

But then there are $\gamma \geq \beta$ and $c \in D\gamma$ such that $\gamma \not\Vdash \exists y Acyb$. We also have $\gamma \Vdash \forall y(Acyb \vee \neg Acyb) \;\&\; \neg\neg \exists y Acyb$. Let $d = j(c,b)$, where j is the standard primitive recursive pairing function with inverses j_1, j_2. Then $\gamma \Vdash \forall y(A(j_1d, y, j_2d) \vee \neg A(j_1d, y, j_2d))$ and $\gamma \Vdash \neg\neg \exists y A(j_1d, y, j_2d)$, whence, applying (iv) to $A'xy : A(j_1x, y, j_2x)$, we have $\gamma \Vdash \exists y A(j_1d, y, j_2d)$, a contradiction.

(iv) $\to$ (iii). Let $\underline{K}$ be a model of (iv), $\alpha \Vdash \forall xy(Axy \vee \neg Axy)$ and $\alpha \not\Vdash \forall x[\neg\neg \exists y Axy \to \exists y Axy]$. Then there are $\beta \geq \alpha$ and $b \in D\beta$ such that $\beta \Vdash \neg\neg \exists y Aby$, $\beta \not\Vdash \exists y Aby$. Now $\beta \Vdash \forall y(Aby \vee \neg Aby)$ and, by (iv), $\beta \Vdash \forall y(Aby \vee \neg Aby) \;\&\; \neg\neg \exists y Aby \to \exists y Aby$, whence $\beta \Vdash \exists y Aby$, a contradiction.

(iii) → (ii). Let $\underline{K}$ be a model of (iii), $\alpha \Vdash \forall xy(Axy \vee \neg Axy)$, and $\alpha \not\Vdash \forall x \neg\neg \exists y Axy \rightarrow \forall x \exists y\, Axy$. Then there are $\beta \geq \alpha$ and $b \in D\beta$ such that $\beta \Vdash \neg\neg\exists y Aby$ and $\beta \not\Vdash \exists y\, Aby$. But, by (iii), $\beta \Vdash \neg\neg \exists y\, Aby \rightarrow \exists y\, Aby$, a contradiction.

(ii) → (i). Trivial. Q. E. D.

Unfortunately, we cannot settle any of the converse implications (simple model-theoretic independence proofs are ruled out - when we replace the operation $(\ ) \rightarrow (\Sigma\ )'$ by one which preserves Markov's schema, we will see that all five schemata are preserved.). However, we can prove the following:

5.4.3. Theorem. The scheme (iv) is derivable in $\underset{\sim}{HA} + RFN((i))$.

Proof. The proof is based on a remark of Kreisel's that the uniform reflection principle allows one to add free variables. We show

$\underset{\sim}{HA} \vdash \forall x \exists z\, Proof_{HA+(i)}(z, \ulcorner \forall y(A\bar{x}y \vee \neg A\bar{x}y)\ \&\ \neg\neg \exists y\, A\bar{x}y \rightarrow \exists y A\bar{x}y \urcorner)$.

Let $B_o(x), \ldots$ be a primitive recursive enumeration of all instances of $\forall y(Axy \vee \neg Axy)\ \&\ \neg\neg \exists y Axy \rightarrow \exists y Axy$.

a) $\underset{\sim}{HA} \vdash \forall w\, Proof_{HA+(i)}(\ulcorner B_w(0) \urcorner, \ulcorner B_w(0) \urcorner)$, i.e. every axiom is its own proof.

b) Let $\forall w \exists z\, Proof_{HA+(i)}(z, \ulcorner B_w(\bar{x}) \urcorner)$. Also, let f be primitive recursive such that

$$\underset{\sim}{HA} \vdash B_w(x+1) \leftrightarrow B_{fw}(x)\ .$$

By well-known properties of Proof,

$$\underset{\sim}{HA} \vdash \exists z\, Proof_{HA+(i)}(z, \ulcorner B_w(\overline{x+1}) \urcorner) \leftrightarrow \exists z\, Proof_{HA+(i)}(z, \ulcorner B_{fw}(x) \urcorner)\ .$$

But $\exists x Proof_{HA+(i)}(z, \ulcorner B_{fw}(\bar{x}) \urcorner)$ and so $\exists z\, Proof_{HA+(i)}(z, \ulcorner B_w(\overline{x+1}) \urcorner)$.

c) Thus

$$\underset{\sim}{HA} \vdash \forall w \exists z\, Proof_{HA+(i)}(z, \ulcorner B_w(\bar{x}) \urcorner) \rightarrow \forall w \exists z\, Proof_{HA+(i)}(z, \ulcorner B_w(\overline{x+1}) \urcorner)\ .$$

This and (a) yields

$$\underset{\sim}{HA} \vdash \forall w\, \forall x \exists z\, Proof_{HA+(i)}(z, \ulcorner B_w(\bar{x}) \urcorner)\ .$$

$RFN'(\underset{\sim}{HA} + (i))$ (which is equivalent to $RFN(\underset{\sim}{HA} + (i))$ by lemma 5.2.18 - the implication $RFN \rightarrow RFN'$, however, is trivial) yields, for w the index of $\forall y(Axy \vee \neg Axy)\ \&\ \neg\neg \exists y\, Axy \rightarrow \exists y Axy$,

$\underset{\sim}{HA} + RFN(\underset{\sim}{HA} + (i)) \vdash \forall x[\ \forall y(Axy \vee \neg Axy)\ \&\ \neg\neg \exists y Axy \rightarrow \exists y Axy]$. Q. E. D.

Thus, if schemata (i) - (iv) are not formally equivalent, they are almost equivalent. Combining this with our model-theoretic inability to distinguish these schemata, we shall allow ourselves to be sloppy and let MP denote any of the schemata (i) - (v) (for the present chapter).

We note that MP may be formulated as a rule of inference (see 3.8.1).

5.4.4 - 5.4.6. <u>The independence of</u> MP.

As remarked above, we have already proven the independence of MP twice. This time, however, we shall be more direct.

5.4.4. <u>Theorem</u>. Let $HA + \Gamma$ be r.e., $\Gamma \in \mathfrak{P}$ (as defined in section 5.2.11). Then, some instance of MP is not derivable in $HA + \Gamma$.

In fact, let $\exists x Ax$ be independent with Ax primitive recursive (so $HA \vdash \forall x(Ax \vee \neg Ax)$). Then

$$HA + \Gamma \nvdash \neg\neg\exists x Ax \to \exists x Ax .$$

<u>Proof</u>. Let $\exists x Ax$ be independent of $HA + \Gamma$ and let $\underline{K}$ be a model of $HA + \Gamma$ with a node β such that $\beta \Vdash \exists x Ax$ and consider $(\underline{K}_\beta)'$:

$$\begin{array}{c} \underline{K}_\beta \\ | \\ \alpha_0 \;\; \omega \end{array} .$$

We will show $\alpha_0 \nVdash \neg\neg\exists x Ax \to \exists x Ax$. Since $\gamma \geq \alpha_0$ implies $\gamma = \alpha_0$ or $\gamma \geq \beta$, $\beta \Vdash \exists x Ax$ implies $\alpha_0 \Vdash \neg\neg\exists x Ax$. But, if $\alpha_0 \Vdash \exists x Ax$, then $\alpha_0 \Vdash An$ for some n. As usual, this means $HA \vdash An$ and so $HA \vdash \exists x Ax$, contradicting independence.

But $\Gamma \in \mathfrak{P}$ and so $HA + \Gamma$ is preserved by the step from $\underline{K}_\beta$ to $(\underline{K}_\beta)'$. Q.E.D.

For example, MP is independent of $HA + RFN(HA)$, $HA + TI(<)$, $HA + CON(HA + MP)$, etc.

In addition to outright independence results, one can obtain results on the form of the axiomatization of MP as follows. First, let us define a measure, m, of the complexity of a formula of number theory. We do this inductively as follows,

(i) if A is atomic, $m(A) = 1$,

(ii) - (iv) $m(A \& B) = m(A \vee B) = m(A \to B) = \max(m(A), m(B))$,

(v) $m(\neg A) = m(A)$,

(vi) $m(\exists x Ax) = \begin{cases} m(Ax), & Ax = \exists y Bxy \text{ for some } B \\ m(Ax) + 1, & \text{otherwise,} \end{cases}$

(vii) $m(\forall x Ax) = \begin{cases} m(Ax), & Ax = \forall y Bxy \text{ for some } B \\ m(Ax) + 1, & \text{otherwise.} \end{cases}$

Observe that, for classical arithmetic, HA^c, this is a reasonable measure in the sense that a truth definition for formulae of bounded complexity can be given. We have observed in 5.3.20 that no such definition can be given in HA. (To obtain one, redefine $m(A \to B) = \max(m(A), m(B)) + 1$ and $m(\neg A) = m(A) + 1$.)

5.4.5. <u>Theorem.</u> $\underset{\sim}{HA} + MP$ is not axiomatized by any restriction of the schema to formulae A for which $m(A) \leq n_o$ for any n_o. I.e. no set of instances of MP of bounded complexity can axiomatize MP (over $\underset{\sim}{HA}$).

<u>Proof.</u> We know from recursion theory that all formulae whose complexity is of measure $\leq n_o$ lie in a certain level of the arithmetical hierarchy. We also know from the hierarchy theorem and the completeness theorem (for classical logic) that there is some non-standard model ω^+ of $\underset{\sim}{HA}^c$ in which the truth of a sentence at or below the given level of the hierarchy agrees with truth in the standard model, but truth at higher levels does not. Thus consider $(\omega^+)'$:

$$\begin{array}{cc} \alpha & \omega^+ \\ & | \\ \alpha_o & \omega \end{array} .$$

We first show by induction on $m(A)$ and on the length of A, that, if $m(A) \leq n_o$, $\alpha, \alpha_o \Vdash A$ iff A is true in ω (written $\omega \models A$). For this, we use the facts that $\alpha \Vdash A$ iff $\omega^+ \models A$

iff $\omega \models A$.

Atomic A are decidable and there is no problem. The connectives & and $\vee$ also offer no difficulty. Consider $B \to C$. If $\alpha_o \Vdash B \to C$, then $\alpha \Vdash B \to C$, whence $\omega^+ \models A$. Since $m(B \to C) \leq n_o$, $\omega \models A$. If $\alpha \not\Vdash B \to C$, then $\alpha_o \not\Vdash B \to C$. So suppose $\alpha_o \not\Vdash B \to C$, whence α or $\alpha_o \Vdash B$, $\not\Vdash C$. $\alpha \Vdash B$ and the length of B is less than that of $B \to C$, whence $\omega \models B$, $\alpha_o \Vdash B$. But $\alpha_o \not\Vdash B \to C$, whence $\alpha_o \not\Vdash C$. Again $\omega \not\models C$, whence $\omega \not\models B \to C$. Also $\alpha \not\Vdash B \to C$.

$\neg B$ is handled similarly.

Now consider the quantified formulae. For convenience, we only exhibit one quantifier, although there may actually be a block of like quantifiers. Thus, consider $\exists x Bx$.

$$\begin{aligned} \alpha_o \Vdash \exists x Bx &\Rightarrow \alpha \Vdash \exists x Bx \\ &\Rightarrow \omega^+ \models \exists x Bx \\ &\Rightarrow \omega \models \exists x Bx, \text{ by choice of } \omega^+. \end{aligned}$$

Conversely,

$$\begin{aligned} \omega \models \exists x Bx &\Rightarrow \underset{\sim}{\exists} n\ \omega \models Bn \\ &\Rightarrow \alpha, \alpha_o \Vdash Bn, \text{ by induction hypothesis.} \end{aligned}$$

For $\forall x Bx$,

$$\begin{aligned} \alpha_o \Vdash \forall x Bx &\Rightarrow \alpha \Vdash \forall x Bx \\ &\Rightarrow \omega^+ \models \forall x Bx \\ &\Rightarrow \omega \models \forall x Bx, \text{ by choice of } \omega^+. \end{aligned}$$

If $\omega \models \forall x Bx$, then $\omega^+ \models \forall x Bx$ and $\alpha \Vdash \forall x Bx$. To conclude $\alpha_0 \Vdash \forall x Bx$, it suffices to show $\alpha_0 \Vdash Bn$ for all numerals n. But

$$\omega \models \forall x Bx \Rightarrow \underset{\sim}{\forall} n \ \omega \models Bn$$
$$\Rightarrow \underset{\sim}{\forall} n \ \alpha_0 \Vdash Bn, \text{ by induction hypothesis.}$$

Thus, we see that, for $m(A) \leq n_0$,

$$\alpha_0 \Vdash A \text{ iff } \alpha \Vdash A$$
$$\text{iff } \omega \models A.$$

But ω^+ is not an elementary extension of ω and, for some prenex sentence A, $\omega^+ \models A$, $\omega \not\models A$. Let A be such a sentence for which $m(A)$ is minimal.

Then A is of the form $\exists x_1 \ldots x_n B$, where $m(B) < m(A)$. To see this, observe that, by minimality of $m(A)$, the above argument holds for all prenex C for which $m(C) < m(A)$. In particular,

$$\alpha_0 \Vdash C \text{ iff } \alpha \Vdash C$$
$$\text{iff } \omega \models C,$$

and C is decidable (since $\alpha_0 \Vdash C \leftrightarrow \alpha \Vdash C$ and $\alpha_0 \Vdash \neg C \leftrightarrow \alpha \Vdash \neg C$).

Suppose A is of the form $\forall x_1 \ldots x_n B(x_1, \ldots, x_n)$, $m(B) < m(A)$. Then

$$\omega^+ \models \forall x_1 \ldots x_n B x_1 \ldots x_n \Rightarrow \alpha \Vdash \forall x_1 \ldots x_n B$$
$$\Rightarrow \underset{\sim}{\forall} a_1 \ldots a_n \in D\alpha (\alpha \Vdash B(a_1, \ldots, a_n))$$
$$\Rightarrow \underset{\sim}{\forall} m_1 \ldots m_n \in \omega \ (\alpha \Vdash B(m_1, \ldots, m_n))$$
$$\Rightarrow \underset{\sim}{\forall} m_1 \ldots m_n \in \omega (\alpha_0 \Vdash B(m_1, \ldots, m_n)),$$

since $m(B) < m(A)$. This last fact, together with the fact that $\alpha \Vdash \forall x_1 \ldots x_n B$ implies $\alpha_0 \Vdash \forall x_1 \ldots x_n B$. But, more importantly, since $m(B) < m(A)$,

$$\alpha_0 \Vdash B(m_1, \ldots, m_n) \Rightarrow \omega \models B(m_1, \ldots, m_n),$$

whence $\omega \models \forall x_1 \ldots x_n B$ and A cannot be of the form suggested.

Hence we have $\omega^+ \models \exists x_1 \ldots x_n B$, $\omega \models \neg \exists x_1 \ldots x_n B$, and $\alpha_0 \Vdash \forall x_1 \ldots x_n (B(x_1, \ldots, x_n) \vee \neg B(x_1, \ldots, x_n))$ (by the fact that $m(B) < m(A)$). But $\alpha \Vdash \exists x_1 \ldots x_n B$, and so $\alpha_0 \Vdash \neg\neg \exists x_1 \ldots x_n B$. But we cannot have $\alpha_0 \Vdash \exists x_1 \ldots x_n B$. Contracting quantifiers, we have an instance,

$$\forall x (B'x \vee \neg B'x) \wedge \neg\neg \exists x B'x \rightarrow \exists x B'x,$$

of (i) which is not forced at α_0.

Finally, for $m(C) < n_0 - 1$, we have $m(\exists x\, C) < n_0$, whence $\exists x C$ is decidable in the model ($m(\exists x C \vee \neg \exists x C) = m(\exists x C) < n_0$). But, whether $\alpha_0 \Vdash \exists x C$ or $\alpha_0 \Vdash \neg \exists x C$, we have

$$\alpha_0 \Vdash \forall x (Cx \vee \neg Cx) \wedge \neg\neg \exists x Cx \rightarrow \exists x Cx.$$

Hence, MP is true in the model for instances of low complexity, but not for high complexity. The fact that we can make the "low complexity" large enough to include n_o yields the theorem. Q. E. D.

5.4.6. Corollary. MP is not derivable from the subscheme,

$$\forall x \neg\neg \exists y Axy \rightarrow \forall x \exists y Axy ,$$

where A is primitive recursive.

The above proof was rather long, but the idea is simple. If we start with a model ω^+ of HA^c which agrees with ω in the truth of formulae of low complexity, but not for formulae of high complexity, then formulae of low complexity are decidable in the model :

$$\begin{array}{c} \omega^+ \\ | \\ \alpha_o \; \omega \end{array} ,$$

whence MP holds for sentences of low complexity. But, sentences of high complexity are not decidable and, in particular, there is some sentence $\exists xBx$ which is not decidable at α_o, but for which. B yields a decidable property. Hence MP fails in some instance of high complexity.

We might also comment on the measure of complexity used. One might object that we should consider an intuitionistically meaningful measure - i.e. one for which the bounded truth definition can be given in HA as well as in HA^c. As observed above, such a measure m' is obtained by defining $m'(A \rightarrow B) = \max(m'(A), m'(B)) + 1$ and $m'(\neg A) = m'(A) + 1$. But then, for any formula A, $m'(A) \geq m(A)$, whence a bound on $m'(A)$ yields one on $m(A)$ and the result follows from theorem 5.4.5. Further, concerning the specific choice of a measure m, theorem 5.4.5 can be shown to hold for any measure for which truth definitions for formulae of bounded complexity can be given in HA^c (and hence for those measures whose bounded truth definitions can be given in HA).

The above theorem 5.4.5 and corollary 5.4.6 easily generalize to any r.e. $HA + \Gamma$, where $\Gamma \in \mathfrak{P}$ and Γ is true in the standard model, e.g. $HA + RFN(HA)$.

5.4.7 - 5.4.9. A comment on proof-theoretic closure properties.

5.4.7. We have used the failure of MP to be preserved by the operation $(\;) \rightarrow (\Sigma\;)'$ to prove its independence and to prove that it cannot be replaced by a bounded set of instances of itself. In 5.4.10-14, we will replace the operation $(\;) \rightarrow (\Sigma\;)'$ by one which will allow us to extend many standard results for HA to HA + MP.

We have also given a derived rule (whose proof was based on this failure to be preserved) :

5.4.8. Theorem. Let A contain only x free and let $\mathbf{HA} \vdash \forall x(Ax \vee \neg Ax)$. Then the following are equivalent :

(i) $\mathbf{HA} \vdash \exists xAx \vee \neg \exists xAx$,

(ii) $\mathbf{HA} \vdash \neg\neg \exists xAx \rightarrow \exists xAx$,

(iii) $\mathbf{HA} \vdash \exists y[\neg\neg \exists xAx \rightarrow Ay]$.

(This is theorem 5.2.6.)

An almost trivial derived rule is

5.4.9. Theorem. Let A contain only x free and let $\mathbf{HA} \vdash \forall x(Ax \vee \neg Ax)$ and $\mathbf{HA} \vdash \neg\neg \exists xAx$. Then $\mathbf{HA} \vdash \exists xAx$.

Proof. Observe that $\mathbf{HA} \vdash \neg\neg \exists xAx$ implies that $\exists xAx$ is true in the standard model and hence An is true for some n. But $\mathbf{HA} \vdash An \vee \neg An$ and the disjunction property yields $\mathbf{HA} \vdash An$ or $\mathbf{HA} \vdash \neg An$. Hence $\mathbf{HA} \vdash An$, i.e. $\mathbf{HA} \vdash \exists xAx$. Q. E. D.

From an earlier chapter, we know that this last result holds when A is allowed to have other free variables. The present proof admits no easy extension to this generalization. Suppose then that we decide to attempt to prove this directly. That is, let $\underline{K}$ be a model with node α such that $\alpha \not\Vdash \forall x \exists y Axy$. Then, for some $\beta \geq \alpha$, $b \in D\beta$, $\beta \not\Vdash \exists y Aby$. Since $\mathbf{HA} \vdash \forall x(Axy \vee \neg Axy)$, we see $\beta \Vdash \neg Abc$ for all $c \in D\beta$. Also, since $\mathbf{HA} \vdash \forall x \neg\neg \exists y Axy$, we can find some $\gamma > \beta$ and $c \in D\gamma$ such that $\gamma \Vdash Abc$. But $c \in D\gamma - D\beta$ and we obtain no contradiction. We cannot pull $D\beta$ out of the model $\underline{K}$ and define a nonstandard model of arithmetic on it, because it does not follow from the fact that an axiom is forced at a given node that it will be true in the classical model determined by the domain and atomic formulae forced at that node. (Note : We have not here done anything that we would not do to show $\mathbf{HA} \vdash \forall xy(Axy \vee \neg Axy) \ \& \ \forall y \neg\neg \exists x Axy \rightarrow \forall y \exists x Axy$. Thus we should not expect to get anywhere. Also, unfortunately, there seems to be no place at which to apply the trick used in proving theorem 5.2.6.)

5.4.10 - 5.4.14. $(\ ) \rightarrow (\Sigma + \omega)'$.

5.4.10. The failure of MP to be preserved under $(\ ) \rightarrow (\Sigma\ )'$ has its applications. But applications such as the ED - theorem required preservation under $(\ ) \rightarrow (\Sigma\ )'$ and, to obtain such results, we must give a similar such operation under which MP is preserved. Fortunately, the solution to this problem is simple : If $\underline{F}$ is a family of models of MP, define an operation on $\underline{F}$ by $\underline{F} \rightarrow (\Sigma\underline{F} + \omega)'$. E.g. let $\underline{F} = \{\underline{K}_1, \underline{K}_2\}$. Then $(\Sigma\underline{F} + \omega)'$ is :

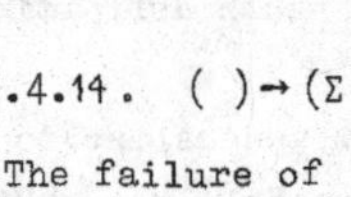

.

5.4.11. Theorem. If $\underset{\sim}{HA} + MP$ is valid in $\underline{F}$, then it is valid in $(\Sigma\underline{F} + \omega)'$ - i.e. the validity of $\underset{\sim}{HA} + MP$ is preserved by the operation $(\ ) \to (\Sigma\ + \omega)'$.

Proof. We consider the schema (i). It being valid in $\Sigma\underline{F} + \omega$, we need only look at α_o. Let $\alpha_o \Vdash \forall x(Ax \vee \neg Ax) \& \neg\neg \exists x Ax$. Let $\alpha > \alpha_o$ be the node corresponding to ω. Then $\alpha \Vdash \neg\neg \exists x Ax$, whence $\alpha \Vdash \exists x Ax$. $D\alpha = \{0, 1, \ldots\}$ and so, for some n, $\alpha \Vdash An$. A is decidable and so $\underset{\sim}{HA} \vdash An$, whence $\alpha_o \Vdash An$, i.e. $\alpha_o \Vdash \exists x Ax$. Thus

$$\alpha_o \Vdash \forall x(Ax \vee \neg Ax) \& \neg\neg \exists x Ax \to \exists x Ax . \qquad \text{Q. E. D.}$$

5.4.12. Corollary. $\underset{\sim}{HA} + MP$ possesses ED, DP.

Proof. We can't quite quote theorem 5.1.20, but we can observe that the proof carries over easily, the additional summand being used only to guarantee the validity of MP. Q. E. D.

Regarding closure properties of the class $\mathfrak{P}^\omega$ of sets Γ such that the validity of $\underset{\sim}{HA} + \Gamma$ is preserved by the operation $(\ ) \to (\Sigma\ + \omega)'$, we get almost exactly the properties corresponding to $(\ ) \to (\Sigma\ )'$ (theorem 5.2.11). The difference is that we must also insist that Γ be true in the standard model.

5.4.13. Theorem. The class $\mathfrak{P}^\omega$ of sets, Γ, such that the validity of $\underset{\sim}{HA} + \Gamma$ is preserved by the operation $(\ ) \to (\Sigma\ + \omega)'$ has the following closure properties:

(i) $\mathfrak{P}^\omega$ is closed under arbitrary union;

(ii) if $\Gamma \in \mathfrak{P}^\omega$ and A is a Harrop-sentence and $\omega \models A$, then $\Gamma \cup \{A\} \in \mathfrak{P}^\omega$;

(iii) if $\Gamma \in \mathfrak{P}^\omega$, A has only the variable x free, and $\underset{\sim}{HA} + \Gamma \vdash An$ for each numeral n, then $\Gamma \cup \{\forall x Ax\} \in \mathfrak{P}^\omega$.

(Note that, in (iii), the fact $\omega \models \forall x Ax$ follows from the facts that $\underset{\sim}{HA} + \Gamma \vdash An$ for all n and $\omega \models \Gamma$.)

Proof. The proof of theorem 5.4.13 is identical to that of theorem 5.2.11 and we omit it.

The results of 5.2.13 - 5.2.23 carry over easily. We leave the verification to the reader.

5.4.14. Remark. The proof of de Jongh's theorem does not carry over: Consider J_3^*:

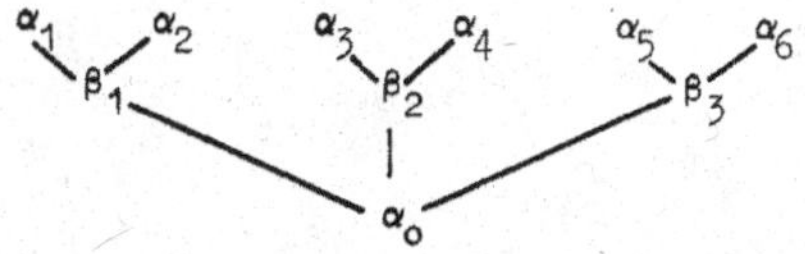

If we wish β_1 to force MP when we turn this into a model of HA, we must associate the standard model with either α_1 or α_2. Similarly, it must be associated with one of α_3 and α_4 and with one of α_5 and α_6. Thus, we end up with something of the form:

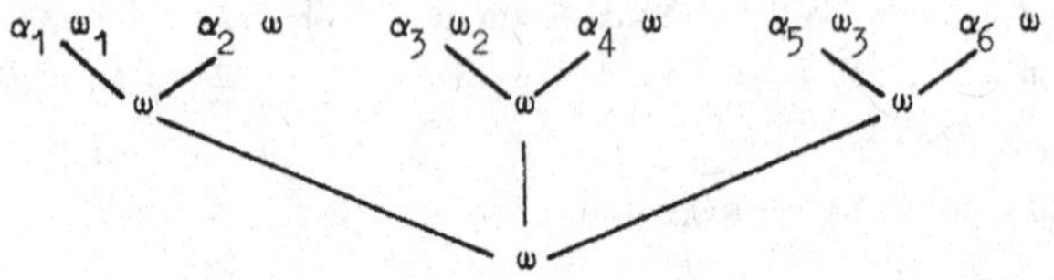

But now the proof of de Jongh's theorem does not go through: Lemma 5.3.9 does not apply since α_2, α_4 and α_6 all behave identically. A sophistication of our technique in section 6 will allow us to get around this problem. Also, it will yield a method of generalizing theorem 5.4.13 to cases where Γ need not be true in the standard model.

§ 5. The schema IP_0^c

5.5.1. In addition to MP, the schema

$$IP_0^c : \forall x(Ax \vee \neg Ax) \& (\forall xAx \to \exists yBy) \to \exists y(\forall xAx \to By),$$

where A, B have only the free variables indicated, is valid under Gödel's interpretation (see 3.5.10). As in section 4, where we considered variants of MP, we may consider variants of this schema. IP_0, however, is simply not as susceptible to study by means of the Kripke models as MP, and, thus, we shall only consider the schema as presented (i.e. with no free variables). The reader may consider variants as he pleases (in particular, IP).

5.5.2 - 5.5.3. Proof theoretic closure results.

Let $\mathfrak{P}$ and $\mathfrak{P}^\omega$ be as defined in sections 5.2.11 and 5.4.13, respectively.

5.5.2. Theorem. Let A have only the variable x free, B only the variable y free. Let $\Gamma \in \mathfrak{P}$. If $\underset{\sim}{HA} + \Gamma \vdash \forall x(Ax \vee \neg Ax)$ and $\underset{\sim}{HA} + \Gamma \vdash \forall xAx \to \exists yBy$, then

$$\underset{\sim}{HA} + \Gamma \vdash \exists y(\forall xAx \to By).$$

Proof. Suppose $\underset{\sim}{HA} + \Gamma \not\vdash \exists y(\forall xAx \to By)$. Then, for each n there is a model $\underset{\sim}{K}_n$ with origin β_n such that $\beta_n \Vdash \forall xAx$, $\beta_n \not\Vdash Bn$. Let α_0 be the origin of $(\Sigma\underset{\sim}{K}_n)'$. By the decidability of A and and the fact that $\beta_n \Vdash \forall xAx$ for all n, $\alpha_0 \Vdash \forall xAx$. Hence $\alpha_0 \Vdash \exists yBy$ and, for some n, $\alpha_0 \Vdash Bn$. But $\beta_n > \alpha_0$ and so $\beta_n \Vdash Bn$, a contradiction. Q. E. D.

5.5.3. Theorem. Let A, B be as in theorem 5.5.2 and let $\Gamma \in \mathfrak{P}^\omega$. If $\underset{\sim}{HA} + \Gamma \vdash \forall x(Ax \vee \neg Ax)$ and $\underset{\sim}{HA} + \Gamma \vdash \forall xAx \to \exists yBy$, then

$$HA + \Gamma \vdash \exists y(\forall xAx \to By).$$

Proof. Replace $(\Sigma\underset{\sim}{K}_n)'$ by $(\Sigma\underset{\sim}{K}_n + \omega)'$. Q. E. D.

For instance, we may let $\Gamma = MP$, $TI(<)$, $RFN(\underset{\sim}{HA})$, $RFN(\underset{\sim}{HA} + MP)$, etc.

5.5.4 - 5.5.7. Mutual independence of MP and IP_0^c.

There is one useful property that IP_0^c has: It is not preserved under $(\ )\to(\Sigma\ )'$ or $(\ )\to(\Sigma\ +\omega)'$. Because of this, we may prove the following

5.5.4. Theorem. Let $\Gamma \in \mathfrak{P}^\omega$ be r.e. Then there is a primitive recursive A Ax and a formula By (each with only the free variables indicated) such that

$$\underset{\sim}{HA} + \Gamma \not\vdash (\forall xAx \to \exists yBy) \to \exists y(\forall xAx \to By).$$

<u>Proof</u>. Let $\exists z Cz$ and $\forall x Ax$ be formally undecidable in $\underset{\sim}{HA}^c + \Gamma$, Ax and Cz primitive recursive. Also, let $\forall x Ax \& \exists z Cz$ be consistent (e.g. use corollary 5.3.12). Define

$$By \equiv (y = 0 \& \exists z Cz) \vee (y = 1 \& \neg \exists z Cz).$$

Let ω^+, ω^{++} be classical models of $\forall x Ax \& \exists z Cz$ and $\neg \forall x Ax$, respectively. Also, observe that ω is a model of $\forall x Ax \& \neg \exists z Cz$. Consider $(\omega + \omega^+ + \omega^{++})'$:

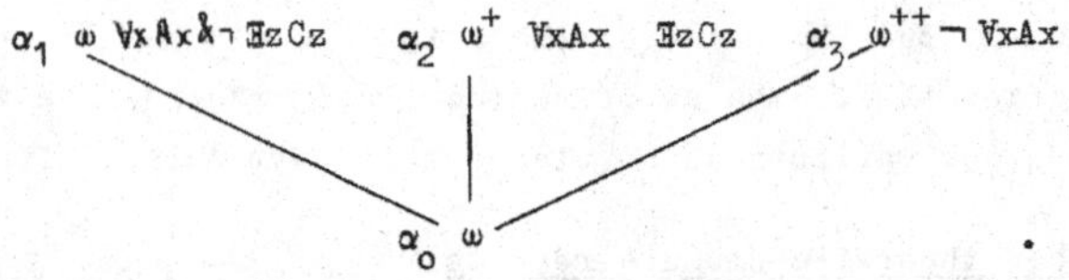

Now $\alpha_0 \Vdash \Gamma$ by theorem 5.4.13. and $\alpha_0 \Vdash \forall x(Ax \vee \neg Ax)$ by the primitive recursiveness of Ax. Also, $\alpha_0 \Vdash \forall x Ax \to \exists y By$, since only $\alpha_1, \alpha_2 \Vdash \forall x Ax$ and $\alpha_1 \Vdash B1$, $\alpha_2 \Vdash B0$.

But $\alpha_0 \nVdash \exists y(\forall x Ax \to By)$, since then $\alpha_0 \Vdash \forall x Ax \to B0$ or $\alpha_0 \Vdash \forall x Ax \to B1$. In the first case, it follows that $\alpha_1 \Vdash \forall x Ax \to B0$ and, in the second case, that $\alpha_2 \Vdash \forall x Ax \to B1$, both implications leading to contradictions. Q. E. D.

It is worth singling out the case $\Gamma = MP$:

5.5.5. <u>Corollary</u>. IP_0^c is not derivable from MP.

However, IP_0^c is preserved under a special case of $(\ )\to(\Sigma\ )'$, under which MP is not preserved:

5.5.6. <u>Theorem</u>. Let ω^+ be a model of $\underset{\sim}{HA}^c$. Then $\underset{\sim}{HA} + IP_0^c$ is valid in $(\omega^+)'$.

<u>Proof</u>. Suppose that, in the model $(\omega^+)'$,

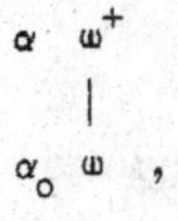

α_0 does not force an instance of IP_0^c:

$$\alpha_0 \nVdash \forall x(Ax \vee \neg Ax) \& (\forall x Ax \to \exists y By) \to \exists y(\forall x Ax \to By).$$

IP_0^c being valid at α, we must have

$$\alpha_0 \Vdash \forall x(Ax \vee \neg Ax) \& (\forall x Ax \to \exists y By),$$
$$\alpha_0 \nVdash \exists y(\forall x Ax \to By).$$

By this last statement, $\alpha_0 \nVdash \forall x Ax \to B0$ and, for some $\beta \geq \alpha_0$, $\beta \Vdash \forall x Ax$, $\beta \nVdash B0$. In particular, $\beta \Vdash \forall x Ax$ and, A being decidable, $\alpha_0 \Vdash An$ for all n. It follows that $\alpha_0 \Vdash \forall x Ax$. But $\alpha_0 \Vdash \forall x Ax \to \exists y By$, whence

$\alpha_0 \Vdash \exists yBy$ and, for some n, $\alpha_0 \Vdash Bn$. Thus $\alpha_0 \Vdash \forall xAx \to Bn$ and so $\alpha_0 \Vdash \exists y(\forall xAx \to By)$, a contradiction. Q. E. D.

The immediate corollary is

5.5.7. Corollary. Let $\underset{\sim}{HA} + \Gamma$ be r.e., $\Gamma \in \mathfrak{P}$, $\underset{\sim}{HA}^c + \Gamma$ consistent. Then some instance of MP is not derivable in $\underset{\sim}{HA} + \Gamma + IP_0^c$.

Proof. $\underset{\sim}{HA} + \Gamma + IP_0$ is valid in $(\omega^+)'$ for any model ω^+ of $\underset{\sim}{HA}^c + \Gamma$. But, as shown in the proof of theorem 5.4.5, MP is not valid in $(\omega^+)'$ unless ω^+ is an elementary extension of ω. Q. E. D.

5.5.8. Final comments on IP_0^c. The non-preservation of IP_0^c under $(\ ) \to (\Sigma\ )'$ allows us to prove for IP_0^c an analogue to theorem 5.4.5. We may also generalize theorem 5.4.5 by using the fact that IP_0^c is preserved under $\omega^+ \to (\omega^+)'$. Also, aside from such results, and formulations of such corollaries as the independence of IP_0^c from $\underset{\sim}{HA} + MP + RFN(\underset{\sim}{HA} + MP) + CON(\underset{\sim}{HA} + IP_0^c) + TI(<)$, etc., we may observe that we can obtain subtler results such as the following:

5.5.9. Theorem. There is a formula By and a primitive recursive Ax (each with only the indicated free variables) such that

(i) $\underset{\sim}{HA} + MP \nvdash (\forall xAx \to \exists yBy) \to \exists y(\forall xAx \to By)$,

(ii) $\underset{\sim}{HA} + IP \nvdash \neg\neg\exists xAx \to \exists xAx$.

(In other words, the same formula A works in both independence proofs.)

Proof. Observe that A may be taken in both independence proofs above to be arbitrary up to the requirement that $\exists xAx$ be independent of $\underset{\sim}{HA}^c$. Q. E. D.

Beyond this, there is little we can do model-theoretically since (i) the only models of IP_0^c we have so far are (except for those given by the completeness theorem) models ω^+ of $\underset{\sim}{HA}^c$ and models of the form $(\omega^+)'$, and (ii) the only models to which we know we can apply the operation $(\ ) \to (\ )'$ and preserve IP_0^c are the models of $\underset{\sim}{HA}^c$.

§ 6. Definability of models of $\underset{\sim}{HA}^c$: applications

5.6.1. The operation $(\ )\to(\Sigma\ )^*$.

The operation $(\ )\to(\Sigma\ )'$, while extremely useful, is too restrictive for certain purposes. In proving de Jongh's theorem, for instance, we had models of the form

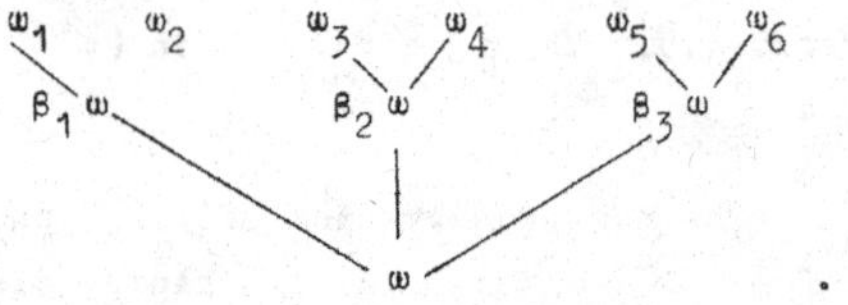

.

This cannot possibly give us Σ_1^0 substitution instances since β_1, β_2 and β_3 must all force the same Σ_1^0 sentences (and similar results for models on the other modified Jaskowski trees). The observant reader will also have noticed that, to apply $(\ )\to(\Sigma+\omega)'$ in proving (say) the ED-property for $\underset{\sim}{HA}+MP+\Gamma$, we had to assume that Γ was true in the standard model.

Let $\underline{F}$ be a family of models and let ω^+ be a non-standard model of $\underset{\sim}{HA}^c$ such that (i) the domain of ω^+ is contained in $D\alpha$ for all $\alpha\in K$, $\underline{K}\in\underline{F}$, and (ii) atomic formulae whose constants name elements in ω^+ are forced at any node α exactly when they are true in ω^+. Then, we can define a model $(\Sigma\underline{F})^*$ in the manner in which we defined $(\Sigma\underline{F})'$. E.g. let $\underline{F}=\{\underline{K}_1,\underline{K}_2\}$. Then $(\Sigma\underline{F})^*$ is

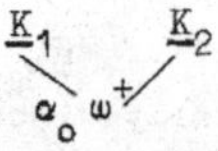

.

Unfortunately, we don't know if $(\Sigma\underline{F})^*$ will always be a model of HA. For, what have we got to guarantee that the induction schema will be forced at α_0? Truth in ω^+ is not convincing - the law of the excluded middle is true in ω^+, but not forced at α_0. In $(\ )\to(\Sigma\ )'$, we did not merely have a model of $\underset{\sim}{HA}^c$ placed at α_0 - we had the natural numbers themselves.

Let us consider how we proved the induction axiom to be valid in $(\Sigma\underline{F})'$. Our "second proof" consisted in observing that, by theorem 5.3.11 (iii), to conclude that induction was valid, we had not to look at the schema without free variables other than x in Ax, but we only had to look at all instances

$$A0\ \&\ \forall xy(Ax\ \&\ S(x,y)\to Ay)\to An\ .$$

This is obviously valid in $(\Sigma\underline{F})^*$ - but, since the domain at α_0 has non-standard integers, we cannot conclude from the fact that $\alpha_0 \Vdash A0\ \&\ \forall xy(Ax\ \&\ S(x,y)\to Ay)$ that $\alpha_0\Vdash Aa$ for all $a\in D\alpha_0$, but only that $\alpha_0\Vdash A0, A1, \ldots$.

The actual proof we gave in proving theorem 5.2.4 was based on the following reasoning: If $\alpha_0 \Vdash A0 \,\&\, \forall xy(Ax \,\&\, S(x,y) \to Ay)$ and $\alpha_0 \nVdash \forall x Ax$, there is a least n such that $\alpha_0 \nVdash An$. There are two cases in which we can guarantee the existence of a least element in ω^+ of which A is not forced: (i) $\omega^+ = \omega$, and (ii) the condition " $\alpha_0 \Vdash A$ " is expressible in the language of ω^+ - in other words, if the truth (or, forcing) definition for a formula in the Kripke model can be given within the classical model ω^+.

5.6.2 - 5.6.7. Definability.

5.6.2. In this subsection, we formally define what we mean by the definability of a Kripke model in a model of $\underset{\sim}{HA}^c$. Suppose $\underline{K}$ is a Kripke model, in which $\underset{\sim}{HA}$ is valid, ω^+ a nonstandard model of arithmetic. Let, for each $\alpha \in K$, $a_\alpha \in D\alpha$, $\bar{a}_\alpha$ denote a number in ω^+ indexing a_α. We assume that we have formulae as follows, with the free variables as indicated:

$K(\alpha)$,
$D(\alpha,x)$,
$\alpha \leq \beta$,
$S(\alpha, x, y)$,
$A(\alpha, x, y, z)$,
$M(\alpha, x, y, z)$.

Also, $\bar{0}, \bar{1}, \ldots$ will denote indices of $0, 1, \ldots$.

Let us suppose that there is a one - to - one correspondence between elements $\alpha \in K$ and elements a of ω^+ for which $\omega^+ \models Ka$, in such a way that, if a, b are associated with α, β, respectively, then

$$\alpha \leq \beta \text{ iff } \omega^+ \models a \leq b,$$

and

$$\omega^+ \models \forall xy(x \leq y \to Kx \,\&\, Ky).$$

Then, obviously, we may identify elements of K with a definable subset of the domain of ω^+ and $\leq$ with the definable partial ordering on this subset. We also assume that

$$\omega^+ \models D(\alpha, \bar{a}_\alpha),$$

and

$$\omega^+ \models D(\alpha,a) \Rightarrow a \text{ is an index } \bar{a}_\alpha \text{ for some } a_\alpha \in D\alpha.$$

We may assume either that the set of constants $\{\bar{a}_\alpha : a_\alpha \in D\alpha\}$ is contained in $\{\bar{b}_\beta : b_\beta \in D\beta\}$ or that there is a definable function $f(x,y,z)$ such that

$$f(\alpha, \beta, \bar{a}_\alpha) \text{ is an index of } a_\alpha \in D\beta \ (\alpha \leq \beta).$$

In what follows, however, we shall ignore this minor distinction.

Finally, if, in addition to all of this, we have

$$\omega^+ \models B(\alpha, \bar{a}_{1\alpha}, \ldots, \bar{a}_{n\alpha}) \text{ iff } \alpha \Vdash B(a_{1\alpha}, \ldots, a_{n\alpha}),$$

for all atomic B, nodes α, and elements $a_{1\alpha}, \ldots, a_{n\alpha} \in D\alpha$, then we say that the model $\underline{K}$ is definable in ω^+.

The definability of a classical model, ω^{++}, in ω^+ can be taken as the definability of the one node Kripke model, or can be taken in the obvious manner.

Three rather obvious lemmas are

5.6.3. Lemma. Let $\underline{K}$ be definable in ω^+. Then, for any formula $A(x_1, \ldots, x_n)$ with free variables as shown, there is a formula $A^*(x, x_1, \ldots, x_n)$ with free variables as shown and parameters from ω^+ such that, for all $\alpha \in K$, $a_{1\alpha}, \ldots, a_{n\alpha} \in D\alpha$,

$$\alpha \Vdash A(a_{1\alpha}, \ldots, a_{n\alpha}) \text{ iff } \omega^+ \models A^*(\alpha, \bar{a}_{1\alpha}, \ldots, \bar{a}_{n\alpha}).$$

5.6.4. Lemma. Let $\underline{F} = \{\underline{K}_1, \ldots, \underline{K}_n\}$ be such that each $\underline{K}_i$ is definable in ω^+, then $\Sigma\underline{F}$ is definable in ω^+.

5.6.5. Lemma. Let $\underline{K}$ be definable in ω^{++} and let ω^{++} be definable in ω^+. Then $\underline{K}$ is definable in ω^+.

These lemmas will be applied shortly in the construction of models. For this, we will need to prove that $(\Sigma\underline{F})^*$ is a model of $\underset{\sim}{HA}$ when each $\underline{K} \in \underline{F}$ is definable in ω^+ ($\underline{F}$ finite). But, before we can do this, we need the following:

5.6.6. Lemma. Let $\underline{K}$ be a model of $\underset{\sim}{HA}$ definable in ω^+ and let $\underline{K}$ have a least node α_0. There is a canonical embedding of ω^+ into the domain of α_0 - i.e. a map of the domain of ω^+ into $D\alpha_0$ which is one - to - one and preserves the atomic formulae.

Proof. Obviously one can match up the 0 of ω^+ with the 0 of $D\alpha_0$, the 1 of ω^+ with the 1 of $D\alpha_0$, etc. But, for non-standard elements, we must observe that, in $\underset{\sim}{HA}$, the relation "y is the result of the x-fold application of the successor function of $\underline{K}$ to 0" (i.e. $y = S^x 0$) is expressible. By the closure of $D\alpha_0$ (in $\underline{K}$) under the successor function, and by induction in ω^+, for all x in ω^+ there is an element in $D\alpha_0$ which is the x-fold application of successor to 0. The truth of atomic formulae $S(a,b)$ is obviously preserved under the map which associates x with the object $S^x 0$ in $D\alpha_0$. The preservation of the truth of other atomic formulae follows from the validity of the recursion equations in $\underline{K}$ and induction in ω^+.

Q. E. D.

(Note: It is in steps like this that the embedding functions $f(\alpha, \beta, x)$ are introduced in the model. We shall, however, suppress further mention of these functions in favor of a more informal approach to the proofs, in much the same way that algebraists avoid mentioning such minor difficulties.)

We may now prove the following

5.6.7. Theorem. Let $\underline{F} = \{\underline{K}_1,\ldots,\underline{K}_n\}$ be definable in ω^+. Then $(\Sigma\underline{F})^*$,

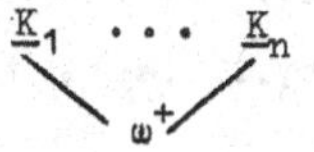

,

is a model of $\underset{\sim}{HA}$. If, in addition, ω^+ is definable in ω^{++}, $(\Sigma\underline{F})^*$ is definable in ω^{++}.

Proof. By lemma 5.6.4, $\Sigma\underline{F}$ is definable in ω^+. Obviously, ω^+ is definable in ω^+. To define $(\Sigma\underline{F})^*$ in ω^+, first recall that, formally, $\Sigma\underline{F} = \underline{K}$ where

$$K = K_1 \times \{1\} \cup K_2 \times \{2\} \cup \ldots \cup K_n \times \{n\}, \text{ etc.}$$

Thus $\alpha_o = (0,0) \notin K$ and we may let α_o be the node for ω^+ and define

$$\alpha \leq^* \beta \text{ iff } \alpha = \alpha_o \vee (\alpha \neq \alpha_o \;\&\; \alpha \leq \beta).$$

Thus K, $\leq$ are definable. Let

$$D^*(\alpha,x) \leftrightarrow (\alpha = \alpha_o \;\&\; x \in \omega^+) \vee (\alpha > \alpha_o \quad D(\alpha,x)).$$

(To explain "$x \in \omega^+$", recall that the elements of ω^+ index elements in all domains and so we must choose special indices to denote elements of ω^+ - that is we have a formula singling out the indices for ω^+. If $a \in \omega^+$, $\bar{a}$ will denote its index. (Recall that ω^+ is definable in ω^+ and consider what we mean by this.).)

To complete the proof that $(\Sigma\underline{F})^*$ is definable in ω^+, we need only show how to define the atomic formulae. Let us assume that ω^+ is defined in ω^+ as a Kripke model (say with node α_o). Let B be atomic: B^Σ denotes its definition in $\Sigma\underline{F}$; B^+ its definition in ω^+. Then

$$B^*(\alpha,x_1,\ldots,x_n) \leftrightarrow (\alpha = \alpha_o \;\&\; B^+(\alpha_o,x_1,\ldots,x_n)) \vee (\alpha > \alpha_o \;\&\; B^\Sigma(\alpha,x_1,\ldots,x_n)).$$

Now, by lemma 5.6.3, for any formula A, there is a formula $A^*(x,x_1,\ldots,x_n)$ such that, for any node α, and elements $a_1,\ldots,a_n \in D\alpha$,

$$\alpha \Vdash A(a_1,\ldots,a_n) \text{ iff } \omega^+ \models A^*(\alpha,\bar{a}_1,\ldots,\bar{a}_n).$$

Suppose $(\Sigma\underline{F})^*$ is not a model of $\underset{\sim}{HA}$. Then, for some $A(x,x_1,\ldots,x_n)$ with only $x,x_1,\ldots,x_n$ free,

$$\alpha_o \nVdash \forall x_1\ldots x_n[A(0,x_1,\ldots,x_n) \;\&\; \forall xy(A(x,x_1,\ldots,x_n) \;\&\; S(x,y) \rightarrow A(y,x_1,\ldots,x_n)) \rightarrow \rightarrow \forall x A(x,x_1,\ldots,x_n)].$$

Then, for some $a_1,\dots,a_n \in \omega^+$,

$$\alpha_o \not\Vdash A(0,a_1,\dots,a_n) \,\&\, \forall xy(A(x,a_1,\dots,a_n) \,\&\, S(x,y) \to A(y,a_1,\dots,a_n)) \to \\ \to \forall x A(x,a_1,\dots,a_n).$$

Then

$$\alpha_o \Vdash A(0,a_1,\dots,a_n),\ \forall xy(A(x,a_1,\dots,a_n) \,\&\, S(x,y) \to A(y,a_1,\dots,a_n)), \\ \alpha_o \not\Vdash A(a,a_1,\dots,a_n)$$

for some $a \in \omega^+$. Letting a be such that $\alpha_o \not\Vdash A(a,a_1,\dots,a_n)$, we see, for A^* defining $\alpha_o \Vdash A(x,x_1,\dots,x_n)$,

$$\omega^+ \models A^*(\alpha_o, \bar{0}, \bar{a}_1,\dots,\bar{a}_n),$$

(1) $\omega^+ \models \forall xy[A^*(\alpha_o,\bar{x},\bar{a}_1,\dots,\bar{a}_n) \,\&\, S^*(\alpha_o,\bar{x},\bar{y}) \to A^*(\alpha_o,\bar{y},\bar{a}_1,\dots,\bar{a}_n)]$,

$$\omega^+ \not\models A^*(\alpha_o, \bar{a}, \bar{a}_1,\dots,\bar{a}_n).$$

Now $S^*(\alpha_o,\bar{x},\bar{y}) \leftrightarrow S(x,y)$, whence (1) becomes

(2) $\omega^+ \models \forall xy[A^*(\alpha_o,\bar{x},\bar{a}_1,\dots,\bar{a}_n) \,\&\, S(x,y) \to A^*(\alpha_o,\bar{y},\bar{a}_1,\dots,\bar{a}_n)]$.

But the map $a \to \bar{a}$ is definable in ω^+, whence there is a least a_o such that

$$\omega^+ \not\models A^*(\alpha_o, \bar{a}_o, \bar{a}_1,\dots,\bar{a}_n).$$

$a_o \neq 0$, whence $a_o = b+1$ for some b. By minimality,

$$\omega^+ \models A^*(\alpha_o, \bar{b}, \bar{a}_1,\dots,\bar{a}_n).$$

By this last statement and (2),

$$\omega^+ \models A^*(\alpha_o, \bar{a}_o, \bar{a}_1,\dots,\bar{a}_n),$$

a contradiction. Thus $(\Sigma \underline{F})^*$ is a model of $\underset{\sim}{HA}$.

The final comment, that $(\Sigma \underline{F})^*$ is definable in ω^{++} if ω^+ is definable in ω^{++} follows from the definability of $(\Sigma \underline{F})^*$ in ω^+ and lemma 5.6.5.

Q. E. D.

5.6.8 - 5.6.9. The Hilbert - Bernays completeness theorem.

5.6.8. In 5.6.2 - 5.6.7, we proved two important results: (i) If $\underline{K}_1,\dots,\underline{K}_n$ are definable in ω^+, then $(\underline{K}_1 + \dots + \underline{K}_n)^*$ is a model of $\underset{\sim}{HA}$: and (ii) if ω^+, as above, is definable in ω^{++}, then $(\underline{K}_1 + \dots + \underline{K}_n)^*$ is also definable in ω^{++}. But, to be able to apply these results, we need a stock of definable Kripke models and definable non-standard models of $\underset{\sim}{HA}^c$. This is obtained by appeal to the Hilbert - Bernays completeness theorem.

5.6.9. Theorem (Hilbert - Bernays completeness theorem). Let $\underset{\sim}{T}$ be a consistent r.e. extension of $\underset{\sim}{HA}^c$. Then, for any model ω^+ of $\underset{\sim}{HA}^c + CON(\underset{\sim}{T})$, there is a non-standard model ω^{++} of $\underset{\sim}{T}$ which is definable in ω^{+}.

For a proof, see Kleene 1952, XIV, Thms 36 - 40, or Feferman 1960, theorem 6.2.

5.6.10 - 5.6.12. The Gödel - Rosser - Mostowski - Kripke - Myhill theorem revisited.

5.6.10. Our first two applications of the above results will be a proof of the existence of Σ_1^0 substitution instances and uniform Π_2^0 substitution instances in de Jongh's theorem. For these results, we need two refinements of theorem 3.3.1 and its corollary. For Σ_1^0 substitution, we need, for r.e. $\underset{\sim}{T}$, $A_1,\ldots,A_m$ such that $\underset{\sim}{T}+A_i$ is consistent and such that $\underset{\sim}{HA}^c \vdash A_i \rightarrow \neg A_j$ for $i \neq j$. We present Kripke's proof:

5.6.11. Theorem. Let $\underset{\sim}{T}$ be a consistent r.e. extension of $\underset{\sim}{HA}^c$. There is an r.e. relation $P(y)$ such that, for every natural number n, $\underset{\sim}{T}+Pn+\exists! xPx$ is consistent.

Proof. (Kripke 1963.) Let $R(e,x,y)$ numeralwise represent the relation $\{e\}(x)=y$. Define a partial recursive function as follows:

$$\varphi(x)=y \quad \text{if} \quad \underset{\sim}{T} \vdash \neg(R(x,x,y) \;\&\; \exists! zR(x,x,z)),$$

(choosing the first theorem of this form if there are more than one). Then φ has an index, e. Let Px be $R(e,e,x)$. We show that, for all n,

$$Pn \;\&\; \exists! xPx$$

is consistent with $\underset{\sim}{T}$.

First, observe that $\varphi(e)$ is undefined. If not, $\varphi(e)=n_o$ for some n_o. Then

$$\underset{\sim}{T} \vdash \neg(R(e,e,n_o) \;\&\; \exists! z\, R(e,e,z)).$$

But clearly, if $\varphi(e)=n_o$,

$$\underset{\sim}{T} \vdash R(e,e,n_o) \;\&\; \exists! z\, R(e,e,z),$$

a contradiction. Hence $\varphi(e)$ is undefined and for no n do we have

$$\underset{\sim}{T} \vdash \neg(R(e,e,n) \;\&\; \exists! x\, R(e,e,x)).$$

Hence, for all n, $\underset{\sim}{T}+R(e,e,n)+\exists! x\, R(e,e,x)$ is consistent. Q.E.D.

Letting A_n be Pn, we have the desired result. One might mention that we have $\underset{\sim}{HA} \vdash A_i \rightarrow \neg A_j$ for $i \neq j$ as well as $\underset{\sim}{HA}^c \vdash A_i \rightarrow \neg A_j$.

For the Π_2^0 substitution, we need the following

5.6.12. Theorem. Let $\underset{\sim}{H}$ denote $\underset{\sim}{HA}^c$ augmented by all true Π_1^0 sentences of arithmetic. If $\underset{\sim}{T} \supseteq \underset{\sim}{H}$ is consistent and has a Σ_2^0 enumeration, then there is an infinite family, $\{A_1,\ldots,A_n,\ldots\}$ of Π_2^0 sentences independent over

$\underset{\sim}{T}$ (in the sense of 5.3.10 that we may choose any subset of them to be true and the rest to be false).

If we observe that the proof predicate is Σ_2^0 and that the Σ_2^0 relations are precisely those numeralwise representable in $\underset{\sim}{T}$, we can mimic the proofs of theorem 5.3.11 and corollary 5.3.12 to obtain an infinite set of independent Σ_2^0 sentences. Replacing these sentences by their negations yields the theorem.

5.6.13 - 5.6.16. Σ_1^0 Substitution instances in de Jongh's theorem.

5.6.13. Recall that the reason we used the modified Jaskowski trees in proving de Jongh's theorem was that every node was determined by the set of terminal nodes not lying beyond it. Thus, if each terminal node was the unique node satisfying a particular sentence, it followed that every node was the least node satisfying a conjunction of negations of sentences corresponding to terminal nodes. Then, any set which could be the set of nodes forcing a propositional variable under a propositional forcing relation was now the set of nodes forcing a disjunction of such conjunctions of negations. In proving the existence of Σ_1^0 substitution instances, we will assign to each node of a tree a Σ_1^0 sentence which is forced only at and above that node. The substitution instances will be disjunctions of these sentences (and will thus be Σ_1^0).

Note that we no longer need to use the special property of the modified Jaskowski trees that every node is determined by a set of terminal nodes. Nonetheless, it will still be convenient to work with them. Consider, e.g., J_3^*:

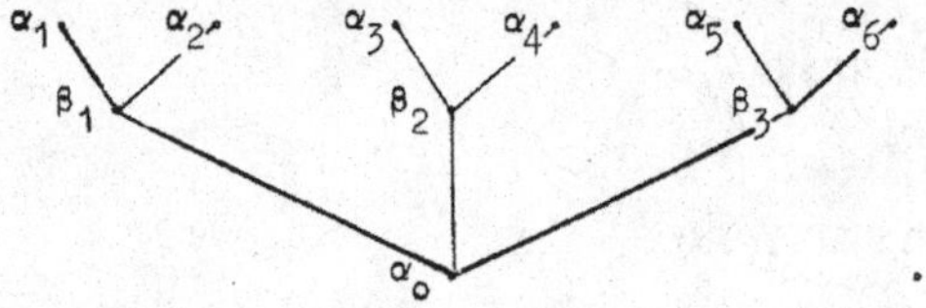

Starting at the terminal nodes and working our way down the tree, we shall assign theories to the nodes. Let $A_1,\ldots,A_6$ be Σ_1^0 mutually independent over $\underset{\sim}{HA}^c$ (or, let them be obtained by theorem 5.6.11). Assign to α_i the theory

$$\underset{\sim}{T}_i = \underset{\sim}{HA}^c + A_i + \bigwedge_{j\neq i} \neg A_j .$$

By the independence of the family $\{A_1,\ldots,A_6\}$, $\underset{\sim}{T}_i$ is consistent. Now choose B_1, B_2, B_3 individually independent over $\underset{\sim}{HA}^c + CON(\underset{\sim}{T}_1) + \ldots + CON(\underset{\sim}{T}_6) + \neg A_1 + \ldots + \neg A_6$ (which is true in ω and hence consistent) such that $\underset{\sim}{HA}^c \vdash B_i \rightarrow \neg B_j$ for $i \neq j$. Assign to β_i the theory

$$\underset{\sim}{T}_i^! = \underset{\sim}{HA}^c + B_i + \overset{6}{\underset{i=1}{\bigwedge}} CON(\underset{\sim}{T}_i) + \overset{6}{\underset{i=1}{\bigwedge}} \neg A_i \, .$$

Again, $\underset{\sim}{T}_i^!$ is consistent. Finally, assign to α_0 the theory

$$\underset{\sim}{HA}^c + CON(\underset{\sim}{T}_1^!) + CON(\underset{\sim}{T}_2^!) + CON(\underset{\sim}{T}_3^!) + \neg B_1 + \neg B_2 + \neg B_3 \, .$$

Having assigned such theories, we now assign models of $\underset{\sim}{HA}^c$ to the nodes. Place ω at α_0. Now $\omega \models CON(\underset{\sim}{T}_1^!) + CON(\underset{\sim}{T}_2^!) + CON(\underset{\sim}{T}_3^!)$, whence there are models ω_1, ω_2, and ω_3 of $\underset{\sim}{T}_1^!$, $\underset{\sim}{T}_2^!$, and $\underset{\sim}{T}_3^!$, respectively, such that each ω_i is definable in ω. Now,

$$\omega_i \models \underset{\sim}{T}_i^! = \underset{\sim}{HA}^c + B_i + \overset{6}{\underset{j=1}{\bigwedge}} CON(\underset{\sim}{T}_j) + \overset{6}{\underset{j=1}{\bigwedge}} \neg A_j \, .$$

Thus, in ω_i there are definable models of $\underset{\sim}{T}_1, \ldots, \underset{\sim}{T}_6$. Let ω_{11}, ω_{12} be models of $\underset{\sim}{T}_1$, $\underset{\sim}{T}_2$, respectively, definable in ω_1; ω_{21}, ω_{22} models of $\underset{\sim}{T}_3$, $\underset{\sim}{T}_4$ definable in ω_2; and ω_{31}, ω_{32} models of $\underset{\sim}{T}_5$, $\underset{\sim}{T}_6$ definable in ω_3.

Thus, we have

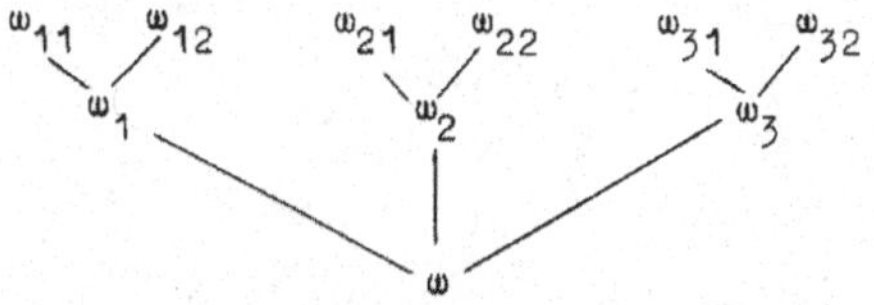

Now, successively apply the lemmas 5.6.3 - 5.6.6 to conclude that the resulting structure is a model of $\underset{\sim}{HA}$. Further, as we shall prove below, A_1 is forced only at the node corresponding to ω_{11}, A_2 at ω_{12}, A_3 at ω_{21}, ... B_1 is forced only at ω_1 and above, B_2 at ω_2 and above, and B_3 at ω_3 and above. Any provable Σ_1^0 sentence is forced at ω. Hence each node is characterized by a Σ_1^0 sentence and we may proceed from here.

For ease in assigning theories and models to nodes in the general case, and for ease in giving the proof, let us use the notation for trees of finite sequences as described in 5.3.3. J_3^*, e.g., will be represented by

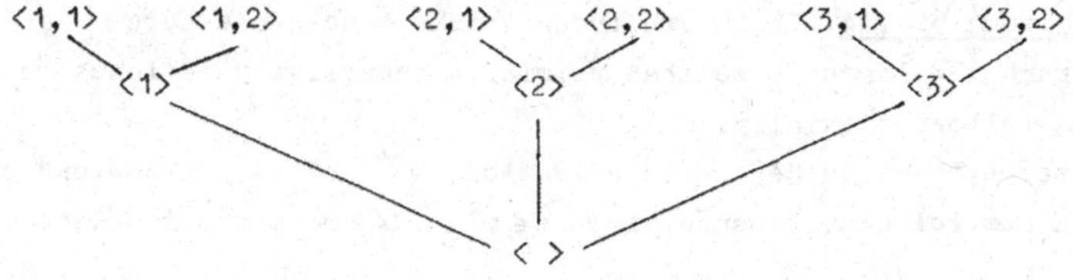

Let J_n^* be given and let $\sigma_1, \ldots, \sigma_{n!}$ be its terminal nodes. Choose $A_{\sigma_1}, \ldots, A_{\sigma_{n!}}$ such that $\underset{\sim}{HA}^c + A_{\sigma_i} + \underset{j \neq i}{\bigwedge} \neg A_{\sigma_j}$ is consistent and let $\underset{\sim}{T}_\sigma = \underset{\sim}{HA}^c + A_{\sigma_i} + \underset{j \neq i}{\bigwedge} \neg A_{\sigma_j}$. Let $\sigma_1, \ldots, \sigma_k$ be the non-terminal nodes of length m and let $\sigma_i*\langle 1\rangle, \dots, \sigma_i*\langle l\rangle$ $(i = 1,\dots,k)$ be the nodes of length $m+1$. Let $A_{\sigma_1}, \dots, A_{\sigma_k}$ be chosen such that $\underset{\sim}{HA}^c \vdash A_{\sigma_i} \to \neg A_{\sigma_j}$ for $i \neq j$ and such that each A_{σ_i} is consistent with

$$\underset{\sim}{T}_{m+1} = \underset{\sim}{HA}^c + \bigwedge_{i=1}^{k} \bigwedge_{j=1}^{l} CON(\underset{\sim}{T}_{\sigma_1*\langle j\rangle}) + \bigwedge_{i=1}^{k} \bigwedge_{j=1}^{l} \neg A_{\sigma_i*\langle j\rangle} .$$

(Observe that ω is a model of this theory and, thus, it is consistent.) Let $\underset{\sim}{T}_{\sigma_i} = \underset{\sim}{T}_{m+1} + A_{\sigma_i}$.

Thus, every node σ gets a theory $\underset{\sim}{T}_\sigma$ assigned to it. Further, if $\sigma * \langle j\rangle$ is a successor of σ, then $CON(\underset{\sim}{T}_{\sigma*\langle j\rangle})$ is provable in $\underset{\sim}{T}_\sigma$. Thus every model ω_σ of $\underset{\sim}{T}_\sigma$ has a definable model of $\underset{\sim}{T}_{\sigma*\langle j\rangle}$. Let $\omega_{\langle\,\rangle}$ be ω and, for each ω_σ and successor $\sigma * \langle j\rangle$, let $\omega_{\sigma*\langle j\rangle}$ be a model of $\underset{\sim}{T}_{\sigma*\langle j\rangle}$ definable in ω_σ. Having defined these models, assign Kripke models $\underset{\sim}{K}_\sigma$ to the nodes as follows:

(i) $\underset{\sim}{K}_\sigma = \omega_\sigma$ for terminal σ,

(ii) let $\sigma * \langle 1\rangle, \dots, \sigma*\langle k\rangle$ be the successors of σ, and let

$\underset{\sim}{K}_\sigma = (\Sigma \underset{\sim}{K}_{\sigma*\langle i\rangle})^*$:

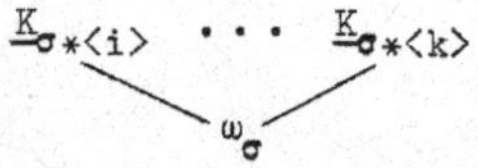

By the lemmas 5.6.3 - 5.6.6, each $\underset{\sim}{K}_\sigma$ is a model of $\underset{\sim}{HA}$.

5.6.14. Lemma. Let A be a Σ^0_1 sentence. $\sigma \Vdash A$ iff $\omega_\sigma \models A$.

If $\sigma \Vdash \exists x Bx$, say B primitive recursive, we know that $\sigma \Vdash Bs$ for some $s \in D\sigma$. Let $\tau \geq \sigma$ be terminal. Then $\omega_\tau \models Bs$. Now, it does not follow trivially that $\omega_\sigma \models Bs$ - model-theoretically, the well-known characterization of recursiveness is preservation under extension and restriction for end extensions (i.e. extensions in which all of the new elements are larger than the old ones. Of course, Matiyasevich 1970 now gives us the result for arbitrary extensions - but this is far from trivial.).

Proof of lemma 5.6.14. There are three tricks we can use here:

(i) Expand the language so that primitive recursive relations are atomic. The lemma follows trivially.

(ii) Observe that the definable extension ω^{++} of ω^+ is an end extension. The lemma now follows, because, for $s \in \omega^+$, $\forall x < s$ means the same in both models.

(iii) Apply the theorem of Matiyasevich 1970 by which, if A is Σ^0_1, $\underset{\sim}{HA} \vdash A \leftrightarrow \exists x_1 \dots x_m D(x_1 \dots x_m)$, where D is quantifier free. The lemma then follows trivially. Q. E. D.

5.6.15. <u>Lemma</u>. $\omega_\sigma \models A_\tau$ iff $\sigma \geq \tau$.

<u>Proof</u>. Clearly $\omega_\tau \models A_\tau$. Then $\tau \Vdash A_\tau$, whence, if $\sigma \geq \tau$, $\sigma \Vdash A_\tau$, whence $\omega_\sigma \models A_\tau$.

Conversely, consider A_τ. First, assume τ is terminal. Then $\omega_\sigma \models A_\tau$ implies that $\omega_{\tau'} \models A_\tau$ for all terminal $\tau' \geq \sigma$. But, the only terminal $\omega_{\tau'} \models A_\tau$ is τ. Hence $\sigma = \tau$, by the property of the modified Jaskowski trees. (To avoid using this property, let $\sigma < \tau$. Then, for some $\sigma < \sigma'$, $\tau = \sigma' * \langle i \rangle$, but $\omega_\sigma \models A_\tau \Rightarrow \sigma \Vdash A_\tau \Rightarrow \sigma' \Vdash A_\tau \Rightarrow \omega_{\sigma'} \models A_\tau$, contradicting the fact that $\omega_{\sigma'} \models \underset{\sim}{T}_{\sigma'} = \underset{\sim}{HA}^c + \bigwedge_{\tau'} \neg A_{\tau'} + \bigwedge_{\tau'} CON(A_{\tau'} + \bigwedge_{\tau'' \neq \tau'} \neg A_{\tau''})$, where τ', τ'' range over terminal nodes.)

Let τ not be terminal and let $\omega_\sigma \models A_\tau$. Assume $\sigma \not\geq \tau$. σ cannot be of length less than τ, because, by choice,

$$\omega_\sigma \models \neg A_\rho,$$

for any ρ of length greater than that of σ. Thus, the length of σ is at least that of τ and $\sigma \geq \sigma'$ for some σ' of length the same as τ. Now, $\omega_{\sigma'} \models A_{\sigma'}$, whence $\sigma' \Vdash A_{\sigma'}$, whence $\sigma \Vdash A_{\sigma'}$, whence $\omega_\sigma \models A_{\sigma'}$. But $\underset{\sim}{HA}^c \vdash A_{\sigma'} \rightarrow \neg A_\tau$, by definition, a contradiction. Q. E. D.

<u>Note</u>. The above proof could have been simplified by unifying the cases - which could have been done by stipulating that theorem 5.6.11 be used in treating the terminal nodes. The non-terminal nodes must be treated by using theorem 5.6.11. - Unless we know $\underset{\sim}{HA}^c \vdash A_{\sigma'} \rightarrow \neg A_{\tau'}$ we have no guarantee that A_τ will be false in extensions σ of σ'. (This observation is due to de Jongh, who found and corrected the corresponding error in our original attempt at proving the existence of Σ_1^0 substitution instances.)

We may now prove

5.6.16. <u>Theorem</u>. Let $P\rho \not\vdash A(p_1,\ldots,p_n)$. Then there are Σ_1^0 sentences $B_1,\ldots,B_n$ such that $\underset{\sim}{HA} \not\vdash A(B_1,\ldots,B_n)$.

<u>Proof</u>. Let J_n^* be given with a forcing relation on it such that $\langle\ \rangle \not\Vdash A(p_1,\ldots,p_n)$. Let, for each p_i,

$$B_i \leftrightarrow \bigvee_{\sigma \Vdash p_i} A_\sigma.$$

(If no σ forces p_i, let B_i be any refutable Σ_1^0 sentence.) Then B_i is Σ_1^0, $\sigma \Vdash B_i$ iff $\sigma \Vdash p_i$, and a simple induction on the length of $C(p_1,\ldots,p_n)$ shows

$$\sigma \Vdash C(p_1,\ldots,p_n) \text{ iff } \sigma \Vdash C(B_1,\ldots,B_n).$$

Q. E. D.

5.6.17 - 5.6.19. Uniform Π_2^0 substitutions in de Jongh's theorem.

5.6.17. The problem with the Σ_1^0 substitutions is that we could not choose the nodes at which we wanted a particular Σ_1^0 sentence to be forced. E.g. consider the tree

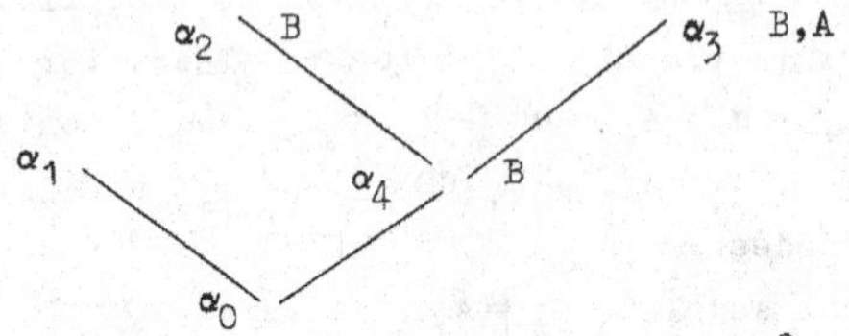

Suppose we want A, B forced as indicated, $A,B \in \Sigma_1^0$. Then, at α_4 we must have a model ω_{α_4} of

$$\underset{\sim}{HA}^c + B + CON(\underset{\sim}{HA}^c + B + \neg A) + CON(\underset{\sim}{HA}^c + B + A).$$

Since B is false in the standard model,

$$\underset{\sim}{HA}^c + \neg B + CON(\underset{\sim}{HA}^c + B + \neg A) + CON(\underset{\sim}{HA}^c + B + A)$$

is also consistent and B must be independent over $\underset{\sim}{HA}^c + CON(\underset{\sim}{HA}^c + B + \neg A) + CON(\underset{\sim}{HA}^c + B + A)$. Larger trees will require larger nested consistency statements and we just don't know if any such Σ_1^0 sentences exist. (Observe that we cannot have as much independence from consistency statements as with Π_2^0 - sentences, since, if $B \in \Sigma_1^0$, then $B + CON(\underset{\sim}{HA}^c + \neg B)$ is inconsistent.)

If, however, $B_1,\ldots,B_n$ are Π_2^0 and mutually independent over $\underset{\sim}{HA}^c$ when augmented by all true Π_1^0 sentences of arithmetic, we can assign nodes to the formulae as desired. This is the basis of the following model-theoretic proof of a result of Friedman A:

5.6.18. Theorem. (Friedman). Let $B_1,\ldots,B_n$ be Π_2^0, independent over $\underset{\sim}{HA}^c$ augmented by all true Π_1^0 sentences, and let $Pp \not\vdash A(p_1,\ldots,p_n)$. Then

$$\underset{\sim}{HA} \not\vdash A(B_1,\ldots,B_n).$$

Proof. Let $(K, \leq, \Vdash)$ be an arbitrary tree model of P and let $\langle\,\rangle \not\Vdash A(p_1,\ldots,p_n)$. Let $B_1,\ldots,B_n$ satisfy the hypothesis of the theorem and assign theories to nodes as follows:

If τ is terminal, $\underset{\sim}{T}_\tau = \underset{\sim}{HA}^c + \bigwedge_{\tau \Vdash p_i} B_i + \bigwedge_{\tau \not\Vdash p_i} \neg B_i$.

If σ has successors $\sigma * \langle 1\rangle, \ldots, \sigma * \langle k\rangle$,

$$\underset{\sim}{T}_\sigma = \underset{\sim}{HA}^c + \bigwedge_{\sigma \Vdash p_i} B_i + \bigwedge_{\sigma \not\Vdash p_i} \neg B_i + \bigwedge_{j=1}^{k} CON(\underset{\sim}{T}_{\sigma * \langle j\rangle}).$$

Each T_σ is obviously consistent and we may define models ω_σ as usual, starting at $\langle\,\rangle$ with an arbitrary model of $T_{\langle\,\rangle}$. Then K_σ is defined and, finally, we have a model of HA. E.g. with the tree featured above, we have

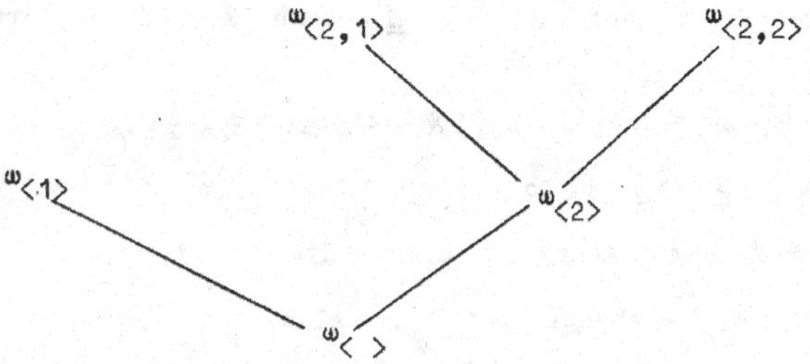

where $\omega_{\langle 2,1\rangle}, \omega_{\langle 2,2\rangle}, \omega_{\langle 2\rangle} \models B$, $\omega_{\langle 2,2\rangle} \models A$.

5.6.19. __Lemma__. Let A be Π^0_2. $\sigma \Vdash A$ iff $\forall \tau \geq \sigma \;\; \omega_\tau \models A$.

__Proof__. Probably the simplest thing to do is to appeal to __Matiyasevich__ 1970 or add new predicate symbols so that A is of the form,

$$\forall x_1 \ldots x_n \; \exists y_1 \ldots y_m C(x_1, \ldots, x_n, y_1, \ldots, y_m),$$

where C is quantifier-free and decidable. Then

$$\begin{aligned}
\sigma \Vdash A &\Rightarrow \forall \tau \geq \sigma \;\; \tau \Vdash A \\
&\Rightarrow \forall \tau \geq \sigma \;\; \forall s_1 \ldots s_n \in D\tau \;\; \exists t_1 \ldots t_m \in D\tau (\tau \Vdash C(s_1, \ldots, s_n, t_1, \ldots, t_m)) \\
&\Rightarrow \forall \tau \geq \sigma \; \forall s_1 \ldots s_n \in D\tau \;\; \exists t_1 \ldots t_n \;\; \omega_\tau \models C(s_1, \ldots, s_n, t_1, \ldots, t_m) \\
&\Rightarrow \forall \tau \geq \sigma \; \omega_\tau \models \forall x_1 \ldots x_n \exists y_1 \ldots y_m C, \text{ i.e. } \omega_\tau \models A.
\end{aligned}$$

The converse is just the definition of forcing for an $\forall\exists$ combination.

Q. E. D.

To finish the proof of the theorem, observe that

$$\begin{aligned}
\sigma \Vdash B_i \;\; &\text{iff} \;\; \forall \tau \geq \sigma \;\; \omega_\tau \models B_i \\
&\text{iff} \;\; \forall \tau \geq \sigma \;\; \tau \Vdash p_i \\
&\text{iff} \;\; \sigma \Vdash p_i.
\end{aligned}$$

The rest is just the usual induction. Q. E. D.

5.6.20 - 5.6.22. __De Jongh's theorem for__ MP .

5.6.20. Just as MP was not preserved by $(\;) \to (\Sigma\;)'$, it is not in general preserved by $(\;) \to (\Sigma\;)^*$. If, however, each element of $F = \{K_1, \ldots, K_n\}$ is definable in ω^+, and if MP is valid in F, then MP is valid in $(\Sigma F + \omega^+)^*$. This is just a variation of the result we will need. A direct verification of this variant is left to the reader. The general lemma we will need is the following:

5.6.21. Lemma. Let K be a tree model of $\mathbf{HA}$ obtained by the process of placing non-standard models of arithmetic at the nodes. Suppose that, for every node σ, there is a terminal node $\tau \geq \sigma$ such that $\omega_\sigma = \omega_\tau$. Then MP is valid in K.

Proof. Assume MP is not valid in K - take MP in the form (iv) of section 4:

$$\langle\,\rangle \nVdash \forall x[\forall y(Axy \vee \neg Axy) \mathbin{\&} \neg\neg\exists yAxy \rightarrow \exists yAxy]\,.$$

Then, for some $\sigma \geq \langle\,\rangle$, $s \in D\sigma$,

$$\alpha \nVdash \forall y(Asy \vee \neg Asy) \mathbin{\&} \neg\neg\exists yAsy \rightarrow \exists yAsy\,.$$

Thus, for some $\rho \geq \sigma$, whence $t \in D_p$. Also,

$$\rho \Vdash \forall y(Asy \vee \neg Asy)\,,\quad \rho \Vdash \neg\neg\exists yAsy\,,\quad \rho \nVdash \exists yAsy\,.$$

Let $\tau \geq \rho$ be terminal with $\omega_\tau = \omega_\rho$. Then $\tau \Vdash \exists yAsy$, say $\tau \Vdash Ast$, $t \in D\tau$. But $D\tau = D\rho$, whence $t \in D\rho$. Also,

$$\rho \Vdash Ast \vee \neg Ast\,,$$

whence $\rho \Vdash Ast$, i.e. $\rho \Vdash \exists yAsy$, a contradiction. Q. E. D.

5.6.22. Theorem. Let Pp $\nvdash A(p_1,\ldots,p_n)$. Then there are sentences $B_1,\ldots,B_n$ such that

$$\mathbf{HA} + MP \nvdash A(B_1,\ldots,B_n)\,.$$

Proof. Let J_n^* be given and define theories as follows: $\mathbf{T}_1 = \mathbf{HA}^c + A_1$, where A_1 is Σ_1^0, independent of $\mathbf{HA}^c$.

$$\mathbf{T}_{m+1} = \mathbf{HA}^c + CON(\mathbf{T}_m) + \neg A_m + A_{m+1}\,,$$

where A_{m+1} is independent of $\mathbf{HA}^c + CON(\mathbf{T}_m) + \neg A_m$. Let $\alpha_1,\ldots,\alpha_{n!}$ be the terminal nodes of J_n^*. Let $\omega_{n!}$ be a model of $\mathbf{T}_{n!}$. Given a model ω_{m+1} of $\mathbf{T}_{m+1}$, let ω_m be a model of $\mathbf{T}_m$ definable in ω_{m+1}. Assign ω_m to the terminal node α_m. In going down the tree, assign to σ the classical model assigned to the right-most successor of σ. J_3^*, e.g., looks like

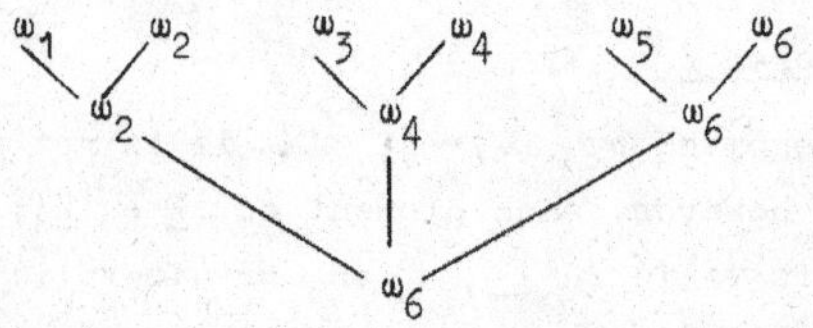

This gives us a Kripke model of $\mathbf{HA}$. By the lemma, it is also a model of MP. Finally, each terminal node is the unique node forcing a particular

sentence . The proof of de Jongh's theorem in section 3 now goes through easily. Q. E. D.

Note that Σ_1^0 substitutions are impossible: If A is Σ_1^0, $\underset{\sim}{HA} + MP \vdash \neg\neg A \rightarrow A$.

5.6.23 - 5.6.25. Other applications.

We first present a lemma.

5.6.23. Lemma. Let $<$ be primitive recursive, $\underset{\sim}{K}_1, \dots, \underset{\sim}{K}_n$ models of $\underset{\sim}{HA} + TI(<)$, each $\underset{\sim}{K}_i$ definable in ω^+, and $\omega^+ \models TI(<)$. Then $TI(<)$ is valid in $(\Sigma\underset{\sim}{F})^*$.

Proof. Use $TI(<)$ in ω^+ applied to $A^*(\alpha_0, \bar{x}, \bar{x}_1, \dots, \bar{x}_n)$ to verify that TI applied to $A(x, x_1, \dots, x_n)$ is forced at α_0. Q.E.D.

By insisting that each theory $\underset{\sim}{T}_\sigma$ used in 5.6.13 - 5.6.22 also contain $TI(<)$, every model $\underset{\sim}{K}_\sigma$ encountered is a model of $TI(<)$. Thus, the Σ_1^0 substitution and uniform Π_2^0 substitution results hold for $\underset{\sim}{HA} + TI(<)$ (where independence over $\underset{\sim}{HA}^c$ ($\underset{\sim}{HA}^c$ + true Π_1^0) is replaced by independence over $\underset{\sim}{HA} + TI(<)$ ($\underset{\sim}{HA}^c + TI(<)$ + true Π_1^0)). Further, the unrefined version of de Jongh's theorem holds for $\underset{\sim}{HA} + MP + TI(<)$.

A similar proof does not work for $RF(\underset{\sim}{T})$ or $RFN(\underset{\sim}{T})$. Recall that, to prove $RF(\underset{\sim}{T})$ was preserved if $\underset{\sim}{T}$ was, when we assumed $\exists x\, Proof_T(x, \ulcorner A \urcorner)$ was forced by α_0 in $(\Sigma\underset{\sim}{F})'$, it followed that $Proof_T(n, \ulcorner A \urcorner)$ was forced for some natural number n. From this it followed that A was indeed provable. We can no longer reason in this manner for $(\Sigma\underset{\sim}{F})^*$. We can, however, appeal to the following result of Kreisel - Levy 1968; (Theorem 12, p. 125):

5.6.24. Theorem. For small ordinals α, $\underset{\sim}{HA}$ together with the scheme $TI(<_{\epsilon_\alpha})$ (transfinite induction on the canonical well-ordering of type ϵ_α) is equivalent to the system obtained from $\underset{\sim}{HA}$ by iterating the process $\underset{\sim}{T} \rightarrow \underset{\sim}{T} + RFN(\underset{\sim}{T})$ $1+\alpha$ times; e.g. $\underset{\sim}{HA} + RFN(\underset{\sim}{HA}) = \underset{\sim}{HA} + TI(<_{\epsilon_0})$.

5.6.25. It follows from theorem 5.6.24 and the above remark that the Σ_1^0 and Π_2^0 results hold for $\underset{\sim}{HA} + RFN(\underset{\sim}{HA})$, $\underset{\sim}{HA} + RFN(\underset{\sim}{HA} + RFN(\underset{\sim}{HA}))$, etc. The results for $\underset{\sim}{HA} + RF(\underset{\sim}{HA})$ follow trivially from the results for the extension $\underset{\sim}{HA} + RFN(\underset{\sim}{HA})$.

It also follows that we get the unrefined version of de Jongh's theorem for $\underset{\sim}{HA} + MP + TI(<)$, $\underset{\sim}{HA} + MP + RFN(\underset{\sim}{HA})$, $\underset{\sim}{HA} + MP + RF(\underset{\sim}{HA})$ - we do not have the result for $\underset{\sim}{HA} + MP + RFN(\underset{\sim}{HA} + MP)$ because it is not known if

$$\underset{\sim}{HA} + MP + RFN(\underset{\sim}{HA} + MP) = \underset{\sim}{HA} + MP + TI(<_{\epsilon_0}).$$

In discussing the operations $(\ )\to(\Sigma\ )'$, $(\ )\to(\Sigma\ +\omega)'$, we gave some closure properties of the classes $\mathfrak{P}$, $\mathfrak{P}^{\omega}$, respectively, of sets Γ preserved by these operations. In disusing the operation $(\ )\to(\Sigma\ )^*$ and the class of models described in the statement of lemma 5.6.21, we should comment on the classes $\mathfrak{P}^*$ of sets of sentences valid in the Kripke model of the lemma provided they are valid in all of the non-standard models of which the Kripke model is composed. We both lose and gain some closure conditions.

In both cases, we lose Friedman's condition (iii) (theorems 5.2.11 and 5.4.13 above) which we restate here for convenience:

Condition (iii) If $\Gamma\in\mathfrak{P}(\mathfrak{P}^{\omega})$, A has only x free, and $\underset{\sim}{HA}+\Gamma\vdash An$ for all n, then $\Gamma\cup\{\forall xAx\}\in\mathfrak{P}$ (resp., $\mathfrak{P}^{\omega}$).

Recall that, if $\Gamma+\forall xAx$ was valid in $\underline{F}$, $\forall xAx$ could only fail to be valid in $(\Sigma\underline{F})'$ or $(\Sigma\underline{F}+\omega)'$ when some instance An was not forced at the node α_o - which is ruled out by the hypothesis. Obviously, this is no longer valid reasoning in the present situation where the new origins have non-standard integers in their domains.

For $(\ )\to(\Sigma\ )^*$, the use of definability does not give us an alternative to condition (iii). For applications to MP, however, we do have a slight rebate. Rather than to try to state an intelligible analogue to theorem 5.4.13, let us consider an example:

5.6.26. Theorem. Let $\mathfrak{P}_1$ be the class obtained by the following:

(i) $\underset{\sim}{HA}+MP\in\mathfrak{P}_1$;

(ii) $\underset{\sim}{HA}+TI(<)\in\mathfrak{P}_1$;

(iii) the union of any r.e. sequence of elements of $\mathfrak{P}_1$ is in $\mathfrak{P}_1$;

(iv) if $\Gamma\in\mathfrak{P}_1$, A is a Harrop-sentence <u>and</u> A is consistent with $\underset{\sim}{HA}^c$ + all true Π^o_1 sentences, then $\Gamma\cup\{A\}\in\mathfrak{P}_1$.

Then, for any $\Gamma\in\mathfrak{P}_1$, $\underset{\sim}{HA}+\Gamma$ has DP.

Proof. First, by (iii) all $\Gamma\in\mathfrak{P}_1$ are consistent with $\underset{\sim}{HA}^c$ + all true Π^o_1 sentences, which includes the consistency statements needed to define models. Also, if $\Gamma\in\mathfrak{P}_1$, Γ is r.e. and, if $\underset{\sim}{HA}+\Gamma\not\vdash A$, $\underset{\sim}{HA}+\Gamma\not\vdash B$, we can find Kripke models $\underline{K}_1$, $\underline{K}_2$, definable in some model ω^+ of $\underset{\sim}{HA}+\Gamma$, in which A, resp. B, fails to be forced. (To see that these are definable Kripke models, use the reduction of the problem to the Hilbert - Bernays completeness theorem outlined in 5.1.26.) Then $(\underline{K}_1+\underline{K}_2+\omega^+)^*$ is a model of $\underset{\sim}{HA}+\Gamma$ in which $A\vee B$ is false. Q. E. D.

We <u>cannot</u> generalize this to obtain ED. First, the domain at the origin would have non-standard integers and $\alpha_o\Vdash\exists xAx$ would not imply $\alpha_o\Vdash An$ for some n. Second, to have the definability of forcing for $(\Sigma\underline{F}+\omega^+)^*$ in ω^+, for an infinite family $\underline{F}$, we must have a uniform forcing definition for all elements in the class. But, the minute we have this, we have models

$\underline{K}_a$ for non-standard a occurring in our description of $\underline{F}$ - i.e. we are not defining the model we want to define.

Such esoteric results as theorem 5.6.26 are of little interest in themselves. They do, however, illustrate the differences in our ability to treat HA and HA + MP (as do such negative results as our inability to prove ED for $\Gamma \in \mathfrak{P}_1$).

§ 7. Other systems

5.7.1 - 5.7.2. Subsystems of Heyting's arithmetic.

5.7.1. In van Dalen - Gordon 1971, van Dalen and Gordon apply Kripke models to settle some independence questions regarding subsystems of $\underset{\sim}{HA}$. For example, consider the system $\underset{\sim}{T}$ with the constant 0, function symbols $'$, $+$, $\cdot$, and axioms (in addition to axioms of the intuitionistic predicate calculus with equality):

$$x' = y' \rightarrow x = y \qquad \neg x' = 0$$
$$x + 0 = x \qquad x + y' = (x + y)'$$
$$x \cdot 0 = 0 \qquad x \cdot y' = x \cdot y + x$$
$$x + y = y + x \qquad x \cdot y = y \cdot x$$
$$(x + y) + z = x + (y + z) \qquad (x \cdot y) \cdot z = x \cdot (y \cdot z)$$
$$\neg x = 0 \rightarrow \exists y (y' = x) .$$

Then, van Dalen and Gordon proved:

5.7.2. **Theorem.** $\underset{\sim}{T} \nvdash \forall xy(x=y \vee \neg x=y)$.

Proof. (van Dalen - Gordon 1971). Let $^*\mathbb{R}$ be a non-standard extension of $\mathbb{R}$, the field of real numbers. Let $^*N$ denote the set of elements of $^*\mathbb{R}$ which are infinitesimally close to some natural number - i.e.

$$^*N = \{x \in {}^*\mathbb{R} : \underset{\sim}{\exists} n \in \omega \ \underset{\sim}{\exists} \text{infinitesimal } \delta (x = n \pm \delta)\} .$$

Now, consider the model,

$$\begin{array}{c} \alpha_1 \ {}^*N_2 \\ | \\ \alpha_0 \ {}^*N_1 \end{array} ,$$

where $D\alpha_1 = D\alpha_0 = {}^*N$, the operations $'$, $+$, $\cdot$ on $^*N_1$, $^*N_2$ are those inherited from $^*\mathbb{R}$, $\alpha_0 \Vdash a=b$ iff a, b actually denote the same element of $^*\mathbb{R}$, and $\alpha_1 \Vdash a=b$ iff a and b are infinitesimally close (i.e. iff a, b are close to the same natural number). The axioms of $\underset{\sim}{T}$ are obviously forced at α_0, but the decidability of equality is not forced, since, if δ is infinitesimal, $\alpha_0 \nVdash n=n+\delta$ and, since $\alpha_1 \Vdash n=n+\delta$, $\alpha_0 \nVdash \neg n=n+\delta$. Thus $\alpha_0 \nVdash n=n+\delta \vee \neg n=n+\delta$. Q. E. D.

Observe that, in the model given in the proof of the theorem, the model at α_1 is, basically, the standard model ω. Thus, every instance of induction is forced at α_1 and the double negation of every instance of induction is forced at α_0. Hence, although equality is provably decidable by induction, it is not provably decidable by the double negation of induction - or by induction on Harrop - formulae, since, as the reader may easily verify, if such a formula is forced at α_1, it is forced at α_0. Thus, as one might expect, one cannot use a negative form of induction to prove the positive result that equality is decidable.

This raises the question:

How much induction is needed to prove the decidability of equality?

By induction on quantifier-free formulae, one can prove, for each n, $\forall x(n=x \vee \neg n=x)$. (Thus, quantifier-free induction fails to hold in the above model and quantifier-free induction is not derivable from the negative formulations of induction mentioned.) Can one use induction on quantifier-free formulae to derive the decidability of equality - $\forall xy(x=y \vee \neg x=y)$?

5.7.3. <u>Extensions of HA : Theory of species.</u>

Aside from some comments on free choice sequences in Kripke's original paper <u>Kripke</u> 1965, the only discussion of Kripke models and higher systems published to date is <u>Prawitz</u> 1970 in which Prawitz proves the completeness of the cut-free rules of the second-order intuitionistic predicate calculus with respect to second-order Kripke models (and also with respect to a proper subclass of these models, namely, the second-order Beth models). Since the induction scheme is given by a single axiom of this second-order language, this yields a completeness theorem for second-order arithmetic plus comprehension with respect to these second-order models.

The simplest way to describe the second-order Kripke models is to say that the domain function D splits into two functions D_1 and D_2, each satisfying the monotonicity condition, and such that there is a binary membership relation, $\in$, between elements of $D_1\alpha$ and $D_2\alpha$ (for given α). Further, when discussing comprehension, one assumes that for every node α and every formula $A(x)$ with parameters from $D_1\alpha$ and $D_2\alpha$, x the only free variable in A, there is an element X of $D_2\alpha$ such that

$$\alpha \Vdash \forall x[Ax \longleftrightarrow x \in X].$$

(In the presence of a little arithmetic, we can restrict ourselves to unary relations in $D_2\alpha$.)

In attempting to construct models of second-order arithmetic, the obvious approach is to mimic our procedure in constructing models of first-order arithmetic - but also insisting that the classical models being used be models of the second-order theory with comprehension. We have not considered this possibility thoroughly enough to say whether or not it will lead anywhere.

Let us consider a simple example. Suppose (ω^{++}, B) is definable in (ω^{+}, A), where B and A are the classes of sets of numbers in the two models. We would define a model:

$$\begin{array}{c} (\omega^{++},\ B) \\ | \\ \alpha_0\ (\omega^{+},\ A^*)\ . \end{array}$$

The choice of B is obvious; but what do we choose for A^*? We cannot simply choose A since, e.g., $\{x \mid C(x)$ is true in $(\omega^+,A)\}$ need not be $\{x \mid C(x)$ is forced at $\alpha_0\}$. (E.g. as long as ω^{++} is not an elementary extension of ω^+, there will be arithmetical $C(x)$ for which these sets will have to differ.) The obvious approach is to start with species X_C for arithmetical C such that

$$\alpha_0 \Vdash \forall x[x \in X_C \leftrightarrow C(x)]\ .$$

We cannot do this for C with species variables since we don't know yet what A^* is. Adding a species at a time, one can handle comprehension for formulae

$$\exists X_1 \ldots X_n\ C(x, X_1, \ldots, X_n)\ ,$$

where C has no bound species variables; but one cannot automatically handle more complex formulae - each new species added changes the domain of species and hence the nature of universal quantification.

Another possibility is the use of ω-models - i.e. models in which the individuals are precisely the natural numbers. The induction scheme, even applied to second-order formulae, will obviously be forced and the only problematic scheme is that of comprehension. A simple way to guarantee comprehension is to guarantee that all possible species are in the domains. Let $\underline{K}$ be a model, with partial order $(K, \leq)$, and let Ax be a formula with only the variable x free. Let $A_\alpha = \{a \in D_1\alpha : \alpha \Vdash Aa\}$. To guarantee comprehension, we need at α_0 a set $X \in D_2\alpha_0$ such that for any α and any $a \in D_1\alpha$, $\alpha \Vdash a \in X$ iff $\alpha \Vdash a \in A_\alpha$. To guarantee this, we simply let $D_2\alpha = D_2\alpha_0$ be the set of all partially ordered systems, $\underline{S} = \{S_\alpha\}_{\alpha \in K}$, of sets of natural numbers indexed by K satisfying $\alpha \leq \beta \Rightarrow S_\alpha \subseteq S_\beta$, and let $\underline{S}$ behave like S_α at node α. For any formula A, the system $\underline{A} = \{A_\alpha\}_{\alpha \in K}$ automatically represents A and comprehension is valid.

This latter type of model can be used to obtain certain formal results, e.g. it is easy to construct a model (the full binary tree) in which

$$\forall x \neg \forall X_1 \ldots \forall X_n A(x \in X_1, \ldots, x \in X_n)$$

is valid for any propositional formula $A(p_1, \ldots, p_n)$ which is not derivable in the intuitionistic propositional calculus. As a second-order counterexample, $A(x \in X_1, \ldots, x \in X_n)$ is as simple as they come - one might hope for an arithmetic counterexample, or at least a version of de Jongh's theorem; but these models will not yield such results, because all true arithmetic formulae are valid in them. Similarly, they cannot be used to prove the explicit definability or disjunction theorems.

5.7.4. Other set-theoretic approaches.

The Kripke models only form one of several classical modellings of intuitionistic systems. Others include the Beth models, interpretations in lattices, and topological interpretations. Their applications to the propositional calculus and the first-order predicate calculus are well-known. For higher systems, they have barely been applied. Prawitz 1970 applies the Beth bodels (and also Kripke models) to the theory of species ; Scott 1968 and Scott 1970 apply the topological interpretation to the theory of the order of the continuum ; and Moschovakis A applies the topological interpretation to second-order arithmetic - i.e. arithmetic with quantification over functions.

In Scott 1970, Scott proved the validity of Kripke's schema in his model. Moschovakis, in Moschovakis A, showed the consistency of this schema with a system of second-order intuitionistic arithmetic. Kripke's schema,

$$\exists\alpha[\exists x(\alpha x \neq 0 \leftrightarrow A(x))],$$

is important in that it contradicts many theorems of classical analysis (see e.g. Hull 1967).

Chapter VI

ITERATED INDUCTIVE DEFINITIONS, TREES AND ORDINALS

J. I. Zucker

§ 1. Introduction.

6.1.1. This chapter contains an investigation of the "provable recursive ordinals" of first-order intuitionistic theories $\underset{\sim}{ID}_2(A)$ of twice iterated inductive definitions. A characterization of the supremum of these ordinals is obtained in terms of recursive functionals on trees of the first three number (or rather tree) classes. Specifically, we describe an intuitionistic theory $\underset{\sim}{T}_2$ of trees of the first three classes and functionals of finite type over them, with a distinct ground type for each class, and a separate functional for (finite or transfinite) recursion on each class.

Although the functionals on trees considered here are less familiar than functionals on ordinals, the word "ordinal" is often used in constructive mathematical literature to denote, not classical ordinals, but in fact (constructive) trees, or well-orderings together with some additional structure, such as distinguished cofinal sequences of previously obtained "ordinals".

One advantage of the present procedure is that certain results, stated in terms of trees, can be interpreted either in a classical or in a constructive sense. An example of this is given in 6.6.3 (a) and (b), where the theory $\underset{\sim}{T}_2$ is given, respectively, a classical and a "constructive" (i.e., recursive) interpretation.

In addition, this work shows that proof-theoretical results can be formulated elegantly and intelligibly in terms of trees (or well-founded relations) rather than ordinals (or well-orderings). (For certain purposes, on the other hand, it may suffice to consider only the ordinals of trees, i.e. to suppress the tree structure.)

6.1.2. Outline of the contents of this chapter.

In § 2 we describe intuitionistic first-order theories $\underset{\sim}{ID}_2(A)$ of (twice) iterated inductively defined sets Q_1 and Q_2 of natural numbers, where the pair of defining formulas $A \equiv (A_1, A_2)$ satisfy a syntactic condition $\mathcal{C}$ which is stricter than positivity, but sufficiently broad to include all the "well-known" inductively defined sets.

With each $a \in Q_1$ is associated an ordinal $|a|_{A_1}$. We then define $|\underset{\sim}{ID}_2(A)|$, the "ordinal of $\underset{\sim}{ID}_2(A)$", as $\sup\{|a|_{A_1} : \underset{\sim}{ID}_2(A) \vdash Q_1\bar{a}\}$, and $|\underset{\sim}{ID}_2|$, the "ordinal of $\underset{\sim}{ID}_2$", as $\sup\{\underset{\sim}{ID}_2(A) : A \in \mathcal{C}\}$. Similarly, the ordinal $|\underset{\sim}{ID}_2^c|$ is defined for the corresponding classical theories $\underset{\sim}{ID}_2^c(A)$; and the ordinals $|\underset{\sim}{ID}_1|$ and $|\underset{\sim}{ID}_1^c|$ are defined analogously for the intuitionistic and classical theories (respectively) of non-iterated inductive definitions.

In § 3 the theory $\underset{\sim}{T}_2$ of trees is described. There are three ground types: 0, for the natural numbers (the "first number class"), 1, for trees of the "second class", and 2, for trees of the "third class".

In § 4 the notions of conversion and reduction for terms of $\underset{\sim}{T}_2$ are defined, uniqueness of normal form is proved, and the normalizability of closed terms of $\underset{\sim}{T}_2$ is proved by a computability argument.

In § 5 we consider <u>strong</u> computability, and hence prove the normalizability of <u>all</u> terms (not necessarily closed) of $\underset{\sim}{T}_2$.

In § 6, well-founded (wf) models of $\underset{\sim}{T}_2$ are discussed. Four specific wf models are considered: (a) the full set-theoretical model $\mathcal{J}_2$, (b) the model HRO_2 (referred to above) of hereditarily recursive operations of finite type over $\mathcal{O}_1$ and $\mathcal{O}_2$, (variants of) the recursive first and second number classes of Kleene and Addison, (c) the extensional variant HEO_2 of (b), and (d) the term model $CTNF_2$.

Let CT_τ be the set of closed terms of $\underset{\sim}{T}_2$ of type τ. With each $t \in CT_\tau$ (for $\tau = 1$ and 2) we can associate an ordinal $|t|_C$ by virtue of its computability, and also the ordinal $|t|_M$ of the tree denoted by t in a given wf model M of $\underset{\sim}{T}_2$. Then for $t \in CT_1$, $|t|_M = |t|_C$ (= $|t|$ say) for any wf model M, but for $t \in CT_2$, $|t|_M$ varies with M, and $|t|_M \geq |t|_C$. This failure of invariance of $|t|_M$ for $t \in CT_2$ points out an interesting difference between non-iterated and iterated inductive definitions (discussed in 6.6.6).

Now we define the "ordinal of $\underset{\sim}{T}_2$":

$$|\underset{\sim}{T}_2| \equiv_{def} \sup\{|t| : t \in CT_1\}.$$

The main result of this chapter can now be stated:

(1) $\qquad |\underset{\sim}{ID}_2| = |\underset{\sim}{T}_2|$.

The one inequality $|\underset{\sim}{ID}_2| \leq |\underset{\sim}{T}_2|$ is proved (6.6.8) by two methods: (a) formalizing the construction of the model HRO_2 in a theory $\underset{\sim}{ID}_2(\mathcal{O})$ of the inductively defined sets $\mathcal{O}_1$ and $\mathcal{O}_2$, and (b) formalizing the proof of computability of terms of $\underset{\sim}{T}_2$ in a theory $\underset{\sim}{ID}_2(C)$ of the inductively defined computability predicates C_1 and C_2 for terms of type 1 and 2 respectively.

In order to prove the reverse inequality, and hence (1), we consider a functional interpretation (modified realizability) of $\underset{\sim}{ID}_2(A)$ in § 7. This extends the $\underset{\approx}{mr}$-interpretation of $\underset{\sim}{HA}$ (chapter III, § 4) by defining: $\alpha^i \ \underset{\approx}{mr}\ Q_i t \equiv P_i(\alpha^i, t)$ $(i = 1,2)$, where α^i is a type i variable, and P_1 and P_2 are two new predicates (not interpretable by recursive relations in HRO_2). So $\underset{\sim}{ID}_2(A)$ is actually interpreted in a theory $\underset{\sim}{E}-\underset{\sim}{T}_2\underset{\sim}{P}$, which is $\underset{\sim}{T}_2$ augmented by the predicates P_1 and P_2, with appropriate axioms (and also extensionality axioms).

By means of this interpretation, it follows that if $\underset{\sim}{ID}_2(A) \vdash Q_1\bar{a}$ for some number a, then from the proof we can find a term $t \in CT_1$ and a proof in $\underset{\sim}{E}-\underset{\sim}{T}_2\underset{\sim}{P}$ of $P_1(t,\bar{a})$. Further, for this t, we can prove: $|t| \geq |a|_{A_1}$, and so (6.7.9) obtain the inequality $|\underset{\sim}{ID}_2| \leq |\underset{\sim}{T}_2|$, and hence (1).

It will also be clear, by a simplification of these arguments, that a corresponding result holds for non-iterated inductive definitions:

$$|\underset{\sim}{ID}_1| = |\underset{\sim}{T}_1|,$$

where $|\underset{\sim}{T}_1|$ is the "ordinal of" a system $\underset{\sim}{T}_1$ of functionals of finite type on trees of the second class only.

In § 8 we turn to the problem of characterizing the ordinals $|\underset{\sim}{ID}_1^c|$ and $|\underset{\sim}{ID}_2^c|$ by means of functional interpretations of the classical theories $\underset{\sim}{ID}_1^c(\mathcal{O})$ and $\underset{\sim}{ID}_2^c(\mathcal{O})$ of the inductively defined sets $\mathcal{O}_1$ and $\mathcal{O}_2$. It turns out that for $\underset{\sim}{ID}_1^c(\mathcal{O})$, a Dialectica interpretation is possible, in a system $\underset{\sim}{T}_1(\mu)$ of functionals, consisting of $\underset{\sim}{T}_1$, together with a (non-constructive) number selection operator μ. Then by means of a "majorizing" technique (essentially due to W.A. Howard) we show that the ordinal $|\underset{\sim}{T}_1(\mu)|$ of this system is no greater than $|\underset{\sim}{T}_1|$, giving the result:

(2) $\quad |\underset{\sim}{ID}_1^c| = |\underset{\sim}{T}_1|.$

However, attempts to prove $|\underset{\sim}{ID}_2^c| \leq |\underset{\sim}{T}_2|$ by a functional interpretation (modified realizability or Dialectica) of $\underset{\sim}{ID}_2^c(\mathcal{O})$, in which $Q_i t$ is translated as $\exists\alpha^i P_i(\alpha^i, t)$, fail. So the problem of whether

$$|\underset{\sim}{ID}_2^c| = |\underset{\sim}{T}_2| \quad \text{or} \quad |\underset{\sim}{ID}_2^c| > |\underset{\sim}{T}_2|$$

remains open.

In § 9 we consider (briefly) extensions of the result (1) for intuitionistic systems $\underset{\sim}{ID}_\nu(A)$ of inductive definitions iterated ν times, for $\nu > 2$, and corresponding systems $\underset{\sim}{T}_\nu$ of trees of the first $1+\nu$ classes. However (unlike the case for $\nu = 1$) it is not known for any $\nu > 1$ whether the result analogous to (2) also holds for the ordinals of classical systems

$\underset{\sim}{ID}^{c}_{\nu}(A)$, i.e. whether $|\underset{\sim}{ID}^{c}_{\nu}| = |\underset{\sim}{T}_{\nu}|$ (as stated above already for $\nu = 2$). Now the proof-theoretical equivalence of these classical systems $\underset{\sim}{ID}^{c}_{\nu}(A)$ with well-known subsystems of classical analysis has been established by Feferman 1970, so a positive answer to the above problem would then also provide an interesting characterization of the "ordinals of" these subsystems of classical analysis (i.e. the suprema of their provably recursive well-orderings).

By contrast (6.9.2), it is known, for (respectively classical and intuitionistic) systems $\underset{\sim}{ID}^{c}_{<\omega}(A)$ and $\underset{\sim}{ID}_{<\omega}(A)$ of finitely iterated inductive definitions, that $|\underset{\sim}{ID}^{c}_{<\omega}| = |\underset{\sim}{ID}_{<\omega}|$; and here we do have an ordinal characterization of a corresponding subsystem of classical analysis.

Finally (6.9.3), we describe a (classical) system of iterated inductive definitions, which is proof-theoretically equivalent to classical analysis with the Π^1_1 - comprehension axiom (modifying a result in Feferman 1970).

6.1.3. Historical note; comparison with other treatments.

Systems (especially intuitionistic) of non-iterated "generalized inductive definitions" were formulated and studied in Kreisel 1963 C.

An intuitionistic system of finitely iterated inductive definitions was formulated in Kreisel 1964 in order to prove the well-foundedness of Takeuti's ordinal diagrams of finite order.

Feferman 1970 (as stated above) considered classical systems of inductive definitions iterated along primitive recursive well-orderings, and established their equivalence with subsystems of classical analysis.

Other known characterizations of $|\underset{\sim}{ID}_1|$ $(= |\underset{\sim}{ID}^{c}_1|)$ in terms of functionals on ordinals (due to Howard and Feferman) and Bachmann's notations (due to Howard and Gerber) are described in 6.8.6. There also we describe the (known) (proof-theoretical) reducibility of $\underset{\sim}{ID}^{c}_1(\mathcal{O})$ to $\underset{\sim}{ID}_1(\mathcal{O})$, and state a conjecture (also made independently by Martin-Löf) giving $|\underset{\sim}{ID}_2|$ in terms of Bachmann - Isles notations.

Martin-Löf 1971 analyzed an intuitionistic theory of finitely iterated inductive definitions by an elegant extension of normalization results, rather than by an extension of functional interpretations, and gave an ordinal characterization of this theory (see 6.9.2). However, the functional interpretation given here seems more amenable to a direct calculation of $|\underset{\sim}{ID}_2|$. In addition, the wf models of $\underset{\sim}{T}_2$ seem to be mathematical structures of independent interest.

As pointed out in Kreisel 1971, § 2, syntactic transformations on derivations are suitable for obtaining derived rules (as can be seen in Martin-Löf

1971, § 9.3), whereas <u>functional</u> <u>interpretations</u> can be useful for independence results, cf. 6.8.8, where it is shown that the formula $\forall n(\neg\neg \mathcal{O}_1 n \rightarrow \mathcal{O}_1 n)$ is independent of $\underset{\sim}{\mathrm{ID}}_2(\mathcal{O})$.

6.1.4. <u>Note to the reader</u>.

Although many of the references can be read with profit, this chapter is intended to be self-contained, or rather, depend only on (parts of) chapters I, II and III. Some paragraphs, of a more or less technical nature, are enclosed in square brackets.

6.1.5. <u>Acknowledgments</u>.

This chapter originally formed part of the author's doctoral dissertation (Stanford University, 1971). My sincere gratitude goes to my dissertation supervisor, Prof. S. Feferman, for his invaluable and unstinting guidance. I also wish to thank Prof. G. Kreisel for his very helpful criticisms and suggestions, and P. Martin-Löf and H. Friedman for many fruitful discussions. I also want to express my appreciation to the Newhouse Foundation for supporting my first year of graduate study at Stanford University.

§ 2. The systems $\underset{\sim}{ID}_2(A)$.

6.2.1. Inductively defined sets of numbers.

Let $\mathcal{L}$ be the language of HA (1.3.2). $\mathcal{L}[X]$ and $\mathcal{L}[X,Y]$ are the languages formed by adjoining to $\mathcal{L}$ unary predicate (or set) variables X and X,Y resp., but with quantification over number variables only.

Notation. a, b, e, k, m, n, ... denote number variables

s, t denote number terms.

Let $A_1(X,a)$ be a formula of $\mathcal{L}[X]$ which is monotonic in X, i.e.

$$\forall X,X' \subset N[A_1(X,a) \,\&\, X \subset X' \to A_1(X',a)]$$

(provably in HAS say), where N is the set of natural numbers.

A set $X \subset N$ is A_1-closed if:

$$\forall a(A_1(X,a) \to a \in X).$$

Let $Q_1 \equiv_{def} \bigcap \{X \subset N : X \text{ is } A_1\text{-closed}\}$.

Then we can prove (in, say, HAS):

(i) Q_1 is A_1-closed (using monotonicity of A_1), and

(ii) $\forall X \subset N(X \text{ is } A_1\text{-closed} \to Q_1 \subset X)$.

(i) and (ii) characterize Q_1 uniquely as the least A_1-closed subset of N.

Now let $A_2(X,Y,a)$ be a formula of $\mathcal{L}[X,Y]$ which is monotonic in Y when X is interpreted as Q_1. We define, similarly, A_2-closed sets to be sets Y such that

$$\forall a(A_2(Q_1,Y,a) \to a \in Y)$$

and $Q_2 \equiv_{def} \bigcap \{Y \subset N : Y \text{ is } A_2\text{-closed}\}$.

Again, Q_2 is the least A_2-closed subset of N.

We define, for ordinals $\nu < \Omega$, sets $Q_{1,\nu}$ by induction on ν:

$$Q_{1,\nu} \equiv_{def} \{a : A_1(\bigcup_{\mu<\nu} Q_{1,\mu}, a)\}.$$

Then $Q_1 = \bigcup_{\nu<\Omega} Q_{1,\nu}$.

In fact, $Q_1 = \bigcup_{\nu<\omega_1} Q_{1,\nu}$

where ω_1 is the first non-recursive ordinal (Spector 1961).

We can associate with each $a \in Q_1$ the ordinal

$$|a|_{A_1} \equiv_{def} \min\{\nu : a \in Q_{1,\nu}\}.$$

In this way the elements of Q_1 can be thought of as ordinal notations.

6.2.2. The theory $\underset{\sim}{ID}_2(A)$; definition of $|\underset{\sim}{ID}_2|$.

In the study of theories of iterated inductive definitions, it seems advisable (at least to start with) to consider conditions on A_1 and A_2 stricter than monotonicity, e.g. positivity. A formula is said to be positive in a set parameter X if it contains only positive occurrences of X, i.e. there are no occurrences of X contained in the antecedent of an implication. (This concept of "positivity" is weaker than the one defined in Kreisel and Troelstra 1970, § 4.4.)

For classical theories the requirement of positivity is not really a restriction, since it turns out that a monotonic inductive definition can be reduced to a positive inductive definition (6.8.1) ; however, it is still unknown whether this holds in the intuitionistic case.

We will consider a syntactic condition $\mathcal{C}$, defined below, which is stricter (even) than positivity, although it is sufficiently broad to include all the "well-known" inductively defined sets.

Classes $\mathcal{C}_0$, $\mathcal{C}_1$ and $\mathcal{C}_2$ of formulas of $\mathcal{L}$, $\mathcal{L}[X]$ and $\mathcal{L}[X,Y]$ resp. are defined as follows.

$\mathcal{C}_0$ is the class of formulas of $\mathcal{L}$ built up from prime formulas $(s = t)$ by &, $\rightarrow$, $\wedge$ and $\forall$ (i.e. the "negative formulas" of $\mathcal{L}$, 1.10.6).

$\mathcal{C}_1$ is defined inductively by :

1. $(s=t) \in \mathcal{C}_1$ for any number terms s, t .
2. $Xt \in \mathcal{C}_1$ for any number term t .
3. If $A,B \in \mathcal{C}_1$ then so are $A \& B$, $\forall nA$, $\exists nA$.
4. If $A \in \mathcal{C}_1$ and $B \in \mathcal{C}_0$ then $(B \rightarrow A) \in \mathcal{C}_1$.

Remark 1. If in clause 4, "$B \in \mathcal{C}_0$" were replaced by "B is a formula of $\mathcal{L}$", then $\mathcal{C}_1$ would be precisely the class of formulas of $\mathcal{L}$ which are positive in X. As it is, $\mathcal{C}_1$ is a subclass of this class. *

$\mathcal{C}_2$ is defined inductively by :

1. $(s=t) \in \mathcal{C}_2$ for number terms s, t .
2. $Xt \in \mathcal{C}_2$ for number terms t .
3. $Yt \in \mathcal{C}_2$ for number terms t .
4. If $A,B \in \mathcal{C}_2$, then so are $A \& B$, $A \vee B$, $\forall nA$, $\exists nA$.
5. If $A \in \mathcal{C}_2$ and $B \in \mathcal{C}_1$ then $(B \rightarrow A) \in \mathcal{C}_2$.

Remark 2. If in clause 5, "$B \in \mathcal{C}_1$" were replaced by "B is a formula of $\mathcal{L}[X]$", then $\mathcal{C}_2$ would be precisely the class of formulas of $\mathcal{L}[X,Y]$ which

----

* On the other hand, $\mathcal{C}_1$ neither includes, nor is included in, "positivity" as defined in Kreisel and Troelstra 1970, § 4.4.

are <u>positive</u> in Y. As it is, $\mathcal{C}_2$ is a subclass of this class.

Finally, $\mathcal{C} \equiv_{def} \mathcal{C}_1 \times \mathcal{C}_2$.

Let $A_1(X,a)$ and $A_2(X,Y,a)$ be formulas of $\mathcal{C}_1$ and $\mathcal{C}_2$ resp. in which $\underline{a}$ is the only free number variable. Let Q_1 and Q_2 be two new unary predicate constants.

$\mathcal{L}[Q_1]$, resp. $\mathcal{L}[Q]$, is $\mathcal{L}$ augmented by Q_1, resp. Q_1 and Q_2.

A denotes the pair of formulas $(A_1,A_2) \in \mathcal{C}$.

<u>The theory</u> $\underset{\sim}{ID}_2(A)$ consists of $\underset{\sim}{HA}$ (1.3.3) in the language $\mathcal{L}[Q]$ (including the equality axioms $a=b \,\&\, Q_i a \to Q_i b$ $(i=1,2)$ and induction for all formulas of $\mathcal{L}[Q]$), and also the following axioms for Q_1 and Q_2:

$Q_1.1)$ $A_1(Q_1,a) \to Q_1 a$

$Q_1.2)$ $\forall a[A_1(F,a) \to F(a)] \to \forall a(Q_1 a \to F(a))$.

$Q_2.1)$ $A_2(Q_1,Q_2 a) \to Q_2 a$

$Q_2.2)$ $\forall a[A_2(Q_1,F,a) \to F(a)] \to \forall a(Q_2 a \to F(a))$,

where $F(a)$ is any formula of $\mathcal{L}[Q]$, and $A_1(F,a)$ is obtained from $A_1(X,a)$ by replacing Xt by $F(t)$.

Axioms $Q_i.1$ (for $i=1$ or 2) state that Q_i is A_i - closed, and $Q_i.2$ express minimality of Q_i among A_i - closed sets (at least those sets which are definable in $\mathcal{L}[Q]$).

Alternatively, $Q_i.2$ can be thought of as expressing (transfinite) <u>induction</u> on Q_i $(i=1,2)$.

The "ordinal of $\underset{\sim}{ID}_2(A)$" is now defined as:

$$|\underset{\sim}{ID}_2(A)| \equiv_{def} \sup\{|a|_{A_1} : \underset{\sim}{ID}_2(A) \vdash Q_1\bar{a}\}$$

and the "ordinal of $\underset{\sim}{ID}_2$" as:

$$|\underset{\sim}{ID}_2| \equiv_{def} \sup\{|\underset{\sim}{ID}_2(A)| : A \in \mathcal{C}\}.$$

(In fact, as we will see, $|\underset{\sim}{ID}_2| = |\underset{\sim}{ID}_2(A)|$ for suitably chosen $A \in \mathcal{C}$.)

For certain pairs of formulas $A \in \mathcal{C}$, to be defined later, we will use special symbols, viz. $\mathcal{O} \equiv (\mathcal{O}_1, \mathcal{O}_2)$ (see below) and $C \equiv (C_1, C_2)$, both for the predicates Q_1, Q_2 of the formal theory and for the corresponding sets in the intended interpretation. In such cases we will write $\underset{\sim}{ID}_2(\mathcal{O})$ (or $\underset{\sim}{ID}_2(C)$) for $\underset{\sim}{ID}_2(A)$, $\mathcal{L}[\mathcal{O}]$ for the language $\mathcal{L}[Q]$, $\mathcal{L}[\mathcal{O}_1]$ for $\mathcal{L}[Q_1]$, etc.

We now define a particular pair of predicates or sets $\mathcal{O} \equiv (\mathcal{O}_1, \mathcal{O}_2)$ which will be used in 6.6.4 and 6.8. They are (simplified versions of) the second and third recursive number classes of <u>Kleene</u> 1955 and <u>Addison and Kleene</u> 1957 resp. They are defined by the formulas:

$$A_1(X,a) \equiv a=0 \ .\vee.\ a=3.5^{(a)_2} \,\&\, \forall n\, X\{(a)_2\}(n),$$

$$A_2(X,Y,a) \equiv a=0 \ .\vee.\ a=3^2.5^{(a)_2} \,\&\, \forall n(Xn \to Y\{(a)_2\}(n)),$$

where $(a)_x$ is the exponent of the xth prime in the prime factor representation of a, and $X\{(a)_2\}(n)$ means $\exists y(T(a)_2 ny \;\&\; X(Uy))$ in Kleene's notation (1.3.9, A) (and similarly for $Y\{(a)_2\}(n)$). *

It is easy to see that $(A_1, A_2) \in \mathcal{C}$.

Some further definitions.

$\underset{\sim}{ID}_1(A)$ is the intuitionistic theory of one inductively defined set, with defining formula $A(X,a)$ (i.e. $\underset{\sim}{HA} + Q_1.1 + Q_1.2$, in $\mathcal{L}[Q_1]$). Again

$$|\underset{\sim}{ID}_1(A)| \equiv_{def} \sup\{|a|_A : \underset{\sim}{ID}_1(A) \vdash Q_1\bar{a}\}$$

and

$$|\underset{\sim}{ID}_1| \equiv_{def} \sup\{|\underset{\sim}{ID}_1(A)| : A \in \mathcal{C}_1\}.$$

$\underset{\sim}{ID}_0$ is just $\underset{\sim}{HA}$.

Finally, $\underset{\sim}{ID}^c_\nu(A)$, $|\underset{\sim}{ID}^c_\nu(A)|$ and $|\underset{\sim}{ID}^c_\nu|$ (for $\nu = 1$ or 2) are defined analogously for the corresponding systems with classical logic.

Remark. For an inductive definition A of the class Γ (chapter 1, § 4), $\underset{\sim}{ID}_1(A)$ is conservative over $\underset{\sim}{HA}$, since the inductively defined set is explicitly definable in $\underset{\sim}{HA}$ (1.4.5).

-----

* Here and elsewhere dots are used to punctuate formulas in a well-known way; so e.g. $A \rightarrow : B . \vee . C \& D$ means $A \rightarrow [B \vee (C \& D)]$.

§ 3. The theory $\underset{\sim}{T}_2$.

This is an intuitionistic theory of trees of the first three number or tree classes.

6.3.1. Type structure. The set $\underset{\approx}{T}_2$ of type symbols of $\underset{\sim}{T}_2$ is defined inductively by the two clauses :

(i) $\underset{\approx}{T}_2$ contains 3 ground types : 0, 1 and 2;

(ii) $\sigma, \tau \in \underset{\approx}{T}_2 \Rightarrow (\sigma)\tau \in \underset{\approx}{T}_2$.

0 is the type of trees of the first class, or natural numbers.

1 is the type of trees of the second class.

2 is the type of trees of the third class. *

$(\sigma)\tau$ is the type of functions from type σ to type τ objects.

6.3.2. Terms of $\underset{\sim}{T}_2$.

Let Tm_τ be the set of terms of type τ, and $Tm \equiv \bigcup \{Tm_\tau : \tau \in \underset{\approx}{T}_2\}$.

Tm is defined inductively by :

(i) Variables of type τ belong to Tm_τ (countably many for each τ).

(ii) Constants (see below) of type τ belong to Tm_τ.

(iii) $s \in Tm_{(\sigma)\tau}$, $t \in Tm_\sigma \Rightarrow st \in Tm_\tau$.

The constants of $\underset{\sim}{T}_2$ are :

$0^0, 0^1, 0^2$: zeroes of type 0, 1, 2.

S_0 : successor, of type $(0)0$.

S_1, S_2 : sup or join for trees of type 1, 2 resp. Their types are $((0)1)1$ and $((1)2)2$ resp.

$R_{0,\tau}, R_{1,\tau}, R_{2,\tau}$: constants for recursion on type 0, 1, 2 resp., for each $\tau \in \underset{\approx}{T}_2$. (Their types can be seen from the axioms for them, below.)

$\Pi_{\sigma,\tau}, \Sigma_{\rho,\sigma,\tau}$: for all $\rho, \sigma, \tau \in \underset{\approx}{T}_2$.

6.3.3. Notational conventions.

$x^\tau, y^\tau, x_1^\tau, \ldots$ (or just $x, y, x_1, \ldots$) denote variables of (any) type τ.

$a, b, e, k, m, n, \ldots$ denote variables of type 0.

α, β or α^1, β^1 denote variables of type 1.

-----

* This notation conflicts with that of chapter I, where 1 and 2 denote the types $(0)0$ and $((0)0)0$ resp. However, this notation will be used only with this meaning in this chapter, so there should be no confusion.

α^2, β^2 denote variables of type 2.
$f^{(0)1}, \ldots, f^{(1)2}, \ldots$ denote variables of type (0)1, (1)2, resp.
$r, s, t, u, v, t_1, t', \ldots$ denote terms.

Type superscripts and subscripts may be dropped: e.g. we write R_i for $R_{i,\tau}$ $(i = 0,1,2)$.

Other conventions stated in chapter I, § 6 are also followed here, e.g. for representing a term as a function of many arguments (1.6.5).

6.3.4. Formulas.

If s and t are terms of the same type, then

$$s = t$$

is a prime formula. Formulas are built up from these using all connectives and quantification at all types.

6.3.5. Axioms and rules.

(a) Axioms and rules for many-sorted intuitionistic predicate logic.

(b) Axioms of equality: the first 4 of 1.6.7 (b), for all $\sigma,\tau \in \underset{\approx}{T}_2$.

(c) Axioms for zero and successor or sup:

$$S_0 n \neq 0^0, \quad S_0 m = S_0 n \to m = n,$$
$$S_1 f^{(0)1} \neq 0^1, \quad S_1 f = S_1 g \to f = g,$$
$$S_2 f^{(1)2} \neq 0^2, \quad S_2 f = S_2 g \to f = g.$$

(d) Axioms for finite and transfinite induction:

$\underline{\mathrm{FI}}$: $F(0^0)\ \&\ \forall n[F(n) \to F(S_0 n)] \to \forall n F(n)$

$\underline{\mathrm{TI}_1}$: $F(0^1)\ \&\ \forall f^{(0)1}[\forall n F(fn) \to F(S_1 f)] \to \forall \alpha F(\alpha)$

$\underline{\mathrm{TI}_2}$: $F(0^2)\ \&\ \forall f^{(1)2}[\forall \beta F(f\beta) \to F(S_2 f)] \to \forall \alpha^2 F(\alpha^2)$,

for arbitrary formulas $F(n)$, $F(\alpha)$, $F(\alpha^2)$.

(e) Axioms for finite and transfinite recursion:

$\underline{\mathrm{FR}}$: finite recursion.

$$\begin{cases} R_{0,\tau} xy0^0 = x, \\ R_{0,\tau} xy(S_0 n) = y(R_{0,\tau} xyn)n, \end{cases}$$

for all $\tau \in \underset{\approx}{T}_2$, where $x \in \tau$, $y \in (\tau)(0)\tau$.

$\underline{\mathrm{TR}_1}$: transfinite recursion on type 1.

$$\begin{cases} R_{1,\tau} xy0^1 = x, \\ R_{1,\tau} xy(S_1 f) = y(\Sigma(\Pi(R_{1,\tau} xy))f)f, \end{cases}$$

for all $\tau \in \underset{\approx}{T}_2$, where $x \in \tau$, $y \in ((0)\tau)((0)1)\tau$, $\Pi \equiv \Pi_{(1)\tau,0}$ and $\Sigma \equiv \Sigma_{0,1,\tau}$. (See remark 6.3.6 (a).)

$\underline{TR_2}$: <u>transfinite recursion on type</u> 2 .

$$\begin{cases} R_{2,\tau}xy0^2 = x , \\ R_{2,\tau}xy(S_2f) = y(\Sigma(\Pi(R_{2,\tau}xy))f)f , \end{cases}$$

for all τ, where $x \in \tau$, $y \in ((1)\tau)((1)2)\tau$, $\Pi \equiv \Pi_{(2)\tau,1}$ and $\Sigma \equiv \Sigma_{1,2,\tau}$. (See remark 6.3.6 (a).)

(f) <u>Axioms for</u> $\Pi_{\rho,\sigma}$ <u>and</u> $\Sigma_{\rho,\sigma,\tau}$: as in 1.6.7 (d).

6.3.6. <u>Remarks.</u>

(a) The λ-operator can be defined in $\underset{\sim}{T}_2$, as in 1.6.8. The axioms TR_1 and TR_2 can then be written (more perspicuously) as:

$$\begin{cases} R_1xy0^1 = x , \\ R_1xy(S_1f) = y(\lambda nR_1xy(fn))f , \end{cases}$$

or (putting $R_1xy = \Phi$, and writing y as a function of two variables):

$$\begin{cases} \Phi\, 0^1 = x , \\ \Phi(S_1f) = y(\Phi \circ f, f) \end{cases}$$

(where $\circ$ denotes composition) ; and similarly:

$$\begin{cases} R_2xy0^2 = x , \\ R_2xy(S_2f) = y(\lambda\alpha R_2xy(f\alpha))f , \end{cases}$$

or (putting $R_2xy = \Phi$):

$$\begin{cases} \Phi\, 0^2 = x , \\ \Phi(S_2f) = y(\Phi \circ f, f) . \end{cases}$$

(b) Let $\underset{\sim}{T}_0$, resp. $\underset{\sim}{T}_1$, be the restriction of $\underset{\sim}{T}_2$, where the only ground types are 0, resp. 0 and 1.

Then $\underset{\sim}{T}_0$ is just $\underset{\sim}{N}$-$\underset{\sim}{HA}^{\omega}$ (chap. I, § 6).

(c) Note that

$$\underset{\sim}{T}_2 \vdash \forall\alpha^i[\alpha^i = 0^i \vee \exists! f(\alpha^i = S_i f)] \qquad (i \equiv 1,2),$$

from the axioms for S_i and TI_i.

§ 4. Computability of closed terms of $\underset{\sim}{T}_2$.

6.4.1. Definition of reduction, normal form, etc.

Let CT_τ be the set of closed terms of type τ, and $CT \equiv \bigcup \{CT_\tau : \tau \in \underset{\sim}{T}_2\}$. *

The relation: t contr t' (for $t, t' \in Tm$), is defined by the clauses 2.2.2 (a) - (d) (with, of course, $R_{0,\tau}$ in (d)), and also (for $R_{1,\tau}$ and $R_{2,\tau}$):

(e) $t \equiv R_i t_1 t_2 0^i$, $\quad t' \equiv t_1$ $\qquad$ (i = 1 or 2),

(f) $t \equiv R_i t_1 t_2 (S_i u)$, $\quad t' \equiv t_2(\Sigma(\Pi(R_i t_1 t_2))u)u$ $\qquad$ (i = 1 or 2).

The relations $t >_1 t'$ and $t \geq t'$ (t reduces to t') and the concepts of normal form (NF) and reduction sequence (red. seq.) are then defined as in 2.2.2.

θ, θ' denote reduction sequences.

The concepts of strict reduction sequence and standard reduction sequence (std red. seq.) are also defined as in 2.2.2.

$t \geq t'$ (std) $\quad \equiv_{def}$ $t \geq t'$ by a std red. seq. †

$t >_1 t'$(std) $\quad \equiv_{def}$ $t >_1 t'$ by a std red. seq. (of length 1).

t is normalizable $\equiv_{def}$ $t > t'$ for some t' in NF.

t is standard (std) normalizable $\equiv_{def}$ $t \geq t'$ (std) for some t' in NF.

6.4.2. Proposition. Every normal closed term of ground type has one of the forms:

$$0^0,\ 0^1,\ 0^2,\ S_0u,\ S_1u,\ S_2u$$

with u normal. In particular, a closed normal term of type 0 is a numeral.

Proof. This is proved simultaneously for all ground types by induction on the number of symbols in the term. So let t be a closed normal term of ground type. If its length is 1, then it must be 0^0, 0^1 or 0^2. If its length is > 1, then it must have the form S_iu, with u normal, or possibly $R_i r s u t_1 t_2 \dots t_n$ $(n \geq 0)$ with u normal. But in the latter case, by induction hypothesis, u has the form 0^j or S_jv, so $R_i rsu$ is not normal.

6.4.3. Uniqueness of normal form.

-----

* Note that in chapters I - III "CTM" is used for "closed terms", and "CT" for "Church's thesis".

† This was denoted by $t \geq' t'$ in chapter II.

Theorem. If $t \geq t'$ and $t \geq t''$ and t', t'' are in NF, then $t' \equiv t''$.
Proof. The proof of theorem 2.2.23 (using 2.2.20 - 2.2.22) extends to the present case.

6.4.4. Definition of (standard) computability.

For each type τ, the set $C_\tau \subset CT_\tau$ of computable (closed) terms of type τ is defined by induction on τ: i.e. first for $\tau = 0, 1$ and 2, and then proceeding from ρ and σ to $(\rho)\sigma$.

$t \in C_0 \equiv_{def} t \in CT_0$ and t is std normalizable.

C_1 is the least subset X of CT_1 such that $\forall t \in CT_1$:

(i) if $t \geq 0^1$ (std) then $t \in X$, and

(ii) if $t \geq S_1 u$ (std) with u normal and $(\forall s \in C_0)(us \in X)$, then $t \in X$.

C_2 is the least subset Y of CT_2 such that $\forall t \in CT_2$:

(i) if $t \geq 0^2$ (std) then $t \in Y$, and

(ii) if $t \geq S_2 u$ (std) with u normal and $(\forall s \in C_1)(us \in Y)$, then $t \in Y$.

For $t \in CT_{(\rho)\sigma}$: $t \in C_{(\rho)\sigma} \equiv_{def} t$ is std normalizable and $(\forall s \in C_\rho)(ts \in C_\sigma)$.

Notes.

(a) Computability is defined here only for closed terms, as opposed to the definition of Comp" in 2.2.5.

(b) The above concept of computability ("standard computability") corresponds to standard reductions (as with Comp" in 2.2.5). We could also define concepts of computability relative to strict reductions, or arbitrary reductions (as with Comp' and Comp resp. in 2.2.5).

(c) Statements about elements of C_τ for $\tau = 1$ or 2 can often be proved by induction corresponding to the inductive definition of C_τ. This will be called: "induction on C_τ" or "induction on $t \in C_\tau$".

6.4.5. Lemma.

(i) $s \geq t$ (std) and $t \in C_\tau \Rightarrow s \in C_\tau$.

(ii) Conversely: $s \geq t$ (std) and $s \in C_\tau \Rightarrow t \in C_\tau$.

(iii) $t \in C_\tau \Rightarrow t$ is std normalizable.

(iv) $t \in C_{(\sigma)\tau}$, $s \in C_\sigma \Rightarrow ts \in C_\tau$.

Proof. (i) Induction on τ, and within that, for $\tau = 1$ and 2, induction on $t \in C_\tau$.

(ii) Induction on τ, and for $\tau = 1$ and 2, induction on $s \in C_\tau$.

(iii) Immediate from the definition of C_τ (and by induction on $t \in C_\tau$ for $\tau = 1$ and 2).

(iv) From the definition of $C_{(\sigma)\tau}$.

Note. The argument for (ii) breaks down if we consider computability relative to strict or arbitrary (rather than standard) reductions.

6.4.6. Lemma. Every constant of $\underset{\sim}{T}_2$ of type τ is in C_τ.
Proof. For 0^0, S_0, R_0, Σ and Π, the proof is as in 2.2.6. Now consider the remaining constants of $\underset{\sim}{T}_2$.

(a) 0^1 and 0^2 are easily seen to be in C_1 and C_2 resp.

(b) S_1 and S_2. Consider S_1. S_1 is normal, so $S_1 \in C_{((0)1)1}$ iff:

(1) $\qquad (\underset{\sim}{\forall} u \in C_{(0)1})(S_1 u \in C_1)$.

So let $u \in C_{(0)1}$. Then u is std normalizable (6.4.5 (iii)).
Let u' be the NF of u. Then $u' \in C_{(0)1}$ (6.4.5 (ii)). So

$$(\underset{\sim}{\forall} s \in C_0)(u's \in C_1).$$

Hence $S_1 u \in C_1$ (by definition of C_1, since $S_1 u \geq S_1 u'$ (std)).
This proves (1).

The proof for S_2 is exactly parallel (replacing throughout subscripts and superscripts "0" and "1" by "1" and "2" resp.).

(c) $R_{1,\tau}$ and $R_{2,\tau}$. Consider $R_{1,\tau}$. $R_{1,\tau}$ is normal, so $R_{1,\tau} \in C_\rho$ (where ρ is the type of $R_{1,\tau}$) iff:

(2) $\qquad (\underset{\sim}{\forall} r \in C_\tau)(\underset{\sim}{\forall} s \in C_\sigma)(\underset{\sim}{\forall} t \in C_1)(R_1 rst \in C_\tau)$

where $\sigma \equiv ((0)\tau)((0)1)\tau$.
(2) is proved for fixed r and s, by induction on $t \in C_1$.
First note that r and s are std normalizable. Let r' and s' be their NF's resp. Then $r' \in C_\tau$, $s' \in C_\sigma$.

(i) Suppose $t \geq 0^1$ (std). Then
$$\begin{aligned} R_1 rst &\geq R_1 r's'0^1 && \text{(std)} \\ &>_1 r' && \text{(std)} \\ &\in C_\tau. \end{aligned}$$

So $R_1 rst \in C_\tau$ (6.4.5 (i)).

(ii) Suppose $t \geq S_1 u$ (std), with u normal and

(3) $\qquad (\underset{\sim}{\forall} v \in C_0)(uv \in C_1)$.

Then $R_1 rst \geq R_1 r's'(S_1 u)$ (std)

(4) $\qquad >_1 s'(\Sigma(\Pi(R_1 r's'))u)u$ (std).

Now we show that

(5) $\qquad \Sigma(\Pi(R_1 r's'))u \in C_{(0)\tau}$.

Let $v \in C_0$. (We must show $\Sigma(\Pi(R_1 r's'))uv \in C_\tau$.) v is std normalizable.
Let v' be the NF of v.

Then $v' \in C_0$, so $uv' \in C_1$ by (3), and

$$R_1rs(uv') \in C_\tau \text{ by \underline{induction hypothesis for} (2),}$$

and hence $R_1r's'(uv') \in C_\tau$.

Also

$$\begin{aligned} \Sigma(\Pi(R_1r's'))uv &\geq \Sigma(\Pi(R_1r's'))uv' \quad \text{(std)} \\ &\succ_1 R_1r's'(uv') \quad \text{(std)} \\ &\in C_\tau, \end{aligned}$$

so

$$\Sigma(\Pi(R_1r's'))uv \in C_\tau .$$

So (5) follows, by definition (since also $\Sigma(\Pi(R_1r's'))u$ is normal). Therefore the term shown in (4) is in C_τ (since also $s' \in C_\sigma$ and $u \in C_{(0)1}$), and hence so is R_1rst. This proves (2).

The argument for $R_{2,\tau}$ is again exactly parallel.

6.4.7. __Theorem.__ Every closed term of $\underset{\sim}{T}_2$ is computable, i.e.

$$C_\tau = CT_\tau \quad \text{for all } \tau .$$

__Proof.__ From 6.4.5 (iv) and 6.4.6.

6.4.8. __Corollary.__ Every closed term of $\underset{\sim}{T}_2$ is std normalizable.

__Proof.__ From 6.4.5 (iii) and 6.4.7.

6.4.9. __Note.__ Theorem 6.4.7 does not (apparently) yield immediately that __every__ term of $\underset{\sim}{T}_2$ is normalizable. However, this will follow (6.5.13) from theorem 6.5.11 below.

6.4.10. __Definition of__ $|t|_C$.

We define for each $t \in CT_i$ ($i = 1$ or 2) an ordinal $|t|_C$ by virtue of its __computability__ (by 6.4.7). The definition is by __induction on__ $t \in C_i$ ($i = 1$ and 2):

(i) If $t \geq 0^i$ (std) then $|t|_C = 0$, and

(ii) if $t \geq S_i u$ (std) with u normal, then $|t|_C = \sup\{|uv|_C + 1 : v \in C_{i-1}\}$.

(__Note.__ Since the only normal closed terms of type 0 are numerals (6.4.2), clause (ii) (for the case $i = 1$) can be changed to:

(ii') if $t \geq S_1 u$ (std) with u normal, then $|t|_C = \sup\{|u\bar{n}|_C + 1 : n \in N\}$,

where $\bar{n} \equiv_{\text{def}} \underbrace{S_0 \ldots S_0}_{n \text{ times}} 0^0$.)

§ 5. Strong computability.

In this section we prove the strong computability of all closed terms of $\underset{\sim}{T}_2$. However, this will not be used elsewhere, except to show that all terms of $\underset{\sim}{T}_2$ (not necessarily closed) are (standard) normalizable (from 6.5.13; cf. 6.4.9).

6.5.1. Definitions.
$t \underline{>} t'$ (strongly) $\equiv_{def}$ every reduction sequence from t contains t'.
t is strongly normalizable $\equiv_{def}$ every reduction sequence from t is finite.

Now we define, for each τ, the set $SC_\tau \subset CT_\tau$ of strongly computable closed terms of type τ. The definition is by induction on τ:

$t \in SC_0 \equiv_{def} t \in CT_0$ and t is strongly normalizable.

SC_1 is the least subset X of CT_1 such that $\underset{\sim}{\forall} t \in CT_1$:

(i) if $t \underline{>} 0^1$ (strongly) then $t \in X$, and

(ii) if for some normal u, $t \underline{>} S_1 u$ (strongly) and $(\underset{\sim}{\forall} s \in SC_0)(us \in X)$, then $t \in X$.

SC_2 is the least subset Y of CT_2 such that $\underset{\sim}{\forall} t \in CT_2$:

(i) if $t \underline{>} 0^2$ (strongly) then $t \in Y$, and

(ii) if for some normal u, $t \underline{>} S_2 u$ (strongly) and $(\underset{\sim}{\forall} s \in SC_1)(us \in Y)$, then $t \in Y$.

For $t \in CT_{(\rho)\sigma}$: $t \in SC_{(\rho)\sigma} \equiv_{def} (\underset{\sim}{\forall} s \in SC_\rho)(ts \in SC_\sigma)$.

Finally, $SC \equiv_{def} \bigcup \{SC_\tau : \tau \in \underset{\approx}{T}_2\}$.

Note. Statements about elements of SC_τ for $\tau = 1$ and 2 can often be proved by induction corresponding to the inductive definition of SC_τ. This will be called: "induction on SC_τ" or "induction on $t \in SC_\tau$".

6.5.2. Lemma. $s \underline{>} t$ and $s \in SC_\tau \Rightarrow t \in SC_\tau$.
Proof. Induction on τ, and for $\tau = 1$ and 2, induction on $s \in SC_\tau$.

6.5.3. Lemma. $t \in SC_{(\sigma)\tau}$ and $s \in SC_\sigma \Rightarrow ts \in SC_\tau$.
Proof. Immediate from definition of $SC_{(\sigma)\tau}$.

6.5.4. Lemma. $t \in SC_i \Leftrightarrow$ every reduction sequence from t contains a term in SC_i ($i = 0, 1$ or 2).
Proof. For $i = 0$: immediate.
For $i = 1$ or 2: $\Rightarrow$ follows immediately by induction on $t \in SC_i$.
For $\Leftarrow$: Suppose every reduction sequence from t contains a term in SC_i.
Let s be a term in SC_i contained in (say) the std reduction sequence

from t. We prove $t \in SC_i$ by <u>induction on</u> $s \in SC_i$:
(i) If $s \geq 0^i$ (strongly): let θ be any reduction sequence from t.
θ contains a term in SC_i, say s'.
s' is strongly normal, so θ ends in a <u>normal</u> term.
By <u>uniqueness of</u> NF of t (6.4.3), this term must be 0^i.
Thus $t \geq 0^i$ (strongly), and so $t \in SC_i$.
(ii) If for some normal u, $s \geq S_i u$ (strongly), and $(\forall v \in SC_{i-1})(uv \in SC_i)$:
then by the same argument, $t \geq S_i u$ (strongly). So $t \in SC_i$.

6.5.5. <u>Note</u>. Every type $\tau \in \underset{\sim}{T}_2$ can be put (uniquely) in the form:

$$(\tau_1)(\tau_2)\ldots(\tau_n)i$$

where $i = 0, 1$ or 2 and $n \geq 0$. (Proof by induction on τ: cf. 1.6.2, remark (iii).)

Now suppose $\tau \equiv (\tau_1)\ldots(\tau_n)i$ and $t \in Tm_\tau$. Then

$$t \in SC_\tau \Leftrightarrow (\forall t_1 \in SC_{\tau_1})\ldots(\forall t_n \in SC_{\tau_n})(tt_1\ldots t_n \in SC_i).$$

If $t \in (\tau_1)\ldots(\tau_n)i$ and $t_j \in \tau_j$ $(1 \leq j \leq n)$, we will use the notation:

	$\underline{\underline{\tau}}$	for $\tau_1\ldots\tau_n$,
	$(\underline{\underline{\tau}})i$	for $(\tau_1)\ldots(\tau_n)i$,
	$\underline{\underline{t}}$	for $t_1\ldots t_n$,
	$t\underline{\underline{t}}$	for $tt_1\ldots t_n$,
	$\underline{\underline{t}} \in \underline{\underline{\tau}}$	for $(t_1 \in \tau_1) \& \ldots \& (t_n \in \tau_n)$,
and	$\underline{\underline{t}} \in SC_{\underline{\underline{\tau}}}$	for $(t_1 \in SC_{\tau_1}) \& \ldots \& (t_n \in SC_{\tau_n})$.

So by lemma 6.5.4, if $t \in \tau \equiv (\underline{\underline{\tau}})i$, then

(1) $$t \in SC_\tau \Leftrightarrow (\forall \underline{\underline{t}} \in SC_{\underline{\underline{\tau}}})(\text{every reduction sequence from } t\underline{\underline{t}} \text{ contains some term in } SC_i).$$

Next, if $t \in (\underline{\underline{\tau}})i$ and $\underline{\underline{t}} \in \underline{\underline{\tau}}$, then <u>a reduction sequence from</u> $t\underline{\underline{t}}$ <u>not affecting</u> t means a reduction sequence of the form:

$$tt_1\ldots t_n >_1 tt'_1\ldots t'_n >_1 \ldots >_1 tt_1^{(k)}\ldots t_n^{(k)} >_1 \ldots$$

(i.e. for each k and $1 \leq j \leq n$, $t_j^{(k)} >_1 t_j^{(k+1)}$ or $t_j^{(k)} \equiv t_j^{(k+1)}$), also written

$$t\underline{\underline{t}} >_1 t\underline{\underline{t}}' >_1 \ldots >_1 t\underline{\underline{t}}^{(k)} >_1 \ldots,$$

with $\underline{\underline{t}} \geq \underline{\underline{t}}^{(k)}$.

Note that

(2) if $t_1, \ldots, t_n$ are all <u>strongly normalizable</u>, then a reduction sequence from $t\underline{\underline{t}}$ not affecting t must be finite.

6.5.6. Definition. A "zero of type τ", 0^τ, is defined for each τ (cf. 2.2.7 (ii)) by induction on τ: 0^τ is the given constant for $\tau = 0$, 1 or 2; and $0^{(\rho)\sigma} \equiv \Pi_{\rho,\sigma} 0^\rho$.

Note that 0^τ is normal.

6.5.7. Lemma. (a) $t \in SC_\tau \Rightarrow t$ is strongly normalizable.
(b) $0^\tau \in SC_\tau$.

Proof. (a) and (b) are proved simultaneously by induction on τ.

If $\tau = 0$, 1 or 2: (a) is immediate for $\tau = 0$, and proved by induction on $t \in SC_\tau$ for $\tau = 1$ or 2.

(b) is immediate.

If $\tau = (\tau_1)\dots(\tau_n)i$, $n \geq 1$:

(a) Suppose $t \in SC_\tau$, and let

$$\theta: \ t \equiv t_0 >_1 t_1 >_1 t_2 >_1 \dots$$

be any reduction sequence from t. Then

$$\theta' : \ t_0 0^{\tau_1} >_1 t_1 0^{\tau_1} >_1 t_2 0^{\tau_1} >_1 \dots$$

is a reduction sequence from $t0^{\tau_1}$, which is $SC_{(\tau_2)\dots(\tau_n)i}$, by induction hypothesis for (b) and 6.5.3.

So by induction hypothesis for (a), θ' is finite; hence so is θ.

(b) Suppose $\underline{t} \in SC_{\underline{\tau}}$. Then by induction hypothesis for (a), $t_1,\dots,t_n$ are strongly normalizable. So (by (2) of 6.5.5) any reduction sequence from $0^\tau \underline{t}$ not affecting 0^τ must be finite. So any reduction sequence from $0^\tau \underline{t}$ has the form:

$$\Pi 0^{\tau_1} t_1 t_2 \dots t_n >_1 \dots >_1 \Pi 0^{\tau_1} t_1^{(k)} t_2^{(k)} \dots t_n^{(k)} >_1 0^{\tau_1} t_2^{(k)} \dots t_n^{(k)} >_1 \dots$$

for some k; and this $(k+1)^{st}$ term is SC_i by induction hypothesis for (b) and 6.5.3. So (by (1) of 6.5.5) $0^\tau \in SC_\tau$.

6.5.8. Corollary. If $t \in (\underline{\tau})i$ and $\underline{t} \in SC_{\underline{\tau}}$ then any reduction sequence from $t\underline{t}$ not affecting t is finite.

Proof. From 6.5.7 (a) and (2) of 6.5.5.

6.5.9. Lemma. For all t in $CT_{(0)1}$ or $CT_{(1)2}$, any reduction sequence from $S_i t$ ($i = 1$ or 2) has the form:

$$S_i t >_1 S_i t' >_1 S_i t'' >_1 \dots$$

where $t >_1 t' >_1 t'' >_1 \dots$

Proof. By inspection of the contraction rules.

6.5.10. Lemma. The constants of $\underset{\sim}{T}_2$ are SC.

Proof. Again, for 0^0, S_0, R_0, Π and Σ, the proof is as in 2.2.19. Now consider the remaining constants of $\underset{\sim}{T}_2$.

(a) 0^1 <u>and</u> 0^2 are easily seen to be SC.

(b) S_1 <u>and</u> S_2. Consider S_1. S_1 is normal, so $S_1 \in SC_{((0)1)}$ iff

(1) $\qquad (\underset{\sim}{\forall} t \in SC_{(0)1})(S_1 t \in SC_1)$.

So suppose $t \in SC_{(0)1}$. Every reduction sequence from $S_1 t$ has the form:

$$S_1 t >_1 S_1 t' >_1 S_1 t'' >_1 \ldots$$

where $t >_1 t' >_1 t'' >_1 \ldots$, by 6.5.9.
Also t is SN, by 6.5.7 (a). Let u be its NF.
Then $S_1 t \geq S_1 u$ (strongly).
Further, $u \in SC_{(0)1}$, by 6.5.2, so $(\underset{\sim}{\forall} v \in SC_0)(uv \in SC_1)$.
Hence $S_1 t \in SC_1$ (by definition of SC_1), proving (1).

The proof for S_2 is parallel.

(c) $R_{1,\tau}$ <u>and</u> $R_{2,\tau}$. Consider $R_{1,\tau}$. $R_{1,\tau}$ is normal, so $R_{1,\tau}$ is SC iff

(2) $\qquad (\underset{\sim}{\forall} r \in SC_\tau)(\underset{\sim}{\forall} s \in SC_\sigma)(\underset{\sim}{\forall} t \in SC_1)(R_1 rst \in SC_\tau)$

(where $\sigma \equiv ((0)\tau)((0)1)\tau)$.

(2) is proved for fixed r and s, by <u>induction on</u> $t \in SC_1$:

(i) Suppose $t \geq 0^1$ (strongly).
Let $\underline{\underline{t}} \in SC_{\underline{\underline{\tau}}}$, where $\tau \equiv (\underline{\underline{\tau}})i$.
Then by 6.5.8, any reduction sequence from $(R_1 rst)\underline{\underline{t}}$, not affecting $R_1 rst$, is finite. Also (by 6.5.7 (a)) any reduction sequence from r or s is finite. So any reduction sequence from $(R_1 rst)\underline{\underline{t}}$ has the form:

$$(R_1 rst)\underline{\underline{t}} >_1 \ldots >_1 (R_1 r's'0^1)\underline{\underline{t}}' >_1 r'\underline{\underline{t}}' >_1 \ldots$$

(where $r \geq r'$, $s \geq s'$, $\underline{\underline{t}} \geq \underline{\underline{t}}'$), with $r'\underline{\underline{t}}' \in SC_i$ (by 6.5.2 and 6.5.3).

So $R_1 rst \in SC_\tau$, by (1) of 6.5.5. Thus (2) is proved in this case.

(ii) Suppose $t \geq S_1 u$ (strongly) where u is normal and

(3) $\qquad (\underset{\sim}{\forall} v \in SC_0)(uv \in SC_1)$.

Let $\underline{\underline{t}} \in SC_{\underline{\underline{\tau}}}$. Then (by the same argument) any reduction sequence from $(R_1 rst)\underline{\underline{t}}$ has the form:

(4) $\qquad (R_1 rst)\underline{\underline{t}} >_1 \ldots >_1 (R_1 r's'(S_1 u))\underline{\underline{t}}) >_1 s'(\Sigma(\Pi(R_1 r's'))u)u >_1 \ldots$

(where $r \geq r'$, $s \geq s'$, $\underline{\underline{t}} \geq \underline{\underline{t}}'$).

Now we show that

(5) $\qquad \Sigma(\Pi(R_1 r's'))u \in SC_{(0)\tau}$.

Let $v \in SC_0$. (We must show $\Sigma(\Pi(R_1 r's'))uv \in SC_\tau$.)

$\underset{\sim}{\forall}\underline{\underline{s}} \in SC_\tau$, any reduction sequence from $\Sigma(\Pi(R_1 r's'))uv\underline{\underline{s}}$ has the form (by 6.5.8 and 6.5.7 (a) again):

$$(6)\quad \begin{array}{|l} \Sigma(\Pi(R_1 r's'))uv\underline{\underline{s}} >_1 \dots >_1 \Sigma(\Pi(R_1 r''s''))uv'\underline{\underline{s}}' \\ \qquad >_1 \Pi(R_1 r''s'')v'(uv')\underline{\underline{s}}' >_1 \dots \\ \qquad \dots >_1 \Pi(R_1 r''s''')\, v''(uv''')\underline{\underline{s}}'' \\ \qquad >_1 R_1 r'''s'''(uv''')s'' >_1 \dots \end{array}$$

(where $r' \geq r'' \geq r'''$, $s' \geq s'' \geq s'''$, $v \geq v'$, $v' \geq v''$, $v' \geq v'''$, $\underline{\underline{s}} \geq \underline{\underline{s}}' \geq \underline{\underline{s}}''$).

Now $uv \in SC_1$ by (3), and $R_1 rs(uv) \in SC_\tau$ by <u>induction hypothesis for</u> (<u>2</u>). Therefore so is $R_1 r'''s'''(uv''')$, and so is the last term shown in (6).

Hence (5) is proved, by (1) of 6.5.5.

So the last term shown in (4) is SC_i (since also s' and u are SC). So (again by (1) of 6.5.5) $R_1 rst \in SC_\tau$, proving (2) again in this case.

The proof for $R_{2,\tau}$ is parallel.

6.5.11. <u>Theorem.</u> All closed terms of $\underset{\sim}{T}_2$ are SC.

<u>Proof.</u> From 6.5.3 and 6.5.10.

6.5.12. <u>Corollary.</u> All closed terms of $\underset{\sim}{T}_2$ are strongly normalizable.

6.5.13. <u>Corollary.</u> <u>All</u> terms of $\underset{\sim}{T}_2$ are strongly normalizable (and hence std normalizable).

<u>Proof.</u> Let $t \equiv t[x_1^{\tau_1}, \dots, x_k^{\tau_k}]$ be a term of $\underset{\sim}{T}_2$, with (only) free variables $x_1^{\tau_1}, \dots, x_k^{\tau_k}$. Consider any reduction sequence from t:

$$\theta:\ t \equiv t_0 >_1 t_1 >_1 t_2 >_1 \dots$$

where $t_i \equiv t_i[x_1^{\tau_1}, \dots, x_k^{\tau_k}]$. Let $t_i^* = t_i[0^{\tau_1}, \dots, 0^{\tau_k}]$. Then

$$t_0^* >_1 t_1^* >_1 t_2^* >_1 \dots$$

is a reduction sequence from the closed term t_0^*, and so is finite, by 6.5.12. Therefore so is θ.

§ 6. Models of $\underset{\sim}{T}_2$; modelling $\underset{\sim}{T}_2$ in $\underset{\sim}{ID}_2(\mathcal{O})$.

6.6.1. Definition. A model of $\underset{\sim}{T}_2$ (cf. 2.4.1) is a structure

$$M = \langle\langle M_\tau\rangle_{\tau\in \underset{\sim}{T}_2}, \langle Ap_M^{\sigma,\tau}\rangle_{\sigma,\tau\in T_2}, Val_M, =_M\rangle$$

which satisfies $\underset{\sim}{T}_2$ when the variables of type τ are interpreted as ranging over the domain M_τ, the function $Ap_M^{\sigma,\tau} : M_{(\sigma)\tau} \times M_\sigma \to M_\tau$ interprets application between terms, $=_M$ interprets $=$, and for each constant C of type τ, $Val_M(C) \in M_\tau$ is its interpretation. If $=_M$ is the identity relation on each M_τ, then M is called normal (as in 2.4.1).

Val_M can be extended to CT by: $Val(st) = Ap(Val(s), Val(t))$.

Notation. We will often use the same notation in discussing M as in the (meta-)language of $\underset{\sim}{T}_2$ (6.3.3), e.g.

$a, b, \ldots, m, n, \ldots$ range over elements of M_0
α, β, or α^1, β^1 range over elements of M_1
α^2, β^2 range over elements of M_2.

For $f \in M_{(\sigma)\tau}$ and $x \in M_\sigma$, we write fx for $Ap_M^{\sigma,\tau}(f,x)$.
For $t \in CT$, we write t_M or just t for $Val_M(t)$.
Type superscripts and subscripts are often dropped.
Also, the subscript M is often dropped from Ap_M, Val_M and $=_M$.

6.6.2. Well-founded models.

M is a wf (well-founded) model of $\underset{\sim}{T}_2$ if it is a model of $\underset{\sim}{T}_2$ and satisfies the following (second-order) conditions:

(1) M_0 is (isomorphic to) N, the set of natural numbers.

(2) M_1 is the least subset X of M_1 compatible with $=_M$* and satisfying:

(i) $0^1 \in X$, and

(ii) if $f \in M_{(0)1}$ and $(\forall n \in M_0)(fn \in X)$ then $S_1 f \in X$.

(3) M_2 is the least subset Y of M_2 compatible with $=_M$ and satisfying:

(i) $0^2 \in Y$, and

(ii) if $f \in M_{(1)2}$ and $(\forall \alpha \in M_1)(f\alpha \in Y)$ then $S_2 f \in Y$.

Note. The above conditions just state that M satisfies the second-order versions of the axioms FI, TI_1 and TI_2 (6.3.5 (d)).

From now on we only consider wf models of $\underset{\sim}{T}_2$.

6.6.3. If M is a wf model of $\underset{\sim}{T}_2$ then statements about elements of M_τ ($\tau = 1$ or 2) can be proved by induction corresponding to the above inductive conditions; this will be called "induction on M_τ". Similarly functions

* (i.e. $\alpha \in X$ and $\beta =_M \alpha \Rightarrow \beta \in X$)

and relations can be defined on M_τ by <u>induction on</u> M_τ.

For example, we associate canonically with each $\alpha^\tau \in M_\tau$ ($\tau = 1$ or 2) an ordinal $|\alpha^\tau|_M$, defined by <u>induction on</u> M_τ (and using remark 6.3.6 (c)):

$$|0^1|_M = 0, \quad |S_1 f|_M = \sup\{|fn|_M + 1 \ : \ n \in M_0\};$$
$$|0^2|_M = 0, \quad |S_2 f|_M = \sup\{|f\alpha|_M + 1 \ : \ \alpha \in M_1\}.$$

We write $|t|_M$ for $|\mathrm{Val}(t)|_M$.

6.6.4. <u>Four examples of well-founded models of</u> $\underset{\sim}{T}_2$: $\mathcal{S}_2$, HRO_2, HEO_2 <u>and</u> CTNF_2.

Four examples of wf models are discussed in this section.

Many definitions and proofs concerning wf models M of $\underset{\sim}{T}_2$ will proceed by <u>induction on</u> τ and, within this, by <u>induction on</u> M_τ for $\tau = 1$ and 2.

(a) $\mathcal{S}_2$, <u>the full set-theoretical model</u>.

This extends the model of $\underset{\sim}{E}$-$\underset{\sim}{HA}^\omega$ in 2.4.6. The domains M_τ are defined (in set theory) by:

$M_0 \equiv N$.

M_1 is the least set X such that:

(i) $0^1 \in X$, and

(ii) if $f : M_0 \to X$ then $S_1 f \in X$.

Here 0^1 is (say) the empty set, f ranges over arbitrary functions from M_0 to X, and $S_1 f$ can be taken as f itself. Similar remarks apply to the definition of M_2:

M_2 is the least set Y such that:

(i) $0^2 \in Y$, and

(ii) if $f : M_1 \to Y$ then $S_1 f \in Y$.

$M_{(\sigma)\tau}$ is the set of all functions from M_σ to M_τ.

Ap and Val are defined in an obvious way.

(b) HRO_2, <u>the hereditarily recursive operations of finite type on</u> $\mathcal{O}_1$ <u>and</u> $\mathcal{O}_2$. <u>Interpreting</u> $\underset{\sim}{T}_2$ <u>in</u> $\underset{\sim}{ID}_2(\mathcal{O})$.

HRO_2 extends the model HRO of $\underset{\sim}{I}$-$\underset{\sim}{HA}^\omega$ (2.4.8). Its construction, given below, can be formalized in $\underset{\sim}{ID}_2(\mathcal{O})$. (The sets $\mathcal{O}_1$ and $\mathcal{O}_2$, and the theory $\underset{\sim}{ID}_2(\mathcal{O})$, were defined in 6.2.2.)

Sets $V_\tau \subset N$ are defined by induction on τ:

$V_0 \equiv N$.

$V_i \equiv \mathcal{O}_i$ for $i = 1$ or 2.

$V_{(\sigma)\tau} \equiv \{e: \forall x \in V_\sigma \exists y(Texy \& Uy \in V_\tau)\}$.

We define the domains M_τ of our model simply by: $M_\tau \equiv V_\tau$. (In 2.4.8, M_τ is defined as $V_\tau \times \{\tau\}$, but this is an unimportant difference.)

For $x \in V_{(\sigma)\tau}$, $y \in V_\sigma$: $Ap^{\sigma,\tau}(x,y) \equiv_{def} \{x\}(y)$.

Equality is interpreted as identity in each V_τ.

Val is defined as follows (writing now $[C]$ for $Val(C)$):

$[0^0] \equiv [0^1] \equiv [0^2] \equiv 0$.

$[\Sigma_{\sigma,\tau}]$ and $[\Pi_{\rho,\sigma,\tau}]$ are the numbers $[\Sigma]$ and $[\Pi]$ resp. defined in 2.4.8.

$[S_i] \equiv \Lambda x(x+1)$, $\Lambda x(3.5^x)$ and $\Lambda x(3^2.5^x)$ resp. for $i = 0, 1, 2$ resp.

$[R_{0,\tau}]$ is the number $[R]$ defined in 2.4.8.

$[R_{1,\tau}] \equiv [R_1]$ is a numeral, found by use of the <u>recursion theorem</u>, such that:

(1) $\begin{cases} \underset{\sim}{HA} \vdash \{[R_1]\}(x,y,0) \simeq x, \\ \underset{\sim}{HA} \vdash \{[R_1]\}(x,y,3.5^z) \simeq \{y\}(\{[\Sigma]\}(\{[\Pi]\}(\{[R_1]\}(x,y)))z,z). \end{cases}$

Similarly, $[R_{2,\tau}] \equiv [R_2]$ is defined to satisfy (1) with $3^2.5^z$ replacing 3.5^z. Then

$$\underset{\sim}{ID}_2(\mathcal{O}) \vdash V_\tau(x) \& V_\sigma(y) \& V_1(z) \to V_\tau(\{[R_1]\}(x,y,z))$$

(where $\sigma \equiv ((0)\tau)((0)1)\tau)$, by <u>induction on</u> $z \in V_1$ $(\equiv \mathcal{O}_1)$.

Hence $\underset{\sim}{ID}_2(\mathcal{O}) \vdash V_\rho([R_1])$

where ρ is the type of $R_{1,\tau}$.

Similarly, $[R_2]$ is (provably in $\underset{\sim}{ID}_2(\mathcal{O})$) in the appropriate V_ρ (by induction on $V_2 \equiv \mathcal{O}_2$).

So for all constants C of type τ,

$$\underset{\sim}{ID}_2(\mathcal{O}) \vdash V_\tau([C]).$$

Hence, more generally, for all $t \in CT_\tau$, writing $[t]$ for (the numeral of) $Val(t)$:

(2) $\underset{\sim}{ID}_2(\mathcal{O}) \vdash V_\tau([t])$

(cf. 2.4.8), since (2) holds for the constants, and further,

$$\underset{\sim}{ID}_2(\mathcal{O}) \vdash V_{(\sigma)\tau}(x) \& V_\sigma(y) \to V_\tau(\{x\}(y)),$$

and for any numbers a, b, c the formula

$$\exists y[T_1(\bar{a}, \bar{b}, y) \& U(y) = \bar{c}]$$

if true, is provable, even in $\underset{\sim}{HA}$ (1.5.10).

For $\tau = 1$, (2) becomes:

(3) $\forall t \in CT_1$, $ID_2(\mathcal{O}) \vdash \mathcal{O}_1([t])$.

This is used in proving theorem 6.6.8.

(c) HEO_2, the hereditarily effective extensional operations over $\mathcal{O}_1$ and $\mathcal{O}_2^!$.

We describe briefly a wf model HEO_2 extending HEO (2.4.11).
Define a binary relation I_τ on N for all τ as follows.

$I_0(x,y) \equiv_{def} x=y$.

I_1 is inductively defined by:

(i) $I_1(0,0)$,

(ii) $\forall n \exists y,y'(Teny \& Te'ny' \& I_1(Uy,Uy')) \rightarrow I_1(3.5^e, 3.5^{e'})$.

I_2 is inductively defined by:

(i) $I_2(0,0)$,

(ii) $\forall n,n'[I_1nn' \rightarrow \exists y,y'(Teny \& Te'n'y' \& I_2(Uy,Uy'))] \rightarrow I_2(3^2.5^e, 3^2.5^{e'})$.

$I_{(\sigma)\tau}(e,e') \equiv_{def} \forall x,x'[I_\sigma xx' \rightarrow \exists y,y'(Texy \& Te'x'y' \& I_\tau(Uy,Uy'))]$.

Finally, $W_\tau \equiv_{def} \{x: I_\tau xx\}$.

(This, in fact, extends the definition of I_τ and W_τ in 2.4.11.)

Now $M = HEO_2$ is defined with $M_\tau \equiv W_\tau$, equality interpreted as I_τ on each M_τ, and Val(C) defined as for HRO_2 for each constant C.

This construction can be formalized in a theory $ID_2(I)$, with the defining formulas for $I \equiv (I_1,I_2)$ in $\mathcal{C}$.

Now write $\mathcal{O}_1^! \equiv W_1$ and $\mathcal{O}_2^! \equiv W_2$.

Then $\mathcal{O}_1^! = \mathcal{O}_1$, but $\mathcal{O}_2^! \subsetneq \mathcal{O}_2$.

(d) $CTNF_2$, the term model of T_2.

This extends the model CTNF (= "closed terms in normal form") of $I\text{-}HA^\omega$ in 2.5.1.

For each $t \in CT$, let $\hat{t}$ be its (unique) normal form.

For each τ, M_τ is the set of normal closed terms of type τ.

$Val(C) \equiv C$ for each constant C.

$Ap(s,t) \equiv \widehat{st}$.

So for each $t \in CT$, $Val(t) \equiv \hat{t}$.

Proposition. $CTNF_2$ is wf.

Proof. By computability of closed terms of type 1 and 2. More precisely (with $M = CTNF_2$):

Let $X \subset M_1$ be such that

(i) $0^1 \in X$, and

(ii) $\forall u \in M_{(0)1}[(\forall n \in M_0)(\widehat{un} \in X) \Rightarrow S_1 u \in X]$.

We show $(\forall t \in CT_1)(\hat{t} \in X)$ by induction on $t \in C_1$.

Similarly for M_2.

Proposition. For $M = CTNF_2$, $t \in CT_1$ or CT_2: $|t|_M = |t|_C$.

Proof. By induction on M_1 and M_2.

Remark. Theorem 2.5.5 can be extended to wf models of $\underset{\sim}{T}_2$: i.e., there is a version of HRO_2 in which $CTNF_2$ can be embedded. The proof extends that of theorem 2.5.5.

This also gives an alternative proof of uniqueness of normal form for the closed terms of $\underset{\sim}{T}_2$ (as in 2.5.6).

6.6.5. Extensionality: some general remarks.

The concepts of homomorphism, embedding and submodel (for wf models of $\underset{\sim}{T}_2$) are defined as in 2.4.3.

The relation $\approx$ of extensional equivalence in wf models M is defined as follows (extending the definition in 2.4.4). It is defined on each M_τ by induction on τ, and, for $\tau = 1$ or 2, by induction on M_τ. (Below, $=$ means $=_M$, and $\approx$ is taken to be compatible with $=_M$*.)

$\tau \equiv 0$: $m \approx n \equiv m = n$.

$\tau \equiv 1$: $\alpha \approx 0^1 \Leftrightarrow \alpha = 0^1$,

$\alpha \approx S_1 f \Leftrightarrow \exists g \in M_{(0)1}[\alpha = S_1 g$ and $(\forall n \in M_0)(fn \approx gn)]$.

$\tau \equiv 2$: $\alpha^2 \approx 0^2 \Leftrightarrow \alpha^2 = 0^2$,

$\alpha^2 \approx S_2 f \Leftrightarrow \exists g \in M_{(1)2}[\alpha^2 = S_2 g$ and $(\forall \beta \in M_1)(f\beta \approx g\beta)]$.

$\tau \equiv (\rho)\sigma$: $f \approx g \equiv (\forall x \in M_\rho)(fx \approx gx)$.

More definitions.

(a) M is pre-extensional if for all σ, τ:

$$(\forall f \in M_{(\sigma)\tau})(\forall x, y \in M_\sigma)(x \approx y \Rightarrow fx \approx fy).$$

(b) M is extensional if for all τ:

$$\forall x, y \in M_\tau (x \approx y \Rightarrow x =_M y).$$

Notes. (1) This is the same as the definition of "extensional model" in 2.4.1. (2) We always have: $x =_M y \Rightarrow x \approx y$, i.e. $\approx$ is a congruence relation w.r.t. $=_M$.

Proposition. (a) M is pre-extensional iff $\approx$ is a congruence relation w.r.t. Ap.

(b) If M is pre-extensional, then M becomes extensional when $=_M$ is re-defined as $\approx$.

(c) If M is extensional, then M can be embedded in $\mathscr{I}_2$.

Proof. In each case, by induction on τ, and for $\tau = 1$ or 2, induction on M_τ.

---

* (i.e. $x \approx y$, $x =_M x'$, $y =_M y' \Rightarrow x' \approx y'$)

Note that $\mathscr{I}_2$ and HEO_2 are extensional, $CTNF_2$ is pre-extensional (corollary below), and HRO_2 is not pre-extensional.

Theorem. If M is a wf model of $\underset{\sim}{T}_2$, then it has a pre-extensional submodel M^E.

Proof. As for 2.4.5. We just mention here the definition of I'_τ for $\tau = 1$ and 2:

$I'_1(\alpha,\beta) \equiv_{def} \alpha \approx \beta$.

I'_2 is the least subset Y of $M_2 \times M_2$ (compatible with $=_M$) such that

(i) $(0^2, 0^2) \in Y$, and

(ii) $\underset{\sim}{\forall}\alpha,\beta[I'_1(\alpha,\beta) \Rightarrow (f\alpha, g\beta) \in Y] \Rightarrow (S_2 f, S_2 g) \in Y$.

The definition of the domains M^E_τ, and the proof, then proceed as in 2.4.5.

Note. M^E is pre-extensional if equality is interpreted in it as the restriction of $=_M$; it is extensional if equality is re-interpreted as I'_τ on each M_τ.

Corollary. $CTNF_2$ is pre-extensional.

Proof. For $M = CTNF_2$, its pre-extensional submodel M^E must be the whole of M, since it includes the constants and is closed under application.

6.6.6. Some distinctions between wf models of $\underset{\sim}{T}_1$ and $\underset{\sim}{T}_2$.

In looking for significant distinctions between theories of non-iterated and iterated inductive definitions, say $\underset{\sim}{ID}_\nu(A)$ for $\nu = 1$ and 2, we may find it useful to compare properties of wf models of the corresponding $\underset{\sim}{T}_\nu$. For example, we notice the following differences between trees of type 1 and 2 in wf models M of $\underset{\sim}{T}_2$:

(a) Relation between $|t|_C$ and $|t|_M$ for t in CT_1 and CT_2. *

For $t \in CT_1$, $|t|_M$ is an invariant of $\underset{\sim}{T}_2$ for any wf model M; in fact

(1) $\qquad |t|_M = |t|_C$.

(Proof by induction on $t \in C_1$.)

However, for $t \in CT_2$, $|t|_M$ varies with M, and

$$|t|_M \geq |t|_C$$

(by induction on $t \in C_2$).

-----

* $|t|_C$ was defined in 6.4.10, and $|t|_M$ in 6.6.3.

In fact, the ordinal $\sup\{|t|_M : t \in CT_2\}$ is uncountable for $M = \mathcal{J}_2$, countable but non-recursive for $M = HRO_2$ and HEO_2, and recursive for $M = CTNF_2$.

(b) Consider the pre-extensional submodel M^E of M (6.6.5). The domain M^E_τ is the whole of M_τ for $\tau = 0$ and 1, but (in general) a proper subset of M_τ for $\tau = 2$.

Remark. We can also make a distinction between $\underset{\sim}{ID}_2(A)$ and $\underset{\sim}{ID}_3(A)$!

Compare the models HRO^E_2 (the pre-extensional submodel of HRO_2) and HEO_2. They are, of course, different since (even) HRO^E and HEO are different (2.4.12). However, their three ground types are the same, for: writing $\mathcal{O}^E_i$ and $\mathcal{O}^!_i$ resp. for the domains of HRO^E_2 and HEO_2 of type i $(i = 1, 2)$, we see that

$$\mathcal{O}^E_1 = \mathcal{O}^!_1 = \mathcal{O}_1, \quad \text{and} \quad \mathcal{O}^E_2 = \mathcal{O}^!_2 \subsetneq \mathcal{O}_2.$$

However, if we define, by analogy, a theory $\underset{\sim}{T}_3$, with models HRO_3, HRO^E_3 and HEO_3, and domains $\mathcal{O}_3$, $\mathcal{O}^E_3$ and $\mathcal{O}^!_3$ of type 3 respectively, then

$$\mathcal{O}^E_3 \neq \mathcal{O}^!_3,$$

since $\mathcal{O}^E_3 \subsetneq \mathcal{O}_3$, but $\mathcal{O}^!_3$ is incomparable with $\mathcal{O}_3$. (Cf. the incomparability of W_4 and $W^!_4$, 2.4.12.)

6.6.7. Definition of the ordinal $|\underset{\sim}{T}_2|$.

We saw in 6.6.6 (a) that $\forall t \in CT_1$,

$$|t|_M = |t|_C \quad \text{for any wf } M.$$

Let us call this ordinal just $|t|$.

Now define the "ordinal of $\underset{\sim}{T}_2$" as

$$|\underset{\sim}{T}_2| \equiv_{def} \sup\{|t| : t \in CT_1\}.$$

6.6.8. Theorem. $|\underset{\sim}{T}_2| \leq |\underset{\sim}{ID}_2|$.

Proof. Two proofs are given here.

(a) Using the formalization of the model HRO_2 in $\underset{\sim}{ID}_2(\mathcal{O})$:
From (3) of 6.6.4 (b), we obtain:

$$|\underset{\sim}{T}_2| \leq |\underset{\sim}{ID}_2(\mathcal{O})| \leq |\underset{\sim}{ID}_2|.$$

(b) Using the formalization of computability in $\underset{\sim}{ID}_2(C)$:
The definition of the computability predicates C_τ (in 6.4.4) and proof of computability of any given closed term of $\underset{\sim}{T}_2$ (theorem 6.4.7) can be formalized in a theory $\underset{\sim}{ID}_2(C)$, with the defining formulas for $C \equiv (C_1, C_2)$ in $\mathcal{C}$, and C_τ defined arithmetically from C_1 and C_2 for each τ. (In fact, this also holds for strong computability.) So for all $t \in CT_\tau$,

$$\underset{\sim}{ID}_2(C) \vdash C_T(\overline{\ulcorner t \urcorner}),$$

where $\ulcorner t \urcorner$ is the gödelnumber of t in the theory.

Further, it is easy to show (by induction on $t \in C_1$) that

$$|\ulcorner t \urcorner|_{A_1} = |t|$$

for $t \in CT_1$, where $A \equiv (A_1, A_2)$ is the pair of defining formulas for $C \equiv (C_1, C_2)$. Hence

$$|\underset{\sim}{T}_2| \leq |\underset{\sim}{ID}_2(C)| \leq |\underset{\sim}{ID}_2|.$$

<u>Remarks</u>. (a) The reverse inequality is proved in 6.7.9.

(b) The above result shows that $|\underset{\sim}{T}_2| < \omega_1$.

§ 7. Functional interpretation of $\underset{\sim}{ID}_2(A)$.

6.7.1. Introduction. In order to obtain the reverse inequality to 6.6.8, we define a functional interpretation of $\underset{\sim}{ID}_2(A)$ which extends the modified realizability interpretation of $\underset{\sim}{HA}$ (chapter III, § 4) by translating $Q_1 t$ and $Q_2 t$ as $\exists\alpha^1 P_1(\alpha^1,t)$ and $\exists\alpha^2 P_2(\alpha^2,t)$ resp., where P_1 and P_2 are two new binary predicate symbols adjoined to the theory $\underset{\sim}{T}_2$. Intuitively, $P_i(\alpha^i,a)$ means: " a is put into Q_i at stage α^i ", or " α^i realizes $Q_i a$ " $(i = 1, 2)$.

So $\underset{\sim}{ID}_2(A)$ is interpreted in a theory $\underset{\sim}{E}-\underset{\sim}{T}_2\underset{\sim}{P}$, which is $\underset{\sim}{T}_2$ augmented by the predicate P_1 and P_2 and appropriate axioms for them (which depend on $A \equiv (A_1,A_2)$), and also axioms of extensionality.

Definitions of the theories $\underset{\sim}{T}_2[P]$, $\underset{\sim}{T}_2\underset{\sim}{P}$, $\underset{\sim}{E}-\underset{\sim}{T}_2$, $\underset{\sim}{E}-\underset{\sim}{T}_2[P]$ and $\underset{\sim}{E}-\underset{\sim}{T}_2\underset{\sim}{P}$.

Let $\mathcal{L}_2$ be the language of $\underset{\sim}{T}_2$.

$\mathcal{L}_2[P]$ is $\mathcal{L}_2$ augmented by the predicate symbols P_1 and P_2, with prime formulas: $s = t$ (as for $\mathcal{L}_2$), and also:

$$P_1(s,t) \quad \text{for } s \in 1,\ t \in 0$$

and

$$P_2(s,t) \quad \text{for } s \in 2,\ t \in 0.$$

$\underset{\sim}{T}_2[P]$ is $\underset{\sim}{T}_2$ in the language $\mathcal{L}_2[P]$, i.e., including the induction axioms for all formulas of $\mathcal{L}_2[P]$, and equality axioms for P_1 and P_2:

$$\alpha^i = \beta^i \ \&\ m = n \rightarrow P_i(\alpha^i,m) \rightarrow P_i(\beta^i,n) \qquad (i = 1, 2).$$

$\underset{\sim}{T}_2\underset{\sim}{P}$ is the theory $\underset{\sim}{T}_2[P]$ together with the axioms for P_1 and P_2 (given below in 6.7.4).

Finally, for $\underset{\sim}{H} \equiv \underset{\sim}{T}_2$, $\underset{\sim}{T}_2[P]$ or $\underset{\sim}{T}_2\underset{\sim}{P}$: $\underset{\sim}{E}-\underset{\sim}{H}$ is the theory $\underset{\sim}{H}$ together with the extensionality axioms:

$$\mathrm{EXT}_{\rho,\sigma}: \quad \forall f^{(\rho)\sigma}, g^{(\rho)\sigma}[\forall x^\rho(fx = gx) \rightarrow f = g]$$

for all $\rho,\sigma \in T_2$.

$\underset{\sim}{ID}_2(A)$ will be interpreted in $\underset{\sim}{E}-\underset{\sim}{T}_2\underset{\sim}{P}$. (The reasons for the extensionality axioms are discussed in 6.7.10. Actually we only need $\mathrm{EXT}_{0,1}$.)

Definition of the translation of $\underset{\sim}{ID}_2(A)$ into $\mathcal{L}_2[P]$.

With each formula F of $\mathcal{L}[Q]$ (the language of $\underset{\sim}{ID}_2(A)$) we associate a formula $\underline{x}\ \underset{\approx}{mr}\ F$ of $\mathcal{L}_2[P]$. The definition is by induction on the complexity of F, and extends the definition for $\underset{\sim}{HA}$ (3.4.2, 3.4.3 (A)) by further defining:

$\underset{\approx}{mr}$ (vii) $\quad \alpha^i\ \underset{\approx}{mr}\ Q_i t \equiv P_i(\alpha^i,t) \qquad (i = 1, 2).$

Another way of saying the same thing is that we define, for each formula F of $\mathcal{L}[Q]$, its $\underset{\approx}{mr}$-translation:

(1) $F^{o} \equiv \exists x F_{o}(x)$

extending the definition in 3.4.2 (see 3.4.3, Notational convention), by further defining

mr (vii) $(Q_i t)^o \equiv \exists \alpha^i P_i(\alpha^i, t) \qquad (i = 1, 2)$.

Outline of this section (§ 7).

Theorem 6.7.5 is the central result of this section. From it (or rather from corollary 6.7.7) and theorem 6.7.8, we immediately obtain (as theorem 6.7.9) the inequality :

$$|ID_2| \leq |T_2| .$$

This, combined with the reverse inequality (6.6.8), yields (as corollary 6.7.10) the main result of this chapter : the characterization of $|ID_2|$ as

$$|ID_2| = |T_2| .$$

In order to prove theorem 6.7.5 we first need lemma 6.7.3, which provides a normal form for the translations of formulas $(A_1, A_2) \in \mathcal{C}$. We can then state the axioms for P_1 and P_2 (6.7.4).

However, in order to prove lemma 6.7.3 and theorem 6.7.5, and even to state the axioms for P_1 and P_2, we need certain pairing functions and their inverses (in $E-T_2$). These are defined in 6.7.2.

Finally, notes 6.7.6 and 6.7.11 discuss, respectively, the reduction of simultaneous to simple transfinite recursion in $E-T_2$, and the rôle of the extensionality axioms.

Notation. In this section, the relation $\cong$ between formulas denotes provable equivalence in $E-T_2[P]$.

Suggestion to the reader. Readers who only want a rough idea of the methods of this section may skip 6.7.3, 6.7.4 and the proof of 6.7.5, and instead look at § 8 (particularly 6.8.1, 6.8.2 and 6.8.7). Although the main aim of § 8 is to consider functional interpretations of classical systems $ID^c_\nu(O)$ ($\nu = 1$ and 2), the subsections indicated also consider briefly the modified realizability interpretation of the corresponding intuitionistic systems $ID_\nu(O)$, which is easier to describe than the general case of $ID_2(A)$ for (any) $A \in \mathcal{C}$.

6.7.2. Pairing functions and their inverses in $E-T_2$.

In order to prove lemma 6.7.3 and theorem 6.7.5, and to state the axioms for P_1 and P_2 (6.7.4), we must show that certain pairing functions and their (left) inverses (i.e. satisfying (2), but not necessarily (3) of 1.6.16), can be defined in $E-T_2$. (Below, "inverse" means left inverse.)

Notation. (i) The λ-notation, used below, is to be understood as defined in 1.6.8 (as stated previously).
(ii) We indicate that f is a function(al) or term of type τ by writing f^τ or $f \in \tau$.

(a) Embeddings of ground types.

First we define embeddings u_{01}, u_{12}, u_{02},
and inverses d_{01}, d_{12}, d_{02}
("u" for up, "d" for down), with $u_{ij} \in (i)j$ and $d_{ij} \in (j)i$, such that

$$\underset{\sim}{E}-\underset{\sim}{T}_2 \vdash d_{ij} u_{ij} x^i = x^i .$$

[The definitions are (using FR, TR_1 and TR_2):

$$\begin{cases} u_{01}0^0 = 0^1, \\ u_{01}(S_0 n) = S_1 \lambda k(u_{01}n), \end{cases} \qquad \begin{cases} d_{01}0^1 = 0^0, \\ d_{01}(S_1 f) = d_{01}(f0) + 1, \end{cases}$$

$$\begin{cases} u_{12}0^1 = 0^2, \\ u_{12}(S_1 f) = S_2 \lambda\alpha u_{12} f(d_{01}\alpha), \end{cases} \qquad \begin{cases} d_{12}0^2 = 0^0, \\ d_{12}(S_2 f) = S_1 \lambda k d_{12} f(u_{01}k). \end{cases}$$

Finally, u_{02} and d_{02} are the composite functions:

$$u_{02} = u_{12} \circ u_{01}, \qquad d_{02} = d_{01} \circ d_{12}.]$$

(b) Pairing of two objects of (the same) ground type as one object of the same type.

For each ground type i there is a pairing function $D_i \in (i)(i)i$ and inverses D_i', D_i'' such that:

$$\underset{\sim}{E}-\underset{\sim}{T}_2 \vdash D_i' D_i x^i y^i = x^i \;\&\; D_i'' D_i x^i y^i = y^i .$$

[The definitions are:

$i = 0$: D_0, D_0', D_0'' can be taken as j, j_1 and j_2 of 1.3.9, B.

$i = 1$: $D_1 \alpha\beta = S_1 f$, where $f \in (0)1$ is defined by FR:

$$\begin{cases} f0^0 = \alpha, \\ f(S_0 n) = \beta. \end{cases}$$

The inverses are defined by TR_1:

$$\begin{cases} D_1' 0^1 = 0^1 \text{ (say)}, \\ D_1'(S_1 f) = f0^0, \end{cases} \qquad \begin{cases} D_1'' 0^1 = 0^1, \\ D_1''(S_1 f) = f1 \quad (\text{where } 1 \equiv S_0 0^0). \end{cases}$$

$i = 2$: $D_2 \alpha^2 \beta^2 = S_2 f$, where $f \in (1)2$ is defined by TR_1:

$$\begin{cases} f0^1 = \alpha^2, \\ f(S_1 g) = \beta^2. \end{cases}$$

The inverses are defined by TR_2 :

$$\begin{cases} D_2'0^2 = 0^2 , & D_2''0^2 = 0^2 , \\ D_2'(S_2 f) = f0^1 , & D_2''(S_2 f) = f(u_{01}1) .] \end{cases}$$

(c) Pairing at all types.

For all $\sigma, \tau \in T_2$ we can define a pairing function $D_{\sigma,\tau}$ with inverses $D'_{\sigma,\tau}$ and $D''_{\sigma,\tau}$ such that (dropping subscripts σ, τ):

$$\underset{\sim}{E} - \underset{\sim}{T}_2 \vdash D'Dx^\sigma y^\tau = x^\sigma \ \& \ D''Dx^\sigma y^\tau = y^\tau .$$

[(i) For σ, τ both ground types: put $i = \sigma$, $j = \tau$, $k = \max(i,j)$.

Functions $D_{i,j} \in (i)(j)k$, $D'_{i,j} \in (k)i$ and $D''_{i,j} \in (k)j$ are defined as follows:

If $i = j$, then the definitions are as in (b).

If $i \neq j$, suppose $i < j$. (The case $j < i$ is symmetrical.) Then by definition:

$$D_{i,j}x^i y^j = D_j(u_{ij}x^i, y^j) ,$$
$$D'_{i,j}z^j = d_{ij}D'_j z^j ,$$
$$D''_{i,j}z^j = D''_j z^j .$$

(ii) In the general case: suppose

$$\sigma \equiv (\sigma_1) \dots (\sigma_m)i , \qquad \tau \equiv (\tau_1) \dots (\tau_n)j$$

(i, j ground types). Then by definition:

$$D_{\sigma,\tau} f^\sigma g^\tau = \lambda x_1^{\sigma_1} \dots x_m^{\sigma_m} y_1^{\tau_1} \dots y_n^{\tau_n} D_{i,j}(fx_1 \dots x_m)(gy_1 \dots y_n)$$

with inverses defined in the obvious way (cf. 1.6.17).]

(d) Pairing functions needed for lemma 6.7.3.

In the proof of lemme 6.7.3, we must code pairs of objects of type τ as single objects of the same type τ, for certain $\tau \in \underset{\approx}{T}_2$. (This is not given, in general, by part (c).) The types for which this is needed are:

$$\tau \equiv 0, \ 1 \ \text{ and } \ (0)\sigma$$

(for certain σ). For each such τ we define a pairing function $\langle \ , \ \rangle_\tau \in (\tau)(\tau)\tau$ with inverses $\pi'_\tau, \pi''_\tau \in (\tau)\tau$ such that (dropping subscripts τ):

$$\underset{\sim}{E} - \underset{\sim}{T}_2 \vdash \pi'\langle x^\tau, y^\tau\rangle = x^\tau \ \& \ \pi''\langle x^\tau, y^\tau\rangle = y^\tau .$$

[The definitions are:

For $\tau \equiv 0$ or 1: they are given by D_τ, D'_τ and D''_τ of (b).

For $\tau \equiv (0)\sigma$: $\langle f^{(0)\sigma}, g^{(0)\sigma}\rangle_{(0)\sigma}$ is defined as the function $h \in (0)\sigma$, where

$$\begin{cases} h(2n) = fn\,, \\ h(2n+1) = gn\,. \end{cases}$$

$\pi'_{(0)\sigma}$ and $\pi''_{(0)\sigma}$ are defined in the obvious way.]

(e) Special pairing functions for the axioms for P_1 and P_2.

In order to state the axioms for P_1 and P_2 (6.7.4) we need the following:

(i) a coding of a pair of functions of type (0)1 and (0)0 as one function of type (0)1:

$$f^{(0)1} = \langle g^{(0)1}, h^{(0)0}\rangle$$

with inverses π_1^1, π_1^0 such that

$$\underset{\sim}{E}-\underset{\sim}{T}_2 \vdash \pi_1^1\langle g, h\rangle = g \ \& \ \pi_1^0\langle g, h\rangle = h\,;$$

(ii) a coding of a triple of functions of type (0)(1)2, (0)(1)1 and (0)(1)0 as one function of type (1)2:

$$f^{(1)2} = \langle g_2^{(0)(1)2}, g_1^{(0)(1)1}, g_0^{(0)(1)0}\rangle$$

with inverses π_2^2, π_2^1, π_2^0 such that

$$\begin{aligned} \underset{\sim}{E}-\underset{\sim}{T}_2 \vdash\ & \pi_2^2\langle g_2, g_1, g_0\rangle = g_2 \\ & \& \ \pi_2^1\langle g_2, g_1, g_0\rangle = g_1 \\ & \& \ \pi_2^0\langle g_2, g_1, g_0\rangle = g_0\,. \end{aligned}$$

Note. The actual form of the definition of π_1^1 (given below), viz. $(\pi_1^1 f)(n) = f(2n)$, is used in the proofs of theorem 6.7.5 (for the interpretation of $Q_1.2$) and theorem 6.7.8 (b). Similarly, the actual form of the definition of π_2^2, viz. $(\pi_2^2 f)(n,\alpha) = fD_{0,1}(3n,\alpha)$, is used in 6.7.5 (for the interpretation of $Q_2.2$).

[The definitions are:

For (i): $\langle g,h\rangle = f$ where $\begin{cases} f(2n) = g(n)\,, \\ f(2n+1) = u_{01}h(n)\,, \end{cases}$

$$(\pi_1^1 f)(n) = f(2n), \quad (\pi_1^0 f)(n) = d_{01}f(2n+1)\,.$$

For (ii): First, from g_2, g_1 and g_0, define $\tilde{f} \in (0)(1)2$ by

$$\begin{aligned} \tilde{f}(3n,\alpha) &= g_2(n,\alpha)\,, \\ \tilde{f}(3n+1, \alpha) &= u_{12}g_1(n,\alpha)\,, \\ \tilde{f}(3n+2, \alpha) &= u_{02}g_0(n,\alpha)\,, \end{aligned}$$

and then define $f \in (1)2$ by

$$f\beta = f(D'_{0,1}\beta,\ D''_{0,1}\beta).$$

Then $(\pi^2_2 f)(n,\alpha) = fD_{0,1}(3n,\alpha)$,

$(\pi^1_2 f)(n,\alpha) = d_{12}f\ D_{0,1}(3n+1,\ \alpha)$,

$(\pi^0_2 f)(n,\alpha) = d_{02}f\ D_{0,1}(3n+2,\ \alpha)$

(with $D_{0,1}$, $D'_{0,1}$ and $D''_{0,1}$ as defined in (c)(i)).]

6.7.3. Lemma. Normal forms for translations of $A \in \mathcal{C}$.

(a) If $A_1(X,a) \in \mathcal{C}_1$ (where a is a list of the free no. variables of A_1), then $A_1(Q_1,a)^o \cong$

$$\exists g^{(0)1} h^{(0)0}\ \forall n B_1(n,\ g,\ h,\ a)$$

where $\forall nB_1$ corresponds to the matrix $F_o(\underline{\underline{x}})$ of (1) of 6.7.1, and B_1 has the form:

$$C(n,\ h,\ a) \rightarrow: D(n,\ h,\ a)\ .\vee.\ E(n,\ h)\ \&\ P_1(gs,\ ts),$$

where: C, D, E are formulas of $\mathcal{L}$ augmented by the variable $h^{(0)0}$ (but with equality only between type 0 terms);

C has quantification over numbers only;

D and E are qf (quantifier free) and hence primitive recursive in h;

$s \in Tm_0$, possibly containing the free variables n, h, a;

and $t \in CT_{(0)0}$ (representing a primitive recursive function).

(b) If $A_2(X,Y,a) \in \mathcal{C}_2$ (where a is a list of the free no. variables of A_1), then $A_2(Q_1,\ Q_2,\ a)^o \cong$

$$\exists g_2^{(0)(1)2} g_1^{(0)(1)1} g_0^{(0)(1)0}\ \forall n,\alpha\ B_2(n,\ \alpha,\ g_2,\ g_1,\ g_0,\ a)$$

where $\forall n,\alpha B_2$ is the matrix, and B_2 has the form:

$$C(n,\alpha,g_0,a) \rightarrow: D(n,\alpha,g_1,g_0,a)\ .\vee.\ E(n,\alpha,g_0)\ \&\ P_2(g_2 rs,tr),$$

where C, D, E contain P_1 positively only, and not P_2 (and equality only between type 0 terms);

C has number quantification only;

D and E are qf;

$r \in Tm_0$, $s \in Tm_1$, possibly containing the free vars. n,α,g_0,a;

and $t \in CT_{(0)0}$ (representing a primitive recursive function).

Proof. By induction on the inductive definitions of $\mathcal{C}_1$ and $\mathcal{C}_2$ resp. (cf. the proof of lemma 4.5 of Kreisel and Troelstra 1970, from which the idea for this lemma came). We give some cases.

(a) If A_1 is $s=t$ or $Q_1 s$, then the form of A_1^o is simpler than that shown, which is O.K., or it may be padded to the required form: e.g. take $(Q_1 s)^o \equiv \exists \alpha P_1(\alpha, s) \cong \exists g^{(0)1} \forall n\, P_1(gs, ts)$, with $t = \lambda k.k$.

Now suppose

$$A^o \cong \exists gh\ \forall n[C(n,h,a) \rightarrow:\ D(n,h,a)\ .\vee.\ E(n,h)\ \&\ P_1(gs,ts)],$$

$$(A')^o \cong \exists gh\ \forall n[C'(n,h,a) \rightarrow:\ D'(n,h,a)\ .\vee.\ E'(n,h)\ \&\ P_1(gs',t's')].$$

Then $(A \& A')^o \cong \exists gg'hh' \forall n[C(n,h,a)\ \&\ C'(n,h',a) \rightarrow$
$\rightarrow:\ D(n,h,a)\ \&\ D'(n,h',a)\ .\vee.\ E(n,h)\ \&\ E'(n,h')\ \&\ P_1(gs,ts)\ \&\ P_1(g's',t's')]$

So put

$$C''(m,n,h,h',a) \equiv m=0\ \&\ C(n,h,a)\ .\vee.\ m\neq 0\ \&\ C'(n,h',a),$$
$$D''(m,n,h,h',a) \equiv m=0\ \&\ D(n,h,a)\ .\vee.\ m\neq 0\ \&\ D'(n,h',a),$$
$$E''(m,n,h,h') \equiv m=0\ \&\ E(n,h)\ .\vee.\ m\neq 0\ \&\ E'(n,h'),$$

and define

$$(1)\qquad \begin{cases} s'' \equiv \langle sg(m),\ \overline{sg}(m).s + sg(m).\ s'\rangle, \\ t'' \equiv \lambda x.[\overline{sg}(\pi' x).t(\pi'' x) + sg(\pi' x).t'(\pi' x).t'(\pi'' x)], \end{cases}$$

so
$$s'' = \begin{cases} \langle 0,s\rangle & \text{if } m=0, \\ \langle 1,s'\rangle & \text{if } m\neq 0, \end{cases}$$

and
$$t''s'' = \begin{cases} ts & \text{if } m=0, \\ t's' & \text{if } m\neq 0. \end{cases}$$

Then $(A \& A') \cong$

$$\exists g''hh'\ \forall mn[C''(m,n,h,h',a) \rightarrow:\ D''(m,n,h,h',a).\ \vee.E(m,n,h,h')\ \&\ P_1(g''s'',t''s'')]$$

(taking g'' as: $g''x = \begin{cases} g(\pi'' x) & \text{if } \pi' x = 0, \\ g'(\pi'' x) & \text{if } \pi' x \neq 0, \end{cases}$

so that, conversely, $gx = g''\langle 0,x\rangle$ and $g'x = g''\langle 1,x\rangle$),

which can be put in the required form by pairing m,n and h,h' respectively as in 6.7.2 (d).

Next, $(A \vee A')^o \cong$

$$\exists mgh[m=0 \wedge \forall n\{C \rightarrow:\ D\ .\vee.\ E\ \&\ P_1(gs,\ ts)\}\ .\vee$$
$$\vee.\ m\neq 0 \wedge \forall n\{C' \rightarrow: D'.\vee.E'\ \&\ P_1(gs',t's')\}].$$

Now define

$$C'''(m,n,h,a) \equiv m=0\ \&\ C\ .\vee.\ m\neq 0\ \&\ C',$$
$$D'''(m,n,h,a) \equiv m=0\ \&\ D\ .\vee.\ m\neq 0\ \&\ D',$$
$$E'''(m,n,h) \equiv m=0\ \&\ E\ .\vee.\ m\neq 0\ \&\ E',$$

and s'',t'' as in (1) again.

Then $(A \vee A')^o \cong \exists mgh\ \forall n[C''' \rightarrow:\ D'''\ .\vee.\ E'''\ \&\ P_1(gs'',t''s'')]$, which again can be put in the required form by coding m and h as one function of type (0)0, which is easy.

Next, let $C_1 \in \mathcal{C}_0$. Then $C_1^0 \equiv C_1$ (by 3.4.4 (i)).

So $(C_1 \rightarrow A)^0 \cong \exists gh[C_1 \rightarrow \forall n\{C \rightarrow: D .\vee. E \& P_1(gs,ts)\}]$

$\cong \exists gh \forall n[(C_1 \& C) \rightarrow: D .\vee. E \& P_1(gs,ts)]$,

which again has the required form.

Next, $(\forall mA)^0 \cong \exists gh \forall mn[C(n,h_m,a) \rightarrow: D(n,h_m,a) .\vee. E(n,h_m) \& P_1(g_m s,ts)]$,
where h_m is defined by : $h_m(k) = h(\langle m,k\rangle)$
and g_m similarly.
So define $s''' \equiv \langle m, s\rangle$,
and $t''' \equiv \lambda k\, t(\pi'' k)$
(so that $t'''s''' = ts$), and we have the required form (after pairing m and n).

Finally, $(\exists mA)^0 \cong \exists mgh \forall n[C \rightarrow: D .\vee. E \& P_1(gs,ts)]$,
which has the required form (after pairing m and n).

(b) The general procedure here is as in (a). We omit details except for the one interesting case : suppose

$$A_2^0 \cong \exists g_2 g_1 g_0 \forall n, \alpha B_2(n,\alpha,g_2,g_1,g_0,a)$$

where $B_2 \equiv C \rightarrow: D .\vee. E \& P_2(g_2 rs,tr)$,
and suppose $A_1 \in \mathcal{C}_1$.
We want to show : $(A_1 \rightarrow A_2)^0$ has the required form.

First define a functional $\Phi \in (1)((0)1)$ by TR_1 :

$$\begin{cases} \Phi(0^1) = \lambda k 0^1 \quad \text{(say)}, \\ \Phi(S_1 f) = f . \end{cases}$$

(So for any $\alpha \neq 0^1$, $\Phi(\alpha)$ is the sequence of immediate predecessors of α.)
Now <u>by part (a)</u> :

$$\begin{aligned} A_1^0 &\cong \exists g^{(0)1} h^{(0)0}\, C'(g,h) \quad \text{(say)} \\ &\cong \exists f^{(0)1}\, C'(\pi_1^1 f, \pi_1^0 f) \quad \text{(see 6.7.2 (e)(i))} \\ &\cong \exists \beta\, C'(\pi_1^1 \Phi\beta, \pi_1^0 \Phi\beta) \\ &\equiv \exists \beta\, C''(\beta) \quad \text{(say)}. \end{aligned}$$

So $(A_1 \rightarrow A_2)^0 \cong \exists g_2 g_1 g_0 \forall n,\alpha,\beta[C''(\beta) \rightarrow B_2(n,\alpha,g_{2,\beta},g_{1,\beta},g_{0,\beta},a)]$,

where $g_{i,\beta}$ is defined by : $g_{i,\beta}(k,\gamma) = g_i(k, \langle\beta,\gamma\rangle)$.

So $(A_1 \rightarrow A_2)^0 \cong \exists g_2 g_1 g_0 \forall n,\alpha,\beta[C''(\beta) \& C(n,\alpha,g_{0,\beta}) \rightarrow$

$\rightarrow: D(n,\alpha,g_{1,\beta},g_{0,\beta},a) .\vee. E(n,\alpha,g_{0,\beta}) \& P_2(g_{2,\beta} rs,tr)]$.

Finally, replace s by $s' \equiv \langle\beta,s\rangle$, and again we have the required form (after pairing α and β).

6.7.4. <u>The axioms for</u> P_1 <u>and</u> P_2 can now be stated.

Let B_1 and B_2 be as in lemma 6.7.3, when A_1 and A_2 are the defining formulas in $\mathbf{ID}_2(A)$.

<u>Axioms for</u> P_1.

$P_1.1)$ $\neg P_1(0^1, a)$

$P_1.2)$ $\forall n B_1(n, g, h, a) \rightarrow P_1(S_1\langle g,h\rangle, a)$

$P_1.3)$ $P_1(S_1 f, a) \rightarrow \forall n B_1(n, \pi^1_1 f, \pi^0_1 f, a)$.

<u>Axioms for</u> P_2.

$P_2.1)$ $\neg P_2(0^1, a)$

$P_2.2)$ $\forall n, \alpha\, B_2(n, \alpha, g_2, g_1, g_0, a) \rightarrow P_2(S_2\langle g_2, g_1, g_0\rangle, a)$

$P_2.3)$ $P_2(S_2 f, a) \rightarrow \forall n, \alpha\, B_2(n, \alpha, \pi^2_2 f, \pi^1_2 f, \pi^0_2 f, a)$

(where $g \in (0)1$, $h \in (0)0$, $g_i \in (0)(1)i$).

6.7.5. <u>Theorem.</u> If $\mathbf{ID}_2(A) \vdash F$, then there is a sequence $\underline{\underline{t}}$ of terms $\mathbf{T}_2$ such that $\mathbf{E}\text{-}\mathbf{T}_2\mathbf{P} \vdash \underline{\underline{t}}\ \mathbf{mr}\ F$ (and the free variables of $\underline{\underline{t}}\ \mathbf{mr}\ F$ are exactly those of F).

<u>Proof.</u> For the axioms and rules of $\mathbf{HA}$ (in $\mathcal{L}[Q]$), the proof proceeds as in the soundness theorem for $\mathbf{mr}$-realizability (3.4.5). It remains to consider the axioms for Q_1 and Q_2.

(Remember, $\underline{\underline{t}}\ \mathbf{mr}\ F \equiv F_0(\underline{\underline{t}})$, where $F^0 \equiv \exists \underline{\underline{x}} F_0(\underline{\underline{x}})$. $\underline{\underline{t}}$ is called a solution for F^0 in $\mathbf{E}\text{-}\mathbf{T}_2\mathbf{P}$.)

$Q_1.1$: By lemma 6.7.3 (a), $(Q_1.1)^0 \cong$

$$\exists X \forall gh[\forall n B_1(n,g,h,a) \rightarrow P_1(Xgh, a)]$$

with $X \in ((0)1)((0)0)1$. A solution is easily obtained in $\mathbf{E}\text{-}\mathbf{T}_2\mathbf{P}$ by setting $X = \lambda gh S_1\langle g,h\rangle$ and using axiom $P_1.2$.

$Q_2.1$: By lemma 6.7.3 (b), $(Q_2.1)^0 \cong$

$$\exists X \forall g_2 g_1 g_0[\forall n, \alpha B_2(n,\alpha,g_2,g_1,g_0) \rightarrow P_2(Xg_2g_1g_0, a)]$$

with $X \in ((0)(1)2)((0)(1)1)((0)(1)0)2$. A solution is again easily obtained, this time by setting $X = \lambda g_2 g_1 g_0 S_2\langle g_2, g_1, g_0\rangle$ and using axiom $P_2.2$.

$Q_1.2$: Let $F(a)$ be any formula of $\mathcal{L}[Q]$, with

(1) $F(a)^0 \equiv \exists \underline{\underline{x}} F_0(\underline{\underline{x}}, a) \cong \exists x^\tau F'_0(x^\tau, a)$,

where the sequence of variables $\underline{\underline{x}}$ is represented as a single variable x^τ by means of the pairing functions of 6.7.2 (c) (cf. 3.4.9, "variant I"). Then, parallel to the proof of lemma 6.7.3 (a), we can prove that

$$A_1(F,a) \cong \exists g^{(0)\tau} h^{(0)0} \forall n\{C \rightarrow: D .\vee. E \,\&\, F'_0(gs, ts)\}$$

with C, D, E, s, t as in lemma 6.7.3 (a) (but now $g \in (0)\tau$).
So $\forall a[A_1(F,a) \rightarrow F(a)] \cong$

$$\exists X \forall a g^{(o)\tau} h^{(o)o}[\forall n\{C \rightarrow: D .\vee. E \,\&\, F'_o(gs,ts)\} \rightarrow F'_o(Xagh, a)]$$

and $\forall a[Q_1 a \rightarrow F(a)] \cong$

$$\exists Y \forall a \alpha[P_1 \alpha a \rightarrow F'_o(Y\alpha a, a)] .$$

So to solve $(Q_1.2)^o$ in $\underset{\sim}{E}-\underset{\sim}{T}_2\underset{\sim}{P}$, we must find Y as a function of α, a and X.
So define by TR_1 on α (suppressing X as an argument of Y):

$$(2) \qquad \begin{cases} Y(0^1,a) = 0^\tau \\ Y(S_1 f,a) = X(a, \lambda k Y(f(2k),tk), \pi^o_1 f) . * \end{cases}$$

This function Y is then proved in $\underset{\sim}{E}-\underset{\sim}{T}_2\underset{\sim}{P}$ to satisfy $(Q_1.2)^o$ by use of TI_1 and the axioms $P_1.1$ and $P_1.3$; i.e. assuming that for X:

$$(3) \qquad \forall a g^{(o)\tau} h^{(o)o}[\forall n\{C \rightarrow: D .\vee. E \,\&\, F'_o(gs,ts)\} \rightarrow F_o(Xagh,a)] ,$$

we prove (in $\underset{\sim}{E}-\underset{\sim}{T}_2\underset{\sim}{P}$):

$$(4) \qquad \forall \alpha, a[P_1(\alpha,a) \rightarrow F_o(Y\alpha a,a)]$$

by TI_1 on α, as follows.
For $\alpha = 0^1$, (4) follows trivially from $P_1.1$.
Now suppose $P_1(S_1 f,a)$. Then by $P_1.3$:

$$(5) \qquad \forall n\{C' \rightarrow: D' .\vee. E' \,\&\, P_1((\pi^1_1 f)s',ts')\}$$

where C', D', E', s' are C, D, E, s resp. with h replaced by $\pi^o_1 f$.
Now $(\pi^1_1 f)s' = f(2s')$ by definition, so

$$P_1((\pi^1_1 f)s',ts') \cong P_1(f(2s'),ts')$$

which implies, from (4) by <u>induction hypothesis</u>, $F'_o(Y(f(2s'),ts'),ts')$.
So (5) implies

$$\forall n\{C' \rightarrow: D' .\vee. E' \wedge F'_o(Y(f(2s'),ts'),ts')\} ,$$

which implies, by (3), taking $g = \lambda k Y(f(2k),tk)$ and $h = \pi^o_1 f$:

$$F'_o(X(a, \lambda k Y(f(2k),tk), \pi^o_1 f), a) ,$$

which implies, by (2): $F'_o(Y(S_1 f,a), a)$.

-----

* Remember, $f(2k) = (\pi^1_1 f)(k)$, and π^1_1 and π^o_1 were defined in 6.7.2 (e)(i). It may be seen more clearly how Y is defined by TR_1 if we write $Y(S_1 f) = \lambda a.X(a, \lambda k(Y \circ f)(2k)(tk), \pi^o_1 f)$.

$Q_2.2$: The argument here follows a similar pattern, but using TR_2 instead of TR_1.

[The following are some of the details. Taking again, for any formula $F(a)$ of $\mathcal{L}[Q]$, $F(a)^o \cong \exists x^\tau F_o^!(x^\tau, a)$, we prove (parallel to the proof of 6.7.3 (b)):

$$A_2(Q_1, F, a) \cong \exists g_2 g_1 g_0 \forall n, \alpha\{C \rightarrow: D . \vee . E \,\&\, F_o^!(g_2 rs, tr)\}$$

with C, D, E, r, s, t as in lemma 6.7.3 (b), $g_1 \in (0)(1)1$, $g_o \in (0)(1)0$, but now $g_2 \in (0)(1)\tau$.

So $\forall a[A_2(Q_1, F, a) \rightarrow F(a)]^o \cong$

$$\exists X \forall a g_2 g_1 g_o[\forall n\alpha\{C \rightarrow: D . \vee . E \,\&\, F_o^!(g_2 rs, tr)\} \rightarrow F_o(Xag_2 g_1 g_0, a)],$$

and $\forall a[Q_2 a \rightarrow F(a)]^o \cong$

$$\exists Y \forall a \alpha^2[P_2(\alpha^2, a) \rightarrow F_o^!(Y\alpha^2 a, a)].$$

So we must find Y as a function of a, α^2 and X.

So define by TR_2 on α^2 (suppressing the argument X):

$$\begin{cases} Y(0^2, a) = 0^\tau \\ Y(S_2 f, a) = X(a, \lambda k\alpha Y(fD_{0,1}(3k, \alpha), tk), \pi_1^1 f, \pi_2^0 f). \end{cases}$$

(Remember, $fD_{0,1}(3k, \alpha) = (\pi_2^2 f)(k, \alpha)$, with π_2^2, π_2^1 and π_2^0 defined in 6.7.2 (e)(ii), and $D_{0,1}$ in 6.7.2 (c).)

Then this function Y is proved (in $\underset{\sim}{E}\text{-}\underset{\sim}{T}_2\underset{\sim}{P}$) to satisfy $(Q_2.2)^o$ by use of the axioms TI_2, $P_2.1$ and $P_2.3$. Details for this are omitted.]

6.7.6. Note. For the solution of $(Q_1.2)^o$, the functional Y (in (2) of 6.7.5) was defined from the constant $R_{1,\tau}$, where τ is the type of the variable x^τ in (1) of 6.7.5, obtained by the use of the pairing functions of 6.7.2 (c). Alternatively, we could have kept $F(a)^o$ in the form $\exists \underline{x} F_o(\underline{x}, a)$ here, and then used simultaneous recursion on type 1 to define a solution of $(Q_1.2)^o$. This could then be reduced to (simple) TR_1 (i.e. (2) of 6.7.5) by means of these same pairing functions (cf. 1.6.16).

Similar remarks apply to $(Q_2.2)^o$.

6.7.7. Corollary. For any number a, if $\underset{\sim}{ID}_2(A) \vdash Q_1\bar{a}$, then there is a closed term t of type 1 such that $\underset{\sim}{E}\text{-}\underset{\sim}{T}_2\underset{\sim}{P} \vdash P_1(t, \bar{a})$.

Proof. Immediate from theorem 6.7.5.

6.7.8. Theorem. (a) In any wf model M of $\underset{\sim}{E}\text{-}\underset{\sim}{T}_2$, relations P_1^M and P_2^M can be defined so as to satisfy the axioms for P_1 and P_2.

(b) Further, if P_1^M is any relation in M which satisfies the axioms for P_1, and $P_1^M(\alpha, a)$ holds for some $\alpha \in M_1$ and $a \in M_0$, then $a \in Q_1$ and $|a|_{A_1} \leq |\alpha|_M$.

<u>Proof</u>. (a) Let M be a wf model of $\underset{\sim}{E}-\underset{\sim}{T}_2$. We can define relations

$$P_1^M \subset M_1 \times M_0 \quad \text{and} \quad P_2^M \subset M_2 \times M_0$$

by <u>induction on</u> M_1 <u>and</u> M_2 respectively, so as to satisfy axioms $P_1.1-3$ and $P_2.1-3$ (and the equality axioms for P_1 and P_2).

For example, P_1^M can be defined inductively by the two clauses

(i) $\forall a \neg P_1^M(0^1, a)$,

(ii) $\forall a[P_1^M(S_1 f, a) \Leftrightarrow \forall n B_1^M(n, \pi_1^1 f, \pi_1^0 f, a)]$

(where, of course, B_1^M is the interpretation of B_1 in M, and P_1^M is taken to be compatible with $=_M$), and similarly for P_2^M.

<u>Notes</u>. 1°) Taking $M = HRO_2$, P_1^M and P_2^M cannot in general be defined (partial) recursively on $\mathcal{O}_1 \times N$ and $\mathcal{O}_2 \times N$.

2°) The axioms do not define P_1^M and P_2^M uniquely. For example, clause (ii) above could be changed to:

(ii)' $\forall a[P_1^M(S_1 f, a) \Leftrightarrow f$ is of the form $\langle g^{(0)1}, h^{(0)0}\rangle$ and $\forall n\, B_1^M(n, g, h, a)]$.

This is not equivalent to (ii).

However, any relation P_1^M in M which does satisfy the axioms for P_1 also satisfies the conclusion of part (b) of the theorem, to which we now turn.

(b) First note that, parallel to the proof of lemma 6.7.3 (a), one can prove that if $A_1(X, a) \in \mathcal{C}_1$, then $A_1(X, a)$ itself is equivalent * to

(1) $\exists h^{(0)0} \forall n[C \rightarrow: D . \vee . E \,\&\, X(ts)]$

with C, D, E, t, s as in 6.7.3 (a) (cf. <u>Kreisel and Troelstra</u> 1970, lemma 4.5).

Now suppose that P_1^M satisfies the axioms for P_1. We must show that for any $\alpha \in M_1$ and $a \in M_0$, if $P_1^M(\alpha, a)$, then $a \in Q_1$ and $|a|_{A_1} \leq |\alpha|_M$, i.e. (using the notation of 6.2.1) $a \in Q_{1,\nu}$. where $\nu = |\alpha|_M$.

The proof is by <u>induction on</u> $\alpha \in M_1$ (or on the ordinal ν).

So suppose $P_1^M(\alpha, a)$.

By $P_1.1$, $\alpha \neq 0^1$, so $\alpha = S_1 f$ for some f.

By $P_1.3$:

(2) $M \models \forall n B_1(n, \pi_1^1 f, \pi_1^0 f, a)$.

Put $g = \pi_1^1 f$, $h = \pi_1^0 f$. Then (2) becomes

-----

* In $\underset{\sim}{EL} + AC_{01} + IP^{\omega} + EXT$ (in the language of $\underset{\sim}{EL}$ augmented by the predicate variable X).

(3) $\quad M \models \forall n[C \to: D.\vee.E \,\&\, P_1(gs,ts)]$

(again in the notation of 6.7.3 (a)).
Further, since $gs = (\pi_1^1 f)s = f(2s)$,

$$P_1^M(gs,ts) \Rightarrow P_1^M(f(2s),ts)$$

which implies by induction hypothesis that

$$ts \in Q_1 \quad \text{and} \quad |ts|_{A_1} \leq |f(2s)|_M < |S_1 f|_M,$$

so $ts \in \bigcup_{\mu<\nu} Q_{1,\mu}$, where $\nu = |S_1 f|_M$.
So (3) $\Rightarrow \forall n[C \to: D.\vee.E \,\&\, (ts \in \bigcup_{\mu<\nu} Q_{1,\mu})]$

$\Rightarrow A_1(\bigcup_{\mu<\nu} Q_{1,\mu}, a)$ by (1)

$\Rightarrow a \in Q_{1,\nu}$.

6.7.9. Theorem. $|\underset{\sim}{ID}_2| \leq |\underset{\sim}{T}_2|$.
Proof. Immediate from corollary 6.7.7 and theorem 6.7.8.

So we finally have our characterization of $|\underset{\sim}{ID}_2|$:

6.7.10. Corollary. $|\underset{\sim}{ID}_2| = |\underset{\sim}{T}_2|$.
Proof. From theorems 6.7.9 and 6.6.8.

6.7.11. A note on the extensionality axioms.

The axioms $EXT_{\rho,\sigma}$ are used to prove the characteristic properties of the pairing functions and their inverses in 6.7.2. These are used in two places:

(i) The pairing functions of 6.7.2 (c) are used in the proof of theorem 6.7.5, for the interpretation of axioms $Q_1.2$ and $Q_2.2$, to reduce simultaneous to simple recursion on types 1 and 2 (see note 6.7.6).

(ii) The pairing functions of 6.7.2 (d), (e) are used in the proof of lemma 6.7.3 and in the axioms for P_1 and P_2 (6.7.4).

Ad (i). The extensionality axioms could be avoided here, by either (a) extending the type structure to include products of types, and thus reducing simultaneous to simple TR_1 and TR_2 (cf. 1.6.16,(B)); or (presumably) (b) introducing constants for simultaneous TR_1 and TR_2 (cf. 1.6.16,(A)). However, it is not known whether either construction results in an expansion of $\underset{\sim}{T}_2$ (as with $\underset{\sim}{N}-\underset{\sim}{HA}^\omega$: cf. 1.8.2 for (a), 1.7.7 for (b)).

Ad (ii). A careful examination shows that the only extensionality axiom really needed here is $EXT_{0,1}$. This is used in proving that $d_{12}u_{12}\alpha^1 = \alpha^1$

(6.7.2 (a)) by TI_1 on σ^1. (The point is that paired functions often occur in formulas "through their values" only, so that one only has to prove statements of the form $(\pi'\langle f,g\rangle)x = fx$, rather than $\pi'\langle f,g\rangle = f$.)

Even the axiom $EXT_{0,1}$ could be avoided by considering a more complicated theory of trees, so as to avoid the codings of trees in the axioms for P_2.

§ 8. Functional interpretations of classical systems $\mathbf{ID}_1^c(\mathcal{O})$ and $\mathbf{ID}_2^c(\mathcal{O})$.

6.8.1. Introduction and definitions.

For convenience, attention is restricted in this section (§ 8) to theories $\mathbf{ID}_\nu(\mathcal{O})$ and $\mathbf{ID}_\nu^c(\mathcal{O})$ $(\nu = 1,2)$ with $\mathcal{O} = (\mathcal{O}_1, \mathcal{O}_2)$ as defined in 6.2.2. This is no real restriction in the study of classical systems, since from the proof that $\mathcal{O}_1$ is complete Π_1^1 (Rogers 1967, chapter 16) and $\mathcal{O}_2$ is complete Π_1^1 in $\mathcal{O}_1$ (Richter 1965), it can be shown that for any A which is monotonic (provably in, say, $\mathbf{HAS}_0^c + \mathrm{EXT}$), each finite subsystem of $\mathbf{ID}_\nu^c(A)$ can be interpreted in $\mathbf{ID}_\nu^c(\mathcal{O})$. (Feferman 1970 gives the proof for $\nu = 1$, and with another complete Π_1^1 set.) Further, it follows that $|\mathbf{ID}^c| = |\mathbf{ID}_\nu^c(\mathcal{O})|$.

It should be clear that (by a simplification of all the preceding work) the analogue of 6.7.10 for $\nu = 1$ holds, i.e.

$$|\mathbf{ID}_1| = |\mathbf{T}_1|.$$

The question we want to consider now is whether $|\mathbf{ID}_\nu^c| = |\mathbf{T}_\nu|$ for $\nu = 1$ and 2.

$\mathbf{ID}_2(\mathcal{O})$ is now conveniently axiomatized as: $\mathbf{HA}$ (in $\mathcal{L}[\mathcal{O}]$) + the following axioms for $\mathcal{O}_1$ and $\mathcal{O}_2$:

$\mathcal{O}_1$. 1a) $\mathcal{O}_1 0$

$\mathcal{O}_1$. 1b) $\forall n \exists y (T\,eny \,\&\, \mathcal{O}_1(Uy)) \rightarrow \mathcal{O}_1(3.5^e)$

$\mathcal{O}_1$. 2) $F(0) \,\&\, \forall e[\forall n \exists y(T\,eny \,\&\, F(Uy)) \rightarrow F(3.5^e)] \rightarrow \forall a(\mathcal{O}_1 a \rightarrow F(a))$

$\mathcal{O}_2$. 1a) $\mathcal{O}_2 0$

$\mathcal{O}_2$. 1b) $\forall n\{\mathcal{O}_1 n \rightarrow \exists y(T\,eny \,\&\, \mathcal{O}_2(Uy))\} \rightarrow \mathcal{O}_2(3^2.5^e)$

$\mathcal{O}_2$. 2) $F(0) \,\&\, \forall e[\forall n\{\mathcal{O}_1 n \rightarrow \exists y(T eny \,\&\, F(Uy))\} \rightarrow F(3^2.5^e)] \rightarrow \forall a(\mathcal{O}_2 a \rightarrow F(a))$,

where $F(a)$ is any formula of $\mathcal{L}[\mathcal{O}]$.

$\mathbf{ID}_1(\mathcal{O})$ is (of course) $\mathbf{HA}$ (in $\mathcal{L}[\mathcal{O}_1]$) + the axioms for $\mathcal{O}_1$ only.

$\mathbf{ID}_\nu^c(\mathcal{O})$ is $\mathbf{ID}_\nu(\mathcal{O})$ with classical logic $(\nu = 1,2)$.

$\mathbf{ID}_2^+(\mathcal{O})$ is $\mathbf{ID}_2(\mathcal{O})$ + the axioms:

$\mathcal{O}_1$. 3) $\forall n(\neg\neg\, \mathcal{O}_1 n \rightarrow \mathcal{O}_1 n)$

$\mathcal{O}_2$. 3) $\forall n(\neg\neg\, \mathcal{O}_2 n \rightarrow \mathcal{O}_2 n)$,

and $\mathbf{ID}_1^+(\mathcal{O})$ is $\mathbf{ID}_1(\mathcal{O}) + \mathcal{O}_1.3$.

Note. $\mathbf{ID}_\nu^c(\mathcal{O})$ can easily be interpreted in $\mathbf{ID}_\nu^+(\mathcal{O})$ by the $\neg\neg$ translation (i.e. $'$ of 1.10.2).

So from now on we will consider $\mathbf{ID}_\nu^+(\mathcal{O})$ instead of $\mathbf{ID}_\nu^c(\mathcal{O})$.

<u>Outline of § 8</u>.

In 6.8.2 - 5, we will consider functional interpretations of $\underset{\sim}{ID}_1(\mathcal{O})$ and $\underset{\sim}{ID}_1^+(\mathcal{O})$; in 6.8.6 we give a historical survey of other (known) methods of characterizing $|\underset{\sim}{ID}_1|$ and $|\underset{\sim}{ID}_1^c|$; and in 6.8.7 - 11 we turn to the problem of functional interpretations of $\underset{\sim}{ID}_2(\mathcal{O})$ and $\underset{\sim}{ID}_2^+(\mathcal{O})$.

First we need some definitions.

<u>The axioms for</u> P_1 <u>and</u> P_2 in the corresponding theory $\underset{\sim}{T}_2\underset{\sim}{P}$ of trees are now more conveniently given in the form :

<u>Axioms for</u> P_1 :

$P_1.1'$) $P_1(0^1,a) \leftrightarrow a=0$

$P_1.2'$) $\forall n[T(e,n,hn) \,\&\, P_1(gn,U(hn))] \to P_1(S_1\langle g,h\rangle,\ 3.5^e)$

$P_1.3'$) $P_1(S_1f,a) \to a = 3.5^{(a)_2} \,\&\, \forall n[T((a)_2,n,\pi_1^o fn) \,\&\, P_1(\pi_1^1 fn,\ U(\pi_1^o fn))]$

with $g \in (0)1$ and $h \in (0)0$; and $\langle\ ,\ \rangle$, π_1^i as defined in 6.7.2 (e)(i).

<u>Axioms for</u> P_2 :

$P_2.1'$) $P_2(0^2,a) \leftrightarrow a=0$

$P_2.2'$) $\forall\alpha,n[P_1\alpha n \to T(e,n,hn\alpha) \,\&\, P_2(gn\alpha,\ U(hn\alpha))] \to P_2(S_2\langle g,h\rangle,\ 3^2.5^e)$

$P_2.3'$) $P_2(S_2f,a) \to a = 3^2.5^{(a)_2}$ &

$\qquad\qquad \& \ \forall\alpha,n[P_1\alpha n \to T((a)_2,n,\pi_2^o fn\alpha) \,\&\, P_2(\pi_2^2 fn\alpha,\ U(\pi_2^o fn\alpha))]$

with $g \in (0)(1)2$ and $h \in (0)(1)0$, and pairing functions $\langle\ ,\ \rangle$ and inverses π_2^2, π_2^o defined similarly to 6.7.2 (e)(ii).

<u>Remark</u>. It may help to clarify these axioms, if we re-state them in a simpler (but less accurate!) form :

for P_1 : $\begin{cases} P_1(0^1,a) \leftrightarrow a=0, \\ P_1(S_1f,\ 3.5^e) \leftrightarrow \forall n P_1(fn,\ \{e\}(n)), \end{cases}$

for P_2 : $\begin{cases} P_2(0^2,a) \leftrightarrow a=0, \\ P_2(S_2f,\ 3^2.5^e) \leftrightarrow \forall\alpha,n[P_1\alpha n \to P_2(f\alpha,\ \{e\}(n))] . \end{cases}$

In this form, $P_1\alpha a$ just means that α is extensionally the same tree as $a \in \mathcal{O}_1$!

However, the axioms for P_1 and P_2, as actually given, have to take into account (for the functional interpretation of $\underset{\sim}{ID}_2(\mathcal{O})$) the quantifier hidden in $\{e\}(n)$. (Also, in the axioms for P_2 , it turns out that f must be taken as a function of n as well as α.)

<u>More definitions</u>.

<u>The theories</u> $\underset{\sim}{T}_1$, $\underset{\sim}{E}-\underset{\sim}{T}_1$ and $\underset{\sim}{E}-\underset{\sim}{T}_1\underset{\sim}{P}$ are defined as the restrictions of the corresponding theories $\underset{\sim}{T}_2$, etc., with 0 and 1 as the only ground types.

The "ordinal of $\mathbf{T}_1$" is defined (as with $\mathbf{T}_2$) by :

$$|\mathbf{T}_1| \equiv_{\mathrm{def}} \sup\{|t| : t \text{ a closed term of } \mathbf{T}_1 \text{ of type } 1\}.$$

The set-theoretical model of $\mathbf{T}_1$ (corresponding to $\mathscr{J}_2$ for $\mathbf{T}_2$) is denoted by $\mathscr{J}_1$.

Functionals in $\mathscr{J}_1$ and $\mathscr{J}_2$.

By "functional", we will mean a functional (in fact, any object) in $\mathscr{J}_1$ or $\mathscr{J}_2$. Functionals are denoted by $\Phi, \Psi, \Psi_1, \ldots$

Further, by $\mathbf{T}_\nu$ ($\nu = 1$ or 2) we will mean, not only the formal theory of trees, but also the collection of functionals in $\mathscr{J}_\nu$ denoted by the closed terms of $\mathbf{T}_\nu$.

Now let $\Psi_1, \Psi_2, \ldots$ be a (finite or infinite) sequence of functionals in $\mathscr{J}_\nu$. Then $\mathbf{T}_\nu(\Psi_1, \Psi_2, \ldots)$ means the system (or collection) of functionals in $\mathscr{J}_\nu$, generated from (the denotations in $\mathscr{J}_\nu$ of) the constants of $\mathbf{T}_\nu$, and also $\Psi_1, \Psi_2, \ldots$, by (repeated) application ; and $|\mathbf{T}_\nu(\Psi_1, \Psi_2, \ldots)|$ is the supremum of the ordinals of the functionals of type 1 in $\mathbf{T}_\nu(\Psi_1, \Psi_2, \ldots)$.

So when the sequence $\Psi_1, \Psi_2, \ldots$ is empty, this is consistent with the previous definition of $|\mathbf{T}_\nu|$ (6.6.7).

6.8.2 - 6.8.5. Functional interpretations of $\mathbf{ID}_1(\mathcal{O})$ and $\mathbf{ID}_1^+(\mathcal{O})$; proof that $|\mathbf{ID}_1^c| = |\mathbf{T}_1|$.

6.8.2. Consider first the modified realizability interpretation of $\mathbf{ID}_1(\mathcal{O})$ obtained by extending the $\mathbf{mr}$ - translation $^\circ$ of $\mathbf{HA}$ (3.4.2 - 3) by defining $(\mathcal{O}_1 t)^\circ \equiv \exists\alpha P_1 \alpha t$ (as in 6.7.1).

Now consider the translations of the axioms for $\mathcal{O}_1$:

$\mathcal{O}_1$. 1a is translated as $\exists\alpha P_1(\alpha, 0)$, which is solved in $\mathbf{E}\text{-}\mathbf{T}_1\mathbf{P}$ by taking $\alpha = 0^1$ (and using axiom $P_1.1'$).

$\mathcal{O}_1$. 1b. The translation proceeds to the stage

(1) $\exists gh\, \forall n\{T(e,n,hn) \,\&\, P_1(gn, U(hn))\} \rightarrow \exists\alpha P_1(\alpha, 3.5^e)$

and then to

$$\exists X\, \forall gh[\forall n\{T(e,n,hn) \,\&\, P_1(gn, U(hn))\} \rightarrow P_1(Xgh, 3.5^e)]$$

which is solved by taking $X = \lambda gh S_1\langle g,h\rangle$ (and using $P_1.2'$).

$\mathcal{O}_1$. 2. The translation is solved (as in the general case, theorem 6.7.5) by use of TR_1 ; and the fact that this does give a solution is proved in $\mathbf{E}\text{-}\mathbf{T}_1\mathbf{P}$ by use of TI_1, $P_1.1'$ and $P_1.3'$. (We omit details.)

Thus we obtain a functional interpretation of $\mathbf{ID}_1(\mathcal{O})$ in $\mathbf{E}\text{-}\mathbf{T}_1\mathbf{P}$, and so we can prove (cf. theorem 6.7.9) :

$$|\mathbf{ID}_1(\mathcal{O})| \leq |\mathbf{T}_1|.$$

Now for an interpretation of $\underset{\sim}{ID}_1^+(\mathcal{O})$, we must also consider $\mathcal{O}_1.3$. The first step in the translation of $\mathcal{O}_1.3$ gives:

$$\forall n[\neg\neg \exists\alpha P_1(\alpha,n) \to \exists\alpha P_1(\alpha,n)],$$

which becomes:

$$\exists X \forall n(\neg\neg \exists\alpha P_1(\alpha,n) \to P_1(Xn,x)].$$

However, a solution for X would mean a functional $\Phi \in (0)1$ such that (in $\mathscr{L}_1$):

$$n \in \mathcal{O}_1 \to P_1(\Phi n,n).$$

Then $S_1\Phi$ would be a tree of type 1, with ordinal <u>at least</u> ω_1, so that

$$|\underset{\sim}{T}_1(\Phi)| \geq \omega_1 > |\underset{\sim}{T}_1|.$$

So the ordinal bound has been "spoilt".

<u>Remark</u>. This argument shows that $\mathcal{O}_1.3$ is independent of $\underset{\sim}{ID}_1(\mathcal{O})$. (See also remark 6.8.8.)

6.8.3. Now let us try rather a <u>Dialectica</u> <u>interpretation</u> of $\underset{\sim}{ID}_1^+(\mathcal{O})$ in $\underset{\sim}{E}$-$\underset{\sim}{T}_1\underset{\sim}{P}$, extending that of $\underset{\sim}{HA}$ in chapter III, § 5.

The Dialectica translation F^D of a formula F is defined as in 3.5.2, d(i) - d(vi). It remains to define $(\mathcal{O}_1 t)^D$. Suppose this is defined, as with the $\underset{\approx}{mr}$-translation, by:

d(vii) $\quad (\mathcal{O}_1 t)^D \equiv \exists\alpha P_1\alpha t$.

Now first we must adjoin to $\underset{\sim}{E}$-$\underset{\sim}{T}_1\underset{\sim}{P}$ the characteristic function of P_1, with appropriate axioms (so that we can construct a term T_B for all quantifier-free formulas B of $\mathscr{L}_2[P]$, as in 1.6.14, to solve the translation of PL 10: see 3.5.4). This will imply the decidability of P_1.

Next, let us consider the translations of the axioms for $\mathcal{O}_1$.

$\mathcal{O}_1.3$ now offers no problems: $(\mathcal{O}_1.3)^D \equiv$

$$\exists X \forall n,\alpha[\neg\neg P_1(\alpha,n) \to P_1(Xn\alpha,n)]$$

which is easily solved by taking $X = \lambda n\alpha.\alpha$.

$\mathcal{O}_1.1a$: $(\mathcal{O}_1.1a)^D \equiv \exists\alpha P_1\alpha 0$, so this is again solved by taking $\alpha = 0^1$.

$\mathcal{O}_1.1b$. The translation proceeds, as with the $\underset{\approx}{mr}$-translation, to stage (1) of 6.8.2, and then to:

$$\exists N,X\forall g,h[T(e,Ngh,h(Ngh)) \,\&\, P_1(g(Ngh),U(h(Ngh))) \to P_1(Xgh, 3.5^e)],$$

i.e. n must also be "pulled out" as a function of g and h. This is solved for X as before, i.e. by taking $X = \lambda ghS_1\langle g,h\rangle$, and for N as a ("non-constructive" and discontinuous) function of g and h, say a total "least number operator", or in fact any number selection operator.

$\mathcal{O}_1.2$ gives no trouble: its translation is solved, as in the modified realizability interpretation, by the use of TR_1.

To sum up: $\underset{\sim}{ID}_1^+(\mathcal{O})$ admits a Dialectica interpretation in the theory $\underset{\sim}{E}-\underset{\sim}{T}_1\underset{\sim}{P}$, augmented by the characteristic function of P_1, and also a number selection operator, i.e. a constant $\mu \in ((0)0)0$, with the axiom

(1) $\quad f^{(0)0}n = 0 \rightarrow f(\mu f) = 0$.

In fact, if we adjoin to the theory $\underset{\sim}{E}-\underset{\sim}{T}_1$ such a functional, then the characteristic function of P_1 can actually be defined by TR_1 so that axioms $P_1.1'-P_1.3'$ are derivable. The point is that μ provides a functional interpretation of number quantification, since (1) implies:

$$\exists n(fn = 0) \leftrightarrow f(\mu f) = 0.$$

We will show that the adjunction to $\underset{\sim}{T}_1$ of any functional μ satisfying (1) does not affect the ordinal bound of $\underset{\sim}{T}_1$, i.e. $|\underset{\sim}{T}_1(\mu)| = |\underset{\sim}{T}_1|$. (In fact we will obtain a more general result.) Hence we obtain the result $|\underset{\sim}{ID}_1^{\mathcal{O}}| = |\underset{\sim}{T}_1|$ (6.8.5 below). First we need:

Definition. A functional in $\mathcal{A}_1$ is type-0-valued if its type has the form $(\tau_1)(\tau_2)\ldots(\tau_n)0$ for some $n \geq 0$. It is type-1-valued if its type has the form $(\tau_1)(\tau_2)\ldots(\tau_n)1$.

Note. Every functional in $\mathcal{A}_1$ is either type-0-values or type-1-valued. The functional μ is type-0-valued. So is any characteristic function, e.g. of equality at any type.

6.8.4. Theorem. If $\Psi_1, \Psi_2, \ldots$ is any sequence of type-0-valued functionals in $\mathcal{A}_1$, then

$$|\underset{\sim}{T}_1(\Psi_1,\Psi_2,\ldots)| = |\underset{\sim}{T}_1|.$$

Proof. We will work in $\mathcal{A}_1$, so τ or M_τ is the domain of objects of type τ in $\mathcal{A}_1$, and $|\alpha|$ is the ordinal canonically associated with $\alpha \in M_1$.

The idea of the proof is this. With each functional Φ of $\underset{\sim}{T}_1(\Psi_1,\Psi_2,\ldots)$ we associate a functional Φ^* of $\underset{\sim}{T}_1$ which "majorizes" it. More precisely, we define a binary relation maj_τ on each M_τ with the following properties (writing maj for maj_τ):

1°) $\forall f^{(\rho)\sigma}, g^{(\rho)\sigma}, x^\rho, y^\rho$ (f maj g and x maj y $\Rightarrow$ fx maj gy)

2°) $\alpha \; maj_1 \; \beta \Rightarrow |\alpha| \geq |\beta|$.

We say "x majorizes y" for x maj y.

Then we show (lemmas 3 and 4 below) that if Φ is one of the constants of

$\underset{\sim}{T}_1$, or one of $\Psi_1,\Psi_2,\ldots$, then there is a functional of $\underset{\sim}{T}_1$ which majorizes Φ .

It follows, by property 1° above, that if Φ is any functional of $\underset{\sim}{T}_1(\Psi_1,\Psi_2,\ldots)$, then there is a functional of $\underset{\sim}{T}_1$ which majorizes Φ (lemma 5). From this and property 2°, the theorem follows.

This "majorizing" technique is a modification of one used by Howard 1963 (section VI and appendix). (See 6.8.6 (b) below. Another example of this technique, as applied to models of $\underset{\sim}{HA}^{\omega}$, is given in the appendix of this volume.)

We now define the relation maj_{τ} by induction on τ .

$\tau = 0$. $\underset{\sim}{\forall}m,n(m \ maj_0 \ n)$. (Note! Any number majorizes any other.)

$\tau = 1$. The definition of $\alpha \ maj_1 \ \beta$ is by induction on $\alpha \in M_1$:

(i) $0^1 \ maj_1 \ 0^1$,

(ii) $\underset{\sim}{\forall}m,n(fm \ maj_1 \ gn) \Rightarrow S_1f \ maj_1 \ S_1g$,

(iii) $\underset{\sim}{\exists}m(fm \ maj_1 \ \beta) \Rightarrow S_1f \ maj_1 \ \beta$.

Note. $\alpha \ maj_1 \ \beta \Rightarrow |\alpha| \geq |\beta|$ (by induction on α or $|\alpha|$), but not conversely.

$\tau = (\rho)\sigma$. $f \ maj_{(\rho)\sigma} \ g \equiv_{def} \underset{\sim}{\forall}x^{\rho},y^{\rho}(x \ maj_{\rho} \ y \Rightarrow fx \ maj_{\sigma} \ gy)$.

Note. It is clear from this that property 1° holds.

Now we define in $\underset{\sim}{T}_1$ a "generalized supremum" $Sup_{\tau} \in ((0)\tau)\tau$ for all τ , such that lemma 1 (below) holds. The definition is by induction on τ :

$Sup_0 \equiv \lambda f^{(0)0}0^0$ (say),

$Sup_1 \equiv S_1$ (the given constant),

$Sup_{(\rho)\sigma} \equiv \lambda f^{(0)(\rho)\sigma}\lambda x^{\rho} \ Sup_{\sigma} \ \lambda n(fnx)$.

Lemma 1. $\underset{\sim}{\forall}f^{(0)\tau}\underset{\sim}{\forall}x^{\tau}[\underset{\sim}{\exists}m(fm \ maj \ x) \Rightarrow Sup_{\tau}f \ maj \ x]$.

Proof. By induction on τ . For $\tau = 1$, we use clause (iii) of the definition of maj_1 .

Next we define in $\underset{\sim}{T}_1$ a "generalized maximum" functional $Max_{\tau} \in (\tau)(\tau)\tau$ for all τ , such that lemma 2 (below) holds. The definition is again by induction on τ :

$Max_0(m,n) = 0$.

$Max_1(\alpha,\beta)$ is defined by TR_1 on β : $\begin{cases} Max_1(\alpha,0) \\ Max_1(\alpha,S_1g) = S_1\lambda n \ Max_1(\alpha,gn) . \end{cases}$

$Max_{(\rho)\sigma}(f,g) = \lambda x^{\rho} \ Max_{\sigma}(fx,gx)$.

Lemma 2. $\underset{\sim}{\forall}x^{\tau},y^{\tau},z^{\tau}[x \ maj \ z \text{ or } y \ maj \ z \Rightarrow Max_{\tau}(x,y) \ maj \ z]$.

Proof. Induction on τ . For $\tau = 1$, use induction on $y \in M_1$, and lemma : $\underset{\sim}{\forall}\alpha(\alpha \ maj_1 \ 0^1)$ (proved by induction on α).

Lemma 3. If Ψ is any type - 0 - valued functional in $\mathcal{J}_1$, then Ψ is majorized by a functional of $\underset{\sim}{T}_1$.

Proof. Suppose $\Psi \in (\tau_1)\ldots(\tau_n)0$. Then

$$\lambda x_1^{\tau_1}\ldots\lambda x_n^{\tau_n}0^o \text{ maj } \Psi .$$

(The whole point is that $\underset{\sim}{\forall} n^o(0^o \text{ maj } n^o)$!)

Lemma 4. For every constant of $\underset{\sim}{T}_1$, there is a functional of $\underset{\sim}{T}_1$ which majorizes it.

(Note. This is not trivial. Not every functional majorizes itself.)

Proof. Consider the constants in turn.

$\Sigma_{\rho,\sigma,\tau}$. $\Sigma_{\rho,\sigma,\tau}$ maj $\Sigma_{\rho,\sigma,\tau}$.

(Proof. Suppose $x,x^* \in (\rho)(\sigma)\tau$, $y,y^* \in (\rho)\sigma$, $z,z^* \in \rho$, with x^* maj x, y^* maj y, z^* maj z. Then, by property 2°, x^*z^* maj xz, y^*z^* maj yz, and so $x^*z^*(y^*z^*)$ maj $xz(yz)$, i.e. $\Sigma x^*y^*z^*$ maj Σxyz.)

$\Pi_{\sigma,\tau}$. Similarly, $\Pi_{\sigma,\tau}$ maj $\Pi_{\sigma,\tau}$.

0°. 0° maj 0°, by definition of maj_o.

0^1. 0^1 maj 0^1, by clause (i) of the definition of maj_1.

S_o. S_o maj S_o, since $\underset{\sim}{\forall} m,n(S_o m \text{ maj } S_o n)$.

S_1. S_1 maj S_1. Proof:

$$f \text{ maj}_{(0)1}\, g \Leftrightarrow \underset{\sim}{\forall} m,n(fm \text{ maj}_1\, gn) \quad \text{by definition,}$$
$$\Rightarrow S_1 f \text{ maj}_1\, S_1 g ,$$

by clause (ii) of the definition of maj_1.

$R_{o,\tau}$. Define $R_o^* \equiv R_{o,\tau}^*$, of the same type as $R_{o,\tau}$, by:

$$R_o^* x^\tau y^{(o)(\tau)\tau} = \lambda m.\text{Sup}_\tau(R_o xy) .$$

Now take $x,x^* \in \tau$ and $y,y^* \in (0)(\tau)\tau$, with x^* maj z and y^* maj y. Then

$$\underset{\sim}{\forall} n(R_o x^*y^*n \text{ maj } R_o xyn)$$

by induction on n. (But this does not imply that $R_o x^*y^*$ maj $R_o xy$!) Then by lemma 1,

$$\underset{\sim}{\forall} n(\text{Sup}_\tau(R_o x^*y^*) \text{ maj } R_o xyn) ,$$

so $\lambda m.\text{Sup}_\tau(R_o x^*y^*) \text{ maj } R_o xy ,$

and hence R_o^* maj R_o.

$R_{1,\tau}$. Define $R_{1,\tau}^*$, of the same type as $R_{1,\tau}$, by TR_1 (with variables $x^* \in \tau$, $y^* \in ((0)\tau)((0)1)\tau)$:

$$R_1^* x^* y^* = \Phi^*,$$

where $\begin{cases} \Phi^* 0^1 = x^* \\ \Phi^*(S_1 g) = \mathrm{Max}_\tau[y^*(\Phi^* \circ g, g),\ \mathrm{Sup}_\tau(\Phi^* \circ g)]. \end{cases}$

Now take $x, x^* \in \tau$ and $y, y^* \in ((0)\tau)((0)1)\tau$, with x^* maj x and y^* maj y.

Let $\Phi = R_1 xy$, $\Phi^* = R_1^* x^* y^*$ as above. Then

$$\alpha \text{ maj } \beta \Rightarrow \Phi^*\alpha \text{ maj } \Phi\beta .$$

The proof is by induction on α, or on the <u>inductive</u> <u>definition</u> <u>of</u> maj_1:

(i) If $\alpha = 0^1$, $\beta = 0^1$; then $\Phi^*\alpha = x^*$, $\Phi\beta = x$, so $\Phi^*\alpha$ maj $\Phi\beta$.

(ii) If $\alpha = S_1 f$, $\beta = S_1 g$, and $\forall m,n(fm \text{ maj } gn)$:

then $\forall m,n\ \Phi^*(fm) \text{ maj } \Phi(gn)$, by <u>induction</u> <u>hypothesis</u>, so

(1) $\Phi^* \circ f \text{ maj } \Phi \circ g$ by definition. Also

(2) $f \text{ maj } g$ by definition,

so $y^*(\Phi^* \circ f, f) \text{ maj } y(\Phi \circ g, g)$, by (1) and (2).

Hence $\Phi^*(S_1 f) \text{ maj } \Phi(S_1 g)$ by lemma 2.

(iii) If $\alpha = S_1 f$ and $\exists m(fm \text{ maj } \beta)$:

then $\Phi^*(fm) \text{ maj } \Phi\beta$ by <u>induction</u> <u>hypothesis</u>.

So $\mathrm{Sup}^\tau(\Phi^* \circ f) \text{ maj } \Phi\beta$ by lemma 1,

and so $\Phi^*(S_1 f) \text{ maj } \Phi\beta$ by lemma 2.

Hence Φ^* maj Φ; and so R_1^* maj R_1.

From lemmas 3 and 4, and property 1° of maj, we immediately obtain:

<u>Lemma 5</u>. If each Ψ_i is type-0-valued, then every functional of $T_1(\Psi_1, \Psi_2, \ldots)$ is majorized by some functional of T_1.

Hence the theorem follows.

6.8.5. <u>Corollary</u>. $|ID_1^c| = |T_1|$.

<u>Proof</u>. From the discussion in 6.8.3 and theorem 6.8.4.

6.8.6. <u>Historical survey: other methods of characterizing</u> $|ID_1^c|$.

(a) Firstly, it is known that

(1) $ID_1^c(\mathcal{O}) \preceq ID_1(\mathcal{O})$

(where $\preceq$ means proof-theoretical reducibility), so that

$$|ID_1^c| = |ID_1|,$$

which gives another proof of 6.8.5 (using $|ID_1| = |T_1|$).

The reduction (1) is highly non-trivial. It is outlined in Kreisel 1968 A, pp. 345 - 6, and also here, for convenience. The steps are:

$$\underset{\sim}{ID}_1^c(\mathcal{O}) \overset{(i)}{\leq} \underset{\sim}{EL}^c + BI_{QF} \overset{(ii)}{\leq} qf - \underset{\sim}{WE} - \underset{\sim}{HA}^\omega + BR_0 \overset{(iii)}{\leq} \underset{\sim}{EL} + BI_0 \overset{(iv)}{\leq}$$

$$\overset{(iv)}{\leq} \underset{\sim}{IDB}_1 \overset{(v)}{\leq} \underset{\sim}{ID}_1(\mathcal{O})$$

where BI_0 can be taken as BI_{QF}, BI_D or BI_M (1.9.20).

Step (i) is accomplished by an explicit definition of $\mathcal{O}_1$ (i.e. saying that there are no infinite descending sequences of a certain kind from elements of $\mathcal{O}_1$).

(ii) is a Dialectica interpretation (Howard 1968: cf. 3.5.19).

(iii): $qf - \underset{\sim}{WE} - \underset{\sim}{HA}^\omega + BR_0$ is modelled in ECF (Tait 1963, Kreisel 1968 A, footnote 33; incidentally, this modelling is extended to BR at all types in 2.9.9).

(iv): by "elimination of choice sequences" (Kreisel and Troelstra 1970, § 7).

(v): by a realizability interpretation of $\underset{\sim}{IDB}_1$ in a theory $\underset{\sim}{ID}_1(K_1)$, where $K_1 \subset N$ is the set of indices of recursive neighbourhood functions representing continuous type 2 functionals (Kreisel and Troelstra 1970, §§ 3.7 and 3.8.1*). Then $\underset{\sim}{ID}_1(K_1)$ can be interpreted directly in $\underset{\sim}{ID}_1(\mathcal{O})$ by explicit definition of K_1, which follows from the proof of the many-one reducibility of K_1 to $\mathcal{O}_1$ (cf. Rogers 1967, exercises 11 - 61 and 16 - 27, where T is used instead of K_1).

(b) Let $\underset{\sim}{\Omega}_1$ be a theory of functionals of finite type over the countable ordinals. There is only one ground type, that of the ordinals, with ordinals less than ω acting as natural numbers. $\underset{\sim}{\Omega}_1$ includes constants for 0 and ω, and transfinite recursion on the ordinals. The exact formulation is not so important here, since Howard 1963 (section VI, appendix I) showed, by a majorizing technique (cf. 6.8.4) that various formulations of $\underset{\sim}{\Omega}_1$ (including e.g. adjoining characteristic functions of predicates or a functional for bounded supremum, or changing the exact form of the recursion functional) lead to the same value for $|\underset{\sim}{\Omega}_1|$, i.e. the supremum of the ordinals denoted by the closed terms of $\underset{\sim}{\Omega}_1$ of ground type.

Howard 1963 (section VI) described a Dialectica interpretation of $\underset{\sim}{IDB}_1$ into a quantifier-free version of $\underset{\sim}{IDB}^\omega$ (say $qf - \underset{\sim}{WE} - \underset{\sim}{IDB}^\omega$, cf. 1.9.25). Now one can associate an ordinal canonically with each element of K, and hence with each closed term of $\underset{\sim}{IDB}^\omega$ of type K, and so define $|\underset{\sim}{IDB}^\omega|$, the "ordinal of $\underset{\sim}{IDB}^\omega$", as the supremum of the ordinals of these closed terms of type K. Then Howard proved, by a majorizing argument between the functionals of $\underset{\sim}{IDB}^\omega$ and those of $\underset{\sim}{\Omega}_1$ (in both directions) that

-----

* Op. cit., § 3.8.1, actually refers to primitive recursive indices, but general recursive indices are more convenient here.

$|\underset{\sim}{IDB}^{\omega}| = |\underset{\sim}{\Omega}_1|$.

From this, and the reduction of $\underset{\sim}{ID}_1^c(0)$ to $\underset{\sim}{IDB}_1$ (described in part (a) above), we obtain :

(2) $|\underset{\sim}{ID}_1^c| \leq |\underset{\sim}{\Omega}_1|$.

We remark that it can also be shown that

$|\underset{\sim}{\Omega}_1| = |\underset{\sim}{T}_1|$,

again by a majorizing argument, this time between $\underset{\sim}{\Omega}_1$ and $\underset{\sim}{T}_1$ (in both directions).

(c) Feferman 1968 gave a direct proof of (2) (in fact with equality : $|\underset{\sim}{ID}_1^c| = |\underset{\sim}{\Omega}_1|$) in the following way.

For the inequality $\leq$, he described a functional interpretation of $\underset{\sim}{ID}_1^c(A)$, for any positive A , into $\underset{\sim}{\Omega}_1$. This proceeds in three stages, as follows.

Let $\underset{\sim}{OR}$ be a first-order, quantified, intuitionistic theory of ordinals (with decidable = and $<$). (We can take the system of Takeuti 1965 without the axiom for cardinals, but with (4) below.) $\underset{\sim}{OR}$ includes constants for 0 and ω, and defining schemata for certain function constants $f, g, \ldots$, including (predicative) transfinite recursion, functions for bounded quantification :

(3) $f(\alpha, \gamma_1, \ldots, \gamma_k) = 0 \leftrightarrow \exists\beta < \alpha\, g(\beta, \gamma_1, \ldots, \gamma_k) = 0$

and an axiom for " ω -upper bounds" :

(4) $\forall\alpha < \omega\, \exists\beta\, (\alpha, \beta, \gamma_1, \ldots) = 0 \leftrightarrow \exists\delta\, \forall\alpha < \omega\, \exists\beta < \delta\; f(\alpha, \beta, \gamma_1, \ldots) = 0$.

$\underset{\sim}{OR}^c$ is $\underset{\sim}{OR}$ with classical logic.

$\underset{\sim}{HA}^c$ can be interpreted directly in $\underset{\sim}{OR}^c$ by translating number quantification as quantification over ordinals bounded by ω. Then (writing $a, b, \ldots$ as variables for ordinals less than ω, and writing the translation of a formula B of $\underset{\sim}{HA}^c$ again as B) : for every formula B of $\underset{\sim}{HA}^c$, there is, by (3), a function constant f of $\underset{\sim}{OR}$ such that

(5) $\underset{\sim}{OR} \vdash B \leftrightarrow f(a_1, \ldots, a_n) = 0$

($a_1, \ldots, a_n$ the free variables of B).

Now $\underset{\sim}{ID}_1^c(A)$, for a positive $A(X,a)$, is interpreted in $\underset{\sim}{OR}^c$ by (further) translating $Q_1 t$ as $\exists\alpha P\alpha t$, where P satisfies

(6) $\underset{\sim}{OR} \vdash P\alpha a \leftrightarrow A(\hat{b}\exists\beta < \alpha P\beta b,\ a)$

and the characteristic function of P can be defined in $\underset{\sim}{OR}$ by transfinite recursion, using (3) to eliminate the number quantifiers in A, as in (5). (Compare this with the use of the μ-operator in 6.8.3 to define the characteristic function of P_1 by TR_1.)

Now the translation of the schema $Q_1.2$ (6.2.2) is proved in $\underset{\sim}{OR}$ by transfinite induction on the ordinals. Finally the translation of $Q_1.1$,

(7) $A(\hat{b}\, \exists \beta\, P\beta b, a) \rightarrow \exists \alpha\, P\alpha a$,

is proved as follows. Using the positivity of $A(X,a)$, we can bring the hypothesis of (7) to prenex normal form:

$$Q_1 c_1 \ldots Q_n c_n\ \exists \beta_1 \ldots \exists \beta_m\ A^*(c_1, \ldots, c_n, \beta_1, \ldots, \beta_m, a)$$

where $Q_i c_i$ denotes quantification over ordinals $< \omega$, and A^* is quantifier-free. This implies by (3) and repeated use of (4):

$$\exists \alpha Q_1 c_1 \ldots Q_n c_n\ \exists \beta_1 < \alpha \ldots \exists \beta_m < \alpha\ A^*(c_1, \ldots, c_n, \beta_1, \ldots, \beta_m, a).$$

But this is equivalent to $\exists \alpha A(\hat{b}\, \exists \beta < \alpha\, P\beta b,\ a)$, i.e., by (6), to the desired conclusion.

The second stage of the interpretation consists in interpreting $\underset{\sim}{OR}^c$ in $\underset{\sim}{OR}$. This is achieved simply by a $\neg\neg$ translation.

The third stage is a Dialectica interpretation of $\underset{\sim}{OR}$ in $\underset{\sim}{\Omega}_1$. Feferman's formulation of $\underset{\sim}{\Omega}_1$ includes a supremum functional:

$$\mathrm{Sup}(\varphi, \alpha) = \sup_{\beta < \alpha} \varphi(\beta),$$

φ of type $(0)0$ (0 = type of ordinals), which solves the Dialectica translation of (4), and a functional of bounded quantification, which takes care of (3).

This proves the inequality (2).

The reverse inequality was proved by modelling $\underset{\sim}{\Omega}_1$ in $\underset{\sim}{ID}_1^c(\mathcal{O})$ as a system of hereditarily hyperarithmetical operations of finite type over $\mathcal{O}_1$ (so as to be able to define a linear order in $\mathcal{O}_1$ interpreting $<$, and also to account for bounded quantification).

It was the present author's (unsuccessful) attempt to extend this method to $\underset{\sim}{ID}_2^c(\mathcal{O})$ that led him to consider a theory of trees.

(d) Howard 1972 considers theories $\underset{\sim}{ID}_1(A)$, with A positive in the sense of Kreisel and Troelstra 1970, § 4.4, and gives another proof of $|\underset{\sim}{ID}_1(A)| \leq |\underset{\sim}{T}_1|$, as follows.

First, $\underset{\sim}{ID}_1(A)$ is interpreted in an intuitionistic first-order theory $\underset{\sim}{U}$ of trees. For this, a normal form theorem for A is used, like that of Kreisel and Troelstra 1970, § 4.5.

Then $\underset{\sim}{U}$ is Dialectica-interpreted in a theory $\underset{\sim}{V}$ which is like a qf (quantifier-free) version of our $\underset{\sim}{T}_1$. $Q_1 t$ is translated as $\exists \alpha\, P\alpha t$, where P is now <u>not</u> qf, but of the form $\forall n P'(n,\alpha,t)$, where P' is qf, with its characteristic function definable by TR_1.

It is interesting to compare this method with the proof of $|\underset{\sim}{ID}_1^c| \le |\underset{\sim}{T}_1|$, by means of another Dialectica interpretation of $\underset{\sim}{ID}_1^+(\mathcal{O})$, given earlier (6.8.3-4).

On the one hand, because of the different translation of $Q_1 t$, Howard's method does not need non-constructive functionals such as a μ-operator.

On the other hand (again because of this different translation), his method applies (apparently) only to intuitionistic $\underset{\sim}{ID}_1(A)$, since it is not clear that the translation of $\mathcal{O}_1.3$ (6.8.1) can be solved without affecting the ordinal bound $|\underset{\sim}{T}_1|$. With our definition of $(Q_1 t)^D$, the translation of $\mathcal{O}_1.3$ (6.8.3) comes for free, so to speak.

In the same paper Howard also gives a characterization of $|\underset{\sim}{T}_1|$ in terms of Bachmann's notations (see below).

(e) <u>Analysis in terms of Bachmann - Isles notations.</u>

We mention that

$$|\underset{\sim}{ID}_1^c| = |\underset{\sim}{ID}_1| = \varphi_{\varepsilon_{\Omega+1}}(1)$$

in the notation of <u>Bachmann</u> 1950. The inequality $|\underset{\sim}{ID}_1| \le \varphi_{\varepsilon_{\Omega+1}}(1)$ follows from <u>Howard</u> 1970 A (or, more simply, <u>Howard</u> 1972), and the reverse inequality from <u>Gerber</u> 1970.

I conjecture that, further,

$$|\underset{\sim}{ID}_2| = \varphi_{F_{\varepsilon_{\omega_2+1}}(1)}(1)$$

in Bachmann's notation (i.e. $F^1(F^2(F^3(2,1),1),1)$ in that of <u>Isles</u> 1970). Martin-Löf conjectured this independently (<u>Martin-Löf</u> 1971).

<u>Note</u>. It is stated, op. cit., that I have proved the above conjecture. This is not so, although it seems that it could be proved by (in one direction) an ordinal analysis of $\underset{\sim}{T}_2$ by means of infinite terms, extending the method of <u>Howard</u> 1972, and (in the other) an extension of the method of <u>Gerber</u> 1970.

<u>Martin-Löf</u> 1971 also gives an ordinal characterization of his system of finitely iterated inductive definitions (op. cit.) in terms of Isles notations. (See 6.9.2.)

6.8.7 - 6.8.11. Functional interpretations of $\underset{\sim}{ID}_2(\mathcal{O})$ and $\underset{\sim}{ID}_2^+(\mathcal{O})$.

6.8.7. Consider, first, the modified realizability ($\underset{\sim}{mr}$-) interpretation of $\underset{\sim}{ID}_2(\mathcal{O})$, defined in 6.7.1. We have already treated the general case of $\underset{\sim}{ID}_2(A)$, for any $A \in \mathcal{C}$, with theorem 6.7.5, but now for convenience we review briefly the special case where $A \equiv (A_1, A_2)$ is the pair of defining predicates for $(\mathcal{O}_1, \mathcal{O}_2)$.

The axioms for $\mathcal{O}_1$ have been dealt with in 6.8.2. Now consider the axioms for $\mathcal{O}_2$.

$\mathcal{O}_2$. 1a is translated as $\exists\alpha^2 P_2(\alpha^2, 0)$, which is solved in $\underset{\sim}{E} - \underset{\sim}{T}_2\underset{\sim}{P}$ by taking $\alpha^2 = 0^2$, and using axiom $P_2.1'$ (6.8.1).

$\mathcal{O}_2$. 1b. The translation proceeds to the stage:

(1) $$\exists g,h\forall\alpha,n[P_1\alpha n \to T(e,n,hn\alpha)\ \&\ P_2(gn\alpha, U(hn\alpha))] \to \exists\beta^2 P_2(\beta^2, 3^2.5^e).$$

The translation is completed, and solved in $\underset{\sim}{E} - \underset{\sim}{T}_2\underset{\sim}{P}$, by taking β^2 as the function $S_2\langle g,h\rangle$ of g,h (and using axiom $P_2.2'$).

$\mathcal{O}_2$. 2. The translation is solved as in the general case (6.7.5) by the use of TR_2 (and the axioms TI_2, $P_2.1'$ and $P_2.3'$).

Thus we obtain a functional interpretation of $\underset{\sim}{ID}_2(\mathcal{O})$ in $\underset{\sim}{E} - \underset{\sim}{T}_2\underset{\sim}{P}$, showing (as a particular case of theorem 6.7.9) that

$$|\underset{\sim}{ID}_2(\mathcal{O})| \le |\underset{\sim}{T}_2|.$$

However, this interpretation is unsuitable for $\underset{\sim}{ID}_2^+(\mathcal{O})$, as shown (already for the axiom $\mathcal{O}_1$. 3 of $\underset{\sim}{ID}_1^+(\mathcal{O})$) in 6.8.2.

6.8.8. Remark. This argument also shows that $\mathcal{O}_1$. 3 is independent of $\underset{\sim}{ID}_2(\mathcal{O})$.

6.8.9. So now let us try a Dialectica interpretation of $\underset{\sim}{ID}_2^+(\mathcal{O})$, with the Dialectica translation defined by: $(\mathcal{O}_i t)^D \equiv \exists\alpha^i P_i(\alpha^i, t)$ $(i = 1,2)$ (i.e. extending the interpretation of $\underset{\sim}{ID}_1^+(\mathcal{O})$ in 6.8.3). Consider again the translations of the axioms for $\mathcal{O}_1$ and $\mathcal{O}_2$:

$(\mathcal{O}_i.\ 3)^D$ $(i = 1,2)$ now gives no trouble (as shown for $i = 1$ in 6.8.3).

$(\mathcal{O}_i.\ 1a)^D$ $(i = 1,2)$ is solved as for the modified realizability interpretation.

$(\mathcal{O}_i.\ 2)^D$ $(i = 1,2)$ is again solved by TR_i.

We are left with $(\mathcal{O}_i.1b)^D$. For $i = 1$, this can be solved as in 6.8.3, by adjoining a number selection operator μ (and then using a majorizing argument for the ordinal analysis).

However, for $i = 2$, the situation is more serious. The translation proceeds to the stage (1) (in 6.8.7), and is then completed by pulling out

n and α (as well as β^2), and solving for them as functions of g, h and e.

For n this can be done again with the functional μ. However, this cannot be done for α (i.e. adjoining a suitable tree selection operator) without affecting the ordinal bound of the system of functionals.

For suppose we could find α (and n) as functions of g, h and e:

(1) $$\begin{cases} \alpha = \Psi ghe, \\ n = N ghe, \end{cases}$$

satisfying (in $\mathcal{J}_2$) for all g, h, e:

$$[P_1\alpha n \to T(e,n,hn\alpha) \; \& \; P_2(gn\alpha, U(hn\alpha))] \to P_2(S_2\langle g,h\rangle, 3^2.5^e),$$

i.e.

(2) $$\neg P_2(S_2\langle g,h\rangle, 3^2.5^e) \to [P_1\alpha n \; \& \; (\neg T(e,n,hn\alpha) \vee \neg P_2(gn\alpha, U(hn\alpha)))],$$

with α, n as in (1).

Then we could define a functional $\Phi \in (0)1$ such that

(3) $$k \in \mathcal{O}_1 \Rightarrow P_1(\Phi k, k)$$

(proved in 6.8.10 below), so that $S_1\Phi$ is a tree of type 1 with ordinal $\geq \omega_1$, and hence

$$|\underset{\sim}{T}_2(\Psi)| \geq \omega_1 > |\underset{\sim}{T}_2|,$$

so that again the ordinal bound has been spoilt.

So both functional interpretations (6.8.7 and 6.8.9) fail to show that $|\underset{\approx}{ID}_2^c| \leq |\underset{\sim}{T}_2|$, and it is still an open problem whether

$$|\underset{\approx}{ID}_2^c| = |\underset{\sim}{T}_2| \quad \text{or} \quad |\underset{\approx}{ID}_2^c| > |\underset{\sim}{T}_2|.$$

[6.8.10. Derivation of (3) of 6.8.9.

Let e_k and $h_{k,n}$ be numbers which depend primitive recursively on k, resp. k, n, such that for all k, n:

(1) $$T(e_k, n, h_{k,n}) \; \& \; U(h_{k,n}) = \delta_{k,n}$$

($\delta_{k,n}$ = Kronecker delta, $= 0$ if $k \neq n$, 1 if $k = n$).

So $\forall k,n \quad \{e_k\}(n) = \delta_{k,n}$.

Now define

(2) $$\begin{cases} g = \lambda n,\alpha.\ 0^2, & \text{and} \\ h_k = \lambda n,\alpha.\ h_{k,n} & \text{(for any } k\text{)}. \end{cases}$$

Finally, define $\Phi \in (0)1$ by

$$\Phi k = \Psi g h_k e_k.$$

We will now show that Φ satisfies (3) of 6.8.9.

From $P_2.2'$, $P_2.3'$ and (1) (with g, h_k as in (2)):

(3) $$P_2(S_2\langle g,h_k\rangle,\ 3^2.5^{e_k}) \Leftrightarrow \forall\alpha,n[P_1\alpha n \Rightarrow P_2(gn\alpha,\ \delta_{k,n})].$$

Now $\forall n,\alpha(gn\alpha = 0^2)$, so

$$\begin{aligned} P_2(gn\alpha,\ \delta_{k,n}) &\Leftrightarrow P_2(0^2,\ \delta_{k,n}) \\ &\Leftrightarrow \delta_{k,n} = 0 \qquad \text{by } P_2.1' \\ &\Leftrightarrow k \neq n. \end{aligned}$$

So (3) becomes:

$$\begin{aligned} P_2(S_2\langle g,h_k\rangle,\ 3^2.5^{e_k}) &\Leftrightarrow \forall\alpha,n[P_1\alpha n \Rightarrow n \neq k] \\ &\Leftrightarrow \forall\alpha\neg P_1\alpha k \\ &\Leftrightarrow k \notin \mathcal{O}_1. \end{aligned}$$

So: $k \in \mathcal{O}_1 \Rightarrow \neg P_2(S_2\langle g,h_k\rangle,\ 3^2.5^{e_k})$

(4) $$\Rightarrow P_1(\Psi gh_k e_k,\ Ngh_k e_k) \text{ and } \neg P_2(0^2,\ \delta_{k,Ngh_k e_k})$$

by (2) of 6.8.9. Now

$$\begin{aligned} \neg P_2(0^2,\ \delta_{k,Ngh_k e_k}) &\Rightarrow \delta_{k,Ngh_k e_k} \neq 0 \qquad \text{by } P_2.1' \\ &\Rightarrow Ngh_k e_k = k. \end{aligned}$$

So (4) $\Rightarrow P_1(\Psi gh_k e_k, k)$,

i.e. $k \in \mathcal{O}_1 \Rightarrow P_1(\Phi k,k)$.]

6.8.11. <u>Remark</u>. There <u>is</u> an easy interpretation of $\underset{\sim}{ID}_2^c(\mathcal{O})$ into an intuitionistic system $\underset{\sim}{ID}_2(\mathcal{O}^{\neg\neg})$ (say) of iterated inductive definitions, namely the $\neg\neg$ interpretation (i.e. ' of 1.10.2); but the inductive definitions of $\underset{\sim}{ID}_2(\mathcal{O}^{\neg\neg})$ do not even satisfy the condition of <u>positivity</u> (let alone $\mathcal{C}$), so this does not seem to help for an ordinal analysis of $\underset{\sim}{ID}_2^c(\mathcal{O})$.

§ 9. Extensions to $\underset{\sim}{ID}_\nu(A)$ and $\underset{\sim}{ID}^c_\nu(A)$ for $\nu > 2$. Equivalences with some subsystems of classical analysis.

6.9.1. The work of this chapter can be extended to systems $\underset{\sim}{ID}_\nu(A)$ of inductive definitions iterated ν times, and corresponding systems $\underset{\sim}{T}_\nu$ of trees of the first $1+\nu$ classes, for $\nu > 2$, even (apparently) for some $\nu \geq \omega$. (See Feferman 1970 for the definitions of the (classical) systems $\underset{\sim}{ID}^c_\nu(A)$ for $\nu \geq \omega$.) For example, we can define a system $\underset{\sim}{ID}_\omega(\mathcal{O})$ of all the recursive finite number classes $(\mathcal{O}_1, \mathcal{O}_2, \mathcal{O}_3, \ldots)$ (cf. Richter 1965), where $\mathcal{O}$ is the binary predicate $\{\langle n,x\rangle : x \in \mathcal{O}_n\}$, and also (it seems) a corresponding theory $\underset{\sim}{T}_\omega$ of trees of all the finite tree classes, with a distinct ground type for each class, and transfinite recursion on each class, such that

(1) $$|\underset{\sim}{ID}_\omega(\mathcal{O})| = |\underset{\sim}{T}_\omega| ,$$

where $|\underset{\sim}{ID}_\omega(\mathcal{O})|$ is the supremum of the ordinals of numbers provably (in $\underset{\sim}{ID}_\omega(\mathcal{O})$) in $\mathcal{O}_1$, and $|\underset{\sim}{T}_\omega|$ is the supremum of the ordinals of the closed terms of $\underset{\sim}{T}_\omega$ of type 1 (i.e. the second tree class).

Further, we can define a system I of ordinal notations up to (apparently) the first recursively inaccessible (something like the systems $\tilde{C}$ of Kreider and Rogers 1961, or F of Richter 1968), and, correspondingly, a theory $\underset{\sim}{ID}_I(\mathcal{O})$ of inductively defined sets $(\mathcal{O}_a : a \in I)$ and a theory $\underset{\sim}{T}_I$ of trees of all classes up to the first (recursively) inaccessible, so that again (it seems):

(2) $$|\underset{\sim}{ID}_I(\mathcal{O})| = |\underset{\sim}{T}_I|$$

(where the two sides are defined analogously to (1)).

The interest of these results lies partly in this. Feferman 1970 established the proof-theoretical equivalence of the classical systems $\underset{\sim}{ID}^c_\nu(A)$ with subsystems of classical analysis, for various ν. For example (with $\simeq$ denoting proof-theoretical equivalence):

(3) $$\underset{\sim}{ID}^c_\omega(\mathcal{O}) \simeq \underset{\sim}{Z}_2 + \Pi^1_1\text{-}CA + BI_0 ,$$

where $\underset{\sim}{Z}_2$ is classical second order arithmetic (i.e. $\underset{\sim}{HAS}^c_0 + EXT$), Π^1_1 - CA is the Π^1_1 - comprehension axiom, and BI_0 can be taken here as BI_D (1.9.20) with $Pn \equiv Xn$ (X a predicate variable) and Qn arbitrary, or as the schema BI in Feferman 1970.

Now let $\underset{\sim}{Z}^-_2$ be $\underset{\sim}{Z}_2$ with induction restricted to the single axiom with induction formula $F(x) \equiv Xx$ (X a predicate variable). For systems of iterated inductive definitions which correspond to $\underset{\sim}{Z}_2^- + \Pi_1^1\text{-CA}$ and $\underset{\sim}{Z}_2 + \Pi_1^1\text{-CA}$, see (resp.) 6.9.2 and 6.9.3 below.

Continuing this, we have (it seems):

(4) $$\underset{\sim}{ID}_I^c(\mathcal{O}) \simeq \underset{\sim}{Z}_2 + \Delta_2^1\text{-CA} + BI_0$$

(where $\underset{\sim}{ID}_I^c(\mathcal{O})$ is $\underset{\sim}{ID}_I(\mathcal{O})$ with classical logic).

Now if we knew that $|\underset{\sim}{ID}_\nu^c(\mathcal{O})| = |\underset{\sim}{ID}_\nu(\mathcal{O})|$ for $\nu = \omega$ and I, we could derive, from (1), (2), (3) and (4), interesting characterizations of the "ordinals of" $\underset{\sim}{Z}_2 + \Pi_1^1\text{-CA} + BI_0$ and $\underset{\sim}{Z}_2 + \Delta_2^1\text{-CA} + BI_0$ (i.e. the suprema of their provable well-orderings): namely,

(5) $$|\underset{\sim}{Z}_2 + \Pi_1^1\text{-CA} + BI_0| = |\underset{\sim}{T}_\omega|, \text{ and}$$

(6) $$|\underset{\sim}{Z}_2 + \Delta_2^1\text{-CA} + BI_0| = |\underset{\sim}{T}_I|.$$

However, it is not known for any $\nu \geq 2$ whether

(7) $$|\underset{\sim}{ID}_\nu^c(\mathcal{O})| = |\underset{\sim}{ID}_\nu(\mathcal{O})| \quad \text{or} \quad |\underset{\sim}{ID}_\nu^c(\mathcal{O})| > |\underset{\sim}{ID}_\nu(\mathcal{O})|,$$

so the truth of (5) and (6) remains an open problem.

The important problem here is to settle (7) for $\nu = 2$, i.e. the truth or falsity of $|\underset{\sim}{ID}_2^c| = |\underset{\sim}{T}_2|$ $(= |\underset{\sim}{ID}_2|)$, since a proof of this, if true, would surely generalize to give proofs of (5) and (6).

6.9.2. One positive result we do have in this direction is the following, pointed out in Martin-Löf 1971.

Let $\underset{\sim}{ID}_n(A)$ be the (intuitionistic) theory of inductive definitions iterated n times, and let $\underset{\sim}{ID}_{<\omega}(A)$ be the union of these theories (for $A \equiv (A_1, A_2, \ldots)$), and $\underset{\sim}{ID}_{<\omega}^c(A)$ the same with classical logic.

Firstly one can see, by adapting the arguments in Feferman 1970, that

$$\underset{\sim}{ID}_{<\omega}^c(A) \simeq \underset{\sim}{Z}_2^- + \Pi_1^1\text{-CA}$$

(for suitable A).

Next, Takeuti 1967 gave an analysis of his system SJNN (essentially $\underset{\sim}{Z}_2^- + \Pi_1^1\text{-CA}$) by means of his system of ordinal diagrams of finite order.

Further, Kreisel 1964 formalized the proof of well-foundedness of these ordinal diagrams in $\underset{\sim}{ID}_{<\omega}(A)$ for suitable A.

Putting this all together, we get

(1) $$|\underset{\sim}{ID}_{<\omega}^c| = |\underset{\sim}{ID}_{<\omega}| = |\underset{\sim}{Z}_2^- + \Pi_1^1\text{-CA}| = \sup_n O(n)$$

where $O(n)$ is the supremum of the ordinal diagrams of order n.

Finally, Levitz 1970 showed that $\sup O(n) = \sup \alpha_n$, where α_n is the ordinal

$$F^1(F^2 \dots (F^n(2,1)) \dots 1)\, 1)$$

in the notation of Isles 1970. So $\sup_n \alpha_n$ is another characterization of the ordinal in (1).

6.9.3. We conclude with a description of a theory $\underset{\sim}{W}-\underset{\sim\sim}{ID}^c_\omega(A)$ ("$\underset{\sim}{W}$" for weak induction), which is equivalent to $\underset{\sim}{Z}_2 + \Pi^1_1-CA$. For convenience, we first repeat the definition of $\underset{\sim\sim}{ID}^c_\omega(A)$ given in Feferman 1970.

Let $A(X,Y,x,y)$ be a formula of $\mathcal{L}[X,Y]$ which is positive in X, with X,Y,x,y its only free variables. Let Q be a new binary predicate symbol. $\underset{\sim\sim}{ID}^c_\omega(A)$ is the theory $\underset{\sim}{HA}^c$ in $\mathcal{L}[Q]$, together with the axioms:

Q.1) $A(Q_y,\ \hat{u}\hat{v}(u < y \ \&\ Quv), x, y) \rightarrow Qyx$

Q.2) $\forall x[A(\hat{z}F(z),\ \hat{u}\hat{v}(u < y \ \&\ Quv), x, y) \rightarrow F(x)] \rightarrow \forall x(Qyx \rightarrow F(x))$

for any $F(x)$ in $\mathcal{L}[Q]$ (where Q_y is the unary predicate $\hat{x}Qyx$).

Definition. $\underset{\sim}{W}-\underset{\sim\sim}{ID}^c_\omega(A)$ is like $\underset{\sim\sim}{ID}^c_\omega(A)$ except that in the induction schema Q.2, $F(x)$ is restricted to being a W-formula (w.r.t. x), which means that in any occurrence of the prime formula Qst in $F(x)$ (where s, t are number terms):

1°) if any variable y occurs in s, then Qst is not in the scope of a quantifier $\forall y$ or $\exists y$ in $F(x)$; and

2°) the variable x does not occur in s.

This just means that $F(x)$ has the form $F(Q_{s_1}, Q_{s_2}, \dots, x)$, where each s_i contains variables only as parameters, i.e. not quantified in F (and distinct from x).

[The proof that $\underset{\sim}{W}-\underset{\sim\sim}{ID}^c_{<\omega}(A) \simeq \underset{\sim}{Z}_2 + \Pi^1_1-CA$ (for suitable A) modifies that in Feferman 1970 (henceforth [Fef]) for (3) of 6.9.1:

(i) To model $\underset{\sim}{W}-\underset{\sim\sim}{ID}_\omega(A)$ in $\underset{\sim}{Z}_2 + \Pi^1_1-CA$: the point is that the extension of a W-formula $\hat{x}F(x)$ can be proved to exist as a set, by Π^1_1-CA.

(ii) Conversely, we model $\underset{\sim}{Z}_2 + \Pi^1_1-CA$ in a theory $\underset{\sim}{W}-\underset{\sim\sim}{ID}_\omega(A)$ like $\underset{\sim\sim}{ID}_{T_\omega}$ of [Fef] (except for having the weakened induction schema, as described). The proof of [Fef, theorem 2.2.1 (ii)] is modified as follows (using the notation there):

First note that the inductive definition of R-securability:

$$R(s, \bar{\beta}_x(Lh(s))) \vee \forall y\, Sec^x_R(s * \langle y \rangle) . \rightarrow Sec^x_R(s),$$

and proof by induction on Sec^x_R:

(1) $\forall s[R(s,\beta_x(Lh(s))) \vee \forall yF(s * \langle y\rangle) . \rightarrow F(s)] \rightarrow \forall s(Sec_R^x(s) \rightarrow F(s))$,

where $F(s)$ is a W-formula w.r.t. s , are both derivable in $\underset{\sim}{W}-\underset{\sim\sim}{ID}_\omega(A)$. (Cf. [Fef, (8) and (9) of § 2.1].)

Now formula (8) of [Fef, § 2.2] can be put in the form

$$\forall \alpha_M[M(x) \ \& \ Sec_R^x(s) \ \& \ \bar{\alpha}(Lh(s)) = s \rightarrow \exists yR(\bar{\alpha}y, \beta_x y)]$$

(i.e. the universal function quantifier is pulled out), and then proved in $\underset{\sim}{W}-\underset{\sim\sim}{ID}_\omega(A)$ by applying the induction schema (1) above to the W-formula

$$\bar{\alpha}(Lh(s)) = s \rightarrow \exists yR(\bar{\alpha}y, \beta_x y) ;$$

and the argument proceeds as in [Fef] to prove (the translation of) Π_1^1-CA.]

Appendix

HEREDITARILY MAJORIZABLE FUNCTIONALS OF FINITE TYPE

W. A. Howard

The purpose of the following is * to show that the Dialectica interpretation (chapter III, § 5) of the simplest nontrivial case $\forall y^2 E_2(y)$ of the axiom of extensionality cannot be carried out by use of a primitive recursive functional (theorem 3.2). To accomplish this we introduce the notion of hereditary majorizability and show that if a functional is hereditarily majorizable, then it does not satisfy the functional interpretation of the axiom of extensionality (§ 2). Then we show that every primitive recursive functional is hereditarily majorizable (§ 3).

The functional interpretation of the next most simple case $\forall y^3 E_3(y)$ of the axiom of extensionality is discussed briefly in § 4. In contrast with the case of $\forall y^2 E_2(y)$, the existence of a functional satisfying the functional interpretation of $\forall y^3 E_3(y)$ depends strongly on the class of functionals over which the variables are taken to range. Indeed, if the variables are taken to range over the class of ordinary set-theoretic functionals, then the existence of a functional providing the functional interpretation of $\forall y\, E_3(y)$ implies the axiom of choice for sets of number-theoretic functionals (theorem 4.1). Hence by known results, there are models of Zermelo - Fraenkel set theory (without the axiom of choice) in which there is no functional satisfying the functional interpretation of $\forall y^3 E_3(y)$.

§ 1. Extensionality.

Supposing $X, W, Z_1, \ldots, Z_s$ to be variables ** such that $XZ_1 \ldots Z_s$ and $WZ_1 \ldots Z_s$ are terms of type 0, let $X =_e W$ denote <u>extensional equality</u>, namely, $(\forall Z_1 \ldots Z_s)(XZ_1 \ldots Z_s = WZ_1 \ldots Z_s)$ (as in 2.7.2). Now let $X_1, \ldots, X_k$

-----

* References with three numbers refer to this volume outside this appendix. References such as 4.1, § 2 etc. refer to this appendix.

** In this appendix also capitals are used as variables for objects of finite type.

be variables of types $\sigma_1,\ldots,\sigma_k$, respectively, and let σ denote $(\sigma_1)(\sigma_2)\ldots(\sigma_k)0$. A functional G of type σ is said to be <u>extensional</u> if

(1.1) $(\forall X_1\ldots X_k)(\forall W_1\ldots W_k)(\forall i[X_i =_e W_i] \to GX_1\ldots X_k = GW_1\ldots W_k)$,

where $\forall i[X_i =_e W_i]$ denotes the conjunction of $X_1 =_e W_1,\ldots,X_k =_e W_k$. Let us abbreviate (1.1) by $E_\sigma(G)$. The <u>axiom of extensionality</u> for functionals of type σ (2.7.2) is here taken in the form $\forall y^\sigma E_\sigma(y)$. The simplest non-trivial case of the axiom of extensionality is $\forall y^2 E_2(y)$; namely,

(1.2) $\forall Y\alpha\beta(\forall u[\alpha u = \beta u] \to Y\alpha = Y\beta)$,

where α and β have type 1 and Y has type 2. A functional F of type (2)(1)(1)0 satisfies the <u>Dialectica functional interpretation</u> of (1.2) if

(1.3) $\forall Y\alpha\beta[Y\alpha \neq Y\beta \to \alpha(FY\alpha\beta) \neq \beta(FY\alpha\beta)]$.

We will work in $\underset{\sim}{HA}^\omega$ (1.6.15). We use the formalism of typed combinators because it simplifies the exposition of § 3. The λ-operator is assumed to be defined by the rules of 1.6.8.

As a point of methodology we note here that the three theorems of § 2, and all instances of the two theorems of § 3, are derivable in $\underset{\sim}{HA}^\omega$ (see remark 3.1). Thus the theorems of §§ 2 - 3 are valid for all models of $\underset{\sim}{HA}^\omega$: in particular (cf. chapter II): the set-theoretical model, the models HRO and HEO based on partial recursive function application, the models ICF and ECF based on continuous function application, the term models CTM and CTNF, and Kleene's general recursive functionals (cf. 2.8.2). Not that the theorems which deal directly with the functional interpretation of the axiom of extensionality (namely, theorems 2.2, 2.3 and 3.2) are of much interest in the case of nonextensional functionals: after all, the negation of the axiom (1.2) of entensionality implies the negation of the functional interpretation of (1.2) trivially. But theorems 2.1 and 3.1 on boundedness and hereditary majorizability appear to be of independent interest. (Also, the intensional continuous functionals of types 1 and 2 are extensional.)

§ 2. Hereditarily majorizable functionals.

A relation F^* maj F (F^* hereditarily majorizes F) will now be defined between functionals F^* and F of the same type σ. The definition is by induction on $\underset{\sim}{T}$. If σ is 0, then F^* and F are numbers n and m: we define n maj m to mean $n \geq m$. If σ is $(\tau)\rho$ then F^* maj F means

$\forall G^* G(G^* \text{ maj } G \rightarrow F^* G^* \text{ maj } FG)$. We say that a functional F is hereditarily majorizable if there exists a functional F^* such that F^* maj F.

The following three remarks are easily verified.

Remark 2.1. If F^* maj F and G^* maj G, then F^*G^* maj FG.

Remark 2.2. If G_r^* maj G_r for $0 \leq r \leq p$, then $G_0^* G_1^* \ldots G_p^*$ maj $G_0 G_1 \ldots G_p$.

Remark 2.3. Suppose $HX_1 \ldots X_p$ has type 0. If $H^* X_1^* \ldots X_p^* \geq HX_1 \ldots X_p$ for all $X_1^*, \ldots, X_p^*, X_1, \ldots, X_p$ such that X_r^* maj X_r for $1 \leq r \leq p$, then H^* maj H.

In the following theorem, X_r and $Z_1, \ldots, Z_{s(r)}$ are variables such that $X_r Z_1 \ldots Z_{s(r)}$ has type 0.

Theorem 2.1. Suppose a functional F of type $(\sigma_1)(\sigma_2)\ldots(\sigma_p)0$ is hereditarily majorizable. Let k be fixed and, for $1 \leq r \leq p$, let $\underline{M}_r$ denote the set of functionals X_r of type σ_r such that $(\forall Z_1 \ldots Z_{s(r)})(X_r Z_1 \ldots Z_{s(r)} \leq k)$. Then

(2.1) $\quad \exists m (\forall X_1 \in \underline{M}_1) \ldots (\forall X_p \in \underline{M}_p)(FX_1 \ldots X_p \leq m)$.

Proof. By assumption there exists F^* such that F^* maj F. For $1 \leq r \leq p$, let G_r^* denote $(\lambda Z_1 \ldots Z_{s(r)}).k$. Then $(\forall X_r \in \underline{M}_r)(G_r^* \text{ maj } X_r)$ by remark 2.3. Hence

$$(\forall X_1 \in \underline{M}_1)\ldots(\forall X_p \in \underline{M}_p)(F^* G_1^* \ldots G_p^* \geq FX_1 \ldots X_p)$$

by remark 2.2. Thus $F^* G_1^* \ldots G_p^*$ is the required number m.

Theorem 2.2. For $r = 1, 2$, let $\underline{N}_r$ denote the set of functionals X of type r such that $\forall Z(XZ \leq 1)$. Let F be a functional of type $(2)(1)(1)0$ such that

(2.2) $\quad \exists m (\forall Y \in \underline{N}_2)(\forall \alpha \in \underline{N}_1)[FY\alpha(\lambda u.0) \leq m]$.

Then F does not satisfy the functional interpretation (1.3) of the axiom of extensionality.

Proof. It is easy to define a primitive recursive functional $\lambda n.Y_n$ such that, for all α,

$$Y_n \alpha = \begin{cases} 1 & \text{if } (\forall u < n)(\alpha u = 0) \text{ and } \alpha n = 1, \\ 0 & \text{otherwise.} \end{cases}$$

Also it is easy to define a functional $\lambda n.\alpha_n$ such that

$$\alpha_n u = \begin{cases} 0 & \text{if } u < n \\ 1 & \text{if } u \geq n. \end{cases}$$

Assume (1.3) and take β to be $\lambda u.0$. Denote $\lambda Y\alpha.FY\alpha(\lambda u.0)$ by G.

Clearly $Y_n(\lambda u.0) = 0$. Hence $Y_n\alpha_n \neq 0 \to \alpha_n(GY_n\alpha_n) \neq 0$ by (1.3). But $Y_n\alpha_n = 1$. Hence $\alpha_n(GY_n\alpha_n) \neq 0$. Hence $GY_n\alpha_n \geq n$ by the definition of α_n. But $Y_n \in \underline{N}_2$ and $\alpha_n \in \underline{N}_1$. Thus the hypothesis (2.2) has been contradicted, so (1.3) has been refuted. Q. E. D.

Theorem 2.3. Suppose a functional of type (2)(1)(1)0 is hereditarily majorizable. Then F does not satisfy the functional interpretation (1.3) of the axiom of extensionality.

Proof. By theorem 2.1 (for $k = 1$) and theorem 2.2.

Remark 2.4. By inspection of theorems 2.1 - 2.3 we see that actually the following sharp form of theorem 2.3 has been proved. There are primitive recursive functionals $\lambda n.Y_n$ and $\lambda n.\alpha_n$ such that, for all F and F^*: if F^* maj F, then

$$\neg[Y_d\alpha_d \neq Y\beta_0 \to \alpha_d(Y_d\alpha_d\beta_0) \neq \beta_0(Y_d\alpha_d\beta_0)],$$

where β_0 is $\lambda u.0$ and $d = F^*(\lambda\alpha.1)(\lambda u.0)(\lambda u.0)$.

Indeed, the above proofs go through for relative hereditary majorization: all that is assumed of the set $\underline{A}$ of majorizing functionals and the set $\underline{B}$ of functionals being majorized is that $\underline{A}$ and $\underline{B}$ are closed under application and contain certain simple primitive recursive functionals.

Construction 2.1. Given F and F^* of type $(0)\sigma$ such that $\forall n(F^*n$ maj $Fn)$, to find H^* such that H^* maj F. Solution: take H^* to be $(\lambda X_1 \ldots X_k).\sum_{m \leq n} F^*mX_1 \ldots X_k$, where $X_1, \ldots, X_k$ are variables of types such that $F^*mX_1 \ldots X_k$ has type 0. We denote this H^* by $(F^*)^+$.

§ 3. Primitive recursive functionals.

We indicate extensional equality of F and G by $F =_e G$ as in § 1. By applying universal quantifiers to the appropriate axioms for $\underset{\sim}{HA}^\omega$ we obtain:

(3.1) $(\forall XY)[\Pi XY =_e X]$,

(3.2) $(\forall XYZ)[\Sigma XYZ =_e XZ(YZ)]$,

(3.3) $(\forall XY)[RXY0 =_e X]$,

(3.4) $(\forall XYu)[RXY(Su) =_e Y(RXYu)u]$.

The set $\underline{P}$ of all primitive recursive functionals is defined with reference to a given notion of functional. In the case of the set-theoretic notion (2.4.6) there is no problem since in this case the equations (3.1) - (3.4) pick out functionals Π, Σ and R of all appropriate types from the supply in a unique way. Hence we define $\underline{P}$ inductively by the following two clauses :

(a) $\underline{P}$ contains zero, the successor function, and the functionals Π, Σ and R of all appropriate types.

(b) If $F \in \underline{P}$ and $G \in \underline{P}$, then $FG \in \underline{P}$.

In the case of the model ICF there is some choice as to which associate is to be named by the constant Π: the equation (3.1) does not pick out an associate from the supply ICF in a unique way. Similarly for the constants Σ and R. Also, the operation of application does not determine uniquely the corresponding operation on associates. Similar remarks apply to the case of HRO. Fortunately these sources of nonuniqueness do not cause any trouble in the present paper. We merely add the following clause to (a) and (b) :

(c) If $F \in \underline{P}$ and $H =_e F$, then $H \in \underline{P}$.

Theorem 3.1. Every primitive recursive functional has a primitive recursive hereditarily majorizing functional.

Proof. We shall indicate, for each of the clauses (a) - (c), how the majorizing functional H^* for H is obtained when H arises by use of the given clause. In the case of clause (b), H is FG so we can take H^* to be F^*G^* by remark 2.1. The case of clause (c) is handled by the observation that if $H =_e F$ and F^* maj F, then F^* maj H. It remains to treat the case of clause (a) ; namely, we must verify theorem 3.1 for each of the generating functionals 0, S, Π, Σ and R. Obviously we can take 0^* and S^* to be 0 and S, respectively. By remarks (2.2) - (2.3) and equations (3.1) - (3.2) we can take Π^* and Σ^* to be Π and Σ, respectively. Similarly, by remarks (2.2) - (2.3), equations (3.3) - (3.4) and induction on n, we find RX^*Y^*n maj $RXYn$ whenever X^* maj X, Y^* maj Y. Hence we can take R^* to be $\lambda XY.(RXY)^+$ as in construction 2.1.

Theorem 3.2. There is no Dialectica interpretation of the axiom of extensionality (1.2) by a primitive recursive functional.

Proof. By theorems 2.3 and 3.1.

Corollary. $\underset{\sim}{E}$ - $\underset{\sim}{HA}^{\omega}$ does not have a Dialectica interpretation in itself.

This corollary follows from the fact that $\underset{\sim}{E}$ - $\underset{\sim}{HA}^{\omega}$ can be axiomatized as $\underset{\sim}{HA}^{\omega} + \forall y E_{\sigma}(y)$ for all σ, and it has a model by primitive recursive functionals. For such a model we can use any of various classes of functionals mentioned in § 1 : the set-theoretic functionals, the extensional continuous functionals, or the extensional effective operations. We can even use the minimal term model consisting of the closed terms of $\underset{\sim}{HA}^{\omega}$ since these terms act extensionally on themselves.

Remark 3.1. To treat theorem 3.1 by use of $\underset{\sim}{HA}^{\omega}$ let us consider first the case of a primitive recursive functional defined by use of clauses (a) and (b) alone. Then the functional will be represented by a closed term F. Inspection of the proof of theorem 3.1 provides a corresponding term F^* together with a derivation, in $\underset{\sim}{HA}^{\omega}$, of the formula F^* maj F. In the case of a primitive recursive functional defined by the use of clause (c) as well as clauses (a) - (b), the functional will be described in $\underset{\sim}{HA}^{\omega}$ by the use of variables. For illustration suppose that clause (c) has been applied only at the last step of the definition. Then the formula to be derived in $\underset{\sim}{HA}^{\omega}$ is $\forall X(X =_e F \rightarrow F^* \text{ maj } X)$. (This shows, by the way, that the majorizing functionals in theorem 3.1 can be chosen from the functionals generated by clauses (a) - (b) alone.) Thus each instance of theorem 3.1 is derivable in $\underset{\sim}{HA}^{\omega}$.

A similar remark applies to theorem 3.2. Also, remark 2.4 and the proof of theorem 3.2 provide an effective procedure which when applied to the definition of F by use of clauses (a) - (c) yields primitive recursive Y_n, α_n and β_0 such that

$$\neg[Y\alpha_n \neq Y\beta_0 \rightarrow \alpha_n(FY_n\alpha_n\beta_0) \neq \beta(FY_n\alpha_n\beta_0)].$$

Thus if a functional H agrees with a primitive recursive functional at all primitive recursive arguments, then H does not satisfy the functional interpretation of the axiom (1.2) of extensionality.

Remark 3.2. There exists a functional F satisfying the functional interpretation (1.3) of the axiom $\forall Y \underline{E}_2(Y)$ of extensionality which is general recursive in the sense of Kleene 1959 (cf. 2.8.2), where it is understood that Y, α and β range over all set-theoretic functionals of types 2, 1 and 1, respectively. Namely, the instructions for calculating $FY\alpha\beta$ are as follows. If $Y\alpha = Y\beta$, take $FY\alpha\beta$ to be 0. If $Y\alpha \neq Y\beta$, examine αn and βn successively for $n = 0,1,2,\ldots$, and take $FY\alpha\beta$ to be the least n such that $\alpha n \neq \beta n$. Since Y is extensional, the required n exists. Thus F is μ-recursive (Kleene 1959, p. 45).

Exactly the same definition yields F satisfying (1.3) when Y, α and β are understood to range over extensional continuous functionals. Hence, by theorem 4 of Kleene 1959A, p. 94, the functional F is itself continuous (which is also easy to see directly).

If Y, α and β are understood to range over (extensional) effective operations, then clearly the above definition yields an effective operation F satisfying (1.3).

§ 4. Discussion of $\forall y\, E_3(y)$.

The axiom $\forall y\, E_3(y)$ is

(4.1) $\forall YXW(\forall\alpha[X\alpha = Y\alpha] \rightarrow YX = YW)$,

where X and W have type 2, and Y has type 3 (i.e., (2)0). The functional interpretation of (4.1) is

(4.2) $\forall YXW[YX \neq YW \rightarrow X(FYXW) \neq W(FYXW)]$.

If Y, X and W are taken to range over extensional continuous functionals, then there exists a Kleene general recursive functional F satisfying (4.2): we merely generalize the definition of F given in remark 3.2, using the fact that the extensional continuous functionals of a given type have a recursively dense base (2.6.16). Let $h_0, h_1, \ldots, h_n, \ldots$ be a recursively dense base for the functionals of type 1. If $YX = YW$, take FYXW to be h_0. If $YX \neq YW$, take FYXW to be h_k, where $k = \min_n(Xh_n \neq Wh_n)$.

These considerations obviously generalize to the case of $\forall y\, E_\sigma(y)$ for arbitrary σ, where all variables are taken to range over extensional continuous functionals.

In the following theorem, Y, X and W are taken to range over set-theoretic functionals.

Theorem 4.1. From a functional F satisfying (4.2) we can construct, in set theory, a function ψ defined on all sets $\underline{M}$ of functions of type 1, such that $\psi\underline{M} \in \underline{M}$ for all nonempty $\underline{M}$.

Proof. Let $\varphi\underline{M}$ denote the characteristic function of $\underline{M}$. That is to say: for all α of type 1, $\varphi\underline{M}\alpha$ is 1 or 0 according as to whether α is in $\underline{M}$ or not. Let H of type 3 be defined by the condition that HX is 0 or 1 according as to whether $\forall\alpha(X\alpha = 0)$ or not. Define $\psi\underline{M}$ to be $FH(\varphi\underline{M})(\lambda\alpha.0)$. Then clearly $\psi\underline{M} \in \underline{M}$ for every nonempty $\underline{M}$.

Let ZF denote Zermelo - Fraenkel set theory and let ZFC denote ZF extended by the addition of the axiom of choice. In Rosser 1969, pp. 113 - 115, it is shown that there are models of ZF in which there is no well-ordering of the real numbers. On the other hand, a well-ordering of the real numbers can be defined in ZF with the help of the choice function ψ of theorem 4.1. Hence we obtain:

Corollary 1. There are models of ZF in which there is no functional satisfying the functional interpretation (4.2) of $\forall y\, E_3(y)$.

Similarly, from the fact that there are models of ZFC in which no formula well-orders the real numbers (Rosser 1969, p. 89), we obtain:

Corollary 2. There are models of ZFC in which no functional definable by a formula of ZF satisfies the functional interpretation (4.2) of $\forall y\, E_3(y)$.

Of course the existence of a functional satisfying the functional interpretation of $\forall y\, E_3(y)$ follows immediately from the axiom of choice.

From corollary 2 and Kleene 1959, p. 32 we obtain:

Corollary 3. There are models of ZFC in which no Kleene general recursive functional satisfies the functional interpretation (4.2) of $\forall y\, E_3(y)$.

BIBLIOGRAPHY

Some abbreviations which are not standard or not self-evident, and which are used in this bibliography are :

J.S.L.	for :	"The Journal of Symbolic Logic" ;
LMPS	for :	"Logic, Methodology and Philosophy of Science".
LMPS III :		B. van Rootselaar, J.F. Staal (editors). Amsterdam (North-Holland Publ. Co.), 1968.
LMPS IV :		Proceedings of the Congress at Bucarest, summer 1971. To appear.
IPT	for :	Intuitionism and proof theory. Proceedings of the summer conference at Buffalo N.Y., 1968. A. Kino, J. Myhill, R.E. Vesley (editors). Amsterdam - London (North-Holland Publ. Co.), 1970.
Oslo Proc.	for :	Proceedings of the Second Scandinavian Logic Symposium. J.E. Fenstad (editor). Amsterdam - London (North-Holland Publ. Co.), 1971.
Cambr. Proc.	for :	The Cambridge Festival of Logic. R.O. Gandy, A.R.D. Matthias (editors). Berlin - Heidelberg - New York (Springer Verlag). To appear.

P.H.G. Aczel

1969 Saturated intuitionistic theories, in :
H.A. Schmidt, K. Schütte, H.-J. Thiele (editors), Contributions to mathematical logic, Amsterdam (North-Holland), pp. 1 - 11.

J.S. Addison and S.C. Kleene

1957 A note on function quantification.
Proc. Am. Math. Soc. 8, pp. 1002 - 1006.

H. Bachmann

1950 Die Normalfunktionen und das Problem der ausgezeichneten Folgen von Ordnungszahlen.
Vierteljahrschr. Naturf. Gesellsch. Zürich 95, pp. 5 - 37.

H.P. Barendregt

1971 Some extensional term models for combinatory logics and λ - calculi :
Thesis, Rijksuniversiteit Utrecht.

A Pairing without conventional restraints (preprint of 1972).

B Combinatorial realizability (preprint of 1972).

C A global representation of the representation of the recursive functions in the λ - calculus.
To appear in : Proceedings of the Logic Conference at Orléans, summer 1972 (G. Sabbagh a.o., editors).

M.J. Beeson

1972 Metamathematics of constructive theories of effective operations.
Thesis, Stanford University.

A Derived rules of inference related to the continuity of effective operations. To appear.

B The non-derivability in constructive formal systems of theorems on the continuity of effective operations. To appear.

C The unprovability in constructive formal systems of the continuity of effective operations on the reals. To appear.

C. Celluci
1971 Operazioni di Brouwer e realizzabilitá formalizzata.
Annali della Scuola Normale Superiore de Pisa 25, pp. 649 - 682.

P.J. Cohen
1966 Set theory and the continuum hypothesis.
New York (W.A. Benjamin).

H.B. Curry and R. Feys
1958 Combinatory logic I. Second edition 1968.
Amsterdam (North-Holland Publ.).

H.B. Curry, J.R. Hindley and J.P. Seldin
1972 Combinatory Logic II.
Amsterdam - London (North-Holland Publ.).

D. van Dalen
A Lectures on intuitionism,
to appear in Cambr. Proc.

D. van Dalen, C.E. Gordon
1971 Independence problems in subsystems of intuitionistic arithmetic.
Indagationes Mathematicae 33, pp. 448 - 456.

J. Diller
1968 Zur Berechenbarkeit primitiv-rekursiver Funktionale endlicher Typen, in:
H.A. Schmidt, K. Schütte, H.-J. Thiele (editors), Contributions to
Mathematical Logic. Amsterdam (North-Holland Publ. Co.), pp. 109 - 120.
1971 Zur Theorie rekursiver Funktionale höherer Typen.
Habilitationsschrift, München 1970.
A A variant to Gödel's interpretation of Heyting arithmetic of finite
types. To appear in LMPS IV.

J. Diller and W. Nahm
A Eine Variante zur Dialectica-Interpretation der Heyting-Arithmetik
endlicher Typen (preprint from 1971).
To appear in: Archiv für mathematische Logik.

J. Diller und K. Schütte
1971 Simultane Rekursionen in der Theorie der Funktionale endlicher Typen.
Archiv für mathematische Logik 14, pp. 69 - 74.

A.G. Dragalin
1968 The computation of primitive recursive terms of finite type, and
primitive recursive realization.
Zap. Naŭcn. Sem. Leningrad. Otdel. Mat. Inst. Steklov (LOM 1) 8,
pp. 32 - 45.

S. Feferman
1960 Arithmetization of metamathematics in a general setting.
Fundamenta Mathematicae 49, pp. 35 - 92.
1962 Transfinite recursive progressions of axiomatic theories.
J.S.L. 27, pp. 259 - 316.
1968 Ordinals associated with theories for one inductively defined set.
Unpublished notes of a lecture on inductive definitions given at the
conference "Intuitionism and Proof Theory". Part of the content
of the lecture is in Feferman 1970.
1970 Formal theories for transfinite iterations of generalized inductive
definitions and some subsystems of analysis, in:
IPT, pp. 303 - 325.

M. Fitting
1969 Intuitionistic logic, model theory and forcing.
Amsterdam (North-Holland).

H. Friedman
A Some applications of Kleene's methods for intuitionistic systems.
To appear in: Cambr. Proc.

D. Gabbay
1969 Applications of trees to intermediate logics, I.
ONR Technical Report No. 31.
1969A Montague type semantics for non-classical logics, I.
Airforce Office Scientific Research. Scientific Report No. 4.

I.L. Gal, J.B. Rosser, D. Scott
1958 Generalization of a lemma of G. Rose.
J.S.L. 23, pp. 137 - 138.

R.O. Gandy
1962 Effective operations and recursive functionals (abstract).
J.S.L. 27, pp. 378 - 379.
1967 Computable functionals of finite type, I, in:
J.N. Crossley (editor), Sets, Models and Recursion Theory.
Amsterdam (North-Holland), pp. 202 - 242.

G. Gentzen
1933 Ueber das Verhältnis zwischen intuitionistischer und klassischer Arithmetik. Galley proof, Mathematische Annalen (received 15th March 1933). First published in English translation in:
The collected papers of Gerhard Gentzen, M.E. Szabo (editor), pp. 53 - 67. Amsterdam (North-Holland).
1935 Untersuchungen über das logische Schliesse.
Mathematische Zeitschrift 39 (1935), pp. 176 - 210, 405 - 431.
English translation in: The collected papers of Gerhard Gentzen, pp. 68 - 131 (cf. under Gentzen 1933).
1938 Neue Fassung des Widerspruchsfreiheitsbeweises für die reine Zahlentheorie.
Forschungen zur Logik und zur Grundlegung der exakten Wissenschaften, New series No. 4. Leipzig (Hirzel), pp. 19 - 44.
English translation in: The collected papers of Gerhard Gentzen, pp. 252 - 286 (cf. under Gentzen 1933).

H. Gerber
1970 Brouwer's bar theorem and a system of ordinal notations.
IPT, pp. 327 - 338.

J.Y. Girard
1971 Une extension de l'interprétation de Gödel à l'analyse, et son application à l'élimination des coupures dans l'analyse et la théorie des types.
Oslo Proc., pp. 63 - 92.
1972 Interprétation fonctionnelle et élimination des coupures de l'arithmétique d'ordre supérieur:
Thèse de doctorat d'état, Université Paris VII.
A Quelques résultats sur les interprétations fonctionnelles, in:
Cambr. Proc.

K. Gödel
1933 Zur intuitionistischen Arithmetik und Zahlentheorie, in: Ergebnisse eines mathematischen Kolloquiums, Heft 4 (for 1931 - 1932, appeared in 1933), pp. 34 - 38.
Translated into English in: The undecidable, M. Davis (editor), pp. 75 - 81, under the title: "On intuitionistic arithmetic and number theory".
For corrections of the translation see review in J.S.L. 31 (1966), pp. 484 - 494.
1958 Ueber eine bisher noch nicht benützte Erweiterung des finiten Standpunktes.
Dialectica 12 (1958), pp. 280 - 287.

N. Goodman
1968 Intuitionistic arithmetic as a theory of constructions. Thesis, Stanford University.
1970 A theory of constructions equivalent to arithmetic, in: IPT, pp. 101 - 120.
1972 A simplification of combinatory logic. J.S.L. 37, pp. 225 - 246.
A The arithmetic theory of constructions. To appear.
B The faithfulness of the interpretation of arithmetic in the theory of constructions. To appear.
C The inductive theory of constructions. To appear.
D The first-order content of the inductive theory of constructions. To appear.
E The theory of the Gödel functionals. To appear.

N. Goodman, J. Myhill
1972 The formalization of Bishop's constructive mathematics, in: F.W. Lawvere (editor), Toposes, Algebraic Geometry and Logic. Lecture Notes in Mathematics, Vol. 274. Berlin - Heidelberg - New York (Springer-Verlag), pp. 83 - 96.

S. Görneman
1971 A logic stronger than intuitionism. J.S.L. 36, pp. 249 - 261.

A. Grzecorczyk
1964 Recursive objects in all finite types. Fundamenta Mathematicae 54, pp. 73 - 93.

Y. Hanatani
1966 Démonstration de l'ω - non-contradiction de l'arithmétique. Annals of the Japan Association for Philosophy of Science 3, pp. 105 - 114.

R. Harrop
1956 On disjunctions and existential statements in intuitionistic systems of logic. Math. Annalen 132, pp. 347 - 361.
1960 Concerning formulas of the types $A \rightarrow B \vee C$, $A \rightarrow (Ex)Bx$ in intuitionistic formal systems. J.S.L. 25, pp. 27 - 32.

A. Heyting
1931 Die intuitionistische Grundlegung der Mathematik.
Erkenntnis 2, pp. 106 - 115.
1934 Mathematische Grundlagenforschung, Intuitionismus, Beweistheorie.
Ergebnisse der Mathematik und ihrer Grenzgebiete, Berlin.
Expanded translation in French as: Les fondements des mathématiques.
Intuitionnisme. Théorie de la démonstration.
Paris - Louvain (Gauthier-Villars, Nauwelaers), 1955.
1956 Intuitionism, an introduction. Amsterdam (North-Holland Publ. Co.).
Second, revised edition 1966. Third, revised edition 1972.
1956A La conception intuitionniste de la logique.
Les études philosophiques 11, pp. 226 - 233.

S. Hinata
1967 Calculability of primitive recursive functionals of finite type.
Science reports of the Tokyo Kyoiku Daigaku, A, 9, pp. 218 -235.

S. Hinata and S. Tugué
1969 A note on continuous functionals.
Annals of the Japan Association for Philosophy of Science 3,
pp. 138 - 145.

J.R. Hindley, B. Lercher and J.P. Seldin
1972 Introduction to combinatory logic.
Cambridge (Cambridge University Press).

W.A. Howard
1963 (i) The axiom of choice (Σ_1^1 - AC_{01}), bar induction and bar recursion
(Section II, with four appendices);
(ii) Transfinite induction and transfinite recursion (Section IV,
with one appendix).
Stanford report on the foundations of analysis. Mimeographed.
Stanford University.
1968 Functional interpretation of bar induction by bar recursion.
Compositio Mathematica 20, pp. 107 - 124.
1970 Assignment of ordinals to terms for primitive recursive functionals
of finite type, in:
IPT, pp. 443 - 458.
1970A Assignment of ordinals to terms for type 0 bar recursive functionals
(abstract). J.S.L. 35, p. 354.
1972 A system of abstract constructive ordinals.
J.S.L. 37, pp. 355 - 374.
A The formulae-as-types notion of construction. 1969,
unpublished manuscript.
B Hereditarily majorizable functionals of finite type.
Appendix, this volume.

W.A. Howard and G. Kreisel
1966 Transfinite induction and bar induction of types zero and one,
and the rôle of continuity in intuitionistic analysis.
J.S.L. 31, pp. 325 - 358.

R. Hull
1969 Counterexamples in intuitionistic analysis using Kripke's schema.
Zeitschrift für mathematische Logik und Grundlagen der Mathematik 15,
pp. 241 - 246.

D. Isles
1970 Regular ordinals and normal forms.
IPT, pp. 339 - 361.

S. Jaskowski
1936 Recherches sur le système de la logique intuitionniste.
Actes du congrès international de philosophie scientifique, VI. Philosophie des mathématiques. Actualités scientifiques et industrielles, 393. Paris (Hermann & Cie), pp. 58 - 61.

H.R. Jervell
1971 A normal form in first order arithmetic, in:
Oslo Proc., pp. 93 - 108.

D.H.J. de Jongh
1968 Investigations on the intuitionistic propositional calculus:
Thesis, University of Wisconsin.
1970 The maximality of the intuitionistic predicate calculus with respect to Heyting's arithmetic (abstract).
J.S.L. 35, p. 606.
1970A A characterization of the intuitionistic propositional calculus.
IPT, pp. 211 - 217.
1971 Disjunction and existence under implication in intuitionistic arithmetic (abstract).
J.S.L. 36, p. 588.
A The maximality of the intuitionistic predicate calculus with respect to Heyting's arithmetic.
To appear in Compositio Mathematica.
B Formulas in one propositional variable in intuitionistic arithmetic.
Report (1973) Mathematical Institute, University of Amsterdam.

D.H.J. de Jongh and A.S. Troelstra
1966 On the connection of partially ordered sets with some pseudo-Boolean algebras.
Indagationes Mathematicae 28, pp. 317 - 329.

S.C. Kleene
1944 On the forms of the predicates in the theory of constructive ordinals.
American Journal of Mathematics 66, pp. 41 - 58.
1945 On the interpretation of intuitionistic number theory.
J.S.L. 10, pp. 109 - 124.
1952 Introduction to metamathematics.
Amsterdam (North-Holland Publ. Co.) , Groningen (P. Noordhoff), and New York, Toronto (D. van Nostrand Comp.).
1955 On the forms of predicates in the theory of constructive ordinals (second paper).
American Journal of Mathematics 77, pp. 405 - 428.
1957 Realizability, in:
Summaries of talks presented at the Summer Institute of Symbolic Logic in 1957 at Cornell University, pp. 100 - 104.
Reprinted in: A. Heyting (editor) Constructivity in Mathematics. Amsterdam (North-Holland Publ. Co.) 1959, pp. 285 - 289.
1959 Recursive functionals and quantifiers of finite types I.
Trans. Amer. Math. Soc. 91, pp. 1 - 52.
1959A Countable functionals, in:
A. Heyting (editor) Constructivity in Mathematics, pp. 81 - 100.
1960 Realizability and Shanin's algorithm for the constructive deciphering of mathematical sentences.
Logique et Analyse 3, pp. 154 - 155.
1962 Disjunction and existence under implication in elementary intuitionistic formalisms.
J.S.L. 27, pp. 11 - 18.
1963 An addendum.
J.S.L. 28, pp. 154 - 156.

(S.C. Kleene continued)

1963A Recursive functionals and quantifiers of finite types II.
Trans. Amer. Math. Soc. 108, pp. 106 - 142.

1965 Classical extensions of intuitionistic mathematics, in:
Y. Bar-Hillel (editor), Proceedings of the 1964 International Congress
Amsterdam (North-Holland Publ. Co.), pp. 31 - 44.

1965A Logical calculus and realizability.
Acta Philosophia Fennica 18, pp. 71 - 80.

1968 Constructive functions, in:
"The foundations of intuitionistic mathematics" in:
LMPS III, pp. 137 - 144.

1969 Formalized recursive functionals and formalized realizability.
Memoirs of the American Mathematical Society, Nr 89.

S.C. Kleene and R.E. Vesley

1965 The foundations of intuitionistic mathematics, especially in relation to recursive functions.
Amsterdam (North-Holland Publ. Co.).

D.L. Kreider and H. Rogers jr

1961 Constructive versions of ordinal number classes.
Trans. Amer. Math. Soc. 100, pp. 325 - 369.

G. Kreisel

1951 On the interpretation of non-finitist proofs, Part I.
J.S.L. 16, pp. 241 - 267.

1958 Mathematical significance of consistency proofs.
J.S.L. 23, pp. 155 - 182.

1958A The non-derivability of $\neg(x)A(x) \rightarrow (Ex)\neg A(x)$, A primitive recursive, in intuitionistic formal systems (abstract).
J.S.L. 23, pp. 456 - 457.

1958B Constructive mathematics. Notes of a course given at Stanford University, 1958 - 1959. (Mimeographed.)

1959 Interpretation of analysis by means of constructive functionals of finite type, in:
A. Heyting (editor), Constructivity in mathematics.
Amsterdam (North-Holland Publ. Co.), pp. 101 - 128.

1959A Reflection principle for subsystems of Heyting's (first order) arithmetic (H) (abstract).
J.S.L. 24, p. 322.

1959B Inessential extensions of Heyting's arithmetic by means of functionals of finite type (abstract).
J.S.L. 24, p. 284.

1959C Proof by transfinite induction and definition by transfinite induction in quantifier-free systems (abstract).
J.S.L. 24, pp. 322 - 323.

1959D Inessential extensions of intuitionistic analysis by functionals of finite type (abstract).
J.S.L. 24, pp. 284 - 285.

1962 On weak completeness of intuitionistic predicate logic.
J.S.L. 27, pp. 139 - 158.

1962A Proof theoretic results on intuitionistic higher order arithmetic (abstract).
J.S.L. 27, p. 380.

1962B Consequences of Brouwer's bar theorem (abstract).
J.S.L. 27, pp. 380 - 381.

1962C Proof theoretic results on intuitionistic first order arithmetic (HA) (abstract).
J.S.L. 27, pp. 379 - 380.

1962D Foundations of intuitionistic logic, in:
E. Nagel, P. Suppes, A. Tarski (editors), LMPS, Stanford
(Stanford University Press), pp. 198 - 210.

(G. Kreisel continued)
1963 The subformula property and reflection principles (abstract).
J.S.L. 28, pp. 305 - 306.
1963A Reflection principle for Heyting's arithmetic (abstract).
J.S.L. 28, pp. 306 - 307.
1963B Reflection principles and ω - consistency (abstract).
J.S.L. 28, pp. 307 - 308.
1963C Generalized inductive definitions.
Section III in the Stanford Report on the foundations of analysis.
Mimeographed. Stanford University.
1964 Review.
Zentrablatt für Mathematik 106, pp. 237 - 238.
1965 Mathematical logic, in:
T.L. Saaty (editor), Lectures on modern mathematics (Vol. III).
New York (Wiley and Sons), pp. 95 - 105.
1968 Functions, ordinals, species, in:
LMPS III.
1968A A survey of proof theory.
J.S.L. 33, pp. 321 - 388.
1968B Lawless sequences of natural numbers.
Compositio Mathematica 20, pp. 222 - 248.
1969 Course notes on functional interpretations. Autumn quarter.
Mimeographed. Stanford University.
1970 Church's thesis: a kind of reducibility axiom for constructive mathematics.
Oslo Proc., pp. 121 - 150.
Review: Zentralblatt für Mathematik 199 (1971), pp. 300 - 301.
1971 A survey of proof theory II.
Oslo Proc., pp. 109 - 170.
1971A Classes of functions of finite type; uniformity properties.
Mimeographed handwritten course notes. Winter 1971 - 1972.
Stanford University.
1972 Which number theoretic problems can be solved in recursive progressions on Π - paths through O?
J.S.L. 37, pp. 311 - 334.

G. Kreisel, D. Lacombe, J.R. Shoenfield
1959 Partial recursive functionals and effective operations, in:
A. Heyting (editor), Constructity in mathematics.
Amsterdam (North-Holland Publ. Co.), pp. 290 - 297.

G. Kreisel and A. Levy
1968 Reflection principles and their use for establishing the complexity of axiomatic systems.
Zeitschrift für Math. Logik 14, pp. 97 - 142.

G. Kreisel and H. Putnam
1957 Eine Unableitbarkeitsbeweismethode für den intuitionistischen Aussagenkalkül.
Archiv für mathematische Logik und Grundlagenforschung 3, pp. 35 - 47.

G. Kreisel and A.S. Troelstra
1970 Formal systems for some branches of intuitionistic analysis.
Annals of mathematical Logic I, pp. 229 - 387.

S. Kripke
1963 "Flexible" predicates of formal number theory.
Proc. Amer. Math. Soc. 13, pp. 647 - 650.
1965 Semantical analysis of intuitionistic logic, I, in:
J.N. Crossley, M.A.E. Dummett (editors), Formal systems and recursive functions.
Amsterdam (North-Bolland Publ. Co.), pp. 92 - 130.

S. Kuroda
1951 Intuitionistische Untersuchungen der formalistischen Logik. Nagoya Math. Journal 2, pp. 35 - 47.

H. Laüchli
1970 An abstract notion of realizability for which intuitionistic predicate calculus is complete, in: IPT, pp. 227 - 234.

H. Levitz
1970 On the relationship between Takeuti's ordinal diagrams O(n) and Schütte's system of ordinal notations Σ(n), in: IPT, pp. 377 - 405.

E.G.K. Lopez-Escobar
1968 An ω-rule in intuitionistic number theory. TR 68 - 63, Technical report, Department of Mathematics, University of Maryland.

H. Luckhardt
1970 Extensionale Funktionalinterpretation der klassischen Analysis. Ein Widerspruchfreiheitsnachweis. Habilitationsschrift, Marburg/Lahn. English translation in Luckhardt 1973.
1971 Anhang: Ueber das bar-rekursive Modell der klassischen Analysis und die allgemeine Barinduktion über Spezies. Manuscript, English translation in Luckhardt 1973.
1973 Extensional Gödel functional interpretation. A consistency proof of classical analysis. Springer Lecture Notes Vol. 306. Berlin, Heidelberg, New York (Springer).

P. Martin-Löf
1971 Hauptsatz for the intuitionistic theory of iterated inductive definitions. Oslo Proc., pp. 179 - 216.
1971A Hauptsatz for the intuitionistic theory of species. Oslo Proc., pp. 217 - 233.
1971B On the strength of intuitionistic reasoning. Report 1971, No. 5 of the Mathematical Institute, University of Stockholm. To appear in LMPS IV.
1971C A theory of types. Report 1971, No. 3 of the Mathematical Institute, University of Stockholm.
1972 Infinite terms and a system of natural deduction. Compositio Mathematica 24, pp. 93 - 103.
1972A About models for intuitionistic type theories and the notion of definitional equality. Report No. 4, 1972, of the Mathematical Institute, University of Stockholm.

Yu.V. Matijasevich
1970 Diofantovost perechislimikh mrozhestv. Doklady ANSSSR 191, pp. 279 - 282; improved English translation in: Soviet Mathematics Doklady 11, pp. 354 - 357.
1971 Diophantine representation of recursively enumerable predicates. Oslo Proc., pp. 171 - 177.

G. Mints
1969 Imbedding operations associated with Kripke's "semantics", in: A.O. Slisenko (editor), Studies in constructive mathematics and mathematical logic, Part I. New York (Consultants Bureau).

J.R. Moschovakis
1967 Disjunction and existence in formalized intuitionistic analysis, in: J.N. Crossley (editor), Sets, models and recursion theory. Amsterdam (North-Holland Publ. Co.).
1969 A survey of intuitionistic logic. Lecture Notes, Spring 1969, University of Bristol.
1971 Can there be no non-recursive functions? J.S.L. 36, pp. 309 - 315.
A A topological interpretation of second-order intuitionistic arithmetic. To appear in Compositio Mathematica.

A. Mostowski
1961 A generalization of the incompleteness theorem. Fundamenta Mathematicae 49, pp. 205 - 232.
1966 Thirty years of foundational studies. New York (Barnes and Noble).

J. Myhill
1967 Notes towards an axiomatization of intuitionistic analysis. Logique et Analyse 35, pp. 280 - 279.
1968 Formal systems of intuitionistic analysis I, in: LMPS III, pp. 161 - 178.
1970 Formal systems of intuitionistic analysis II: the theory of species, in: IPT, pp. 151 - 162.
1972 An absolutely independent set of Σ_1^0 sentences. Zeitschr. für Math. Logik und Grundl. der Mathematik 18, pp. 107 - 109.

D. Nelson
1947 Recursive functions and intuitionistic number theory. Trans. Amer. Math. Soc. 61, pp. 307 - 368.

I. Nishimura
1960 On formulas in one variable in intuitionistic propositional calculus. J.S.L. 25, pp. 327 - 331.

H. Osswald
1972 Vollständigkeit und Schnittelimination in der intuitionistischen Typenlogik. Manuscripta Mathematica 6, pp. 17 - 31. (Preprint under the title: Ein Berechnungsverfahren für die Zulässigkeit der Schnittregel im Kalkül von Schütte für die intuitionistische Typenlogik.)
1973 Ein syntaktischer Beweis für die Zulässigkeit der Schnittregel für die intuitionistische Typenlogik. Manuscripta Mathematica 8, pp. 243 - 249.

Ch. Parsons
1970 On a number theoretic choice schema and its relation to induction. IPT, pp. 459 - 473.
1972 On n - quantifier induction. J.S.L. 37, pp. 466 - 482.

W. Pohlers
1973 Ein starker Normalisationssatz für die intuitionistische Typentheorie. Manuscripta Mathematica 8, pp. 371 - 387.

D. Prawitz
1965 Natural deduction, a proof-theoretical study. Stockholm, Göteborg, Uppsala (Almqvist & Wiksell).
1970 Some results for intuitionistic logic with second order quantification rules. IPT, pp. 259 - 269.

(D. Prawitz continued)
1971 Ideas and results in proof theory.
Oslo Proc., pp. 235 - 307.
A Towards a foundation of a general proof theory.
LMPS IV.

W. Richter
1965 Extensions of the constructive ordinals.
J.S.L. 30, pp. 193 - 211.
1968 Constructively accessible ordinal numbers.
J.S.L. 33, pp. 43 - 55.

T.T. Robinson
1965 Interpretations of Kleene's metamathematical predicate $\Gamma|A$ in intuitionistic arithmetic.
J.S.L. 30, pp. 140 - 154.

H. Rogers jr
1958 Gödel numberings of partial recursive functions.
J.S.L. 23, pp. 331 - 341.
1967 Theory of recursive functions and effective computability.
New York etc. (McGraw-Hill).

G.F. Rose
1953 Propositional calculus and realizability.
Trans. Amer. Math. Soc. 75, pp. 1 - 19.

J.B. Rosser
1935 A mathematical logic without variables.
Annals of Mathematics (2) 36, pp. 127 - 150,
Duke Mathematical Journal 1, pp. 328 - 355.
1969 Simplified independence proofs.
New York (Academic Press).

L.E. Sanchis
1967 Functionals defined by recursion.
Notre Dame Journal of Formal Logic 8, pp. 161 - 174.

B. Scarpellini
1969 Some applications of Gentzen's second consistency proof.
Math. Annalen 181, pp. 325 - 344.
1970 On cut elimination in intuitionistic systems of analysis.
IPT, pp. 271 - 285.
1970A A model of intuitionistic analysis.
Commentarii Mathematici Helvetici 45, pp. 440 - 471.
1971 Proof theory and intuitionistic systems.
Lecture Notes in Mathematics, Vol. 212.
Berlin, Heidelberg, New York (Springer-Verlag).
1971A A model for bar recursion of higher types.
Compositio Mathematica 23, pp. 123 - 153.
1972 Induction and transfinite induction in intuitionistic systems.
Annals of Math. Logic 4 , pp. 173 - 227.
1972A A formally constructive model for bar recursion of higher types.
Zeitschr. für Math. Logik und Grundl. der Mathematik 18, pp. 321 - 383.
1973 On bar induction of higher types for decidable predicates.
Annals of Math. Logic 5, pp. 77 - 163.
A Disjunctive properties of intuitionistic systems.
To appear.

D.S. Scott
1957 Completeness proofs for the intuitionistic sentential calculus, in: Summaries of talks presented at the Summer Institute of Symbolic Logic in 1957 at Cornell University.
1968 Extending the topological interpretation to intuitionistic analysis. Compositio Mathematica 20, pp. 194 - 210.
1970 Extending the topological interpretation to intuitionistic analysis, II, in: IPT, pp. 235 - 255.

H. Schwichtenberg
A On impredicative definitions of primitive recursive functionals. (Preprint 1972.)

J.R. Shoenfield
1967 Mathematical logic. Reading, Menlo Park, London, Don Mills (Ont.) (Addison-Wesley).

C. Smorynski
A Some remarks on measures of complexity and unboundedness theorems (unpublished).
B Peano's arithmetic is an essentially unbounded extension of Heyting's arithmetic (unpublished).
C PA / HA is essentially unbounded (unpublished).

C. Spector
1961 Inductively defined sets of natural numbers, in: Infinitistic methods. Proceedings of the Symposium on Foundations of Mathematics, Warsaw, 2 - 9 September 1959. Oxford, London, New York, Paris (Pergamon Press), Warszawa (Państwowe Wydawnictwo Naukowe), pp. 97 - 102.
1962 Provably recursive functionals of analysis: a consistency proof of analysis by an extension of principles formulated in current intuitionistic mathematics, in: J.C.E. Dekker (editor), Proceedings of Symposia in Pure Mathematics V. Providence (R.I.) (American Mathematical Society), pp. 1 - 27.

S. Stenlund
1971 Introduction to combinatory logic. Filosofiska Studier Nr 11. Uppsala (Philosophical Soc. and Dept. of Philosophy, University of Uppsala).
1972 Combinators, λ - terms, and proof theory. Dordrecht (D. Reidel).

W.W. Tait
1959 A characterization of ordinal recursive functions (abstract). J.S.L. 24, p. 325.
1963 A second order theory of functionals of higher type, with two appendices. Appendix A: Intentional functionals. Appendix B: An interpretation of functionals by convertible terms (reworked in Tait 1967). Appeared as: Section V, with two appendices, in: Stanford report on the foundations of analysis. Mimeographed. Stanford University.
1965 Infinitely long terms of transfinite type, in: J. Crossley and M.A.E. Dummett (editors), Formal systems and recursive functions. Amsterdam (North-Holland Publ. Co.), pp. 176 - 185.
1965A Functionals defined by transfinite recursion. J.S.L. 30, pp. 155 - 174.
1967 Intensional interpretations of functionals of finite type I. J.S.L. 32, pp. 198 - 212.
1968 Constructive reasoning, in: LMPS III, pp. 185 - 199.

(W.W. Tait, continued)
1971 Normal form theorem for barrecursive functions of finite type, in: Oslo Proc., pp. 353 - 367.

G. Takeuti
1965 A formalization of the theory of ordinal numbers. J.S.L. 30, pp. 295 - 317.
1967 Consistency proofs of subsystems of classical analysis. Annals of Mathematics 86, pp. 299 - 348.

R.H. Thomason
1968 On the strong semantical completeness of the intuitionistic predicate calculus. J.S.L. 33, pp. 1 - 7.

R.R. Tompkins
1968 On Kleene's recursive realizability as an interpretation for intuitionistic elementary number theory. Notre Dame Journal of Formal Logic 9, pp. 289 - 293.

A.S. Troelstra
1968 The theory of choice sequences, in: LMPS III, pp. 201 - 223. Some errata: page 221, line 12, replace " $\Lambda\alpha$ " by " $\Lambda\alpha \in a$ "; lines -6, -9, -10 replace " $en \dot{-} 1$ " by " $fn \dot{-} 1$ ".
1969 Principles of intuitionism. Lecture Notes in Mathematics, Vol. 95. Berlin - Heidelberg - New York (Springer-Verlag). Some corrections: page 41, line 13, replace $\leq$ by $\geq$; line 15 must read: $\Lambda n(1th(n) + k = en \dot{-} 1 \wedge en \neq 0 \rightarrow Yn)$; line 16, replace K by N. Page 51: formula (10) is not correct as it stands. See instead the techniques developed in Kreisel-Troelstra 1970, § 5 for dealing with schemata with parameters. Page 79, the definition 14.2.4 is misstated and should read: φ_u is a mapping defined as follows. Let $u = \langle x_o, \ldots, x_n \rangle$, $v = \langle y_o, \ldots, y_m \rangle$, $w = (\overline{\lambda x.1})(n+1)$. Then (a) If $v \leq u$ we take $\varphi_u v = w$; (b) If $y_o > x_o$ we take $\varphi_u v = \langle y_o - x_o, y_1, \ldots, y_m \rangle$; (c) If $y_o = x_o, \ldots, y_i = x_i$, $y_{i+1} > x_{i+1}$ we take $\varphi_u v = \langle 0, \ldots, 0, y_{i+1} - x_{i+1}, y_{i+2}, \ldots, y_m \rangle$; (d) In all other cases we take e.g. $\varphi_u v = w$. φ_u is an order-isomorphism when restricted to species $F[u]$, $F[u,v]$; φ_u maps $F[u]$, $F[u,v]$ onto elements of WO.
1969A Notes on the intuitionistic theory of sequences, I. Indagationes Mathematicae 31, pp. 430 - 440.
170 Notes on the intuitionistic theory of sequences, III. Indagationes Mathematicae 32, pp. 245 - 252.
1971 Notions of realizability for intuitionistic arithmetic and intuitionistic arithmetic in all finite types, in: Oslo Proc., pp. 369 - 405. Some errata are listed in Troelstra 1972A. Some further errata: p. 372, line -5, read: " $\underset{\sim}{E}$ - $\underline{HA}^{\omega}$ "; page 381, line -8, add formula number (1) at the end of the line; page 383, line 11, replace "3.16" by "3.15"; page 384, line 8, replace " $\vee$ " by " & "; page 396, line 6, replace " IP " by " IPR "; page 397, line -4, add ") " at end; page 398, line 4, replace " $\forall x$ " by " $\forall x \in V_\sigma$ ";

(A.S. Troelstra, continued)
page 398, line 6, replace "HRE" by "HEO"; line -3, read "8.2" for "7.2".
1971A Computability of terms and notions of realizability for intuitionistic analysis.
Report 71 - 02 of the Department of Mathematics, University of Amsterdam (mimeographed).
1972 Review of Yasugi 1963.
J.S.L. 37, p. 404.
1972A Self review of Troelstra 1971.
Zentralblatt für Mathematik 227, Nr 02015.
A Notes on intuitionistic second order arithmetic.
To appear in: Cambr. Proc.
Correction: in the proof of 2.7, under "Extensions", the analysis should not be applied to a path, but to a spine as defined in § 4.2 of this volume. (The reason is that our permutative reductions did not reckon with permutation with UP, IP - rules.)
The change does not affect the applications.
A preprint of this paper was issued as Report 71 - 05 of the Department of Mathematics, University of Amsterdam (1971).

R.E. Vesley
1970 A palatable substitute for Kripke's schema, in:
IPT, pp. 197 - 207.
1972 Choice sequences and Markov's principle.
Compositio Mathematica 24, pp. 33 - 53.

M. Yasugi
1963 Intuitionistic analysis and Gödel's interpretation.
Journal of the Math. Society of Japan 15, pp. 101 - 112.
Review with corrections in:
J.S.L. 37 (1972), p. 104.

J.I. Zucker
1971 Proof theoretic studies of systems of iterated inductive definitions and subsystems of analysis.
Thesis, Stanford University (mimeographed).

INDEX

Notions and notations are listed only when they have more than local significance in the text. The references given refer to definitions or special conventions regarding the notion or notations listed.

I. List of symbols

A) Formal systems, arranged primarily in alphabetic order

For some general conventions see 1.1.2 (viii). For any formal system $\mathbf{H}$ based on intuitionistic logic, $\mathbf{H}^c$ indicates the corresponding classical system.

B) Schemata and rules, arranged primarily in alphabetic order

C) Syntactical variables (in order of appearance)

There are many local deviations in the use of variables. Often new variables are made by adding sub- or super-scripts reserved for variables of a certain category.

x, y, z, u, v, w 1.1.2 (ii); cf. 6.2.1, 6.3.3

a, b, c 1.1.2 (ii); cf. 6.2.1

$A, B, C, \ldots$ 1.1.2 (ii)

t, s 1.1.2 (iii), 6.2.1.; cf. 1.6.5, 6.3.3

$\bar{n}, \bar{m}, \bar{x}, \bar{y}, \bar{z}, \bar{u}, \bar{v}, \bar{w}$ 1.3.9 D; cf. 5.2.3

n, m 5.2.3

$x^\sigma, y^\sigma, z^\sigma, u^\sigma, v^\sigma, w^\sigma$ 1.6.3

$\underline{\underline{x}}, \underline{\underline{y}}, \underline{\underline{z}}, \underline{\underline{u}}, \underline{\underline{v}}, \underline{\underline{w}}$ 1.6.5

$\underline{\underline{X}}, \underline{\underline{Y}}, \underline{\underline{Z}}, \underline{\underline{U}}, \underline{\underline{V}}, \underline{\underline{W}}$ 1.6.5

s, t, T 1.6.5

s^σ, t^σ 1.6.5

$\underline{\underline{s}}, \underline{\underline{t}}, \underline{\underline{T}}$ 1.6.5, 6.5.5

X^n, Y^n, Z^n 1.9.3

$x^1, y^1, z^1, u^1, v^1, w^1$ 1.9.10

$\alpha, \beta, \gamma, \ldots$ 1.9.10; cf. 6.3.3

$e, f, e', e'', e_1, \ldots, f', f'', \ldots$ 1.9.25; cf. 6.2.1

$\Pi, \Pi', \Pi'', \Pi_1, \ldots$ 4.1.2

$\Sigma, \Sigma', \Sigma'', \ldots, \Sigma_0, \Sigma_1, \ldots$ 4.1.2

a, b, e, k, m, n 6.2.1

$\alpha, \beta, \alpha^1, \beta^1$ 6.3.3

α^2, β^2 6.3.3

$f^{(0)1}, f^{(1)2}$ 6.3.3

r, s, t, u, v, t_1, t_1' 6.3.3

Θ, Θ' 6.4.1

$\Phi, \Psi, \Psi', \ldots$ 6.8.1.

D) Other symbols

The symbols are primarily listed in order of appearance, some very similar ones are grouped together.

$\&, \vee, \exists, \forall, \rightarrow, \curlywedge$ 1.1.2 (i)

$\Rightarrow, \Leftrightarrow, \underset{\sim}{\forall}, \underset{\sim}{\exists}, \in, \subseteq$ 1.1.2 (i)

$\equiv_{\text{def}}$ 1.1.2 (i)

$\equiv$ 1.1.2 (i)

$t[x], t[x,y]$ 1.1.2 (iii)

$\neg$ 1.1.2 (vi)

$\leftrightarrow$ 1.1.2 (vi)

$[x/t]E$ 1.1.2 (vii)

II. List of notions

Lecture Notes in Mathematics Vol. 344
ISBN 978-3-540-06491-6

Anne S. Troelstra (Ed.)

Metamathematical Investigation of Intuitionistic Arithmetic and Analysis

Errata

This text contains corrections and additions to "Metamathematical Investigation of Intuitionistic Arithmetic and Analysis" which appeared in 1973 as number 344 in the "Springer Lecture Notes in Mathematics" series.

The original edition ran out of print many years ago. A small run of a corrected edition has been produced in 1993 as a report (X-93-05) of the Institute of Logic, Language and Information (ILLC) of the University of Amsterdam, but this is now also out of print.

In the mean time Springer Verlag has decided to make the volumes of the series Lecture Notes in Mathematics which are out of print available as "publish-on-demand" editions.

The text has been typeset in Latex. Wavy underlining in the original text is now interpreted as boldface, underlining as italics. Double wavy underlining has been interpreted by a sans serif fount. However, we have retained double underlining and did not replace it by Fraktur.

A first list of Errata appeared in 1974 as a report of the Mathematical Institute of the University of Amsterdam; many more errata have been discovered since then. In particular I should like to thank Marc Bezem, Susumu Hayashi, Jane Bridge Kister, Jaap van Oosten and Jeffery Zucker.

The counting of lines includes the lines in displayed formulas; for indications, e.g. a name or a number for a group of displayed lines, which are between lines so to speak, an ad hoc indication will be chosen.

Underlining in the original text has been rendered as italics in these correction; double underlining has been rendered as such, but a double wavy underlining corresponds to a sans serif letter in these corrections.

XIII Add below the summary of §6:

§7 *Applications: proof theoretic closure properties* 258
List of rules (3.7.1) — closure under ED, DP, CR_0, ECR_0, ED′, ACR, IPR'^{ω} (3.7.2–5) — closure under CR (3.7.6) — closure under ECR_1 (3.7.7–8) — extensions to analysis (3.7.9)

6 In 1.1.7, interchange "$\vee I_r$" and "$\vee I_l$".

7_{13} Read " $\exists E$ " for " $\exists I$ ".

8^5 Read "essentially".

8 In the first proof tree, "$B \to \curlywedge$" should be "$B \to \curlywedge$ (2)" and "(2)" should be repeated at the lowest horizontal line.

16^{13} Read "A" for "F".

18 In 1.3.3 the axiom

$$x_i = x_i' \to \phi(x_1, \ldots, x_i, \ldots, x_n) = \phi(x_1, \ldots, x_i', \ldots, x_n)$$

(for any n-ary function constant ϕ, $1 \leq i \leq n$) can be replaced by the corresponding axiom for S only:

$$x = y \rightarrow Sx = Sy,$$

since the general case can be established by induction (since all ϕ except S are introduced by schemas for primitive recursive functions).

18_{11} Read "*Defining*" for "*Definining*".

19 Add at the bottom of the page rules expressing the functional character of the F_k:

$$\frac{F_k t_1 \ldots t_{n-1} t_n \qquad F_k t_1 \ldots t_{n-1} t'_n}{t_n = t'_n}.$$

20^8 Replace " F'_{k_m} " by " F_{k_m} ".

25 Addition to second paragraph of (D) : "Canonical" essentially means that the arithmetization provably satisfies the "same" inductive closure conditions as the predicate itself.

26^3 Read " $\ulcorner t \urcorner$ " for " t ".

26^5 Read "that $\ulcorner A(\bar{x}_1, \ldots, \bar{x}_n) \urcorner$ stands for ...".

27^8 Read " $\simeq$ " for " $=$ ".

27 Add at the bottom of the page a paragraph:

We follow *Kleene 1952* and use $\Lambda x.t$, t a p-term, to indicate a gödelnumber for t as partial recursive function of x; if t contains, besides the free variables $x, x_1, \ldots, x_n$, $\Lambda x.t$ is a (primitive) recursive function of $x_1, \ldots, x_n$.

29_2 Read " y_0 " for " y ".

29_1 Read " y_1 " for " y ".

33_{10} Read "We put $\ulcorner \xi t_1 \ldots t_n \urcorner = \ldots$ ".

41_2 Read " $t' \not\equiv x^\sigma$ " for " $t' \neq x^\sigma$ ".

44^{17} Replace "slightly ... is" by "seemingly stronger (but in fact equivalent) variant is".

44^{19-21} Delete "In ... EXT-R′."

44_6–45^5 Replace these lines by the following:

The following two propositions are due to M. Bezem (Equivalence of Bar Recursors in the Theory of Functionals of Finite Type, *Archive for Mathematical Logic* 27 (1988), 149–160).

PROPOSITION. The rule EXT-R′ is derivable in qf-**WE-HA**$^\omega$.

PROOF. Assume EXT-R, and let $\vdash P \to s_1 = s_2$, $\vdash Q[x/t_1]$ Here $s_1 = s_2$ as usual is shorthand for an equation between terms of type 0 $s_1 x_1 x_2 \dots x_n = s_2 x_1 x_2 \dots x_n$, where $x_1, x_2, \dots, x_n$ are variables not free in P, s_1, s_2. Without loss of generality we can assume $P \equiv (t_1 = 0)$, $Q[x] \equiv (t[x] = 0)$ (x not free in P, s_1, s_2. Below we shall abbreviate $t[x/s]$, for arbitrary s, as $t[s]$. So we have

(1) $\quad \vdash t_1 = 0 \to s_1 = s_2, \quad \vdash t[s_1] = 0.$

Define

$$s_i' := \mathbf{R}_\sigma s_i 0^{(\sigma)(0)\sigma}, \quad s_i \in \sigma.$$

Then, with $x \notin \mathrm{FV}(s_i), i = 1, 2$:

$$\vdash x = 0 \to s_i'x = s_i, \qquad \vdash x \neq 0 \to s_i'x = 0^\sigma.$$

Applying EXT-R to $s_i'0 = s_i$ yields $\vdash t[s_i] = t[s_i'0]$. By replacement (i.e. $x = y \to t[x] = t[y]$) we obtain

$$\vdash t_1 = 0 \to t[s_i'0] = t[s_i't_1].$$

Since also (1) holds, and $t_1 = 0$ is decidable ,$\vdash s_1't_1 = s_2't_1$,, so again using EXT-R

$$\vdash t[s_1't_1] = t[s_2't_1],$$

hence

$$\vdash t_1 = 0 \to t[s_2] = t[s_2'0] = t[s_2't_1] = t[s_1't_1] = t[s_1'0] = t[s_1] = 0.$$

Q.e.d.

PROPOSITION. The deduction theorem holds for qf-**WE-HA**$^\omega$ + EXT-R$'$, hence also for qf-**WE-HA**$^\omega$.

PROOF. It suffices to prove the deduction theorem for the system with EXT-R$'$, and in this case the deduction theorem is easy.

$45_{16,17}$ Delete these lines.

55_1 Read " $z < x$ " for " $z < z$ ".

56_{12} Read " $Q(x, \underline{\underline{v}})$ " for " $Q(0, \underline{\underline{v}})$ ".

56_6 Read " T " for " T ".

58_{12} Read " T " for " T ".

59 Add at the bottom: "Cf. also *Luckhardt 73*, pp. 66–67."

63^{10} Delete the first equation.

67_{11} At end of line we add "assumed to be provably linear in **HA**".

71^{15} Read " $\sigma((Q_2 V_n)A) \equiv (Q_1 v_{m+n+1})\sigma(A), \ldots$ ".

74 In comparing section 1.9.14 with more recent literature (such as A.S. Troelstra, D. van Dalen, *Constructivism in Mathematics*, Amsterdam 1988), it is to be noted that definedness of a term containing functions and numbers with partial application is here supposed to be defined in the sense of *Kleene* 1969, that is to say a function applied to an argument is defined if we can sufficiently many values of the function to find its value at the argument; this convention does not agree with the logic of partial terms with its strictness condition.

75^{4} Read " $U(\mathrm{lth}(u)) = y$ " for " $U(\mathrm{lth}(u) \dot{=} y)$ ".

75_{5} Read " x " for " α ".

80_{13} Addition to 1.9.23: "Cf. also *Kreisel 1967*, page 249, where the role of generalized bar induction in proving the continuous functionals to be a model of bar recursion is mentioned."

$83_{18,13}$ Read " 0_σ " for " 0^σ ".

83_{10} Read " $B_\sigma yzu(c * \hat{v}))c$ " for " $B_\sigma yzuc(c * \hat{v}))c$ ".

91^{12} Add " In *Friedman B* it is shown that for r.e. axiomatizable extensions of **HA**, DP implies ED. ".

$94_{4,5}$ Replace by: "In *Luckhardt A* it is shown that the principle is equivalent to M. "

95 Add at the end of 1.11.6:

It has been noted by C.A. Smorynski that, for theories with decidable prime formulas, IP + M together amount to the principle of the excluded third. E.g. for **HA**, **HA** + IP + M = $\mathbf{HA}^c$, which is seen as follows. Assume $A \vee \neg A$ to be proved already in **HA** + IP + M, and consider $\exists x Ax$. By M, $\forall x(Ax \vee \neg Ax) \& \neg\neg\exists x Ax \rightarrow \exists x Ax$; by the induction hypothesis and IP, this implies $\exists x Ax \vee \neg\exists x Ax$. Application of propositional operators preserves decidability, and $\forall x Ax \leftrightarrow \neg\exists\neg Ax$ by the decidability of A, hence $(\forall x Ax \vee \neg\forall x Ax) \leftrightarrow (\neg\exists x \neg Ax \vee \neg\neg\exists x \neg Ax) \leftrightarrow \neg\exists x \neg Ax \vee \exists x \neg Ax$, hence $\forall x Ax \vee \neg\forall x Ax$.

95_{4} Replace " $\alpha x = Uz$ " by " $\alpha y = Uz$ ".

98^{6} Read " $x_1^{(\sigma)\tau}$ " for " $x^{(\sigma)\tau}$ ".

111^{9} Read " $t_1 \equiv t_1'$ " for " $t_1 \equiv t'$ ".

113_{4} Read " $\mathbf{H} \vdash t = \bar{n}$ ".

$114^{21,22}$ Delete the sentence beginning "For yet another ...".

117_{6} Read "and t is" for "and τ is".

119^{4} Read "... representing αn).".

125_1 Read " $(\sigma)(\tau)\sigma$ " for " $(\sigma)(\tau), \sigma$ ".

126^{10} Read "of" for "if".

$128^{12,13}$ The open problem has been solved by M. Bezem, in the sense that the two structures are isomorphic: J.S.L. 50 (1985), pp. 359–371.

129 Add between lines 6 and 7:

If we replace in the right hand side of this equivalence A by a predicate letter X, we have the inductive condition $B(X, x, y)$ characterizing A.

$132^{18,20}$ Replace " $\mathbf{I\text{-}HA}^\omega$ " by " $\mathbf{I\text{-}HA}^\omega + \mathrm{IE}_0$ ".

133^4 Delete ", CTM′, CTNF′ ".

$133^{10,11}$ Replace final comma by stop in line 10 and delete " CTM′, CTNF′ " in line 11.

$133_{5,4}$ Read " CTNF′ " for " CTNF ".

141^{10} Read " W^1_σ " for " I^1_σ ".

144^8 Read " (4) " for " (1) ".

147_1,148^1 Read " p_q " for " p_k ".

148^1 Insert before comma "& $\{z\}(p_q) = \{z\}(y)$ (since $z \in E(V)$, $y \in V^*$) ".

158^{13} Read " $t =_e s$ " for " $t = s$ ".

158^{16} Read " CTNF " for " CTNF′ ".

158_{11} Read " $x^2[\Sigma(\Pi x^2)\Pi]$ ".

158_9 Read ", that $\Sigma(\Pi x^2_1)\Pi$ and ".

158_8 Read " $\Sigma(\Pi x^2_2)\Pi$ in the model ".

159^{12} Read "...which $x^1\bar{n}_i$ contr $\overline{\alpha n_i}$ has".

159_{14} Read " $t' \in 2$ " for " $t' \in z$ ".

$159_{1,2}$ – 160^{1-4} Replace these lines by:

$$t^2\alpha = \begin{cases} 0 \text{ if } \exists i(1 \le i \le k \;\&\; \bar{\alpha}_1(k+1) = \bar{\alpha}(k+1) \\ m+1 \text{ otherwise, where } m = \max\{\alpha_i(y) \mid 1 \le i \le k,\ 0 \le y \le k\}. \end{cases}$$

Now $Ft^2 \neq 0$; for, $\overline{(\pi_{0,0}0)}(k+1)$, ..., $\overline{(\pi_{0,0}k)}(k+1)$ are all distinct, hence one of them, say $\overline{(\pi_{0,0}k_0)}(k+1)$ $(0 \le k_0 \le k)$ is distinct from all $\bar{\alpha}_1(k+1)$, ..., $\bar{\alpha}_k(k+1)$, and therefore $t^2(\pi_{0,0}k_0) = m+1$; but then $\overline{(\lambda x^0.t^2(\pi_{0,0}x))}(k+1)$ differs from all $\bar{\alpha}_1(k+1)$, ..., $\bar{\alpha}_k(k+1)$, and thus Ft^2 takes the value $m+1$.

160_1 Read " = " for " ≡ ".

161^{12} Delete "not".

164 Remark to be added at the end of 2.8.5: J.M.E. Hyland showed in his thesis that Scarpellini's model coincides with the model ECF ".

167_7 Read "establishing" for "establish".

173 Remark to be added in 2.9.10:

If a coding by functions is given for the elements of $\sigma^{\smile}$, such that there are continuous Φ_0, Φ with $\Phi_0\xi$ the length of the sequence coded by ξ, $\Phi(n, \xi)$ the n-th component extracted from ξ, then one can construct a bijection between two codings of this kind."

179^{17} Read " $\mathbf{HA}^c + \mathrm{G}_1$.

180^6 Read " $|F$ holds " for " $|$ holds ".

180_{13} read " P " for " D ".

183_2 Read " PCA). " for " P A). ".

184^{17} Add after " terms ": "satisfying ($t \in V$ and $t' = t \Rightarrow t' \in V$) ".

188^{17} Read "mathematical".

189^{12} Read " $\& P(B(j_1 x))$ " for " $\& P(A(j_1 x))$ ".

190_{17} Delete ", and $P(F_1^*), \ldots, P(F_n^*)$ ".

190_{12} Delete " Also $\forall x P(Bx)$ ".

190_4 Delete "It follows that ... hence $P(C)$.", and replace "Also" by "Then".

192^1 Add after "hence": " $!t \,\&\, t \,\mathsf{r}_P\, A$ is an abbreviation for $(\exists x(t \simeq x \,\&\, x \,\mathsf{r}_P\, A).)$".

192_{1-7} The argument given in the first edition is not correct. The result is a consequence of the unprovability in **HA** of the DP, which has been proved by J. Myhill (A note on indicator functions, Proc. Amer. math. Soc. 29 (1973), 181–183) and by Friedman in a stronger form (On the derivability of instantiation properties, J.S.L. 42 (1977), 506–514).

194_{10} Insert " $\mathbf{HA} \vdash$ " between " (ii) " and " $A(\underline{\underline{a}}) \to !\psi(\underline{\underline{a}})\& \ldots$ ".

194_9 Replace this line by

Proof. The "only if" part is established as follows. Assume $\vdash A\underline{\underline{a}} \leftrightarrow B\underline{\underline{a}}$, B almost negative. Then there is a recursive ϕ such that $\vdash \forall u(u \,\mathsf{r} A\underline{\underline{a}} \to !\{j_1\phi\underline{\underline{a}}\}(u) \,\&\, \{j_1\phi\underline{\underline{a}}\}(u) \,\mathsf{r} B\underline{\underline{a}})$, and $\vdash \forall u(u \,\mathsf{r} B\underline{\underline{a}} \to !\{j_2\phi\underline{\underline{a}}\}(u) \,\&\, \{j_2\phi\underline{\underline{a}}\}(u) \,\mathsf{r} A\underline{\underline{a}})$, which together with 3.2.11 for B readily yields the desired conclusion.

194_2 Read " $Uv \,\mathsf{r}\, A\underline{\underline{a}}$ " for " $v \,\mathsf{r}\, A\underline{\underline{a}}$ ".

198 Add after 3.2.22:

Remark. In the writings of the Russian constructivist school (cf. e.g. Dragalin 1969) one finds the following extension of CT_0:

$$\mathrm{CT}' \qquad \forall x(\neg Ax \to \exists y Bxy) \to \exists u \forall x(\neg Ax \to \exists v(Tuxv \,\&\, B(x, Uv))).$$

However, in the presence of M this is equivalent to ECT_0, i.e.

$$\mathbf{HA} + \mathrm{ECT}_0 + M = \mathbf{HA} + \mathrm{CT}' + M.$$

To see this, let us first assume CT′, M, and let Ax be almost negative. then by M $Ax \leftrightarrow A'x$, A' negative, and hence $\neg\neg A'x \leftrightarrow Ax$ (1.10.8); thus an instance of ECT_0 can be interpreted as an instance of CT′.

Conversely, if ECT_0 and M are assumed, and we let $\forall x(\neg Ax \to \exists y Bxy)$, then by ECT_0, 3.2.8 $\neg Ax \leftrightarrow \exists z(z \,\mathsf{r}\, \neg Ax) \leftrightarrow \forall z(z \,\mathsf{r}\, \neg Ax) \leftrightarrow 0 \,\mathsf{r}\, \neg Ax$; $0 \,\mathsf{r}\, \neg Ax$ is almost negative. Replacing $\neg Ax$ by $0 \,\mathsf{r}\, \neg Ax$ we have $\forall x(0 \,\mathsf{r}\, \neg Ax \to \exists y Bxy)$ to which we can apply ECT_0 etc.

200–201 The claim of 3.2.26 is correct, but the proof given is incorrect. The attempted proof in list of Errata from 1974 is also flawed, but J. van Oosten presented me with a proof that IP_0 is derivable from ECT_0 and MP.

Here follows the proof: Assume

$$\forall x(Ax \vee \neg Ax), \quad \forall x\, Ax \to \exists y\, B.$$

By ECT_0 there is a number n such that

$$\forall x(\{n\}x = 0 \leftrightarrow Ax), \quad \text{hence } \forall x(\{n\}x = 0) \to \exists y\, B.$$

Again by ECT_0 there is a number m such that

$$\forall x(\{n\}x = 0) \to !\{m\}0 \wedge B(\{m\}0).$$

Let k be a number such that

$$\{k\}j(a, b) = \min_x[\{a\}x \neq 0 \vee Tb0x.]$$

Since

$$\neg\neg(\exists x(\{n\}x \neq 0) \vee \forall x(\{n\}x = 0)), \text{ it follows that}$$
$$\neg\neg(\exists x(\{n\}x \neq 0) \vee !\{m\}0, \text{ therefore}$$
$$\neg\neg!\{k\}j(n, m); \text{ hence with MP } !\{k\}j(n, m).$$

From this we see, that by the definition of k

$$\{n\}(\{k\}j(n, m)) \neq 0 \to \neg\forall x\, Ax,$$
$$T(m, 0, \{k\}j(n, m)) \to (\forall x\, Ax \to B(U(\{k\}j(n, m))).$$

Hence

$$\exists y(\forall x\, Ax \to By).$$

202^{6-13} The alternative argument for the non-realizability of formula (1) is erroneous.

203 Add after 3.2.29: "Friedman has shown (*Friedman B*) how to extend q-realizability by a similar trick.".

203_4 Read "*Cellucci*".

214_5 Replace "negative" by " $\exists$-free (i.e. not containing $\vee, \exists$) ".

$214_{4,3}$ Delete "on the convention ... omitted,".

215^1 Replace "negative" by " $\exists$-free".

215^9 Add " **N-HA**$^\omega$," after " **HA**$^\omega$, ".

215^{11} Read "... sequence $\underline{t}$ of ... ".

215_{15} Replace "negative" by " $\exists$-free".

216^{20} Read " $\underline{\underline{y}}\,\mathsf{mr}_P\,A$ " for " $y\,\mathsf{mr}_P\,A$ ".

$217^{1,2,17}$ Read "$\underline{\underline{T}}$" for "$T$" (4 times).

217^{13} Replace this line by:

$$\mathrm{IP}^- \qquad (A \to \exists y^\sigma B) \to \exists y^\sigma (\neg A \to B)$$

(y^σ not free in A, A $\exists$-free, i.e. not containing $\vee, \exists$)

$217_{16,15}$ Delete ", taking for ... into account". Add after 3.4.7:
Remark. The schema

$$\mathrm{IP}^\omega \qquad (\neg A \to \exists y^\sigma B) \to \exists y^\sigma (\neg A \to B) \;\; (y^\sigma \text{ not free in } B).$$

is readily seen to be modified-realizable, hence $\mathbf{H} + \mathrm{IP}^- + AC \vdash \mathrm{IP}^\omega$. Since in systems with decidable prime formulae negative and $\exists$-free formulas coincide, and for negative A $\neg\neg A \leftrightarrow A$, we have in such cases also that IP^ω implies IP^-.

$217_{10,9,4,2,1}$ Replace "IP^ω " by " IP^- ".

217_{10} Read " $\mathbf{H} +$ " for " $\mathbf{H} \vdash$ ".

217_8 Add after line: "For $\mathbf{H} = \mathbf{HA}^\omega$, $\mathbf{I\text{-}HA}^\omega$, $\mathbf{HRO}^-$, $\mathbf{E\text{-}HA}^\omega$, IP^- may be replaced by IP^ω.".

217_5 Read "$\exists$-free" for "negative".

221_7 Read " $\mathrm{M_{PR}}$ " for " $\mathrm{MP_{PR}}$ ".

221_2 Read "3.4.14" for "3.4.4".

222 Add after 3.4.14:

Remark. V.A. Lifschitz has shown (Proceedings of the American Mathematical Society 73 (1979), 101–106) that also **HA** + $\mathrm{CT}_0! \nvdash \mathrm{CT}_0$, where

$$\mathrm{CT}_0! \qquad \forall x \exists ! y A(x,y) \to \exists u \forall x \exists v (Tuxv \,\&\, A(x, Uv)). \text{ ".}$$

222_2 Read " $\forall\alpha\neg\neg\exists x$ " for " $\forall\alpha\neg\neg\exists z$.

223^1 Read "... was suggested by results contained in".

224_1,225^1 Read " IP^- " for " IP^ω ".

226_{16} Replace " for " by " . For ".

226_8 Insert after "... numbers" "(provably linear in **HA**)".

227 Add "($\prec$ provably linear in **HA**)".

228_{16} Read " $U^1_{j(n,i)}x$ " for " $U^1_{j(n,i)}$ ".

$228_{7,6}$ These lines must read respectively " $\ldots \equiv \forall X^* \forall D_X (x \,\mathsf{mr}\, A(X))$ " and " $\ldots \equiv \exists X^* \exists D_X (x \,\mathsf{mr}\, A(X))$ ".

229_{11} Read " of s^1 in HRO ".

230^5 Read "eliminating" for elementary".

233_5 Read " Π^0_2 " for " Π^0_z ".

$238^{4,6,7}$ Delete " $]^{\mathrm{D}}$ ".

239_4 Read " $(\underline{\underline{x}}, \underline{\underline{v}}, \underline{\underline{Zv}})$ " for " $(\underline{\underline{x}}, \underline{\underline{vZ}}, \underline{\underline{v}})$ ".

240^3 Read " $\underline{\underline{y}}$ " for " $\underline{\underline{Yx}}$ ".

242^{12} Read " $\vdash F^{\mathrm{D}}$ " for " $+F^{\mathrm{D}}$ ".

242^{17} Read "now" for "not".

$244_{7,6}$ Replace these lines by:

If we take everywhere X to be identically 1, we obtain the Dialectica interpretation.

245^{10} "Shoenfield" should be underlined.

251_4 Delete " $\mathbf{N\text{-}HA}^\omega$, " and add " ; $(\mathbf{N\text{-}HA}^\omega + \mathrm{IP}^- + \mathrm{AC}) \cap \Gamma_1 = \mathbf{N\text{-}HA}^\omega \cap \Gamma_1$ " (cf. the corrections to page 217).

255_6 Read " $\ldots \,\&\, A^*y)]$ ".

264^{16} Read " **HA** " for " HA ".

264_8 Read " $\to \exists x^\sigma A$ " for " $\to \neg\exists x^\sigma A$ ".

265^1 Read " $(z \neq 0 \rightarrow \neg A)$ " for " $(z \neq 0 \rightarrow A)$ ".

267_2 Read " $\exists y(y\,\mathsf{p}$ " for " $\exists(y\,\mathsf{p}$ ".

273_{17} Read " 3.9.13 " for " 3.9.11 ".

274^1 Read " 3.9.14 " for " 3.9.12 ".

274_4 Read " 3.9.15 " for " 3.9.13 ".

275_6 Delete " (".

278 Second proof tree under 4), read " Π " for the highest " Π_i ".

279 Replace in the first four proof trees exhibited the occurrences of "A" (but not the A in "A_1", "A_2", "Aa" or "$\exists x Ax$") by "B" (7 occurrences).

280 Replace under " 13) " " $A0$ " by " Aa ".

282^7 Read " $\Pi' \succ_1 \Pi''$, (without ... ".

282^{23} Add " of ⅄$_{\mathrm{I}}$ " at the end.

282 In the display at the bottom of the page, the first two lines should be

$$\begin{array}{cc} \Pi & \\ \underline{A} & \underline{0 = 0} \end{array} \qquad\qquad \begin{array}{cc} \Pi & \\ \underline{A} & \underline{sa = sa} \end{array}$$

283 In the displayed prooftree read " $\&_{\mathrm{l}}$E " for " &E ".

284^1 read "form (*Prawitz*" for "(from *Prawitz*".

285 Immediately above the paragraph starting with "This makes it ..." read

$$\begin{array}{c} \Pi' \\ [A] \\ \Pi \\ B \end{array} \quad \text{for} \quad \begin{array}{c} \Pi \\ [A] \\ \Pi \\ B \end{array}$$

286 Replace last paragraph of 4.1.7. by:

For applications, we need only a normalization theorem (not a strong normalization theorem) relative to $\mathcal{R}_{\mathrm{c}⅄}$; so if the reader wishes, he may use the preceding remark and delete everything in the proof below referring to ⅄-contractions.

287^{16} Read " $\mathrm{Prd}_1(\Pi), \mathrm{Prd}_2(\Pi), \ldots$ ".

287_5 Read " $\mathrm{SV}(\mathrm{Sub}(A, \Pi, \mathrm{Prd}(\Pi), \mathrm{Ass}(\Pi)))$ ".

287_2 Read " Π " for " Π_1 ".

288^9 Insert ", $\mathrm{Ass}(\Pi)$ " after " $\mathrm{Prd}_2(\Pi)$ ".

288 Directly above footnote, read

$$\frac{\Pi'_i}{A} \quad \text{for} \quad \frac{\Pi_i}{A}.$$

290[17] Read " $\Pi' \in \mathrm{PRD}(\Pi)$ ".

290 In the first displayed prooftree, replace " At " by " At' ".

291 The second displayed prooftree should read:

$$\frac{\Pi'_1 \ \dots \ \Pi'_n}{A}.$$

292[1] Read "condition IV " for " condition II ".

293 Second line of paragraph starting with "Condition IV for Π'...", read "for Π " instead of "for $\exists$ ".

294 Replace in the second display " Π_3 " by " Π_4 ", and in the line under this display, replace "Π_4" by "Π_5".

294 Replace in the third display "Π_3" by "Π_4". In the line under the third display, insert after "reduces to": "the left subdeduction of".

295[6] Read " Π_6 " for " Π_3 ".

295 In the third display, place in the second proof tree " Π_0 " above " $\exists x Bx$ ".

295_{12} Insert between "condition" and "I": "IV, and hence".

295 In the line below the third display, insert before "is SV":
"and also

$$\begin{matrix} \Pi_4 \\ [B_1 t'] \\ \Pi'_1(t') \\ [Dt] \\ \Pi_1(t) \\ A \end{matrix}$$

"

297[17] Read " 2.2.25 " for " 2.2.31 ".

299[15] Add " (major) " at the end.

300[6] Add after the comma "which may be empty,".

300[9,10] Replace "preceding" by "succeeding".

301[2] Read "were" for "would be".

301^{10} Add before ") ": " ; also, A_i cannot be discharged by IND, since no application of IND lies below A_1 ".

301_6 read " 4.2.8 " for " 2.8 ".

302^5 Read "normal" for "formal".

302^{14} Add " σ " at the end.

304^3 Insert "(by 4.2.7)" before " ; ".

304^{10} Read "were" for "would be".

304^{11} Read "or IND-application occurring" for "occurring".

304^{14} Replace " IND-application " by " begin with a conclusion of an IND-application".

304_{19} Read "Let Φ denote the ".

304 The proof in subsection 4.2.16 in the first edition contains a gap. Much simpler is the following argument:

PROOF. Note that Ψ is equivalent to a set of Harrop formulas: if $\exists x Px \in \Psi$, then we may replace this formula by $P\bar{n}$ for some $\bar{n}$ such that $\mathbf{HA} \vdash P\bar{n}$. Then we can apply 4.2.12.

305_1 delete " , or A_1 is a basic axiom ".

305^5 Read " $A_1 \in \Psi$, i.e. B is prime ".

305 Remark at 4.2.17: instead of referring to 3.6.7(ii), it suffices to note that only true closed Σ^0_1-formulae are provable in **HA** and $\mathbf{HA}^c$.

306, proof of 4.2.18. This proof is incorrect as it stands, since the conclusion of an IND-application is not necessarily atomic, only quantifier-free. The proof is correct if we replace in the statement of the theorem **H**, qf-**HA** by the corresponding systems with induction for atomic formulas only.

To establish the theorem as stated, we can e.g. proceed as follows: define a path of order 0 to be a path $A_1, \ldots, A_n$ with A_n conclusion of the deduction, and define a path of order $m+1$ to be a path $A_1, \ldots, A_n$ such that either A_n is minor premiss of an $\rightarrow$E-application the major premiss of which belongs to a path of order m, or premiss of an IND-application the conclusion of which belongs to a path of order m.

In a strictly normal derivation, every formula occurrence belongs to some path of order m, for suitable m (since redundant applications of $\vee$E, $\exists$E have been removed). Then one readily proves, by induction on m, that for a strictly normal derivation of a quantifier-free formula in **H** all formula occurrences on a path of order m are quantifier-free. (Note that here normalization also w.r.t. permutative reductions is necessary, in contrast to other applications. This could have been avoided by reduction of qf-**HA** to a logic-free calculus, which is not a very elegant solution, however.)

306_{16} Read " 2.5.7 " for " 2.5.6 ".

307, line 2 below second display. Read " $\mathcal{R}_c$ " for " $\mathcal{R}_C$ ".

307_2 Read " 4.1.16 " for " 4.1.15 ".

308 Last line of first display, replace in proof tree Π'' " $\neg A \to Bx$ " by " $\neg A \to \exists x Bx$ ".

309 Replace second sentence of the statement of 4.3.5 by:

Then a spine of Π not ending with an introduction does not contain IP-applications, and ends with an application of a basic rule or an atomic instance of $\curlywedge_I$.

Delete the third sentence.

The proof of 4.3.5. should be reformulated as follows:

Let $A_1, \ldots, A_n$ be a spine of Π not ending with an introduction. Then, by 4.3.4(iii) there are two cases:

(1°) A_1 is a basic axiom. So the spine coincides with its minimum part, hence A_n is atomic.

(2°) A_1 is of the form $\neg B$, to be discharged by $\to$I, followed by IP. this case is excluded, for the sort of inference following A_1 can be (not IP, or $\to$I + IP, but) $\to$E only, leaving us with $A_2 \equiv \curlywedge$, and a minimum part $A_2, \ldots, A_n$.

311^8 read "any one".

311^{17} Read "Red_1" for "Red".

311_{19} Read "of" for "from".

311_{12} Read " $\vee\mathrm{I_r}$ " for " $\vee\mathrm{I_l}$ ".

311_{11} Read " $\vee\mathrm{I_l}$ " for " $\vee\mathrm{I_r}$ ".

311_7 Read " $\mathrm{SV_{d-1}}(\mathrm{Subst}(\mathrm{Param}(\Pi), x, \mathrm{Prd}_1(\Pi)))]$ ".

313^1 Read " (ii) " for " (iii) ".

313_{12} Read " 4.4.3 " for " 4.4.2 ".

313_{10} Insert "(1.5.6)" before " : ".

314^4 Read " Proof_n " for " Proof ".

314_5 Read " 4.4.3 " for " 4.4.2 ".

314_3 Read " $\mathbf{HA} \vdash \forall x \exists yz(\mathrm{Proof}_n(y, \ulcorner A(\bar{x}, \bar{z})\urcorner \,\&\, A(x,z))$ ".

321 As observed by S.Hayashi, (On derived rules of intuitionistic second order arithmetic, Commentarii Mathematici Universitatis Sancti Pauli 26 (1977), 77–103), the proof of 4.5.8 indicated in the text of the first edition establishes a result which is too weak, e.g.

$$\forall n \forall A \in \mathrm{Fm}^{(n)}(\vdash \mathrm{Sat}^{(n)}(X, \ulcorner \forall x Ax \urcorner) \leftrightarrow \forall x \mathrm{Sat}^{(n)}(X, \ulcorner A(\bar{x}) \urcorner))$$

instead of

$$\forall n(\vdash \forall A \in \mathrm{Fm}^{(n)}(\mathrm{Sat}^{(n)}(X, \ulcorner \forall x A x \urcorner) \leftrightarrow \forall x \mathrm{Sat}^{(n)}(X, \ulcorner A(\bar{x}) \urcorner))).$$

Following Hayashi, the desired stronger conclusion can be established as follows. We first define the notion of a formation sequence of a formula A in $\mathrm{Fm}^{(n)}$.

DEFINITION. A *formation sequence* (fs) of $A \in \mathrm{Fm}^{(n)}$ is a finite sequence of quadruples $\langle a_0, b_0, c_0, t_0 \rangle, \ldots, \langle a_m, b_m, c_m, t_m \rangle$ such that

(1) $t_m = \ulcorner A \urcorner$; t_0, and c_i for $1 \leq i \leq m$ are codes of formulas of complexity $\leq n$.

(2) $a_i \in \mathbb{N}$ for $0 \leq i \leq m$, $a_{i+1} \leq i$ for $0 \leq i < m$.

(3) $b_i, c_i \in \mathbb{N}$ for $0 \leq i \leq m$; t_{i+1} is the code of the term which is the result of substituting the term with code $t_{a_{i+1}}$ for the second-order variable $V^p_{b_{i+1}}$ in the formula (with index) c_{i+1} and logical complexity $\leq n$, where p is the number of free variables in $t_{a_{i+1}}$ (end of definition).

Now $\mathrm{Sat}_n(X, \ulcorner A \urcorner)$ is constructed as before. Let f, g, h range over formation sequences. We then define, similar to $\mathrm{Sat}^{(n)}(X, \ulcorner A \urcorner)$ of the text, and with help of Sat_n, the formula $\mathrm{Sat}^{(n)}_f(X, \ulcorner A \urcorner)$, where f is an fs for A with $t_m = \ulcorner A \urcorner$, and $\mathrm{Sat}^{(n)}_f$ is constructed parallel to the substitutions of f. Then one proves

LEMMA. In **HAS**

(i) $\forall f \forall A, B \in \mathrm{Fm}^{(n)} \exists g, h \forall X(\mathrm{Sat}^{(n)}_f(X, \ulcorner A \circ B \urcorner)$

$$\leftrightarrow \mathrm{Sat}^{(n)}_g(X, \ulcorner A \urcorner) \circ \mathrm{Sat}^{(n)}_h(X, \ulcorner B \urcorner))$$

for $\circ \in \{\rightarrow, \&, \vee\}$.

(ii) $\forall f \forall A \in \mathrm{Fm}^{(n)} \exists g \forall X(\mathrm{Sat}^{(n)}_f(X, \ulcorner Q v_i A(v_i) \urcorner) \leftrightarrow (Q v_i)\mathrm{Sat}^{(n)}_g(X, \ulcorner A(\bar{v}_i) \urcorner))$ for $Q \in \{\forall_1, \exists_1\}$.

(iii) $\forall f \forall A \in \mathrm{Fm}^{(n)} \exists g \forall X(\mathrm{Sat}^{(n)}_f(X, \ulcorner Q V^p_i A(V^p_i) \urcorner) \leftrightarrow$
$(Q Y^p) \forall Z^1(\forall y_1, y_2(j(y_1, y_2) \neq j(p, i) \rightarrow Z^1_{(y_1, y_2)} = X_{(y_1, y_2)}) \wedge Z^1_{(p,i)} = Y \rightarrow$
$\mathrm{Sat}^{(n)}_g(Z, \ulcorner A(V^p_i) \urcorner)$

for $Q \in \{\forall_2, \exists_2\}$.

(iv) $\forall X, f, g, n(\mathrm{FS}(f, n) \wedge \mathrm{FS}(g, n) \rightarrow \mathrm{Sat}^{(n)}_f(X, n) \leftrightarrow \mathrm{Sat}^{(n)}_g(X, n))$, where $\mathrm{FS}(f, n)$ expresses "f is a formation sequence of a formula A with $\ulcorner A \urcorner = n$".

Proof. The proof of (i)–(iii) by induction on the length of f; the proof of (iv) uses (i)–(iii) and induction on n.

We may then put

$$\mathrm{Sat}^{(n)}(X, \ulcorner A \urcorner) \leftrightarrow \exists f \mathrm{Sat}^{(n)}_f(X, \ulcorner A \urcorner)$$

and can then establish a stronger version of 4.5.8, namely

$$\forall n(\mathbf{HAS} \vdash \forall A \in \mathrm{Fm}^{(n)}(\mathrm{Sat}^{(n)}(X, \ulcorner \forall x A x \urcorner) \leftrightarrow \forall x \mathrm{Sat}^{(n)}(X, \ulcorner A \bar{x} \urcorner)))$$

etc. etc.

322^{13} read "choice" for "hoice".

375_{17} Read "$\alpha > \alpha_0$ &" for "$\alpha > \alpha_0$".

389 Subsection 5.7.3: more information about Kripke models for second-order intuitionistic arithmetic may be found in: D.H.J. de Jongh, C.A. Smoryński, Kripke models and the intuitionistic theory of species, Annals of Mathematical Logic 9 (1977), 157–186.

391_3 read "$(\alpha x \neq 0)$" for "$(\alpha x \neq 0$".

391_1 Add after ")" "in the presence of continuity axioms".

398_{12} Read "$\mathcal{L}[x]$" for "$\mathcal{L}$".

414_{11} Read "$S_2 f \in Y$" for "$S_1 f \in Y$".

422_{10} Read "$\mathbf{ID}^{\mathrm{c}}_{\nu}$" for "$\mathbf{ID}^{\mathrm{C}}_{\nu}$".

435^{10} Read "$|\mathbf{ID}^{\mathrm{c}}_{\nu}|$" for "$|\mathbf{ID}^{\mathrm{c}}|$".

438^5 Read "$P_1(Xn, n)]$" for "$P_1(Xn, x)]$".

439_{16} Read "type-0-valued".

440_6 Read "$\mathrm{Max}_1(\alpha, 0) = \alpha$" for "$\mathrm{Max}_1(\alpha, 0)$".

$448_{13,14}$ The equality in (7) of 6.9.1 was proved for all recursive ν by 1977, independently by Buchholz, Pohlers and Sieg, using various sophisticated proof-theoretic techniques (see W. Buchholz, S. Feferman, W. Pohlers, and W. Sieg, Iterated Inductive Definitions and Subsystems of Analysis: Recent Proof-Theoretical Studies. Springer Verlag, Berlin 1977). Hence the equalities

$$|\mathbf{ID}^{\mathrm{c}}_2| = |\mathbf{ID}_2| = |\mathbf{T}_2|$$

hold (end of 6.8.9). Hence also the equalities (5) and (6) of 6.9.1 are true.

451^{12-17} See the remark to page 448.

457_{14} Read "$(\lambda n.X_1 \ldots X_k)$" for "$(\lambda X_1 \ldots X_k)$".

462^1 Insert before the second line "Corrections in the bibliography consist sometimes in replacements, sometimes in added information between square brackets."

$462^{8,9}$ Replace by "LMPS IV: P.Suppes, L.Henkin, Gr.C. Moisil, A. Joja (eds.), North-Holland Publ. Co., Amsterdam 1973."

462^{17} Read "Cambridge Summer School in Mathematical Logic"

462^{18} Read: H.Rogers, A.R.D. Mathias (eds.), 1973.

462^{19} Add at the end ", 1973".

462^{20} Delete.

462_{15} Add "[Zeitschrift für mathematische Logik und Grundlagen der Mathematik 20 (1974), 289–306.]"

462_{14} Add: "[cf. H.P. Barendregt, Combinatory logic and the axiom of choice, Indagationes Mathematicae 35(1973), 203–221.]"

$462_{10,11}$ Replace by: " Theoretical Computer Science 3 (1977), 225–242.

462_5 Add "[J.S.L. 41 (1976), 328– 336]".

462_3 Add "[J.S.L. 40 (1975), 321– 346]".

462_1 Add "[J.S.L. 41 (1976), 18–24]".

463^1 Read " Cellucci".

463^{15} Read "Cambr. Proc. 1–94."

463_{25} Add "[Did not appear]"

463_{21} Read "Archiv für mathematische Logik 16 (1974), 49–66."

464^5 Read "Cambr. Proc. 113–170."

$464_2$3 Read "Schliessen."

464_1 Add ", 232–252".

465^{21} Read "Cambr. Proc. 274–298."

465^{24} Read "J.S.L. 38 (1973), 453–459."

465^{26} Add "[Did not appear]"

465^{28} Add "[Did not appear]"

465^{30} Read "J.S.L. 41 (1976), 574–582."

466^{26} Read "Section VI" for "Section IV".

466_{14} Add: "[Appeared in: J.P.Seldin, J.R.Hindley (eds.), To H.B. Curry: Essays on Combinatory Logic, Lambda Calculus and Formalism. Academic Press, New York 1980, 480–490.]"

467^{22} Add: "[Never published]"

467^{24} Add: "[Appeared in: A.S. Troelstra, D. van Dalen (eds.), The L.E.J Brouwer Centenary Symposium. North-Holland Publ. Co., Amsterdam 1982, 51–64.]"

468^8 Read "Philosophica".

469^9 Read "in" for "in:".

469^{12} Read "Zentralblatt".

469^{26} Read "IPT" for Oslo Proc."

469_{24} Read " Π_1^1" for "Π".

469_8 Read "1" for "I".

470_{12} Add: "[Cf. paper under this title in: S. Kanger (ed.), Proceedings of the 3rd Scandinavian Logic Symposium, North-Holland Publ. Co., Amsterdam 1975, 81–109.]"

471^{10} Read: "Compositio Mathematica 26 (1973), 261–275."

472^5 Insert before stop: ", 225–250".

472_1 Replace by "Archiv für Mathematische Logik 16 (1974), 147–158."

473^{11} Add "[Unpublished]".

474_{11} Read "1970" for "170".

475^{13} Read "Cambr. Proc. 171–205."